# Steel Structures Design for Lateral and Vertical Forces

## About the International Code Council

The **International Code Council (ICC),** a membership association dedicated to building safety, fire prevention, and energy efficiency, develops the codes and standards used to construct residential and commercial buildings, including homes and schools. The mission of the ICC is to provide the highest quality codes, standards, products, and services for all concerned with the safety and performance of the built environment. Most U.S. cities, counties, and states choose the International Codes, building safety codes developed by the ICC. The International Codes also serve as the basis for construction of federal properties around the world, and as a reference for many nations outside the United States. The ICC is also dedicated to innovation and sustainability, and a subsidiary, ICC Evaluation Service, issues Evaluation Reports for innovative products and reports of Sustainable Attributes Verification and Evaluation (SAVE).

Headquarters: 500 New Jersey Avenue NW, 6th Floor, Washington, DC 20001-2070

District Offices: Birmingham, AL; Chicago, IL; Los Angeles, CA

1-888-422-7233; www.iccsafe.org

# Steel Structures Design for Lateral and Vertical Forces

Alan Williams

**Second Edition**

McGraw Hill Education

New York Chicago San Francisco
Athens London Madrid
Mexico City Milan New Delhi
Singapore Sydney Toronto

**Library of Congress Control Number: 2016934571**

McGraw-Hill Education books are available at special quantity discounts to use as premiums and sales promotions or for use in corporate training programs. To contact a representative, please visit the Contact Us page at www.mhprofessional.com.

**Steel Structures Design for Lateral and Vertical Forces, Second Edition**



1 2 3 4 5 6 7 8 9 QVS 21 20 19 18 17 16

ISBN 978-1-25-958801-3
MHID 1-25-958801-7

This book is printed on acid-free paper.

**Sponsoring Editor**
Lauren Poplawski

**Editorial Supervisor**
Stephen M. Smith

**Production Supervisor**
Pamela A. Pelton

**Acquisitions Coordinator**
Lauren Rogers

**Project Managers**
Ishita Bhatnagar and Yashmita Hota, Cenveo® Publisher Services

**Copy Editor**
Surendra Nath Shivam, Cenveo Publisher Services

**Proofreader**
Cenveo Publisher Services

**Indexer**
Robert Swanson

**Art Director, Cover**
Jeff Weeks

**Composition**
Cenveo Publisher Services

## About the Author

**Alan Williams, Ph.D., S.E., F.I.C.E., C.Eng.,** is a registered structural engineer in California who has had extensive experience in the practice and teaching of structural engineering. In California, he has worked as a Senior Transportation Engineer in the Department of Transportation and as Principal for Structural Safety in the Division of the State Architect. His prior positions include Professor of Structural Analysis at Ahmadu Bello University, Nigeria, and consulting structural engineer in South Africa and the United States. Dr. Williams's practical experience includes the design and construction of bridges, schools, and commercial and industrial structures.

He obtained his bachelor of science degree and doctorate from Leeds University and has published 13 papers and nine books on structural engineering topics. Dr. Williams is a member of the Structural Engineers Association of Southern California, a Fellow and Life Member of the Institution of Civil Engineers, and a Chartered Engineer in the United Kingdom.

# Contents

# Preface

For the second edition of this book, material has been revised and expanded to include the most current engineering codes, specifications, and manuals. The text and examples have been checked and updated so as to be consistent with the new design standards.

The first edition was based on the final draft of the 2010 *Specification for Structural Steel Buildings* (AISC 360-10) as the final specification was not yet available. This second edition provides the opportunity to correct the text for organizational and technical changes that were introduced in the final specification.

The first edition was based on the 13th edition of the *Steel Construction Manual* (AISC 325-05) while this second edition uses the 14th edition (AISC 325-11). The *Steel Construction Manual* provides tables and design aids to expedite implementation of AISC 360-10 regulations. The 14th edition of the manual has been extensively revised to conform to the requirements of the 2010 specification.

A new edition of the *Seismic Design Manual* (AISC 327-12) was introduced in 2012 and provides tables and design aids to expedite implementation of the *Seismic Provisions for Structural Steel Buildings* (AISC 341-10) and the *Prequalified Connections for Special and Intermediate Steel Moment Frames for Seismic Applications* (AISC 358-10). The manual is utilized in the new Chap. 14 that has been added to the second edition of this book.

Finally, this new edition conforms to the requirements of the 2015 *International Building Code* which replaced the 2012 *International Building Code.* The above-mentioned AISC provisions are adopted by reference in IBC-15. In addition, IBC-15 adopts by reference *Minimum Design Loads for Buildings and Other Structures* (ASCE 7-10) and this ASCE standard is covered extensively in the second edition of this book.

New topics have been added to the original text and existing topics have been expanded, particularly in the following chapters:

- Chapter 2, Sec. 2.9, Seismic Loads, includes new material on design response spectrum, overstrength factor, deflection amplification factor, seismic force-resisting systems, overturning, story drift, *P*-delta effects, stability coefficient, building separation, and structural irregularities.
- Chapter 3, Sec. 3.3, includes new material on design of collectors.
- Chapter 14 is a new chapter on the design of steel structures for seismic loads and covers the topics ordinary concentrically braced frames, special concentrically braced frames, eccentrically braced frames, and special moment frames.

The purpose of this edition is to provide comprehensive coverage of the design of steel structures for both vertical and lateral forces. In this book, the theoretical background and fundamental basis of steel design are introduced followed by the detailed design of members and their connections for gravity and seismic loads. This book provides detailed interpretations of the requirements of AISC *Specification for Structural Steel Buildings*, which covers the routine design of steel structures for vertical loads. The new Chap. 14 deals with the design of lateral load-resisting systems and assists in the interpretation and application of AISC *Seismic Provisions for Structural Steel Buildings.* The ASCE standard, *Minimum Design Loads for Buildings and Other Structures*, provides live, dead, wind, seismic, and snow design loads and their load combinations. The examples in this text provide extensive coverage of this standard.

As in the previous edition, this book covers both ASD and LRFD methods and presents design problems and solutions side-by-side in both formats. This allows the reader to readily distinguish the similarities and differences between the two methods. The design principles are illustrated with 190 examples with concise step-by-step solutions. Each example focuses on a specific issue and provides a clear and concise solution to the problem.

This publication provides a comprehensive resource document, reference, and guide for practicing engineers, teachers, professional engineering examination candidates, and undergraduate and graduate students requiring a comprehensive textbook on structural steel design. It fully covers the NCEES syllabus requirements for the Civil PE exam and the Structural SE exam.

*Alan Williams*

# Nomenclature

| | |
|---|---|
| $a$ | Clear distance between transverse stiffeners, in; shortest distance from edge of pin hole to edge of member measured parallel to direction of force, in; width of pressure coefficient zone, ft |
| $a_i$ | Acceleration at level $i$ obtained from a modal analysis, ft/s |
| $a_p$ | Amplification factor related to the response of a system or component as affected by the type of seismic attachment |
| $A$ | Effective wind area, ft$^2$ |
| $A_c$ | Area of concrete, in$^2$; area of concrete slab within effective width, in$^2$ |
| $A_e$ | Effective net area, in$^2$ |
| $A_{eff}$ | Summation of the effective areas of the cross section based on the reduced effective width, $b_e$, in$^2$ |
| $A_{fc}$ | Area of compression flange, in$^2$ |
| $A_{ft}$ | Area of tension flange, in$^2$ |
| $A_g$ | Gross area of member, in$^2$; gross area of composite member, in$^2$ |
| $A_{gv}$ | Gross area subject to shear, in$^2$ |
| $A_{lw}$ | Link web area (excluding flanges), in$^2$ |
| $A_n$ | Net area of member, in$^2$ |
| $A_{nt}$ | Net area subject to tension, in$^2$ |
| $A_{nv}$ | Net area subject to shear, in$^2$ |
| $A_o$ | Total area of openings in a wall that receives positive external pressure, ft$^2$ |
| $A_{oi}$ | Sum of the areas of openings in the building envelope not including $A_o$, ft$^2$ |
| $A_{og}$ | Total area of openings in the building envelope, ft$^2$ |
| $A_{pb}$ | Projected bearing area, in$^2$ |
| $A_s$ | Area of steel cross section, in$^2$ |
| $A_{sc}$ | Cross-sectional area of stud shear connector, in$^2$ |
| $A_{sf}$ | Shear area on the failure path, in$^2$ |
| $A_{sr}$ | Area of continuous reinforcing bars, in$^2$ |
| $A_{st}$ | Stiffener area, in$^2$ |
| $A_{st}$ | Horizontal cross-sectional area of the link stiffener, in$^2$ |
| $A_T$ | Tributary area, ft$^2$ |
| $A_w$ | Web area, the overall depth times the web thickness, $d_{tw}$, in$^2$; effective area of the weld, in$^2$ |

| | |
|---|---|
| $A_{wi}$ | Effective area of weld throat of any *i*th weld element, in$^2$ |
| $A_x$ | Torsional amplification factor for seismic loads |
| $A_1$ | Area of steel concentrically bearing on a concrete support, in$^2$ |
| $A_2$ | Maximum area of the portion of the supporting surface that is geometrically similar to and concentric with the loaded area, in$^2$ |
| $b$ | Width of unstiffened compression element; width of stiffened compression element, in |
| $b_{cf}$ | Width of column flange, in |
| $b_e$ | Reduced effective width, in; effective edge distance for calculation of tensile rupture strength of pin-connected members, in |
| $b_f$ | Flange width, in |
| $b_{fc}$ | Compression flange width, in |
| $b_{ft}$ | Width of tension flange, in |
| $B$ | Factor for lateral-torsional buckling in tees and double angles; horizontal dimension of building measured normal to wind direction, ft |
| $B_1$, $B_2$ | Factors used in determining $M_u$ for combined bending and axial forces when first-order analysis is employed |
| $C$ | HSS torsional constant |
| $C_a$ | Ratio of required strength to available strength |
| $C_b$ | Lateral-torsional buckling modification factor for nonuniform moment diagrams when both ends of the unsupported segment are braced |
| $C_d$ | Coefficient relating relative brace stiffness and curvature; deflection amplification factor for seismic loads |
| $C_e$ | Exposure factor for snow load |
| $C_f$ | Constant based on stress category |
| $C_m$ | Coefficient assuming no lateral translation of the frame |
| $C_{net}$ | Net-pressure coefficient based on $Kd\,[(G)(C_p) - (GC_{pi})]$ |
| $C_p$ | External pressure coefficient to be used in determination of wind loads for buildings |
| $C_r$ | Coefficient for web sidesway buckling |
| $C_s$ | Slope factor for snow load; seismic response coefficient |
| $C_T$ | Building period coefficient |
| $C_t$ | Thermal factor for snow load |
| $C_v$ | Web shear coefficient |
| $C_{vx}$ | Vertical distribution factor for seismic loads |
| $C_w$ | Warping constant, in$^6$ |
| $d$ | Nominal fastener diameter, in; nominal bolt diameter, in; full nominal depth of the section, in; diameter, in; pin diameter, in |
| $d_b$ | Beam depth, in; nominal diameter (body or shank diameter), in |
| $d_c$ | Column depth, in |
| $D$ | Dead load; outside diameter of round HSS member, in; outside diameter, in |
| $D$ | effect of dead load |
| $D_u$ | In slip-critical connections, a multiplier that reflects the ratio of the mean installed bolt pretension to the specified minimum bolt pretension |
| $e$ | EBF link length, in |
| $e_{mid\text{-}ht}$ | Distance from the edge of stud shank to the steel deck web, measured at mid-height of the deck rib, and in the load-bearing direction of the stud, in |

| | |
|---|---|
| $E$ | Modulus of elasticity of steel = 29,000 ksi; earthquake load; effect of horizontal and vertical earthquake induced forces |
| $E_c$ | Modulus of elasticity of concrete, ksi |
| $EI_{eff}$ | Effective stiffness of composite section, kip-in$^2$ |
| $E_{mh}$ | Horizontal seismic load effect including overstrength factor, kips |
| $E_s$ | Modulus of elasticity of steel, ksi |
| $f'_c$ | Specified minimum compressive strength of concrete, ksi |
| $f_{ra}$ | Required axial stress at the point of consideration using LRFD or ASD load combinations, ksi |
| $f_{rb(w,z)}$ | Required flexural stress at the point of consideration (major axis, minor axis) using LRFD or ASD load combinations, ksi |
| $f_{rv}$ | Required shear strength per unit area, ksi |
| $F$ | Load due to fluids with well-defined pressures and maximum heights |
| $F_a$ | Short-period site coefficient (at 0.2 s-period), s |
| $F_{cr}$ | Critical stress, ksi |
| $F_{cre}$ | Critical stress calculated from AISC 360 Chapter E using expected yield stress, ksi |
| $F_e$ | Elastic critical buckling stress, ksi |
| $F_{ex}$ | Elastic flexural buckling stress about the major axis, ksi |
| $F_{EXX}$ | Electrode classification number, ksi |
| $F_{ey}$ | Elastic flexural buckling stress about the minor axis, ksi |
| $F_{ez}$ | Elastic torsional buckling stress, ksi |
| $F_n$ | Nominal stress, ksi; nominal tensile stress $F_{nt}$, or shear stress, $F_{nv}$, from Table J3.2, ksi |
| $F_{nt}$ | Nominal tensile stress from Table J3.2, ksi |
| $F_{nv}$ | Nominal shear stress from Table J3.2, ksi |
| $F_p$ | Seismic force acting on a component of a structure, lb |
| $F_{SR}$ | Design stress range, ksi |
| $F_{TH}$ | Threshold fatigue stress range, maximum stress range for indefinite design life, ksi |
| $F_u$ | Specified minimum tensile strength, ksi |
| $F_v$ | Long-period site coefficient (at 1.0-s period) |
| $F_w$ | Nominal strength of the weld metal per unit area, ksi |
| $F_{wi}$ | Nominal stress in any $i$th weld element, ksi |
| $F_x$ | Portion of the seismic base shear, $V$, induced at level $x$, lb |
| $F_y$ | Specified minimum yield stress, ksi |
| $F_{yf}$ | Specified minimum yield stress of the flange, ksi |
| $F_{yr}$ | Specified minimum yield stress of reinforcing bars, ksi |
| $F_{yst}$ | Specified minimum yield stress of the stiffener material, ksi |
| $F_{yw}$ | Specified minimum yield stress of the web, ksi |
| $g$ | Transverse center-to-center spacing (gage) between fastener gage lines, in; acceleration due to gravity |
| $G$ | Shear modulus of elasticity of steel, ksi; gust effect factor |
| $G_f$ | Gust effect factor for MWFRS of flexible buildings |
| $GC_p$ | Product of external pressure coefficient and gust effect factor |
| $GC_{pf}$ | Product of the equivalent external pressure coefficient and gust-effect factor to be used in determination of wind loads for MWFRS of low-rise buildings |

| | |
|---|---|
| $GC_{pi}$ | Product of internal pressure coefficient and gust effect factor |
| $h$ | Average roof height of structure with respect to the base; width of stiffened compression element, in; height of shear element, in; mean roof height of a building, except that eave height shall be used for roof angle $\theta$ of less than or equal to 10°, ft |
| $h_b$ | Height of balanced snow load determined by dividing $p_s$ by $\gamma$, ft |
| $h_c$ | Clear height from top of balanced snow load to (1) closest point on adjacent upper roof, (2) top of parapet, or (3) top of a projection on the roof, ft; twice the distance from the centroid to the following: the inside face of the compression flange less the fillet or corner radius, for rolled shapes; the nearest line of fasteners at the compression flange or the inside faces of the compression flange when welds are used, for built-up sections, in |
| $h_d$ | Height of snow drift, ft |
| $h_f$ | Factor for fillers |
| $h_o$ | Distance between flange centroids, in; height of obstruction above the surface of the roof, for snow load, ft |
| $h_p$ | Twice the distance from the plastic neutral axis to the nearest line of fasteners at the compression flange or the inside face of the compression flange when welds are used, in |
| $h_r$ | Nominal height of ribs, in |
| $h_x$ | Height above the base to level $x$ |
| $H$ | Height of hill or escarpment, ft; load due to lateral earth pressure; story shear produced by the lateral forces used to compute $\Delta_H$, kips; overall height of rectangular HSS member, measured in the plane of the connection, in |
| $I$ | Moment of inertia in the plane of bending, in$^4$; moment of inertia about the axis of bending, in$^4$; importance factor |
| $I_c$ | Moment of inertia of the concrete section, in$^4$ |
| $I_p$ | Component importance factor for seismic loads |
| $I_s$ | Moment of inertia of steel shape, in$^4$ |
| $I_{sr}$ | Moment of inertia of reinforcing bars, in$^4$ |
| $I_x$, $I_y$ | Moment of inertia about the principal axes, in$^4$ |
| $I_z$ | Minor principal axis moment of inertia, in$^4$ |
| $I_{yc}$ | Moment of inertia of the compression flange about $y$-axis, in$^4$ |
| $J$ | Torsional constant, in$^4$ |
| $k$ | Distance from outer face of flange to the web toe of fillet, in; distribution exponent for seismic loads |
| $k_c$ | Coefficient for slender unstiffened elements, in |
| $k_{sc}$ | Slip-critical combined tension and shear coefficient |
| $k_v$ | Web plate buckling coefficient |
| $K$ | Effective length factor |
| $K_d$ | Wind directionality factor |
| $K_h$ | Velocity pressure exposure coefficient evaluated at height $z = h$ |
| $K_{LL}$ | Live load element factor |
| $K_z$ | Effective length factor for torsional buckling; velocity pressure exposure coefficient evaluated at height $z$ |
| $K_{zt}$ | Topographic factor |

| | |
|---|---|
| $K_1$ | Effective length factor in the plane of bending, calculated based on the assumption of no lateral translation |
| $K_2$ | Effective length factor in the plane of bending, calculated based on a sidesway buckling analysis |
| $K_1, K_2, K_3$ | Multipliers to obtain $K_{zt}$ |
| $l$ | Actual length of end-loaded weld, in; length of connection in the direction of loading, in |
| $l_b$ | Length of bearing, in |
| $l_c$ | Length of channel anchor, in; clear distance, in the direction of the force, between the edge of the hole and the edge of the adjacent hole or edge of the material, in |
| $l_u$ | Length of the roof upwind of the snow drift, ft |
| $L$ | Story height, in; length of the member, in; occupancy live load; laterally unbraced length of a member, in; span length, in; reduced design live load per ft² of area supported by the member, psf; horizontal dimension of a building measured parallel to the wind direction, ft |
| $L$ | effect of live load |
| $L_b$ | Length between points that are either braced against lateral displacement of compression flange or braced against twist of the cross section, in |
| $L_c$ | Length of channel shear connector, in |
| $L_h$ | Distance upwind of crest of hill or escarpment to where the difference in ground elevation is half the height of hill or escarpment, ft |
| $L_h$ | Distance between plastic hinge locations, in |
| $L_o$ | Unreduced design live load per ft² of area supported by the member, psf |
| $L_p$ | Limiting laterally unbraced length for the limit state of yielding, in |
| $L_r$ | Limiting laterally unbraced length for the limit state of inelastic lateral-torsional buckling, in; roof live load; reduced roof live load per ft² of horizontal projection, psf |
| $L_v$ | Distance from maximum to zero shear force, in |
| $M_a$ | Required flexural strength, using ASD load combinations, kip-in |
| $M_A$ | Absolute value of moment at quarter point of the unbraced segment, kip-in |
| $M_B$ | Absolute value of moment at centerline of the unbraced segment, kip-in |
| $M_C$ | Absolute value of moment at three-quarter point of the unbraced segment, kip-in |
| $M_{cx}, M_{cy}$ | Available flexural strength, kip-in |
| $M_{cx}$ | Available flexural-torsional strength for strong axis flexure, kip-in |
| $M_{lt}$ | First-order moment under LRFD or ASD load combinations caused by lateral translation of the frame only, kip-in |
| $M_{max}$ | Absolute value of maximum moment in the unbraced segment, kip-in |
| $M_n$ | Nominal flexural strength, kip-in |
| $M_{nt}$ | First-order moment using LRFD or ASD load combinations assuming there is no lateral translation of the frame, kip-in |
| $M_p$ | Plastic bending moment, kip-in |
| $M^*_{pb}$ | Moment at the intersection of the beam and column centerlines determined by projecting the beam maximum developed moments from the column face, kip-in |

| | |
|---|---|
| $M^*_{pc}$ | Moment at beam and column centerline determined by projecting the sum of the nominal column plastic moment strength, reduced by the axial stress $P_{uc}/A_g$, from the top and bottom of the beam moment connection, kip-in |
| $M_r$ | Required second-order flexural strength under LRFD or ASD load combinations, kip-in; required flexural strength using LRFD or ASD load combinations, kip-in |
| $M_t$ | Torsional moment resulting from eccentricity between the locations of center of mass and the center of rigidity |
| $M_{ta}$ | Accidental torsional moment |
| $M_u$ | Required flexural strength, using LRFD load combinations, kip-in |
| $M_y$ | Yield moment about the axis of bending, kip-in |
| $M_1$ | Smaller moment, calculated from a first-order analysis, at the ends of that portion of the member unbraced in the plane of bending under consideration, kip-in |
| $M_2$ | Larger moment, calculated from a first-order analysis, at the ends of that portion of the member unbraced in the plane of bending under consideration, kip-in |
| $n$ | Threads per inch |
| $n_b$ | Number of bolts carrying the applied tension |
| $n_s$ | Number of slip planes |
| $n_{sr}$ | Number of stress range fluctuations in design life |
| $N$ | Standard penetration resistance; number of stories |
| $N_i$ | Additional lateral load; notional lateral load applied at level $i$, kips |
| $p$ | Pitch, in per thread; design pressure to be used in determination of wind loads for buildings, psf |
| $p_d$ | Maximum intensity of snow drift surcharge load, psf |
| $p_f$ | Snow load on flat roofs, psf |
| $p_g$ | Ground snow load, psf |
| $p_i$ | Ratio of element $i$ deformation to its deformation at maximum stress |
| $p_{net}$ | Net design wind pressure, psf |
| $p_s$ | Net design wind pressure, psf; sloped roof snow load, psf |
| $p_W$ | Wind pressure acting on windward face, psf |
| $P_a$ | Required axial strength of a column using ASD load combinations, kips |
| $P_c$ | Available axial strength, kips |
| $P_{cy}$ | Available compressive strength out of the plane of bending, kip |
| $P_e$ | Elastic critical buckling load, kips |
| $P_{e1}$ | Elastic critical buckling strength of the member in the plane of bending, kips |
| $P_{lt}$ | First-order axial force using LRFD or ASD load combinations as a result of lateral translation of the frame only, kips |
| $P_n$ | Nominal axial strength, kips |
| $P_{net}$ | Design wind pressure to be used in determination of wind loads on buildings or other structures or their components and cladding, psf |
| $P_{no}$ | Nominal axial compressive strength without consideration of length effects, kips |
| $P_{nt}$ | First-order axial force using LRFD or ASD load combinations, assuming there is no lateral translation of the frame, kips |
| $P_p$ | Nominal bearing strength, kips |

| | |
|---|---|
| $P_r$ | Required second-order axial strength using LRFD or ASD load combinations, kips; required axial compressive strength using LRFD or ASD load combinations, kips; required axial strength using LRFD or ASD load combinations, kips; required strength, kips |
| $P_u$ | Required axial strength in compression, kips |
| $P_u$ | Required axial strength using LRFD load combinations, kips |
| $P_y$ | Axial yield strength, kips |
| $q$ | Velocity pressure, psf |
| $q_h$ | Velocity pressure evaluated at height $z = h$, psf |
| $q_i$ | Velocity pressure for internal pressure determination, psf |
| $q_s$ | Wind stagnation pressure, psf |
| $q_z$ | Velocity pressure evaluated at height $z$ above ground, psf |
| $Q$ | Full reduction factor for slender compression elements |
| $Q_a$ | Reduction factor for slender stiffened compression elements |
| $Q_E$ | Effect of horizontal seismic forces |
| $Q_n$ | Nominal strength of one stud shear connector, kips |
| $Q_s$ | Reduction factor for slender unstiffened compression elements |
| $r$ | Radius of gyration, in |
| $r_{cr}$ | Distance from instantaneous center of rotation to weld element with minimum $\Delta_u/r_i$ ratio, in |
| $r_i$ | Minimum radius of gyration of individual component in a built-up member, in; distance from instantaneous center of rotation to $i$th weld element, in |
| $\bar{r}_o$ | Polar radius of gyration about the shear center, in |
| $r_t$ | Radius of gyration of the flange components in flexural compression plus one-third of the web area in compression due to application of major axis bending moment alone, in |
| $r_{ts}$ | Effective radius of gyration, in |
| $r_x$ | Radius of gyration about $x$-axis, in |
| $r_y$ | Radius of gyration about $y$-axis, in |
| $r_z$ | Radius of gyration about the minor principal axis, in |
| $R$ | Load due to rainwater; seismic response modification coefficient; response modification coefficient |
| $R_a$ | Required strength using ASD load combinations |
| $R_g$ | Coefficient to account for group effect |
| $R_M$ | Coefficient to account for the influence of P-δ on P-Δ |
| $R_n$ | Nominal strength; available slip resistance, kips |
| $R_{nwl}$ | Total nominal strength of longitudinally loaded fillet welds |
| $R_{nwt}$ | Total nominal strength of transversely loaded fillet welds, |
| $R_p$ | Position effect factor for shear studs; component response modification factor for seismic loads |
| $R_t$ | Ratio of the expected tensile strength to the specified minimum tensile strength $F_u$ |
| $R_u$ | Required strength using LRFD load combinations |
| $R_y$ | Ratio of the expected yield stress to the specified minimum yield stress, $F_y$ |
| $R_1, R_2$ | Reduction factor for roof live load |
| $s$ | Longitudinal center-to-center spacing (pitch) of any two consecutive holes, in; separation distance between buildings, for snow load, ft |

| | |
|---|---|
| $s_h$ | Distance from center of column to centroid of plastic hinge |
| $S$ | Elastic section modulus, in$^3$; snow load; roof slope run for a rise of one |
| $S$ | Effect of snow load |
| $S_{DS}$ | Design, 5 percent damped, spectral response acceleration parameter at short periods |
| $S_{D1}$ | Design, 5 percent damped, spectral response acceleration parameter at a period of 1 s |
| $S_e$ | Effective section modulus about major axis, in$^3$ |
| $S_h$ | Distance from face of column to centroid of plastic hinge |
| $S_{min}$ | Lowest elastic section modulus relative to the axis of bending, in$^3$ |
| $S_S$ | Mapped 5 percent damped, spectral response acceleration parameter at short periods |
| $S_x$ | Elastic section modulus taken about the $x$-axis, in$^3$ |
| $S_{xc}$ | Elastic section modulus referred to compression flange, in$^3$ |
| $S_{xt}$ | Elastic section modulus referred to tension flange, in$^3$ |
| $S_y$ | Elastic section modulus taken about the $y$-axis, in$^3$ |
| $S_1$ | Mapped 5 percent damped, spectral response acceleration parameter at a period of 1 s |
| $t$ | Thickness of element, in; wall thickness, in; angle leg thickness, in; thickness of connected material, in; thickness of plate, in; design wall thickness of HSS, in; total thickness of fillers, in |
| $t_{cf}$ | Thickness of the column flange, in |
| $t_f$ | Thickness of the flange, in; |
| $t_f$ | Flange thickness of channel shear connector, in |
| $t_{fc}$ | Compression flange thickness, in |
| $t_p$ | Thickness of plate, in |
| $t_w$ | Web thickness of channel shear connector, in |
| $t_w$ | Web thickness, in |
| $T$ | Self-straining force; fundamental period of the building |
| $T_a$ | Approximate fundamental period of the building; tension force due to ASD load combinations, kips |
| $T_b$ | Minimum fastener tension |
| $T_L$ | Long-period transition period for seismic loads |
| $T_p$ | Fundamental period of the component and its attachment |
| $T_S$ | $S_{D1}/S_{DS}$ |
| $T_u$ | Tension force due to LRFD load combinations, kips |
| $T_0$ | $0.2S_{D1}/S_{DS}$ |
| $U$ | Shear lag factor |
| $U_{bs}$ | Reduction coefficient, used in calculating block shear rupture strength |
| $V$ | Basic wind speed, mi/h; total design lateral force or shear at the base |
| $V'$ | Nominal shear force between the steel beam and the concrete slab transferred by steel anchors, kips |
| $V_a$ | Required shear strength using ASD load combinations, kips |
| $V_c$ | Available shear strength, kips |
| $V_h$ | Shear force at the plastic hinge, kips |
| $V_n$ | Nominal shear strength, kips |
| $V_n$ | Link nominal shear strength, kips |
| $V_p$ | Nominal shear strength of an active link, kips |
| $V_r$ | Required shear strength using LRFD or ASD load combinations, kips |

$V'_r$ Required longitudinal shear force to be transferred to the steel or the concrete, kips

$V_u$ Required shear strength using LRFD load combinations, kips

$V_x$ Seismic design shear in story $x$

$w$ Width of cover plate, in; weld leg size, in; plate width, in; width of snow drift, ft

$w_c$ Weight of concrete per unit volume, pcf

$w_r$ Average width of concrete rib or haunch, in

$W$ Wind load; width of building, ft; horizontal distance from eave to ridge, ft; effective seismic weight of the building, lb

$x$ Subscript relating symbol to strong axis; distance upwind or downwind of crest, ft

$\bar{x}$ Eccentricity of connection, in

$x_i$ $x$ component of $r_i$

$x_o$, $y_o$ Coordinates of the shear center with respect to the centroid, in

$y$ Subscript relating symbol to weak axis

$y_i$ $y$ component of $r_i$

$Y_i$ Gravity load from the LRFD load combination or 1.6 times the ASD load combination applied at level i, kips

$z$ Subscript relating symbol to minor principal axis bending; height in structure of point of attachment of component with respect to the base, ft; height above ground level, ft

$Z$ Plastic section modulus about the axis of bending, $in^3$

$Z_x$ Plastic section modulus about the $x$- axis, $in^3$

$Z_y$ Plastic section modulus about the $y$-axis, $in^3$

## Greek Symbols

$\alpha$ ASD/LRFD force level adjustment factor

$\delta_x$ Total drift of the frame, elastic plus plastic, in

$\delta_{xe}$ Theoretical displacement, caused by the design level earthquake forces, as determined by an elastic analysis, in

$\Delta$ First-order interstory drift due to the design loads, in; design story drift, in

$\Delta_a$ Allowable story drift, in

$\Delta_p$ Plastic story drift, in

$\Delta_H$ First-order interstory drift due to lateral forces, in

$\Delta_i$ Deformation of weld elements at intermediate stress levels, linearly proportioned to the critical deformation based on distance from the instantaneous center of rotation, $r_i$, in

$\Delta_{mi}$ Deformation of weld element at maximum stress, in

$\Delta_{ui}$ Deformation of weld element at ultimate stress (rupture), usually in element furthest from the instantaneous center of rotation, in

$\Delta_{avg}$ Average of the displacements at the extreme points of the structure at level $x$, in

$\Delta_{max}$ Maximum displacement at level $x$, considering torsion, in

$\Delta_x$ Deflection of level $x$ at the center of the mass at and above level $x$, in

$\Delta_{xe}$ Deflection of level $x$ at the center of the mass at and above level $x$ determined by an elastic analysis, in

| | |
|---|---|
| $\kappa$ | Ratio of expected yield stress to elastic critical buckling stress |
| $\gamma$ | Snow density, pcf |
| $\lambda$ | Slenderness parameter |
| $\lambda$ | Adjustment factor for building height and exposure |
| $\lambda_{hd}$ | Limiting slenderness parameter for highly ductile compression elements |
| $\lambda_{md}$ | Limiting slenderness parameter for moderately ductile compression elements |
| $\lambda_p$ | Limiting slenderness parameter for compact element |
| $\lambda_{pd}$ | Limiting slenderness parameter for plastic design |
| $\lambda_{pf}$ | Limiting slenderness parameter for compact flange |
| $\lambda_{pw}$ | Limiting slenderness parameter for compact web |
| $\lambda_r$ | Limiting slenderness parameter for noncompact element |
| $\lambda_{rf}$ | Limiting slenderness parameter for noncompact flange |
| $\lambda_{rw}$ | Limiting slenderness parameter for noncompact web |
| $\mu$ | Mean slip coefficient for class A or B surfaces, as applicable, or as established by tests |
| $\varphi$ | Resistance factor |
| $\varphi_b$ | Resistance factor for flexure |
| $\varphi_B$ | Resistance factor for bearing on concrete |
| $\varphi_c$ | Resistance factor for compression; resistance factor for axially loaded composite columns |
| $\varphi_{sf}$ | Resistance factor for shear on the failure path |
| $\varphi_t$ | Resistance factor for tension |
| $\varphi_T$ | Resistance factor for torsion |
| $\varphi_v$ | Resistance factor for shear; resistance factor for steel headed stud anchor in shear |
| $\Omega$ | Safety factor |
| $\Omega_b$ | Safety factor for flexure |
| $\Omega_B$ | Safety factor for bearing on concrete |
| $\Omega_c$ | Safety factor for compression; safety factor for axially loaded composite columns |
| $\Omega_s$ | Safety factor for steel headed stud anchor in tension |
| $\Omega_{sf}$ | Safety factor for shear on the failure path |
| $\Omega_t$ | Safety factor for tension |
| $\Omega_v$ | Safety factor for shear; safety factor for steel headed stud anchor in shear |
| $\Omega_0$ | Overstrength factor |
| $\rho$ | Redundancy factor based on the extent of structural redundancy present in a building |
| $\rho_{sr}$ | Minimum reinforcement ratio for longitudinal reinforcing |
| $\theta$ | Angle of loading measured from the weld longitudinal axis, degrees; angle of loading measured from the longitudinal axis of *i*th weld element, degrees; angle of plane of roof from horizontal, degrees; roof slope on the leeward side, degrees |
| $\theta_p$ | Plastic story drift angle |
| $\tau_b$ | Stiffness reduction parameter |

# Steel Structures Design for Lateral and Vertical Forces

# CHAPTER 1

# Steel Buildings and Design Criteria

## 1.1 Introduction

Steel is widely used as a building material. This is because of a number of factors including its mechanical properties, availability in a variety of useful and practical shapes, economy, design simplicity, and ease and speed of construction.

Steel can be produced with a variety of properties to suit different requirements. The principal requirements are strength, ductility, weldability, and corrosion resistance. Figure 1.1 shows the stress-strain curves for ASTM A36 mild steel and a typical high-strength steel. Until recently, mild steel was the most common material for hot-rolled shapes but has now been superceded by higher strength steels for a number of shapes. ASTM A242 and A588 are corrosion resistant low-alloy steels. These are known as weathering steels and they form a tightly adhering patina on exposure to the weather. The patina consists of an oxide film that forms a protective barrier on the surface, thus preventing further corrosion. Hence, painting the steelwork is not required, resulting in a reduction in maintenance costs.

The stress-strain curve for mild steel indicates an initial elastic range, with stress proportional to strain, until the yield point is reached at a stress of 36 ksi. The slope of the stress-strain curve, up to this point, is termed the modulus of elasticity and is given by

$$\begin{aligned} E &= \text{stress}/\text{strain} \\ &= 29{,}000 \text{ ksi} \end{aligned}$$

Loading and unloading a mild steel specimen within the elastic range produces no permanent deformation and the specimen returns to its original length after unloading. The yield point is followed by plastic yielding with a large increase in strain occurring at a constant stress. Elongation produced after the yield point is permanent and non-recoverable. The plastic method of analysis is based on the formation of plastic hinges in a structure during the plastic range of deformation. The increase in strain during plastic yielding may be as much as 2 percent. Steel with a yield point in excess of 65 ksi does not exhibit plastic yielding and may not be used in structures designed by plastic design methods. At the end of the plastic zone, stress again increases with strain because of strain hardening. The maximum stress attained is termed the tensile strength of the steel and subsequent strain is accompanied by a decrease in stress.

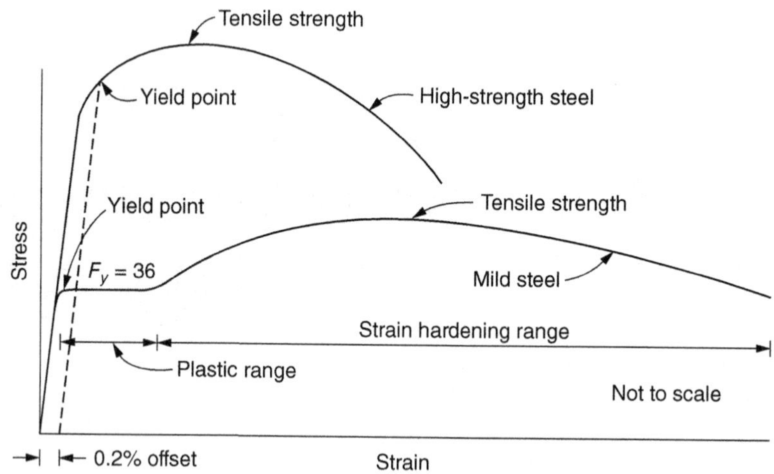


**FIGURE 1.1** Stress-strain curves for steel.

The stress-strain curve for high-strength steel does not exhibit a pronounced yield point. After the elastic limit is reached, the increase in stress gradually decreases until the tensile strength is reached. For these steels, a nominal yield stress is defined as the stress that produces a permanent strain of 0.2 percent.

Rolled steel sections are fabricated in a number of shapes, as shown in Fig. 1.2 and listed in Table 1.1.

Dimensions, weights, and properties of these sections are given by American Institute of Steel Construction, *Steel Construction Manual* (AISC Manual)[1] Part 1. The W-shape is an I-section with wide flanges having parallel surfaces. This is the most commonly used shape for beams and columns and is designated by nominal depth and weight per foot. Thus a W24 × 84 has a depth of 24.1 in and a weight of 84 lb/ft. Columns are loaded primarily in compression and it is preferable to have as large a radius of gyration about the minor axis as possible to prevent buckling. W12 and W14 sections are fabricated with the flange width approximately equal to the depth so as to achieve this. For example, a W14 × 132 has a depth of 14.7 in and a flange width of 14.7 in. The radii of gyration about

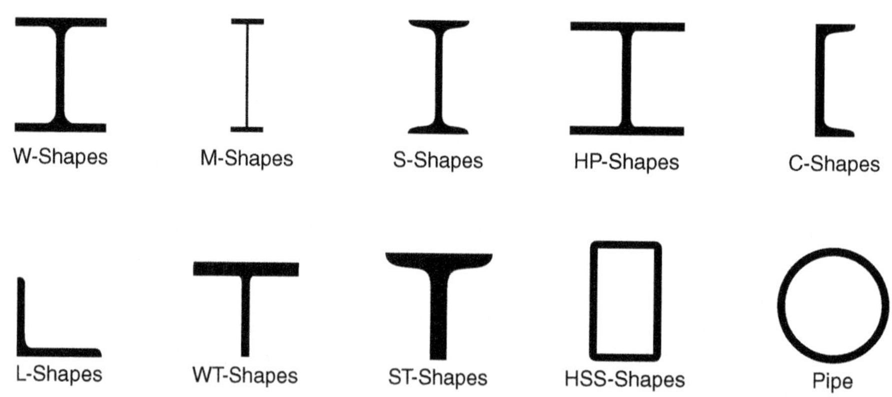


**FIGURE 1.2** Standard rolled shapes.

| Shape | Designation |
|---|---|
| Wide flanged beams | W |
| Miscellaneous beams | M |
| Standard beams | S |
| Bearing piles | HP |
| Standard channels | C |
| Miscellaneous channels | MC |
| Angles | L |
| Tees cut from W-shapes | WT |
| Tees cut from M-shapes | MT |
| Tees cut from S-shapes | ST |
| Rectangular hollow structural sections | HSS |
| Square hollow structural sections | HSS |
| Round hollow structural sections | HSS |
| Pipe | Pipe |

**TABLE 1.1** Rolled Steel Sections

the major and minor axes are 6.28 in and 3.76 in, respectively. Both S-shapes and M-shapes are I-sections with tapered flanges that are narrower than comparable W-shapes and provide less resistance to lateral torsional buckling. M-shapes are available in small sizes up to a depth of 12.5 in. S-shapes are available up to a depth of 24 in and have thicker webs than comparable W-shapes making them less economical.

The HP-shape is also an I-section and is used for bearing piles. To withstand piling stresses, they are of robust dimensions with webs and flanges of equal thickness and with depth and flange width nominally equal. The HP-shape is designated by nominal depth and weight per foot. Thus, an HP14 × 117 has a depth of 14.2 in and a weight of 117 lb/ft.

The C-shape is a standard channel with a slope of 2 on 12 to the inner flange surfaces. The MC-shape is a miscellaneous channel with a nonstandard slope on the inner flange surfaces. Channels are designated by exact depth and weight per foot. Thus, a C12 × 30 has a depth of 12 in and a weight of 30 lb/ft.

Angles have legs of equal thickness and either equal or unequal length. They are designated by leg size and thickness with the long leg specified first and the thickness last. Thus, an L8 × 6 × 1 is an angle with one 8-in leg, one 6-in leg, and with each leg 1 in thickness.

T-sections are made by cutting W-, M-, and S-shapes in half and they have half the depth and weight of the original section. Thus, a WT15 × 45 has a depth of 14.8 in and a weight of 45 lb/ft and is split from a W30 × 90.

There are three types of hollow structural sections: rectangular, square, and round. Hollow structural sections are designated by out side dimensions and nominal wall thickness. Thus, an HSS12 × 12 × ½ is a square hollow structural section with overall outside dimensions of 12 in by 12 in and a design wall thickness of 0.465 in. An HSS14.000 × 0.250 is a round hollow structural section with an outside dimension of 14 in and a design wall thickness of 0.233 in. Hollow structural sections are particularly suited for members that require high torsional resistance.

There are three classifications of pipes: standard, extra strong, and double-extra strong. Pipes are designated by nominal out side dimensions. Thus, a pipe 8 Std. is a pipe with an outside diameter of 8.63 in and a wall thickness of 0.322 in. A pipe 8 xx-Strong is a pipe with an outside diameter of 8.63 in and a wall thickness of 0.875 in.

Dimensions and properties of double angles are also provided in the AISC Manual Part 1. These are two angles that are interconnected through their back-to-back legs along the length of the member, either in contact for the full length or separated by spacers at the points of interconnection. Double angles are frequently used in the fabrication of open web joists. They are designated by specifying the size of angle used and their orientation. Thus, a 2L8 × 6 × 1 LLBB has two 8 × 6 × 1 angles with the 8 in (long) legs back-to-back. A 2L8 × 6 × 1 SLBB has two 8 × 6 × 1 angles with the 6 in (short) legs back-to-back.

Dimensions and properties of double channels are also provided in the AISC Manual Part 1. These are two channels that are interconnected through their back-to-back webs along the length of the member, either in contact for the full length or separated by spacers at the points of interconnection. Double channels are frequently used in the fabrication of open web joists. They are designated by specifying the depth and weight of the channel used. Thus, a 2C12 × 30 consists of two C12 × 30 channels each with a depth of 12 in and a weight of 30 lb/ft.

The types of steel commonly available for each structural shape are listed by Anderson and Carter[2] and are summarized in Table 1.2.

| Shape | Steel Type | | |
|---|---|---|---|
| | ASTM Designation | $F_y$, ksi | $F_u$, ksi |
| Wide flanged beams | A992 | 50 | 65 |
| Miscellaneous beams | A36 | 36 | 58–80 |
| Standard beams | A36 | 36 | 58–80 |
| Bearing piles | A572 Gr. 50 | 50 | 65 |
| Standard channels | A36 | 36 | 58–80 |
| Miscellaneous channels | A36 | 36 | 58–80 |
| Angles | A36 | 36 | 58–80 |
| Ts cut from W-shapes | A992 | 50 | 65 |
| Ts cut from M-shapes | A36 | 36 | 58–80 |
| Ts cut from S-shapes | A36 | 36 | 58–80 |
| Hollow structural sections, rectangular | A500 Gr. B | 46 | 58 |
| Hollow structural sections, square | A500 Gr. B | 46 | 58 |
| Hollow structural sections, round | A500 Gr. B | 42 | 58 |
| Pipe | A53 Gr. B | 35 | 60 |

Note: $F_y$ = specified minimum yield stress; $F_u$ = specified minimum tensile strength

**TABLE 1.2** Type of Steel Used

## 1.2 Types of Steel Buildings

Steel buildings are generally framed structures and range from simple one-story buildings to multistory structures. One of the simplest type of structure is constructed with a steel roof truss or open web steel joist supported by steel columns or masonry walls, as shown in Fig. 1.3.

An alternative construction technique is the single bay rigid frame structure shown in Fig. 1.4.

Framed structures consist of floor and roof diaphragms, beams, girders, and columns as shown in Fig. 1.5. The building may be one or several stories in height.

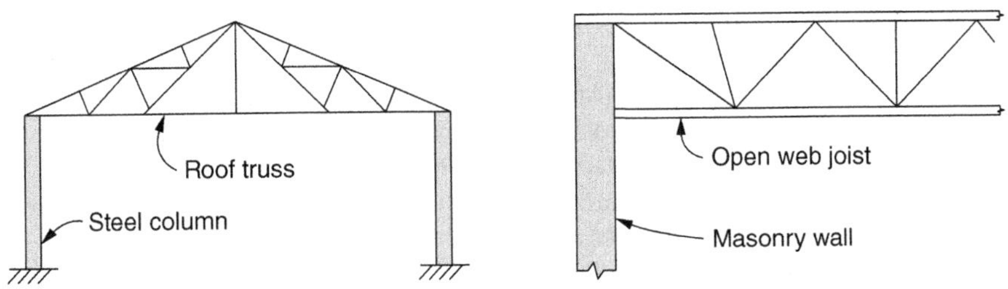


**FIGURE 1.3** Steel roof construction.

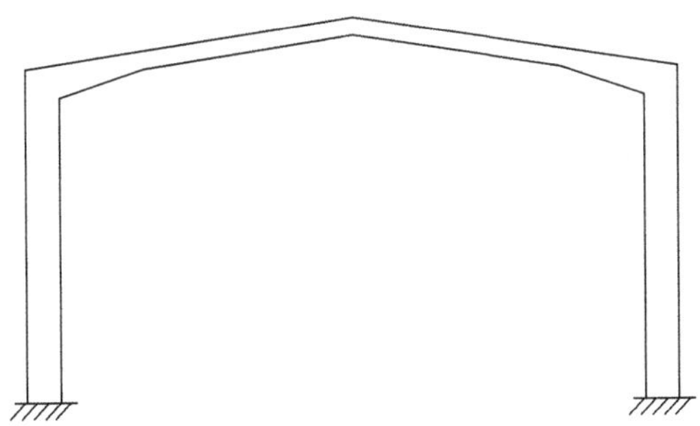

**FIGURE 1.4** Single bay rigid frame.

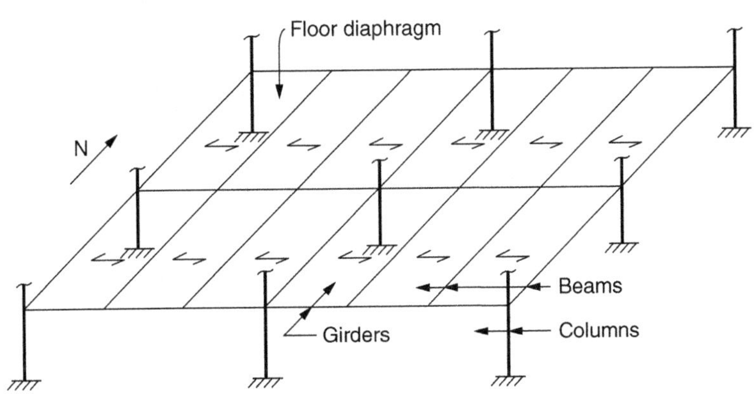


**FIGURE 1.5** Framed building.

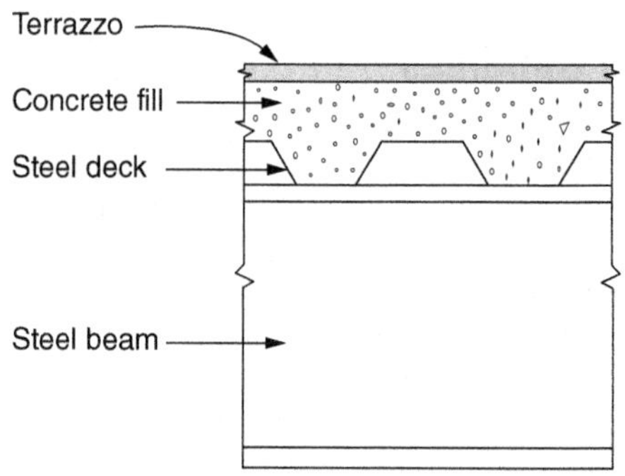


**Figure 1.6** Beam detail.

Figure 1.5 illustrates the framing arrangements at the second floor of a multistory building. The floor diaphragm spans east-west over the supporting beams and consists of concrete fill over formed steel deck as shown in Fig. 1.6.

The beams span north-south and are supported on girders, as shown in Fig. 1.7.

The girders frame into columns as shown in Fig. 1.8.

As well as supporting gravity loads, framed structures must also be designed to resist lateral loads caused by wind or earthquake. Several techniques are used to provide lateral

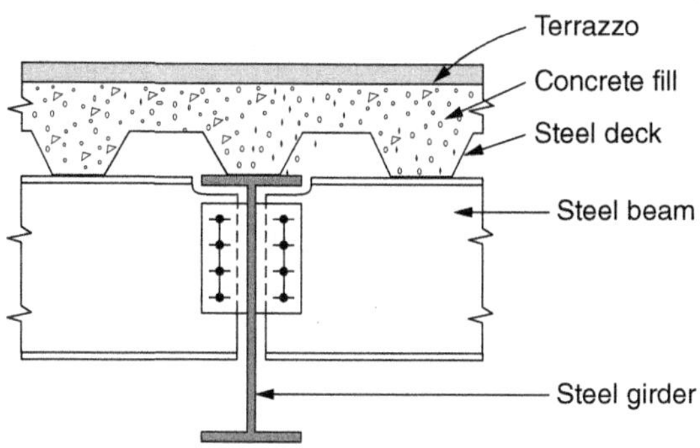


**Figure 1.7** Girder detail.

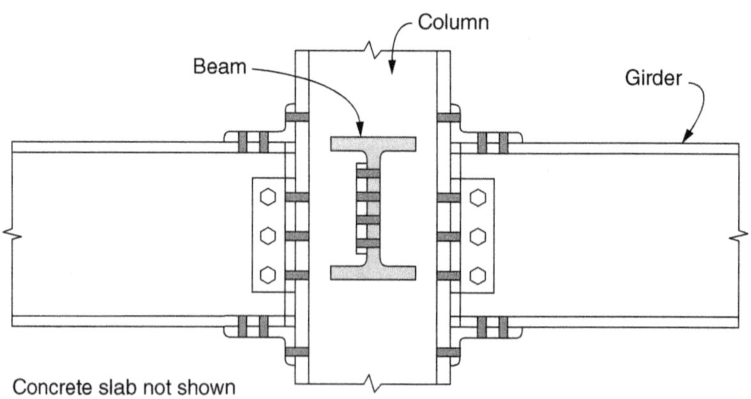


**Figure 1.8** Column detail.

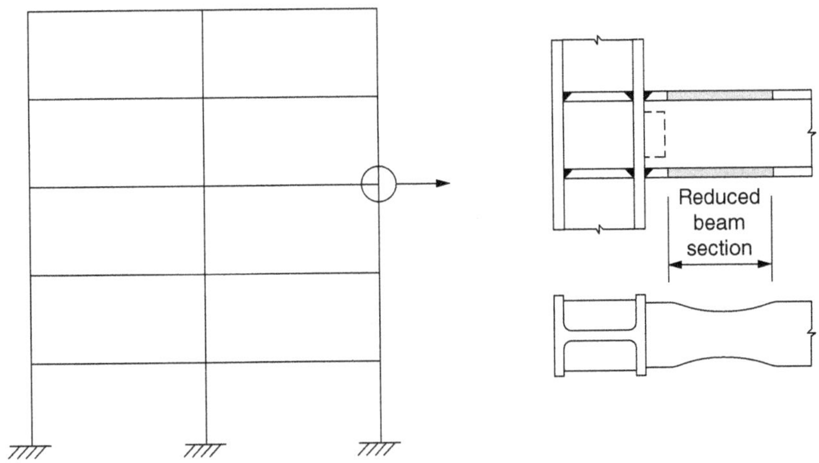


**Figure 1.9** Moment-resisting frame.

resistance including special moment-resisting frames, braced frames, and shear walls. Moment-resisting frames resist lateral loads by means of special flexural connections between the columns and beams. The flexural connections provide the necessary ductility at the joints to dissipate the energy demand with large inelastic deformations. A number of different methods is used to provide the connections and these are specified in American Institute of Steel Construction, *Prequalified Connections for Special and Intermediate Steel Moment Frames for Seismic Applications* (AISC 358-10).[3] A typical moment-resisting frame building is shown in Fig. 1.9 with a reduced beam section connection detailed.

Moment-resisting frames have the advantage of providing bays free from obstructions. However, special detailing is required for finishes and curtain walls to accommodate, without damage, the large drifts anticipated.

Concentrically braced frames, described by Cochran and Honeck,[4] and eccentrically braced frames, described by Becker and Ishler,[5] are illustrated in Fig. 1.10. These systems

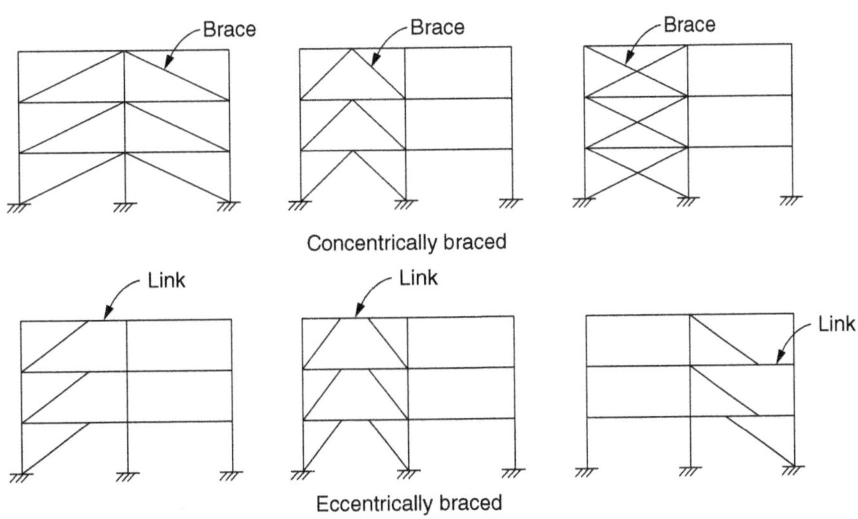


**Figure 1.10** Braced frames.

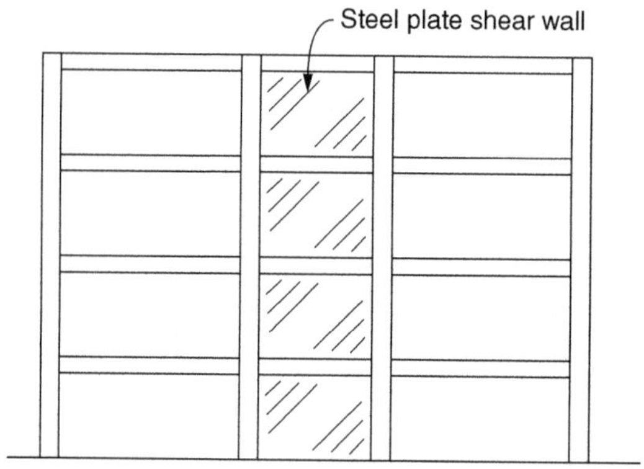


FIGURE 1.11 Steel plate shear wall building.

have the advantage over moment-resisting frames of less drift and simpler connections. In addition, braced frames are generally less expensive than moment-resisting frames. Their disadvantages are restrictions on maximum building height and architectural limitations.

A building with a steel plate shear wall lateral force-resisting system is shown in Fig. 1.11 and is described by Sabelli.[6] This system provides good drift control but lacks redundancy.

## 1.3 Building Codes and Design Criteria

The building code adopted by most jurisdictions throughout the United States is the International Code Council, *International Building Code* (IBC).[7] Some states and some cities publish their own code and this is usually a modification of the IBC to conform to local customs and preferences. The IBC establishes minimum regulations for building systems using prescriptive and performance-related provisions. When adopted by a local jurisdiction, it becomes a legally enforceable document.

The code provides requirements to safeguard public health, safety, and welfare through provisions for structural strength, sanitation, light, ventilation, fire, and other hazards. To maintain its relevance to changing circumstances and technical developments, the code is updated every 3 years. The code development process is an open consensus process in which any interested party may participate.

The requirements for structural steelwork are covered in IBC Chap. 22. In IBC Sec. 2205, two specifications of the American Institute of Steel Construction are adopted by reference. These are, *Specification for Structural Steel Buildings* (AISC 360)[8] and *Seismic Provisions for Structural Steel Buildings* (AISC 341).[9] The *Specification for Structural Steel Buildings* is included in AISC Manual Part 16. The *Seismic Provisions for Structural Steel Buildings* is included in AISC *Seismic Design Manual* (SDM)[10] Part 6. The Specification and the Provisions provide complete information for the design of buildings. Both include a Commentary that provides background information on the derivation and application of the specifications and provisions.

AISC 360 provides criteria for the design, fabrication, and erection of structural steel buildings and structures similar to buildings. It is specifically intended for low-seismic applications where design is based on a seismic response modification coefficient $R$ of 3 or less. This is permissible in buildings assigned to seismic design category A, B, or C and ensures a nominally elastic response to the applied loads. When design is based on a seismic response modification coefficient $R$ greater than 3, the design, fabrication, and erection of structural steel buildings and structures similar to buildings must comply with the requirements of the Seismic Provisions, AISC 341. This is mandatory in buildings assigned to seismic design category D, E, or F. In situations where wind effects exceed seismic effects, the building elements must still be detailed in accordance with AISC 341 provisions. These provisions provide the design requirements for structural steel seismic force-resisting systems to sustain the large inelastic deformations necessary to dissipate the seismic induced demand. The *Seismic Manual* provides guidance on the application of the provisions to the design of structural steel seismic force-resisting systems.

## 1.4 ASD and LRFD Concepts

The traditional method of designing steel structures has been by the allowable stress design method. The objective of this method was to ensure that a structure was capable of supporting the applied working loads safely. Working loads, also referred to as nominal or service loads, are the dead loads and live loads applied to a structure. Dead load includes the self-weight of the structure and permanent fittings and equipment. Live load includes the weight of the structure's occupants and contents and is specified in American Society of Civil Engineers, *Minimum Design Loads for Buildings and Other Structures* (ASCE 7-10)[11] Table 4-1. The allowable stress design method specified that stresses produced in a structure by the working loads must not exceed a specified allowable stress. The method was based on elastic theory to calculate the stresses produced by the working loads. The allowable stress, also known as working stress, was determined by dividing the yield stress of the material by an appropriate factor of safety. Hence:

$$
\begin{aligned}
F &= F_y/\Omega \\
&\geq f
\end{aligned}
$$

where $F$ = allowable stress
$F_y$ = yield stress
$\Omega$ = factor of safety
$f$ = actual stress in a member, subjected to working loads, as determined by elastic theory

The advantages of the allowable stress method are its simplicity and familiarity.

In 1986, American Institute of Steel Construction introduced the load and resistance factor design (LRFD) method. In this method, the working loads are factored before being applied to the structure. The load factors are given by ASCE 7 Sec. 2.3.2 and these are used in the strength design load combinations. The load factors are determined by probabilistic theory and account for

- Variability of anticipated loads
- Errors in design methods and computations
- Lack of understanding of material behavior

The force in a member, caused by the factored load combination, may be determined by elastic, inelastic, or plastic analysis methods and this is the required strength of the member. The nominal strength of the member, also known as the ultimate capacity, is determined according to AISC 360 or AISC 341 provisions. The design strength, is determined by multiplying the nominal strength of the member by an appropriate resistance factor. The resistance factors are determined by probabilistic theory and account for

- Variability of material strength
- Poor workmanship
- Errors in construction

Hence, in accordance with AISC 360 Eq. (B3-1)

$$R_u \le \varphi R_n$$

where $R_u$ = required strength of a member subjected to strength design load combinations (LRFD)
$\varphi$ = resistance factor
$R_n$ = nominal strength of the member as determined by the specifications or provisions
$\varphi R_n$ = design strength

In 2005, American Institute of Steel Construction issued the unified specification, AISC 360. In accordance with AISC 360 Sec. B3, structural steel design must be done by either load and resistance factor design (LRFD) or by allowable *strength* design (ASD). In the ASD method, the members in a structure are proportioned so that the required strength, as determined by the appropriate ASD load combination, does not exceed the designated allowable strength of the member. The ASD load combinations are given by ASCE 7 Sec. 2.4.1. The allowable strength is determined as the nominal strength of the member divided by a safety factor. The nominal strength of the member is determined according to AISC 360 or AISC 341 provisions. The nominal strength is identical for both the LRFD and ASD methods. Hence, in accordance with AISC 360 Eq. (B3-2):

$$R_a \le R_n/\Omega$$

where $R_a$ = required strength of a member subjected to allowable stress design load combinations (ASD)
$\Omega$ = safety factor
$R_n$ = nominal strength of the member as determined by the specifications or provisions
$R_n/\Omega$ = allowable strength

The relationship between safety factor and resistance factor is

$$\Omega = 1.5/\varphi$$

**Example 1.1** Relationship between Safety Factor and Resistance Factor
Assuming a live load to dead load ratio of $L/D = 3$, derive the relationship between safety factor and resistance factor.

Consider a simply supported beam of length $\ell$ supporting a uniformly distributed dead load of $D$ and a uniformly distributed live load of $L$. The required nominal flexural strength determined using both the LRFD and ASD methods is as follows:

| LRFD | ASD |
|---|---|
| Load combination from ASCE 7 Sec. 2.3.2 is | Load combination from ASCE 7 Sec. 2.4.1 is |
| $w_u = 1.2D + 1.6L$ | $w_a = D + L$ |
| Substituting $L = 3D$ gives | Substituting $L = 3D$ gives |
| $w_u = 6D$ | $w_a = 4D$ |
| The required flexural strength is | The required flexural strength is |
| $M_u = w_u\ell^2/8$ | $M_a = w_a\ell^2/8$ |
| $= 3D\ell^2/4$ | $= D\ell^2/2$ |
| The required nominal flexural strength is | The required nominal flexural strength is |
| $M_n = M_u/\varphi$ | $M_n = M_a\Omega$ |
| $= 3D\ell^2/4\varphi$ | $= D\ell^2\Omega/2$ |

Equating the nominal strength for both design methods

$$3D\ell^2/4\varphi = D\ell^2\Omega/2$$

Hence:
$$\Omega = 1.5/\varphi$$

Allowable *strength* design is similar to allowable *stress* design in that both utilize the ASD load combinations. However, for strength design, the specifications are formatted in terms of force in a member rather than stress. The stress design format is readily derived from the strength design format by dividing allowable strength by the appropriate section property, such as cross-sectional area or section modulus, to give allowable stress.

**Example 1.2** Relationship between Allowable Strength Design and Allowable Stress Design
For the limit state of tensile yielding, derive the allowable tensile stress from the allowable strength design procedure.

For tensile yielding in the gross section, the nominal tensile strength is given by AISC 360 Eq. (D2-1) as

$$\boxed{P_n = F_yA_g}$$

where $A_g$ = gross area of member

The safety factor for tension is given by AISC 360 Sec. D2 as

$$\boxed{\Omega_t = 1.67}$$

The allowable tensile strength is given by AISC 360 Sec. D2 as

$$\begin{aligned} P_c &= P_n/\Omega_t \\ &= F_yA_g/1.67 \\ &= 0.6F_yA_g \end{aligned}$$

The allowable tensile stress for the limit state of tensile yielding is

$$\begin{aligned} F_t &= P_c/A_g \\ &= 0.6F_y \end{aligned}$$

## References


1. American Institute of Steel Construction (AISC). 2011. *Steel Construction Manual.* 14th edition, AISC, Chicago, IL.
2. Anderson, M. and Carter, C. J. 2015. "Are You Properly Specifying Materials?," *Modern Steel Construction*, February 2015.
3. American Institute of Steel Construction (AISC). 2010. *Prequalified Connections for Special and Intermediate Steel Moment Frames for Seismic Applications* (AISC 358-10) AISC, Chicago, IL.
4. Cochran, M. and Honeck, W. C. 2004. *Design of Special Concentric Braced Frames.* Structural Steel Educational Council, Moraga, CA.
5. Becker, R. and Ishler, M. 1996. *Seismic Design Practice for Eccentrically Braced Frames.* Structural Steel Educational Council, Moraga, CA.
6. Sabelli, R. 2006. *Steel Plate Shear Walls.* Design Guide No. 20. AISC, Chicago, IL.
7. International Code Council (ICC). 2015. *International Building Code*, 2015 edition, ICC, Falls Church, VA.
8. American Institute of Steel Construction (AISC). 2010. *Specification for Structural Steel Buildings* (AISC 360-10), AISC, Chicago, IL.
9. American Institute of Steel Construction (AISC). 2010. *Seismic Provisions for Structural Steel Buildings* (AISC 341-10), AISC, Chicago, IL.
10. American Institute of Steel Construction (AISC). 2012. AISC *Seismic Design Manual.* 2nd edition, AISC, Chicago, IL.
11. American Society of Civil Engineers (ASCE). 2010. *Minimum Design Loads for Buildings and Other Structures* (ASCE 7-10), ASCE, Reston, VA.


## Problems

**1.1** *Given:* American Institute of Steel Construction, *Steel Construction Manual*

*Find:* Using the manual

*a.* The differences between W-, M-, S-, and HP-shapes
*b.* The uses of each of these shapes

**1.2** *Given:* American Institute of Steel Construction, *Steel Construction Manual*

*Find:* Using the manual the meaning of

*a.* W16 × 100
*b.* WT8 × 50
*c.* 2MC13 × 50
*d.* HSS8.625 × 0.625
*e.* 2L4 × 3 × ½ LLBB
*f.* Pipe 6 xx-Strong
*g.* HSS6 × 4 × ½

**1.3** *Given:* American Institute of Steel Construction, *Steel Construction Manual*

*Find:* Using the manual

*a.* The meaning of "Unified Code"
*b.* How the unified code developed

**1.4** *Given:* American Institute of Steel Construction, *Steel Construction Manual*

*Find:* Using the manual the distinction between
*a.* Safety factor and resistance factor
*b.* Nominal strength and required strength
*c.* Design strength and allowable strength

**1.5** *Given:* American Institute of Steel Construction, *Steel Construction Manual*

*Find:* Using the manual
*a.* Four different types of steel that may be used for rectangular HSS-shapes
*b.* The preferred type of steel for rectangular HSS-shapes

**1.6** *Given:* A building to be designed to resist seismic loads and three different lateral force-resisting methods are to be evaluated.

*Find:* *a.* Three possible methods that may be used
*b.* The advantages of each method
*c.* The disadvantages of each method

**1.7** *Given:* American Institute of Steel Construction, *Steel Construction Manual* and International Code Council, *International Building Code*

*Find:* Describe the purpose of each document and their interrelationship.

# CHAPTER 2
# Design Loads

## 2.1 Introduction

A structure must be designed and constructed so as to safely resist the applied load. The applied load consists of dead loads, live loads, and environmental loads. Dead load includes the self-weight of the structure and permanent fittings and equipment. Live load includes the weight of the structure's occupants and contents. Environmental loads include the effects of snow, wind, earthquake, rain, and flood. Additional loads may also be imposed by the self-straining forces caused by temperature changes, shrinkage, or settlement.

The design of a structure must take into consideration the different combinations of loads that may be applied to the structure and also the variable nature of each load. To allow for this, load factors are applied to the nominal loads and several different combinations are checked. In each combination, one variable load is taken at its maximum lifetime value and the other variable loads assume arbitrary point-in-time values.

A steel structure may be designed by either the allowable *strength* design (ASD) method or by the load and resistance factor design (LRFD) method. In the ASD method, the members in a structure are proportioned so that the required strength, as determined by the appropriate ASD load combination, does not exceed the designated allowable strength of the member. The allowable strength is determined as the nominal strength of the member divided by a safety factor. Traditionally, the acronym ASD has meant allowable *stress* design with the requirement that stress induced by the ASD load combination does not exceed the designated allowable stress for the member. In general, the allowable stress is determined as the yield stress of the member divided by a safety factor. The stress design format is readily derived from the strength design format by dividing allowable strength by the appropriate section property, such as cross-sectional area or section modulus, to give allowable stress. In the LRFD method, the members in a structure are proportioned so that the force induced in the member by the appropriate strength design load combination does not exceed the member capacity multiplied by a resistance factor. To account for this difference in approach, different load factors are assigned, in the two methods, to the nominal loads in the load combinations.

In addition to designing a structure for strength limit states, serviceability limit states must also be met. Deflection of members must be limited to ensure that damage does not occur in supported elements. This is particularly important in the case of beams supporting plastered ceilings and sensitive partitions and cladding. Vibrations must also be controlled to ensure that annoyance is not caused to the occupants of a building.

## 2.2 Dead Loads

Dead loads are defined in *International Building Code* (IBC)[1] Sec. 1606 and in American Society of Civil Engineers, *Minimum Design Loads for Buildings and Other Structures* (ASCE 7-10)[2] Sec. 3.1. Dead loads consist of the permanent loads imposed on a structure. These include the self-weight of the structure, architectural features, fixed service equipment such as heating and air-conditioning systems, sprinkler systems, and utility services. An allowance should also be included for future additional wearing surfaces. The actual dead loads cannot be calculated accurately until after the structure is designed and the size of all members is known. Hence, an estimate of the dead load must be determined in advance and checked after the design is completed. A comprehensive list of the weight of building materials is given in ASCE 7 Table C3-1. A comprehensive list of the density of building materials is given in ASCE 7 Table C3-2.

### Tributary Area

In order to determine the dead load applied to a structural member, use is made of the tributary area concept. As shown in Fig. 2.1, the second-floor slab is supported on beams which, in turn, are supported on either girders or columns. The girders carry the dead load to columns that transfer the total load to the foundations. It may be assumed that all beams have the same section with a weight of $w_B$ lb/ft and all girders have the same section with a weight of $w_G$ lb/ft.

### Slab Supports

Each slab panel is supported on its periphery by either a beam or a girder. For the situation shown, the aspect ratio of a panel exceeds two, and the slab resists moments in the direction of the shorter span between beams essentially as a one-way slab. Beams are supported at each end by either a girder or a column. A further assumption is made that beams are simply supported at each end. Then the tributary area of a beam is defined as the area of the slab that is directly supported by the beam.

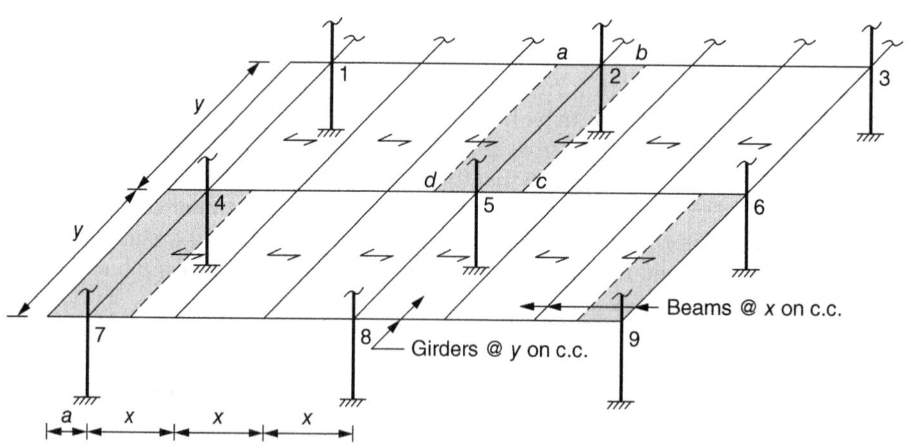


**Figure 2.1** Beam tributary area.

## Dead Load Applied to Beams

For a typical interior beam 25, the tributary area is the shaded area *abcd*, shown in Fig. 2.1, which extends over its full length $y$ and a distance of $x/2$ on either side. The tributary area is

$$A_T = xy$$

The tributary width is

$$B_T = x$$

For a slab with a weight of $q$ lb/ft$^2$ and a beam with a self-weight of $w_B$ lb/ft, the uniformly distributed dead load on beam 25 is

$$\begin{aligned} w_{D25} &= qB_T + w_B \\ &= qx + w_B \end{aligned}$$

The total dead load on beam 25 is

$$\begin{aligned} W_{D25} &= qA_T + w_B y \\ &= y(qx + w_B) \end{aligned}$$

The dead load reaction at end 2 of beam 25 is

$$\begin{aligned} V_{D25} &= y(qx + w_B)/2 \\ &= V_{D52} \end{aligned}$$

The loading on beam 25 is shown in Fig. 2.2.

As shown in Fig. 2.1, beam 47 supports a cantilevered portion of the slab and the tributary area is shown shaded. The corresponding loads on beam 47 are

$$\begin{aligned} w_{D47} &= qB_T + w_B \\ &= q(a + x/2) + w_B \\ W_{D47} &= qA_T + w_B y \\ &= y[q(a + x/2) + w_B] \\ V_{D47} &= y[q(a + x/2) + w_B]/2 \\ &= V_{D74} \end{aligned}$$

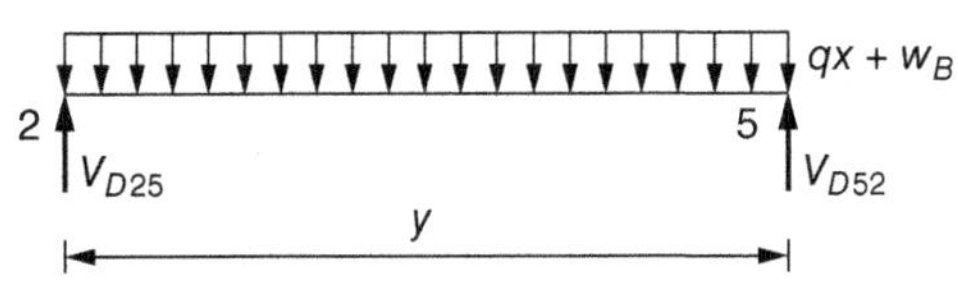


**FIGURE 2.2** Dead load on beam 25.

Similarly for edge beam 69, the corresponding loads are

$$\begin{aligned} w_{D69} &= qB_T + w_B \\ &= qx/2 + w_B \\ W_{D69} &= qA_T + w_B y \\ &= y(qx/2 + w_B) \\ V_{D69} &= y(qx/2 + w_B)/2 \\ &= V_{D96} \end{aligned}$$

**Example 2.1.** Dead Load Applied to Beams
The second-floor layout of an office facility is shown in Fig. 2.1 and a detail of the floor construction and of a typical interior beam is shown in Fig. 2.3. Dimension $x = 10$ ft and $y = 30$ ft. The floor consists of composite steel-concrete construction with a 3-in concrete fill over a 3-in-high formed steel deck. The lightweight concrete fill has a weight of 110 lb/ft$^3$ and the formed steel deck is of 20 gage material. A 1-in terrazzo finish, and a suspended acoustic ceiling are provided. All beams are W14 × 22 and ribs of the steel deck are perpendicular to the beams. Determine the dead load acting on a typical interior beam 25.

The steel deck thickness of 20 gage is selected to support the concrete fill over a span of 10 ft without requiring propping during construction. From the manufacturer's catalogue, the weight of the steel deck is obtained as 2 lb/ft$^2$ and the weight of the lightweight concrete fill as 41 lb/ft$^2$.

The total distributed load on the floor is

| | |
|---|---|
| 1-in terrazzo | = 13 lb/ft$^2$ |
| Concrete fill | = 41 lb/ft$^2$ |
| Steel deck | = 2 lb/ft$^2$ |
| Acoustical ceiling and supports | = 3 lb/ft$^2$ |
| Mechanical and electrical services | = 3 lb/ft$^2$ |
| Total, $q$ | = 62 lb/ft$^2$ |

The tributary area of beam 25 is

$$\begin{aligned} A_T &= xy \\ &= 10 \times 30 \\ &= 300 \text{ ft}^2 \end{aligned}$$

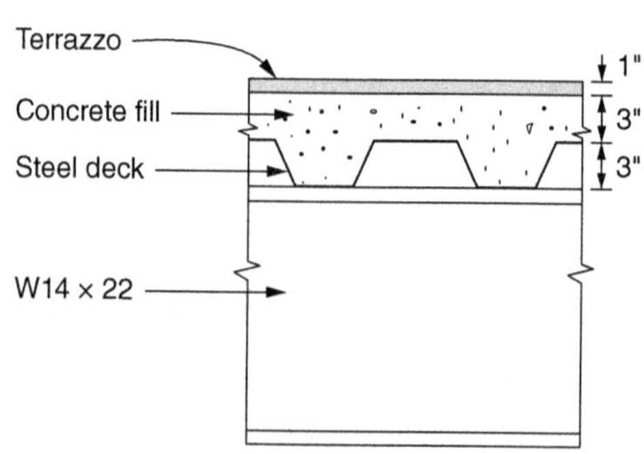


FIGURE 2.3 Beam detail.

The total dead load on beam 25 is

$$\begin{aligned} W_{D25} &= qA_T + w_B y \\ &= (62 \times 300 + 22 \times 30)/1000 \\ &= 19.26 \text{ kips} \end{aligned}$$

The dead load reaction at end 2 of beam 25 is

$$\begin{aligned} V_{D25} &= W_{D25}/2 \\ &= 19.26/2 \\ &= 9.63 \text{ kips} \\ &= V_{D52} \end{aligned}$$

## Dead Load Applied to Girders

As shown in Fig. 2.4, the tributary area of a typical interior girder 56 is the shaded area efgh. The girder supports its own weight $w_G$ and also the end reactions of the beams framing into each side of the girder. Girders are supported at each end by a column.

Since all beams are identical, the concentrated loads acting on the girder at third points of the span are

$$2V_{D25} = y(qx + w_B)$$

The dead load reaction at end 5 of girder 56 is

$$\begin{aligned} V_{D56} &= y(qx + w_B) + 3xw_G/2 \\ &= V_{D65} \end{aligned}$$

The loading on girder 56 is shown in Fig. 2.5.

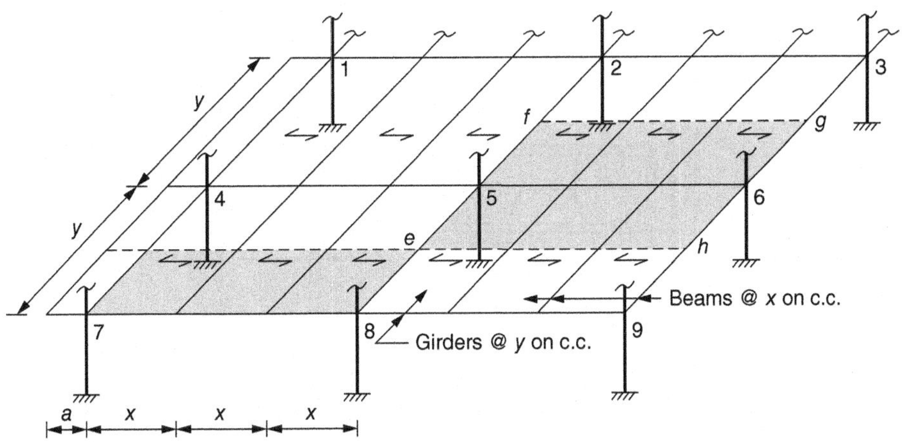


**FIGURE 2.4** Girder tributary area.

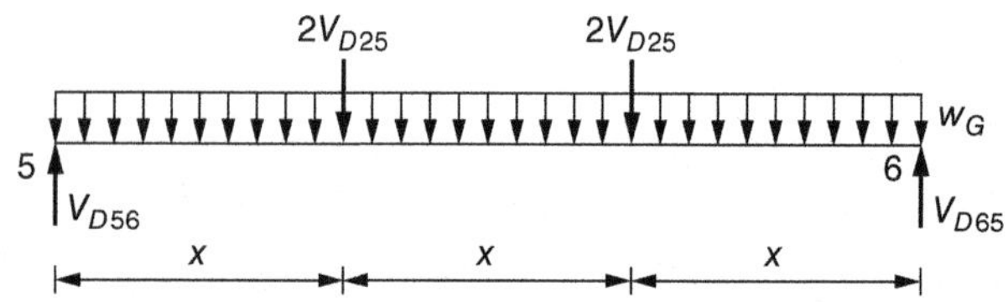


**FIGURE 2.5** Dead load on girder 56.

The tributary area of girder 78 is shown shaded in Fig. 2.4. This is an edge girder and beams frame into only one side of the girder. Since all beams are identical, the concentrated loads acting on the girder at third points of the span are

$$V_{D25} = y(qx + w_B)/2$$

The dead load reaction at end 7 of girder 78 is

$$V_{D78} = y(qx + w_B)/2 + 3xw_G/2$$
$$= V_{D87}$$

**Example 2.2.** Dead Load Applied to Girders
The second-floor layout of an office facility is shown in Fig. 2.1 and a detail of the floor construction and of a typical girder 56 is shown in Fig. 2.6. Dimension $x = 10$ ft and $y = 30$ ft. The floor consists of composite steel-concrete construction with a 3-in concrete fill over a 3-in-high formed steel deck. The lightweight concrete fill has a weight of 110 lb/ft³ and the formed steel deck is of 20 gage material. A 1-in terrazzo finish, and a suspended acoustic ceiling are provided. All beams are W14 × 22. All girders are W18 × 40 and ribs of the steel deck are parallel to the girders. Determine the dead load acting on a typical girder 56.

The total distributed load on the floor is obtained in Example 2.1 as

$$q = 62 \text{ lb/ft}^2$$

Since all beams are identical, the concentrated loads acting on the girder at third points of the span are

$$2V_{D25} = 2 \times 9.63$$
$$= 19.26 \text{ kips}$$

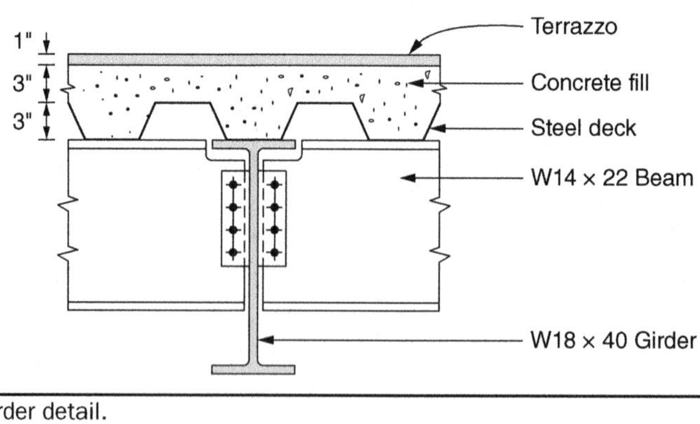


**FIGURE 2.6** Girder detail.

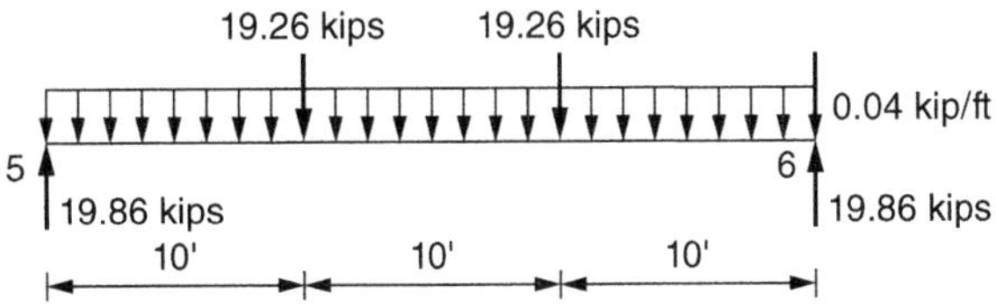


**FIGURE 2.7** Dead load on girder.

The dead load reaction acting at end 5 of girder 56 is

$$V_{D56} = 19.26 + 0.04 \times 15$$
$$= 19.86 \text{ kips}$$
$$= V_{D65}$$

The dead load acting on the girder is shown in Fig. 2.7.

## Dead Load Applied to Columns

As shown in Fig. 2.8, the tributary area of column 5 is the shaded area *ijkl*. Framing into the column are beams 52 and 58 and girders 54 and 56, and the column supports the end reactions from these members.

Hence, the total dead load applied to column 5 at the second floor is

$$V_{D5} = V_{D52} + V_{D58} + V_{D54} + V_{D56}$$

The end reaction from beam 52 is

$$V_{D52} = y(qx + w_B)/2$$

The end reaction from beam 58 is

$$V_{D58} = y(qx + w_B)/2$$

The end reaction from girder 54 is

$$V_{D54} = y(qx + w_B) + 3xw_G/2$$

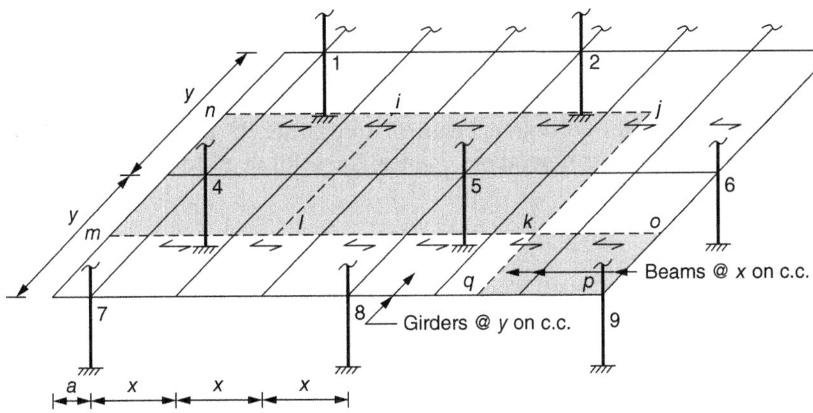


**FIGURE 2.8** Column tributary area.

The end reaction from girder 56 is

$$V_{D56} = y(qx + w_B) + 3xw_G/2$$

Alternatively, the weight of the beams and girders may be included in the distributed floor load, since all beams are identical and all girders are identical. Thus, the total distributed load on the floor is

$$q_D = q + w_B/x + w_G/y$$

The area tributary to column 5 is

$$A_T = 3xy$$

The dead load applied to column 5 at the second floor is, then

$$V_{D5} = q_D(3xy)$$

The tributary area of column 4 is the shaded area *lmni* shown in Fig. 2.8. This is a side column that supports a cantilevered slab, and has beams framing into two sides and a girder framing into only one side. Framing into the column are beams 41 and 47 and girder 45, and the column supports the end reactions from these members. Hence, the total dead load applied to column 4 at the second floor is

$$V_{D4} = V_{D41} + V_{D47} + V_{D45}$$

The end reaction from beam 41 is

$$V_{D41} = y[q(a + x/2) + w_B]/2$$

The end reaction from beam 47 is

$$V_{D47} = y[q(a + x/2) + w_B]/2$$
$$= V_{D41}$$

The end reaction from girder 45 is

$$V_{D45} = y(qx + w_B) + 3xw_G/2$$

Because of the cantilevered slab, the alternative method of calculating the column load using the tributary area $A_T$ and the total distributed load $q_d$ does not apply.

The tributary area of column 9 is the shaded area *opqk* in Fig. 2.8. This is a corner column that has a beam framing into only one side and a girder framing into only one side. Framing into the column is beam 96 and girder 98, and the column supports the end reactions from these members. Hence, the total dead load applied to column 9 at the second floor is

$$V_{D9} = V_{D96} + V_{D98}$$

The end reaction from beam 96 is

$$V_{D96} = y(qx/2 + w_B)/2$$

The end reaction from girder 98 is

$$V_{D98} = y(qx + w_B)/2 + 3xw_G/2$$

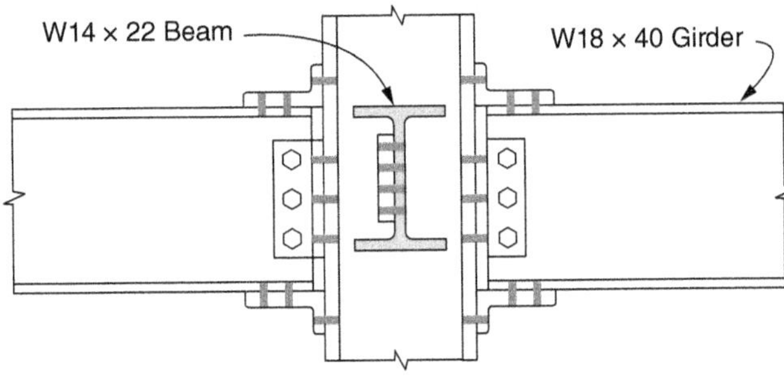


**FIGURE 2.9** Column detail.

**Example 2.3.** Dead Load Applied to Columns
The floor layout of an office facility is shown in Fig. 2.1 and a detail of the column framing is shown in Fig. 2.9. Dimension $x = 10$ ft and $y = 30$ ft. The floor consists of composite steel-concrete construction with a 3-in concrete fill over a 3-in-high formed steel deck. The lightweight concrete fill has a weight of 110 lb/ft$^3$ and the formed steel deck is of 20 gage material. A 1-in terrazzo finish, and a suspended acoustic ceiling are provided. All beams are W14 × 22. All girders are W18 × 40 and ribs of the steel deck are parallel to the girders. Determine the dead load acting on column 5 at each floor.

The total dead load applied to column 5 at each floor is

$$P_{DF} = V_{D52} + V_{D58} + V_{D54} + V_{D56}$$

From Example 2.1, the end reaction from beam 52 is

$$V_{D52} = 9.63 \text{ kips}$$
$$= V_{D58}$$

From Example 2.2, the end reaction from girder 54 is

$$V_{D54} = 19.86 \text{ kips}$$
$$= V_{D56}$$

Hence, the dead load acting on column 5 at each floor is

$$P_{DF} = 2(9.63 + 19.86)$$
$$= 58.98 \text{ kips}$$

Alternatively, the total distributed load on the floor is

$$q_D = q + w_B/x + w_G/y$$
$$= 62 + 22/10 + 40/30$$
$$= 65.53 \text{ lb/ft}^2$$

The area tributary to column 5 is

$$A_T = 3xy$$
$$= 3 \times 10 \times 30$$
$$= 900 \text{ ft}^2$$

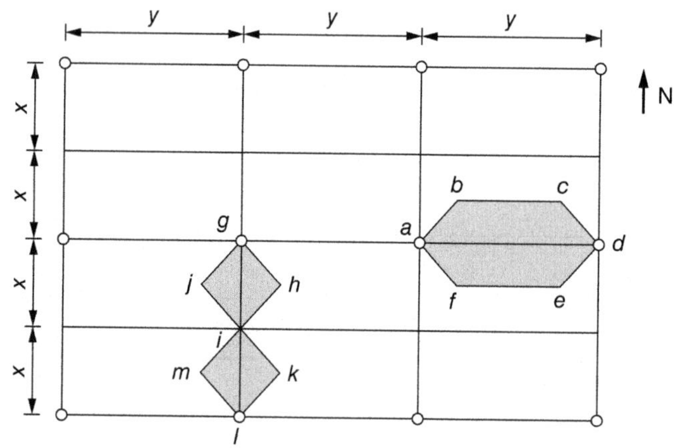


**FIGURE 2.10** Two-way slab tributary areas.

The dead load applied to column 5 at each floor is, then

$$P_{DF} = q_D A_T$$

$$= 65.53 \times 900/1000$$

$$= 58.98 \text{ kips}$$

## Two-Way Slabs

When the aspect ratio of a slab is not more than two, the slab resists moments essentially as a two-way slab. The tributary areas for the supporting beams are bounded by 45° lines drawn from the corners of the panels and by the center lines of the panels parallel to the long sides. A plan view of floor framing is shown in Fig. 2.10 with beams spanning east-west and girders spanning north-south. The beams are spaced at $x$ on centers and the girders at $y$ on centers and the aspect ratio is

$$y/x = 2$$

For a typical interior beam spanning east-west, the tributary area is the trapezoidal area *abcdef* shown shaded in Fig. 2.10. The dead load acting on the beam is shown in Fig. 2.11, where $q$ is the uniformly distributed weight of the floor and $y$ is $2x$.

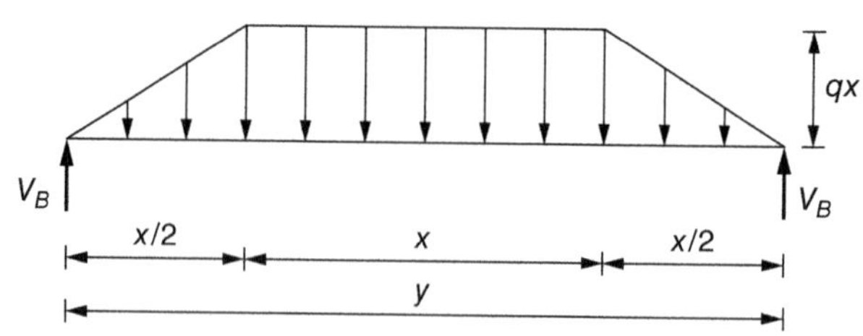


**FIGURE 2.11** Dead load supported by beam.

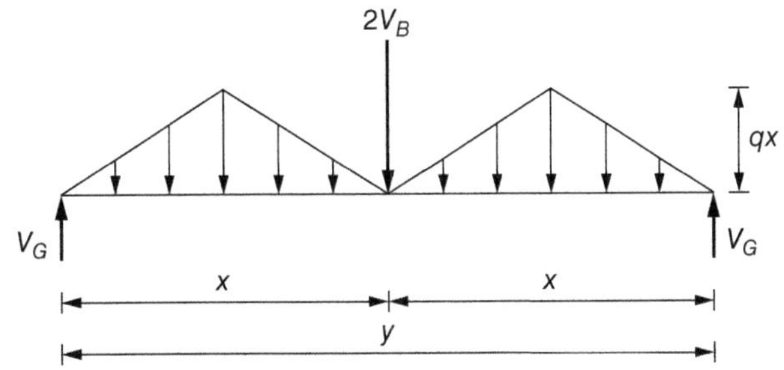


**FIGURE 2.12** Dead load supported by girder.

The dead load reaction at each end of the beam, including its own weight $w_B$, is

$$V_B = 0.75qx^2 + w_B x$$

For a typical interior girder spanning north-south, the tributary area is the double triangular area *ghij* plus the area *iklm* shown shaded in Fig. 2.10. The dead load acting on the girder is shown in Fig. 2.12.

The dead load reaction at each end of the girder, including its own weight $w_G$, is

$$V_G = 0.5qx^2 + V_B + w_G x$$

## 2.3 Live Loads

Live loads are defined in ASCE 7 Sec. 4.1 as loading on a floor produced by the occupancy or use of the building that does not include construction loads, dead loads, or environmental loads. Partition walls, used to subdivide large floor areas into smaller offices and cubicles, may be reconfigured and modified during the life of the building. Hence, partitions are designated as live loads and are applied, in addition to the specified floor live load, irrespective of whether or not partitions are shown on the building plans. As required by ASCE 7 Sec. 4.3.2, moveable partitions are considered a nonreducible live load with a magnitude of 15 lb/ft$^2$. However, a partition live load is not required where the floor live load exceeds 80 lb/ft$^2$. Roof live load is defined as loading on a roof caused by maintenance work or by moveable planters or other decorative features. The minimum values of the uniformly distributed unit loads required for design are given in ASCE 7 Table 4-1. These values are statistically derived from the maximum load that can reasonably be expected to occur during the life of the building, normally 50 years.

### Continuous Beam Systems

Partial loading or "checkerboard" loading conditions that produce more critical loading on a member must also be considered.

A continuous beam with two partial loading conditions is shown in Fig. 2.13. In loading condition 1, alternate spans are loaded. This produces maximum positive moments at the center of the loaded spans 12, 34, and 56. This also produces maximum negative moments at the center of the unloaded spans 23 and 45. In loading condition 2, two adjacent spans are loaded with alternate spans loaded beyond these. This produces maximum negative moment at support 4 and maximum beam shears. As it is unlikely

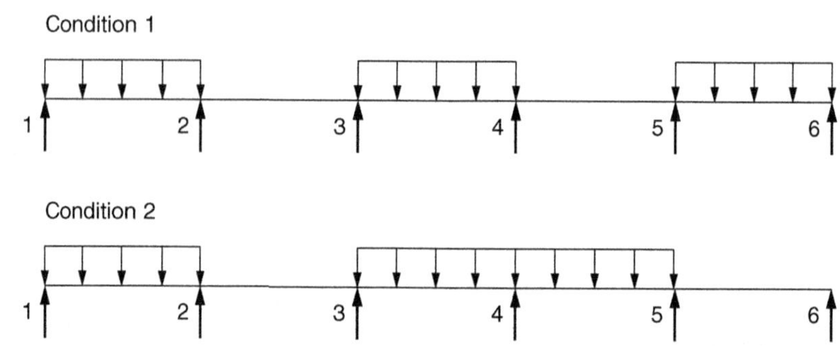


**Figure 2.13** Partial loading conditions.

that the maximum design live load will occur over the whole of a large area, it is permissible to reduce live loads for this situation. The allowable reduction increases with the tributary area of the floor supported by a member. Different methods are specified for reducing floor loads and roof loads.

## Influence Area

In the case of floor loads, the concept of influence area $A_I$ is introduced. Influence area is defined in ASCE 7 Sec. C4.7.1 as that floor area over which the influence surface for structural effects is significantly different from zero. The live load element factor $K_{LL}$ is defined as the ratio of the influence area of a member to its tributary area. Thus:

$$K_{LL} = A_I / A_T$$

As shown in Fig. 2.14, the influence area for a typical interior beam 25 is the shaded area *abcd* and the live load element factor is

$$\begin{aligned} K_{LL} &= A_I / A_T \\ &= 2xy / xy \\ &= 2 \end{aligned}$$

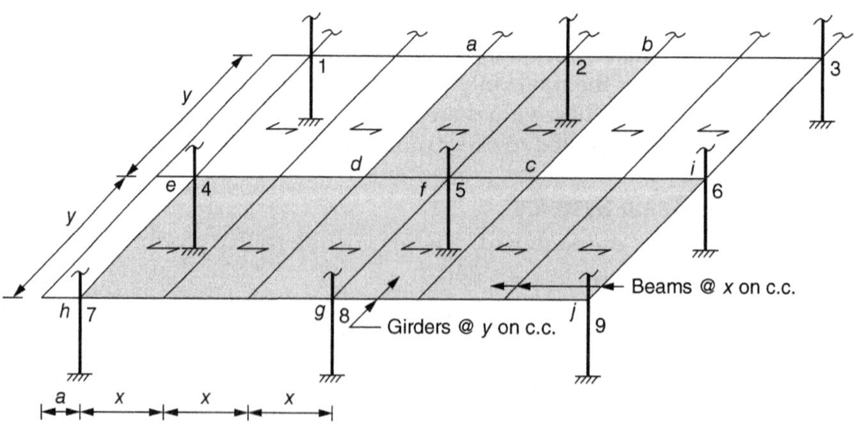


**Figure 2.14** Influence areas.

The influence area for an edge girder 78 is the shaded area *efgh*, and the live load element factor is

$$K_{LL} = A_I/A_T$$
$$= 3xy/1.5xy$$
$$= 2$$

The influence area for a corner column 9 is the shaded area *ijgf*, and the live load element factor is

$$K_{LL} = A_I/A_T$$
$$= 3xy/0.75xy$$
$$= 4$$

Values of $K_{LL}$ are given in ASCE 7 Table 4-2 and an abbreviated listing is given in the following Table 2.1.

## Reduction in Floor Live Load

It is unlikely that all floors in a multistory building will be subjected to the full design live load simultaneously. Similarly, a large floor area is unlikely to be subjected to as high a loading intensity as a smaller area. For members that have a value of $K_{LL}A_T \geq 400\ \text{ft}^2$, floor live loads may be reduced in accordance with ASCE 7 Eq. (4.7-1) which is

$$L = L_o[0.25 + 15/(K_{LL}A_T)^{1/2}]$$
$$\geq 0.5L_o \text{ ... for members supporting only one floor}$$
$$\geq 0.4L_o \text{ ... for members supporting two or more floors}$$

where $L$ = reduced design live load of area supported by member
$L_o$ = unreduced design live load
$K_{LL}$ = live load element factor
$A_T$ = tributary area

| Element | $K_{LL}$ |
|---|---|
| Interior columns | 4 |
| Exterior columns without cantilever slabs | 4 |
| Edge columns with cantilever slabs | 3 |
| Corner columns with cantilever slabs | 2 |
| Edge beams without cantilever slabs | 2 |
| Interior beams | 2 |
| All other members | 1 |

**TABLE 2.1** Live Load Element Factor $K_{LL}$

There are a number of exceptions to this requirement. Heavy live loads exceeding 100 lb/ft$^2$ may not be reduced when a member is supporting only one floor. These loads usually occur in storage facilities and it is possible that the maximum design live load will occur over the whole area of a floor. However, when a member supports two or more floors, it is unlikely that all floors will be fully loaded simultaneously and the heavy live load may be reduced by not more than 20 percent. The same provision is applied to garages for passenger cars. In facilities used for public assembly, live loads not exceeding 100 lb/ft$^2$ shall not be reduced because of the likely possibility that all floors of the facility may be fully loaded. Live loads may be reduced for the design of one-way slabs, but the tributary area of the slab is restricted to the product of the slab span and a tributary width of 1.5 times the slab span. This compensates for the lack of redundancy of a one-way slab compared with a two-way slab.

**Example 2.4.** Live Load Applied to Beams
The second-floor layout of an office facility is shown in Fig. 2.1 and a detail of the floor construction and of a typical interior beam is shown in Fig. 2.3. Dimension $x$ = 10 ft and $y$ = 30 ft. Determine the live load acting on a typical interior beam 25.

From ASCE 7 Table 4-1, the unreduced live load is

$$L_o = 50 \text{ lb/ft}^2$$

From ASCE 7 Table 4-2, the live load element factor for an interior beam is

$$K_{LL} = 2$$

From Example 2.1, the area tributary to beam 25 is

$$\begin{aligned} A_T &= 300 \text{ ft}^2 \\ K_{LL}A_T &= 2 \times 300 \\ &= 600 \text{ ft}^2 \\ &> 400 \text{ ft}^2 \text{ ... ASCE 7 Eq. (4.7-1) is applicable} \end{aligned}$$

The reduced design live load for beam 25 is

$$\begin{aligned} L &= L_o[0.25 + 15/(K_{LL}A_T)^{1/2}] \\ &= 50[0.25 + 15/(600)^{1/2}] \\ &= 43.12 \text{ lb/ft}^2 \text{ ... satisfactory} \\ &> 0.5L_o \text{ ... minimum does not govern} \end{aligned}$$

Hence, the minimum permitted value for a member supporting only one floor of $L = 0.5L_o$ is not applicable.

In accordance with ASCE 7 Sec. 4.3.2, an additional 15 lb/ft$^2$ must be added to allow for the weight of moveable partitions. Hence, the total live load intensity is

$$\begin{aligned} L &= 43.12 + 15 \\ &= 58.12 \text{ lb/ft}^2 \end{aligned}$$

The total live load is

$$\begin{aligned} P &= LA_T \\ &= 58.12 \times 300/1000 \\ &= 17.44 \text{ kips} \end{aligned}$$

**Example 2.5.** Floor Live Load Applied to Columns
The floor layout of a four-story office facility is shown in Fig. 2.8. Dimension $x = 10$ ft and $y = 30$ ft. Determine the floor live load produced on a typical interior column 5 at each story.

From ASCE 7 Table 4-1, the unreduced live load is

$$L_o = 50 \text{ lb/ft}^2$$

From ASCE 7 Table 4-2, the live load element factor for an interior column is

$$K_{LL} = 4$$

The area tributary to column 5 at each floor level is

$$\begin{aligned} A_T &= 3xy \\ &= 3 \times 10 \times 30 \\ &= 900 \text{ ft}^2 \end{aligned}$$

The design live load on the column, in each story, must account for the tributary floor area supported by that story.

**Top story**
There is no floor live load acting on the column in the top story.

**Third story**
The column supports the floor live load from the fourth floor.

$$\begin{aligned} K_{LL}A_T &= 4 \times 900 \\ &= 3600 \text{ ft}^2 \\ &> 400 \text{ ft}^2 \text{ ... ASCE 7 Eq. (4.7-1) is applicable} \end{aligned}$$

The reduced design live load for column 5 is

$$\begin{aligned} L &= L_o[0.25 + 15/(K_{LL}A_T)^{1/2}] \\ &= 50[0.25 + 15/(3600)^{1/2}] \\ &= 25 \text{ lb/ft}^2 \text{ ... satisfactory} \\ &= 0.5L_o \text{ ... minimum does not govern} \end{aligned}$$

In accordance with ASCE 7 Sec. 4.3.2, an additional 15 lb/ft$^2$ must be added to allow for the weight of moveable partitions. Hence the total live load intensity is

$$\begin{aligned} L &= 25 + 15 \\ &= 40 \text{ lb/ft}^2 \end{aligned}$$

The total live load is

$$\begin{aligned} P_L &= LA_T \\ &= 40 \times 900/1000 \\ &= 36 \text{ kips} \end{aligned}$$

**Second story**

The column supports the floor live load from the third and fourth floor.

$$\begin{aligned} K_{LL}A_T &= 4 \times 900 \times 2 \\ &= 7200 \text{ ft}^2 \\ &> 400 \text{ ft}^2 \text{ ... ASCE 7 Eq. (4.7-1) is applicable} \end{aligned}$$

The reduced design live load for column 5 is

$$\begin{aligned} L &= L_o[0.25 + 15/(K_{LL}A_T)^{1/2}] \\ &= 50[0.25 + 15/(7200)^{1/2}] \\ &= 21.34 \text{ lb/ft}^2 \text{ ... satisfactory} \\ &> 0.4L_o \text{ ... minimum does not govern} \end{aligned}$$

In accordance with ASCE 7 Sec. 4.3.2, an additional 15 $\text{lb/ft}^2$ must be added to allow for the weight of moveable partitions on the third and fourth floor. Hence, the total live load intensity is

$$\begin{aligned} L &= 21.34 + 15 \\ &= 36.34 \text{ lb/ft}^2 \end{aligned}$$

The total live load is

$$\begin{aligned} P_L &= LA_T \\ &= 36.34 \times 2 \times 900/1000 \\ &= 65.41 \text{ kips} \end{aligned}$$

**Bottom story**

The column supports the floor live load from the second, third, and fourth floor.

$$\begin{aligned} K_{LL}A_T &= 4 \times 900 \times 3 \\ &= 10{,}800 \text{ ft}^2 \\ &> 400 \text{ ft}^2 \text{ ... ASCE 7 Eq. (4.7-1) is applicable} \end{aligned}$$

The reduced design live load for column 5 is

$$\begin{aligned} L &= L_o[0.25 + 15/(K_{LL}A_T)^{1/2}] \\ &= 50[0.25 + 15/(10{,}800)^{1/2}] \\ &= 19.72 \text{ lb/ft}^2 \\ &< 0.4L_o \text{ ... minimum governs} \end{aligned}$$

Hence, use the minimum permitted value for a column supporting three floors of

$$\begin{aligned} L &= 0.4L_o \\ &= 0.4 \times 50 \\ &= 20 \text{ lb/ft}^2 \end{aligned}$$

In accordance with ASCE 7 Sec. 4.3.2, an additional 15 lb/ft², on each floor, must be added to allow for the weight of moveable partitions. Hence, the total live load intensity is

$$\begin{aligned} L &= 20 + 15 \\ &= 35 \text{ lb/ft}^2 \end{aligned}$$

The total live load is

$$\begin{aligned} P_L &= LA_T \\ &= 35 \times 3 \times 900/1000 \\ &= 94.50 \text{ kips} \end{aligned}$$

## Reduction in Roof Live Load

Roof live loads account for loads imposed during construction of the roof and subsequently, by maintenance and re-roofing operations. Roof loads are assumed to act on the horizontal projection of the roof surface. Snow, rain, and wind loads are considered on an individual basis. Roof live loads are specified in ASCE 7 Table 4-1 and the normal value is $L_o = 20$ lb/ft². Roofs used for roof gardens have a specified value of 100 lb/ft² and roofs used for promenade purposes have a specified value of 60 lb/ft².

In the case of roof loads, the reduced design live load depends on the tributary area supported by the member and on the roof slope. For normal roof live loads, the reduced load is given by ASCE 7 Eq. (4.8-1) as

$$\begin{aligned} L_r &= L_o R_1 R_2 \\ &\geq 12 \text{ lb/ft}^2 \\ &\leq 20 \text{ lb/ft}^2 \end{aligned}$$

where $R_1 = 1.0 \ldots$ for $A_T \leq 200$ ft²
$= 1.2 - 0.001A_T \ldots$ for 200 ft² $< A_T <$ 600 ft²
$= 0.6 \ldots$ for $A_T \geq 600$ ft²
$R_2 = 1.0 \ldots$ for $F \leq 4$
$= 1.2 - 0.05F \ldots$ for $4 < F < 12$
$= 0.6 \ldots$ for $F \geq 12$
$F$ = number of inches of rise per foot of a pitched roof

The live load is applied per square foot of horizontal projection of a pitched roof.

Roofs used for roof gardens or promenade purposes may have their loads reduced in accordance with the provisions for floor loads.

**Example 2.6.** Roof Live Load Applied to Beams
The roof framing layout of an office facility is similar to the second-floor layout shown in Fig. 2.1. Dimension $x = 10$ ft and $y = 30$ ft and the roof is nominally flat. Determine the live load acting on a typical interior roof beam 25.

The tributary area of beam 25 is

$$\begin{aligned} A_T &= xy \\ &= 10 \times 30 \\ &= 300 \text{ ft}^2 \\ &> 200 \text{ ft}^2 \\ &< 600 \text{ ft}^2 \end{aligned}$$

The reduction factor is then

$$\begin{aligned} R_1 &= 1.2 - 0.001A_T \\ &= 1.2 - 0.001 \times 300 \\ &= 0.9 \end{aligned}$$

For a flat roof, the rise per foot is

$$F = 0$$

$$R_2 = 1.0$$

The roof live load intensity is given by ASCE 7 Eq. (4.8-1) as

$$\begin{aligned} L_r &= L_o R_1 R_2 \\ &= 20 \times 0.9 \times 1.0 \\ &= 18 \text{ lb/ft}^2 \text{ ... satisfactory} \\ &> 12 \text{ lb/ft}^2 \text{ ... minimum does not govern} \end{aligned}$$

The total live load is

$$\begin{aligned} P_R &= L_r A_T \\ &= 18 \times 300/1000 \\ &= 5.40 \text{ kips} \end{aligned}$$

**Example 2.7.** Roof Live Load Applied to Columns
The roof framing layout of an office facility is similar to the second-floor layout shown in Fig. 2.1. Dimension $x = 10$ ft and $y = 30$ ft and the roof is nominally flat. Determine the roof live load acting on a typical interior column 5.

The area tributary to column 5 is

$$\begin{aligned} A_T &= 3xy \\ &= 3 \times 10 \times 30 \\ &= 900 \text{ ft}^2 \\ &> 600 \text{ ft}^2 \end{aligned}$$

The reduction factor is then

$$R_1 = 0.6$$

For a flat roof, the rise per foot is

$$F = 0$$

$$R_2 = 1.0$$

The roof live load intensity is given by ASCE 7 Eq. (4.8-1) as

$$
\begin{aligned}
L_r &= L_o R_1 R_2 \\
&= 20 \times 0.6 \times 1.0 \\
&= 12 \text{ lb/ft}^2 \text{ ... satisfactory} \\
&\not< 12 \text{ lb/ft}^2 \text{ ... minimum does not govern}
\end{aligned}
$$

The total live load is

$$
\begin{aligned}
P_R &= L_r A_T \\
&= 12 \times 900/1000 \\
&= 10.80 \text{ kips}
\end{aligned}
$$

## Combined Dead and Live Load

In the design of an element, nominal dead loads and live loads must be combined in accordance with the LRFD or ASD requirements. The following example is an illustration.

**Example 2.8.** Total Loads Applied to Columns
The roof framing layout of a four-story office facility is similar to the floor layout shown in Fig. 2.1. Dimension $x = 10$ ft and $y = 30$ ft and the roof is nominally flat. The dead load of the roof, including the weight of framing members is 61 lb/ft$^2$. The column weight is 70 lb/ft in all stories and the height of each story is 12 ft. Determine the total dead load, floor live load, and roof live load acting on a typical interior column 5 and the design load on the column footing.

The area tributary to column 5 is

$$
\begin{aligned}
A_T &= 3xy \\
&= 3 \times 10 \times 30 \\
&= 900 \text{ ft}^2 \\
P_{DR} &= \text{roof dead load} \\
&= 0.061 \times 900 \\
&= 54.90 \text{ kips}
\end{aligned}
$$

The column weight over the total height of the building is

$$
\begin{aligned}
P_C &= w_C \times h \times n \\
&= 0.070 \times 12 \times 4 \\
&= 3.36 \text{ kips}
\end{aligned}
$$

From previous examples

$$
\begin{aligned}
P_R &= \text{roof live load} \\
&= 10.80 \text{ kips} \\
P_{DF} &= \text{dead load of each floor} \\
&= 58.98 \text{ kips} \\
P_L &= \text{total floor live load on column footing} \\
&= 94.50 \text{ kips}
\end{aligned}
$$

Total dead load on the column footing is

$$P_D = P_{DR} + P_C + 3P_{DF}$$
$$= 54.90 + 3.36 + 3 \times 58.98$$
$$= 235.20 \text{ kips}$$

In accordance with ASCE 7 Sec. 2.3.2 Exception 1, the load factor on $P_L = 0.5$ in LRFD combinations 3, 4, and 5. Applying ASCE 7 Secs. 2.3 and 2.4 gives

| LRFD | ASD |
|---|---|
| From ASCE 7 Sec. 2.3.2 combination 3: | From ASCE 7 Sec. 2.4.1 combination 4: |
| $P_u = 1.2P_D + 1.6P_R + 0.5P_L$ | $P_a = P_D + 0.75P_R + 0.75P_L$ |
| $= 1.2 \times 235.2 + 1.6 \times 10.8 + 0.5 \times 94.5$ | $= 235.2 + 0.75 \times 10.8 + 0.75 \times 94.5$ |
| = 347 kips | = 314 kips |
| From ASCE 7 Sec. 2.3.2 combination 2: | From ASCE 7 Sec. 2.4.1 combination 2: |
| $P_u = 1.2P_D + 1.6P_L + 0.5P_R$ | $P_a = P_D + P_L$ |
| $= 1.2 \times 235.2 + 1.6 \times 94.5 + 0.5 \times 10.8$ | $= 235.2 + 94.5$ |
| = 439 kips ... governs | = 330 kips ... governs |
| = required strength | = required strength |

## 2.4 Snow Loads

In accordance with IBC Sec. 1608.1, design snow loads shall be determined by ASCE 7 Chap. 7. Provisions are provided in Chap. 7 for calculating snow loads on flat roofs, sloped roofs, and the effects of sliding snow, ice dams, rain-on-snow, snow drifts, unbalanced loads, and partial loading. As specified in ASCE 7 Sec. C7.3, the live load reductions in ASCE 7 Sec. 4.8 based on tributary areas are not applied to snow loads. The following factors are used in the determination of snow loads:

### Flat Roof

A flat roof is defined in ASCE 7 Sec. 7.1 as a roof with a slope ≤ 5°.

### Ground Snow Load

A value for the ground snow load $p_g$ at any locality may be obtained from ASCE 7 Fig. 7-1. The values given in the figure represent the 2 percent annual probability of being exceeded in a 50-year recurrence interval. In some areas, the ground snow load is too variable to accurately estimate and a site-specific case study is required.

### Flat Roof Snow Load

Less snow accumulates on a roof than on the ground and the design value for a flat roof is calculated from ASCE 7 Eq. (7.3-1) as

$$p_f = 0.7C_eC_tI_sp_g$$

where $C_e$ = exposure factor
$C_t$ = thermal factor
$I_s$ = importance factor

The factor of 0.7 is a basic exposure factor to convert ground snow loads to roof snow loads. A minimum value for the roof snow load is specified in ASCE 7 Sec. 7.3.4 for low-slope roofs. When the value of the ground snow load is 20 lb/ft$^2$ or less, the minimum value of the roof snow load is

$$p_m = I_s p_g$$

When the value of the ground snow load exceeds 20 lb/ft$^2$, the minimum value of the roof snow load is

$$p_m = 20 I_s$$

A monoslope, hip, or gable roof with a slope less than 15° is defined in ASCE 7 Sec. 7.3.4 as a low-slope roof.

### Exposure Factor

Roofs in fully exposed, windswept locations tend to accumulate less snow than roofs in more sheltered areas since wind tends to blow snow off the roof. This is reflected in the values for the exposure factor given in ASCE 7 Table 7-2. The three categories of exposure are described as fully exposed, partially exposed, and sheltered. A sheltered location is defined as a roof surrounded by conifers. An exposed location is defined as a roof without any shelter provided by higher structures or trees. A building or other obstruction within a distance of $10h_o$ is considered to provide shelter, where $h_o$ is the height of the obstruction above the roof level. A partially exposed location is one that does not fall into the category of either fully exposed or sheltered. The exposure factor also depends on the terrain category which is defined in ASCE 7 Sec. 26.7.2 as type B, C, or D. Category B is applicable to urban, suburban, and mixed wooded areas. Category C is applicable to open terrain with scattered obstructions less than 30 ft in height. Category D is applicable to flat, unobstructed areas and to wind blowing over open water.

### Thermal Factor

More snow accumulates on a cold roof than on a warm roof. Values of the thermal factor are given in ASCE 7 Table 7-3 and vary from 0.85 for continuously heated greenhouses to 1.1 for roofs kept just above freezing and 1.3 for structures intentionally kept below freezing. For all other structures, the thermal factor is taken as 1.0.

### Importance Factor

Importance factors are listed in ASCE 7 Table 1.5-2 and are an indication of the degree of protection against failure required for the structure. The importance factor provides enhanced performance for those facilities that constitute a substantial public hazard because of high levels of occupancy or because of the essential nature of their function. The importance factor ensures that these facilities are designed for higher loads so as to reduce possible structural damage. Four risk categories are listed in ASCE 7 Table 1.5-1 and the corresponding importance factors are listed in ASCE 7 Table 1.5-2. Category I structures are low-hazard structures such as agricultural facilities and minor storage buildings and are allocated an importance factor of 0.8. Category III structures are structures with a high occupant load which pose a substantial risk to human life or have a potential to cause a substantial economic impact in the event of failure. These include facilities such as schools, colleges, and health care facilities where large numbers of people assemble, power stations and jails. Category III structures are allocated an importance factor of 1.1. Category IV structures are essential facilities such as hospitals, police stations, fire stations, and emergency centers and these facilities are allocated an importance factor of 1.2. This higher importance factor ensures minimal damage without disruption to the continued operation of the facility. Category II structures consist of all other facilities not listed in categories I, III, and IV such as residential, office, and commercial facilities and are allocated an importance factor of 1.0.

## Rain-on-Snow Surcharge Load

In accordance with ASCE 7 Sec. 7.10, in locations where the ground snow load is 20 lb/ft$^2$ or less, all roofs with a slope in degrees of less than $W/50$ shall have a 5 lb/ft$^2$ rain-on-snow surcharge imposed. This surcharge is not applied in combination with drift, sliding, unbalanced, or partial snow loads.

**Example 2.9.** Rain-on-Snow Surcharge and Minimum Value of the Roof Snow Load
A single-story storage facility, with a maintained temperature of just above freezing, is located in a suburban area of Charleston, Missouri and is considered partially exposed. The monoslope roof slopes at 1 on 48 and is 80 ft wide. Determine the design rain-on-snow surcharge load on the roof and the minimum value of the roof snow load.

Ground snow load for Charleston, Missouri, from ASCE 7 Fig. 7-1 is

$$p_g = 15 \text{ lb/ft}^2$$

$$\begin{aligned}\theta &= \text{roof slope}\\ &= 1.0/48\\ &= 1.19° \text{ ... qualifies as a flat roof}\\ &< W/50\\ &= 80/50\\ &= 1.6°\end{aligned}$$

Since the ground snow load does not exceed 20 lb/ft$^2$ and the slope of the roof is less than $W/50$, a rain-on-snow surcharge of 5 lb/ft$^2$ is required by ASCE 7 Sec. 7.10.

$$\begin{aligned}C_t &= \text{thermal factor}\\ &= 1.1 \text{ ... from ASCE 7 Table 7-3 for a structure kept just above freezing}\\ C_e &= \text{exposure factor}\\ &= 1.0 \text{ ... from ASCE 7 Table 7-2, for terrain category B, partially exposed}\\ I_s &= \text{importance factor}\\ &= 1.0 \text{ ... from ASCE 7 Table 1.5-2, for risk category II}\end{aligned}$$

Since the roof slope is less than 15°, the roof qualifies as a low-slope roof in accordance with ASCE 7 Sec. 7.3.4. The flat roof snow load is calculated from ASCE 7 Eq. (7.3-1) as

$$\begin{aligned}p_f &= 0.7C_eC_tI_sp_g\\ &= 0.7 \times 1.0 \times 1.1 \times 1.0 \times 15\\ &= 11.6 \text{ lb/ft}^2\end{aligned}$$

The total flat roof snow load, including the rain-on-snow surcharge, is

$$p_f = 11.6 + 5.0$$
$$= 16.6 \text{ lb/ft}^2$$

For a low-slope roof with the value of the ground snow load not exceeding 20 lb/ft$^2$, the minimum value of the roof snow load is specified by ASCE 7 Sec. 7.3.4 as

$$p_m = I_s p_g$$
$$= 1.0 \times 15$$
$$= 15 \text{ lb/ft}^2 \text{ ... does not govern}$$
$$< 16.6 \text{ lb/ft}^2$$

Hence, the design flat roof snow load is 16.6 lb/ft$^2$.

## Ice Dams and Icicles along Eaves

The design requirements for ice dams and icicles along eaves are covered in ASCE 7 Sec. 7.4.5. Warm roofs that drain water over their eaves, that are unventilated and have an R-value less than 30 ft$^2$ h °F/Btu, shall be designed for a uniformly distributed load of two times the flat roof snow load on all overhanging portions. A similar surcharge is required for warm roofs that drain water over their eaves, that are ventilated and have an R-value less than 20 ft$^2$ h °F/Btu. The surcharge is applied only to the overhanging portion of the structure, not to the entire building.

**Example 2.10.** Flat Roof Snow Load

A single-story heated commercial structure is located in a suburban area of Madison, Wisconsin and is considered partially exposed. The roof slopes at 1 on 48 and is without overhanging eaves. Determine the design snow load on the roof.

Ground snow load for Madison, Wisconsin, from ASCE 7 Fig. 7-1 is

$$p_g = 30 \text{ lb/ft}^2$$
$$> 20 \text{ lb/ft}^2 \text{ ... rain-on-snow surcharge is not required}$$
$$\theta = \text{roof slope}$$
$$= 1.0/48$$
$$= 1.19°$$
$$< 15° \text{ ... qualifies as a low-slope roof in accordance with ASCE 7 Sec. 7.3.4}$$
$$C_t = \text{thermal factor}$$
$$= 1.0 \text{ ... from ASCE 7 Table 7-3 for a heated structure}$$
$$C_e = \text{exposure factor}$$
$$= 1.0 \text{ ... from ASCE 7 Table 7-2, for terrain category B, partially exposed}$$
$$I_s = \text{importance factor}$$
$$= 1.0 \text{ ... from ASCE 7 Table 1.5-2, for risk category II}$$

The flat roof snow load is calculated from ASCE 7 Eq. (7.3-1) as

$$p_f = 0.7C_eC_tI_sp_g$$
$$= 0.7 \times 1.0 \times 1.0 \times 1.0 \times 30$$
$$= 21 \text{ lb/ft}^2 \text{ ... governs}$$
$$> 20 \times I_s = 20 \text{ lb/ft}^2 \text{ ... } p_f \text{ exceeds minimum specified in ASCE 7 Sec. 7.3.4}$$

Hence, the design flat roof snow load is 21 lb/ft$^2$.

## Snow Drifts on Lower Roofs

Snow drifts accumulate on roofs in the wind shadow of a higher roof or parapet, as described by O'Rourke.[3] Adjacent structures or higher terrain features within 20 ft may also produce snow drifts. Drift size is related to the upwind roof length and ground snow load.

## Leeward Snow Drifts

As shown in Fig. 2.15, wind blowing off a high roof, upwind of a low roof, forms a leeward step drift in the aerodynamic shade region at the change in elevation.

The size of the drift increases as the length of the roof upwind of the drift, or upwind fetch, increases and as the ground snow load increases. Drift loads are superimposed on the existing balanced snow load on the lower roof. As shown in Fig. 2.16, drift loads

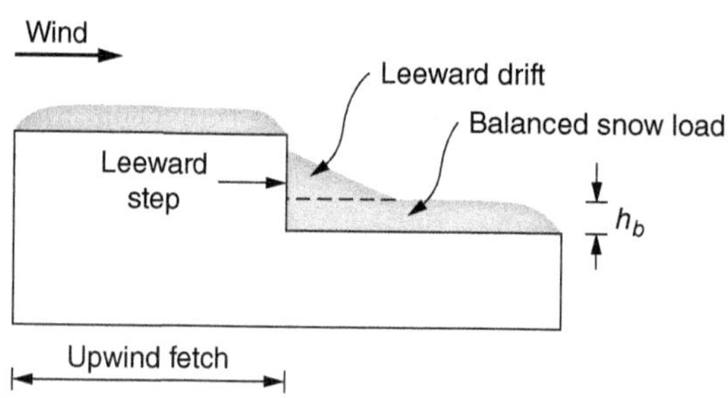


**FIGURE 2.15** Leeward snow drifts.

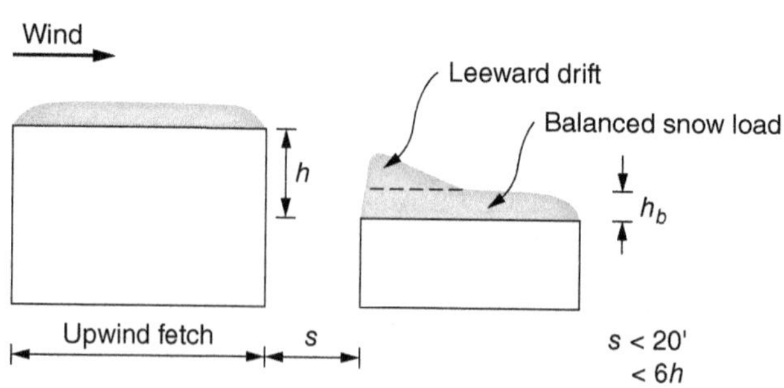


**FIGURE 2.16** Drift formed on adjacent low structure.

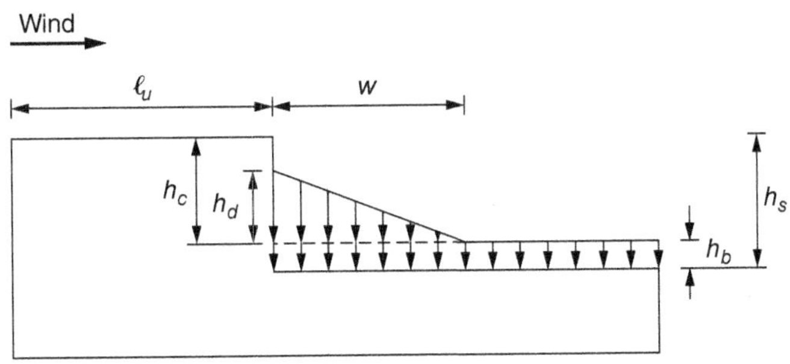


FIGURE 2.17 Configuration of snow drift on lower roof.

caused by a higher adjacent structure must also be considered when the horizontal separation distance is

$$s < 20 \text{ ft}$$
$$< 6h$$

where $h$ is vertical separation distance.

As specified in ASCE 7 Sec. 7.7.1, the geometry of the surcharge load is approximated by a triangle as shown in Fig. 2.17.

Snow density is calculated from ASCE 7 Eq. (7.7-1) as

$$\gamma = 0.13p_g + 14$$
$$\leq 30.0 \text{ lb/ft}^3$$

The height of the flat roof snow load is

$$h_b = p_s/\gamma$$

where $p_s$ = sloped roof balanced snow load
$= C_s p_f$
$C_s$ = slope factor

Clear height from top of flat roof snow load to upper roof level is given by Fig. 2.17 as

$$h_c = h_s - h_b$$

Drift loads need not be considered when

$$h_c/h_b < 0.2$$

The drift height is determined from ASCE 7 Fig. 7-9 as

$$h_d = 0.43(\ell_u)^{1/3}(p_g + 10)^{1/4} - 1.5$$
$$\leq h_c$$

where $\ell_u$ is upwind fetch.

When $h_d \le h_c$ the drift width is

$$w = 4h_d$$
$$\le 8h_c$$

When $h_d > h_c$ the drift width is

$$w = 4h_d^2/h_c$$
$$\le 8h_c$$

and

$$h_d = h_c$$

The maximum intensity of the drift surcharge load at the change in elevation is

$$p_d = h_d\gamma$$

If the drift width, $w$, exceeds the length of the lower roof, the drift shall be truncated at the far edge of the roof and not reduced to zero.

For drift loads caused by a higher adjacent structure, the height of the drift shall be based on the length of the upwind fetch and determined from ASCE 7 Fig. 7-9 with a maximum value of

$$h_d = (6h - s)/6$$

The drift width is

$$w = 6h_d$$
$$< 6h - s$$

For drift loads caused by roof projections, the height of a drift is three-quarters the drift height determined from ASCE 7 Fig. 7-9 and is given by

$$h_d = 0.75[0.43(\ell_u)^{1/3}(p_g + 10)^{1/4} - 1.5]$$

The upwind fetch is the length of roof upwind of the projection. A drift load is not required for projections less than 15 ft long.

**Example 2.11.** Leeward Snow Drift (After Ref. 4)
The single-story heated industrial structure shown in Fig. 2.18*a* is located in a suburban area of Madison, Wisconsin. The roof slopes at 1 on 48 and is stepped as indicated. Determine the design snow load on the low roof.

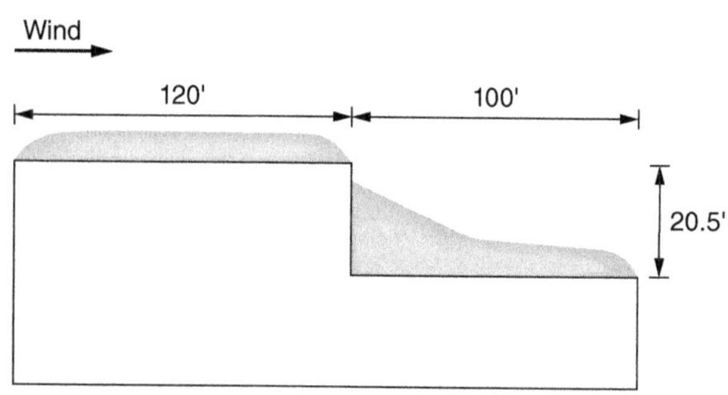


**FIGURE 2.18*a*** Details for Example 2.11.

The relevant parameters determined in Example 2.10 are

$$p_g = 30 \text{ lb/ft}^2$$

$$> 20 \text{ lb/ft}^2 \text{ ... rain-on-snow surcharge is not required}$$

$$\theta = \text{roof slope}$$

$$= 1.19°$$

$$< 15° \text{ ... qualifies as a low-slope roof in accordance with ASCE 7 Sec. 7.3.4}$$

$$p_f = 21 \text{ lb/ft}^2 \text{ ... governs}$$

$$> 20 \times I_s = 20 \text{ lb/ft}^2 \text{ ... } p_f \text{ exceeds minimum specified in ASCE 7 Sec. 7.3.4}$$

Snow density is calculated from ASCE 7 Eq. (7.7-1) as

$$\gamma = 0.13p_g + 14$$

$$= 0.13 \times 30 + 14$$

$$= 17.9 \text{ lb/ft}^3$$

$$< 30.0 \text{ lb/ft}^3 \text{ ... satisfactory}$$

The height of the flat roof snow load is

$$h_b = p_f/\gamma$$

$$= 21/17.9$$

$$= 1.2 \text{ ft}$$

ASCE 7 Sec. 7.7.1 is used to calculate drift loads at changes in elevation. Clear height from top of flat roof snow load to upper roof level is given by ASCE 7 Fig. 7-8 as

$$h_c = h_s - h_b$$

$$= 20.5 - 1.2$$

$$= 19.3 \text{ ft}$$

Since

$$h_c/h_b = 19.3/1.2$$

$$= 16.1$$

$$> 0.2 \text{ ... drift loads must be considered}$$

For wind blowing from the high roof to the low roof, a leeward step drift forms on the low roof at the change in elevation. From ASCE 7 Fig. 7-9 with $\ell_u = 120$ ft and $p_g = 30 \text{ lb/ft}^2$

$$h_d = 0.43(\ell_u)^{1/3}(p_g + 10)^{1/4} - 1.5$$

$$= 0.43(120)^{1/3}(30 + 10)^{1/4} - 1.5$$

$$= 3.8 \text{ ft}$$

$$< h_c = 19.3 \text{ ft ... satisfactory}$$

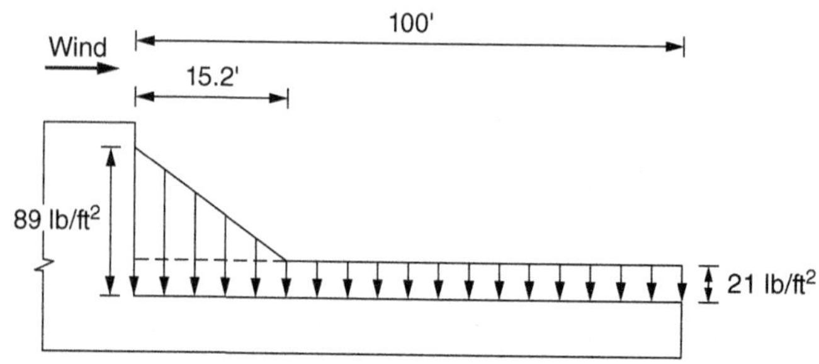


**FIGURE 2.18b** Design snow load on low roof.

Since $h_d < h_c$ the drift width is

$$
\begin{aligned}
w &= 4h_d \\
&= 4 \times 3.8 \\
&= 15.2 \text{ ft} \\
&< 8h_c = 8 \times 19.3 = 154 \text{ ft} \ldots \text{ satisfactory}
\end{aligned}
$$

The maximum intensity of the drift surcharge load at the change in elevation is

$$
\begin{aligned}
p_d &= h_d\gamma \\
&= 3.8 \times 17.9 \\
&= 68.0 \text{ lb/ft}^2
\end{aligned}
$$

The loading diagram for the low roof is shown in Fig. 2.18*b*.

## Windward Snow Drifts

As shown in Fig. 2.19, a windward step drift is formed by wind blowing off a low roof, upwind of a high roof. A windward drift may also form at a parapet, as shown in Fig. 2.19.

The height of a windward drift is three-quarters the height of a leeward drift in accordance with ASCE 7 Sec. 7.7.1 and is given by

$$h_d = 0.75[0.43(\ell_u)^{1/3}(p_g + 10)^{1/4} - 1.5]$$

The larger of the windward and the leeward drifts at a step shall be used in design.

Windward drift loads caused by roof projections and adjacent structures are obtained similarly.

**Example 2.12.** Windward Snow Drift

A single-story heated industrial structure is located in a suburban area of Madison, Wisconsin. The roof slopes at 1 on 48 and is 120 ft long between 5 ft parapets. Determine the design snow load on the roof.

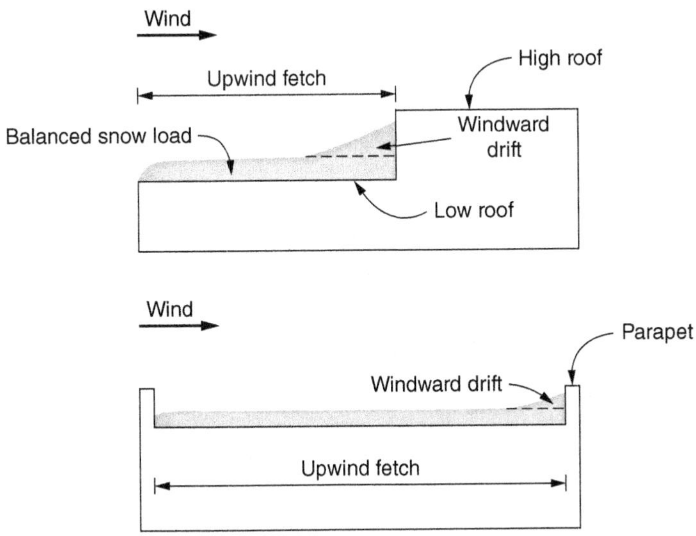


**FIGURE 2.19** Windward snow drifts.

The relevant parameters determined in Examples 2.10 and 2.11 are

$$p_g = 30 \text{ lb/ft}^2$$
$$> 20 \text{ lb/ft}^2 \text{ ... rain-on-snow surcharge is not required}$$
$$\theta = \text{roof slope}$$
$$= 1.19°$$
$$< 15° \text{ ... qualifies as a low-slope roof in accordance with ASCE 7 Sec. 7.3.4}$$
$$p_f = 21 \text{ lb/ft}^2 \text{ ... governs}$$
$$> 20 \times I_s = 20 \text{ lb/ft}^2 \text{ ... } p_f \text{ exceeds minimum specified in ASCE 7 Sec. 7.3.4}$$
$$\gamma = 17.9 \text{ lb/ft}^3$$
$$< 30.0 \text{ lb/ft}^3 \text{ ... satisfactory}$$
$$h_b = 1.2 \text{ ft}$$

For wind blowing along the roof, a windward step drift forms at the parapet at the end of the roof. Clear height from top of flat roof snow load to top of parapet is given by ASCE 7 Fig. 7-8 as

$$h_c = h_p - h_b$$
$$= 5.0 - 1.2$$
$$= 3.8 \text{ ft}$$

Since

$$h_c/h_b = 3.8/1.2$$
$$= 3.2$$
$$> 0.2 \text{ ... drift loads must be considered}$$

From ASCE 7 Fig. 7-9 with $\ell_u = 120$ ft and $p_g = 30$ lb/ft², and taking the height of the windward drift as three-quarters the height of a leeward drift in accordance with ASCE 7 Sec. 7.7.1

$$h_d = 0.75[0.43(\ell_u)^{1/3}(p_g + 10)^{1/4} - 1.5]$$
$$= 0.75[0.43(120)^{1/3}(30 + 10)^{1/4} - 1.5]$$
$$= 2.9 \text{ ft}$$
$$< h_c = 3.8 \text{ ft ... satisfactory}$$

Since $h_d < h_c$ the drift width is

$$w = 4h_d$$
$$= 4 \times 2.9$$
$$= 11.6 \text{ ft}$$
$$< 8h_c = 8 \times 3.8 = 30 \text{ ft ... satisfactory}$$

The maximum intensity of the drift surcharge load at the parapet is

$$p_d = h_d\gamma$$
$$= 2.9 \times 17.9$$
$$= 52 \text{ lb/ft}^2$$

The loading diagram for the roof is shown in Fig. 2.20.

## Sloped Roof Snow Load

Less snow accumulates on a sloped roof than on a flat roof because of the effects of wind action and by shedding of some of the snow by sliding. The design value is calculated from ASCE 7 Eq. (7.4-1) as

$$p_s = C_s p_f$$

where $C_s$ is slope factor.

The snow load on a sloping surface is assumed to act on the horizontal projection of that surface and is designated the balanced snow load.

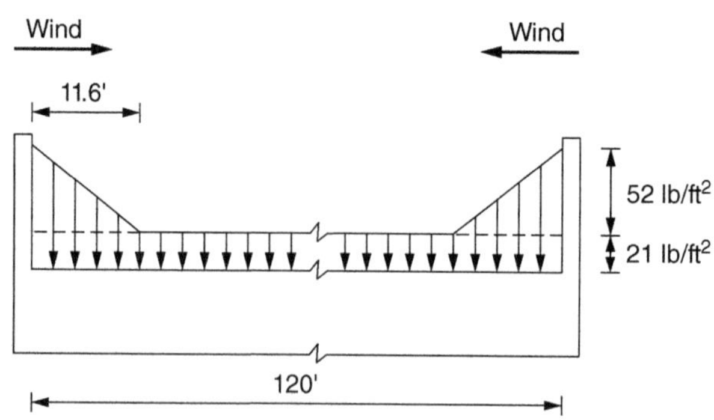


**FIGURE 2.20** Design snow load at parapet.

### Slope Factor

The slope factor depends on the pitch of the roof, the temperature of the roof, the slipperiness of the roof surface, whether or not the surface is obstructed, and whether sufficient space is available below the eaves to accept all of the sliding snow. Snow loads decrease as the roof slope increases and for slopes of 70° or more is assumed to be zero. Roof materials that are considered slippery are defined in ASCE 7 Sec. 7.4 as metal, slate, glass, bituminous, rubber, and plastic membranes with a smooth surface. Membranes with an imbedded aggregate or mineral granule surface are not considered smooth. Asphalt shingles, wood shingles, and shakes are not considered smooth.

### Warm Roof Slope Factor

A warm roof is defined in ASCE 7 Sec. 7.4.1 as a roof with a value for the thermal factor of $C_t \leq 1.0$. For a warm roof with an unobstructed slippery surface, the roof slope factor is determined using the dashed line in ASCE 7 Fig. 7-2*a*, provided that for nonventilated roofs the R-value is not less than 30 ft$^2$ h °F/Btu and that for ventilated roofs the R-value is not less than 20 ft$^2$ h °F/Btu. For warm roofs that do not meet these conditions, the roof slope factor is determined using the solid line in ASCE 7 Fig. 7-2*a*.

### Cold Roof Slope Factor

A cold roof is defined in ASCE 7 Sec. 7.4.2 as a roof with a value for the thermal factor of $C_t > 1.0$. For a cold roof with $C_t = 1.1$ and an unobstructed slippery surface, the roof slope factor is determined using the dashed line in ASCE 7 Fig. 7-2*b*. For all other cold roofs with a thermal factor of $C_t = 1.1$, the roof slope factor is determined using the solid line in ASCE 7 Fig. 7-2*b*.

For a cold roof with $C_t = 1.2$ and an unobstructed slippery surface, the roof slope factor is determined using the dashed line in ASCE 7 Fig. 7-2*c*. For all other cold roofs with a thermal factor of $C_t = 1.2$, the roof slope factor is determined using the solid line in ASCE 7 Fig. 7-2*c*.

**Example 2.13.** Balanced Snow Load
A single-family home is located in a suburban area of Madison, Wisconsin and is considered partially exposed. The building has a wood shingle, gable roof, with a slope of 6 on 12 and an overall width of 40 ft. Determine the balanced snow load on the roof.

Ground snow load for Madison, Wisconsin, from ASCE 7 Fig. 7-1 is

$$p_g = 30 \text{ lb/ft}^2$$

$$> 20 \text{ lb/ft}^2 \text{ ... rain-on-snow surcharge does not apply}$$

$$\theta = \text{roof slope}$$

$$= 6/12$$

$$= 26.6°$$

$$C_t = \text{thermal factor}$$

$$= 1.0 \text{ ... from ASCE 7 Table 7-3 for a heated structure}$$

$$C_e = \text{exposure factor}$$

$$= 1.0 \text{ ... from ASCE 7 Table 7-2, for terrain category B, partially exposed}$$

$$I_s = \text{importance factor}$$

$$= 1.0 \text{ ... from ASCE 7 Table 1.5-2, for risk category II}$$

The flat roof snow load is calculated from ASCE 7 Eq. (7.3-1) as

$$\begin{aligned} p_f &= 0.7C_eC_tI_sp_g \\ &= 0.7 \times 1.0 \times 1.0 \times 1.0 \times 30 \\ &= 21 \text{ lb/ft}^2 \\ \theta &= \text{roof slope} \\ &= 26.6° \\ &> 15° \end{aligned}$$

Hence, in accordance with ASCE 7 Sec. 7.3.4, the minimum flat roof load does not apply and the flat roof snow load is

$$p_f = 21 \text{ lb/ft}^2$$

For a heated roof with a slope of 26.6° and a nonslippery surface, the slope factor is determined from the solid line in ASCE 7 Fig. 7-2*a* as

$$C_s = 1.0$$

The balanced sloped roof snow load is given by ASCE 7 Eq. (7.4-1) as

$$\begin{aligned} p_s &= C_sp_f \\ &= 1.0 \times 21 \\ &= 21 \text{ lb/ft}^2 \end{aligned}$$

The balanced snow load is shown in Fig. 2.21.

## Unbalanced Snow Load for Hip and Gable Roofs

Unbalanced snow loads develop on gable roofs due to the effects of wind, as described by O'Rourke et al.[5] Wind blowing over the roof removes snow from the windward side

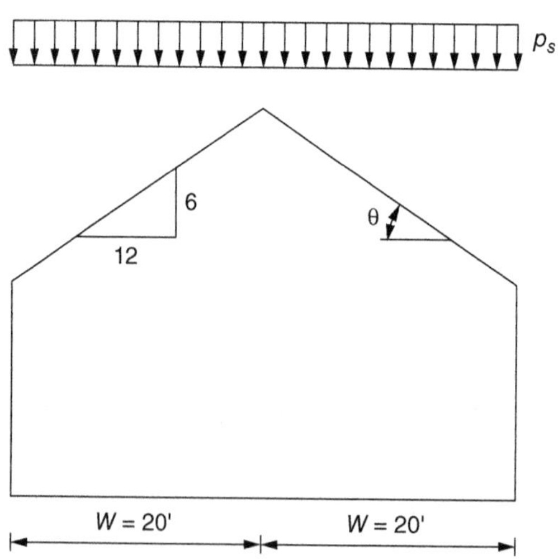


**FIGURE 2.21** Balanced snow load on sloped roof.

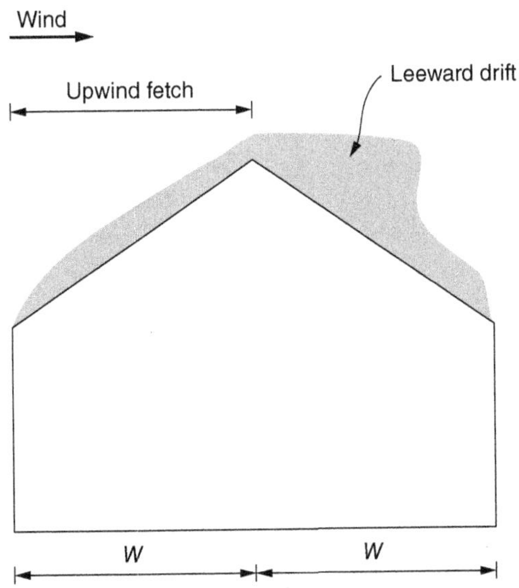


**FIGURE 2.22** Gable roof drift.

of the gable and deposits it in the aerodynamic shade region on the leeward side. The drift formed on the leeward side of the gable is nominally triangular in shape with a roughly horizontal upper surface as shown in Fig. 2.22.

The size of the drift is comparable to a leeward step drift with an upwind fetch equal to the length of the slope on the windward side of the gable. As specified in ASCE 7 Sec. 7.6.1, analysis for unbalanced snow loads is not required for gable roof slopes exceeding 30.2° or with a slope less than 2.38°. In accordance with ASCE 7 Sec. 7.6, balanced and unbalanced loading conditions shall be analyzed separately to determine the most critical situation. As indicated in Ref. 5, the governing load on exterior walls for values of *W* exceeding 20 ft is produced by the balanced snow condition. For values of *W* not exceeding 20 ft, the unbalanced snow condition governs.

## Unbalanced Snow Load for Gable Roof with $W \leq 20$ ft

For simple structural systems, a simplified representation of the unbalanced loading is used, as detailed in ASCE 7 Sec. 7.6.1. With a simply supported rafter spanning from eave to ridge and with a horizontal eave to ridge distance not exceeding 20 ft, a uniformly distributed load of $I_s p_g$ is applied to the leeward slope. The windward slope is considered to be unloaded, as shown in Fig. 2.23.

**Example 2.14.** Unbalanced Snow Load for Residential Roof Rafter System
A single-family home is located in a suburban area of Madison, Wisconsin. The building has a wood shingle, gable roof, with a slope of 6 on 12 and an overall width of 40 ft. The roof rafters are simply supported at the eave and the ridge. Determine the unbalanced snow load on the roof.

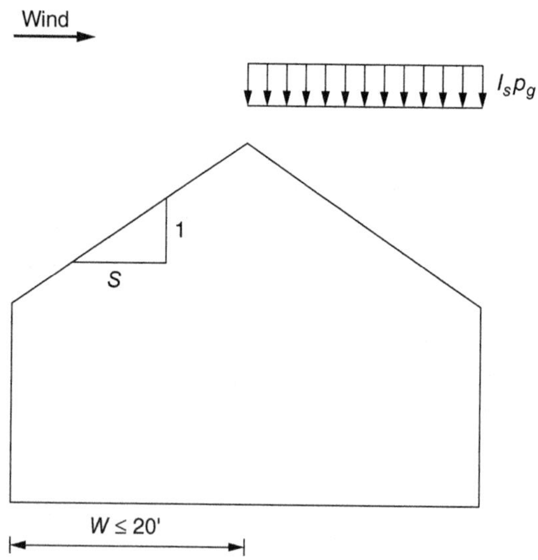


**FIGURE 2.23** Unbalanced snow load for a roof rafter system.

The relevant parameters determined in Example 2.13 are

$$
\begin{aligned}
p_g &= \text{ground snow load} \\
&= 30 \text{ lb/ft}^2 \\
&> 20 \text{ lb/ft}^2 \text{ ... rain-on-snow surcharge does not apply} \\
\theta &= \text{roof slope} \\
&= 26.6^\circ \\
&> 2.38^\circ \\
&< 30.2^\circ
\end{aligned}
$$

Hence, analysis for unbalanced snow loads is required.

$$
\begin{aligned}
I_s &= \text{importance factor} \\
&= 1.0 \\
p_L &= \text{load on leeward slope} \\
&= I_s p_g \text{ ... from ASCE 7 Sec. 7.6.1} \\
&= 1.0 \times 30 \\
&= 30 \text{ lb/ft}^2 \\
p_W &= \text{load on windward slope} \\
&= 0 \text{ ... from ASCE 7 Sec. 7.6.1}
\end{aligned}
$$

The unbalanced loading condition is shown in Fig. 2.24.

## Unbalanced Snow Load for Gable Roof with $W$ > 20 ft

When the horizontal eave to ridge distance exceeds 20 ft or nonprismatic roof members are used, the unbalanced loading is specified in ASCE 7 Sec. 7.6.1 and shown in Fig. 2.25.

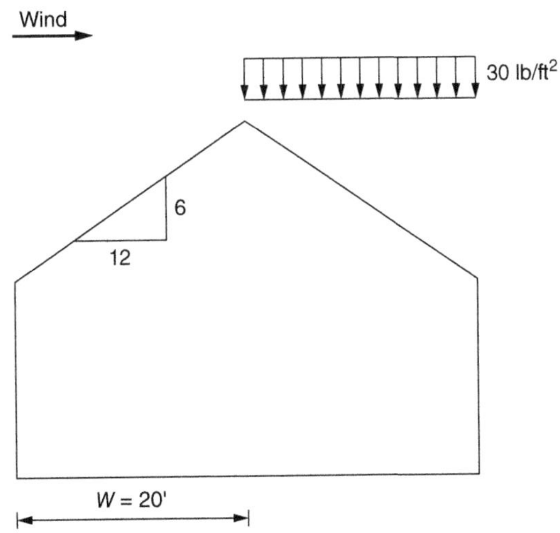


**Figure 2.24** Unbalanced snow load for a simple roof rafter system.

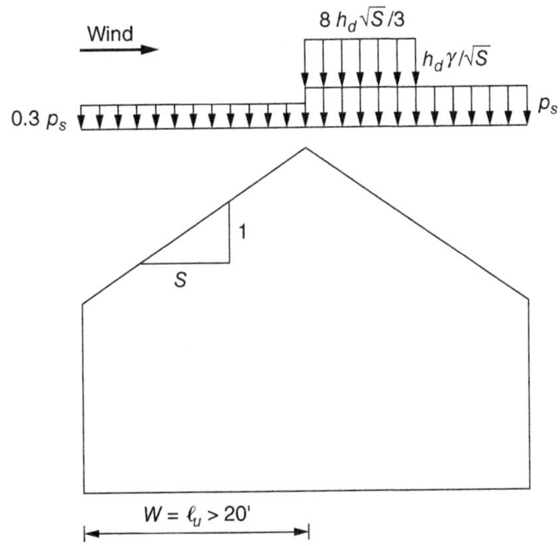


**Figure 2.25** Unbalanced snow load for a roof with $W > 20$ ft.

For this situation, the load on the windward slope is $0.3p_s$. The load on the leeward side consists of a distributed load of $p_s$ plus a rectangular surcharge of $h_d\gamma/(S)^{1/2}$ that extends a horizontal distance of $8(S)^{1/2}\,h_d/3$ from the ridge where

$h_d$ = height of snow drift determined for a value of $\ell_u = W$

$\ell_u$ = length of upwind fetch

$= W$

$W$ = horizontal distance, in ft, from eave to ridge of windward slope of roof

$S$ = roof slope run for a rise of one

$\gamma$ = snow density

**Example 2.15.** Unbalanced Snow Load for Gable Roof with $W > 20$ ft
An office building is located in a suburban area of Madison, Wisconsin. The building has a wood shingle, gable roof, with a slope of 6 on 12 and an overall width of 60 ft. Determine the unbalanced snow load on the roof.

The relevant parameters determined in Example 2.13 are

$$p_g = \text{ground snow load}$$
$$= 30 \text{ lb/ft}^2$$
$$> 20 \text{ lb/ft}^2 \text{ ... rain-on-snow surcharge does not apply}$$
$$\theta = \text{roof slope}$$
$$= 26.6°$$
$$> 2.38°$$
$$< 30.2°$$

Hence, analysis for unbalanced snow loads is required.

$$p_s = \text{balanced sloped roof snow load}$$
$$= 21 \text{ lb/ft}^2$$
$$p_W = \text{load on windward slope}$$
$$= 0.3p_s \text{ ... from ASCE 7 Sec. 7.6.1}$$
$$= 0.3 \times 21$$
$$= 6.3 \text{ lb/ft}^2$$
$$p_{LD} = \text{distributed load on leeward slope}$$
$$= p_s \text{ ... from ASCE 7 Sec. 7.6.1}$$
$$= 21 \text{ lb/ft}^2$$
$$\ell_u = \text{length of upwind fetch}$$
$$= W$$
$$= 30 \text{ ft}$$
$$h_d = \text{height of leeward step drift}$$
$$= 0.43(\ell_u)^{1/3}(p_g + 10)^{1/4} - 1.5 \text{ ... from ASCE 7 Fig. 7-9}$$
$$= 0.43(30)^{1/3}(30 + 10)^{1/4} - 1.5$$
$$= 1.86 \text{ ft}$$
$$S = \text{roof slope run for a rise of one}$$
$$= 12/6$$
$$= 2.0$$
$$\gamma = \text{snow density}$$
$$= 0.13p_g + 14 \text{ ... from ASCE 7 Eq. (7-3)}$$
$$= 0.13 \times 30 + 14$$
$$= 17.9 \text{ lb/ft}^3$$
$$< 30.0 \text{ lb/ft}^3 \text{ ... satisfactory}$$

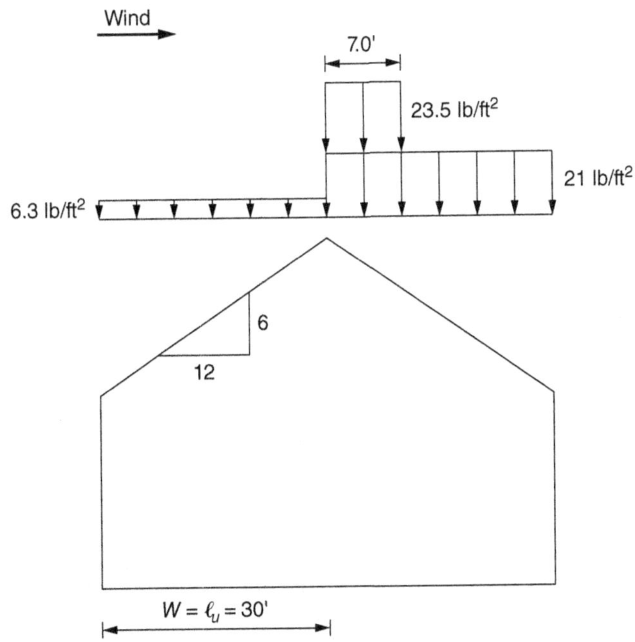


**FIGURE 2.26** Unbalanced snow load for a gable roof with $W > 20$ ft.

$$\begin{aligned} p_{LS} &= \text{surcharge load on leeward slope} \\ &= h_d\gamma/(S)^{1/2} \\ &= 1.86 \times 17.9/(2.0)^{1/2} \\ &= 23.5\ \text{lb/ft}^2 \end{aligned}$$

$$\begin{aligned} d_S &= \text{horizontal extent of the surcharge} \\ &= 8(S)^{1/2}h_d/3 \\ &= 8 \times (2.0)^{1/2} \times 1.86/3 \\ &= 7.0\ \text{ft} \end{aligned}$$

The unbalanced loading condition is shown in Fig. 2.26.

## Sliding Snow

The load caused by snow sliding off a sloped roof onto a lower roof is specified by ASCE 7 Sec. 7.9. As shown in Fig. 2.27, the sliding snow is uniformly distributed over a distance of 15 ft on the lower roof. The intensity of loading is

$$p_{sliding} = 0.4p_f W/15$$

where $p_f$ is flat roof snow load for the sloping roof and $W$ is horizontal distance from eave to ridge of the sloping roof.

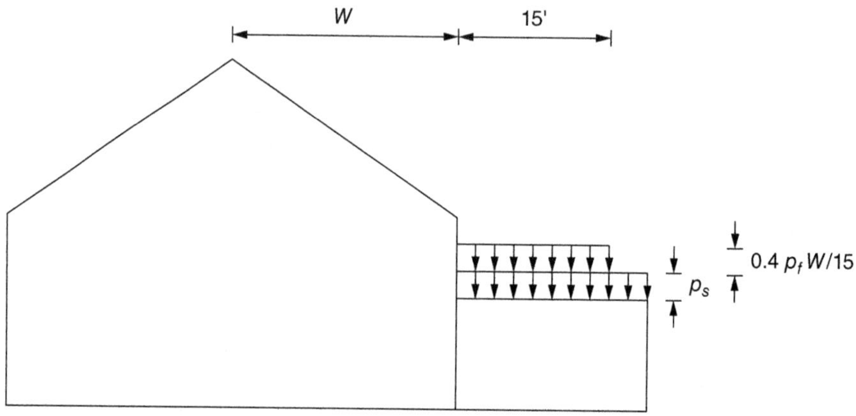


**FIGURE 2.27** Sliding snow load.

Sliding snow loads are superimposed on the existing balanced snow load on the lower roof. Where the lower roof is less than 15 ft wide, the surcharge extends over the full width of the lower roof. It is assumed that sliding will occur on slippery roofs with a slope greater than 1/4 on 12 and on nonslippery slopes with a slope greater than 2 on 12.

**Example 2.16.** Sliding Snow Load on Garage Roof
A heated single-family home is located in a suburban area of Madison, Wisconsin. The building has a wood shingle, gable roof, with a slope of 6 on 12 and an overall width of 60 ft. An unheated garage with a roof slope of 1 on 48 is adjacent to the home, as shown in Fig. 2.28*a*. Determine the snow load on the garage roof.

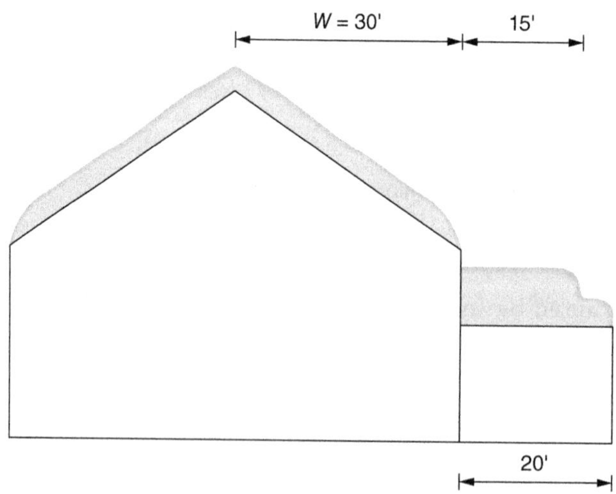


**FIGURE 2.28*a*** Details for Example 2.16.

The relevant parameters for the sloping roof are determined in Example 2.13 and are

$$
\begin{aligned}
p_g &= \text{ground snow load} \\
&= 30 \text{ lb/ft}^2 \\
&> 20 \text{ lb/ft}^2 \ldots \text{ rain-on-snow surcharge does not apply} \\
p_f &= \text{flat roof snow load} \\
&= 21 \text{ lb/ft}^2 \\
\theta &= \text{roof slope} \\
&= 26.6^\circ \\
&> 2.38^\circ \\
&< 30.2^\circ
\end{aligned}
$$

Hence, in accordance with ASCE 7 Sec. 7.3.4, the minimum flat roof load does not apply and the flat roof snow load is

$$p_f = 21 \text{ lb/ft}^2$$

The sloping roof is classified as a nonslippery slope in ASCE 7 Sec. 7.4 and has a slope greater than 2 on 12. Hence, it is necessary to consider snow sliding off the sloped roof onto the garage roof. For the unheated garage roof

$$
\begin{aligned}
\theta &= \text{roof slope} \\
&= 1.0/48 \\
&= 1.19^\circ \\
&< 15^\circ \ldots \text{ qualifies as a low-slope roof in accordance with ASCE 7 Sec. 7.3.4} \\
C_t &= \text{thermal factor} \\
&= 1.2 \ldots \text{ from ASCE 7 Table 7-3 for an unheated structure} \\
C_e &= \text{exposure factor} \\
&= 1.0 \ldots \text{ from ASCE 7 Table 7-2, for terrain category B, partially exposed} \\
C_s &= \text{slope factor} \\
&= 1.0 \ldots \text{ from ASCE 7 Fig. 7-2, for a low-slope roof} \\
I_s &= \text{importance factor} \\
&= 0.8 \ldots \text{ from ASCE 7 Table 1.5-2 for risk category I}
\end{aligned}
$$

The balanced roof snow load is calculated from ASCE 7 Eq. (7.3-1) as

$$
\begin{aligned}
p_s &= 0.7C_sC_eC_tI_sp_g \\
&= 0.7 \times 1.0 \times 1.0 \times 1.2 \times 0.8 \times 30 \\
&= 20.2 \text{ lb/ft}^2
\end{aligned}
$$

The surcharge caused by snow sliding from the sloped roof is

$$
\begin{aligned}
p_{sliding} &= 0.4p_fW/15 \\
&= 0.4 \times 21 \times 30/15 \\
&= 16.8 \text{ lb/ft}^2
\end{aligned}
$$

The loading condition on the garage roof is shown in Fig. 2.28*b*.

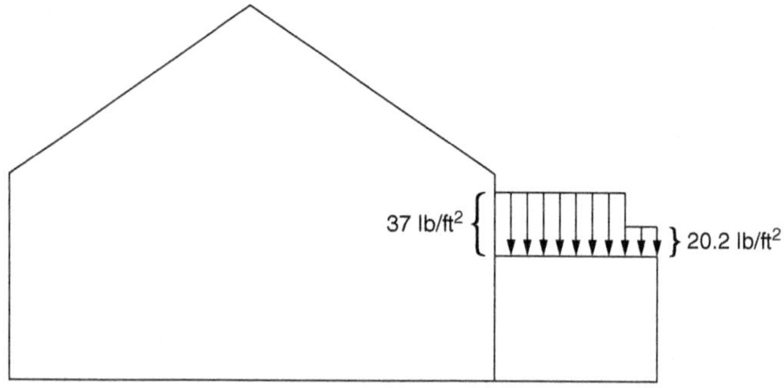


**FIGURE 2.28*b*** Sliding snow load on garage roof.

## Snow Load on Continuous Beam Systems

Partial loading on some spans of a continuous beam system may produce more critical stresses in the members than full loading on all spans. The three partial loading conditions that require consideration are detailed in ASCE 7 Sec. 7.5.1 and shown in Fig. 2.29. The three loading conditions are

1. Full balanced snow load on either exterior span and half the balanced snow load on all other spans.
2. Half balanced snow load on either exterior span and full balanced snow load on all other spans.
3. All possible combinations of full balanced snow load on any two adjacent spans and half the balanced snow load on all other spans.

In all of these loading conditions, the end span may be replaced by a cantilever.

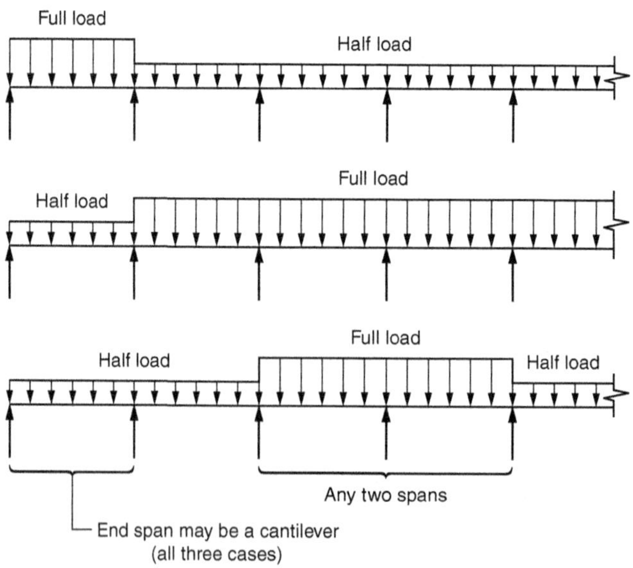


**FIGURE 2.29** Partial loading conditions for continuous beams.

Partial loading conditions need not be considered for members spanning perpendicular to the ridge in gable roofs where the slope is 2.38° or greater.

## 2.5 Soil Lateral Load

Lateral loads caused by soil pressure are addressed in ASCE 7 Sec. 3.2 and in IBC Sec. 1610. Basement and foundation walls are designed for the lateral pressure of adjacent soil. Soil loads may be specified in a soil investigation report. Alternatively, design active pressure, for various soils, are given in ASCE 7 Table 3.2-1 and IBC Table 1610.1. When adjacent soil is below the free-water table, design must be based on the submerged soil pressure plus hydrostatic pressure. Allowance is also required for the additional lateral pressure produced by surcharge loads adjacent to the basement wall. Design lateral pressure must also be increased if soil with expansion potential is present at the site.

### Earth Pressure at Rest

Basement walls free to move and rotate at the top are designed for active earth pressure. Basement walls braced by floors that restrict movement at the top must be designed for at-rest earth pressure. For sand and gravel type soils, this amounts to 60 lb/ft$^2$/ft. For silt and clay type soils, the equivalent value is 100 lb/ft$^2$/ft. In accordance with IBC Sec. 1610.1, basement walls extending not more than 8 ft below grade and supporting flexible floor systems may be designed for active pressure.

## 2.6 Flood Loads

Loads caused by flood conditions are covered in ASCE 7 Chap. 5 and in IBC Sec. 1612. Buildings constructed in flood hazard zones must be built to resist the effects of flood hazards and flood loads. A flood hazard zone is defined as an area subject to a 1 percent or greater chance of flooding in any given year, or as an otherwise legally designated area.

### Loads during Flooding

Water loads are the loads or pressures on surfaces of structures caused by the presence of flood waters. These loads are either hydrostatic or hydrodynamic in nature. Wave loads are a special type of hydrodynamic load, and debris transported by flood waters produce impact loads on striking a structure.

### Hydrostatic Loads

Hydrostatic loads are caused by a depth of water to the level of the design flood elevation. The water is either stagnant or moves at a velocity less than 5 ft/s. The resulting pressure is applied over all surfaces involved, both above and below ground level. The design load is equal to the product of the water pressure multiplied by the surface area on which it acts.

### Hydrodynamic Loads

Hydrodynamic loads are loads induced by the flow of water moving above ground level. These are lateral loads caused by the impact of the moving water on the surface area of a structure. Where water velocity does not exceed 10 ft/s, dynamic effects may be converted into equivalent hydrostatic loads by increasing the design flood elevation by an equivalent surcharge depth given by ASCE 7 Eq. (5.4-1) of

$$d_h = aV^2/2g$$

where $a$ = shape factor or drag coefficient
$V$ = average velocity of water, in ft/s
$g$ = acceleration due to gravity, 32.2 ft/s$^2$

### Wave Loads

Wave loads result from the water waves propagating over the water surface and striking the structure. Design procedures are described in ASCE 7 Sec. 5.4.4.

### Impact Loads

Impact loads result from debris transported by flood waters striking a structure. Design procedures are discussed in ASCE 7 Sec. C5.4.5.

## 2.7 Rain Loads

Design rain loads are addressed in ASCE 7 Chap. 8 and in IBC Sec. 1611. A primary drainage system is provided to cope with the anticipated rainfall intensity produced during short, intense rainfall events. In the event that the primary drainage system becomes blocked by debris or ice, a secondary overflow drain must be provided with a capacity not less than that of the primary system. A free discharge system is the preferred method of emergency drainage.

### Design Rain Loads

The roof must be capable of resisting the maximum water depth that occurs when the primary drainage system is blocked. As shown in Fig. 2.30, the maximum depth is the sum of the static head developed at the inlet of the overflow drain plus the hydraulic head that develops above the inlet at its design flow.

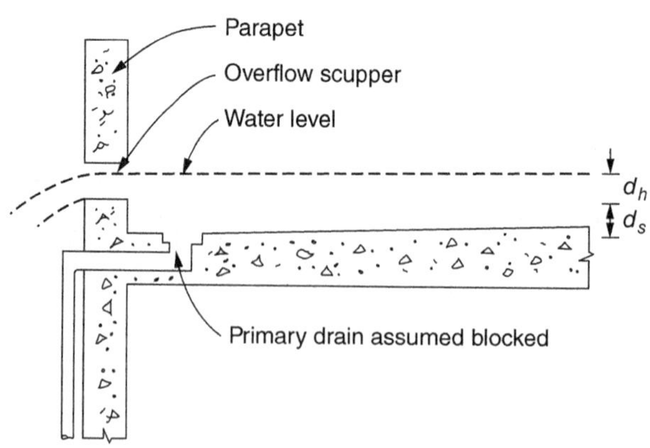


**Figure 2.30** Roof drainage.

The rain load on the roof, in lb/ft², is given by ASCE 7 Eq. (8.3-1) as

$$R = 5.2(d_s + d_h)$$

where $d_s$ = depth of water up to the inlet of the overflow drain when the primary system is blocked, in inches
$d_h$ = additional depth of water, or hydraulic head, above the inlet of the overflow drain at its design flow, in inches
5.2 = lb/ft² per inch of water

In applying this equation, it is assumed that ponding has not occurred and the roof is undeflected. The hydraulic head may be determined from ASCE 7 Table C8-1 which relates hydraulic head to flow rate for several types of overflow drain. The required flow rate is given by ASCE 7 Eq. (C8-1) as

$$Q = 0.0104Ai$$

where $A$ is area of roof being drained and $i$ is design rainfall intensity as specified by the local jurisdiction.

**Example 2.17.** Design Rain Loads
A 2000 ft² roof with a pitch of 1/4 in/ft is enclosed by parapet walls and is provided with a single overflow scupper in the parapet wall. The closed scupper is 6 in wide and 4 in high and is set 2 in above the roof surface. The specified design rainfall intensity is 2 in/h. Determine the design rain load.

The required flow rate for the overflow scupper is given by ASCE 7 Eq. (C8-1) as

$$\begin{aligned} Q &= 0.0104Ai \\ &= 0.0104 \times 2000 \times 2 \\ &= 41.6 \text{ gal/min} \end{aligned}$$

The hydraulic head is obtained from ASCE 7 Table C8-1 as

$$d_h = 1.74 \text{ in}$$

The rain load on the roof is given by ASCE 7 Eq. (8.3-1) as

$$\begin{aligned} R &= 5.2(d_s + d_h) \\ &= 5.2(2 + 1.74) \\ &= 19.4 \text{ lb/ft}^2 \end{aligned}$$

### Ponding Instability

Ponding is defined in ASCE 7 Sec. 8.4 as the retention of water due solely to the deflection of relatively flat roofs. Water may accumulate as ponds on roofs with a slope of less than 1/4 in/ft. As additional water flows into the area, the roof deflects more, allowing a deeper pond to form. The roof must possess adequate stiffness to resist this progression and prevent overloading.

## 2.8 Wind Loads

When wind strikes an enclosed building the wind flows around the sides and over the roof and either a pressure or a suction is produced on the external surfaces of the building. As shown in Fig. 2.31, the windward wall that is perpendicular to the wind direction

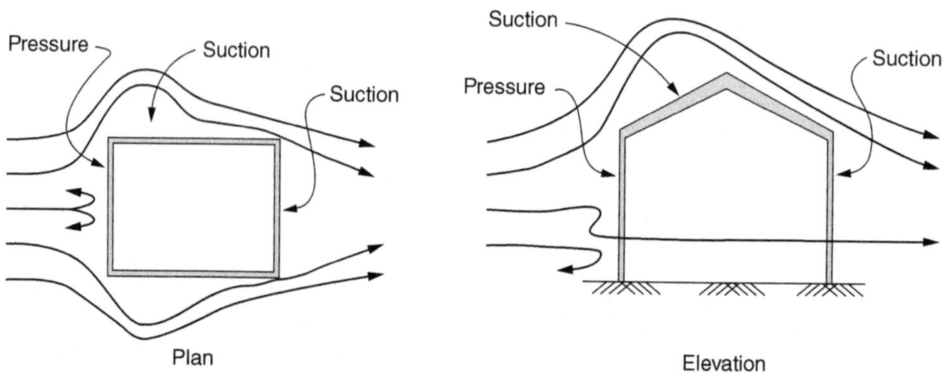


FIGURE 2.31 Wind pressure.

experiences an inward, positive pressure. As wind flows round the corners of the windward wall, the turbulence produced separates the air flow from the walls and causes an outward, negative pressure or suction on the side walls and the leeward wall. As wind flows over a high-sloping gable roof, a positive pressure is produced on the windward side of the ridge and a suction on the leeward side of the ridge. However, for gable roofs with shallow slopes, suction also develops on the windward side of the ridge and for flat roofs, suction develops over the whole roof.

Procedures are provided in the ASCE Standard for determining pressures on the main wind-force resisting system (MWFRS) and on components and cladding. The main wind-force resisting system is defined in ASCE 7 Sec. 26.2 as the structural elements that transfer wind loads to the ground and provide support and stability for the whole structure. Components and cladding are defined as elements of the building envelope that do not qualify as part of the main wind-force resisting system. The cladding of a building receives wind loading directly. Examples of cladding include wall and roof sheathing, windows, and doors. Components receive wind loading from the cladding and transfer the load to the main wind-force resisting system. Components include purlins, studs, girts, fasteners, and roof trusses. Some elements, such as roof trusses and sheathing may also form part of the main wind-force resisting system and must be designed for both conditions. Because of local turbulence, which may occur over small areas at ridges and corners of buildings, components and cladding are designed for higher wind pressures than the main wind force resisting system.

The design procedures consist of two basic approaches:

- The directional procedure determines the wind loads on buildings for specific wind directions, in which the external pressure coefficients are based on wind tunnel testing of prototypical building models for the corresponding direction of wind.
- The envelope procedure determines the wind load cases on buildings, in which pseudo external pressure coefficients are derived from wind tunnel testing of prototypical building models successively rotated through 360°, such that the pseudo pressure cases produce key structural actions (uplift, horizontal shear, bending moments, etc.) that envelope their maximum values among all possible wind directions.

In accordance with IBC Sec. 1609.1.1, wind loads on buildings shall be determined by the provisions of ASCE 7 or by the alternate method of IBC Sec. 1609.6. Five different procedures for determining wind loads on the main wind-force resisting system of buildings are specified in ASCE 7 and these are

- The analytical directional design method of ASCE 7 Chap. 27 Part 1 Sec. 27.4. This is applicable to enclosed, partially enclosed, and open buildings of all heights and roof geometry. Wind pressure is calculated using specific wind pressure equations applicable to each building surface. The method uses the directional procedure to separate applied wind loads onto the windward, leeward, and side walls of the building to correctly assess the forces in the members.
- The simplified method of ASCE 7 Chap. 27 Part 2 Sec. 27.5. This is based on the analytical method of Chap. 27 Part 1 and is applicable to enclosed, simple diaphragm buildings of any roof geometry with a height not exceeding 160 ft. Wind pressures are obtained directly from a table.
- The envelope design method of ASCE 7 Chap. 28 Part 1 Sec. 28.4. This is applicable to enclosed, partially enclosed, and open low-rise buildings having a flat, gable, or hip roof with a height not exceeding 60 ft. Wind pressure is calculated using specific wind pressure equations applicable to each building surface. The method uses the envelope procedure to separate applied wind loads onto the windward, leeward, and side walls of the building to correctly assess the forces in the members.
- The simplified method of ASCE 7 Chap. 28 Part 2 Sec. 28.6. This is based on the envelope procedure of Chap. 28 Part 1 and is applicable to enclosed, simple diaphragm low-rise buildings having a flat, gable, or hip roof with a height not exceeding 60 ft. Wind pressures are obtained directly from a table and applied to vertical and horizontal projected surfaces of the building.
- The wind tunnel procedure of ASCE 7 Chap. 31, that may be used for any structure. This is a procedure for determining wind loads on buildings and other structures, in which pressures, forces, and moments may be determined for each wind direction considered, from a model of the building or other structure and its surroundings. The wind tunnel procedure must be used when the limiting conditions of the previous methods are not satisfied.

In order to apply these methods, a number of prerequisites must be determined. These include exposure category, wind speed, low-rise building designation, velocity pressure exposure coefficient, site topography, wind direction, and velocity pressure.

## Exposure Category

Exposure category accounts for the effect of terrain roughness on wind speed and is defined and illustrated in ASCE-7 Sec. C26.7. The exposure category is dependent on surface roughness category and the upwind fetch distance. The exposure category assigned to each surface roughness category is listed in Table 2.2.

## Basic Wind Speed

Wind speed *V* is determined from the wind speed maps ASCE 7 Fig. 26.5-1A, B, and C. The values given are based on the 3-second-gust wind speed, in miles per hour, adjusted

| Exposure Category | Surface Roughness Category |
|---|---|
| B | *Roughness category B:* Applicable to urban, suburban, and wooded areas with numerous closely spaced obstructions the size of single-family dwellings or larger. The minimum specified upwind fetch distance is (i) the greater of 2600 ft or 20$h$, for $h > 30$ ft or (ii) 1500 ft, for $h \leq 30$ ft. |
| C | *Roughness category C:* Applicable to open terrain with scattered obstructions having heights generally less than 30 ft. This category includes flat open country and grasslands. Exposure C applies for all cases where exposures B or D do not apply. |
| D | *Roughness category D:* Applicable to flat, unobstructed areas and water surfaces. This category includes smooth mud flats, salt flats, and unbroken ice. The minimum specified upwind fetch distance is (i) surface roughness category D the greater of 5000 ft or 20$h$ or (ii) surface roughness category B or C for a distance $d_2$ followed by surface roughness category D for a distance $d_1$. |

Note: $h$ = building height, $d_1$ = greater of 5000 ft or 20$h$, $d_2$ = greater of 600 ft or 20$h$.

**Table 2.2** Exposure Categories

to a reference height of 33 ft and for exposure category C. Drag effects retard wind flow close to the ground and wind speed increases with height above ground level until the gradient height is reached and the speed becomes constant. The gradient heights $z_g$ for different exposure conditions are given in ASCE 7 Table 26.9-1.

The wind speed is given at the strength level value. This differs from previous editions of the ASCE Standard where service level values were used. ASCE 7-05 design wind speeds are multiplied by $(1.6)^{0.5}$ to convert to ASCE 7-10 design wind speeds. ASCE 7 Table C26.5-6 provides a listing of design wind speeds and their converted values.

Also, previous editions of the ASCE Standard used an importance factor to provide enhanced performance for those facilities assigned to a high risk category. ASCE 7-10 achieves the same objective by using a probabilistic approach with three wind speed maps provided for buildings with different risk categories. An increased return period provides enhanced performance for those facilities that constitute a substantial public hazard because of high levels of occupancy or because of the essential nature of their function. The design wind speed return period for each map is based on the risk category assigned to the building and the importance factor is eliminated. This ensures that high risk facilities are designed for higher loads so as to reduce possible structural damage. The three wind speed maps provided are

- ASCE 7 Fig. 26.5-1A gives basic wind speeds for risk category II buildings and provides a return period of 700 years.
- ASCE 7 Fig. 26.5-1B gives basic wind speeds for risk category III and IV buildings and provides a return period of 1700 years.
- ASCE 7 Fig. 26.5-1C gives basic wind speeds for risk category I buildings and provides a return period of 300 years.

| Risk Category | Nature of Occupancy | Return Period | Wind Speed Map |
|---|---|---|---|
| I | Low hazard structures | 300 | 26.5-1C |
| II | Standard occupancy structures | 700 | 26.5-1A |
| III | Assembly structures | 1700 | 26.5-1B |
| IV | Essential or hazardous structures | 1700 | 26.5-1B |

**TABLE 2.3** Risk Category and Return Period

Details of the different occupancy categories and corresponding risk categories and return periods are given in Table 2.3.

### Low-Rise Building

A low-rise building is defined in ASCE 7 Sec. 26.2 as an enclosed or partially enclosed building that satisfies both the following conditions:

- Mean roof height $h$ is less than or equal to 60 ft.
- Mean roof height $h$ does not exceed least horizontal dimension.

Applying the analytical method to low-rise buildings, requires the use of specific velocity pressure exposure coefficients.

### Regular Building

A regular building is defined in ASCE 7 Sec. 26.2 as a building having no unusual geometrical irregularity in spatial form.

### Simple Diaphragm Building

A simple diaphragm building is defined in ASCE 7 Sec. 26.2 as a building in which both windward and leeward wind loads are transmitted by vertically spanning wall elements through continuous roof and floor diaphragms to the main wind-force resisting system.

### Velocity Pressure Exposure Coefficient

Wind speed increases with height and also as the exposure changes from category B to category D. The velocity pressure exposure coefficient $K_z$ reflects this and values are listed in ASCE 7 Tables 27.3-1 and 28.3-1.

### Site Topography

Structures sited on the upper half of an isolated hill or escarpment experience a significant increase in the wind speed. To account for this, the velocity pressure exposure coefficient is multiplied by the topographic factor $K_{zt}$. The topographic factor is a function of the three criteria:

- Slope of the hill
- Distance of the building from the crest
- Height of the building above the local ground surface

These three criteria are represented by the multipliers $K_1$, $K_2$, and $K_3$ that are tabulated in ASCE 7 Fig. 26.8-1. The topographic factor is given by ASCE 7 Eq. (26.8-1) as

$$K_{zt} = (1 + K_1K_2K_3)^2$$

When no topographic effect is to be considered, the topographic factor is given by

$$K_{zt} = 1.0$$

## Directionality Factor

The directionality factor $K_d$ is obtained from ASCE 7 Table 26.6-1 and for buildings is given as 0.85. The directionality factor accounts for the reduced probability of

- Extreme winds occurring in any specific direction
- The peak pressure coefficient occurring for a specific wind direction

## Velocity Pressure

The basic wind speed is converted to a velocity pressure at height $z$ by ASCE 7 Eq. (28.3-1) which is

$$q_z = 0.00256K_zK_{zt}K_dV^2$$

The constant 0.00256 reflects the mass density of air at a temperature of 59°F and a pressure of 29.92 in of mercury and this value should be used unless sufficient data is available to justify a different value. The velocity pressure varies with the height above ground level since the value of the velocity pressure exposure coefficient also varies with the height above ground level.

For the directional design procedure of ASCE 7 Chap. 27 Part 1 Sec. 27.4 applied to buildings of all heights, the velocity pressure on the windward wall is evaluated at each floor and roof level and increases toward the top of the building. For leeward walls, side walls, and roofs, the velocity pressure is evaluated at mean roof height only and is a constant value over the height of the building.

For the envelope design procedure of ASCE 7 Chap. 28 Part 1 Sec. 28.4 applied to low-rise buildings, the velocity pressure is evaluated at mean roof height for all walls and roof and is a constant value over the height of the building.

**Example 2.18.** Wind Velocity Pressure

The two-story office building, shown in Fig. 2.32, is located in a suburban area with a wind speed $V$ of 115 mi/h. Determine the wind velocity pressure at roof height for the main wind-force resisting system.

The height to minimum width ratio is

$$\begin{aligned} h/L &= 20/32 \\ &= 0.63 \\ &< 1 \end{aligned}$$

The mean roof height is

$$\begin{aligned} h &= 20 \text{ ft} \\ &< 60 \text{ ft} \end{aligned}$$

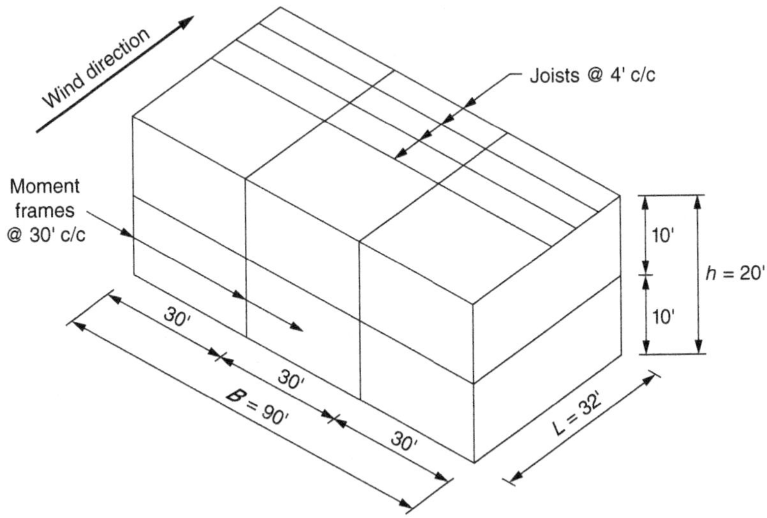


**FIGURE 2.32** Details for Example 2.18.

Hence, the building qualifies as a low-rise building and ASCE 7 Table 28.3-1 values for the velocity pressure exposure coefficients are applicable for a building designed using ASCE 7 Fig. 28.4-1.

For a suburban area, the exposure is category B and the relevant parameters are obtained as

$K_z$ = velocity pressure exposure coefficient
= 0.70 ... from ASCE 7 Table 28.3-1 for a height of 20 ft for the main wind-force resisting system and exposure category B

$K_{zt}$ = topographic factor
= 1.0 ... from ASCE 7 Sec. 26.8.2

$K_d$ = wind directionality factor
= 0.85 ... from ASCE 7 Table 26.6-1

The velocity pressure $q_h$ at the roof height of 20 ft above the ground is given by ASCE 7 Eq. (28.3-1) as

$$\begin{aligned} q_h &= 0.00256 K_z K_{zt} K_d V^2 \\ &= 0.00256 \times 0.70 \times 1.0 \times 0.85 \times 115^2 \\ &= 20.14 \text{ lb/ft}^2 \end{aligned}$$

## ASCE 7 Chap. 28 Part 1—Envelope Procedure

This procedure is outlined in ASCE 7 Sec. 28.4 and is applicable to low-rise buildings that meet the following requirements:

- The structure is a regular-shaped building without irregularities such as projections or indentations.
- The structure does not have response characteristics making it subject to across wind loading, vortex shedding, and instability due to galloping or flutter.
- The structure is not located at a site subject to channeling effects or buffeting in the wake of upwind obstructions.

In order to determine the design wind pressures on a structure, it is necessary to convert the wind velocity pressures to design pressures and the following prerequisites must be determined:

- Rigidity of the structure
- Gust effect factor
- Enclosure classification

## Rigidity of the Structure

A rigid structure is defined in ASCE 7 Sec. 26.2 as a structure with a fundamental frequency greater than or equal to 1 Hz. Most structures, according to ASCE 7 Sec. C26.2, having a height to minimum width ratio less than four qualify as rigid. Where necessary, the fundamental frequency may be determined using the procedures given in ASCE 7 Sec. 26.9.2. A low-rise building is permitted to be considered rigid. A structure with a fundamental frequency less than 1 Hz is considered flexible. A flexible structure exhibits a significant dynamic resonant response to wind gusts.

## Gust Effect Factor

The gust effect factor accounts for along-wind loading effects caused by dynamic amplification in flexible structures and for wind turbulence-structure interaction. For a rigid structure, the gust effect factor may be taken as 0.85. Alternatively, the gust effect factor may be calculated using the procedures given in ASCE 7 Secs. 26.9.4 and 26.9.5.

## Enclosure Classifications

The internal pressure produced in a structure by wind depends on the size and location of openings in the external walls of the structure. As shown in Fig. 2.33, an opening in the windward wall of a structure produces an internal pressure. An opening in the leeward wall of a structure produces an internal suction.

Glazing that is breached by missiles must be treated as openings, as this may result in the development of high internal pressures. In accordance with ASCE 7 Sec. 26.10.3.1, in a wind-borne debris region, glazing in the lower 60 ft of structures shall be assumed to be openings unless such glazing is impact resistant or protected with an impact resistant covering. The same requirement applies to glazing that is less than 30 ft above

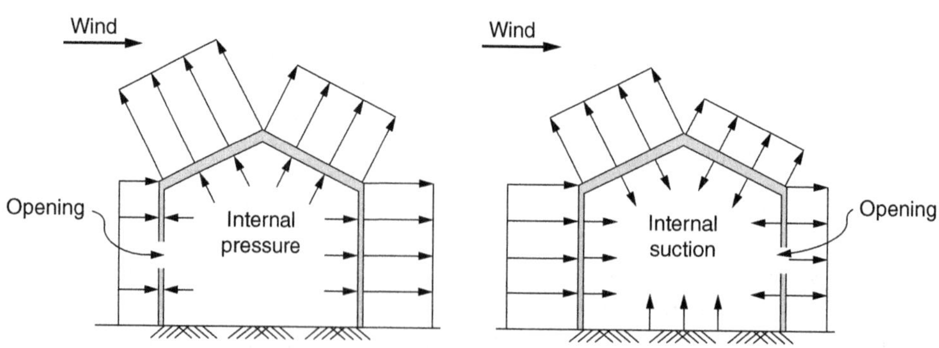


**FIGURE 2.33** Effect of openings on internal pressure.

aggregate surface roofs located within 1500 ft of the structure. A wind-borne debris area is defined in ASCE 7 Sec. 26.10.3.1 as being a hurricane prone area:

- Within 1 mi of the coastal mean high water line where the basic wind speed is equal to or greater than 130 mi/h.
- Within a region where the basic wind speed is not less than 140 mi/h.

An *open building* is defined in ASCE 7 Sec. 26.2 as a building having each wall at least 80 percent open. This is given for each wall by the expression

$$A_o \geq 0.8A_g$$

where $A_o$ is total area of openings in a wall that receives positive external pressure and $A_g$ is the gross area of the wall in which $A_o$ is identified.

A *partially enclosed building* is defined as satisfying both of the following requirements:

- The total area of openings in a wall that receives positive external pressure exceeds the sum of the areas of openings in the balance of the building envelope (walls and roof) by more than 10 percent.
- The total area of openings in a wall that receives positive external pressure exceeds the smaller of 4 ft² or 1 percent of the area of the wall, and the percentage of openings in the balance of the building envelope does not exceed 20 percent.

These requirements are given by the following expressions

$$A_o > 1.10A_{oi}$$

$$A_{oi}/A_{gi} \leq 0.20$$

and either

$$A_o > 0.01A_g$$

or

$$> 4 \text{ ft}^2$$

where $A_{oi}$ is sum of the areas of openings in the building envelope (walls and roof) not including $A_o$ and $A_{gi}$ is sum of the gross surface area of the building envelope (walls and roof) not including $A_g$.

An *enclosed building* is defined as one that does not comply with the requirements for open or partially enclosed buildings.

## Design Wind Pressure on MWFRS for Low-Rise, Rigid Buildings

For the envelope procedure of ASCE 7 Chap. 28 Part 1, the gust effect factor is combined with the external and internal pressure coefficients. The design wind pressure on the main wind-force resisting system is given by ASCE 7 Eq. (28.4-1) as

$$p = q_h[(GC_{pf}) - (GC_{pi})]$$

where $q_h$ = wind velocity pressure at mean roof height $h$ for the applicable exposure category

$(GC_{pf})$ = product of the equivalent external pressure coefficient and gust effect factor as given in ASCE 7 Fig. 28.4-1

$(GC_{pi})$ = product of the internal pressure coefficient and gust effect factor as given in ASCE 7 Table 26.11-1

In accordance with ASCE 7 Sec. 28.4.4, the wind load to be used in design shall not be less than 16 lb/ft$^2$ multiplied by the wall area of the building and 8 lb/ft$^2$ multiplied by the roof area of the building projected on a plane normal to the wind direction as shown in ASCE 7 Fig. C27.4.1.

**Example 2.19.** Design Wind Pressure for Main Wind-Force Resisting System
For the transverse wind direction, determine the design wind pressure acting on the end frames of the two-story office building analyzed in Example 2.18. The building may be considered enclosed and the roof and floor diaphragms are flexible. Consider only load case A.

From Example 2.18, the velocity pressure at mean roof height is obtained as

$$q_h = 20.14 \text{ lb/ft}^2$$

The height to minimum width ratio is

$$h/L = 20/32$$
$$= 0.63$$
$$< 4 \text{ ... rigid structure as defined by ASCE 7 Sec. 26.2}$$

The mean roof height is

$$h = 20 \text{ ft}$$
$$< 60 \text{ ft ... low-rise building as defined by ASCE 7 Sec. 26.2}$$

Hence, the low-rise building analytical method of ASCE 7 Sec. 28.4 is applicable and values of $(GC_{pf})$ may be obtained from ASCE 7 Fig. 28.4-1.

For a two-story building with flexible diaphragms, ASCE 7 Fig. 28.4-1 Note 5 specifies that torsional load cases may be neglected. To design the end frame of the building, the pressures on surfaces 1E, 2E, 3E, and 4E must be determined. For an enclosed building, the product of the internal pressure coefficient and gust effect factor is

$$(GC_{pi}) = \pm 0.18 \text{ ... from ASCE 7 Table 26.11-1}$$

For surface 1E, the product of the equivalent external pressure coefficient and gust effect factor is

$$(GC_{pf}) = 0.61 \text{ ... from ASCE 7 Fig. 28.4-1}$$

The design wind pressure is given by ASCE 7 Eq. (28.4-1) as

$$p = q_h[(GC_{pf}) - (GC_{pi})]$$
$$= 20.14[(0.61) - (\pm 0.18)]$$
$$= 15.91 \text{ lb/ft}^2 \text{ for negative internal pressure (suction); adopt 16 lb/ft}^2 \text{ minimum}$$
$$= 8.66 \text{ lb/ft}^2 \text{ for positive internal pressure; adopt 16 lb/ft}^2 \text{ minimum}$$

For surface 2E, the product of the equivalent external pressure coefficient and gust effect factor is

$$(GC_{pf}) = -1.07 \text{ ... from ASCE 7 Fig. 28.4-1}$$

The design wind pressure is given by ASCE 7 Eq. (28.4-1) as

$$p = q_h[(GC_{pf}) - (GC_{pi})]$$
$$= 20.14[(-1.07) - (\pm 0.18)]$$
$$= -17.93 \text{ lb/ft}^2 \text{ for negative internal pressure (suction)}$$
$$= -25.18 \text{ lb/ft}^2 \text{ for positive internal pressure}$$

For surface 3E, the product of the equivalent external pressure coefficient and gust effect factor is

$$(GC_{pf}) = -0.53 \text{ ... from ASCE 7 Fig 28.4-1}$$

The design wind pressure is given by ASCE 7 Eq. (28.4-1) as

$$\begin{aligned} p &= q_h[(GC_{pf}) - (GC_{pi})] \\ &= 20.14[(-0.53) - (\pm 0.18)] \\ &= -7.05 \text{ lb/ft}^2 \text{ for negative internal pressure (suction); adopt 16 lb/ft}^2 \text{ minimum} \\ &= -14.30 \text{ lb/ft}^2 \text{ for positive internal pressure} \end{aligned}$$

For surface 4E, the product of the equivalent external pressure coefficient and gust effect factor is

$$(GC_{pf}) = -0.43 \text{ ... from ASCE 7 Fig. 28.4-1}$$

The design wind pressure is given by ASCE 7 Eq. (28.4-1) as

$$\begin{aligned} p &= q_h[(GC_{pf}) - (GC_{pi})] \\ &= 20.14[(-0.43) - (\pm 0.18)] \\ &= -5.04 \text{ lb/ft}^2 \text{ for negative internal pressure (suction); adopt 16 lb/ft}^2 \text{ minimum} \\ &= -12.29 \text{ lb/ft}^2 \text{ for positive internal pressure; adopt 16 lb/ft}^2 \text{ minimum} \end{aligned}$$

The wind pressure diagrams for both cases, internal suction and internal pressure, are shown in Fig. 2.34.

## Design Wind Pressure on Components and Cladding

ASCE 7 Chap. 30 provides six separate procedures for the determination of wind pressure on components and cladding. All procedures require compliance with the following conditions:

- The structure is a regular-shaped building without irregularities such as projections or indentations.
- The structure does not have response characteristics making it subject to across wind loading, vortex shedding, and instability due to galloping or flutter.
- The structure is not located at a site subject to channeling effects or buffeting in the wake of upwind obstructions.

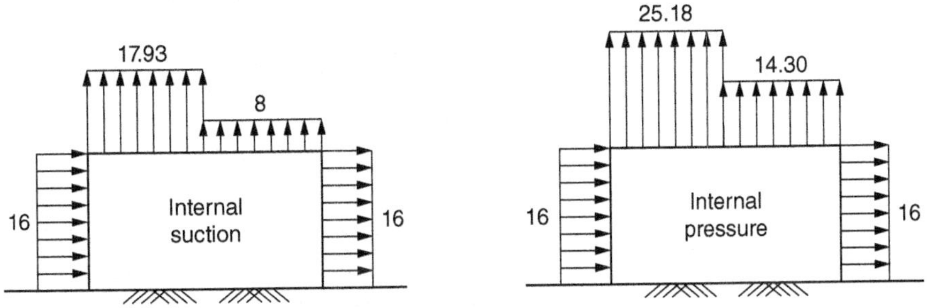


**FIGURE 2.34** Wind pressure diagrams, in lb/ft², for Example 2.19.

The six different procedures are

- The analytical envelope design method of ASCE 7 Chap. 30 Part 1 Sec. 30.4. This is applicable to enclosed and partially enclosed low-rise buildings and buildings with $h \leq 60$ ft having flat roofs, gable roofs, multispan gable roofs, hip roofs, monoslope roofs, stepped roofs, and saw tooth roofs.
- The simplified envelope design method of ASCE 7 Chap. 30 Part 2 Sec. 30.5. This is applicable to enclosed low-rise buildings and buildings with $h \leq 60$ ft having flat roofs, gable roofs, and hip roofs. This method is based on the procedure of Part 1. Wind pressures are determined from a table and adjusted where necessary.
- The analytical directional design method of ASCE 7 Chap. 30 Part 3 Sec. 30.6. This is applicable to enclosed and partially enclosed buildings with $h > 60$ ft having flat roofs, pitched roofs, gable roofs, hip roofs, mansard roofs, arched roof, and domed roof. Wind pressures are determined from the specified equation applicable to each building surface.
- The simplified directional design method of ASCE 7 Chap. 30 Part 4 Sec. 30.7. This is applicable to enclosed buildings with $h \leq 160$ ft having flat roofs, gable roofs, hip roofs, monoslope roofs, and mansard roofs. This method is based on the procedure of Part 3. Wind pressures are determined from a table and adjusted where necessary.
- The analytical directional design method of ASCE 7 Chap. 30 Part 5 Sec. 30.8. This is applicable to open buildings of all heights having pitched free roofs, monoslope free roofs, and troughed free roofs. Wind pressures are determined from the specified equation applicable to each roof surface.
- The analytical directional design method of ASCE 7 Chap. 30 Part 6 Sec. 30.9. This is applicable to building appurtenances such as roof overhangs and parapets. Wind pressures are determined from the specified equation applicable to each roof overhang or parapet surface.

## Design of Components and Cladding Using ASCE 7 Sec. 30.4

The design wind pressure on components and cladding for low-rise building and buildings with a height not exceeding 60 ft is given by ASCE 7 Eq. (30.4-1) as

$$p = q_h[(GC_p) - (GC_{pi})]$$

where $q_h$ = wind velocity pressure at mean roof height $h$ for the applicable exposure category

$(GC_p)$ = product of the equivalent external pressure coefficient and gust effect factor as given in ASCE 7 Figs. 30.4-1 through 30.4-7

$(GC_{pi})$ = product of the internal pressure coefficient and gust effect factor as given in ASCE 7 Fig. 26.11-1

In accordance with ASCE 7 Sec. 30.2.2, the design wind pressure shall not be less than a net pressure of 16 lb/ft$^2$ applied in either direction normal to the surface.

The velocity pressure exposure coefficients $K_z$ are given in ASCE 7 Table 30.3-1.

Local turbulence at corners and at the roof eaves produces an increase in pressure in these areas. Hence, as shown in ASCE 7 Figs. 30.4-1 and 30.4-2, walls are divided into two zones and roofs are divided into three zones with a different wind pressure coefficient assigned to each. The zone width is given by ASCE 7 Fig. 30.4-1 Note 6 as the lesser of

$$a = 0.1 \times (\text{least horizontal dimension})$$

or

$$a = 0.4h$$

but not less than either

$$a = 0.04 \times (\text{least horizontal dimension})$$

or

$$a = 3 \text{ ft}$$

The values of $(GC_p)$ depend on the effective area attributed to the element considered. Because of local turbulence that may occur over small areas of buildings, components and cladding are designed for higher wind pressures than the main wind-force resisting system. An effective wind area is used to determine the external pressure coefficient. This is defined in ASCE 7 Sec. 26.2 as

$$A = b_e \ell$$

where $\ell$ = element span length
$b_e$ = effective tributary width
$\geq \ell/3$

For cladding fasteners, the effective wind area shall not be greater than the area that is tributary to an individual fastener. In accordance with ASCE 7 Fig. 30.4-1 Note 5, the values of $(GC_p)$ may be reduced by 10 percent for the walls of buildings with a roof slope of 10° or less.

**Example 2.20.** Design Wind Pressure for Components

The roof framing of the building analyzed in Example 2.18 consists of open web joists spaced at 4 ft centers and spanning 30 ft parallel to the long side of the building. For the transverse wind direction, determine the design wind pressure acting on a roof joist in interior zone 1 of the building and determine the width of the eave zone. The building may be considered enclosed.

From Example 2.18, the velocity pressure at mean roof height using Case 1 values for $K_z$ is obtained as

$$q_h = 20.14 \text{ lb/ft}^2$$

The mean roof height is

$$h = 20 \text{ ft}$$
$$< 60 \text{ ft}$$

Hence, the low-rise building method of ASCE 7 Sec. 30.4 is applicable.

The product of the internal pressure coefficient and the gust effect factor is obtained from Example 2.19 as

$$(GC_{pi}) = \pm 0.18$$

The width of the eave zone 2 is given by ASCE 7 Fig. 30.4-1 Note 6 as the lesser of

$$a = 0.1 \times L$$
$$= 0.1 \times 32$$
$$= 3.2 \text{ ft}$$

or

$$a = 0.4h$$
$$= 0.4 \times 20$$
$$= 8.0 \text{ ft}$$

but not less than either

$$a = 0.04 \times L$$
$$= 0.04 \times 32$$
$$= 1.28 \text{ ft}$$

or

$$a = 3 \text{ ft}$$

Hence,

$$a = 3.2 \text{ ft} \ldots \text{ governs}$$

The effective tributary width of a roof joist is defined in ASCE 7 Sec. 26.2 as the larger of

$$b_e = \text{joist spacing}$$
$$= 4 \text{ ft}$$

or

$$b_e = \ell/3$$
$$= 30/3$$
$$= 10 \text{ ft} \ldots \text{ governs}$$

The effective wind area attributed to the roof joist is then

$$A = b_e\ell$$
$$= 10 \times 30$$
$$= 300 \text{ ft}^2$$

The negative external pressure coefficient for roof interior zone 1, reduced by 10% for $\theta = 0$, is obtained from ASCE 7 Fig. 30.4-2A as

$$(GC_p) = -0.81$$

The negative design wind pressure on a roof joist for interior zone 1 is obtained from ASCE 7 Eq. (30.4-1 A) as

$$p = q_h[(GC_p) - (GC_{pi})]$$
$$= 20.14[(-0.81) - (0.18)]$$
$$= -19.94 \text{ lb/ft}^2$$

The upward load on the roof joist over interior zone 1 is

$$w = ps$$
$$= -19.94 \times 4$$
$$= -79.76 \text{ lb/ft}$$

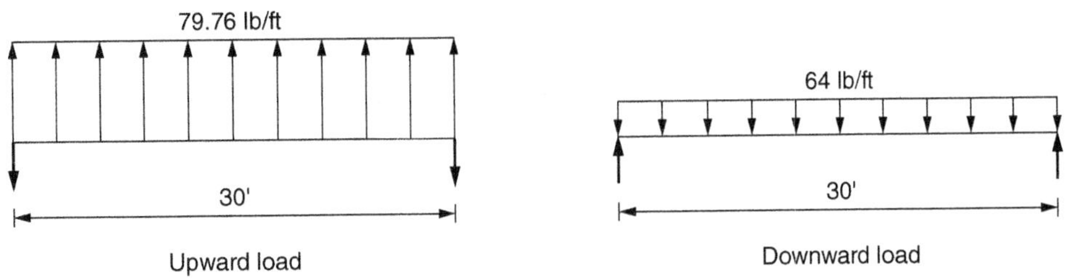


FIGURE 2.35 Wind loading on roof joist.

The positive external pressure coefficient for roof interior zone 1, reduced by 10% for θ = 0, is obtained from ASCE 7 Fig. 30.4-2A as

$$(GC_p) = 0.18$$

The positive design wind pressure on a roof joist for interior zone 1 is obtained from ASCE 7 Eq. (30.4-1) as

$$\begin{aligned} p &= q_h[(GC_p) - (GC_{pi})] \\ &= 20.14[(0.18) - (-0.18)] \\ &= 7.25 \text{ lb/ft}^2\text{; adopt 16 lb/ft}^2 \text{ minimum} \end{aligned}$$

The downward load on the roof joist over interior zone 1 is

$$\begin{aligned} w &= ps \\ &= 16 \times 4 \\ &= 64 \text{ lb/ft} \end{aligned}$$

The wind loading acting on the roof joist is shown in Fig. 2.35.

## IBC Alternate All-Heights Method

The alternate method of IBC Sec. 1609.6 is a simplified procedure based on ASCE 7 Chap. 27 Part 1 Sec. 27.4. The derivation of the procedure is described by Barbera and Scott[6] and Huston[7] and is applicable to buildings that meet the following conditions:

- The building does not exceed 75 ft in height, with a height to least width ratio not exceeding 4, or the building is of any height and has a fundamental frequency not less than 1 Hz. This is consistent with the definition of a rigid structure given in ASCE 7 Sec. 26.2.
- The building is not sensitive to dynamic effects.
- The structure is not located at a site subject to channeling effects or buffeting in the wake of upwind obstructions.
- The building shall meet the requirements of a simple diaphragm building as defined in ASCE 7 Sec. 26.2.

In order to apply the alternate method, a number of prerequisites must be determined. These include velocity pressure exposure coefficient, topography factor, wind stagnation pressure, and net pressure coefficient.

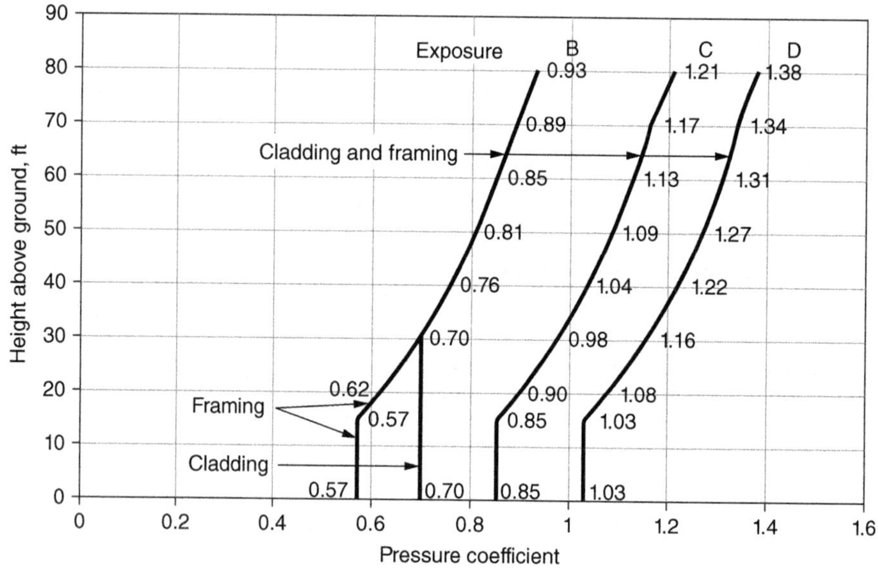


**FIGURE 2.36** Velocity pressure exposure coefficients.

## Velocity Pressure Exposure Coefficient

Velocity pressure exposure coefficients $K_z$ values are listed in ASCE 7 Tables 27.3-1 and 30.3-1 for exposure categories B, C, and D. For the windward wall, $K_z$ is based on the actual height above ground level of each floor of the building. For leeward walls, side walls, and roofs, $K_z = K_h$ is evaluated at mean roof height only and is a constant value over the height of the building. The values of $K_z$ listed in ASCE 7 Tables 27.3-1 and 30.3-1 are shown graphed in Fig. 2.36.

## Topography Factor

The topography factor $K_{zt}$ accounts for wind speedup in the vicinity of hills. The topography factor is given by ASCE 7 Eq. (26.8-1) as

$$K_{zt} = (1 + K_1K_2K_3)^2$$

Values of $K_1$, $K_2$, and $K_3$ are tabulated in ASCE 7 Fig. 26.8-1. When no topography effect is to be considered, the topography factor is given by

$$K_{zt} = 1.0$$

## Wind Stagnation Pressure

The basic wind speed is converted to a stagnation pressure at a standard height of 33 ft by the expression

$$q_s = 0.00256V^2$$

Values of $q_s$ are shown in Table 2.4.

| Basic wind speed $V$ mph | 85 | 90 | 100 | 105 | 110 | 120 | 125 | 130 | 140 | 150 | 160 |
|---|---|---|---|---|---|---|---|---|---|---|---|
| Pressure $q_s$ lb/ft$^2$ | 18.5 | 20.7 | 25.6 | 28.2 | 31.0 | 36.9 | 40.0 | 43.3 | 50.2 | 57.6 | 65.5 |

**Table 2.4** Wind Velocity Pressure

## Wind Importance Factor

An importance factor was included in the original IBC equations. However, this may now be omitted since the three wind speed maps provided for buildings assigned to different risk categories ensure enhanced performance for those facilities that constitute a substantial public hazard.

## Net-Pressure Coefficient

The net-pressure coefficient is given by IBC Sec. 1609.6.2 as

$$C_{net} = K_d[(G)(C_p) - (GC_{pi})]$$
$$= 0.85[0.85(C_p) - (GC_{pi})]$$

where $K_d$ = wind directionality factor
= 0.85 ... from ASCE 7 Table 26.6-1 for buildings
$G$ = gust effect factor
= 0.85 ... from ASCE 7 Sec. 26.9.1 for a rigid structure
$C_p$ = external pressure coefficient from ASCE 7 Fig. 17.4-1
$(GC_{pi})$ = product of internal pressure coefficient and gust-effect factor from ASCE 7 Table 26.11-1

The expression effectively provides net pressures on a building by adding internal and external pressures. This is appropriate for a simple diaphragm structure where the internal pressures on the windward and leeward walls cancel each other out. Values of $C_{net}$ are provided in IBC Table 1609.6.2 for the design of main wind-force resisting frames and components and cladding. Values are given for enclosed and partially enclosed buildings and for roofs of varying slopes.

## Design Wind Pressure on MWFRS: IBC Alternate All-Heights Method

The analytical directional design method of ASCE 7 Chap. 27 Part 1 Sec. 27.4 is essentially based on two expressions, the velocity pressure equation ASCE 7 Eq. (27.3-1) which is

$$q_z = 0.00256K_zK_{zt}K_dV^2$$

and the design wind pressure equation ASCE 7 Eq. (27.4-1) which is

$$p = qGC_p - q_i(GC_{pi})$$

By combining these two equations and rearranging terms, as illustrated by Lai,[8] the IBC alternate design wind pressure expression is derived as IBC Eq. (16.35) which is

$$P_{net} = q_sK_zC_{net}[K_{zt}]$$

In accordance with ASCE 7 Sec. 27.1.5, the wind load to be used in design shall not be less than 16 lb/ft$^2$ multiplied by the wall area of the building and 8 lb/ft$^2$ multiplied by the roof area of the building projected on a plane normal to the wind direction.

**Example 2.21.** Main Wind-Force Resisting System: Alternate Design Method
The two-story office building, shown in Fig. 2.32, meets the requirements of a simple diaphragm building and is located in a suburban area with a wind speed of $V = 115$ mi/h. For the transverse wind direction, determine the design wind pressure acting on the windward wall, leeward wall, and roof. For the windward roof, consider condition 1 only. The building may be considered enclosed and the roof and floor diaphragms are flexible. The building is not sensitive to dynamic effects and is not located on a site at which channeling or buffeting occurs.

The height to minimum width ratio is

$$h/L = 20/32$$
$$= 0.63$$
$$< 4$$

The mean roof height is

$$h = 20 \text{ ft}$$
$$< 75$$

Hence, the alternate method of IBC Sec. 1609.6 is applicable and values of $C_{net}$ may be obtained from IBC Table 1609.6.2.

The value of the wind stagnation pressure for a wind speed $V$ of 115 mi/h is obtained from Table 2.4 as

$$q_s = 34.0 \text{ lb/ft}^2$$

For a suburban area, the exposure is category B and the relevant parameters are obtained as

$K_h$ = velocity pressure exposure coefficient at roof height

= 0.62 ... from ASCE 7 Table 27.3-1 for a height of 20 ft for the main wind-force resisting system and exposure category B

$K_z$ = 0.57 ... from ASCE 7 Table 27.3-1 for a height of 10 ft for the main wind-force resisting system and exposure category B

$K_{zt}$ = topography factor

= 1.0 ... from ASCE 7 Fig. 26.8-1

For a two-story building with flexible diaphragms, ASCE 7 App. D Sec. D1.1 specifies that torsional load cases may be neglected.

**For the windward wall,** IBC Table 1609.6.2 gives the values of the net-pressure coefficient as

$C_{net}$ = 0.43 ... for positive internal pressure

= 0.73 ... for negative internal pressure (suction)

The design wind pressure at roof height is given by IBC Eq. (16-35) which is

$$P_{net} = q_s K_h C_{net}[K_{zt}]$$
$$= 34.0 \times 0.62 \times 0.73 \times 1.0$$

= 15.39 lb/ft$^2$ for negative internal pressure (suction); adopt 16 lb/ft$^2$ minimum

$$= 34.0 \times 0.62 \times 0.43 \times 1.0$$

= 9.06 lb/ft$^2$ for positive internal pressure; adopt 16 lb/ft$^2$ minimum

The design wind pressure at first floor height is

$$P_{net} = q_s K_z C_{net}[K_{zt}]$$
$$= 34.0 \times 0.57 \times 0.73 \times 1.0$$
$$= 14.15 \text{ lb/ft}^2 \text{ for negative internal pressure (suction); adopt 16 lb/ft}^2 \text{ minimum}$$
$$= 34.0 \times 0.57 \times 0.43 \times 1.0$$
$$= 8.33 \text{ lb/ft}^2 \text{ for positive internal pressure; adopt 16 lb/ft}^2 \text{ minimum}$$

**For the leeward wall,** IBC Table 1609.6.2 gives the values of the net-pressure coefficient as

$$C_{net} = -0.51 \text{ ... for positive internal pressure}$$
$$= -0.21 \text{ ... for negative internal pressure (suction)}$$

The design wind pressure is given by IBC Eq. (16-35) which is

$$P_{net} = q_s K_h C_{net}[K_{zt}]$$
$$= 34.0 \times 0.62 \times -0.21 \times 1.0$$
$$= -4.43 \text{ lb/ft}^2 \text{ for negative internal pressure (suction); adopt 16 lb/ft}^2 \text{ minimum}$$
$$= 34.0 \times 0.62 \times -0.51 \times 1.0$$
$$= -10.75 \text{ lb/ft}^2 \text{ for positive internal pressure; adopt 16 lb/ft}^2 \text{ minimum}$$

**For the windward roof,** IBC Table 1609.6.2 gives the values of the net-pressure coefficient as

$$C_{net} = -1.09 \text{ ... for positive internal pressure}$$
$$= -0.79 \text{ ... for negative internal pressure (suction)}$$

The design wind pressure is given by IBC Eq. (16-35) which is

$$P_{net} = q_s K_h C_{net}[K_{zt}]$$
$$= 34.0 \times 0.62 \times -0.79 \times 1.0$$
$$= -16.65 \text{ lb/ft}^2 \text{ for negative internal pressure (suction)}$$
$$= 34.0 \times 0.62 \times -1.09 \times 1.0$$
$$= -22.98 \text{ lb/ft}^2 \text{ for positive internal pressure}$$

**For the leeward roof,** IBC Table 1609.6.2 gives the values of the net-pressure coefficient as

$$C_{net} = -0.66 \text{ ... for positive internal pressure}$$
$$= -0.35 \text{ ... for negative internal pressure (suction)}$$

The design wind pressure is given by IBC Eq. (16-35) which is

$$P_{net} = q_s K_h C_{net}[K_{zt}]$$
$$= 34.0 \times 0.62 \times -0.35 \times 1.0$$
$$= -7.38 \text{ lb/ft}^2 \text{ for negative internal pressure (suction); adopt 8 lb/ft}^2 \text{ minimum}$$
$$= 34.0 \times 0.62 \times -0.66 \times 1.0$$
$$= -13.91 \text{ lb/ft}^2 \text{ for positive internal pressure}$$

The wind pressure diagrams for both cases, internal suction and internal pressure, are shown in Fig. 2.37.

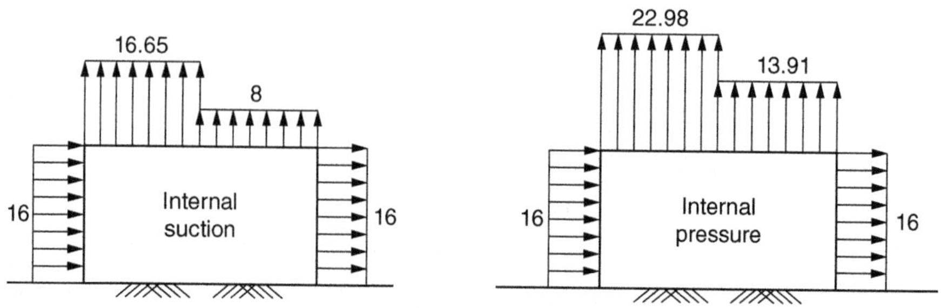


**FIGURE 2.37** Wind pressure diagrams, in lb/ft², for Example 2.21.

## Design Wind Pressure on Components and Cladding: IBC Alternate All-Heights Method

The alternate method of IBC Sec. 1609.6 is a simplified procedure based on ASCE 7 Chap. 30 Part 3 Sec. 30.6. The design wind pressure for components and cladding is determined using IBC Eq. (16-35) which is

$$P_{net} = q_s K_z C_{net} [K_{zt}]$$

In accordance with ASCE 7 Sec. 30.2.2, the design wind pressure shall not be less than a net pressure of 16 lb/ft² applied in either direction normal to the surface.

Velocity pressure exposure coefficients $K_z$ values are listed in ASCE 7 Table 30.3-1 for exposure categories B, C, and D. For the windward wall, $K_z$ is based on the actual height above ground level of the component. For leeward walls, side walls, and roofs, $K_z = K_h$ is evaluated at mean roof height only and is a constant value over the height of the building. The values of $K_z$ listed in ASCE 7 Table 30.3-1 are shown graphed in Fig. 2.36.

Values of $C_{net}$ are provided in IBC Table 1609.6.2 for the design of components and cladding. Values are given for enclosed and partially enclosed buildings and for roofs of varying slopes. In addition, to account for local turbulence at corners and eaves, appropriate values are provided for the five zones shown in Fig. 2.38.

**Example 2.22.** Design Wind Pressure for Components

The roof framing of the building analyzed in Example 2.21 consists of open web joists spaced at 4 ft centers and spanning 30 ft parallel to the long side of the building. For the transverse wind direction, determine the design wind pressure acting on a roof joist in interior zone 1 of the building. The building may be considered enclosed.

The effective tributary width of a roof joist is defined in ASCE 7 Sec. 26.2 as the larger of

$$b_e = \text{joist spacing}$$
$$= 4 \text{ ft}$$

or

$$b_e = \ell/3$$
$$= 30/3$$
$$= 10 \text{ ft ... governs}$$

The effective wind area attributed to the roof joist is then

$$A = b_e \ell$$
$$= 10 \times 30$$
$$= 300 \text{ ft}^2$$

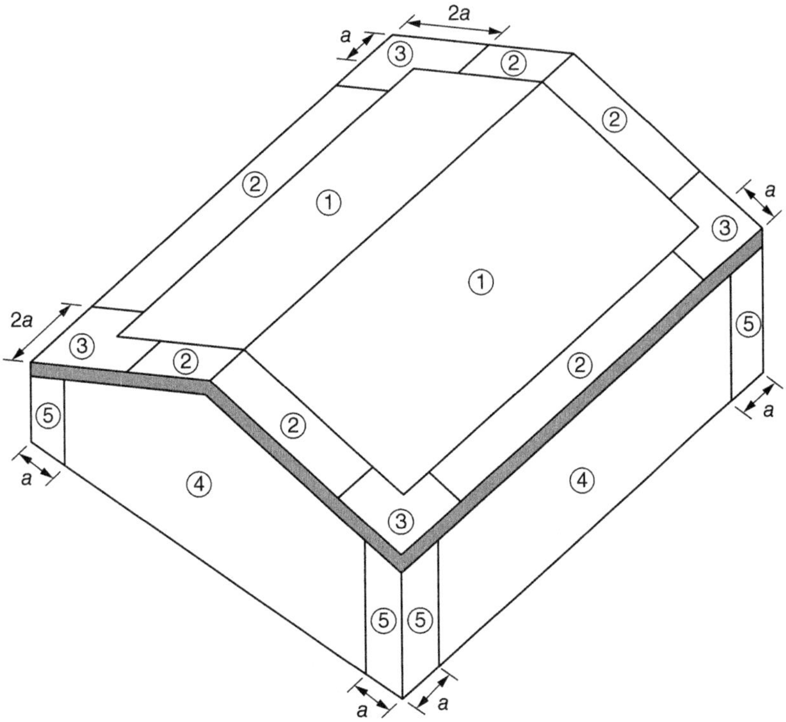


**FIGURE 2.38** Pressure zones on walls and roofs.

The value of the wind stagnation pressure for a wind speed $V$ of 115 mi/h is obtained from Table 2.4 as

$$q_s = 34.0 \text{ lb/ft}^2$$

For a suburban area, the exposure is category B and the relevant parameters are obtained as

$K_h$ = velocity pressure exposure coefficient at roof height from ASCE 7 Table 30.3-1 for a height of 20 ft for components and cladding for exposure category B

= 0.70

$K_{zt}$ = topography factor from ASCE 7 Sec. 26.8.2

= 1.0

The net-pressure coefficient for interior zone 1 is obtained from IBC Table 1609.6.2(2) for a tributary area of 300 ft$^2$ and a slope of < 27° as

$$C_{net} = -0.92 \text{ ... for negative pressure}$$

The net negative design wind pressure on a roof joist for interior zone 1 is obtained from IBC Eq. (16-35) which gives

$$\begin{aligned} P_{net} &= q_s K_h C_{net} [K_{zt}] \\ &= 34.0 \times 0.70 \times -0.92 \times 1.0 \\ &= -21.90 \text{ lb/ft}^2 \end{aligned}$$

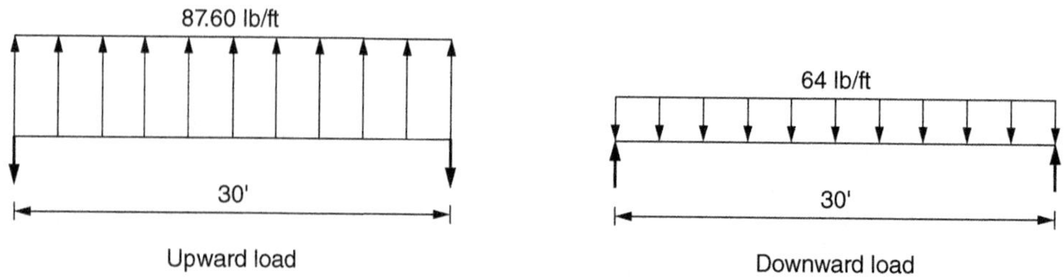


**FIGURE 2.39** Wind loading on roof joist.

**The upward load** on the roof joist in interior zone 1 is

$$\begin{aligned} w &= P_{net}s \\ &= -21.90 \times 4 \\ &= -87.60 \text{ lb/ft} \end{aligned}$$

The net-pressure coefficient for interior zone 1 is obtained from IBC Table 1609.6.2(2) for a tributary area of 300 ft$^2$ and a slope of < 27° as

$$C_{net} = 0.41 \text{ ... for positive pressure}$$

The net positive design wind pressure on a roof joist for interior zone 1 is obtained from IBC Eq. (16-35) which gives

$$\begin{aligned} P_{net} &= q_s K_h C_{net} [K_{zt}] \\ &= 34.0 \times 0.70 \times 0.41 \times 1.0 \\ &= 9.76 \text{ lb/ft}^2\text{; adopt 16 lb/ft}^2 \text{ minimum} \end{aligned}$$

**The downward load** on the roof joist in interior zone 1 is

$$\begin{aligned} w &= P_{net}s \\ &= 16 \times 4 \\ &= 64 \text{ lb/ft} \end{aligned}$$

The wind loading acting on the roof joist is shown in Fig. 2.39.

## 2.9 Seismic Loads

The objective of the seismic provisions in ASCE 7 is to preclude structural collapse in a major earthquake. Hence, fatalities and economic loss are minimized. Because of economic considerations and the degree of protection required for a structure, the probability of collapse varies with the risk associated with unacceptable performance. Anticipated probabilities of collapse are given in ASCE 7 Table C1.3.1b as 10, 6, and 3 percent probability, under maximum considered earthquake accelerations, for risk category II, III, and IV structures.

Seismic loads on a structure are generated by the effects that an earthquake has on the structure. Earthquakes are the result of an abrupt rupture along a fault zone below the earth's surface. The ground vibrations created as a result of this rupture produce inertial forces in a structure that may cause severe damage unless the structure is appropriately designed and constructed. In accordance with Newton's second law of motion,

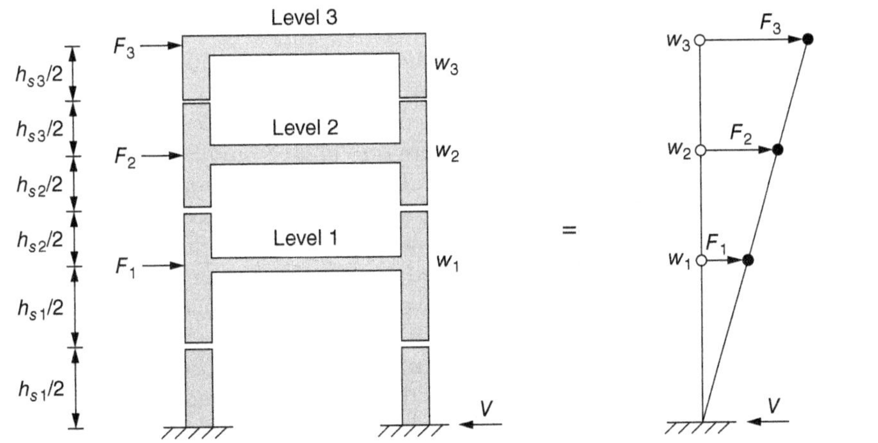


**FIGURE 2.40** Seismic base shear.

the inertial force produced equals the mass of the structure multiplied by the imposed acceleration. The seismic load must be accurately estimated in order to design the structure to withstand an earthquake.

In accordance with IBC Sec. 1613.1, seismic loads on structures shall be determined by the provisions of ASCE 7, excluding Chap. 14 and App. 11A. Seismic loads are dynamic in nature and require a complex dynamic analysis for a complete solution. An alternative method of analysis, the *equivalent lateral force procedure*, is presented in ASCE 7 Sec. 12.8 and this provides a simple and direct approach when a sophisticated dynamic analysis is not warranted. The procedure consists of replacing the dynamic seismic force with a single static force at the base of the building, as shown in Fig. 2.40. This static force, $V$, is termed the *seismic base shear*. After the seismic base shear is determined, the forces acting on the individual structural elements in the building may be calculated. The equivalent lateral force procedure is applicable to regular structures, defined as structures without irregular features that have a reasonably uniform distribution of stiffness, strength, and mass over the height of the structure. The structure mass is assumed to be concentrated at floor and roof levels as shown. Determination of the seismic base shear depends on a number of factors including the ground accelerations, soil profile, fundamental period of the building, mass of the building, classification of the structural system, ductility of the structural elements, and the function and occupancy of the building.

In accordance with IBC Sec. 1613.1 and ASCE 7 Sec. 11.1.2, the following buildings are exempt from seismic design requirements:

- Detached one- and two-family dwellings that are located where the mapped, short period, spectral response acceleration parameter, $S_S$, is less than 0.4 or where the Seismic Design Category determined in accordance with Sec. 11.6 is A, B, or C.
- Dwellings of wood-frame construction satisfying the limitations of and constructed in accordance with the International Residential Code.[9]
- Buildings of wood-frame construction satisfying the limitations of and constructed in accordance with IBC Sec. 2308.
- Agricultural storage structures that are intended only for incidental human occupancy.

## Ground Motion Parameters

The ground motion parameters $S_S$ and $S_1$ are defined in ASCE 7 Sec. 11.4.1 and are mapped in ASCE 7 Figs. 22-1 through 22-6. Alternatively, the ground motion parameters may be obtained from the U.S. Geological Survey website for locations with known latitude and longitude. The two values provided, $S_S$ and $S_1$, represent the risk-adjusted maximum considered earthquake ($MCE_R$) response accelerations at periods of 0.2 second and 1.0 second for a site class B soil profile and 5 percent damping. The primary cause of earthquake injuries and deaths is the collapse of a building in the event of a major earthquake, the maximum considered earthquake. The ground motion parameters are risk-adjusted to provide a uniform probability of structural collapse for the entire United States.

For a seismically active region such as coastal California, a probabilistic approach results in much higher accelerations than that of the characteristic earthquakes in the region. Hence, for this region, the values represent the deterministic event defined as the median estimate of the accelerations of the characteristic earthquakes increased by 50 percent. The characteristic earthquake is defined as the maximum acceleration capable of occurring in the region but not less than the largest acceleration that has been recorded in the region.

## Site Classification Characteristics

The ground motion produced by an earthquake is modified by the type of soil through which the vibrations pass. Soft soil amplifies ground vibrations to a greater extent than stiffer soil. Long period ground vibration is amplified in soft soil to a greater extent than short period vibration. The soil profile is classified by measuring the shear wave velocity in the upper 100 ft of material as specified in ASCE 7 Chap. 20. Alternatively for site classification C, D, and E, the classification may be made by measuring the standard penetration resistance or the undrained shear strength of the material. Six different soil profile types are identified in ASCE 7 Sec. 11.4.2 and are shown in Table 2.5.

For site classification F, a special site response analysis is required. When soil properties are unknown, in accordance with ASCE 7 Sec. 11.4.2, site classification type D may be assumed, unless the authority having jurisdiction determines that site class E or F soils are present at the site.

## Site Coefficients

Site coefficients, $F_a$ and $F_v$, account for the amplification of ground accelerations by the soil profile at a specific site. Site coefficients are a function of the site classification

| Site Classification | Description | Shear Wave Velocity (ft/s) |
|---|---|---|
| A | Hard rock | > 5000 |
| B | Rock | 2500–5000 |
| C | Very dense soil and soft rock | 1200–2500 |
| D | Stiff soil | 600–1200 |
| E | Soft clay soil | < 600 |
| F | Peat, sensitive clay | – |

**Table 2.5** Soil Profiles

| Site Classification | Response Acceleration, $S_S$ | | | | | Response Acceleration, $S_1$ | | | | |
|---|---|---|---|---|---|---|---|---|---|---|
| | ≤ 0.25 | 0.50 | 0.75 | 1.00 | ≥ 1.25 | ≤ 0.1 | 0.2 | 0.3 | 0.4 | ≥ 0.5 |
| A | 0.8 | 0.8 | 0.8 | 0.8 | 0.8 | 0.8 | 0.8 | 0.8 | 0.8 | 0.8 |
| B | 1.0 | 1.0 | 1.0 | 1.0 | 1.0 | 1.0 | 1.0 | 1.0 | 1.0 | 1.0 |
| C | 1.2 | 1.2 | 1.1 | 1.0 | 1.0 | 1.7 | 1.6 | 1.5 | 1.4 | 1.3 |
| D | 1.6 | 1.4 | 1.2 | 1.1 | 1.0 | 2.4 | 2.0 | 1.8 | 1.6 | 1.5 |
| E | 2.5 | 1.7 | 1.2 | 0.9 | 0.9 | 3.5 | 3.2 | 2.8 | 2.4 | 2.4 |
| F | (a) | (a) | (a) | (a) | (a) | (a) | (a) | (a) | (a) | (a) |

Note: (a) Site-specific geotechnical investigation and dynamic site response analysis required.

**TABLE 2.6** Site Coefficients

characteristics and ground motion parameters. Values are given in ASCE 7 Tables 11.4-1 and 11.4-2 and are combined in Table 2.6. $F_a$ is the short period or acceleration based amplification factor corresponding to response acceleration $S_S$, and $F_v$ is the long period or velocity based amplification factor corresponding to response acceleration $S_1$. Since the maximum considered earthquake response parameters are derived for site classification type B profiles, the site coefficient is 1.0 for site class B. The table also indicates a 20 percent reduction in the ground response for a hard rock site classification type A. In general, as the soil profile becomes softer, the ground response increases. However, the value of $F_a$ for a value of $S_S = 1.0$ reduces for site classification type E reflecting the tendency for the ground response to attenuate as the seismicity increases in soil profile type E. Linear interpolation may be used to obtain intermediate values.

## Adjusted Earthquake Response Accelerations

The maximum considered response accelerations are modified by the site coefficients at a specific site to give the adjusted response accelerations. Hence, the adjusted ground response accelerations, $S_{MS}$ and $S_{M1}$, at a specific site are obtained from ASCE 7 Sec. 11.4.3 as

$$S_{MS} = F_a S_S$$

and

$$S_{M1} = F_v S_1$$

## Design Response Acceleration Parameters

The objective of the design provisions is to provide a uniform margin against collapse, in all regions, for ground motions in excess of the design levels. It is estimated that a structure subjected to a risk-targeted maximum considered earthquake ground motion 1.5 times the level of the ground motion for which it is designed will be unlikely to collapse because of the inherent overstrength of the structure. Hence, in accordance with ASCE 7 Sec. 11.4.4, the design response accelerations are

$$S_{DS} = 2S_{MS}/3$$

and

$$S_{D1} = 2S_{M1}/3$$

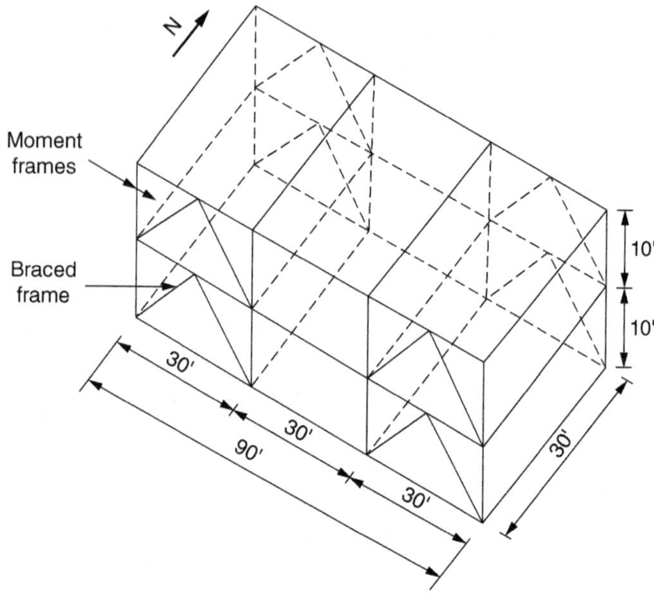


FIGURE 2.41 Details for Example 2.23.

**Example 2.23.** Calculation of Design Accelerations
The two-story steel-framed building shown in Fig. 2.41 is located on a site in Orange County, California, with a soil profile of stiff soil having a shear wave velocity of 700 ft/s. The maximum considered response accelerations for the location are $S_S = 1.5g$ and $S_1 = 0.6g$. Determine the design response accelerations $S_{DS}$ and $S_{D1}$.

From ASCE 7 Table 20.3-1, the applicable site classification for stiff soil with a shear wave velocity of 700 ft/s is site classification D.

The site coefficients for site classification D and for the given values of $S_S$ and $S_1$ are obtained from ASCE 7 Tables 11.4-1 and 11.4-2 and are

$$F_a = 1.0$$
$$F_v = 1.5$$

From ASCE 7 Eqs. (11.4-1) and (11.4-2), the adjusted response accelerations at short periods and at a period of 1 second are

$$\begin{aligned} S_{MS} &= F_a S_S \\ &= 1.0 \times 1.5 \\ &= 1.5g \end{aligned}$$

and

$$\begin{aligned} S_{M1} &= F_v S_1 \\ &= 1.5 \times 0.6 \\ &= 0.9g \end{aligned}$$

The corresponding design response accelerations are

$$\begin{aligned} S_{DS} &= 2S_{MS}/3 \\ &= 2 \times 1.5/3 \\ &= 1.0g \end{aligned}$$

and $$S_{D1} = 2S_{M1}/3$$
$$= 2 \times 0.9/3$$
$$= 0.6g$$

## Design Response Spectrum

When a design response spectrum is required, this is developed as indicated in ASCE 7 Sec. 11.4.5. The design response spectrum shown in ASCE 7 Fig. 11.4-1 consists of several segments. The constant-acceleration segment covers the period from $T_0$ to $T_s$ and response accelerations in this band are constant and equal to $S_{DS}$. The constant-velocity segment covers the period from $T_s$ to $T_L$ and the response accelerations in this band are proportional to $1/T$ with the response acceleration at 1 second equal to $S_{D1}$. The constant-displacement segment commences at $T_L$ and the response accelerations in this band are proportional to $1/T^2$. Values of $T_L$ are provided by ASCE 7 Figs. 22-12 through 22-16.

Applying ASCE 7 Sec. 11.4.5 to Example 2.23 provides the following data:

The corresponding design spectral response accelerations are

$$S_{DS} = 2S_{MS}/3$$
$$= 1.0g$$
$$S_{D1} = 2S_{M1}/3$$
$$= 0.6g$$

The response spectrum parameters are

$$T_S = S_{D1}/S_{DS}$$
$$= 0.6g/1.0g$$
$$= 0.6$$
$$T_0 = 0.2S_{D1}/S_{DS}$$
$$= 0.2 \times 0.6g/1.0g$$
$$= 0.12 \text{ second}$$

At $T = 0$, the design spectral response acceleration is given by ASCE 7 Eq. (11.4-5) as

$$S_a = S_{DS}(0.4 + 0.6\,T/T_0)$$
$$= 0.4S_{DS}$$
$$= 0.40g$$

For periods greater than $T_S$ and less than or equal to $T_L$, the design spectral response acceleration is given by ASCE 7 Eq. (11.4-6) as

$$S_a = S_{D1}/T$$
$$= 0.6g/T$$

For a building located on a site in Orange County, California, the long-period transition period is given on ASCE 7 Fig. 22-12 as

$$T_L = 8 \text{ seconds}$$

At $T = 8$ sec, the design spectral response acceleration is given by ASCE 7 Eq. (11.4-6) as

$$\begin{aligned} S_a &= S_{D1}/T \\ &= 0.6g/8 \\ &= 0.075g \end{aligned}$$

For periods greater than $T_L$, the design spectral response acceleration is given by ASCE 7 Eq. (11.4-7) as

$$\begin{aligned} S_a &= S_{D1}T_L/T^2 \\ &= 0.6g \times 8/T^2 \\ &= 0.48g/T^2 \end{aligned}$$

The response spectrum is shown in Fig. 2.42.

The effect of soil classification on the response spectrum is shown in Fig. 2.43 for the building analyzed in Example 2.23. As the site classification changes from type D (stiff soil) to type A (hard rock), the spectral acceleration in the constant velocity segment of the response curve reduces.

## Risk Category and Importance Factors

The importance factor relates design loads to the consequences of failure and is a measure of the degree of protection required for a building. An increase in the importance factor results in an increase in the design base shear for a building, with a consequent reduction in the damage caused by the design earthquake. The importance factor ensures that essential facilities are designed for higher loads so as to reduce possible

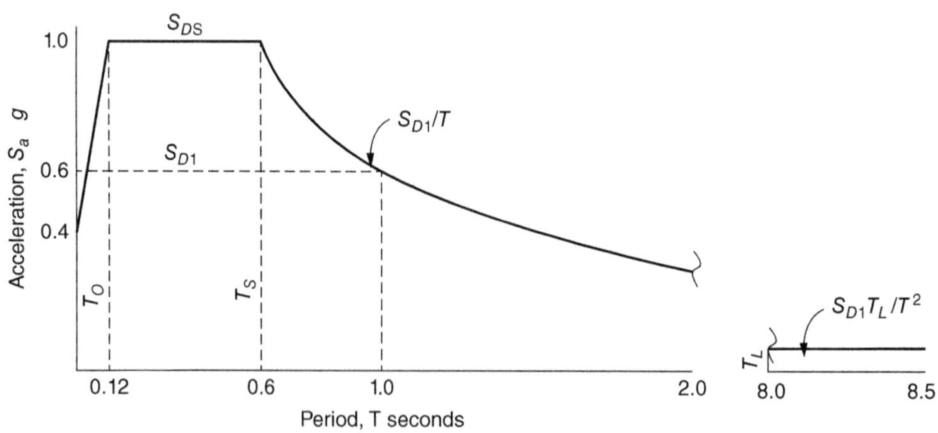


**FIGURE 2.42** Design response spectrum for Example 2.23.

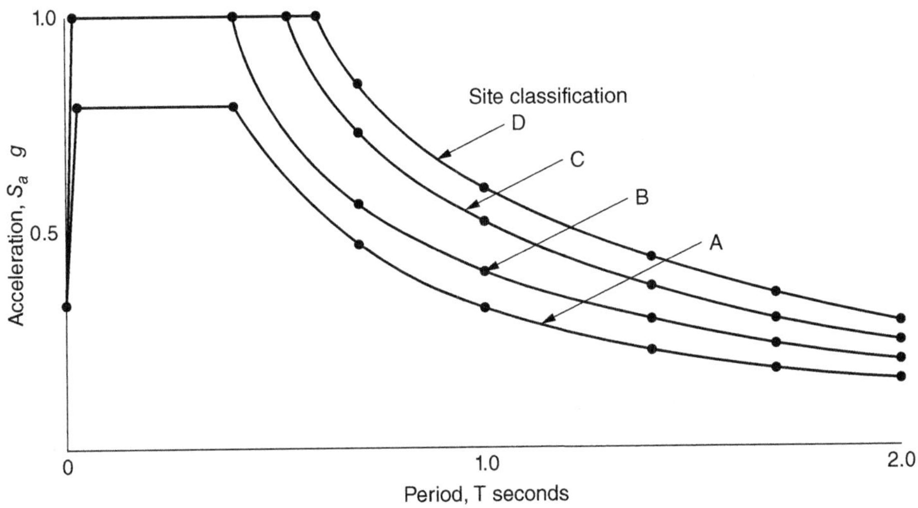


**FIGURE 2.43** Design response spectra for site classifications A, B, C, and D.

structural damage. In ASCE 7 Table 1.5-1, risk categories are assigned to buildings based on the nature of their occupancies. The importance factors corresponding to the risk categories are given in ASCE 7 Table 1.5-2. Details of the risk category, nature of occupancy, and importance factors are given in Table 2.7.

Risk category IV buildings are essential facilities such as hospitals, fire and police stations, post earthquake recovery centers and buildings housing equipment for these facilities. Also included in risk category IV are buildings housing toxic materials that will endanger the safety of the public if released. Risk category IV buildings are allocated an importance factor of 1.5. This higher importance factor ensures minimal damage without disruption to the continued operation of the facility after an earthquake. Risk category III buildings are facilities with a high occupant load such as buildings where more than 300 people congregate, schools with a capacity exceeding 250, colleges with a capacity exceeding 500, health care facilities with a capacity of 50 or more, jails and power stations. Risk category III buildings are allocated an importance factor of 1.25. Risk category I buildings are low hazard structures such as agricultural facilities, minor storage buildings, and temporary facilities. Risk category I buildings are allocated an importance factor of 1.00. Risk category II buildings are standard occupancy structures that consist of all other types of facilities and are also allocated an importance factor of 1.00.

| Risk Category | Nature of Occupancy | Importance Factor, $I_e$ |
|---|---|---|
| I | Low hazard structures | 1.00 |
| II | Standard occupancy structures | 1.00 |
| III | Assembly structures | 1.25 |
| IV | Essential or hazardous structures | 1.50 |

**TABLE 2.7** Occupancy Category and Importance Factor

| $S_{DS}$ | $S_{D1}$ | Risk Category | |
|---|---|---|---|
| | | I, II, or III | IV |
| $S_{DS} < 0.167g$ | $S_{D1} < 0.067g$ | A | A |
| $0.167g \le S_{DS} < 0.33g$ | $0.067g \le S_{D1} < 0.133g$ | B | C |
| $0.33g \le S_{DS} < 0.50g$ | $0.133g \le S_{D1} < 0.20g$ | C | D |
| $0.50g \le S_{DS}$ | $0.20g \le S_{D1}$ | D | D |
| Response acceleration at 1 second period, $S_1 \ge 0.75g$ | | E | F |

**TABLE 2.8** Seismic Design Category

## Seismic Design Category

The seismic design category determines the following design requirements for a building:

- Maximum allowable height
- Permissible structural system
- Allowable types of irregularity
- Permissible analysis procedure
- Necessary seismic detailing procedure

Six design categories are defined in ASCE 7 Sec. 11.6 and a building is assigned a category based on its design response accelerations and risk category. The seismic design category is determined twice, first as a function of the design response acceleration at short periods, using ASCE 7 Table 11.6-1, and then as a function of the design response acceleration at a period of 1 second, using ASCE 7 Table 11.6-2. The most severe seismic design category governs. ASCE 7 Tables 11.6-1 and 11.6-2 are combined in Table 2.8.

The significance of the seismic design category is summarized in Table 2.9.

| Seismic Design Category | Design Requirements |
|---|---|
| A | Minimal ground movements anticipated. A nominal amount of structural integrity provided in accordance with ASCE 7 Sec. 11.7. |
| B | Low seismicity anticipated. Equivalent lateral force procedure required. |
| C | Moderate seismicity anticipated. Some structural systems are restricted. Some nonstructural components must be designed for seismic resistance. Detached one- and two-story family dwellings are exempt from these requirements. |
| D | High seismicity anticipated. Some structural systems are restricted. Irregular structures must be designed by dynamic analysis methods. |
| E or F | Very high seismicity anticipated. Severe restrictions are placed on the use of some structural systems, irregular structures, and analysis methods. |

**TABLE 2.9** Design Requirements

**Example 2.24.** Calculation of Seismic Design Category
The two-story, moment-resisting, structural steel frame shown in Fig. 2.41 forms part of the lateral force-resisting system of an office building. Calculate the importance factor and the applicable seismic design category of the building.

The building is used as an office building which is a standard occupancy structure with a risk category of II and an importance factor of

$$I_e = 1.00$$

The design response acceleration at short periods is obtained in Example 2.23 as

$$S_{DS} = 1.0g$$
$$> 0.5g$$

Hence, for a risk category of II, the seismic design category is determined from ASCE 7 Table 11.6-1 as

$$\text{SDC} = D$$

The design response acceleration at a period of 1 second is obtained in Example 2.23 as

$$S_{D1} = 0.6g$$
$$> 0.2g$$

Hence, for a risk category of II, the seismic design category is determined from ASCE 7 Table 11.6-2 as

$$\text{SDC} = D$$

Hence, the seismic design category is *D*.

## Seismic Force-Resisting System

ASCE 7 Table 12.2-1 lists the following seismic force-resisting systems:

**Bearing walls:** ASCE 7 Sec. 11.2 defines a bearing wall system as bearing walls providing support for all or a major portion of the vertical loads. Braced frames or shear walls resist all seismic loads and may also support gravity loads. These systems reduce deformations and limit damage under seismic loads. They have poor ductility and lack redundancy since they support both gravity and lateral loads. ASCE 7 Sec. 11.2 defines a bearing wall as a metal or wood stud wall that supports more than 100 lb/ft of vertical load or a concrete or masonry wall that supports more than 200 lb/ft of vertical load. Figure 2.44 shows the roof framing of a tilt-up concrete building with open web joists spanning in the north-south direction and supported on bearing walls on the north and south faces. The roof area tributary to the bearing walls is shown shaded and is given by

$$A_{\text{trib}} = 5\text{a} \times 3\text{a}$$
$$= 15\text{a}^2$$

The total roof area is

$$A_{\text{total}} = 6\text{a} \times 3\text{a}$$
$$= 18\text{a}^2$$

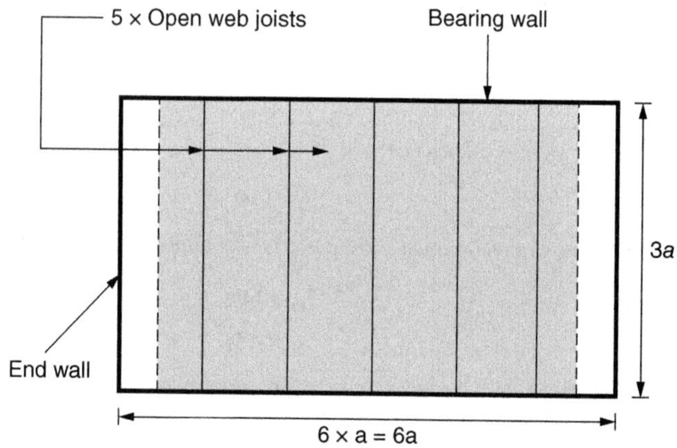


FIGURE 2.44 Bearing wall system.

The ratio of tributary area to total area is

$$A_{trib}/A_{total} = 15a^2/18a^2$$
$$= 0.83 > 0.5$$

Hence, this is a bearing wall structure since the bearing walls support the major portion of the gravity load[10].

**Building frames:** ASCE 7 Sec. 11.2 defines a building frame system as a complete space frame providing support for vertical loads with independent shear walls or braced frames resisting all seismic loads. This system provides better ductility than a bearing wall system. Figure 2.45 shows the roof framing of a tilt-up concrete building with interior pipe columns, W-shape joists spanning in the north-south direction and supported on bearing walls on the north and south faces, and

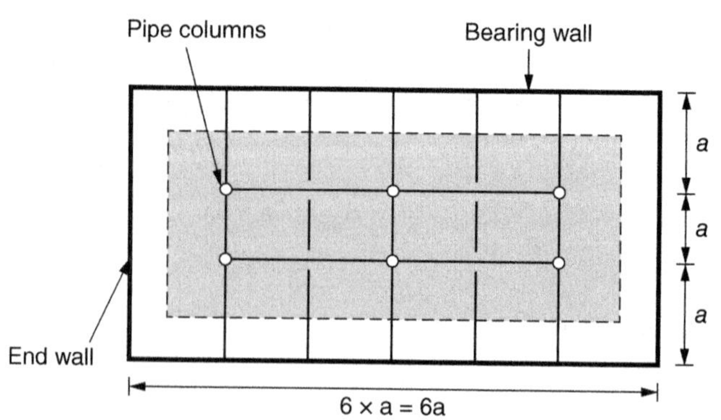


FIGURE 2.45 Building frame system.

W-shape beams spanning east-west. The roof area tributary to the pipe columns is shown shaded and is given by

$$A_{trib} = 5a \times 2a$$
$$= 10a^2$$

The total roof area is

$$A_{total} = 6a \times 3a$$
$$= 18a^2$$

The ratio of tributary area to total area is

$$A_{trib}/A_{total} = 10a^2/18a^2$$
$$= 0.56 > 0.5$$

Hence, this is a building frame structure since the pipe columns support the major portion of the gravity load.

**Moment-resisting frames:** Moment-resisting frames support gravity loads and also resist all seismic loads by flexure in the beams and columns. Moment frames are categorized as ordinary moment frames, intermediate moment frames, and special moment frames. These systems provide good ductility and redundancy and provide better free access than the two previous systems. However, large deformations under seismic loads may damage finishes, and the elements require special detailing to ensure integrity.

**Dual systems with special moment-resisting frames:** Shear walls or braced frames provide the primary seismic support system and special moment frames comprise an essentially complete space frame to support gravity loads and also provide a minimum 25 percent of the lateral force resisting system. These systems provide excellent redundancy and seismic safety.

**Dual systems with intermediate moment frames:** Shear walls or braced frames provide the primary seismic support system and intermediate moment frames comprise an essentially complete space frame to support gravity loads and also provide a minimum 25 percent of the lateral force-resisting system. These systems may be satisfactorily used in regions of moderate seismic risk.

**Shear wall-frame interactive system:** Shear walls of ordinary reinforced concrete must provide a shear strength of at least 75% of the design story shear at each story. Moment frames of ordinary reinforced concrete must be capable of resisting at least 25% of the design story shear at each story. This system is used for structures assigned to seismic design categories A and B. Construction is entirely of reinforced concrete and there is no equivalent structural steel system.

**Cantilevered column structure:** In a cantilevered column system, seismic forces are resisted entirely by columns acting as cantilevers from the base. The system has limited redundancy. In an inverted pendulum structure more than 50 percent of its mass is concentrated at the top of a cantilevered structure. Stability of the mass at the top of the structure relies on rotational restraint at the top of the cantilever.

**Steel systems not specifically detailed for seismic resistance:** Steel building structures assigned to seismic design categories B or C, with the exception of cantilever column systems, may be designed and detailed as specified by AISC 360 using a response modification factor, *R*, of 3 or less. AISC 360 provides criteria for the design of steel buildings intended for low-seismic applications where design is based on a seismic response modification coefficient of 3 or less. Steel building structures assigned to seismic design categories D, E, or F, must be designed and detailed as specified by AISC 341. AISC 341 provides criteria for the design of steel buildings intended for high-seismic applications where design is based on a seismic response modification coefficient greater than 3. Steel building structures assigned to seismic design categories B or C may also be designed using the value of *R* given in ASCE 7 Table 12.2-1. In this case, the structure must be designed and detailed as specified by AISC 341.

Using a response modification coefficient of R = 3 ensures a nominally elastic response to the applied loads and this may result in a more economical structure as the requirements of AISC 341 are not necessary. Steel building structures assigned to seismic design category A are designed and detailed as specified by AISC 360 and special seismic detailing is not required. In locations where wind effects exceed seismic effects, building members must still be detailed in accordance with AISC 341 provisions so that the members can sustain the large inelastic deformations produced by the seismic loads. .

Figure 2.46 illustrates the various structural systems.

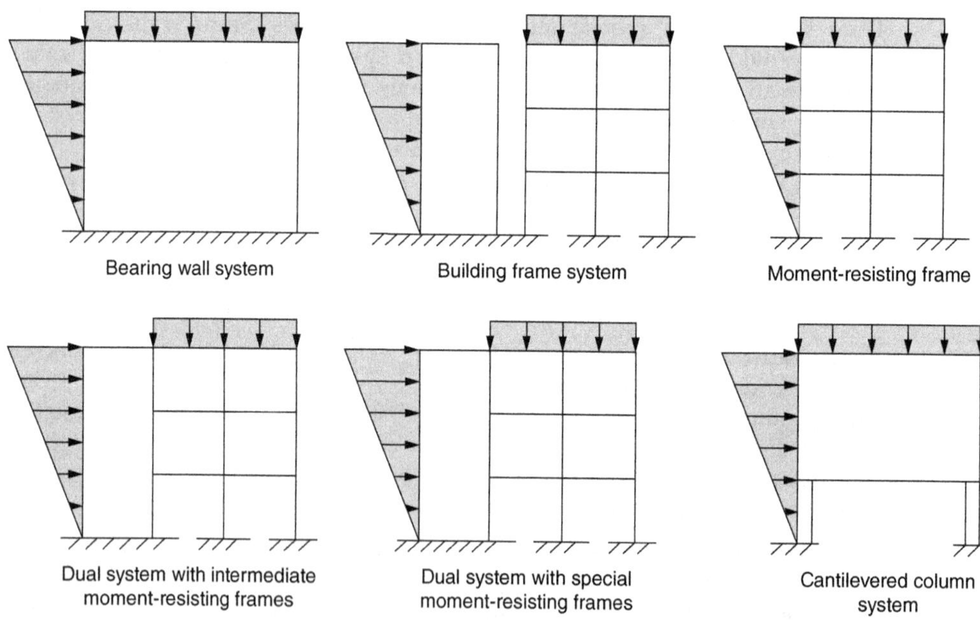


**FIGURE 2.46** Structural systems.

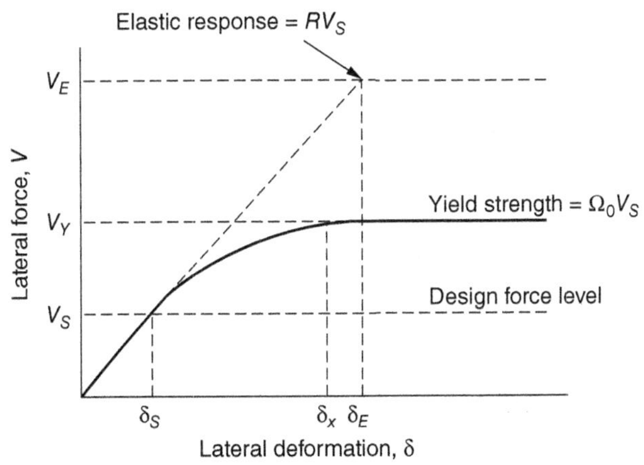


**FIGURE 2.47** Force-deformation relationship.

## Response Modification Coefficient

In ASCE 7 Table 12.2-1, structural systems are assigned a response modification coefficient $R$ corresponding to their perceived ability to resist a major seismic event. A structure with good hysteretic behavior, sufficiently ductile to sustain several cycles of inelastic deformation, adequate redundancy, and material overstrength, will dissipate imposed seismic forces without significant loss of strength. Hence, a structure may be designed for a significantly smaller force than is predicted by a linear-elastic analysis without collapse, provided that inelastic deformations are accommodated by careful detailing. The inelastic deformations produced after yield result in an increase in the natural period of the structure with a corresponding reduction in the seismic demand. As shown in Fig. 2.47, the response modification coefficient is given by

$$R = V_E / V_S$$

where $V_E$ is theoretical base shear in an elastic structure and $V_S$ is design base shear.

The more ductility and redundancy that a structural system possesses, the greater the energy dissipation capacity, the higher the value of $R$, and the lower the design seismic force.

## Overstrength Factor

The overstrength factor is the ratio of the actual capacity of a building to the design value and results from the inherent over design of members and the overstrength of members. When the weakest member in a structure yields, the structure continues to support additional load until sufficient members have failed to produce a collapse mechanism. The overstrength factor is given by

$$\Omega_0 = V_Y / V_S$$

The overstrength factor is used to determine an amplified design load for members critical to the stability of the structure.

| Structural System | $R$ | $\Omega_0$ | $C_d$ | System and height limitations | | | | |
|---|---|---|---|---|---|---|---|---|
| | | | | Seismic design category | | | | |
| | | | | B | C | D | E | F |
| **Bearing wall** | | | | | | | | |
| Light-framed walls sheathed with steel sheets | 6.5 | 3.0 | 4.0 | NL | NL | 65 | 65 | 65 |
| **Building frame** | | | | | | | | |
| Steel eccentrically braced frame | 8.0 | 2.0 | 4.0 | NL | NL | 160[a] | 160[a] | 100[b] |
| Steel special concentrically braced frames | 6.0 | 2.0 | 5.0 | NL | NL | 160[a] | 160[a] | 100[b] |
| Steel ordinary concentrically braced frames | 3.25 | 2.0 | 3.25 | NL | NL | 35 | 35 | NP |
| Light frame walls sheathed with steel sheets | 7.0 | 2.5 | 4.5 | NL | NL | 65 | 65 | 65 |
| Steel buckling-restrained braced frame | 8.0 | 2.5 | 5.0 | NL | NL | 160[a] | 160[a] | 100[b] |
| Steel special plate shear walls | 7.0 | 2.0 | 6.0 | NL | NL | 160[a] | 160[a] | 100[b] |
| **Moment-resisting frame** | | | | | | | | |
| Steel special moment frames | 8.0 | 3.0 | 5.5 | NL | NL | NL | NL | NL |
| Steel special truss moment frames | 7.0 | 3.0 | 5.5 | NL | NL | 160 | 100 | NP |
| Steel intermediate moment frames | 4.5 | 3.0 | 4.0 | NL | NL | 35[c] | NP[c] | NP[c] |
| Steel ordinary moment frames | 3.5 | 3.0 | 3.0 | NL | NL | NP[d] | NP[d] | NP[d] |
| **Dual system with special moment frames** | | | | | | | | |
| Steel eccentrically braced frames | 8.0 | 2.5 | 4.0 | NL | NL | NL | NL | NL |
| Steel special concentrically braced frames | 7.0 | 2.5 | 5.5 | NL | NL | NL | NL | NL |
| Steel buckling-restrained braced frames | 8.0 | 2.5 | 5.0 | NL | NL | NL | NL | NL |
| Steel special plate shear walls | 8.0 | 2.5 | 6.5 | NL | NL | NL | NL | NL |
| **Dual system with intermediate moment frames** | | | | | | | | |
| Steel special concentrically braced frames | 6.0 | 2.5 | 5.0 | NL | NL | 35 | NP | NP |
| Composite special concentrically braced frames | 5.5 | 2.5 | 4.5 | NL | NL | 160 | 100 | NP |
| **Shear wall-frame interactive system with ordinary reinforced concrete moment frames and ordinary reinforced shear walls** | 4.5 | 2.5 | 4.0 | NL | NP | NP | NP | NP |
| **Cantilevered column** | | | | | | | | |
| Steel special cantilever column systems | 2.5 | 1.25 | 2.5 | 35 | 35 | 35 | 35 | 35 |
| Steel ordinary cantilever column systems | 1.25 | 1.25 | 1.25 | 35 | 35 | NP[d] | NP[d] | NP[d] |
| **Steel systems not specifically detailed for seismic resistance, excluding cantilever column systems** | 3.0 | 3.0 | 3.0 | NL | NL | NP | NP | NP |

Note: NL = not limited, NP = not permitted

In accordance with ASCE 7 Sec. 12.2.5.4, the indicated heights may be increased to [a] 240 ft [b] 160 ft provided that the structure does not have an extreme torsional irregularity as defined in ASCE 7 Table 12.3-1 (horizontal structural irregularity Type 1b), and the braced frames or shear walls in any one plane do not resist more than 60 percent of the total seismic forces in each direction, neglecting accidental torsional effects.

[c]Addition height limitations are given in ASCE 7 Sec. 12.2.5.7.

[d]Addition height limitations are given in ASCE 7 Sec. 12.2.5.6.

**Table 2.10** Seismic Design Factors

## Deflection Amplification Factor

The deflection amplification factor is used to determine the actual displacements produced by the actual seismic load and is listed in ASCE 7 Table 12.2-1. As shown in Fig. 2.47, the deflection amplification factor is

$$C_d = \delta_x / \delta_S$$

where $\delta_x$ is the actual displacement caused by the actual seismic load and $\delta_S$ is the theoretical displacement caused by the design seismic loads.

Allowing for the importance factor, $I_e$, IBC Eq. (12.8-15) gives the value of the actual displacement as

$$\delta_x = C_d \delta_S / I_e$$

A listing of $R$ values, $\Omega_0$ values, structural system limitations, and building height limits for structural steel lateral force-resisting systems is provided in Table 2.10.

## Combinations of Framing Systems

When several different lateral force-resisting systems are used in a building, ASCE 7 Sec. 12.2 introduces controls to ensure that an appropriate value for the response modification coefficient is adopted. For the situation, shown in Fig. 2.48, where different lateral force-resisting systems are used along each of the two orthogonal axes of a building, the respective values of $R$ shall apply to each system.

The building shown in Fig. 2.48 has special steel moment-resisting frames in the north-south direction providing a value of $R = 8$ in that direction. Special concentrically braced frames provide the lateral force-resisting system in the east-west direction giving a value of $R = 6$ in that direction.

When different structural systems are used over the height of a building and the upper system has a response modification coefficient higher than that of the lower system, both systems are designed using their individual response modification coefficients. Forces transferred from the upper system to the lower system are increased by multiplying by the ratio of the higher response modification coefficient to the lower response modification coefficient as shown in Fig. 2.49*a*. When different structural systems are used over the height of a building and the upper system has a response modification coefficient lower than that of the lower system, the lower response modification coefficient is used for both systems as shown in Fig. 2.49*b*. Penthouses and one- and two-family dwellings are exempt from this requirement.

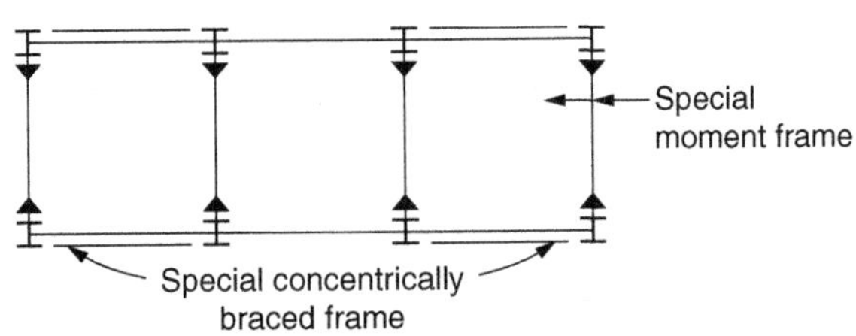


**Figure 2.48** Different systems in two orthogonal directions.

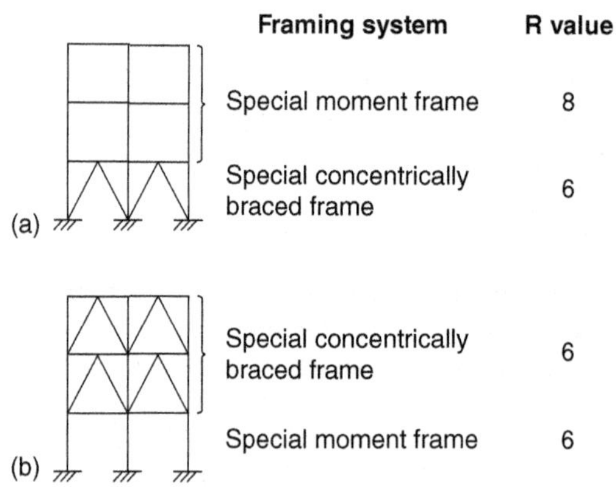


**FIGURE 2.49** Vertical combinations of framing systems.

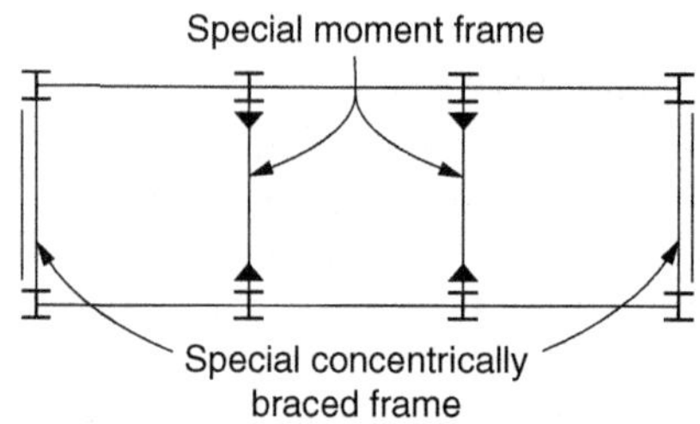


**FIGURE 2.50** Different systems in same direction.

In accordance with ASCE 7 Sec. 12.2.3.3, when different seismic-force-resisting systems are used in the same direction, as shown in Fig. 2.50, the least value of $R$ for any of the systems shall be used for that direction. The values of $C_d$ and $\Omega_0$ must be consistent with the $R$ value in that direction. For the building in Fig. 2.50, both special concentrically braced frames and special moment frames provide lateral resistance in the transverse direction. The special concentrically braced frames have an $R$ value of 6 and the special moment frames an $R$ value of 8. The $R$ value of 6 governs for the transverse direction.

Alternatively, one- and two-story structures assigned to risk category I or II, with flexible diaphragms or of light-frame construction, are exempted from this requirement. For this condition, resisting elements may be designed using the least value of $R$ for the different structural systems found in each independent line of resistance. Diaphragms must be designed for the least value of $R$ for any of the systems utilized in that direction.

**Example 2.25.** Calculation of Response Modification Coefficient
For the two-story steel-framed building shown in Fig. 2.41, determine the response modification coefficients in the north-south and east-west directions.

The values of the response modification coefficients are obtained from ASCE 7 Table 12.2-1 as

$$R = 8.0 \text{ ... for steel moment-resisting frames, north-south direction}$$

$$= 6.0 \text{ ... for special concentrically braced steel frames, east-west direction}$$

## Fundamental Period of Vibration

*(a) General approximate method:* The natural, or fundamental, period of a building depends on its height and stiffness and can vary from 0.1 second for a single-story building to 2.0 second for a 20-story building. The approximate fundamental period is given by ASCE 7 Sec. 12.8.2.1 as:

$T_a = 0.028(h_n)^{0.8}$ ... for steel moment-resisting frames

$T_a = 0.016(h_n)^{0.9}$ ... for reinforced concrete moment-resisting frames

$T_a = 0.030(h_n)^{0.75}$ ... for eccentrically braced steel frames and buckling-restrained frames

$T_a = 0.020(h_n)^{0.75}$ ... for all other structural systems

where $h_n$ is height, in feet, of the roof above the base, not including the height of penthouses or parapets.

**Example 2.26.** Calculation of Fundamental Period

For the two-story steel-framed building shown in Fig. 2.41, determine the approximate fundamental periods of vibration in both the north-south and east-west directions.

The approximate fundamental period is given by ASCE 7 Eq. (12.8-7) as

$T_a = 0.028(h_n)^{0.8}$ ... for steel moment-resisting frames

$= 0.020(h_n)^{0.75}$ ... for concentrically braced steel frames (all other systems)

where $h_n$ = roof height
= 20 ft

Then, the fundamental period is

$T_a = 0.028(20)^{0.8}$

$= 0.31$ second ... for steel moment-resisting frames, north-south direction

$= 0.020(20)^{0.75}$

$= 0.19$ second ... for concentrically braced steel frames, east-west direction

*(b) Approximate method based on number of stories:* For buildings not exceeding 12 stories above the base and with an average story height not less than 10 feet, where the seismic force-resisting system consists entirely of moment-resisting frames, the approximate fundamental period may be determined by ASCE 7 Eq. (12.8-8)

$$T_a = 0.1N$$

where $N$ is the number of stories above the base.

**Example 2.27.** Calculation of Fundamental Period

Determine the approximate fundamental period of vibration for the four-story special steel moment frame building shown in Fig. 2.51 using ASCE 7 Eq. (12.8-8)

The story height is

$h_s = 15$ ft

> 10 ft ... satisfactory

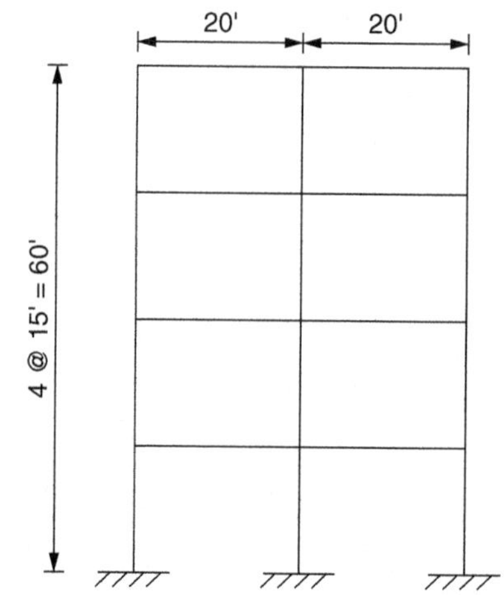


**Figure 2.51** Details for Example 2.27.

The number of stories is

$$N = 4$$
$$< 12 \text{ ... satisfactory}$$

Then, for a special steel moment frame structure, ASCE 7 Eq. (12.8-8) specifies a value for the building period of

$$T_a = 0.1N$$
$$= 0.1 \text{ H } 4$$
$$= 0.40 \text{ sec}$$

*(c) Analytical method:* In accordance with ASCE 7 Sec. 12.8.2, the fundamental period may be determined by a "properly substantiated analysis." The Rayleigh procedure may be used and the fundamental period is

$$T_r = 2\pi(\Sigma w_i \delta_i^2 / g\Sigma f_i \delta_i)^{1/2}$$
$$= (0.32)(\Sigma w_i \delta_i^2 / \Sigma f_i \delta_i)^{1/2}$$

where $\delta_i$ = elastic deflection at level $i$ due to the forces $f_i$
$f_i$ = lateral force at level $i$
$w_i$ = seismic weight located at level $i$
$g$ = acceleration due to gravity

The procedure consists of applying to the structure the lateral forces $f_i$ shown in Fig. 2.52. These forces increase approximately uniformly with height and correspond to the distribution of base shear that is assumed in ASCE 7. The weight of each story is assumed concentrated at the floor and roof levels, lateral displacements are calculated, and values inserted into the Rayleigh equation. All significant elements of the lateral force resisting system must be included in determining the stiffness of the structure. If the stiffness is underestimated, the calculated deflections and fundamental periods are

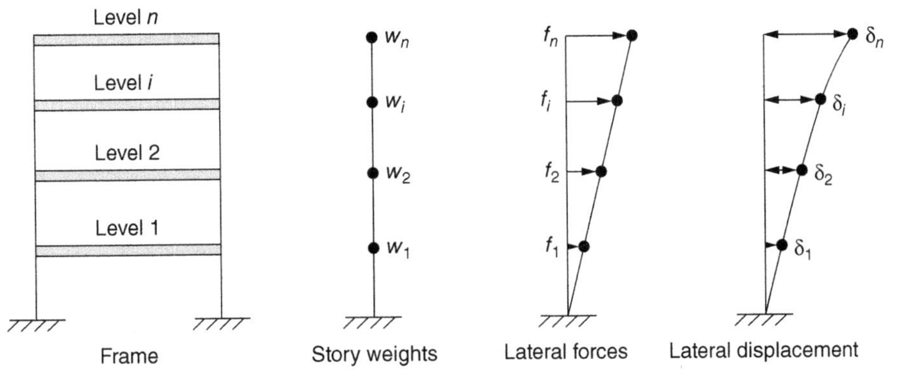


**FIGURE 2.52** Rayleigh procedure.

overestimated, giving a value for the base shear which is too low. To compensate for this, ASCE 7 Sec. 12.8.2 specifies that the value of the fundamental period determined by the Rayleigh method may not exceed

$$T = C_u T_a$$

where $T_a$ = approximate fundamental period determined by ASCE 7 Eq. (12.8-7)
$C_u$ = coefficient for upper limit on calculated period

Values of $C_u$ are given in ASCE 7 Table 12.8-1 and are shown in Table 2.11. As indicated, the values of $C_u$ are dependent on $S_{D1}$ the design, five percent damped, spectral response acceleration at a period of one second.

**Example 2.28.** Application of the Rayleigh Method
Using Rayleigh's method, determine the fundamental period of vibration of the two-story frame shown in Fig. 2.53 which is located in an area with a value for $S_{D1}$ exceeding 0.40$g$. The lateral force system shown in Fig. 2.53 may be utilized. The seismic dead weight at the second floor is 20 kips and at the roof is 10 kips. The total lateral stiffness of each story is 10 kips/in.

Applying the given force system, the displacements at each level are:

$$\text{at the second floor, } \delta_1 = (f_2 + f_1)/k_1$$
$$= (4 + 2)/10$$
$$= 0.6 \text{ in}$$
$$\text{at the roof, } \delta_2 = f_2/k_2 + \delta_1$$
$$= 4/10 + 0.6$$
$$= 1.0 \text{ in}$$

Details are given in Table 2.12.

| $S_{D1}$ | ≥0.40 | 0.30 | 0.20 | 0.15 | ≤0.10 |
|---|---|---|---|---|---|
| $C_u$ | 1.4 | 1.4 | 1.5 | 1.6 | 1.7 |

**TABLE 2.11** Coefficient for Upper Limit on the Calculated Period

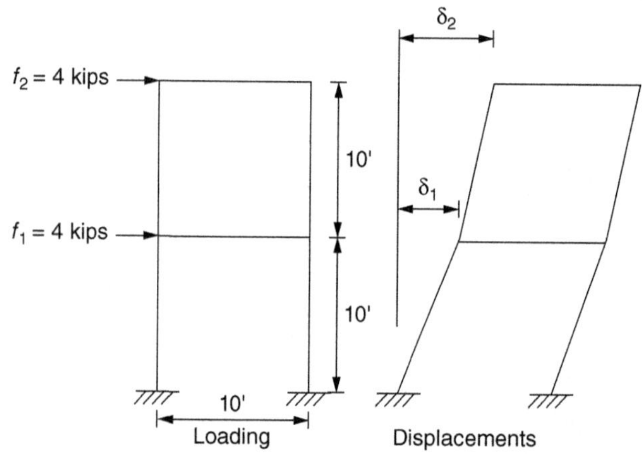


**FIGURE 2.53** Details for Example 2.28.

| Level | $w_i$ | $f_i$ | $\delta_i$ | $w_i\delta_i^2$ | $f_i\delta_i$ |
|---|---|---|---|---|---|
| 2 | 10 | 4 | 1.0 | 10.0 | 4.0 |
| 1 | 20 | 2 | 0.6 | 7.2 | 1.2 |
| Total | 30 | - | - | 17.2 | 5.2 |

**TABLE 2.12** Rayleigh procedure for Ex. 2.28

The Rayleigh fundamental period is

$$T_r = (0.32)(\Sigma w_i\delta_i^2/\Sigma f_i\delta_i)^{1/2}$$
$$= 0.32(17.2/5.2)^{1/2}$$
$$= 0.58 \text{ sec}$$

In a location with a value for $S_{D1} > 0.4$, the value of the fundamental period is limited by Table 2.11 to $1.4T_a$ where $T_a$ is given by ASCE 7 Eq. (12.8-7) as

$$T_a = 0.028(h_n)^{0.8}$$
$$= 0.028(20)^{0.8}$$
$$= 0.31 \text{ second}$$

Hence the fundamental period, in accordance with ASCE 7 Sec. 12.8.2, may not exceed

$$C_uT_a = 1.4T_a$$
$$= 1.4 \text{ H } 0.31$$
$$= 0.43 \text{ second}$$
$$< T_r$$

Hence the maximum permitted value of the fundamental period is

$$T = 0.43 \text{ second}$$

## Seismic Response Coefficient

As noted in *NEHRP Recommended Seismic Provisions for New Buildings and Other Structures* (FEMA)[11] Sec. 8.2, determination of the seismic response coefficient forms the basis of the equivalent lateral force (ELF) procedure. This provides a simple design approach when a complex analysis is not required and is permitted for most structures. The procedure is valid only for structures without significant discontinuities in mass and stiffness over the height. The ELF procedure has three basic steps:

- Determine the seismic response coefficient and the base shear.
- Distribute the shear vertically along the height of the structure.
- Distribute the shear horizontally across the width and breadth of the structure.

The seismic response coefficient $C_S$ is defined in ASCE 7 Sec. 12.8.1.1. For the longer period, velocity-governed region of the spectrum ASCE 7 Eq. (12.8-3) gives the value as

$$C_s = S_{D1}I_e/RT$$

where $S_{D1}$ = design response acceleration at a period of 1 second
$I_e$ = importance factor
$R$ = response modification coefficient
$T$ = fundamental period of the structure

The maximum value of the seismic response coefficient for the short period, constant acceleration region of the spectrum, is given by ASCE 7 Eq. (12.8-2) as

$$C_s = S_{DS}I_e/R$$

where $S_{DS}$ is design response acceleration at short periods.

At a period of $T = T_S$, the transition from one equation to the other occurs and $T_S$ is given by

$$T_S = S_{D1}/S_{DS}$$

For the constant-displacement region of the spectrum, at periods exceeding the value $T = T_L$, the value of the seismic response coefficient is given by ASCE 7 Eq. (12.8-4) as

$$C_s = S_{D1}T_LI_e/RT^2$$

where $T_L$ is long-period transition period given in ASCE 7 Figs. 22-12 through 22-15.

The code related response spectrum is illustrated in Fig. 2.54.

To prevent too low a value of the seismic response coefficient being used for tall buildings, where $S_1 < 0.6g$, the minimum permitted value of $C_s$ is given by ASCE 7 Eq. (12.8-5) as

$$C_s = 0.044S_{DS}I_e$$
$$\geq 0.01$$

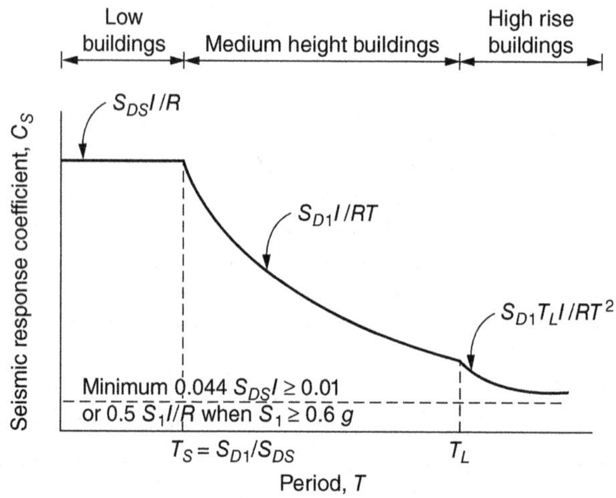


**FIGURE 2.54** Seismic response coefficient.

In addition, for structures located in areas of high seismicity where $S_1 \geq 0.6g$, the seismic response coefficient must not be less than that given by ASCE 7 Eq. (12.8-6) as

$$C_s = 0.5S_1I_e/R$$

The equivalent lateral force procedure is permitted for all buildings assigned to seismic design categories B and C. The equivalent lateral force procedure is not permitted:

- For a regular structure assigned to seismic design categories D, E, and F with a height exceeding 160 ft and a fundamental period exceeding $3.5T_S$.
- For buildings assigned to seismic design categories D, E, and F having one or more horizontal irregularities of Type 1a or 1b in ASCE 7 Table 12.3-1 or vertical irregularities of Type 1, 2, 3 in ASCE 7 Table 12.3-2.

When the equivalent lateral force procedure is not permitted, a modal response spectrum analysis, as described by Paz[12] and SEAOC,[13] may be used.

The equivalent lateral force procedure is also permitted for buildings assigned to seismic design category D, E, or F provided that the building

- Is of light-frame construction
- Has a risk category of I or II and is of any construction not exceeding two stories in height
- Has no structural irregularities with a height not exceeding 160 feet
- Does not exceed a height of 160 feet and has neither horizontal irregularities Type 1a (torsional) or Type 1b (extreme torsional) nor vertical irregularities

Type 1a (soft story), Type 1b (extreme soft story), Type 2 (mass), or Type 3 (geometric) as defined in ASCE 7 Tables 12.3-1 and 12.3-2

- Has a height exceeding 160 feet with no structural irregularities and a fundamental period $T < 3.5T_S$

As specified in ASCE 7 Table 12.6-1, a modal analysis is necessary for all other structures assigned to seismic design categories D, E, and F.

In seismic design category B, members are designed for the seismic forces applied independently in each of two orthogonal directions and orthogonal interaction effects are neglected in conformance with ASCE 7 Sec. 12.5.2.

In seismic design category C, interaction effects must be considered in a structure with vertical lateral force-resistant elements aligned in two or more directions that are not orthogonal to each other as defined by horizontal irregularity Type 5 (nonparallel system) in ASCE 7 Table 12.3-1. For this situation, ASCE 7 Sec. 12.5.3a requires members to be designed for 100% of the seismic force applied along one primary axis simultaneously with 30% of the seismic force along the orthogonal direction.

In seismic design category D, E, and F, ASCE 7 Sec. 12.5.4 requires interaction effects to be considered in a member forming part of two or more intersecting seismic force-resisting systems when the axial force in the member due to the seismic force applied along either principal orthogonal axis equals or exceeds 20% of the axial design strength of the member. The member must be designed for 100% of the seismic force applied along one primary axis simultaneously with 30% of the seismic force along the orthogonal direction.

In accordance with ASCE 7 Sec. 12.3.3.1, structures in seismic design categories E and F with horizontal irregularities Type 1b as defined in ASCE 7 Table 12.3-1 or vertical irregularities Types 1b, 5a, and 5b as defined in ASCE 7 Table 12.3-2 are not permitted. Structures in seismic design category D with vertical irregularity Type 5b are not permitted.

**Example 2.29.** Calculation of Seismic Response Coefficient

Determine the seismic response coefficient for the two-story, moment-resisting, structural steel frame shown in Fig. 2.41.

From previous examples the relevant parameters are

$$S_{D1} = 0.6g$$

$$S_{DS} = 1.0g$$

$$S_1 = 0.6g$$

$$I_e = 1.0$$

$$\text{SDC} = \text{D}$$

$T_a = 0.31$ second ... for steel moment-resisting frames

$= 0.19$ second ... for concentrically braced steel frames

$R = 8.0$ ... for steel moment-resisting frames

$= 6.0$ ... for special concentrically braced steel frames

The value of the response spectrum parameter is obtained from ASCE 7 Sec. 11.4.5 as

$$T_S = S_{D1}/S_{DS}$$
$$= 0.6/1.0$$
$$= 0.6 \text{ second}$$
$$> T_a \text{ ... for both north-south and east-west directions}$$

Hence, ASCE 7 Eq. (12.8-2) governs for both directions and the seismic response coefficient is given by

$$C_s = S_{DS}I_e/R$$
$$= 1.0 \times 1.0/8.0$$
$$= 0.125 \text{ ... for steel moment-resisting frames}$$
$$= 1.0 \times 1.0/6.0$$
$$= 0.167 \text{ ... for special concentrically braced steel frames}$$

## Effective Seismic Weight

The effective seismic weight $W$ consists of the total dead load of the building above the base and the following additional loads specified by ASCE 7 Sec. 12.7.2:

- Twenty-five percent of the floor live load for storage and warehouse occupancies. Storage loads at any level adding no more than 5 percent to the effective seismic weight is excepted as is also floor live load in public parking garages and open parking structures.
- An allowance of 10 lb/ft$^2$ for moveable partitions or the actual weight whichever is greater.
- Where the flat roof snow load $p_f$ exceeds 30 lb/ft$^2$, 20 percent of the uniform design snow load, regardless of roof slope.
- The total operating weight of permanent equipment.
- Weight of landscaping and other materials at roof gardens and similar areas.

Roof and floor live loads, except as noted above, are not included in the value of $W$. The effective seismic weight is assumed concentrated at the roof and at each floor level.

**Example 2.30.** Calculation of Effective Seismic Weight
Determine the effective seismic weight of the two-story structural steel frame shown in Fig. 2.41.

The two-story, steel-framed building shown in Fig. 2.41 forms part of the construction of an office building. The component weights, including all framing, are

| | |
|---|---|
| Roof | 60 lb/ft$^2$ |
| Second floor | 65 lb/ft$^2$ |
| Walls | 25 lb/ft$^2$ |

No allowance is required for permanent equipment or snow loads and the end walls may be neglected. Determine the effective seismic weight, corresponding to the north-south direction, at the roof and second-floor levels.

The relevant dead load tributary to the roof in the north-south direction is due to the roof dead load and half the story height of the north wall, south wall, and partition load and is given by

| | | |
|---|---|---|
| Roof | $= 60 \times 30$ | $= 1800$ lb/ft |
| North wall | $= 25 \times 10/2$ | $= 125$ lb/ft |
| South wall | | $= 125$ lb/ft |
| Partitions | $= 10 \times 30/2$ | $= 150$ lb/ft |
| Total at roof | | 2200 lb/ft |

The relevant dead load tributary to the second floor in the north-south direction is due to the floor dead load and the full story height of the north wall, south wall, and partition load and is given by

| | | |
|---|---|---|
| Floor | $= 65 \times 30$ | $= 1950$ lb/ft |
| North wall | $= 25 \times 10$ | $= 250$ lb/ft |
| South wall | | $= 250$ lb/ft |
| Partitions | $= 10 \times 30$ | $= 300$ lb/ft |
| Total at second floor | | 2750 lb/ft |

The total effective seismic weight for the building is

$$W = 2200 + 2750$$
$$= 4950 \text{ lb/ft}$$

## Seismic Base Shear

The base shear produced in a building by the ground motion may be determined by the equivalent lateral force procedure. This utilizes Newton's second law of motion and the seismic base shear is determined by ASCE 7 Eq. (12.8-1) as

$$V = C_s W$$

Forces and displacements in the structural elements are determined from the base shear assuming linear elastic behavior and relying on the dissipation of earthquake energy by inelastic behavior. However, critical elements in a building are designed to remain nominally elastic by designing for an amplified force consisting of the design seismic force multiplied by the overstrength factor $\Omega_0$.

**Example 2.31.** Calculation of Seismic Base Shear
The two-story steel-framed building shown in Fig. 2.41 forms part of the construction of an office building. Determine the seismic base shear, corresponding to the north-south direction.

The total effective seismic weight for the building is derived in Example 2.30 as

$$W = 4950 \text{ lb/ft}$$

The value of the seismic response coefficient is derived in Example 2.29 as

$$C_s = 0.125 \text{ ... for steel moment-resisting frames}$$

Hence, the base shear is given by ASCE 7 Eq. (12.8-1) as

$$V = C_s W$$
$$= 0.125 \times 4950$$
$$= 619 \text{ lb/ft}$$

## Vertical Distribution of Seismic Forces

For a multistory building, the distribution of the base shear over the height of the building depends on the fundamental period of the building. In a building with a fundamental period of not more than 0.5 second, the fundamental mode predominates and the base

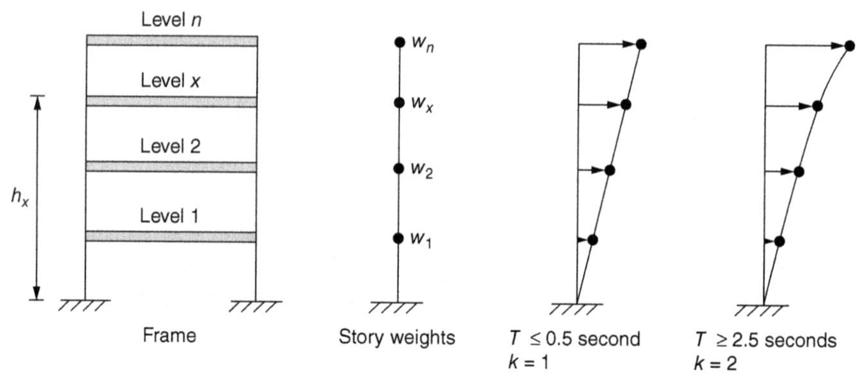


**FIGURE 2.55** Vertical force distribution.

shear is distributed linearly over the height, varying from zero at the base to a maximum at the top. The distribution is shown in Fig. 2.55. In a building with a fundamental period of 2.5 seconds or more, higher mode effects must be considered and the distribution is parabolic in shape as indicated in FEMA Sec. C12.8.3.

The forces are calculated at each floor level and at roof level and are used in the design of the lateral force-resisting systems consisting of shear walls, braced frames, and moment-resisting frames. The distribution of base shear over the height of a building is obtained from ASCE 7 Sec. 12.8.3 and the design lateral force at level $x$ is given by ASCE 7 Eqs. (12.8-11) and (12.8-12) as

$$F_x = Vw_x h_x^k / \Sigma w_i h_i^k$$

where $V$ = base shear
$F_x$ = design lateral force at level $x$
$w_i$ = portion of the total effective seismic weight located at any level $i$
$w_x$ = portion of the total effective seismic weight located at a specific level $x$
$h_i$ = height above the base to any level $i$
$h_x$ = height above the base to a specific level $x$
$\Sigma w_i h_i^k$ = summation, over the whole structure, of the product of $w_i$ and $h_i^k$
$k$ = distribution exponent
= 1.0 ... for $T \leq 0.5$ second
= 2.0 ... for $T \geq 2.5$ seconds

For intermediate values of $T$, a linear variation of $k$ may be assumed.

This method is appropriate for a building with a uniform distribution of floor mass and story stiffness over its height. For a single-story structure, this reduces to

$$F_x = V$$

**Example 2.32.** Vertical Force Distribution
Determine the vertical force distribution, in the north-south direction, for the two-story steel-framed building shown in Fig. 2.41.

The fundamental period of vibration was derived in Example 2.26 as

$$T = 0.31 \text{ second}$$
$$< 0.5 \text{ second}$$

| Level | $w_x$ | $h_x$ | $w_x h_x$ | $F_x$ |
|---|---|---|---|---|
| Roof | 2200 | 20 | 44,000 | 381 |
| 2nd floor | 2750 | 10 | 27,500 | 238 |
| Total | 4950 | – | 71,500 | 619 |

**TABLE 2.13** Details for Example 2.32

The value of the distribution exponent factor is obtained from ASCE 7 Sec. 12.8.3 as

$$k = 1.0$$

Hence, the expression for $F_x$ reduces to

$$F_x = Vw_x h_x / \Sigma w_i h_i$$

The effective seismic weights located at second-floor and roof levels are obtained from Example 2.30 and the relevant values are given in Table 2.13.
From Example 2.31, the base shear is given by

$$V = 619 \text{ lb/ft}$$

The design lateral force at level $x$ is

$$F_x = Vw_x h_x / \Sigma w_i h_i$$
$$= 619(w_x h_x)/71{,}500$$
$$= 0.00866 w_x h_x$$

The values of $F_x$, in lb/ft, are shown in Table 2.13.

## Overturning

In accordance with ASCE 7 Sec. 12.8.5, buildings must be designed to resist the overturning effects caused by the seismic forces at each level. In Fig. 2.56, the overturning moment at level $x$ is given by

$$M_x = \Sigma F_i(h_i - h_x)$$

where: $F_i$ = design lateral force at level $i$ as determined by ASCE 7 Eq. (12.8-11)
$h_i$ = height above the base to level $i$
$h_x$ = height above the base to a specific level $x$

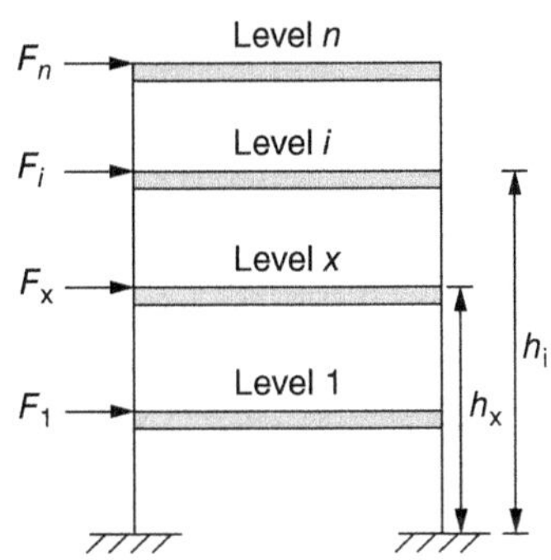


**FIGURE 2.56** Overturning moment.

In accordance with ASCE 7 Sec. 12.4.2.2, the vertical effect of the seismic forces is neglected when determining demands on the soil-structure interface of foundations. Hence, in ASCE 7 Sec. 12.4.2.3, ASD load combination no. 8 becomes $(0.6D + 0.7\rho Q_E)$.

The horizontal forces determined using ASCE 7 Eq. (12.8-11) represent values that envelop the maximum values that can occur. Hence, overturning moments based on these values are conservative and, in accordance with ASCE 7 Sec. 12.13.4, overturning moments at the soil-foundation interface are permitted to be reduced by 25% provided that both of the following conditions are satisfied:

a. The structure is designed by the equivalent lateral force procedure

b. The structure is not a cantilever column structure

No overturning reduction is allowed in the above-grade portion of the structure.

**Example 2.33.** Calculation of Overturning Moment
Determine the factor of safety against overturning in the north-south direction for the two-story structural steel frame building shown in Fig. 2.41. The length of the building is $l = 90$ ft and the width is $b = 30$ ft. The weight of the end walls may be neglected. The redundancy factor is $\rho = 1.0$ and the seismic design category is D.

The effective seismic weight determined in Ex. 2.30 includes an allowance for partitions and this is used to calculate the seismic base shear as required by ASCE 7 Sec. 12.7.2. For all other procedures, ASCE 7 Sec. 4.3.2 requires partitions to be considered a live load. Hence, the dead load, not including partitions, at level 2 is obtained from Ex. 2.30 as

$$\begin{aligned} W_2 &= (1800 + 125 + 125)l/1000 \\ &= (1800 + 125 + 125)90/1000 \\ &= 184.5 \text{ kips} \end{aligned}$$

The dead load, not including partitions, at level 1 is obtained from Ex. 2.30 as

$$\begin{aligned} W_1 &= (1950 + 250 + 250)l/1000 \\ &= (1950 + 250 + 250)90/1000 \\ &= 220.5 \text{ kips} \end{aligned}$$

The lateral forces acting on the building are obtained from Example 2.32 as

$$\begin{aligned} F_2 &= 381 \times l/1000 \\ &= 381 \times 90/1000 \\ &= 34.3 \text{ kips} \\ F_1 &= 238 \times l/1000 \\ &= 238 \times 90/1000 \\ &= 21.4 \text{ kips} \end{aligned}$$

The factor of safety against overturning is determined using the load factors given in ASCE 7 Sec. 12.4.2.3 ASD load combination no. 8. The load factors are 0.6 for dead load and 0.7 for seismic load. The effects of vertical seismic forces are neglected and the factored overturning moment is given by

$$\begin{aligned} M_O &= 0.7(F_2 \times h_2 + F_1 \times h_1) \\ &= 0.7(34.3 \times 20 + 21.4 \times 10) \\ &= 630 \text{ kip-ft} \end{aligned}$$

A reduction of 25% is allowed to give

$$M_{OR} = 0.75 \times 630$$
$$= 473 \text{ kip-ft}$$

The factored restoring moment is given by

$$M_R = 0.6 \times b(W_1 + W_2)/2$$
$$= 0.6 \times 30(220.5 + 184.5)/2$$
$$= 3645 \text{ kip-ft}$$

The factor of safety against overturning is

$$M_R/M_{OR} = 3645/473$$
$$= 7.7$$

## Diaphragm Loads

The method used to calculate the vertical distribution of forces can underestimate the forces applied to diaphragms, particularly at the lower levels, as indicated by Sabelli et al.[14] This is due to the higher modes of excitation being underestimated in ASCE 7 Sec. 12.8.3 which represents primarily the first mode of vibration. However, the higher modes of vibration can impose significantly larger forces at lower levels. ASCE 7 Sec. 12.10.1.1 requires diaphragms to be designed for the maximum of the two following conditions:

(a) The value of the design lateral force $F_x$ at each level calculated from ASCE 7 Eq. (12.8-11) and (12.8-12) which give

$$F_x = Vw_x h_x^k / \Sigma w_i h_i^k$$

(b) The value of the diaphragm design force $F_{px}$ calculated from ASCE 7 Eq. (12.10-1) as

$$F_{px} = w_{px} \Sigma F_i / \Sigma w_i$$
$$\geq 0.2 S_{DS} I_e w_{px} \quad \text{Eq. (12.10-2)}$$
$$\leq 0.4 S_{DS} I_e w_{px} \quad \text{Eq. (12.10-3)}$$

where $S_{DS}$ = design response acceleration coefficient at short periods
$I_e$ = occupancy importance factor
$F_i$ = lateral force at level $i$ calculated by ASCE 7 Sec. 12.8.3
$\Sigma F_i$ = total shear force at level $i$
$w_i$ = total seismic weight located at level $i$
$\Sigma w_i$ = total seismic weight at level $i$ and above
$w_{px}$ = seismic weight tributary to the diaphragm at level $x$, including walls normal to the direction of the seismic load

For a single story structure, this equation reduces to

$$F_{px} = Vw_{px}/W$$
$$= C_s w_{px}$$

| Level | $\Sigma w_i$ | $\Sigma F_i$ | $\Sigma F_i/\Sigma w_i$ | Max | Min | $w_{px}$ | $F_{px}$ |
|---|---|---|---|---|---|---|---|
| Roof | 2200 | 381 | 0.173 | 0.400 | 0.200 | 2200 | 440 |
| 2nd floor | 4950 | 619 | 0.125 | 0.400 | 0.200 | 2750 | 550 |

**Table 2.14** Details for Example 2.34

Diaphragm loads determined using ASCE 7 Eq. (12.10-1) are not subject to the redundancy factor ρ. In calculating the value of $w_{px}$, the weight of shear walls parallel to the direction of the seismic load are not included. These walls do not contribute to the force on the diaphragm but rather support the diaphragm and transfer the lateral load to the foundations.

**Example 2.34.** Diaphragm Loads
Determine the diaphragm loads, in the north-south direction, for the two-story steel-framed building shown in Fig. 2.41.

ASCE 7 Sec. 12.10.1.1 is applicable and the diaphragm loads are given by

$$F_{px} = w_{px}\Sigma F_i/\Sigma w_i$$

The values of $w_{px}$ are determined in Example 2.30 and are shown in Table 2.14. In Example 2.30, end walls of the building were neglected hence, in this case, values of $w_i$ and $w_{px}$ are identical.

From previous examples, the relevant parameters are

$$S_{DS} = 1.0g$$

$$I_e = 1.0$$

The maximum applicable value for the diaphragm load is given by

$$\begin{aligned} F_{px} &= 0.4S_{DS}I_e w_{px} \\ &= 0.4 \times 1.0 \times 1.0w_{px} \\ &= 0.40w_{px} \end{aligned}$$

The minimum applicable value for the diaphragm load is given by

$$\begin{aligned} F_{px} &= 0.2S_{DS}I_e w_{px} \\ &= 0.2 \times 1.0 \times 1.0w_{px} \\ &= 0.20w_{px} \text{ ... governs at both levels} \end{aligned}$$

The values of the diaphragm loads, in lb/ft, are given in Table 2.14 and are greater than the values of $F_x$ derived in Example 2.32. Hence, the values for $F_{px}$ govern.

## Story Drift

Story drift is defined in ASCE 7 Sec. 12.8.6 as the difference of the inelastic deflections at the centers of mass at the top and bottom of the story under consideration. The maximum allowable story drift $\Delta_a$ is given in ASCE 7 Table 12.12-1 and shown in Table 2.15.

The effect of drift in a two-story structure is shown in Fig. 2.57. The drift in the top story is $\Delta_2$ and the drift in the bottom story is $\Delta_1$. Excessive drift in a structure may cause damage to cladding and finishes and to nonstructural walls and partitions. In addition, the secondary stresses, or *P*-delta effects, introduced into the columns may produce instability and collapse.

| Building Type | Risk Category | | |
|---|---|---|---|
| | I or II | III | IV |
| One-story buildings with fittings designed to accommodate drift | no limit | no limit | no limit |
| Buildings, other than masonry shear wall buildings, of four stories or less with fittings designed to accommodate drift[a] | $0.025h_{sx}$ | $0.020h_{sx}$ | $0.015h_{sx}$ |
| Masonry cantilever shear wall buildings | $0.010h_{sx}$ | $0.010h_{sx}$ | $0.010h_{sx}$ |
| Other masonry shear wall buildings | $0.007h_{sx}$ | $0.007h_{sx}$ | $0.007h_{sx}$ |
| All other buildings[a] | $0.020h_{sx}$ | $0.015h_{sx}$ | $0.010h_{sx}$ |

$h_{sx}$ = story height below level $x$.
[a]For moment frames in structures assigned to seismic design categories D through F, these values are divided by ρ.

**TABLE 2.15** Maximum Allowable Story Drift $\Delta_a$

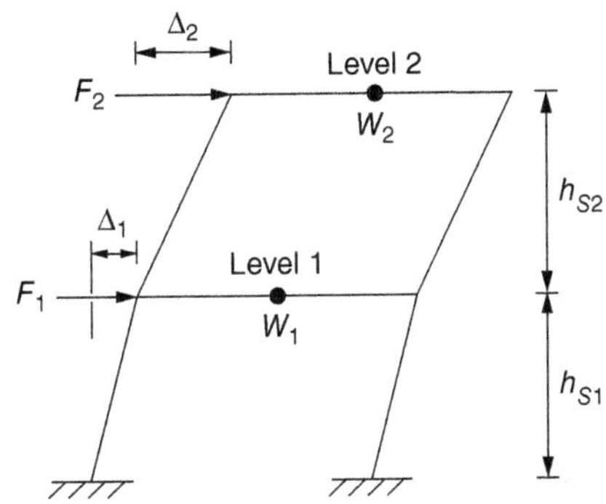


**FIGURE 2.57** Story drift.

Deflections in the structure are calculated by an elastic analysis and multiplied by the deflection amplification factor to determine the inelastic deflections. The amplified deflection at level $x$ is given by ASCE 7 Eq. (12.8-15) as

$$\delta_x = C_d \delta_{xe} / I_e$$

where: $\delta_{xe}$ = horizontal deflection at level $x$ determined by an elastic analysis using the code-prescribed design-level forces
$I_e$ = seismic importance factor given in Table 2.7
$C_d$ = deflection amplification factor given in Table 2.10

When allowable stress design methods are used, ASCE 7 Sec. 12.8.6 requires drift to be calculated using the code-prescribed strength level forces without multiplying by the load factor 0.7.

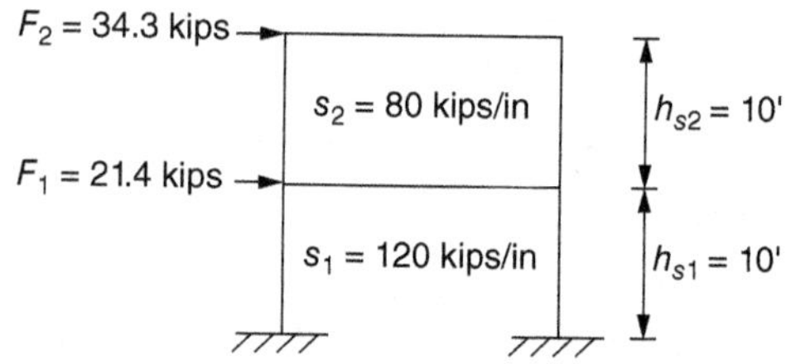


**FIGURE 2.58** Details for Example 2.35.

In accordance with ASCE 7 Sec. 12.8.6.2, the fundamental period of vibration determined by the Rayleigh method may be used to calculate the seismic base shear and the drift. The upper bound limitation imposed by ASCE 7 Sec. 12.8.2 is not applicable.

In accordance with ASCE 7 Sec. 12.3.4.1 the redundancy factor $\rho = 1.0$ when calculating drift and in accordance with ASCE 7 Sec. 12.8.7 *P*-delta effects need not be included in the calculation of drift when the stability coefficient $\theta \leq 0.10$.

**Example 2.35.** Calculation of Drift

Determine the drift in the north-south direction in the top and bottom stories of the two-story structural steel frame building shown in Fig. 2.41. The redundancy factor is $\rho = 1.0$, the seismic design category is D, and the building is assigned to risk category II with fittings designed to accommodate drift. The lateral stiffness of the top story is $s_2 = 80$ kips/in, the lateral stiffness of the bottom story is $s_1 = 120$ kips/in, and the relevant details are shown in Fig. 2.58. *P*-delta effects need not be considered.

The value of the deflection amplification factor for a steel special moment frame is obtained from Table 2.10 as

$$C_d = 5.5$$

The value of the importance factor, for a standard occupancy structure, is obtained from Table 2.7 as

$$I_e = 1.0$$

For each story, the maximum allowable inelastic story drift for a two-story structure with $\rho = 1.0$, assigned to risk category II, with fittings designed to accommodate drift is given by Table 2.15 as

$$\begin{aligned}\Delta_a &= 0.025h_{sx}\\ &= 0.025 \times 10 \times 12\\ &= 3.0 \text{ in}\end{aligned}$$

The elastic displacements at level 1 and level 2 are given by

$$\begin{aligned}\delta_{1e} &= (F_2 + F_1)/s_1\\ &= (34.3 + 21.4)/120\\ &= 0.46 \text{ in}\end{aligned}$$

$$\begin{aligned}\delta_{2e} &= F_2/s_2 + \delta_{1e}\\ &= 34.3/80 + 0.46\\ &= 0.89 \text{ in}\end{aligned}$$

The inelastic displacements at level 1 and level 2 are given by ASCE 7 Eq. (12.8-15) as

$$\delta_1 = C_d\delta_{1e}/I_e$$
$$= 5.5 \times 0.46/1.0$$
$$= 2.53 \text{ in}$$
$$\delta_2 = C_d\delta_{2e}/I_e$$
$$= 5.5 \times 0.89/1.0$$
$$= 4.90 \text{ in}$$

The drift in the top story is

$$\Delta_2 = (\delta_2 - \delta_1)$$
$$= 4.90 - 2.53$$
$$= 2.37 \text{ in}$$
$$< \Delta_a = 3.0 \text{ in} \ldots \text{satisfactory}$$

The drift in the bottom story is

$$\Delta_1 = \delta_1$$
$$= 2.53 \text{ in}$$
$$< \Delta_a = 3.0 \text{ in} \ldots \text{satisfactory}$$

## *P*-Delta Effects and Stability Coefficient

The primary moment in a story is defined as the product of the seismic shear in the story and the height of that story. The primary moments in the top and bottom stories of the two-story frame shown in Fig. 2.57 are

$$M_{P2} = F_2 h_{s2}$$
$$M_{P1} = (F_2 + F_1)h_{s1}$$

The *P*-delta effects in a given story are caused by the eccentricity of the gravity load above the story which produces secondary moments augmenting the primary moment. The secondary moment in a story is defined as the product of the total unfactored gravity load above the story and the *elastic* drift of that story. The secondary moments in the top and bottom stories of the two-story frame shown in Fig. 2.57 are

$$M_{S2} = W_2\Delta_2 I_e/C_d$$
$$M_{S1} = (W_2 + W_1)\Delta_1 I_e/C_d$$

where: $W_x$ = total unfactored dead plus floor live plus roof snow load acting at and above level $x$

$\Delta_x$ = story drift at level $x$

$\Delta_x I_e/C_d$ = elastic drift at level $x$

The stability coefficient is defined as the ratio of the secondary moment at a story to the primary moment at that story and is an indication of how sensitive the structure is to *P*-delta effects. The stability coefficient is given by

$$\theta = M_{Sx}/M_{Px}$$

The stability coefficient is also defined by ASCE 7 Eq. (12.8-16) as

$$\theta = P_x \Delta I_e / V_x h_{sx} C_d$$

where: $P_x$ = total unfactored gravity load at and above level $x$
$\Delta$ = design story drift occurring simultaneously with $V_x$
$V_x$ = seismic shear force acting between levels $x$ and $(x - 1)$
$h_{sx}$ = story height below level $x$
$C_d$ = deflection amplification factor given in Table 2-10

When the stability coefficient $\theta \geq 1.0$, the structure is unstable. In accordance with ASCE 7 Eq. (12.8-17), the stability coefficient in any story must not exceed the value

$$\theta_{max} = 0.5/\beta C_d$$
$$\leq 0.25$$

Where $\beta$ is the ratio of the shear demand to the shear capacity for the story between levels $x$ and $(x - 1)$ and may conservatively be considered equal to 1.0. Where the stability coefficient in any story exceeds 0.1, the effects of secondary moments must be included in the determination of story drifts and element forces for the whole structure. In accordance with ASCE 7 Sec. 12.8.7, the revised story drift, allowing for $P$-delta effects, is obtained as the product of the calculated drift and the factor $1/(1 - \theta)$. Where $\theta > \theta_{max}$, the structure is potentially unstable and must be redesigned.

The stability coefficients in the top and bottom stories of the two-story frame shown in Fig. 2.57 are

$$\theta_2 = M_{S2}/ M_{P2}$$
$$\theta_1 = M_{S1}/ M_{P1}$$

**Example 2.36.** Calculation of Stability Coeficient
Determine the stability coefficient in the north-south direction in the bottom story of the two-story office building shown in Fig. 2.41. The redundancy factor is $\rho = 1.0$, the seismic design category is D, the importance factor is $I_e = 1.0$. Snow load on the roof is not applicable and the relevant details are shown in Fig. 2.58. $P$-delta effects need not be considered.

Live load on the second floor is obtained from ASCE 7 Table 4-1 as 50 lb/ft$^2$. To this must be added a 15 lb/ft$^2$ allowance for partitions as specified in ASCE 7 Sec. 4.3.2. The total live load on the second floor is

$$W_L = (0.05 + 0.015)30 \times 90$$
$$= 175.5 \text{ kips}$$

The dead loads at level 1 and level 2 are determined in Example 2.33 as

$$W_{D1} = 220.5 \text{ kips}$$
$$W_{D2} = 184.5 \text{ kips}$$

The total dead plus floor live plus roof snow load on the structure is

$$W = W_{D1} + W_{D2} + W_L$$
$$= 220.5 + 184.5 + 175.5$$
$$= 580.5 \text{ kips}$$

The drift in the bottom story is derived in Example 2.35 as

$$\Delta_1 = 2.53 \text{ in}$$

The seismic loads acting on the building are derived in Example 2.33 as

$$F_2 = 34.3 \text{ kips}$$

$$F_1 = 21.4 \text{ kips}$$

The primary moment in the bottom story is

$$\begin{aligned} M_{p1} &= (F_2 + F_1)h_{s1}/I_e \\ &= (34.3 + 21.4) \times 10 \times 12/1.0 \\ &= 6684 \text{ kip-in} \end{aligned}$$

$$\begin{aligned} M_{S1} &= W\Delta_1/C_d I_e \\ &= 580.5 \times 2.53/(5.5 \times 1.0) \\ &= 267 \text{ kip-in} \end{aligned}$$

The stability coefficient in the bottom story is

$$\begin{aligned} \theta_1 &= M_{S1}/M_{P1} \\ &= 267/6684 \\ &= 0.04 \end{aligned}$$

$< 0.1$ . . . the effects of secondary moments may be neglected

## Building Separation

In accordance with ASCE 7 Sec. 12.12.3, all portions of a structure must be separated a sufficient distance to permit independent seismic motion without impact between adjacent portions. The required separation is given by AISC 7 Eq. (12.12-2) as

$$\delta_{MT} = [(\delta_{M1})^2 + (\delta_{M2})^2]^{½}$$

where: $\delta_{M1}$ = maximum inelastic displacement of building 1
$\delta_{M2}$ = maximum inelastic displacement of building 2

The displacements of both buildings are determined at the same height.

Similarly, where a structure adjoins a property line not common to a public way, the structure must be set back from the property line by at least the distance $\delta_M$ of that structure.

**Example 2.37.** Calculation of Building Separation
Determine the separation required for the two steel special moment frames shown in Fig. 2.59. The drift in each story of building 1 is $0.020h_s$ and the drift in each story of building 2 is $0.015h_s$.

The maximum inelastic displacement at the roof level of building 1 is

$$\begin{aligned} \delta_{M1} &= 2 \times 0.020h_s \\ &= 2 \times 0.020 \times 10 \times 12 \\ &= 4.80 \text{ in} \end{aligned}$$

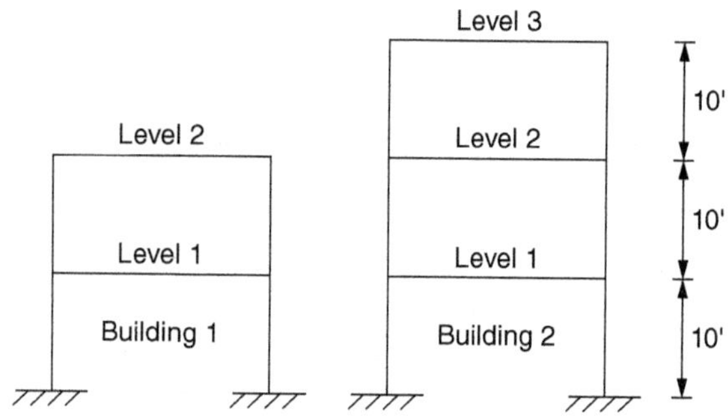


**FIGURE 2.59** Details for Example 2.37.

The maximum inelastic displacement at level 2 of building 2 is

$$\begin{aligned}\delta_{M2} &= 2 \times 0.015h_s \\ &= 2 \times 0.015 \times 10 \times 12 \\ &= 3.60 \text{ in}\end{aligned}$$

The required separation is

$$\begin{aligned}\delta_{MT} &= [(\delta_{M1})^2 + (\delta_{M2})^2]^{½} \\ &= [(4.80)^2 + (3.60)^2]^{½} \\ &= 6.0 \text{ in}\end{aligned}$$

## Flexible Diaphragms

A diaphragm is classified as flexible, in accordance with ASCE 7 Sec. 12.3.1.3, when the maximum displacement of the diaphragm under lateral load, exceeds twice the average story drift of adjoining vertical elements of the lateral force-resisting system. This is shown in Fig. 2.60 and the diaphragm is flexible if

$$\delta_M > 2\delta_A$$

where $\delta_M$ is maximum displacement of the diaphragm and $\delta_A$ is average story drift.

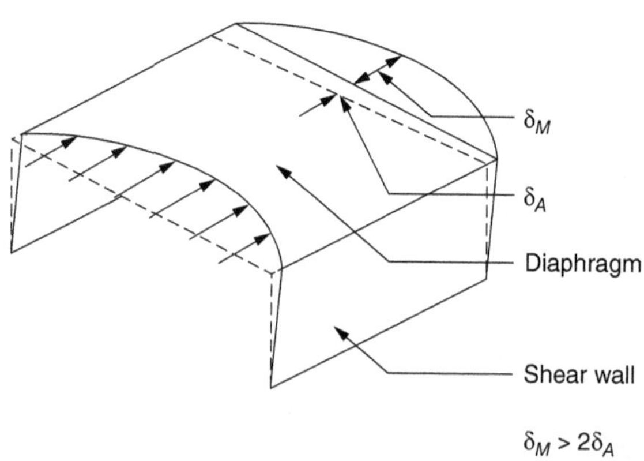


**FIGURE 2.60** Flexible diaphragm.

In accordance with ASCE 7 Sec. 12.3.1.1, the following types of diaphragms may be considered flexible:

- Untopped steel decking or wood structural panels supported by vertical elements of steel or composite braced frames, or concrete, masonry, steel, or composite shear walls.
- Untopped steel decking or wood structural panels in one- and two-family residential buildings.
- In structures of light-frame construction provided all of the following conditions are met:
  - Toppings of concrete or similar materials are not placed over wood structural panel diaphragms except for nonstructural toppings no greater than 1.5 in thick.
  - Each line of the lateral force-resisting system complies with the allowable story drift of ASCE 7 Table 12.12-1.

## Horizontal Distribution of Seismic Forces

The inertial forces developed in a building by an earthquake must be transferred by a suitable seismic force-resisting system to the foundation. This system consists of two parts, horizontal diaphragms that transfer the seismic forces at each floor to the vertical seismic force-resisting elements and the vertical elements that transfer the lateral forces to the foundation. As shown in Fig. 2.61, a flexible diaphragm is assumed to act as a simply supported beam between vertical seismic force-resisting elements. Hence, lateral force is distributed to the vertical elements based on tributary mass, without producing any torsional effects.

**Example 2.38.** Loads on Vertical Seismic Force-Resisting Elements
For the north-south direction, determine the lateral loads acting on an internal frame of the two-story steel-framed building shown in Fig. 2.41. The diaphragms may be considered flexible.

The length of diaphragm tributary to an interior frame is obtained from Fig. 2.41 as

$$L = 30 \text{ ft}$$

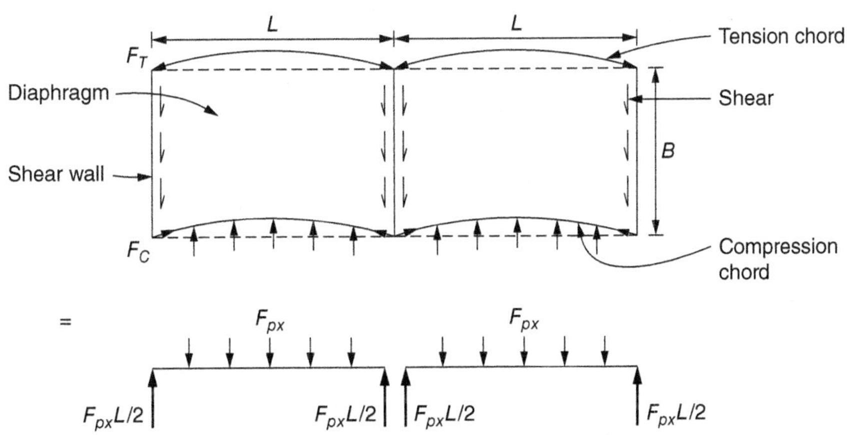


**FIGURE 2.61** Flexible diaphragm.

| Level | $L$, ft | $F_x$, lb/ft | $F$, kips |
|---|---|---|---|
| Roof | 30 | 381 | 11.43 |
| 2nd floor | 30 | 238 | 7.14 |

**TABLE 2.16** Details for Example 2.38

The lateral forces $F_x$ at the roof and first floor are obtained from Example 2.32 and are shown in Table 2.16.

The lateral forces acting on an interior frame at floor and roof levels are given by

$$F = F_x L$$

Values of $F$ are shown in Table 2.16.

A flexible diaphragm is analogous to a deep beam with the floor deck acting as the web to resist shear and the boundary members, normal to the load, acting as the flanges or chords to resist flexural effects. Design techniques are discussed in the APA publications.[15,16] Shear stresses are assumed uniformly distributed across the depth of the diaphragm. As shown in Fig. 2.61, the chords develop axial forces and provide a couple to resist the applied moment. The chord force is given by

$$\begin{aligned} F_T &= F_C \\ &= M/B \\ &= F_{px}L^2/8B \end{aligned}$$

The boundary unit shear is given by

$$q = F_{px}L/2B$$

where $M$ = applied bending moment
$F_{px}$ = uniformly distributed seismic load
$B$ = distance between chord centers
= depth of diaphragm
$L$ = span of diaphragm

**Example 2.39.** Diaphragm Forces

For the north-south direction, determine the chord force and boundary shear produced in the second-floor diaphragm of the two-story steel-framed building shown in Fig. 2.41. The diaphragms may be considered flexible.

The span length and depth of the diaphragm are obtained from Fig. 2.41 as

$$L = 30 \text{ ft}$$

$$B = 30 \text{ ft}$$

The seismic load on the second-floor diaphragm is obtained from Example 2.34 as

$$F_{px} = 550 \text{ lb/ft}$$

The chord force is given by

$$
\begin{aligned}
F_t &= F_c \\
&= F_{px}L^2/8B \\
&= 550 \times 30^2/(8 \times 30 \times 1000) \\
&= 2.06 \text{ kips}
\end{aligned}
$$

The unit shear at the diaphragm boundaries parallel to the applied load is

$$
\begin{aligned}
q &= F_{px}L/2B \\
&= 550 \times 30/(2 \times 30) \\
&= 275 \text{ lb/ft}
\end{aligned}
$$

## Anchorage of Structural Walls to Diaphragms

During past earthquakes, a major cause of failure has been the separation of flexible diaphragms from supporting walls. This is due to diaphragm flexibility amplifying out-of-plane accelerations. To prevent separation occurring, anchorage ties must be provided to tie the diaphragms and walls together. The specified design force depends on the seismic design category of the building, as reported by Henry.[17] These forces apply only to design of the ties and not to the overall wall design. Anchorage forces are not subject to the redundancy factor ρ.

*For buildings in seismic design category A:* ASCE 7 Sec. 11.7 requires structures to be provided with a continuous load path as specified in ASCE 7 Sec. 1.4 with the lateral force-resisting system designed to resist the following notional loads $N$:

- Any smaller element of a structure must be tied to the remainder of the structure with a connection capable of resisting a lateral force of $N = 5$ percent of its weight.
- The lateral force-resisting system must be designed to resist lateral forces, applied simultaneously at each level of the structure, given by $N = 0.01W_x$, where $W_x$ is the dead load assigned to level $x$.
- For each beam, girder, or truss, a connection must be provided to resist a lateral force, acting parallel to the member, of $N = 0.05W_R$, where $W_R$ is the reaction due to dead plus live load.
- Structural walls must be anchored to the roof and floors of the structure that provide lateral support for the wall or that are supported by the wall. The connection must be capable of resisting a horizontal force perpendicular to the wall of $N = 20$ percent of the weight of the wall tributary to the connection but not less than 5 lb/ft$^2$.

ASCE 7 Sec. 1.4.1 requires the following load combinations to be applied when using the notional loads:

- For strength design load combinations:

$$1.2D + 1.0N + L + 0.2S$$

$$0.9D + 1.0N$$

- For allowable stress design load combinations:

$$D + 0.7N$$

$$D + 0.75(0.7N) + 0.75L + 0.75(L_r \text{ or } S \text{ or } R)$$

$$0.6D + 0.7N$$

*For buildings with flexible diaphragms in seismic design category B through F:* Diaphragm flexibility can amplify out-of-plane accelerations, and ASCE 7 Sec. 12.11.2.1 requires anchors of flexible diaphragms to be designed for the force

$$F_p = 0.4S_{DS}k_a I_e W_p$$

$$\geq 0.2k_a I_e W_p$$

where $I_e$ = occupancy importance factor
$S_{DS}$ = design response acceleration, for a period of 0.2 second
$W_p$ = weight of the wall tributary to the anchor
$k_a$ = amplification factor for diaphragm flexibility
$= 1.0 + L_f/100$
$\leq 2$
$L_f$ = span, in ft, of a flexible diaphragm measured between vertical elements that provide lateral support to the diaphragm in the direction considered
$= 0$ . . . for rigid diaphragm

**Example 2.40.** Wall Anchor Forces for Flexible Diaphragms
For the north-south direction, determine the required seismic design force for the anchors in the second-floor diaphragm of the two-story steel-framed building shown in Fig. 2.41. The diaphragms may be considered flexible and the anchorages are at $s = 2$ ft centers.

From previous examples, the relevant parameters are

$I_e$ = occupancy importance factor

$= 1.0$

$S_{DS}$ = design response acceleration, for a period of 0.2 second

$= 1.0g$

$w$ = weight of the wall

$= 25 \text{ lb/ft}^2$

SDC = seismic design category

= D

$h_s$ = story height

$= 10$ ft

The equivalent area of wall tributary to each anchor is

$$A_w = sh_s$$

$$= 2 \times 10$$

$$= 20 \text{ ft}^2$$

The weight of wall tributary to each anchor is

$$W_p = wA_w$$
$$= 25 \times 20$$
$$= 500 \text{ lb}$$

$$L_f = \text{span, in ft, of the flexible diaphragm}$$
$$= 30 \text{ ft}$$

$$k_a = \text{amplification factor for diaphragm flexibility}$$
$$= 1.0 + L_f/100$$
$$= 1.0 + 30/100$$
$$= 1.3$$

For seismic design category D, the seismic lateral force on an anchor is given by ASCE 7 Eq. (12.11-1) as

$$F_p = 0.4S_{DS}k_aI_eW_p$$
$$= 0.40 \times 1.0 \times 1.3 \times 1.0 \times 500$$
$$= 260 \text{ lb}$$

The minimum permissible force on one anchor is

$$F_p = 0.2k_aI_eW_p$$
$$= 0.20 \times 1.3 \times 1.0 \times 500$$
$$= 130 \text{ lb}$$

The required seismic design force for the anchors is

$$F_p = 260 \text{ lb}$$

*For buildings with rigid diaphragms in seismic design category B through F:* In accordance with IBC Sec. 1604.4, a diaphragm is classified as rigid for the purpose of distribution of story shear and torsional moment when the lateral deformation of the diaphragm is less than or equal to twice the average story drift. Diaphragms of concrete slabs or concrete filled metal deck with span-to-depth ratios of 3 or less in structures that have no horizontal irregularities are considered by ASCE 7 Sec. 12.3.1.2 to be rigid. ASCE 7 Sec. 12.11.2.1 permits anchors of rigid diaphragms, with the exception of roof diaphragms, to be designed for the force

$$F_p = (0.4S_{DS}I_e)(1 + 2z/h)W_p/3$$
$$\geq 0.2I_eW_p$$

For a rigid diaphragm, ASCE 7 Sec. 12.11.2.1 gives $K_a = 1$ and

$$F_p = 0.4S_{DS}I_eW_p$$
$$\geq 0.2I_eW_p$$

where $F_p$ = seismic design force on the anchor
$I_e$ = occupancy importance factor
$S_{DS}$ = design response acceleration, for a period of 0.2 second
$W_p$ = weight of wall tributary to the anchor
$h$ = height of roof above the base
$z$ = height of the anchor above the base

Walls shall be designed to resist bending between anchors where the anchor spacing exceeds 4 ft.

**Example 2.41.** Wall Anchor Forces for Rigid Diaphragms
For the north-south direction, determine the required seismic design force for the anchorages in the second-floor diaphragm of the two-story steel-framed building shown in Fig. 2.41. The diaphragms may be considered rigid and the anchorages are at $s = 2$ ft centers.

From previous examples, the relevant parameters are

$$I_e = \text{occupancy importance factor}$$
$$= 1.0$$
$$S_{DS} = \text{design response acceleration, for a period of 0.2 second}$$
$$= 1.0g$$
$$w = \text{weight of the wall}$$
$$= 25 \text{ lb/ft}^2$$
$$\text{SDC} = \text{seismic design category}$$
$$= \text{D}$$
$$h_s = \text{story height}$$
$$= 10 \text{ ft} = z$$

The area of wall tributary to each anchor is

$$A_w = sh_s$$
$$= 2 \times 10$$
$$= 20 \text{ ft}^2$$
$$h = \text{height of roof above the base} = 20 \text{ ft}$$

The weight of wall tributary to each anchor is

$$W_p = wA_w$$
$$= 25 \times 20$$
$$= 500 \text{ lb}$$

For seismic design category D, the seismic lateral force on an anchor is given by ASCE 7 Sec. 12.11.2.1 as

$$F_p = (0.4S_{DS}I_e)(1 + 2z/h)W_p/3$$
$$= (0.4 \times 1.0 \times 1.0)[1 + (2 \times 10)/(20)]500/3$$
$$= 133 \text{ lb}$$

The minimum permissible force on one anchor is

$$F_p = 0.2I_eW_p$$
$$= 0.20 \times 1.0 \times 500$$
$$= 100 \text{ lb}$$

The required seismic design force for the anchors is

$$F_p = 133 \text{ lb}$$

*Additional requirements for buildings in seismic design category C through F:* To transfer anchorage forces across the complete depth of the diaphragm and to prevent the walls and diaphragm from separating, ASCE 7 Sec. 12.11.2.2.1 requires the provision of continuous ties across the complete depth of the diaphragm. To reduce the number of full depth ties required, subdiaphragms and added chords are used to span between the full depth ties. The maximum permitted length-to-width ratio of the subdiaphragm is 2.5 to 1.

In accordance with ASCE 7 Secs. 12.11.2.2.3 and 12.11.2.2.4, neither plywood sheathing nor metal deck may be considered effective as providing the ties. In addition, anchorage may not be accomplished by use of toenails or nails subject to withdrawal nor may wood ledgers be used in cross-gain bending.

In accordance with ASCE 7 Sec. 12.11.2.2.2, steel elements of the structural wall anchorage system must be designed for a force of 1.4 times the calculated force.

**Example 2.42.** Subdiaphragms and Crossties
For the north-south direction, determine a suitable subdiaphragm arrangement for the second floor of the two-story steel-framed building shown in Fig. 2.41. The diaphragm may be considered flexible with joists spaced at 5 ft centers and subpurlins spaced at 2 ft centers, as shown in Fig. 2.62.

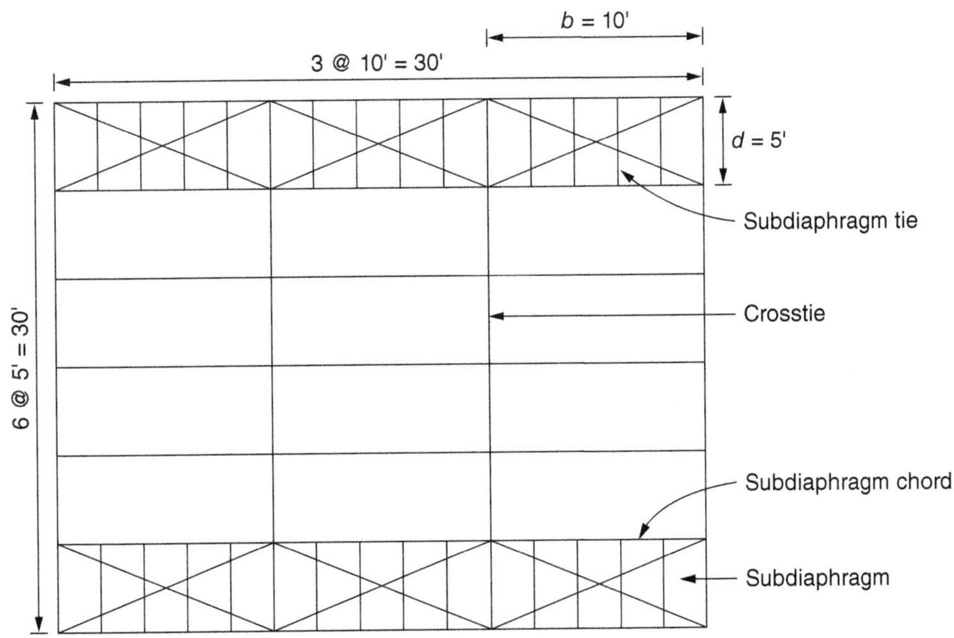


**FIGURE 2.62** Details for Example 2.42.

From Example 2.40, the strength level design force on each anchor spaced at $s = 2$ ft centers is

$$F_p = 260 \text{ lb}$$

Hence, the pull-out force along the wall is

$$\begin{aligned} p &= F_p/s \\ &= 260/2 \\ &= 130 \text{ lb/ft} \end{aligned}$$

Provide three subdiaphragms, with dimension $b = 10$ ft and $d = 5$ ft, in each bay of the building, with crossties at 10 ft centers as shown in Fig. 2.62.

The aspect ratio of each subdiaphragm is

$$\begin{aligned} b/d &= 10/5 \\ &= 2.0 < 2.5 \text{ ... complies with ASCE 7 Sec. 12.11.2.2.1} \end{aligned}$$

The subpurlins provide the subdiaphragm ties and the force in each is

$$F_p = 260 \text{ lb}$$

The unit shear stress in the subdiaphragm is

$$\begin{aligned} q &= pb/2d \\ &= 130 \times 10/(2 \times 5) \\ &= 130 \text{ lb/ft} \end{aligned}$$

The subdiaphragm chord force is

$$\begin{aligned} P_c &= pb^2/8d \\ &= 130 \times 10^2/(8 \times 5) \\ &= 325 \text{ lb} \end{aligned}$$

The force in the crossties is

$$\begin{aligned} P_t &= pb \\ &= 130 \times 10 \\ &= 1300 \text{ lb} \end{aligned}$$

## Rigid Diaphragms

In accordance with IBC Sec. 1604.4, a diaphragm is classified as rigid for the purpose of distribution of story shear and torsional moment when the lateral deformation of the diaphragm is less than or equal to twice the average story drift. Lateral force is distributed to the vertical seismic-load-resisting elements based on the relative rigidity of these elements and the torsional displacements produced by the rigid-body rotation of the diaphragm and vertical elements.

## Torsional Effects

As shown in Fig. 2.63, computation of the forces in the vertical elements requires the determination of the centers of mass and rigidity for each story. The center of rigidity is the point about which a structure rotates when subjected to a torsional moment. For the

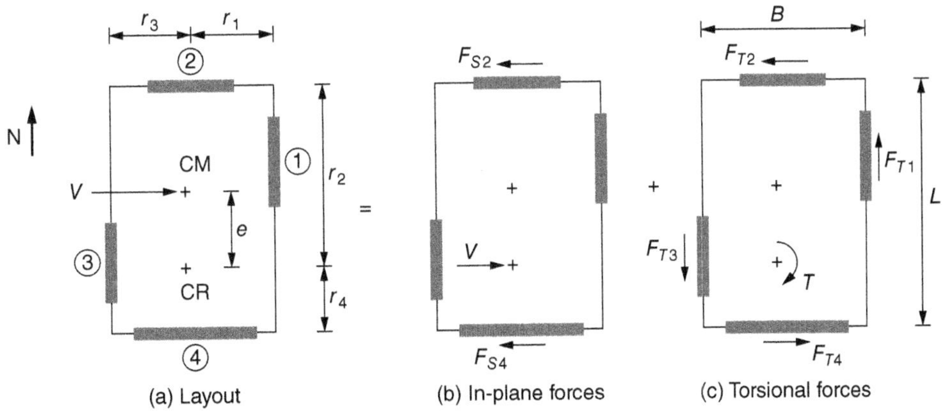


**FIGURE 2.63** Rigid diaphragm.

single story building shown in Fig. 2.63, the east-west seismic base shear $V$ acts at the center of mass of the building. The torsional moment acting is

$$T = Ve$$

where $e$ eccentricity of the center of mass with respect to the center of rigidity.

The displacement of the building consists of an east-west translation and a clockwise rotation about the center of rigidity. As shown in Fig. 2.63*b*, the translation produces in-plane forces in vertical elements 2 and 4 proportional to their relative translational stiffness. No forces are produced in vertical elements 1 and 3 by this translation. The clockwise rotation produces forces in all four walls, proportional to their torsional stiffness, as shown in Fig. 2.63*c*.

In a perfectly symmetric building, the centers of mass and rigidity coincide and torsion is not produced. However, the centers of mass and rigidity may not be located exactly because of uncertainties in determining the mass and stiffness distribution in the building. In addition, torsional components of the ground motion may also cause torsion to develop. Hence, accidental eccentricity may in fact exist even in a nominally symmetric structure. Torsion resulting from this accidental eccentricity is referred to as accidental torsion. To account for accidental torsion, ASCE 7 Sec. 12.8.4.2 specifies that the center of mass is assumed displaced each way from its actual location by a distance equal to 5 percent of the building dimension perpendicular to the direction of the applied force.

When a building, assigned to seismic design category C through F, has a torsional irregularity as defined in ASCE 7 Table 12.3-1 (horizontal structural irregularity Type 1a or 1b), the accidental torsion is amplified as specified in ASCE 7 Sec. 12.8.4.3. This is to account for the possibility of unsymmetrical yielding of the vertical seismic force-resisting elements resulting in a large increase in torsional effects. The amplification factor is given by ASCE 7 Sec. 12.8.4.3 as

$$A_x = (\delta_{max}/1.2\delta_{avg})^2$$
$$\leq 3.0$$

where $\delta_{max}$ is maximum displacement at level $x$ computed assuming $A_x = 1$ and $\delta_{avg}$ is average of displacements at extreme points of the structure at level $x$ computed assuming $A_x = 1$. Structures assigned to seismic design categories E or F having horizontal irregularity type 1b are not permitted.

**Example 2.43.** Loads on Vertical Seismic Force-Resisting Elements with Rigid Diaphragms
For the east-west direction of the seismic force, determine the lateral force acting on vertical element 2 of the building shown in Fig. 2.63. The building dimensions and the relative vertical element stiffness are

$$L = 40 \text{ ft}$$
$$B = 20 \text{ ft}$$
$$s_4 = 3$$
$$s_1 = s_2 = s_3 = 1$$

The base shear is $V = 20$ kips. The diaphragm may be considered rigid and the building mass is symmetrically disposed about the centerline of the building. The building is assigned to seismic design category D.

From the symmetry of the structure, for an east-west seismic load, the center of mass is located midway between wall 2 and wall 4 and its distance from wall 4 is

$$x = 40/2 = 20 \text{ ft}$$

The position of the center of mass in the orthogonal direction is not relevant to the question.

In locating the center of rigidity for an east-west seismic load, wall 1 and wall 3, which have no stiffness in the east-west direction, are omitted. Taking moments about wall 4, the distance of the center of rigidity from wall 4 is given by

$$\begin{aligned} r_4 &= \Sigma s_x y/\Sigma s_x \\ &= (s_1 L + s_4 \times 0)/(s_1 + s_4) \\ &= (1 \times 40 + 3 \times 0)/(1 + 3) \\ &= 10 \text{ ft} \end{aligned}$$

The distance of the center of rigidity from wall 2 is

$$\begin{aligned} r_2 &= 40 - 10 \\ &= 30 \text{ ft} \end{aligned}$$

In locating the center of rigidity for a north-south seismic load, wall 2 and wall 4, which have no stiffness in the north-south direction are omitted. Due to the symmetry of the vertical seismic force-resisting elements 1 and 3, the center of rigidity is located midway between wall 1 and wall 3 and

$$r_1 = r_3 = 10 \text{ ft}$$

The polar moment of inertia of the walls is

$$\begin{aligned} J &= \Sigma r^2 s \\ &= r_1^2 \times s_1 + r_2^2 \times s_2 + r_3^2 \times s_3 + r_4^2 \times s_4 \\ &= 10^2 \times 1 + 30^2 \times 1 + 10^2 \times 1 + 10^2 \times 3 \\ &= 1400 \text{ ft}^2 \end{aligned}$$

The sum of the wall rigidities for a seismic load in the east-west direction is

$$\begin{aligned}\Sigma s_x &= s_2 + s_4\\ &= 1 + 3\\ &= 4\end{aligned}$$

For a seismic load in the east-west direction, the eccentricity is

$$\begin{aligned}e_y &= x - r_4\\ &= 20 - 10\\ &= 10 \text{ ft}\end{aligned}$$

Accidental eccentricity, in accordance with Sec. 12.8.4.2 is

$$\begin{aligned}e_a &= \pm 0.05 \times L\\ &= \pm 0.05 \times 40\\ &= \pm 2 \text{ ft}\end{aligned}$$

An accidental displacement of the center of mass to the north gives the maximum eccentricity of

$$\begin{aligned}e &= e_y + e_a\\ &= 10 + 2.0\\ &= 12 \text{ ft}\end{aligned}$$

The maximum eccentricity governs for the force in wall 2 since the torsional force and the in-plane force are of the same sense and are additive.

The maximum torsional moment acting about the center of rigidity is

$$\begin{aligned}T &= Ve\\ &= 20 \times 12\\ &= 240 \text{ kip-ft}\end{aligned}$$

The force produced in a wall by the base shear acting in the east-west direction is the algebraic sum of the in-plane shear force and the torsional shear force.

The in-plane shear force is

$$\begin{aligned}F_S &= Vs_x/\Sigma s_x\\ &= 20s_x/4\\ &= 5s_x\end{aligned}$$

The maximum torsional shear force is

$$\begin{aligned}F_T &= Trs/J\\ &= 240rs/1400\\ &= 0.171rs\end{aligned}$$

The total force in a wall is

$$F = F_S + F_T$$

with a negative value for $F_T$ indicating that the torsional force is opposite in sense to the in-plane force.

The total forces produced in wall 2 and wall 4, by the maximum torsional moment, are

$$\begin{aligned}F_2 &= 5 \times s_2 + 0.171 \times r_2 \times s_2\\ &= 5 \times 1 + 0.171 \times 30 \times 1\\ &= 5 + 5.13\\ &= 10.13 \text{ kips}\end{aligned}$$

$$\begin{aligned}F_4 &= 5 \times s_4 - 0.171 \times r_4 \times s_4\\ &= 5 \times 3 - 0.171 \times 10 \times 3\\ &= 15 - 5.13\\ &= 9.87 \text{ kips}\end{aligned}$$

To determine if amplification of the torsional moment is necessary, the displacements of wall 1 and wall 4 must be determined.

The relative displacement of a wall is given by

$$\delta = F/s$$

The relative displacements of wall 2 and wall 4 are

$$\delta_2 = 10.13/1 = 10.13$$

$$\delta_4 = 9.87/3 = 3.29$$

The ratio of the maximum displacement of wall 2 to the average displacement of wall 2 and wall 4 is

$$\begin{aligned}\mu &= 2\delta_2/(\delta_2 + \delta_4)\\ &= 2 \times 10.13/13.42\\ &= 1.51\\ &> 1.40\end{aligned}$$

This constitutes an extreme torsional irregularity as defined in ASCE 7 Table 12.3-1 and, for a structure assigned to seismic design category D, the accidental eccentricity must be amplified, as specified in ASCE 7 Sec. 12.8.4.3 by the factor

$$\begin{aligned}A_x &= (\mu/1.2)^2\\ &= (1.51/1.2)^2\\ &= 1.58\\ &< 3.00 \text{ ...satisfactory}\end{aligned}$$

The revised accidental eccentricity is

$$\begin{aligned}e' &= \pm A_x e_a\\ &= \pm 1.58 \times 2.00\\ &= \pm 3.16 \text{ ft}\end{aligned}$$

The revised maximum eccentricity for a displacement of the center of mass to the north is

$$\begin{aligned}e'' &= e_x + e'\\ &= 10 + 3.16\\ &= 13.16 \text{ ft}\end{aligned}$$

The amplified torsional moment is

$$T' = Ve''$$
$$= 20 \times 13.16$$
$$= 263.20 \text{ kip-ft}$$

The revised total force produced in wall 2 by the maximum amplified torsional moment is

$$F_{2A} = F_S + F_{TA}$$
$$= Vs_2/\Sigma s_x + T'r_2 s_2/J$$
$$= 20 \times 1/4 + 263.20 \times 30 \times 1/1400$$
$$= 5 + 5.64$$
$$= 10.64 \text{ kips}$$

## Out-of-Plane Force on Structural Walls

ASCE 7 Sec. 12.11.1 requires structural walls to be designed for a force normal to the surface of

$$F_p = 0.4S_{DS}I_eW_p$$
$$\geq 0.1W_p$$

where $I_e$ = occupancy importance factor
$S_{DS}$ = design response acceleration, for a period of 0.2 second
$W_p$ = weight of the wall

Out-of-plane forces on walls are not subject to the redundancy factor ρ.

**Example 2.44.** Lateral Force on Structural Wall
A tilt-up concrete office building has 15-ft-high walls that are hinged at top and bottom. The walls are of normal weight concrete 6 in thick. The design response acceleration, for a period of 0.2 second, is $S_{DS} = 1.0g$. Determine the design lateral seismic force on the wall.
Weight of the wall is

$$W_p = 150 \times 6/12$$
$$= 75 \text{ lb/ft}^2$$

The seismic lateral force on the wall is given by ASCE 7 Sec. 12.11.1 as

$$F_p = 0.4I_eS_{DS}W_p$$
$$= 0.4 \times 1.0 \times 1.0 \times 75$$
$$= 30 \text{ lb/ft}^2$$

## Out-of-Plane Force on Parapets

ASCE 7 Sec. 13.3.1 requires parapets to be designed as architectural components using ASCE 7 Eqs. (13.3-1) through (13.3-3) which are

$$F_p = (0.4a_pS_{DS}I_p/R_p)(1 + 2z/h)W_p$$
$$\leq 1.6S_{DS}I_pW_p$$
$$\geq 0.3S_{DS}I_pW_p$$

where $F_p$ = seismic design force on the parapet
$I_p$ = component importance factor given in ASCE 7 Sec. 13.1.3
$S_{DS}$ = design response acceleration, for a period of 0.2 second
$W_p$ = weight of parapet
$a_p$ = component amplification factor from ASCE 7 Table 13.5-1
= 2.5
$h$ = height of roof above the base
$z$ = height of point of attachment of parapet above the base
= $h$
$R_p$ = component response modification factor from ASCE 7 Table 13.5-1
= 2.5

Because of the lack of redundancy of a parapet, a high value of 2.5 is assigned to the component amplification factor. The lateral force is considered uniformly distributed over the height of the parapet.

**Example 2.45.** Lateral Force on Parapet
A tilt-up concrete office building with a roof height of 20 ft has a 3-ft-high parapet that cantilevers above the roof. The parapet is of normal weight concrete 6 in thick. The design response acceleration, for a period of 0.2 second, is $S_{DS} = 1.0g$. Determine the design lateral seismic force on the parapet.

Weight of the parapet is

$$W_p = 150 \times 6/12$$
$$= 75 \text{ lb/ft}^2$$

The seismic lateral force on the parapet is given by ASCE 7 Sec. 13.3.1 as

$$F_p = (0.4a_p S_{DS} I_p / R_p)(1 + 2z/h)W_p$$

where $I_p = 1.0$
$S_{DS} = 1.0\,g$
$W_p = 75 \text{ lb/ft}^2$
$a_p = 2.5$
$h$ = height of roof above the base
= 20 ft
$z$ = height of point of attachment of parapet above the base
= 20 ft
$R_p = 2.5$

Then $F_p = (0.4 \times 2.5 \times 1.0 \times 1.0/2.5)(1 + 2 \times 20/20)W_p$
$= 1.2W_p$
$= 90 \text{ lb/ft}^2$

Neither ASCE 7 Eq. (13.3-2) nor Eq. (13.3-3) govern.

## Redundancy Factor

The redundancy factor ρ penalizes buildings that lack redundancy in their lateral force-resisting systems and is applicable to buildings assigned to seismic design category D, E, and F. Without adequate redundancy, structural failure may result when failure of an element in the lateral force-resisting system results in an excessive loss of shear strength or development of an extreme torsional irregularity. The redundancy factor is determined for each orthogonal direction of the building and is assigned a value of 1.0 for a building with adequate redundancy or 1.3 for a building with inadequate redundancy. Intermediate values are not permitted. In accordance with ASCE 7 Sec. 12.3.4.1, a redundancy factor of 1.0 may be adopted when any of the following conditions exist:

- The building is assigned to seismic design category B or C.
- When drift and P-delta effects are calculated.
- For design of nonstructural components.
- When the overstrength factor $\Omega_0$ is required in design of an element.
- For determination of diaphragm loads.
- For design of structural walls for out-of-plane forces including their anchorage.
- For buildings with damping devices.

A building in seismic design category D, E, or F may be assigned a redundancy factor of 1.0 provided that, at each story resisting more than 35 percent of the base shear, one of the following two conditions is satisfied:

- The building is regular in plan at all levels and has at least two bays of seismic force-resisting perimeter framing on each side of the building in each orthogonal direction for the story. The number of bays for a shear wall is calculated as the length of the shear wall divided by the story height. For light-framed construction, the number of bays is calculated as twice the length of the shear wall divided by the story height.
- Removal of an element of the lateral force-resisting system does not result in more than a 33 percent reduction in story strength, nor does the resulting system have an extreme torsional irregularity Type 1b of ASCE 7 Table 12.3-1.

Removal of an element of the lateral force-resisting system is defined in ASCE 7 Table 12.3-3 as

- Removal of an individual brace, or connection thereto, for a braced frame
- Loss of moment resistance at the beam-to-column connections at both ends of a single beam, for a moment resisting frame
- Removal of a shear wall or wall pier with a height-to-length ratio greater than 1.0 within any story, or collector connections thereto, for a shear wall system
- Loss of moment resistance at the base connection of any single cantilever column, for a cantilever column system

**Example 2.46.** Redundancy Factor
For the north-south direction, determine the redundancy factor on the second floor of the two-story steel-framed building shown in Fig. 2.41. Lateral resistance in the north-south direction is provided by special steel moment frames with identical stiffness. Lateral resistance in the east-west direction is provided by concentrically braced frames with identical stiffness. The stiffness of a braced frame is ten times the stiffness of a moment frame. The diaphragms may be considered rigid and the building is assigned to seismic design category D.

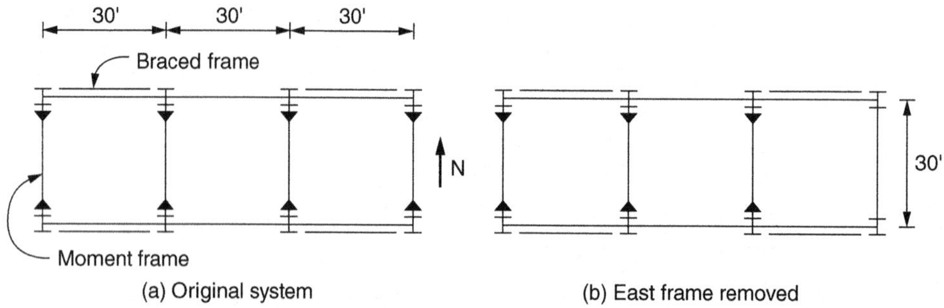


FIGURE 2.64 Lateral force-resisting elements.

From Example 2.32

$$V = \text{base shear}$$
$$= 619 \text{ lb/ft}$$
$$F_{x2} = \text{shear at second story}$$
$$= 238 \text{ lb/ft}$$
$$F_{x2}/V = 0.38$$
$$> 0.35 \ldots \text{ASCE 7 Sec. 12.3.4.2 must be applied}$$

The plan view of the second story, showing the lateral force-resisting elements, is shown in Fig. 2.64*a*. Only one bay of special steel moment frames is provided at each end of the building and, thus, the building does not comply with case (b) of ASCE 7 Sec. 12.3.4.2.

To investigate case (a) of ASCE 7 Sec. 12.3.4.2, the special steel moment frame at the east end of the building is removed as shown in Fig. 2.64*b*.

**Check reduction in story strength**

This leaves three moment frames in the north-south direction. At the ultimate limit state, it may be assumed that the three moment frames have each reached their ultimate capacity of $F_M$ and are providing restraint to lateral displacement. The total story shear capacity in the north-south direction is now $3F_M$. Prior to removal of this moment frame, the story shear capacity in the north-south direction was $4F_M$. Hence, the reduction in story strength is

$$F_M/4F_M = 0.25$$
$$< 0.33 \ldots \text{satisfactory}$$

**Check for extreme torsional irregularity**

After removal of the moment frame at the east end of the building, the ultimate story shear is

$$V_u = 3F_M$$

Torsional restraint is provided by the four braced frames each with a force of $F_B$. This is shown in Fig. 2.65*a*.

For forces in the north-south direction, the center of rigidity is located at a distance from the west wall given by

$$x_R = 30 \text{ ft}$$

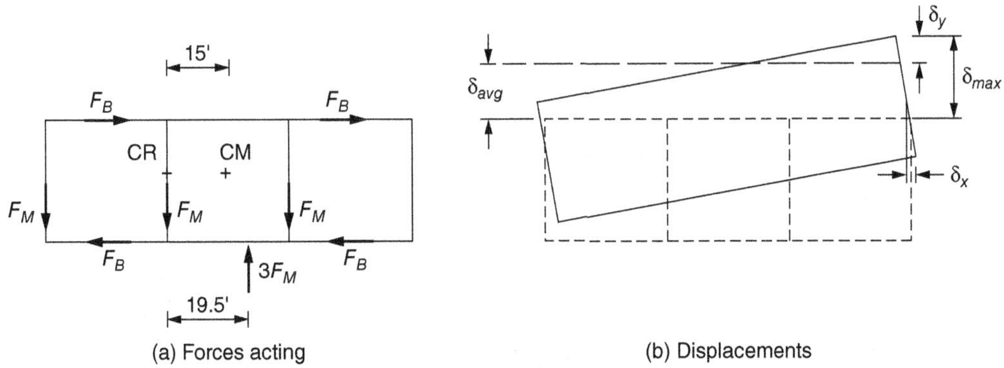


**FIGURE 2.65** Forces acting at ultimate limit state.

From symmetry, the center of mass is located 45 ft from the west wall. Accidental eccentricity is given by ASCE 7 Sec. 12.8.4.2 as

$$e_a = 0.05 \times 90$$
$$= 4.5 \text{ ft}$$

The displaced center of mass, allowing for accidental eccentricity, is located a distance from the west wall given by

$$x_M = 45 + 4.5$$
$$= 49.5 \text{ ft}$$

The maximum eccentricity is

$$e = x_M - x_R$$
$$= 49.5 - 30$$
$$= 19.5 \text{ ft}$$

The torsion acting on the building is

$$T_u = eV_u$$
$$= 19.5 \times 3F_M$$
$$= 58.5F_M$$

The couple created by the force in the braced frames produces an equal and opposite torsion which is

$$T_u = 2F_B \times 30$$
$$= 60F_B$$

Hence
$$58.5F_M = 60F_B$$
$$F_B = 0.98F_M$$

The displacement produced in the story is shown in Fig. 2.65*b*. The average story drift is

$$\delta_{avg} = F_M/s$$

where *s* is stiffness of one moment frame.

The east-west deflection of the south-east corner of the story is

$$\delta_x = F_B/10s$$
$$= 0.098F_M/s$$

where $10s$ is stiffness of one braced frame.

From the similar triangles

$$\delta_y/45 = \delta_x/15$$
$$\delta_y = 3\delta_x$$
$$= 0.29F_M/s$$

The maximum story drift is

$$\delta_{max} = \delta_y + \delta_{avg}$$
$$= 1.29F_M/s$$

The ratio of the maximum drift to the average drift is

$$\delta_{max}/\delta_{avg} = 1.29$$
$$< 1.4 \ldots \text{extreme torsional irregularity does not exist}$$

Removing an interior moment frame is less critical than the removal of an end frame.

Hence, the redundancy factor is

$$\rho = 1.0$$

## Structural Irregularities

A regular structure is one in which mass and stiffness are uniformly distributed throughout the structure and the inelastic response produced by earthquake ground motion tends to be uniformly distributed throughout the structure. Under these circumstances, the equivalent lateral force procedure (ELF) provides a reasonably accurate representation of the forces generated in the structure. When irregularities are introduced into the configuration, the assumptions inherent in the ELF procedure become invalid and it may be necessary to adopt a dynamic analysis to determine an accurate distribution of the seismic forces.

Past earthquakes have demonstrated that structures with irregularities have suffered greater damage than structures with a regular configuration. This is because inelastic behavior tends to concentrate in the location of an irregularity resulting in major damage and failure in this location. Configurations prone to damage include those with a weak story and those susceptible to torsional effects due to the eccentricity between the centers of mass and rigidity. ASCE 7 Sec. 12.3.2 lists the types of irregularity that occur and specifies when a dynamic analysis must be used or when the ELF procedure must be augmented to ensure satisfactory behavior.

Structural irregularities are classified into two main types; horizontal irregularities which are defined in ASCE 7 Table 12.3-1 and vertical irregularities which are defined in ASCE Table 12.3-2. The criteria for determining the presence of each irregularity is detailed in the tables and the additional requirements that must be satisfied, if the irregularity is present, are given. Restrictions are placed on the use of the ELF procedure and these are detailed in ASCE 7 Table 12.6-1.

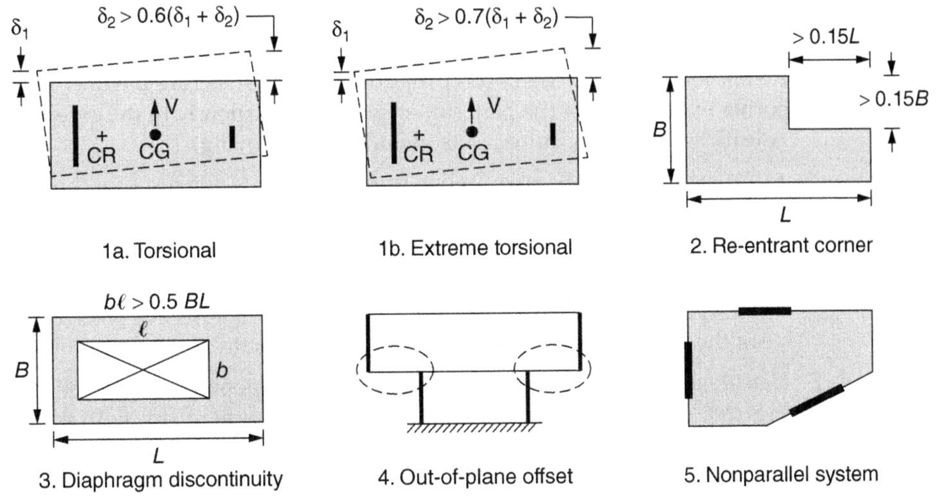


**FIGURE 2.66** Horizontal irregularities.

Horizontal irregularities are shown in Fig. 2.66 and consist of the following types:

1a. *Torsional Irregularity*: when the maximum story drift at one end of a rigid diaphragm, including accidental torsion with amplification factor neglected, exceeds 1.2 times the average story drift. (Applies only to structures with rigid or semirigid diaphragms). Additional design requirements include the following:

- A three dimensional analysis is required in all seismic design categories (ASCE 7 Sec. 12.7.3)
- An increase of 25% is required in the calculated design force for the connection of collectors to diaphragms and vertical elements and for the connection of diaphragms to vertical elements in seismic design categories D through F (ASCE 7 Sec. 12.3.3.4)
- In seismic design categories D through F, restrictions are placed on the use of the ELF procedure as detailed in ASCE 7 Table 12.6-1

1b. *Extreme Torsional Irregularity*: when the maximum story drift at one end of a rigid diaphragm, including accidental torsion with amplification factor neglected, exceeds 1.4 times the average story drift. (Applies only to structures with rigid or semirigid diaphragms). Additional design requirements include the following:

- A three dimensional analysis is required in seismic design categories B through D (ASCE 7 Sec. 12.7.3)
- Structures assigned to seismic design categories E and F are not permitted (ASCE 7 Sec. 12.3.3.1)
- An increase of 25% is required in the calculated design force for the connection of collectors to diaphragms and vertical elements and for the connection of diaphragms to vertical elements in seismic design categories D through F (ASCE 7 Sec. 12.3.3.4)

- In seismic design category D, restrictions are placed on the use of the ELF procedure as detailed in ASCE 7 Table 12.6-1

2. *Re-Entrant Corners*: when both projections of the structure beyond a re-entrant corner exceed 15% of the plan dimension of the structure in the given direction. Additional design requirements include the following:
   - An increase of 25% is required in the calculated design force for the connection of collectors to diaphragms and vertical elements and for the connection of diaphragms to vertical elements in seismic design categories D through F (12.3.3.4)
   - In seismic design categories D through F, restrictions are placed on the use of the ELF procedure as detailed in ASCE 7 Table 12.6-1
3. *Diaphragm Discontinuity*: when the area of an opening exceeds 50 percent of the area of the diaphragm or when the diaphragm stiffness from one story to the next changes by more than 50%. Additional design requirements include the following:
   - An increase of 25% is required in the calculated design force for the connection of collectors to diaphragms and vertical elements and for the connection of diaphragms to vertical elements in seismic design categories D through F (12.3.3.4)
   - In seismic design categories D through F, restrictions are placed on the use of the ELF procedure as detailed in ASCE 7 Table 12.6-1
4. *Out-Of-Plane Offsets*: when there is a discontinuity in the lateral-force-resistance path such as an out-of-plane offset of at least one of the vertical elements. Additional design requirements include the following:
   - A three dimensional analysis is required in all seismic design categories except for structures with flexible diaphragms (ASCE 7 Sec. 12.7.3)
   - An increase of 25% is required in the calculated design force for the connection of collectors to diaphragms and vertical elements and for the connection of diaphragms to vertical elements in seismic design categories D through F (12.3.3.4)
   - Structural elements supporting discontinuous walls or frames must be designed using the overload factor $\Omega_0$ (ASCE 7 Sec. 12.3.3.3)
   - In seismic design categories D through F, restrictions are placed on the use of the ELF procedure as detailed in ASCE 7 Table 12.6-1
5. *Nonparallel Systems*: when the vertical lateral-force-resisting elements are not parallel to the major orthogonal axes of the lateral-force-resisting system. Additional design requirements include the following:
   - A three dimensional analysis is required in seismic design categories B through F (ASCE 7 Sec. 12.7.3)
   - For seismic design categories C through F, design is required for 100% of the forces for one direction plus 30% of the forces for the perpendicular direction ((ASCE 7 Sec. 12.5.3)
   - In seismic design categories D through F, restrictions are placed on the use of the ELF procedure as detailed in ASCE 7 Table 12.6-1

Vertical irregularities are shown in Fig. 2.67 and consist of the following types:

1a. *Stiffness-Soft Story*: when the stiffness of one story is less than 70 percent of the stiffness of the story above or less than 80 percent of the average stiffness of the three stories above. Additional design requirements include the following:

- In seismic design categories D through F, restrictions are placed on the use of the ELF procedure as detailed in ASCE 7 Table 12.6-1
- Irregularity does not exist if no story drift ratio, excluding the top two stories, is greater than 1.3 times the drift ratio of the story above

1b. *Stiffness-Extreme Soft Story*: when the stiffness of one story is less than 60 percent of the stiffness of the story above or less than 70 percent of the average stiffness of the three stories above. Additional design requirements include the following:

- Structures assigned to seismic design categories E and F are not permitted (ASCE 7 Sec. 12.3.3.1)
- In seismic design category D, restrictions are placed on the use of the ELF procedure as detailed in ASCE 7 Table 12.6-1
- Irregularity does not exist if no story drift ratio, excluding the top two stories, is greater than 1.3 times the drift ratio of the story above

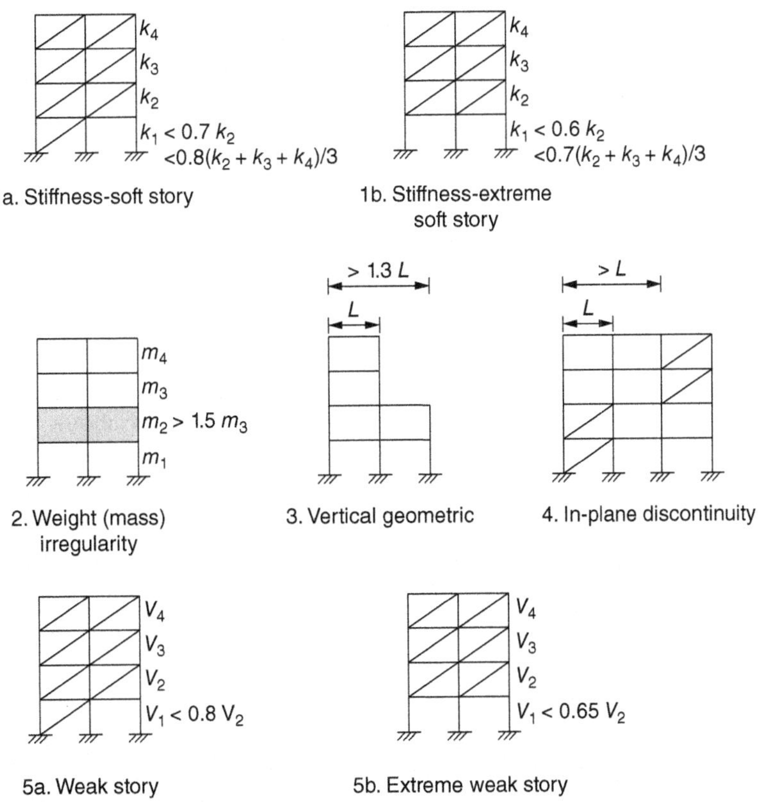


**FIGURE 2.67** Vertical irregularities.

2. *Weight (Mass) Irregularity*: when the mass of any story is more than 150 percent of the mass of an adjacent story. A roof that is lighter than the floor below need not be considered. Additional design requirements include the following:
    - In seismic design categories D through F, restrictions are placed on the use of the ELF procedure as detailed in ASCE 7 Table 12.6-1
    - Irregularity does not exist if no story drift ratio, excluding the top two stories, is greater than 1.3 times the drift ratio of the story above
3. *Vertical Geometric Irregularity*: when the horizontal dimension of the lateral-force-resisting system is more than 130 percent of that in an adjacent story. Additional design requirements include the following:
    - In seismic design categories D through F, restrictions are placed on the use of the ELF procedure as detailed in ASCE 7 Table 12.6-1
4. *In-Plane Discontinuity*: when an in-plane offset of the lateral-force-resisting elements is greater than the length of those elements. Additional design requirements include the following:
    - An increase of 25% is required in the calculated design force for the connection of collectors to diaphragms and vertical elements and for the connection of diaphragms to vertical elements in seismic design categories D through F (12.3.3.4)
    - Structural elements supporting discontinuous walls or frames must be designed using the overload factor $\Omega_0$ (ASCE 7 Sec. 12.3.3.3)
    - In seismic design categories D through F, restrictions are placed on the use of the ELF procedure as detailed in ASCE 7 Table 12.6-1

5a. *Discontinuity In Lateral Strength-Weak Story*: where the lateral strength of a story is less than 80 percent of that in the story above. Additional design requirements include the following:
- Structures assigned to seismic design categories E and F are not permitted (ASCE 7 Sec. 12.3.3.1)
- In seismic design category D, restrictions are placed on the use of the ELF procedure as detailed in ASCE 7 Table 12.6-1

5b. *Discontinuity In Lateral Strength-Extreme Weak Story*: when the lateral stength of a story is less than 65 percent of that in the story above. Additional design requirements include the following:
- Structures assigned to seismic design categories D through F are not permitted (ASCE 7 Sec. 12.3.3.1)
- Structures assigned to seismic design categories B or C shall not exceed two stories or 30 ft in height unless designed using the overload factor $\Omega_0$ (ASCE 7 Sec. 12.3..3.2)

## 2.10 Load Combinations

Load combinations are applied in both ASD and LRFD design methods to determine the most critical effect produced by the loads acting on a building. The individual nominal loads that may act on a building are multiplied by specific load factors determined by probabilistic analysis. As indicated in Sec. 2.1 of this chapter, different load factors are assigned, in the two methods, to the nominal loads in the load combinations. The factors have been selected so that both design methods produce comparable results. Load combinations in ASCE 7 are applicable to all conventional structural materials.

Design loads are comprised of both permanent loads, consisting of the self-weight of the building, and transient loads, commonly consisting of floor live loads, roof live loads, wind, earthquake, and snow loads. During the life of a building, several of the loads may act simultaneously and a load combination represents this situation. The concept of a load combination process is that, in addition to the permanent dead load, one of the transient loads is assigned its maximum lifetime value while the other transient loads are assigned arbitrary point-in-time values. Each of the applicable load combinations is evaluated in turn and elements in a building are designed for the combination that results in the most critical effect. In some cases, this may occur when one or more loads are not acting. As shown in Fig. 2.13, when alternate spans of a continuous beam are loaded, maximum positive moments are produced at the center of the loaded spans.

When several transient loads are considered, it is unlikely that all will attain their maximum value simultaneously and a reduction in the combined loads is justified. Wind and earthquake loads are assumed not to act simultaneously. However, even when the effects of wind load exceed those of earthquake load, member detailing may be governed by earthquake requirements in order to provide adequate ductility for the inelastic deformations. The seismic and wind loads specified in ASCE 7 are at the strength design level in contrast to all other loads that are at the service or nominal level. Hence, in the ASD load combinations, seismic loads are multiplied by the factor 0.7 and wind loads are multiplied by the factor 0.6 to reduce them to service level values.

In addition to externally applied loads, the load combinations also account for self-straining forces caused by temperature changes, shrinkage, or settlement.

## Strength Design Load Combinations

The basic requirement in strength design is to ensure that the design capacity of a member is not less than the demand produced by the applied loads. The following seven strength design load combinations are given in ASCE 7 Sec. 2.3.2:

The governing loading condition for dead load is

1. $1.4D$

   The governing loading condition when the structure is subjected to maximum values of occupancy live load is

2. $1.2D + 1.6L + 0.5(L_r \text{ or } S \text{ or } R)$

   The governing loading condition when the structure is subjected to maximum values of roof live load, rainwater, or snow load is

3. $1.2D + 1.6(L_r \text{ or } S \text{ or } R) + (0.5L \text{ or } 0.5W)$ ... where $0.5L$ is replaced with $L$ for garages, places of public assembly, and areas where $L > 100 \text{ lb/ft}^2$.

   The governing loading condition when the structure is subjected to maximum values of wind load increasing the effects of dead load is

4. $1.2D + 1.0W + 0.5L + 0.5(L_r \text{ or } S \text{ or } R)$ ... where $0.5L$ is replaced with $L$ for garages, places of public assembly, and areas where $L > 100 \text{ lb/ft}^2$.

   The governing loading condition when the structure is subjected to maximum values of seismic load increasing the effects of dead load is

5. $1.2D + 1.0E + 0.5L + 0.2S$ ... where $0.5L$ is replaced with $L$ for garages, places of public assembly, and areas where $L > 100\ \text{lb/ft}^2$.

   The governing loading condition when the structure is subjected to maximum values of wind load opposing the effects of dead load is

6. $0.9D + 1.0W$

   The governing loading condition when the structure is subjected to maximum values of seismic load opposing the effects of dead load is

7. $0.9D + 1.0E$

   where $D$ = dead load
   $E$ = earthquake load
   $L$ = live load
   $L_r$ = roof live load
   $R$ = rain load
   $S$ = snow load
   $W$ = wind load

The earthquake load $E$ is defined in ASCE 7 Sec. 12.4.2 and comprises both horizontal and vertical components. When the effects of gravity and seismic loads are additive, $E$ is given by ASCE 7 Eqs. (12.4-1), (12.4-3), and (12.4-4) as

$$E = \rho Q_E + 0.2S_{DS}D$$

where $Q_E$ = effect of horizontal seismic forces
$S_{DS}$ = design response acceleration, for a period of 0.2 second
$D$ = effect of dead load
$\rho$ = redundancy factor

Load combination 5 may now be defined as

$(1.2 + 0.2S_{DS})D + \rho Q_E + 0.5L + 0.2S$ ... where $0.5L$ is replaced with $L$ for garages, places of public assembly, and areas where $L > 100\ \text{lb/ft}^2$.

When the effects of gravity and seismic loads counteract, $E$ is given by ASCE 7 Eqs. (12.4-2), (12.4-3), and (12.4-4) as

$$E = \rho Q_E - 0.2S_{DS}D$$

Load combination 7 may now be defined as

$$(0.9 - 0.2S_{DS})D + \rho Q_E$$

**Example 2.47.** Strength Design Load Combinations

The nominal applied loads acting on a two-story steel frame are shown in Fig. 2.68. The frame forms part of the structure of an office building. Determine the maximum and minimum strength design loads acting on the column footings. The redundancy factor is $\rho = 1$ and the design response acceleration, for a period of 0.2 second, is $S_{DS} = 1g$.

The force produced on a column footing by the dead load is

$$D = (60 + 80)/2$$

$$= 70 \text{ kips}$$

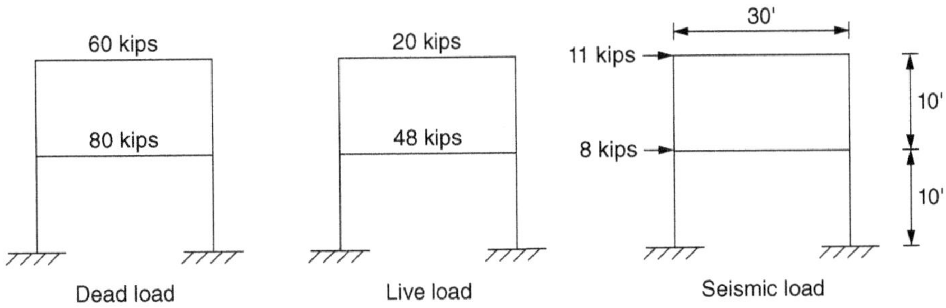


**FIGURE 2.68** Details for Example 2.47.

The force produced on a column footing by floor live load is

$$L = 48/2 = 24 \text{ kips}$$

The force produced on a column footing by roof live load is

$$L_r = 20/2 = 10 \text{ kips}$$

The force produced on a column footing by horizontal seismic load is

$$\rho Q_E = \pm\rho(\Sigma F_x h_x)/B$$

where $B$ = width of frame
$F_x$ = design lateral force at level $x$
$h_x$ = height above base to level $x$

$$\rho Q_E = \pm 1.0(11 \times 20 + 8 \times 10)/30 = \pm 10 \text{ kips}$$

The force produced in a column footing by vertical seismic load is

$$Q_V = \pm 0.2 S_{DS} D = \pm 0.2 \times 1 \times 70 = \pm 14 \text{ kips}$$

Applying load combination 5 of ASCE 7 Sec. 2.3.2 gives the strength design load as

$$F_C = (1.2 + 0.2S_{DS})D + \rho Q_E + 0.5L = 1.2 \times 70 + 14 + 10 + 0.5 \times 24 = 120 \text{ kips}$$

Applying load combination 2 of ASCE 7 Sec. 2.3.2 gives the strength design load as

$$F_C = 1.2D + 1.6L + 0.5L_r = 1.2 \times 70 + 1.6 \times 24 + 0.5 \times 10 = 127 \text{ kips ... governs for maximum load}$$

Applying load combination 7 of ASCE 7 Sec. 2.3.2 gives the strength design load as

$$F_C = (0.9 - 0.2S_{DS})D + \rho Q_E = 0.9 \times 70 - 14 - 10 = 39 \text{ kips ... governs for minimum load}$$

### Allowable Stress Load Combinations

The seismic and wind loads specified in ASCE 7 are at the strength design level in contrast to all other loads that are at the service or nominal level. Hence, in the ASD load combinations, seismic loads are multiplied by the factor 0.7 and wind loads are multiplied by the factor 0.6 to reduce them to service level values. The following nine load combinations are given in ASCE 7 Sec. 2.4.1 for allowable stress design:

The governing loading condition for dead load is

1. $D$

The governing loading condition when the structure is subjected to maximum values of occupancy live load is

2. $D + L$

The governing loading condition when the structure is subjected to maximum values of roof live load, rainwater, or snow load is

3. $D + (L_r \text{ or } S \text{ or } R)$

The governing loading condition when the structure is subjected to simultaneous values of occupancy live load and roof live load, or rainwater, or snow load is

4. $D + 0.75L + 0.75(L_r \text{ or } S \text{ or } R)$

The governing loading condition when the structure is subjected to maximum values of seismic load or wind load increasing the effects of dead load is

5. $D + (0.6W \text{ or } 0.7E)$

The governing loading condition when the structure is subjected to simultaneous values of occupancy live load, roof live load, or rain water, or snow load plus wind load increasing the effects of dead load is

6a. $D + 0.75(0.6W) + 0.75L + 0.75(L_r \text{ or } S \text{ or } R)$

The governing loading condition when the structure is subjected to simultaneous values of occupancy live load, or snow load plus seismic load increasing the effects of dead load is

6b. $D + 0.75(0.7E) + 0.75L + 0.75S$

The governing loading condition when the structure is subjected to maximum values of wind load opposing the effects of dead load is

7. $0.6D + 0.6W$

The governing loading condition when the structure is subjected to maximum values of seismic load opposing the effects of dead load is

8. $0.6D + 0.7E$

The earthquake load E is defined in ASCE 7 Sec. 12.4.2 and comprises both horizontal and vertical components. When the effects of gravity and seismic loads are additive, $E$ is given by ASCE 7 Eqs. (12.4-1), (12.4-3), and (12.4-4) as

$$E = \rho Q_E + 0.2S_{DS}D \text{ ... at the strength level}$$

Load combination 5 may now be defined as

$$(1 + 0.14S_{DS})D + 0.7\rho Q_E \text{ or } D + 0.6W$$

Load combination 6b may now be defined as

$$(1 + 0.105S_{DS})D + 0.525\rho Q_E + 0.75L + 0.75S$$

When the effects of gravity and seismic loads counteract, $E$ is given by ASCE 7 Eqs. (12.4-2), (12.4-3), and (12.4-4) as

$$E = \rho Q_E - 0.2S_{DS}D$$

Load combination 8 may now be defined as

$$(0.6 - 0.14S_{DS})D + 0.7\rho Q_E$$

**Example 2.48.** Allowable Stress Design Load Combinations
The nominal applied loads acting on a two-story steel frame are shown in Fig. 2.68. The frame forms part of the structure of an office building. Determine the maximum and minimum allowable stress design loads acting on the column footings. The redundancy factor is $\rho = 1$ and the design response acceleration, for a period of 0.2 second, is $S_{DS} = 1g$.

The loads acting on the frame are shown in Fig. 2.68 and the forces acting on a column footing are determined in Example 2.47.

The force produced on a column footing by the dead load is

$$D = 70 \text{ kips}$$

The force produced on a column footing by floor live load is

$$L = 24 \text{ kips}$$

The force produced on a column footing by roof live load is

$$L_r = 10 \text{ kips}$$

The strength level force produced in a column footing by horizontal seismic load is

$$\rho Q_E = \pm 10 \text{ kips}$$

The strength level force produced in a column footing by vertical seismic load is

$$Q_V = \pm 14 \text{ kips}$$

Applying load combination 2 of ASCE 7 Sec. 2.4.1 gives the allowable stress design load as

$$\begin{aligned} F_C &= D + L \\ &= 70 + 24 \\ &= 94 \text{ kips} \end{aligned}$$

Applying load combination 5 of ASCE 7 Sec. 2.4.1 gives the allowable stress design load as

$$\begin{aligned} F_C &= (1 + 0.14S_{DS})D + 0.7\rho Q_E \\ &= 70 + 0.7 \times 14 + 0.7 \times 10 \\ &= 87 \text{ kips} \end{aligned}$$

Applying load combination 6b of ASCE 7 Sec. 2.4.1 gives the allowable stress design load as

$$\begin{aligned} F_C &= (1.0 + 0.105S_{DS})D + 0.525\rho Q_E + 0.75L \\ &= 1.0 \times 70 + 0.75 \times 0.7 \times 14 + 0.525 \times 10 + 0.75 \times 24 \\ &= 101 \text{ kips ... governs for maximum load} \end{aligned}$$

Applying load combination 8 of ASCE 7 Sec. 2.4.1 gives the allowable stress design load as

$$F_C = (0.6 - 0.14S_{DS})D + 0.7\rho Q_E$$
$$= 0.6 \times 70 - 0.7 \times 14 - 0.7 \times 10$$
$$= 25 \text{ kips ... governs for minimum load}$$

## Strength Design Special Load Combinations

The special load combinations incorporate the overstrength factor to protect critical elements in a building. These critical elements must be designed with sufficient strength to remain elastic during seismic loading. Seismic loads multiplied by the overstrength factor are an approximation of the maximum load an element will experience.

The load combinations are modified by replacing the factor $E$ with $E_m$. When the effects of gravity and seismic loads are additive, ASCE 7 Eqs. (12.4-5) and (12.4-7) give

$E_m$ = maximum effect of horizontal and vertical earthquake forces that can be developed in an element

$= \Omega_0 Q_E + 0.2S_{DS}D$

$\Omega_0$ = structure overstrength factor given in ASCE 7 Table 12.2-1

= amplification factor to account for the overstrength of the structure in the inelastic range

$Q_E$ = effect of horizontal seismic forces

$S_{DS}$ = design response acceleration, for a period of 0.2 second

Load combination 5 of ASCE 7 Sec. 2.3.2 may now be defined as

$(1.2 + 0.2S_{DS})D + \Omega_0 Q_E + 0.5L + 0.2S$ ... where $0.5L$ is replaced with $L$ for garages, places of public assembly, and areas where $L > 100\ \text{lb/ft}^2$

When the effects of gravity and seismic loads counteract, ASCE 7 Eqs. (12.4-6) and (12.4-7) give

$$E_m = \Omega_0 Q_E - 0.2S_{DS}D$$

Load combination 7 of ASCE 7 Sec. 2.3.2 may now be defined as

$$= (0.9 - 0.2S_{DS})D + \Omega_0 Q_E$$

## Allowable Stress Design Special Load Combinations

Load combinations are modified by replacing the factor $E$ with $E_m$. When the effects of gravity and seismic loads are additive, load combinations 5 and 6b of ASCE 7 Sec. 2.4.1 become

$$(1 + 0.14S_{DS})D + 0.7\Omega_0 Q_E \text{ or } D + 0.6W$$

and

$$(1 + 0.105S_{DS})D + 0.525\Omega_0 Q_E + 0.75L + 0.75S$$

When the effects of gravity and seismic loads counteract, load combination 8 of ASCE 7 Sec. 2.4.1 becomes

$$(0.6 - 0.14S_{DS})D + 0.7\Omega_0 Q_E$$

## 2.11 Serviceability Criteria

Serviceability concerns that impair the functioning of a building are excessive deflections, drift, vibrations, and deterioration. The serviceability limit state is defined as the condition beyond which a structure or member becomes unfit for service and is judged to be no longer useful for its intended function. Structural systems and members are required by IBC Sec. 1604.3 to be designed to have adequate stiffness to limit deflections and lateral drift. Floor systems must be designed with due regard for vibrations caused by pedestrian traffic and mechanical equipment. Buildings must be designed to tolerate long-term environmental effects or be protected from such effects.

### Deflection

The deflection of beams under load must be compatible with the degree of movement that can be tolerated by the supported elements in order that the serviceability of the structure is not impaired. Common deflection limits given by American Institute of Steel Construction, *Specification for Structural Steel Buildings* (AISC 360-10)[18] Commentary Sec. L3 are 1/360 of the span for floors subjected to reduced live load and 1/240 of the span for roof members. In long-span floors, it may be necessary to impose a limit on the maximum deflection, independent of span length, to prevent damage to adjacent non-structural elements. Damage to non-load-bearing partitions may occur when deflections exceed 3/8 in.

Deflection limitations are specified in IBC Sec. 1604 and listed in IBC Table 1604.3 and are summarized in Table 2.17.

Excessive vertical deflections may be caused by loads, temperature, creep, differential settlement, and construction tolerances and errors. Deflections in excess of 1/300 of

| Construction | $L$ | $S$ or $W$ | $D + L$* |
|---|---|---|---|
| **Roof members:** | | | |
| Supporting plaster ceiling | 1/360 | 1/360 | 1/240 |
| Supporting nonplaster ceiling | 1/240 | 1/240 | 1/180 |
| Not supporting ceiling | 1/180 | 1/180 | 1/120 |
| **Floor members** | 1/360 | – | 1/240 |
| **Exterior walls and interior partitions:** | | | |
| With plaster or stucco finishes | – | 1/360 | – |
| With other brittle finishes | – | 1/240 | – |
| With flexible finishes | – | 1/120 | – |

*For steel structural members, the dead load is taken as zero.

**Table 2.17** Deflection Limits

the span may be visually objectionable and cause architectural damage, cladding leakage, and impair the operation of doors and windows.

### Drift

Excessive drift of walls and frames may cause damage to cladding and nonstructural walls and partitions. Common drift limits given by AISC 360-05 Commentary Sec. L3 are 1/600 to 1/400 of the building or story height. Smaller drift limits may be necessary if the cladding is brittle. Damage to cladding and glazing may occur when interstory drift exceeds 3/8 in unless special detailing methods are implemented.

### Vibration

Vibration of floors or of the building as a whole may cause discomfort to the building occupants. Stiffness, mass distribution, and damping all influence the degree of vibration produced. Long-span beams with insufficient damping may be subject to perceptible transient vibrations caused by pedestrian traffic. This is particularly significant for beams with a span exceeding 20 ft and a natural frequency of less than 8 Hz. To determine the degree of vibration anticipated, analysis techniques as described by Murray et al.[19] and Allen et al.[20] may be employed.

To prevent resonance from human activities, a floor system must be tuned so that its natural frequency is at least twice the frequency of the steady-state excitation to which it is exposed. Common human activities impart dynamic forces to a floor at frequencies in the range of 2 to 6 Hz. The relationship between fundamental frequency and deflection is given by ASCE 7 App. C Commentary Sec. CC.1.3 as

$$f_o \approx 18/(\delta)^{1/2}$$

where $\delta$ is maximum deflection under uniform load, in mm.

Hence, the deflection due to uniform dead load and participating live load must be limited to 5 mm if the fundamental frequency of the floor system is to be not less than 8 Hz.

### Durability

Buildings may deteriorate in inhospitable environments. Visible staining may be caused by weathering and corrosion. Deterioration of the structural members may be caused by water infiltration and corrosion. Design must include adequate protection measures or planned maintenance to prevent these problems occurring.

## References

1. International Code Council (ICC). 2015. *International Building Code*, 2015 ed., ICC, Falls Church, VA.
2. American Society of Civil Engineers (ASCE). 2010. *Minimum Design Loads for Buildings and Other Structures* (ASCE 7-10), ASCE, Reston, VA.
3. O'Rourke, M. 2008. "Structural Collapse from Snow Loads," *Structure*, November 2008.
4. Kusek, L. 2008. *Trial Design Problem 01/08*. Structural Engineering Institute (SEI), Reston, Virginia.
5. O'Rourke, M., Mazikins, P. J., and Showalter J. 2006. *Snow Provisions in ASCE 7-05*. American Forest and Paper Association (AF&PA), Washington, DC.

6. Barbera, J. J. and Scott, D. R. 2007. "Simplifying Wind Provisions," *Structural Engineer*, November 2007.
7. Huston, E. 2007. "SEAW's Handbook of a Rapid Solutions Methodology for Wind Design," *Structure*, November 2007.
8. Lai, J. S. 2009. *2009 IBC Alternate Wind Design Approach*. Structural Engineers Association of Southern California (SEAOSC) Design Seminar, Long Beach, CA, SEAOSC, Whittier, CA.
9. International Code Council (ICC). 2006. *International Residential Code*, 2006 ed., ICC, Falls Church, VA.
10. Charney, F. A. 2014. *Seismic Loads: Guide to the Seismic Load Provisions of ASCE 7-10*. ASCE Press, Reston, VA.
11. Building Seismic Safety Council (BSSC). 2009. *NEHRP Recommended Seismic Provisions for New Buildings and Other Structures*, FEMA P-750/2009 ed., Federal Emergency Management Agency, Washington, DC.
12. Paz, M. 1991. *Structural Dynamics*. Van Nostrand Reinhold, New York, NY.
13. Structural Engineers Association of California (SEAOC). 1999. *Recommended Lateral Force Requirements and Commentary*. SEAOC, Sacramento, CA.
14. Sabelli, R., Pottebaum, W., and Dean, B. 2009. "Diaphragms for Seismic Loading," *Structural Engineer*, January 2009.
15. APA-The Engineered Wood Association(APA). 2007. *Diaphragms and Shear Walls: Design/Construction Guide*. APA, Tacoma, WA.
16. Tissell, J. R. and Elliot, J. R. 2000 Reprint. *Plywood Diaphragms: Research Report 138*. APA, Tacoma, WA.
17. Henry, J. R. 2008. "Structural Wall Anchorage Requirements in 2006 IBC and ASCE/SEI 7-05," *Structural Engineer*, November 2008.
18. American Institute of Steel Construction (AISC). 2010. *Specification for Structural Steel Buildings* (AISC 360-10), AISC, Chicago, IL.
19. Murray, T. M., Allen, D. E., and Ungar, E. E. 1997. *Floor Vibrations due to Human Activity*. Design Guide No. 11. AISC, Chicago, IL.
20. Allen, D. E., Onysko, D. M., and Murray, T. M. 1999 *ATC Design Guide 1: Minimizing Floor Vibration*. Applied Technology Council, Redwood City, CA.


## Problems

**2.1** *Given:* Figure 2.69 shows the roof framing plan of a single-story concrete tilt-up factory building. The roof is nominally flat with a dead load of 20 lb/ft$^2$.

*Find:* Using allowable stress level (ASD) load combinations the design loads on

*a.* Girder 12
*b.* Beam 23
*c.* Column 2

**2.2** *Given:* Figure 2.69 shows the roof framing plan of a single-story concrete tilt-up factory building. The roof is nominally flat with a dead load of 20 lb/ft$^2$.

*Find:* Using strength level (LRFD) load combinations the design loads on

*a.* Girder 12
*b.* Beam 23
*c.* Column 2

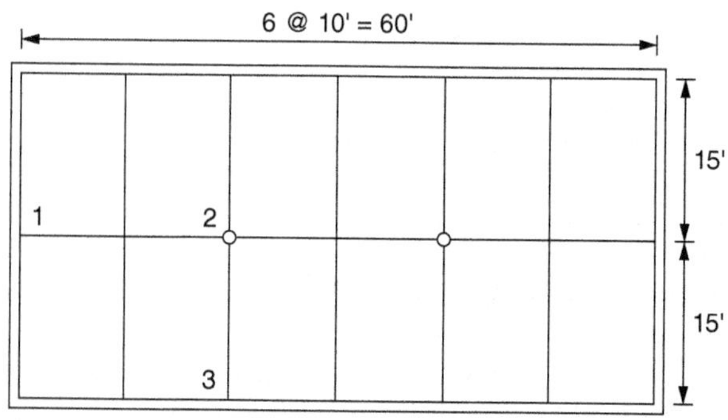


**FIGURE 2.69** Details for Problems 2.1 to 2.6.

**2.3** *Given:* Figure 2.69 shows the roof framing plan of a single-story concrete tilt-up factory building. The roof is pitched with a rise of 5 in/ft from the outer walls to the central girder. The roof dead load is 20 lb/ft$^2$.

*Find:* Using allowable stress level (ASD) load combinations the design loads on
*a.* Girder 12
*b.* Beam 23
*c.* Column 2

**2.4** *Given:* Figure 2.69 shows the roof framing plan of a single-story concrete tilt-up factory building. The roof is pitched with a rise of 5 in/ft from the outer walls to the central girder. The roof dead load is 20 lb/ft$^2$.

*Find:* Using strength level (LRFD) load combinations the design loads on
*a.* Girder 12
*b.* Beam 23
*c.* Column 2

**2.5** *Given:* Assume Fig. 2.69 is the second-floor framing plan of a heavy storage warehouse. The floor dead load is 80 lb/ft$^2$.

*Find:* Using allowable stress level (ASD) load combinations the design loads on
*a.* Girder 12
*b.* Beam 23

**2.6** *Given:* Assume Fig. 2.69 is the second-floor framing plan of a heavy storage warehouse. The floor dead load is 80 lb/ft$^2$.

*Find:* Using strength level (LRFD) load combinations the design loads on
*a.* Girder 12
*b.* Beam 23

**2.7** *Given:* Figure 2.70 shows the elevation of a steel-framed office building with moveable partitions. Frames are at 25 ft on center. The roof dead load of 60 lb/ft$^2$ and the floor dead load of 80 lb/ft$^2$ include the weight of the columns.

*Find:* Using allowable stress level (ASD) load combinations the design load at the base of an interior column.

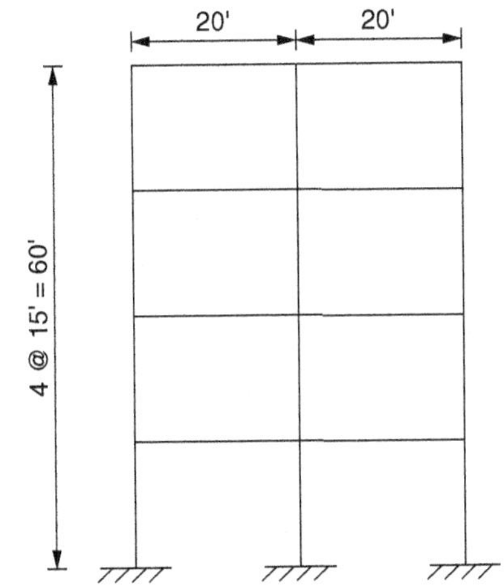


**FIGURE 2.70** Details for Problems 2.7 and 2.8.

**2.8** *Given:* Figure 2.70 shows the elevation of a steel-framed office building with moveable partitions. Frames are at 25 ft on center. The roof dead load of 60 lb/ft$^2$ and the floor dead load of 80 lb/ft$^2$ include the weight of the columns.

*Find:* Using strength level (LRFD) load combinations the design load at the base of an interior column.

**2.9** *Given:* Figure 2.71 shows a factory building having a ventilated roof with a thermal factor of $C_t = 0.9$ and a thermal resistance of $R = 15$ ft$^2$-$h$ °F/Btu. The roof has an unobstructed smooth surface. The ground snow load for the locality is $p_g = 40$ lb/ft$^2$.

*Find:* The balanced snow load on the roof.

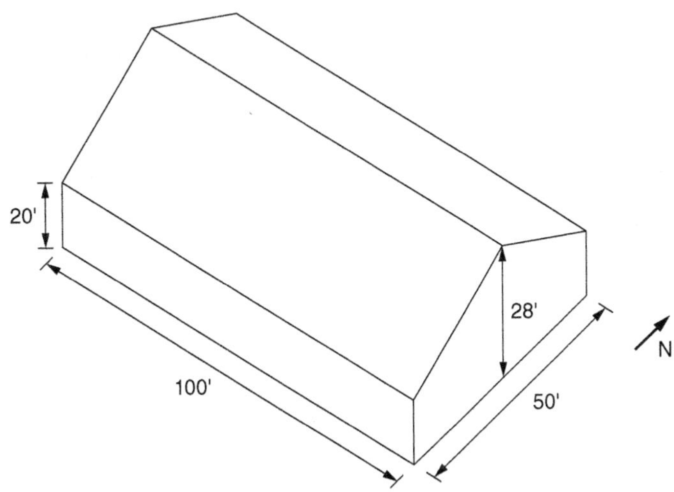


**FIGURE 2.71** Details for Problems 2.9 to 2.12.

**2.10** *Given:* Figure 2.71 shows a factory building having a ventilated roof with a thermal factor of $C_t = 0.9$ and a thermal resistance of $R = 15$ ft²-h °F/Btu. The roof has an unobstructed smooth surface. The ground snow load for the locality is $p_g = 40$ lb/ft².

*Find:* The unbalanced snow load on the roof.

**2.11** *Given:* Figure 2.71 shows a factory building located in a suburban area with a wind speed of $V = 100$ mi/h. The building may be considered enclosed and the roof diaphragm is flexible. The building is not sensitive to dynamic effects and is not located on a site at which channeling or buffeting occurs.

*Find:* Using the IBC alternate method, the wind loads on the main wind-force resisting system.

**2.12** *Given:* Figure 2.71 shows a factory building located in a suburban area with a wind speed of $V = 100$ mi/h. The building may be considered enclosed and the roof diaphragm is flexible. The building is not sensitive to dynamic effects and is not located on a site at which channeling or buffeting occurs.

*Find:* Using the ASCE 7 low-rise building analytical method, the wind loads on the main wind-force resisting system.

**2.13** *Given:* The office building shown in Fig. 2.72.

*Find:* The fundamental period of the building assuming it is constructed with the following structural systems.
*a.* Steel moment-resisting frames
*b.* Reinforced concrete shear walls

**2.14** *Given:* The office building shown in Fig. 2.72 located on soil profile type site class C. The maximum considered response accelerations for the location are $S_S = 1.2g$ and $S_1 = 0.5g$.

*Find:* The design response accelerations $S_{DS}$ and $S_{D1}$.

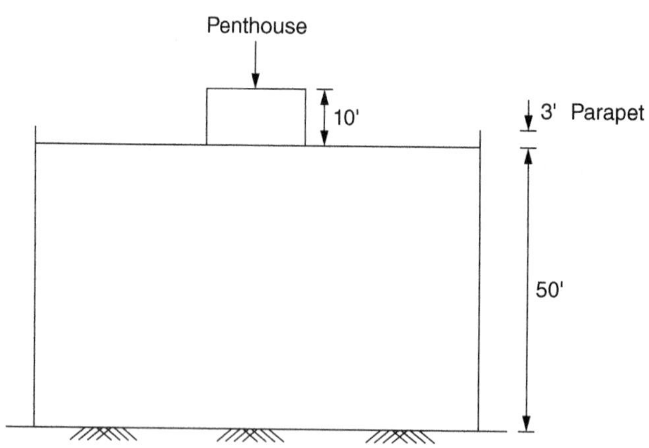


**FIGURE 2.72** Details for Problems 2.13 to 2.17.

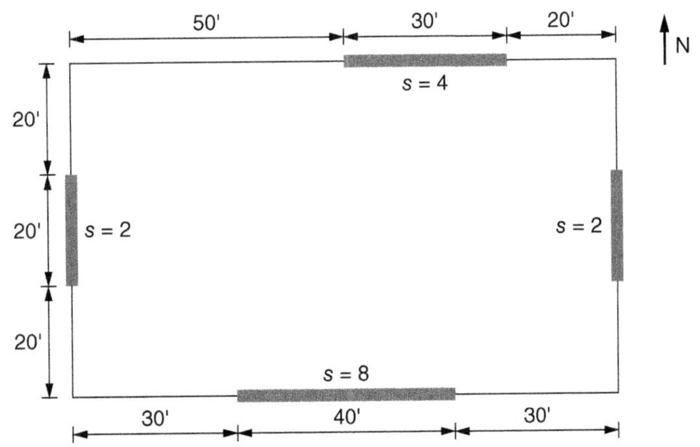


**FIGURE 2.73** Details for Problems 2.18 and 2.19.

**2.15** *Given:* The office building shown in Fig. 2.72.

*Find:* *a.* The occupancy category
*b.* The importance factor
*c.* The seismic design category

**2.16** *Given:* The office building shown in Fig. 2.72.

*Find:* The response modification coefficient of the building assuming it is constructed with the following structural systems.
*a.* Steel moment-resisting frames
*b.* Reinforced concrete shear walls

**2.17** *Given:* The office building shown in Fig. 2.72.

*Find:* The seismic response coefficient of the building assuming it is constructed with the following structural systems.
*a.* Steel moment-resisting frames
*b.* Reinforced concrete shear walls

**2.18** *Given:* Figure 2.73 shows the plan view of a single-story masonry bearing wall building with the relative stiffness s of each wall indicated. The height of each wall is 20 ft. The weight of the roof is 60 lb/ft$^2$ and the weight of the walls is 80 lb/ft$^2$.

*Find:* *a.* The distance of the center of mass from the west wall
*b.* The distance of the center of rigidity from the west wall
*c.* The distance of the center of rigidity from the south wall

**2.19** *Given:* The building in Fig. 2.73. The seismic base shear in the north-south direction is 120 kips.

*Find:* *a.* The maximum torsion in the north-south direction (ignore torsional amplification)
*b.* The maximum force in the east wall

# CHAPTER 3
# Behavior of Steel Structures under Design Loads

## 3.1 Introduction

The forces applied to a building must be transferred through the structural framework to the building foundation and the supporting ground. This applies to both vertical forces, caused by building self-weight and occupancy loads, and lateral forces, caused by wind and earthquake. To provide a stable structure, a complete and continuous load path is required from the point of application of the forces to the foundation.

In accordance with the International Code Council, *International Building Code* (IBC)[1] Sec. 1604.4 design of the structural framework shall provide a complete load path capable of transferring loads from their point of origin to the load-resisting elements. In addition, the total lateral force shall be distributed to the vertical elements of the lateral force-resisting system in proportion to their rigidities. In buildings with rigid diaphragms, provision must be made for the increased forces resulting from torsion caused by the eccentricity between the center of mass and the center of rigidity. The building must also be designed to resist the overturning effects caused by the lateral forces.

## 3.2 Gravity Load-Resisting Systems

As indicated in Fig. 1.5, a steel-framed building consists of floor and roof slabs, beams, girders, and columns. The floor slab is supported by beams that are supported by girders that frame into columns. The structural behavior of the frame is governed by the connection of the girder ends to the columns. American Institute of Steel Construction, *Specification for Structural Steel Buildings* (AISC 360)[2] Sec. B3.6 classifies connections into three basic types. These are, simple connections, fully restrained (FR) moment connections, and partially restrained (PR) connections.

### Simple Connections

Simple connections are defined in AISC 360 Glossary as connections that transmit negligible bending moment between connected members. The connection must have enough rotation capacity to accommodate the required end rotation of the girder

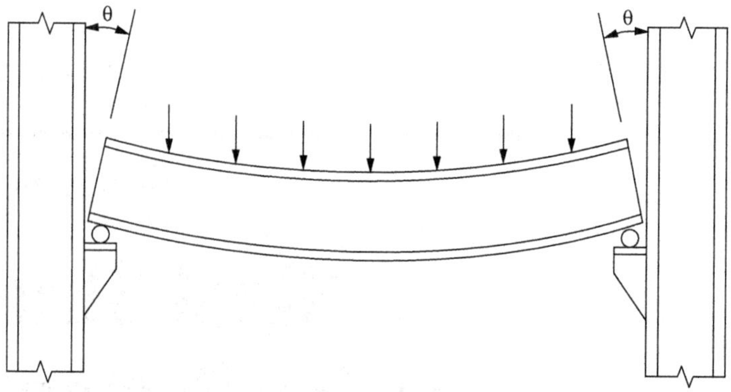


**FIGURE 3.1** Simply supported girder.

without imparting significant moment to the column. In addition, the connection must have sufficient strength to transfer the end shear reaction of the girder to the column. To provide the necessary rotation, inelastic rotation in the connection is permitted.

An ideal simple connection is to support girder ends on roller bearings as shown in Fig. 3.1. This is neither feasible nor practical.

Testing of a girder end connection, as described by Kennedy,[3] provides the typical moment-rotation response shown in Fig. 3.2. The secant stiffness of the connection is

$$K_S = \text{rotational stiffness of the connection at service load}$$
$$= M_S/\theta_S$$

where $M_S$ is bending moment at service load and $\theta_S$ is rotation at service load.

In AISC 360 Commentary Sec. B3.6, the limits on simple, PR, and FR connections are defined as

$$K_S L/EI \leq 2 \text{ ... simple connection}$$
$$K_S L/EI \geq 20 \text{ ... FR connection}$$
$$2 < K_S L/EI < 20 \text{ ... PR connection}$$

where $L$ is girder length and $EI$ is flexural rigidity of the girder.

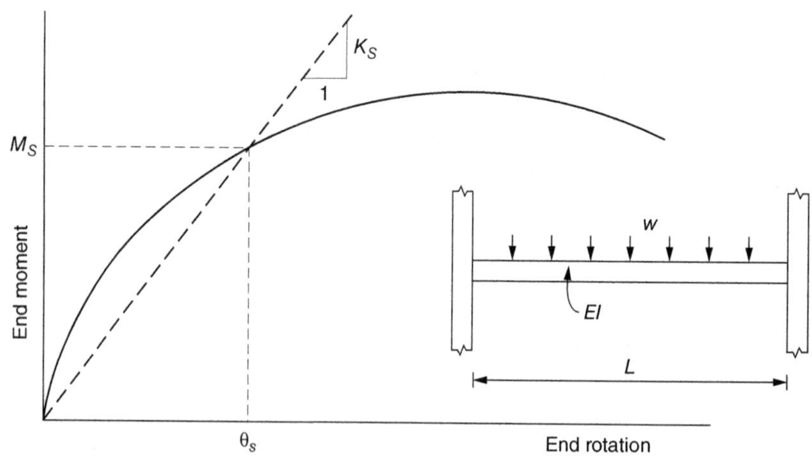


**FIGURE 3.2** Moment-rotation response.

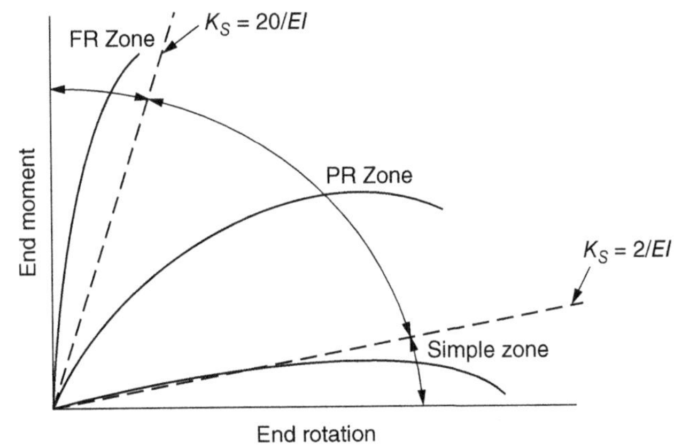


**FIGURE 3.3** Moment-rotation response limits.

The moment-rotation response limits are shown in Fig. 3.3. In accordance with AISC 360 Commentary Sec. B3.6, connections that transmit less than 20 percent of the fully plastic moment of the girder at a rotation of 0.02 radian may be considered simple connections.

Design details for *simple shear connections* are provided in American Institute of Steel Construction, *Steel Construction Manual* (AISC Manual)[4] Part 10. The different types covered include

- *One-sided connections*

  Single-plate (shear tab), single-angle, and T connections
- *Two-sided connections*

  Double-angle connections
- *Seated connections*

  Unstiffened- and stiffened-seated connections

The *single-plate connection* consists of a plate shop-welded to the column at one edge and bolted in the field to the girder web with one vertical row of bolts. Figure 3.4 shows a typical application of a single-plate shear connection as detailed by Astaneh et al.[5]

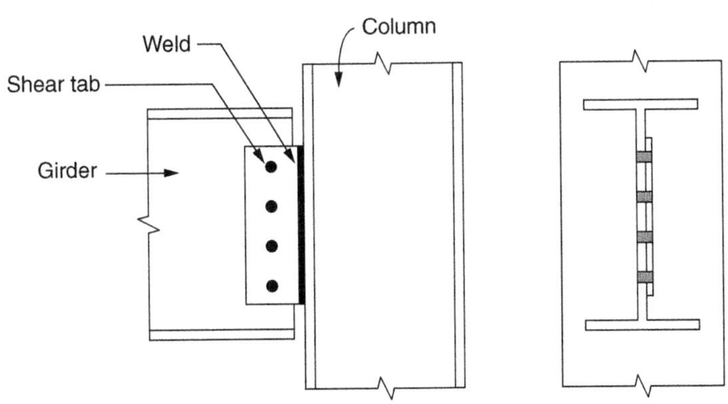


**FIGURE 3.4** Shear tab connection.

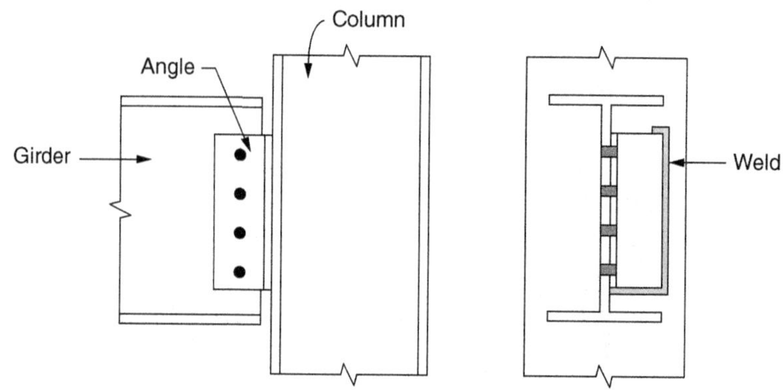


**Figure 3.5** Single-angle connection.

This connection has the advantage of facilitating side erection of the girder, with ample clearance, and short-slotted holes allow adjustments for tolerance in the length of the girder.

The *single-angle connection* consists of one leg of an angle field-bolted to one side of the web of the girder. The other leg of the angle is shop-bolted or welded to the column. The welded alternative is shown in Fig. 3.5. To ensure sufficient flexibility in the connection, welding is placed only along the toe and across the bottom of the leg. A return of length $2w$ is required at the top of the leg, where $w$ is the nominal weld size. Welding across the entire top of the leg must be avoided as this will make the connection too rigid.

The *T connection* consists of the web of a T shape field-bolted to one side of the web of the girder. The flange of the T is shop-bolted or welded to the column. The welded alternative, as discussed by Astaneh and Nader,[6] is shown in Fig. 3.6. This single-sided connection is suitable for use when the end reaction from the girder is large, since two rows of bolts may be provided in the web of the T. To provide the required flexibility, when the T is welded to the column, welding is placed only along the toes of the T flange with a return of length $2w$ at the top.

The *double-angle connection* consists of two angles, one on each side of the web of the girder as shown in Fig. 3.7, and is suitable when the end reaction is large. The angles may be bolted or welded to the girder as well as to the column flange. The angles are

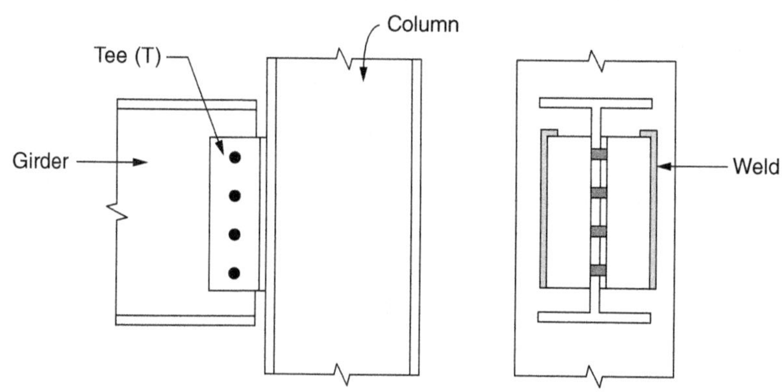


**Figure 3.6** T connection.

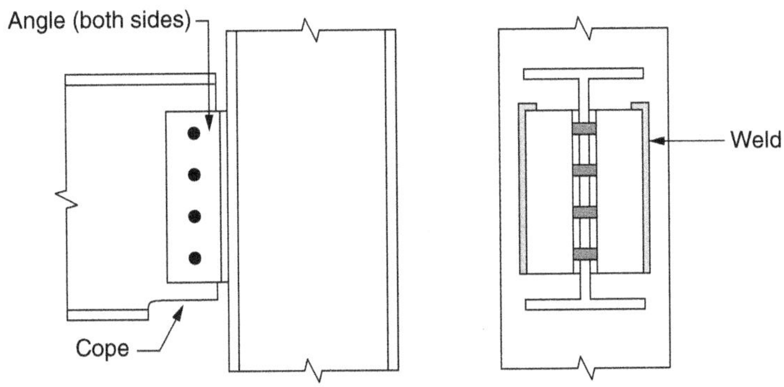


**Figure 3.7** Double-angle connection.

usually shop-attached to the column flange and this requires the bottom flange of the girder to be coped, as shown, to allow the girder to be knifed into position in the field. When the angles are welded to the column, welding is placed only along the toe of the angles to ensure sufficient flexibility in the connection. A return of length $2w$ is required at the top of each leg, where $w$ is the nominal weld size. Welding across the entire top of the legs or across the bottom of the legs must be avoided as this will make the connection too rigid.

In an *unstiffened-seated connection* the girder sits on an unstiffened angle that is shop attached to the column. The entire end reaction of the girder is supported by this angle seat. A clip angle is located on the top flange and connected to the column to provide stability to the girder. The two angles may be bolted or welded to the girder as well as to the column flange. The connection has been analyzed by Garrett and Brockenbrough[7] and Yang et al.[8] and is shown in Fig. 3.8. With this connection, erection is simplified as the seat provides a stable erection platform for the girder while the installation is completed. An additional advantage over a framed connection is that larger fabrication and erection tolerances can be accommodated.

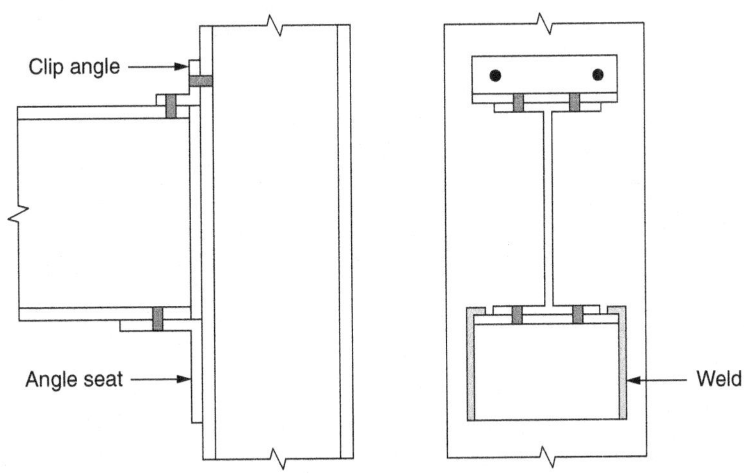


**Figure 3.8** Unstiffened-seated connection.

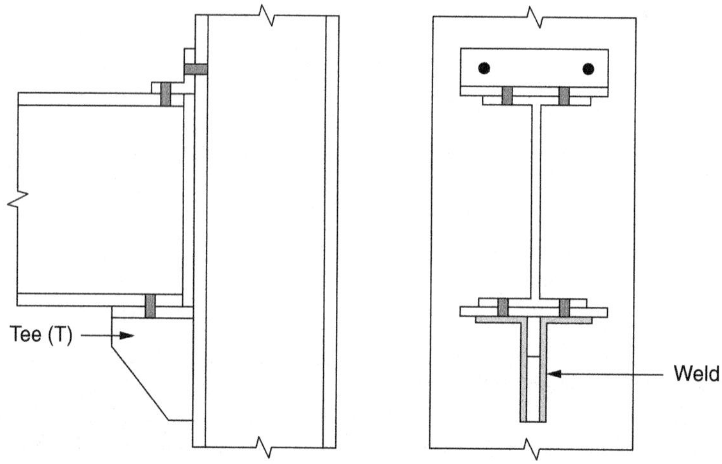


**Figure 3.9** Stiffened-seated connection.

The *stiffened-seated connection* can support heavier loads than the *unstiffened seated connection*. The stiffened seat may be fabricated from two angles back-to-back, a T section cut to shape, or from plates. Design details and experimental data are available from Ellifritt and Sputo[9] and Roeder and Dailey.[10] Figure 3.9 shows a stiffened-seated connection fabricated from a T section.

**Example 3.1.** Two-Story Frame with Simple Connections
Figure 3.10 shows a two-story frame. The girders are attached to the columns with simple connections. The distributed load on each girder is 1.2 kips/ft. Determine the bending moments produced in the frame.

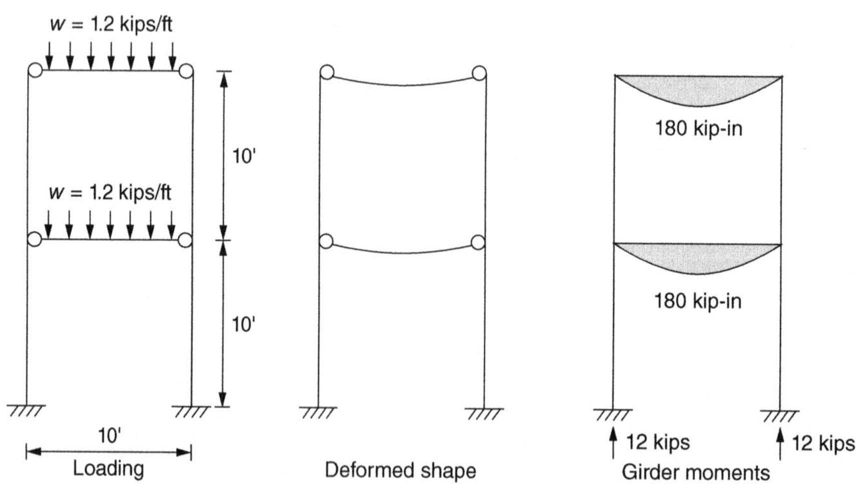


**Figure 3.10** Details for Example 3.1.

The girders are considered simply supported and the maximum bending moment produced at midspan of each girder is

$$\begin{aligned} M &= wL^2/8 \\ &= 1.2 \times 10^2/8 \\ &= 15 \text{ kip-ft} \\ &= 180 \text{ kip-in} \end{aligned}$$

No bending moments are produced at the ends of the girders or in the columns. The bending moments produced in the frame are indicated in Fig. 3.10 with the moments drawn on the tension side of the members.

## Fully Restrained Moment Connections

Fully restrained (FR) moment connections are defined in AISC 360 Glossary as connections that are capable of transmitting bending moment and shear force with negligible rotation between connected members. The bending moment at the end of a girder causes axial forces in the flanges of the girder. As shown in Fig. 3.11, a tensile force is developed in the top flange and a compressive force in the bottom flange when gravity loads are applied to the girder. These concentrated forces are applied to the column flange, and stiffener plates, as shown, may be required to prevent local flange bending and local web yielding in the column. The end reaction in the girder is transferred through a shear tab to the column flange. The angle between the connected members is maintained under load.

Design details for fully restrained connections are provided in AISC Manual Part 12. The different types covered are

- Flange-plated moment connections
- Directly welded moment connections
- Extended end-plate connections

These connections are suitable for buildings assigned to seismic design categories A, B, and C that are designed for a response modification coefficient of $R = 3$ as specified in ASCE 7 Table 12.2-1 Part H. They are not permitted in seismic design categories D

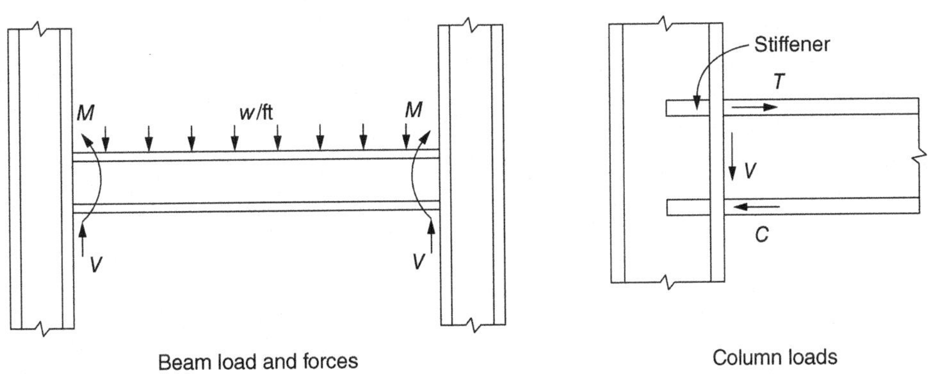


**FIGURE 3.11** Fully restrained moment connection.

through F for moment-resisting frames required for seismic force-resisting systems as they cannot accommodate the large inelastic deformation demand that is required. As specified by IBC Sec. 2205.2.2, steel structures assigned to seismic design categories D through F shall be designed and detailed in accordance with the American Institute of Steel Construction, *Seismic Provisions for Structural Steel Buildings* (AISC 341).[11] In seismic design categories A, B, and C, the designer has a choice of adopting an $R$ factor of 3 and the provisions of AISC 360 or of adopting the appropriate higher value of $R$ and the provisions of AISC 341. In seismic design categories D through F, design must be in accordance with AISC 341. Where applicable, using an $R$ value of 3 and designing to AISC 360 is generally more economical than designing and detailing to the more stringent requirements of AISC 341.

A *flange-plated moment connection* consists of a shear tab, welded to the column flange and bolted to the girder web, and top and bottom flange plates, as shown in Fig. 3.12. The flange plates are fillet welded to the column flange and either bolted or welded to the girder flanges. Details of flange-plated moment connections are given by Astaneh.[12] Because of possible mill tolerances in the column flange producing out-of-squareness with the web, it may be desirable to shop-attach the plates to the column to prevent assembly problems in the field. Alternatively, the plates may be shipped loose for field attachment to the column.

The *directly welded moment connection*, shown in Fig. 3.13, consists of a shear tab connection to the girder web and complete-joint-penetration groove welds connecting the top and bottom flanges of the girder to the supporting column. Allowance is required in this type of connection for the shrinkage that occurs in the groove weld when it cools

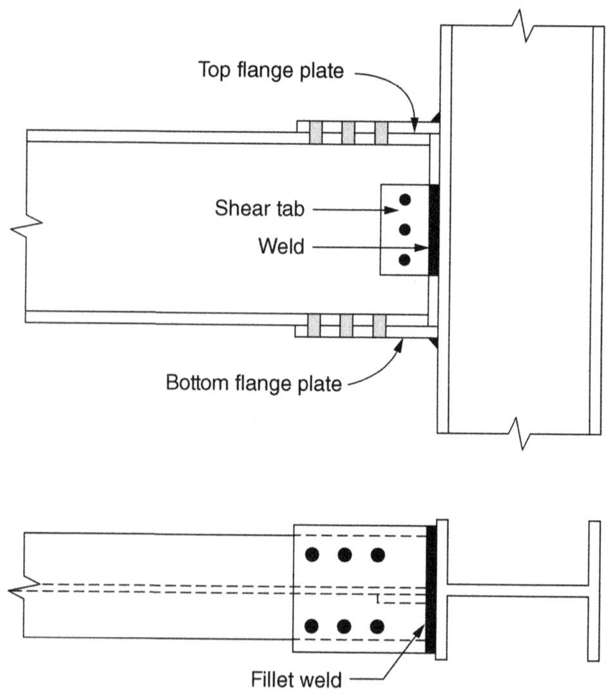


**FIGURE 3.12** Flange-plated moment connection.

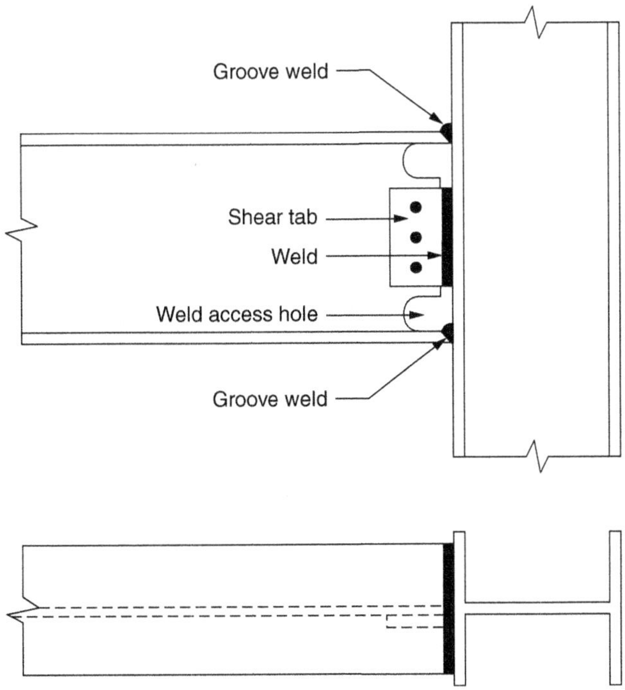


**FIGURE 3.13** Directly welded moment connection.

and contracts. This amounts to approximately $\frac{1}{16}$ in and can cause erection problems along long lines of continuous girders. To compensate for this, girders may be fabricated longer or the weld-joint root opening can be increased.

The *extended end-plate connection* consists of an end-plate shop-welded to the web and flanges of a girder and bolted to a column flange. As shown in Fig. 3.14, the length of the end-plate is greater than the girder depth. Details of the application and design of this connection are given by Murray and Sumner.[13] This connection is readily erected and is economical in use.

**Example 3.2.** Two-Story Frame with Fully Restrained Connections
Figure 3.15 shows a two-story frame. The flexural rigidity of all members in the frame is identical. The girders are attached to the columns with fully restrained connections. The distributed load on each girder is 1.2 kips/ft. Determine the bending moments produced in the frame.

The girders are considered rigidly connected to the columns and moments produced in the frame may be conveniently determined by the moment distribution procedure. As described in Williams[14] Part 2, Sec. 7.4, advantage may be taken of the symmetry in the structure and loading to simplify the procedure. For all members

$$EI/L = k$$

Allowing for symmetry, the stiffness of the members are

$$s_{21} = s_{23} = s_{32} = 4k$$
$$s_{25} = s_{36} = 2k$$

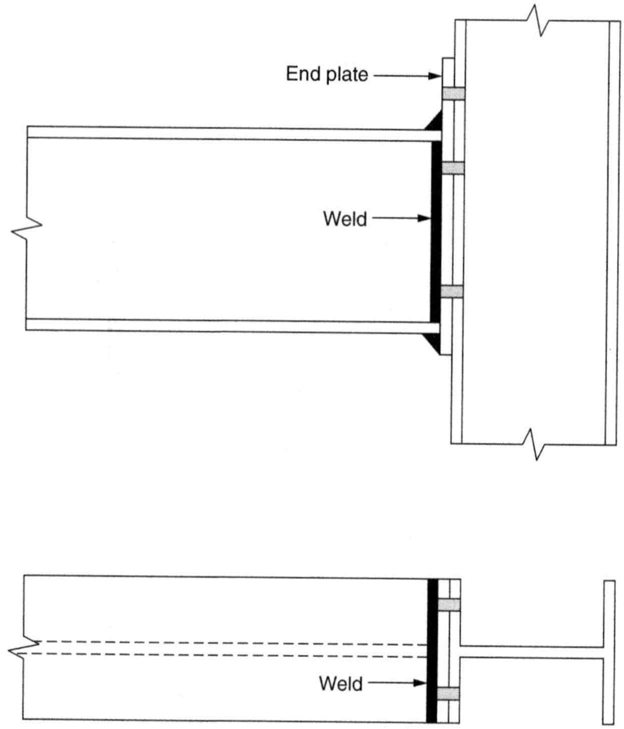


**FIGURE 3.14** Extended end-plate connection.

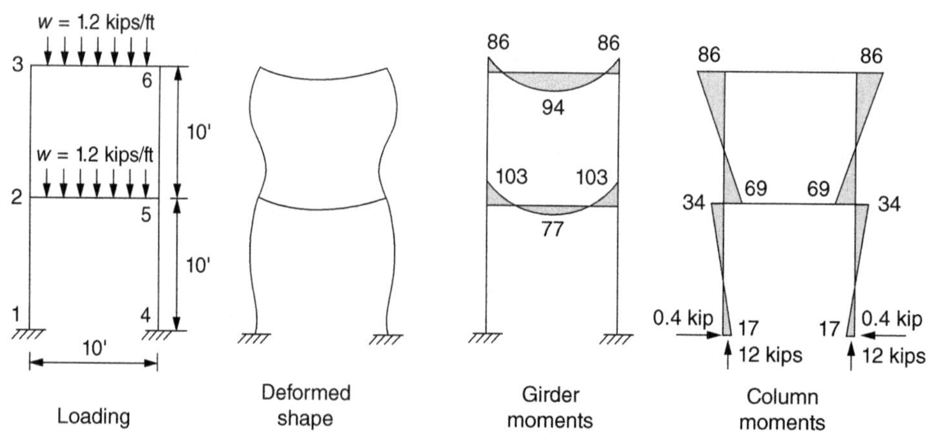


**FIGURE 3.15** Details for Example 3.2.

The distribution factors at joint 2 are

$$\begin{aligned} d_{25} &= s_{25}/(s_{25} + s_{21} + s_{23}) \\ &= 2k/10k \\ &= 1/5 \\ d_{21} &= d_{23} = 2/5 \end{aligned}$$

The distribution factors at joint 3 are

$$\begin{aligned} d_{36} &= s_{36}/(s_{36} + s_{32}) \\ &= 2k/6k \\ &= 1/3 \\ d_{32} &= 2/3 \end{aligned}$$

Allowing for symmetry, the carry-over factors in the columns are ½ and there is no carry-over between beam ends. The fixed-end moments are

$$\begin{aligned} M_{F25} = M_{F36} &= -wL^2/12 \\ &= -1.2 \times 10^2/12 \\ &= -10 \text{ kip-ft} \\ &= -120 \text{ kip-in} \end{aligned}$$

The distribution procedure is shown in Table 3.1 with distribution occurring in the left half of the frame only. For convenience, the fixed-end moments are multiplied by 100.

The final bending moments, in kip-in units, produced in the members, are indicated in Fig. 3.15 with the moments drawn on the tension side of the members. The moments produced at midspan of the fully restrained girders are much smaller than those produced in the simply supported girders. Significant moments are developed at the girder ends and also in the columns.

| Member | 21 | 25 | 23 | 32 | 36 |
|---|---|---|---|---|---|
| Distribution Factor | 2/5 | 1/5 | 2/5 | 2/3 | 1/3 |
| Fixed-end Moments | | −12000 | | | −12000 |
| Distribution | 4800 | 2400 | 4800 | 8000 | 4000 |
| Carry-over | | | 4000 | 2400 | |
| Distribution | −1600 | −800 | −1600 | −1600 | −800 |
| Carry-over | | | −800 | −800 | |
| Distribution | 320 | 160 | 320 | 533 | 267 |
| Carry-over | | | 267 | 160 | |
| Distribution | −107 | −53 | −107 | −107 | −53 |
| Carry-over | | | −53 | −53 | |
| Distribution | 21 | 11 | 21 | 35 | 18 |
| Final Moments | 3434 | −10282 | 6848 | 8568 | -8568 |

**TABLE 3.1** Distribution of Moments in Example 3.2

## Partially Restrained Moment Connections

Partially restrained (PR) moment connections are defined in AISC 360 Glossary as connections that are capable of transmitting bending moment and shear force with rotation between connected members that is not negligible. To design a framework using partially restrained moment connections, the moment-rotation relationship of the connection must be evaluated and included in the analysis of the framework.

The characteristics of girder end connections may be summarized in terms of the restraint factor $R$ defined by Blodgett[15] Sec. 5.1. The degree of restraint $R$ is the ratio of the actual end moment in a girder to the end moment in a fully fixed-ended girder. Simple connections provide sufficient ductility to accommodate girder end rotation while resisting less than 20 percent of the fixed-end moment and

$$\begin{aligned} R &= M/M_F \\ &< 0.2 \end{aligned}$$

where $M$ is moment produced at the end connection of the girder supporting a uniformly distributed load $w$ and $M_F$ is moment produced at the end of the fixed-end girder supporting a uniformly distributed load $w$.

Figure 3.16 shows the bending moment produced in a simply supported girder with a restraint factor $R = 0$. The maximum moment of $wL^2/8$ occurs at midspan and the remainder of the girder is underutilized.

Fully restrained moment connections have sufficient restraint to resist 90 percent of the fixed-end moment with negligible rotation and

$$\begin{aligned} R &= M/M_F \\ &> 0.9 \end{aligned}$$

Assuming rigid columns, Fig. 3.17 shows the bending moment produced in a fixed-ended girder with a restraint factor $R = 1$. The maximum moment of $wL^2/12$ occurs at

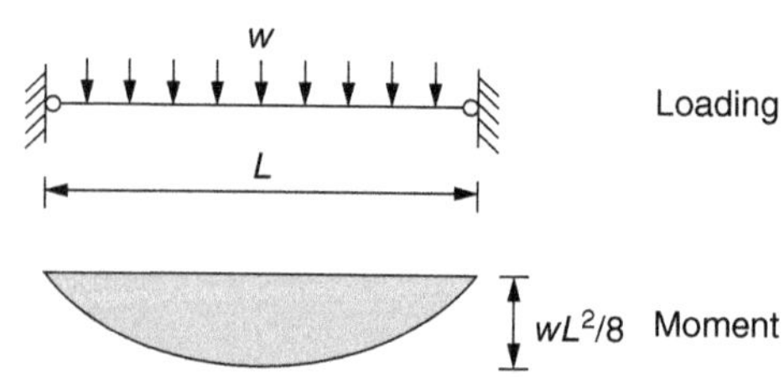


**FIGURE 3.16** Simply supported girder.

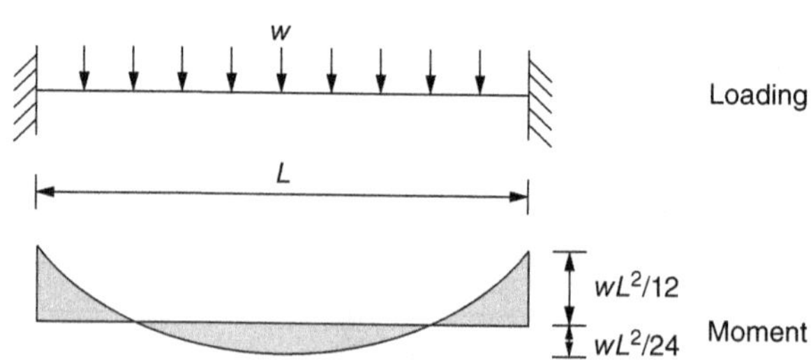


**FIGURE 3.17** Fixed-ended girder.

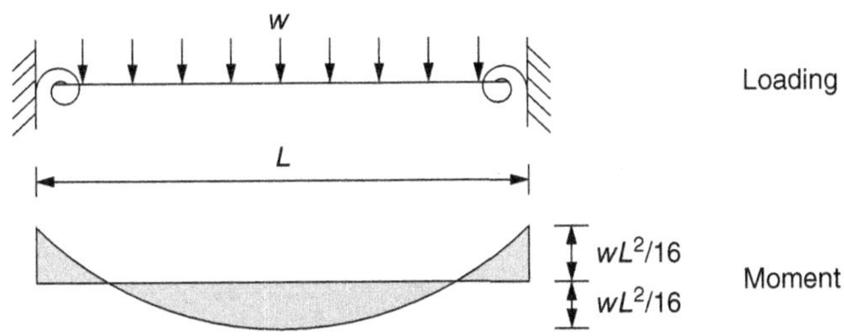


**FIGURE 3.18** Girder with partially restrained moment connections.

the end connections and the remainder of the girder is underutilized. The maximum moment in this case is 2/3 of the maximum moment in the simply supported girder resulting in an economy in the size of the girder compared with a simply supported girder.

Assuming rigid columns, and using an end connection with a restraint factor $R = 0.75$ results in the moment diagram shown in Fig. 3.18. The bending moment produced at each end of the girder is

$$\begin{aligned} M_e &= 0.75 \times wL^2/12 \\ &= wL^2/16 \end{aligned}$$

The bending moment produced at midspan is

$$\begin{aligned} M_c &= wL^2/8 - wL^2/16 \\ &= wL^2/16 \end{aligned}$$

The maximum moments of $wL^2/16$ occur at both midspan and at the ends of the girder. The maximum moment in this case is ½ the maximum moment in the simply supported girder resulting in a further economy in the size of the girder.

A determination of the design moment and end rotation of a partially restrained girder may be obtained by the method of Geschwindner.[16] This entails superimposing the beam line on the moment-rotation response curve for the connection. For a fully fixed-ended girder, the end rotation is zero and the fixed-end moment is

$$M_F = wL^2/12$$

This provides a point on the moment-rotation graph shown in Fig. 3.19. For a simply supported girder, the end moment is zero and the end rotation is

$$\theta_0 = wL^3/24EI$$

This provides an additional point on Fig. 3.19. The beam line connects these two extreme conditions. The point at which the beam line intersects the moment-rotation response curve of the girder gives the resulting end moment and rotation under the given load. These are shown on Fig. 3.19 as $M$ and $\theta$, respectively.

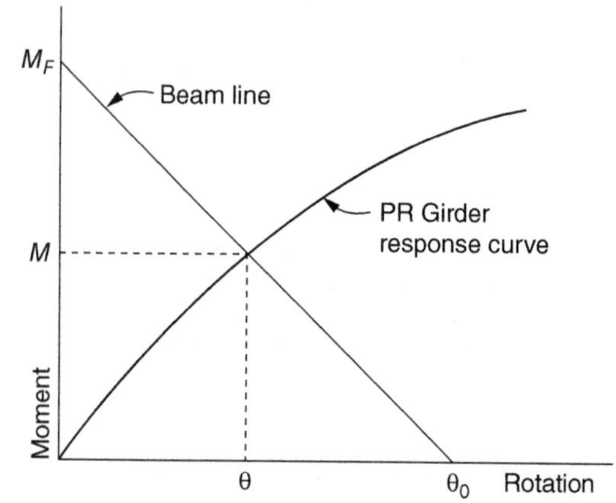


**FIGURE 3.19** Moment-rotation response curve and beam line.

Design details for partially restrained moment connections are given in AISC Manual Part 11 and in Blodgett Sec. 5.1 and, for steel-concrete composite connections, in Leon et al.[17] A flange-plated partially restrained moment connection is shown in Fig. 3.20. This consists of a shear tab and a bottom flange plate shop welded to the column flange. The top flange plate is field groove welded to the column flange and fillet welded, over a portion of its length as shown, to the girder flange.

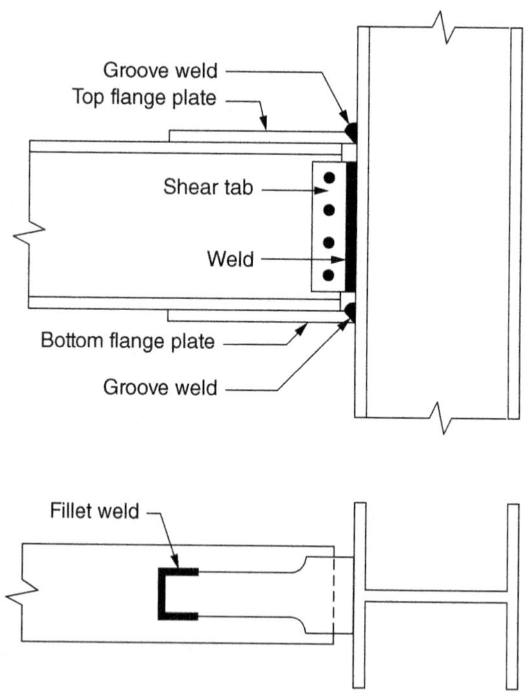


**FIGURE 3.20** Flange-plated partially restrained moment connection.

The flexible moment connection (FMC) method, described by Geschwindner and Disque,[18] is a simplified and conservative version of the partially restrained moment connection method. Using this technique, a frame is designed for gravity loads assuming the girders are simply supported. For lateral loads, the frame is designed assuming that the girders are moment connected to the columns.

**Example 3.3.** Two-Story Frame with Partially Restrained Connections
Figure 3.21 shows a two-story frame. The $EI/L$ values for all members are identical with a value of $k = 600$ kip-ft. The girders are attached to the columns with partially restrained moment connections with a moment/rotation ratio of $\eta = 1200$ kip-ft/rad at each end of the girders. The distributed load on each girder is 1.2 kips/ft. Determine the bending moments produced in the frame.

The moments produced in the frame may be determined by the moment distribution procedure. The stiffness, carry-over factors, and fixed-end moments are modified by the elastic connections at the beam ends and are derived as described in Williams Part 2, Sec. 6.7. For all members

$$EI/L = k$$

$$\begin{aligned}\beta &= 2k/\eta \\ &= 2 \times 600/1200 \\ &= 1.0\end{aligned}$$

Allowing for symmetry, the modified stiffness of the members are

$$\begin{aligned}s_{21} = s_{23} = s_{32} &= 4k \\ &= 4 \times 600 \\ &= 2400\end{aligned}$$

$$\begin{aligned}s_{25} = s_{36} &= 2k/(1+\beta) \\ &= 2 \times 600/(1+1) \\ &= 600\end{aligned}$$

The distribution factors at joint 2 are

$$\begin{aligned}d_{25} &= s_{25}/(s_{25} + s_{21} + s_{23}) \\ &= 600/(600 + 2400 + 2400) \\ &= 1/9\end{aligned}$$

$$d_{21} = d_{23} = 4/9$$

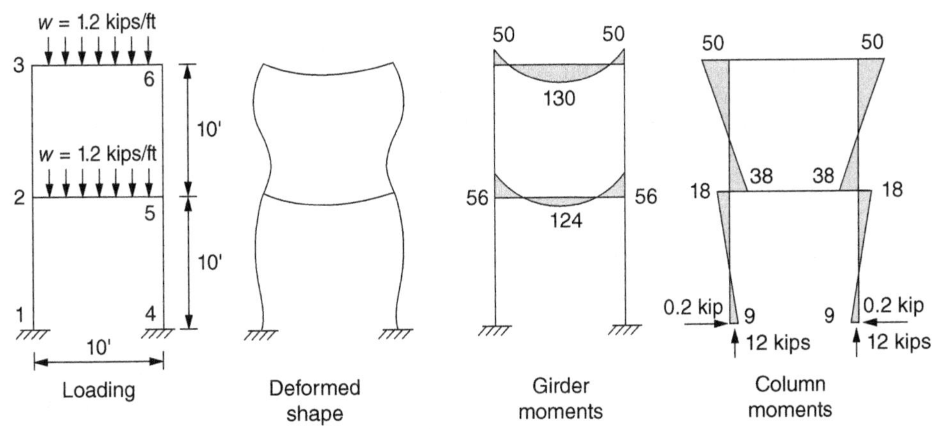


**FIGURE 3.21** Details for Example 3.3.

| Member | 21 | 25 | 23 | 32 | 36 |
|---|---|---|---|---|---|
| Distribution Factor | 4/9 | 1/9 | 4/9 | 4/5 | 1/5 |
| Fixed-end Moments | | –6000 | | | –6000 |
| Distribution | 2667 | 666 | 2667 | 4800 | 1200 |
| Carry-over | | | 2400 | 1334 | |
| Distribution | –1067 | –266 | –1067 | –1067 | –267 |
| Carry-over | | | –534 | –534 | |
| Distribution | 237 | 59 | 237 | 427 | 107 |
| Carry-over | | | 214 | 119 | |
| Distribution | –95 | –24 | –95 | –95 | –24 |
| Carry-over | | | –48 | –48 | |
| Distribution | 21 | 5 | 21 | 38 | 10 |
| Final Moments | 1763 | –5560 | 3795 | 4974 | –4974 |

**Table 3.2** Distribution of Moments in Example 3.3

The distribution factors at joint 3 are

$$
\begin{aligned}
d_{36} &= s_{36}/(s_{36} + s_{32}) \\
&= 600/(600 + 2400) \\
&= 1/5 \\
d_{32} &= 4/5
\end{aligned}
$$

Allowing for symmetry, the carry-over factors in the columns are ½ and there is no carry-over between beam ends. The fixed-end moments are

$$
\begin{aligned}
M_{F25} = M_{F36} &= -(wL^2/12)(1 + 3\beta)/(1 + 4\beta + 3\beta^2) \\
&= -(1.2 \times 10^2/12)(4)/(8) \\
&= -5 \text{ kip-ft} \\
&= -60 \text{ kip-in}
\end{aligned}
$$

The distribution procedure is shown in Table 3.2 with distribution occurring in the left half of the frame only. For convenience, the fixed-end moments are multiplied by 100.

The final bending moments, in kip-in units, produced in the members, are indicated in Fig. 3.21 with the moments drawn on the tension side of the members. The moments produced at midspan of the partially restrained girders are smaller than those produced in the simply supported girders and larger than those produced in the fully restrained girders. The moments produced in the columns of the frame with the partially restrained girders are smaller than those produced in the columns of the frame with the fully restrained girders.

## 3.3 Lateral Load-Resisting Systems

### Diaphragms

As shown in Fig. 3.22, a typical lateral load-resisting system consists of horizontal and vertical elements connected together so as to transfer lateral forces from the top of a building to the foundations. Forces caused by wind or seismic effects, acting on the east and west walls of the building, are transferred to the horizontal roof diaphragm.

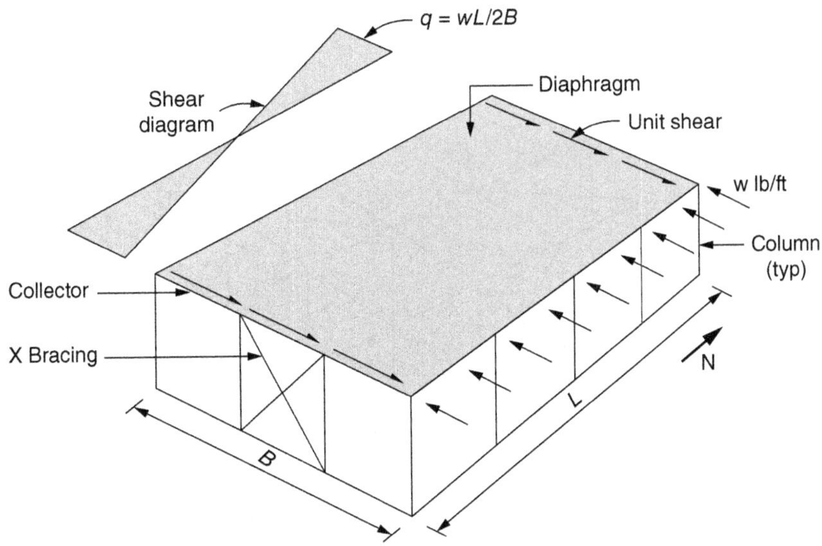


**FIGURE 3.22** Lateral load transfer by horizontal diaphragm.

A diaphragm is defined in IBC Sec. 202 as a horizontal or sloped system acting to transmit lateral forces to the vertical-resisting elements. The diaphragm may be considered flexible or rigid as defined in American Society of Civil Engineers, *Minimum Design Loads for Buildings and Other Structures* (ASCE 7)[19] Sec. 12.3 and may consist of plywood panels, topped or untopped steel decking, or reinforced concrete. A diaphragm is analogous to a deep beam. The roof deck acts as a web to resist shear and the boundary members, normal to the load, act as flanges or chords to resist flexural effects. A chord is defined in IBC Sec. 202 as a diaphragm boundary element perpendicular to the applied load that is assumed to take axial stresses due to the diaphragm moment. The diaphragm need not necessarily be horizontal, a pitched roof may also act as a diaphragm. Shear stress in the diaphragm is uniformly distributed across the north and south edges of the diaphragm and is transferred to the braced frames and collector beams at the north and south ends of the building. The braced frames transfer the diaphragm shear to the foundation. Detailed design procedures are given in Prasad et al.[20] As shown in Fig. 3.22, the shear in a diaphragm decreases uniformly from the end supports to midspan. Hence, the strength of the diaphragm may be progressively decreased toward the center of the span.

As an alternative to a horizontal diaphragm, bracing in the plane of the roof can also be used to transfer lateral loads to the vertical braced frames. This is shown in Fig. 3.23.

## Collectors

A collector, also known as a drag strut, is defined in IBC Sec. 202 as a horizontal diaphragm element, parallel and in line with the applied force that collects and transfers diaphragm shear forces to the vertical elements of the lateral force-resisting system. Collectors are required where shear walls or braced frames terminate along the boundary of a diaphragm in order to ensure deformation compatibility and prevent localized tearing of the diaphragm. As shown in Fig. 3.24, the east-west lateral load of

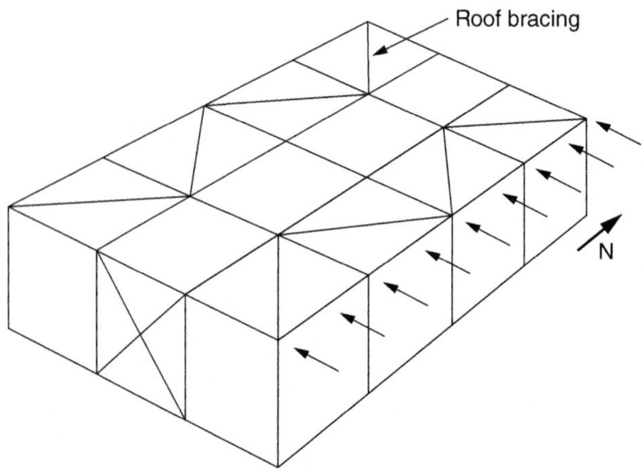


FIGURE 3.23 Lateral load transfer by horizontal-braced frame.

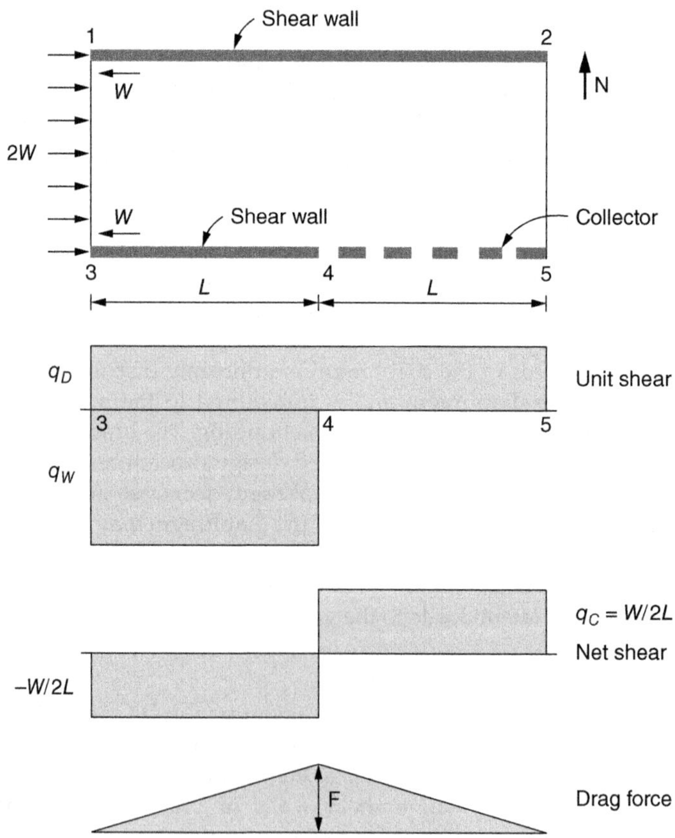


FIGURE 3.24 Collector force.

2$W$ produces a shear force on the north and south diaphragm boundaries of $W$, assuming that the diaphragm is flexible. The unit shear at the diaphragm south boundary is given by

$$q_D = W/2L$$

where 2$L$ is length of diaphragm.

This is plotted on the shear diagram as shown. The shear wall on the south side of the building ends at point 4 and a collector is required from 4 to 5 to collect the shear from the diaphragm. The unit shear in shear wall 34 is

$$q_W = -W/L$$

This is also plotted on the shear diagram and is opposite in sense to the unit shear in the diaphragm. The net shear acting along the south diaphragm boundary is also shown. The net shear over the length of the collector is

$$q_C = \pm W/2L$$

The drag force in the collector at end 5 is zero. The maximum drag force occurs at end 4, where the collector is connected to the shear wall, and is given by

$$F = q_C \times L$$
$$= W/2$$

A collector is also provided at a reentrant corner of a building, as shown in Fig. 3.25. The shear wall at the reentrant corner, 23, terminates at end 2 and a collector is provided

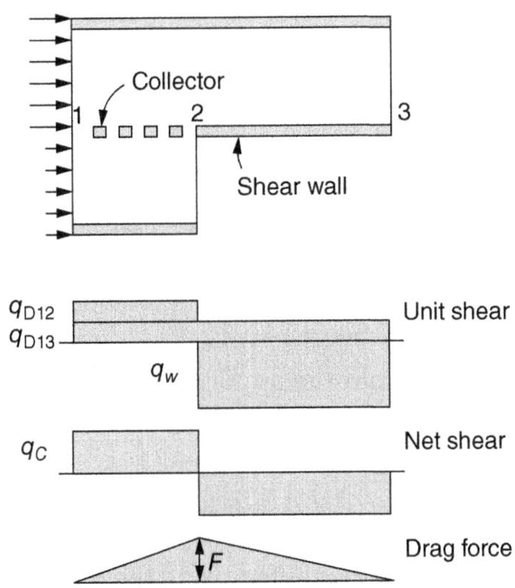


**Figure 3.25** Reentrant corner.

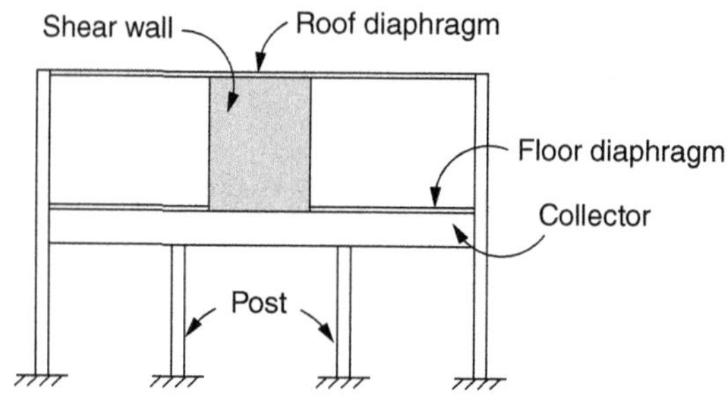


**FIGURE 3.26** Offset shear walls.

to transfer the diaphragm shear from the west end wall of the building to end 2 of the shear wall. The compression force in the collector reaches a maximum at its connection with the shear wall at end 2. When the seismic load reverses, the force in the collector becomes tensile.

Figure 3.26 shows a shear wall located between the roof diaphragm and the second floor diaphragm of a two-story building, and this shear wall is offset from the shear walls in the bottom story. The shear wall transfers shear from the roof diaphragm to the collector on the second floor. The collector distributes this shear to the second floor diaphragm which is required to transfer the shear to the offset shear walls below the diaphragm. The transfer forces are subject to the redundancy factor $\rho$, applicable to the structural system, and are added to the diaphragm forces determined in accordance with ASCE 7 Sec. 12.10.1.1.

## Collector Design Forces

For structures assigned to seismic design categories C through F, collector elements and their connections are designed in accordance with ASCE 7 Sec. 12.10.2. The collector design forces are the maximum of the three following conditions:

(a) Forces resulting from application at each level of the design lateral force $F_x$ calculated from ASCE 7 Eq. (12.8-11) and (12.8-12) which give

$$F_x = Vw_x h_x^k / \Sigma w_i h_i^k$$

The value of $F_x$ is determined using the load combinations with overstrength factor $\Omega_0$ of ASCE 7 Sec. 12.4.3.2.

(b) Forces resulting from the application at each level of the diaphragm design force $F_{px}$ calculated from ASCE 7 Eq. (12.10-1) as

$$F_{px} = w_{px} \Sigma F_i / \Sigma w_i$$

The value of $F_{px}$ is determined using the load combinations with overstrength factor $\Omega_0$ of ASCE 7 Sec. 12.4.3.2.

(c) Forces resulting from the application at each level of the minimum diaphragm design force $F_{px}$ given by ASCE 7 Eq. (12.10-2) as

$$F_{px} = 0.2S_{DS}I_e w_{px}$$

The value of $F_{px}$ is determined using the basic load combinations of ASCE 7 Sec. 12.4.2.3.

The following two exceptions are permitted by ASCE 7 Sec. 12.10.2.1:

(d) The collector forces determined from cases (a) through (c) need not exceed the forces resulting from the application at each level of the maximum diaphragm design force $F_{px}$ given by ASCE 7 Eq. (12.10-3) as

$$F_{px} = 0.4S_{DS}I_e w_{px}$$

The value of $F_{px}$ is determined using the basic load combinations of ASCE 7 Sec. 2.4.2.3.

(e) In structures braced entirely by light-frame shear walls, collector elements and their connections need only be designed for forces resulting from the application at each level of the diaphragm design force $F_{px}$ calculated from ASCE 7 Eq. (12.10-1) as

$$F_{px} = w_{px}\Sigma F_i/\Sigma w_i$$

The value of $F_{px}$ is determined using the basic load combinations of ASCE 7 Sec. 12.4.2.3.

**Example 3.4.** Drag Force Calculation with Flexible Diaphragm

Figure 3.27 shows a single-story, bearing wall structure. The east-west lateral load acting on the roof is $W = 144$ kips. The roof diaphragm may be considered flexible and all masonry shear walls are 20 ft long. Determine the maximum force in collector 78 and the chord force in member 24.

The diaphragm is flexible and diaphragms 1243 and 3465 may be treated as independent simply supported beams. The shear force at the boundaries of each diaphragm is

$$\begin{aligned} Q_D &= W/4 = 144{,}000/4 \\ &= 36{,}000 \text{ lb} \end{aligned}$$

The unit shear at the boundaries of each diaphragm is

$$\begin{aligned} q_D &= Q_D/L \\ &= 36{,}000/60 \\ &= 600 \text{ lb/ft} \end{aligned}$$

The unit shears at the south side of diaphragm 1243 and the north side of diaphragm 3465 are plotted on the shear diagram. The two shear walls 37 and 84 each support a shear force of

$$\begin{aligned} Q_W &= W/4 \\ &= 144{,}000/4 \\ &= 36{,}000 \text{ lb} \end{aligned}$$

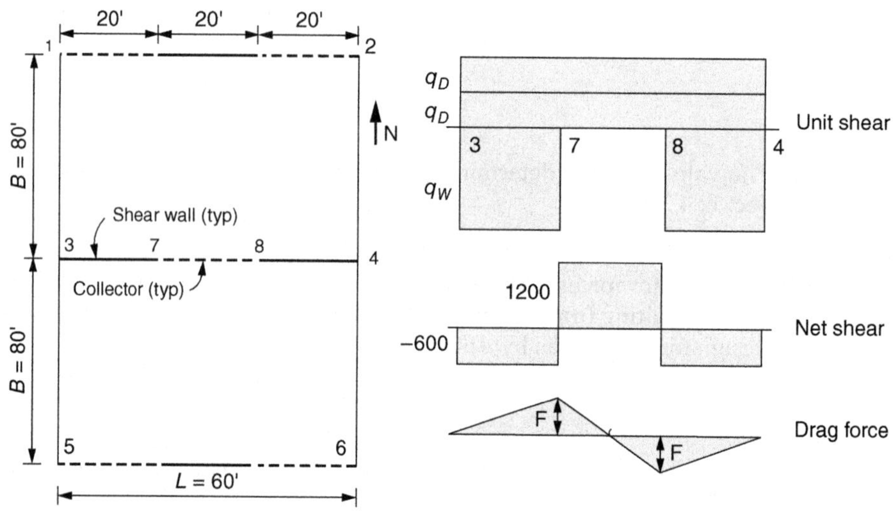


**FIGURE 3.27** Details for Example 3.4.

The unit shear in each shear wall is

$$q_W = Q_W/l = 36{,}000/20$$
$$= 1800 \text{ lb/ft}$$

This is plotted on the shear diagram. The net shear along the diaphragm interface is also shown. The maximum drag force occurs at each end of the collector at the connection to the shear wall, and is given by

$$F = 600 \times 20/1000$$
$$= 12 \text{ kips}$$

The drag force diagram is shown.

The bending moment acting on diaphragm 1243 is

$$M = (W/2)B/8$$
$$= 72 \times 80/8$$
$$= 720 \text{ kip-ft}$$

The chord force in 24 is

$$F_C = M/L$$
$$= 720/60$$
$$= 12 \text{ kips}$$

When the vertical lateral load-resisting element is a braced frame, the drag force diagram must be modified to account for the force in the brace members.

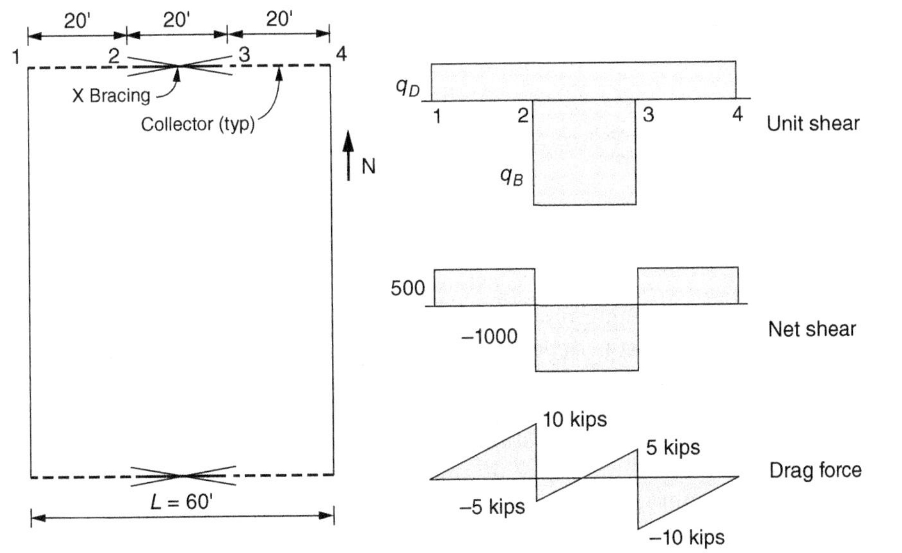


**FIGURE 3.28** Details for Example 3.5.

**Example 3.5.** Drag Force Calculation with Braced Frame
Figure 3.28 shows a single-story building frame structure. The east-west lateral load acting on the roof is 60 kips. The roof diaphragm may be considered flexible and the bracing in the north and south walls consists of a cross-braced frame. Determine the maximum force in the collectors.

The diaphragm is flexible and the unit shear at the boundary of the diaphragm is

$$\begin{aligned} q_D &= Q/L \\ &= 30{,}000/60 \\ &= 500 \text{ lb/ft} \end{aligned}$$

The total force applied to each braced frame is 30 kips and the unit shear in each braced frame is

$$\begin{aligned} q_B &= 30{,}000/20 \\ &= 1500 \text{ lb/ft} \end{aligned}$$

These are plotted on the shear diagram and the net shear along the diaphragm is shown. The horizontal component of the force in each brace is

$$\begin{aligned} Q_B &= 30/2 \\ &= 15 \text{ kips} \end{aligned}$$

The drag force at point 2 is

$$\begin{aligned} F_2 &= 500 \times 20/1000 \\ &= 10 \text{ kips} \end{aligned}$$

and

$$\begin{aligned} &= 10 - Q_B \\ &= 10 - 15 \\ &= -5 \text{ kips} \end{aligned}$$

The drag force at point 3 is

$$F_3 = -500 \times 20/1000$$
$$= -10 \text{ kips}$$

and

$$= -10 + Q_B$$
$$= -10 + 15$$
$$= 5 \text{ kips}$$

The drag force diagram is shown.

When the diaphragm is rigid, the lateral force distributed to the vertical lateral load-resisting elements depends on their relative stiffness. For the case of symmetrical loading on a symmetrical structure, the diaphragm displaces laterally as a rigid body and the forces produced in the vertical lateral load-resisting elements is directly proportional to their relative stiffness. When either the loading or the structure is unsymmetrical, additional forces are produced in the vertical lateral load-resisting elements due to the diaphragm rotation.

**Example 3.6.** Drag Force Calculation with Rigid Diaphragm
Figure 3.29 shows a single-story building frame structure. The east-west lateral load acting on the roof is 50 kips. The roof diaphragm may be considered rigid and the stiffness of the shear walls in the north and south ends of the building are shown ringed. Ignoring accidental eccentricity, determine the maximum force in collector 23.

The structure and the loading are symmetrical and, neglecting accidental eccentricity, rotation does not occur in the rigid diaphragm. The unit shear at the boundary of the diaphragm is

$$q_D = Q/L$$
$$= 25{,}000/50$$
$$= 500 \text{ lb/ft}$$

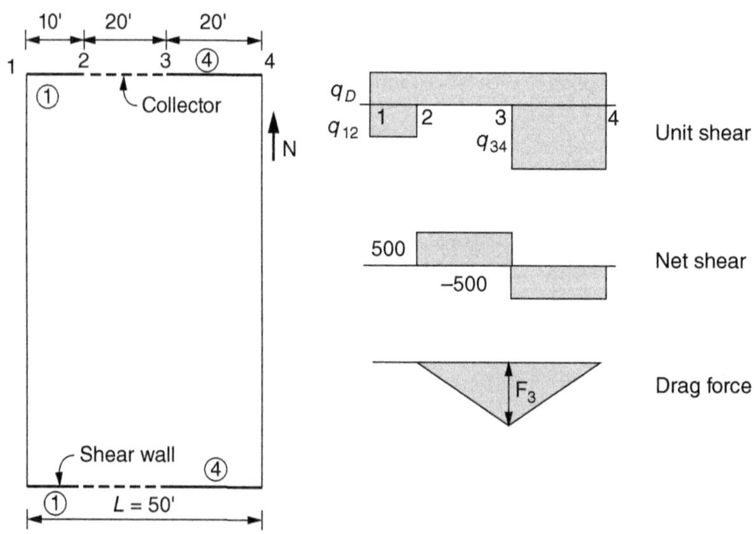


**FIGURE 3.29** Details for Example 3.6.

The two shear walls in the north wall support a shear force of

$$Q = 25{,}000 \text{ lb}$$

Shear wall 12 supports a shear force of

$$\begin{aligned} Q_{12} &= Qs_{12}/(s_{12} + s_{34}) \\ &= 25{,}000 \times 1/(1+4) \\ &= 5000 \text{ lb} \end{aligned}$$

The unit shear in shear wall 12 is

$$\begin{aligned} q_{12} &= -5000/10 \\ &= -500 \text{ lb/ft} \end{aligned}$$

Shear wall 34 supports a shear force of

$$\begin{aligned} Q_{34} &= Qs_{34}/(s_{12} + s_{34}) \\ &= 25{,}000 \times 4/(1+4) \\ &= 20{,}000 \text{ lb} \end{aligned}$$

The unit shear in shear wall 34 is

$$\begin{aligned} q_{34} &= -20{,}000/20 \\ &= -1000 \text{ lb/ft} \end{aligned}$$

The unit shears are plotted on the shear diagram and the net shear along the diaphragm is shown.

The drag force at point 3 is

$$\begin{aligned} F_3 &= -500 \times 20/1000 \\ &= -10 \text{ kips} \end{aligned}$$

The drag force at point 2 is

$$\begin{aligned} F_2 &= F_3 + 500 \times 20/1000 \\ &= -10 + 10 \\ &= 0 \text{ kip} \end{aligned}$$

The drag force diagram is shown.

## Steel Deck Diaphragms

Steel deck diaphragms are a common means of providing a diaphragm to steel-framed structures. Untopped steel deck may be used for roof diaphragms and concrete filled steel deck may be used for floor diaphragms as described by Luttrell.[21] The typical profile of the decking used without topping is shown in Fig. 3.30 and is normally provided in thicknesses of 22, 20, 18, and 16 gage. Three profiles are available, wide rib deck (WR),

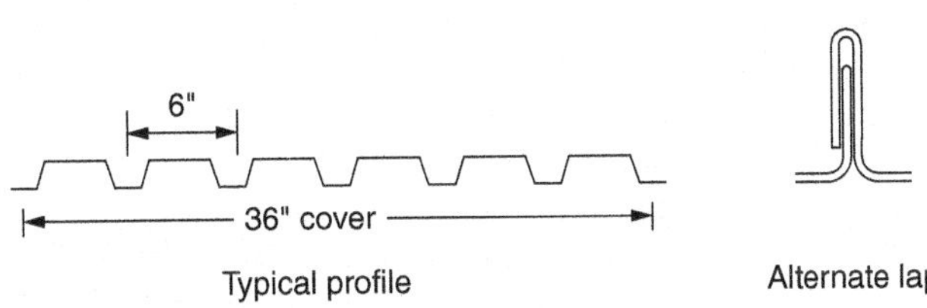


**Figure 3.30** Steel deck profile.

intermediate rib deck (IR), and narrow rib deck (NR). The widths at the bottom of the flutes are, respectively, 1⁹⁄₁₆-in minimum, ½-in minimum, ⅜-in minimum.

The strength of a steel deck in horizontal in-plane shear is governed by the thickness of the deck, the connection of the deck to the supporting beams, the span between beams, the connection of the deck boundary to the perimeter structure, and the side-lap or stitch connection between deck panels. The horizontal in-plane shear strength of the deck is the same in both perpendicular directions and is unaffected by the orientation of the corrugations. Attachment of the deck to supporting beams provides stiffness to the diaphragm to prevent general buckling of the deck. Fastening to the beams may be by puddle welding, pneumatically driven pins, or powder-actuated fasteners.

Puddle welds are formed by striking an arc with an electrode on the decking. Through the hole formed in the deck, the base metal is raised to fusion temperature and the process continues until the hole is filled with electrode metal. For deck thicknesses less than 22 gage, a weld washer with a prepunched hole is required. The washer acts as a heat sink and prevents excessive burnout in the deck.

Pneumatically driven fasteners or pins are made from high-carbon steel and have a ballistic point for penetration into the supporting steel beam. The pins are applied by means of a handheld, compressed air or gas actuated tool.

Powder-actuated tools rely on a controlled explosion, created by a small chemical charge, to drive the fastener into the support. The fastener consists of a hardened steel pin with a knurled shaft.

Side-lap connections between adjacent panels may be made by puddle welding, by screws, or by button punching. Screw connections are usually made with self-drilling screws. Some deck panels are supplied with an upstanding single element on one side and a downstanding folded-over double element on the other side, as shown in Fig. 3.30. In placing the deck, the single element from one panel is knifed into the double element on the adjacent panel. A crimping tool is then used to punch cone shaped indentations into the connection to interlock the two panels.

Steel roof decks are required by IBC Sec. 2210.1.1.2 to be designed and constructed in accordance with Steel Deck Institute *Standard for Steel Roof Deck* (SDI RD1.0).[22] In accordance with SDI RD1.0 Sec. 1.2B, the design of roof decks shall comply with the applicable provisions of Steel Deck Institute, *Diaphragm Design Manual* (SDI DDM03).[23]

The nominal shear values for steel deck diaphragms for various combinations of span length, type of deck, type of support and side-lap connections, and number of support and side-lap fasteners are given in the *Diaphragm Design Manual*. Also provided are the resistance factors, $\varphi$, for use with LRFD load combinations, and safety factors, $\Omega$, for use with ASD load combinations. These factors account for the different coefficients of variation in strength of the different types of connections and for the different types of loading applied—whether wind, seismic, or other. Using the LRFD design method, the design strength is determined by multiplying the nominal strength of the deck by the appropriate resistance factor. The design strength must not be less than the required strength determined by the LRFD load combinations. Using the ASD design method, the allowable strength is determined by dividing the nominal strength of the deck by the appropriate safety factor. The allowable strength must not be less than the required strength determined by the ASD load combinations. A summary of the applicable safety and resistance factors is given in Table 3.3.

The SDI Tables indicate the number of fasteners required along the sides of a panel by specifying the number of fasteners required in each span. The number of fasteners required at supports is indicated by specifying the panel width and the number of

| Load Type | Connection Type | φ | Ω |
|---|---|---|---|
| Seismic | Welds | 0.55 | 3.00 |
| | Screws | 0.65 | 2.50 |
| Wind | Welds | 0.70 | 2.35 |
| | Screws | 0.70 | 2.35 |
| Other | Welds | 0.60 | 2.65 |
| | Screws | 0.65 | 2.50 |

**TABLE 3.3** List of φ and Ω Values

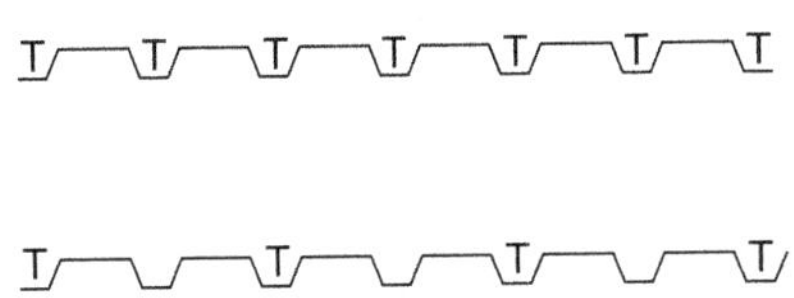

36/7 Pattern

36/4 Pattern

**FIGURE 3.31** Fastener layout.

fasteners required. As shown in Fig. 3.31, a 36/7 pattern represents 7 fasteners in a panel 36 in wide, a 36/4 pattern represents 4 fasteners in a 36-in-wide panel.

In addition to providing nominal shear strength values for the steel deck, the SDI Tables also indicate the nominal strength based on panel buckling. The resistance factor and safety factor for buckling are 0.80 and 2.00, respectively. It is necessary to determine whether in-plane shear or panel buckling governs the design.

As a result of wind uplift, fasteners are subjected to tensile forces and this reduces the shear capacity. To determine if the fasteners are satisfactory under the combined stresses, the following interaction equations must be applied

| LRFD | ASD |
|---|---|
| $(T_u/\varphi_u T_n)^{1.5} + (Q_u/\varphi Q_f)^{1.5} < 1.0$ | $(\Omega_u T/T_n)^{1.5} + (\Omega Q/Q_f)^{1.5} < 1.0$ |
| where<br>$T_u$ = fastener required tensile strength from LRFD load combinations<br>$\varphi_u$ = resistance factor for fastener in tension<br>$T_n$ = fastener nominal tensile strength<br>$Q_u$ = fastener required shear strength from LRFD load combinations<br>$\varphi$ = resistance factor for fastener in shear<br>$Q_f$ = fastener nominal shear strength<br>If $(T_u/\varphi_u T_n)^{1.5} \le 0.15$ ... no interaction check is required | where<br>$T$ = fastener required tensile strength from ASD load combinations<br>$\Omega_u$ = safety factor for fastener in tension<br>$T_n$ = fastener nominal tensile strength<br>$Q$ = fastener required shear strength from ASD load combinations<br>$\Omega$ = safety factor for fastener in shear<br>$Q_f$ = fastener nominal shear strength<br>If $(\Omega_u T/T_n)^{1.5} \le 0.15$ ... no interaction check is required |

The SDI tables rate the thickness of deck panels in terms of the design thickness rather than the gage size. A comparison of the two classifications is given in Table 3.4.

| Steel deck gage | 22 | 20 | 18 | 16 |
|---|---|---|---|---|
| Design thickness, in | 0.0295 | 0.0358 | 0.0474 | 0.0598 |

**TABLE 3.4** List of Steel Deck Thickness

**Example 3.7.** Steel Deck Diaphragm Design
Figure 3.32 shows a single-story industrial structure with a seismic force of 700 lb/ft acting in the north-south direction on the roof diaphragm. The lateral force resisting system consists of concentrically braced frames and the roof is supported at 6 ft centers on open web steel joists. Design the roof diaphragm using 1½ deep, 20 gage, wide rib untopped steel decking. The deck panels are 3 × 18 ft and are continuous over 3 spans. Adopt a 36/7 pattern for the support fastening with Buildex BX-14 connectors and No. 10 self-drilling screws for side-lap fastening. Determine the number of side-lap connections required at the diaphragm boundaries and at the quarter points of the span.

The diaphragm is flexible in accordance with ASCE 7 Sec. 12.3.1.1 and may be treated as a simply supported beam for lateral loads. Then

| LRFD | ASD |
|---|---|
| From ASCE 7 Sec. 2.3.2, the factored shear force at the diaphragm boundary is $Q_u = 1.0 \times 700 \times 120/2 = 42{,}000$ lb | From ASCE 7 Sec. 2.4.1, the factored shear force at the diaphragm boundary is $Q_a = 0.7 \times 700 \times 120/2 = 29{,}400$ lb |
| The unit shear at the diaphragm boundary is $q_u = Q_u/L = 42{,}000/60 = 700$ lb/ft | The unit shear at the diaphragm boundary is $q_a = Q_a/L = 29{,}400/60 = 490$ lb/ft |
| The resistance factor is $\varphi = 0.65$ | The safety factor is $\Omega = 2.50$ |
| Required nominal strength of diaphragm is $q_n = q_u/\varphi = 700/0.65 = 1077$ lb/ft | Required nominal strength of diaphragm is $q_n = q_a \times \Omega = 490 \times 2.5 = 1225$ lb/ft ... governs |

A portion of SDI Table AV-26 is reproduced, by permission, in Fig. 3.33.

From SDI Table AV-26, 6 side-lap connections in each span provide a nominal strength of

$$q_n = 1310 \text{ lb/ft}$$
$$> 1225 \text{ lb/ft ...satisfactory}$$

From SDI Table AV-26, the nominal strength of wide rib profile sheeting based on panel buckling is

$$q_n = 2005 \text{ lb/ft}$$
$$> 1225 \text{ lb/ft ...satisfactory}$$

At a distance of 30 ft from the diaphragm boundary, the required nominal strength is

$$q_n = 1225/2$$
$$= 613 \text{ lb/ft}$$

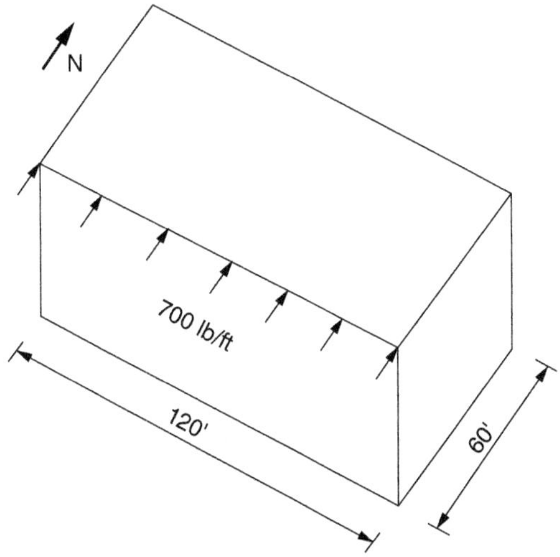


**Figure 3.32** Details for Example 3.7.

**1.5 (WR, IR, NR)**
**t = design thickness = 0.0358"**
**SUPPORT FASTENING: Builder BX-14**
**SIDE-LAP FASTENING: #10 screws**

**φ (EQ): 0.65 Ω (EQ): 2.50**
**φ (WIND): 0.70 Ω (WIND): 2.35**
**φ (Other): 0.65 Ω (Other): 2.50**

| FASTENER LAYOUT | SIDE-LAP CONN./SPAN | NOMINAL SHEAR STRENGTH, PLF | | | | | | | | | |
|---|---|---|---|---|---|---|---|---|---|---|---|
| | | SPAN, FT | | | | | | | | | |
| | | 4.0 | 4.5 | 5.0 | 5.5 | 6.0 | 6.5 | 7.0 | 7.5 | **8.0** | K1 |
| 36/9 | 0 | 1470 | 1310 | 1170 | 1055 | 965 | 880 | 815 | 755 | 700 | 0.775 |
| | 1 | 1625 | 1470 | 1325 | 1195 | 1090 | | | | | 0.547 |
| | 2 | 1775 | 1605 | 1465 | 1335 | 1220 | 1120 | 1035 | 960 | 895 | 0.423 |
| | 3 | 1915 | 1740 | 1590 | 1465 | 1345 | 1235 | 1145 | 1060* | 990` | 0.345 |
| | 4 | 2050 | 1865 | 1710 | 1580 | 1460 | 1355 | 1255* | 1165* | 1085` | 0.291 |
| | 5 | 2175 | 1990 | 1830 | 1690 | 1565 | 1460 * | 1360 * | 1265 * | 1180 " | 0.251 |
| | 6 | 2295 | 2105 | 1940 | 1795 | 1670 ' | 1560 " | 1460 * | 1370 * | 1280 * | 0.221 |
| 36/7 | 0 | 910 | 805 | 720 | 650 | 595 | 545 | 505 | 470 | 435 | 1.163 |
| | 1 | 1095 | 975 | 875 | 790 | 725 | | | | | 0.716 |
| | 2 | 1255 | 1130 | 1025 | 930 | 850 | 780 | 725 | 675 | 630 | 0.517 |
| | 3 | 1410 | 1275 | 1160 | 1065 | 980 | 900 | 835 | 775 | 725 | 0.404 |
| | 4 | 1560 | 1415 | 1290 | 1185 | 1095 | 1020 | 945 | 880 | 820 | 0.332 |
| | 5 | 1695 | 1545 | 1415 | 1300 | 1205 | 1120 | 1050 | 980 | 915 | 0.282 |
| | 6 | 1825 | 1670 | 1530 | 1415 | 1310 | 1220 | 1145 | 1075` | 1010* | 0.245 |
| 36/5 | 0 | 830 | 745 | 665 | 605 | 550 | 505 | 465 | 430 | 400 | 1.395 |
| | 1 | 985 | 890 | 810 | 740 | 675 | | | | | 0.797 |
| | 2 | 1125 | 1025 | 935 | 865 | 800 | 740 | 685 | 635 | 595 | 0.558 |
| | 3 | 1255 | 1145 | 1055 | 975 | 905 | 845 | 790 | 740 | 690 | 0.429 |
| | 4 | 1370 | 1260 | 1165 | 1080 | 1005 | 940 | 880 | 830 | 785 | 0.349 |
| | 5 | 1475 | 1365 | 1265 | 1175 | 1100 | 1030 | 970 | 915 | 865 | 0.294 |
| | 6 | 1565 | 1455 | 1355 | 1270 | 1190 | 1115 | 1050 | 995 | 940* | 0.254 |

* NOMINAL SHEAR SHOWN ABOVE MAY BE LIMITED BY SHEAR BUCKLING. SEE TABLE BELOW.

THE SHADED VALUES DO NOT COMPLY WITH THE MINIMUM SPACING REQUIREMENTS FOR SIDE-LAP CONNECTIONS AND SHALL NOT BE USED EXCEPT WITH PROPERLY SPACED SIDE-LAP CONNECTIONS.

**φ (Buckling): 0.80 Ω (Buckling): 2.00**

| DECK PROFILE | I | NOMINAL SHEAR DUE TO PANEL BUCKLING ($S_n$), PLF / SPAN, FT | | | | | | | | |
|---|---|---|---|---|---|---|---|---|---|---|
| | $in^4$/ ft | 4.0 | 4.5 | 5.0 | 5.5 | 6.0 | 6.5 | 7.0 | 7.5 | 8.0 |
| NR | 0.128 | 3255 | 2570 | 2085 | 1720 | 1445 | 1230 | 1060 | 925 | 810 |
| IR | 0.139 | 3465 | 2735 | 2215 | 1830 | 1540 | 1310 | 1130 | 985 | 865 |
| WR | 0.198 | 4515 | 3570 | 2890 | 2390 | 2005 | 1710 | 1475 | 1285 | 1125 |

NOTE:
ASO Required Strength (Service Applied Load) <= Minimum (Nominal Shear Strength / Ω (ED or WIND), Nominal Buckling Strength $S_n$ / Ω (Buckling)
LRFD Required Strength (Factored Applied Load) ,<= Minimum (φ (EQ or WIND) x Nominal Shear Strength, φ (Buckling) x Nominal Buckling Strength $S_n$I

**Figure 3.33** Portion of SDI Table AV-26.

From SDI Table AV-26, 1 side-lap connection in each span provides a nominal strength of

$$q_n = 725 \text{ lb/ft}$$
$$> 613 \text{ lb/ft}$$

However, the deck span exceeds 5 ft and, in accordance with SDI RD1.0 Sec. 3.2, connections are required at a maximum spacing of 36 in. Hence, use two connections in each span at a spacing of 36 in.

## Frames Subjected to Lateral Forces

The two-story frame previously analyzed for gravity loads will now be analyzed for lateral loads. The following example considers the girders rigidly attached to the columns.

**Example 3.8.** Rigid Frame with Lateral Loads
Figure 3.34 shows a two-story frame. The flexural rigidity of all members in the frame are identical and the girders are attached to the columns with fully restrained connections. Determine the bending moments produced in the frame by the lateral loads indicated.

The girders are considered rigidly connected to the columns and moments produced in the frame may be conveniently determined by the moment distribution procedure. As described in Williams Part 2, Sec. 7.9, advantage may be taken of the skew symmetry in the structure to allow automatically for sidesway. For all members

$$EI/L = k$$

Allowing for skew symmetry, the stiffness of the members are

$$s_{21} = s_{23} = s_{32} = k$$
$$s_{25} = s_{36} = 6k$$

The distribution factors at joint 2 are

$$d_{25} = s_{25}/(s_{25} + s_{21} + s_{23})$$
$$= 6k/8k$$
$$= 3/4$$
$$d_{21} = d_{23} = 1/8$$

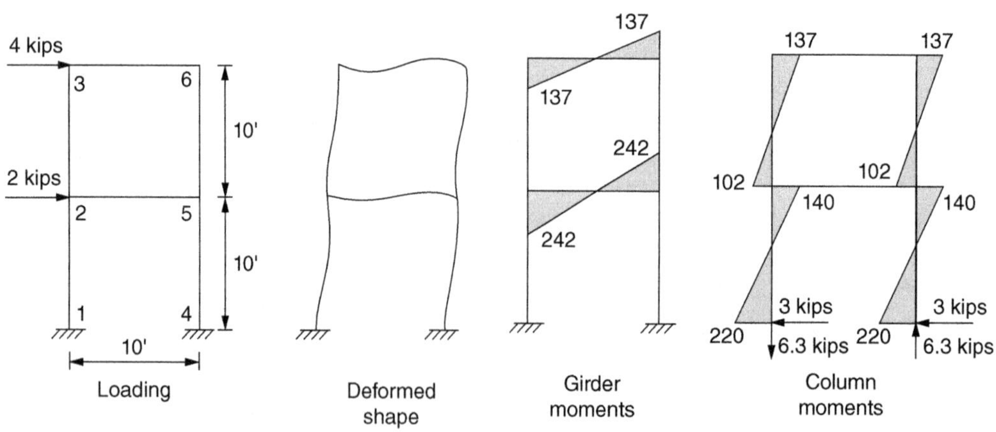


**FIGURE 3.34** Details for Example 3.8.

The distribution factors at joint 3 are

$$d_{36} = s_{36}/(s_{36} + s_{32})$$
$$= 6k/7k$$
$$= 6/7$$
$$d_{32} = 1/7$$

Allowing for skew symmetry, the carry-over factors in the columns are –1 and there is no carry-over between girder ends. The initial fixed-end moments are obtained by imposing unit virtual sway displacement on each story. For the upper story

$$M_{F23} = M_{F32} = -V_3 h_{23}/4$$
$$= -4 \times 10/4$$
$$= -10 \text{ kip-ft}$$
$$= -120 \text{ kip-in}$$

For the lower story

$$M_{F12} = M_{F21} = -(V_3 + V_2)h_{12}/4$$
$$= -6 \times 10/4$$
$$= -15 \text{ kip-ft}$$
$$= -180 \text{ kip-in}$$

The distribution procedure is shown in Table 3.5 with distribution occurring in the left half of the frame only. For convenience, the fixed-end moments are multiplied by 100.

The final bending moments, in kip-in units, produced in the members, are indicated in Fig. 3.34 with the moments drawn on the tension side of the members.

| Member | 12 | 21 | 25 | 23 | 32 | 36 |
|---|---|---|---|---|---|---|
| Distribution Factor | 0 | 1/8 | 3/4 | 1/8 | 1/7 | 6/7 |
| Fixed-end Moments | –18000 | –18000 | | –12000 | –12000 | |
| Distribution | | 3750 | 22500 | 3750 | 1714 | 10286 |
| Carry-over | –3750 | | | –1714 | –3750 | |
| Distribution | | 214 | 1286 | 214 | 536 | 3214 |
| Carry-over | –214 | | | –536 | –214 | |
| Distribution | | 67 | 402 | 67 | 31 | 183 |
| Carry-over | –67 | | | –31 | –67 | |
| Distribution | | 4 | 23 | 4 | 10 | 57 |
| Final Moments | –22031 | –13965 | 24211 | –10246 | –13740 | 13740 |

**Table 3.5** Distribution of Moments in Example 3.8

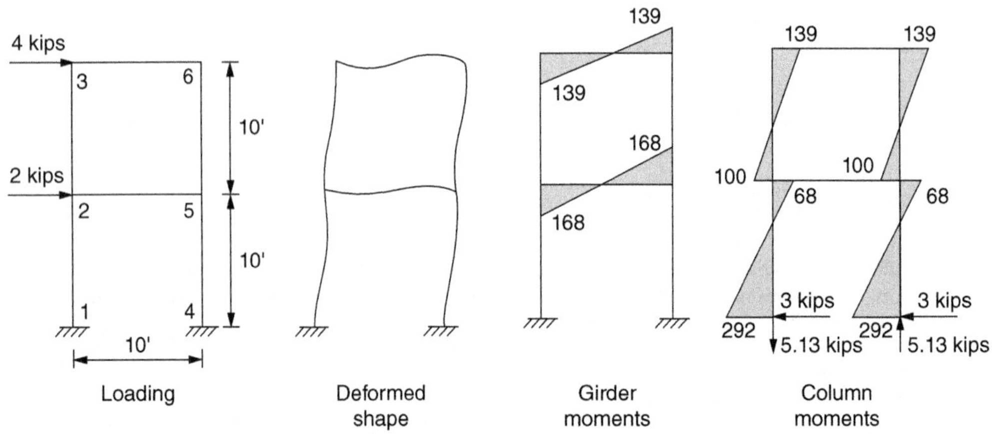


**FIGURE 3.35** Details for Example 3.9.

The following example considers the girders partially restrained.

**Example 3.9.** Laterally Loaded Two-Story Frame with Partially Restrained Connections
Figure 3.35 shows a two-story frame. The $EI/L$ values for all members are identical with a value of $k = 600$ kip-ft. The girders are attached to the columns with partially restrained moment connections with a moment/rotation ratio of $\eta = 1200$ kip-ft/rad at each end of the girders. Determine the bending moments produced in the frame by the lateral loads shown.

The moments produced in the frame may be determined by the moment distribution procedure. The girder stiffness is modified by the elastic connections at the girder ends and is derived as described in Williams Part 2, Sec. 6.7. For all members

$$
\begin{aligned}
EI/L &= k \\
\beta &= 2k/\eta \\
&= 2 \times 600/1200 \\
&= 1.0
\end{aligned}
$$

Allowing for elastic restraint at the beam supports, the stiffness of the beams are

$$
\begin{aligned}
s_{25} = s_{36} &= 6k/(1 + 3\beta) \\
&= 3k/2
\end{aligned}
$$

Allowing for skew symmetry, the stiffness of the columns are

$$s_{21} = s_{23} = s_{32} = k$$

The distribution factors at joint 2 are

$$
\begin{aligned}
d_{25} &= s_{25}/(s_{25} + s_{21} + s_{23}) \\
&= 1.5/(1.5 + 1 + 1) \\
&= 3/7 \\
d_{21} = d_{23} &= 2/7
\end{aligned}
$$

The distribution factors at joint 3 are

$$
\begin{aligned}
d_{36} &= s_{36}/(s_{36} + s_{32}) \\
&= 1.5/(1.5 + 1) \\
&= 3/5 \\
d_{32} &= 2/5
\end{aligned}
$$

Allowing for skew symmetry, the carry-over factors in the columns are –1 and there is no carry-over between girder ends. The fixed-end moments are obtained by imposing unit virtual sway displacement on each story. For the upper story

$$M_{F23} = M_{F32} = -V_3 h_{23}/4$$
$$= -4 \times 10/4$$
$$= -10 \text{ kip-ft}$$
$$= -120 \text{ kip-in}$$

For the lower story

$$M_{F12} = M_{F21} = -(V_3 + V_2)h_{12}/4$$
$$= -6 \times 10/4$$
$$= -15 \text{ kip-ft}$$
$$= -180 \text{ kip-in}$$

The distribution procedure is shown in Table 3.6 with distribution occurring in the left half of the frame only. For convenience, the fixed-end moments are multiplied by 100.

The final bending moments, in kip-in units, produced in the members, are indicated in Fig. 3.35 with the moments drawn on the tension side of the members. Comparing these results to the frame with rigidly connected beams, it is clear that the major difference is at node 2. The moment in column 21 is 50 percent less than in the rigidly connected frame. The moment in beam 25 is 30 percent less than in the rigidly connected frame.

| Member | 12 | 21 | 25 | 23 | 32 | 36 |
|---|---|---|---|---|---|---|
| Distribution Factor | 0 | 2/7 | 3/7 | 2/7 | 2/5 | 3/5 |
| Fixed-end Moments | –18000 | –18000 | | –12000 | –12000 | |
| Distribution | | 8571 | 12857 | 8571 | 4800 | 7200 |
| Carry-over | –8571 | | | –4800 | –8571 | |
| Distribution | | 1371 | 2057 | 1371 | 3428 | 5142 |
| Carry-over | –1371 | | | –3428 | –1371 | |
| Distribution | | 979 | 1469 | 979 | 548 | 823 |
| Carry-over | –979 | | | –548 | –979 | |
| Distribution | | 157 | 235 | 157 | 392 | 587 |
| Carry-over | –157 | | | –392 | –157 | |
| Distribution | | 112 | 168 | 112 | 63 | 94 |
| Carry-over | –112 | | | –63 | –112 | |
| Distribution | | 18 | 27 | 18 | 45 | 67 |
| Carry-over | –18 | | | –45 | –18 | |
| Distribution | | 13 | 19 | 13 | 7 | 11 |
| Final Moments | –29208 | –6779 | 16832 | –10055 | –13925 | 13924 |

**TABLE 3.6** Distribution of Moments in Example 3.9

## 3.4 Approximate Methods for Laterally Loaded Frames

Approximate methods may be useful in a preliminary design to obtain an initial estimate of forces and member sizes. Two such methods, the **portal method** and the **cantilever method**, are described by Grinter[24] Chap. 11, Sec. 286.

### Portal Method

The portal method assumes that each bay of a frame acts independently of the adjacent bays as an individual portal. As shown Fig. 3.36, this results in an interior column receiving a compressive force as the leeward column of the left bay and an equal tensile force as the windward column of the right bay. Hence, the resultant axial force is zero. Similarly, the shear and moment in an interior column is the sum of the shear and moment in the leeward column of the left bay and an equal shear and moment in the windward column of the right bay. Hence, the resultant shear and moment is twice that of an exterior column. The fundamental assumptions are

- A point of inflection occurs at the midheight of each column.
- A point of inflection occurs at the midpoint of each girder.
- The shear and moment in each interior column is twice that of an exterior column.
- The axial force in an interior column is zero.

The assumption that points of inflection occur at the midpoints of girders and columns is equivalent to inserting a hinge in the member at the midpoint. As shown in Fig. 3.37, a single bay portal with fixed column bases is 3 degrees indeterminate. Inserting a hinge in each member has the effect of introducing a moment release in each member and the frame is now statically determinate.

In a multistory, multibay frame, introducing a hinge into all columns and all girders will not make the frame statically determinate. However, in conjunction with the other assumptions made in the portal method, the frame can be solved by the principles of statics.

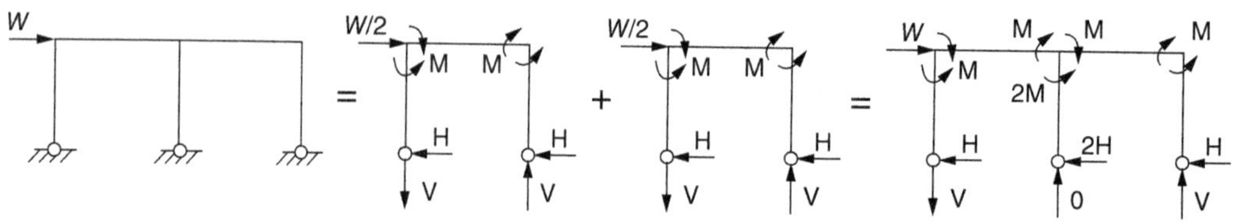


**FIGURE 3.36** Portal method assumptions.

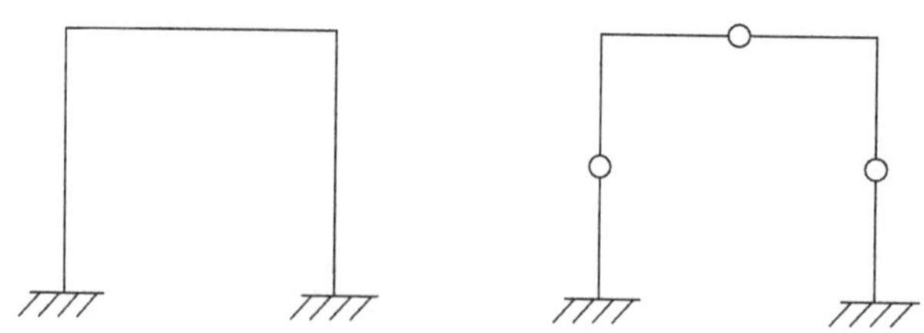


**FIGURE 3.37** Inserting hinges in a fixed base portal.

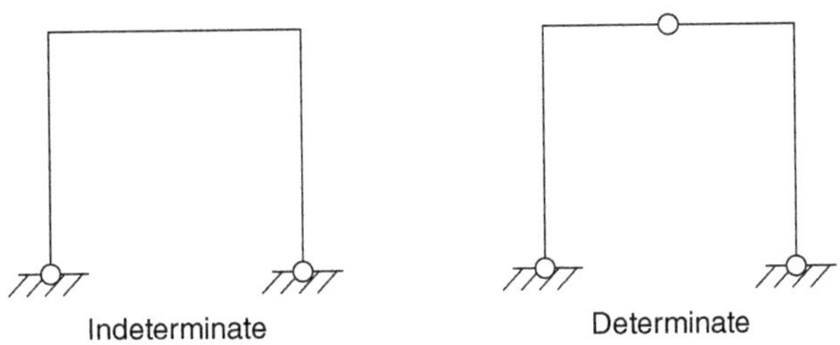


FIGURE 3.38 Inserting a hinge in a pinned base portal.

As shown in Fig. 3.38, a single bay portal with pinned column bases is 1 degree indeterminate. Inserting a hinge in the girder has the effect of introducing a moment release in the girder and the frame is now statically determinate.

In a multistory, multibay frame, with pinned column bases, introducing an additional hinge into columns in the bottom story is unnecessary.

**Example 3.10.** Analysis by the Portal Method
Use the portal method to determine the forces in the members of the frame shown in Fig. 3.39.

Taking moments about hinge 1 gives

$$V_{10} \times 30 + V_7 \times 20 + V_4 \times 10 = 40 \times 20 + 20 \times 10$$
$$= 1000 \text{ kip-ft}$$

Since
$$V_4 = V_7 = 0$$
$$V_{10} = 1000/30$$
$$= 33 \text{ kips ... up}$$

and
$$V_1 = 33 \text{ kips ... down}$$

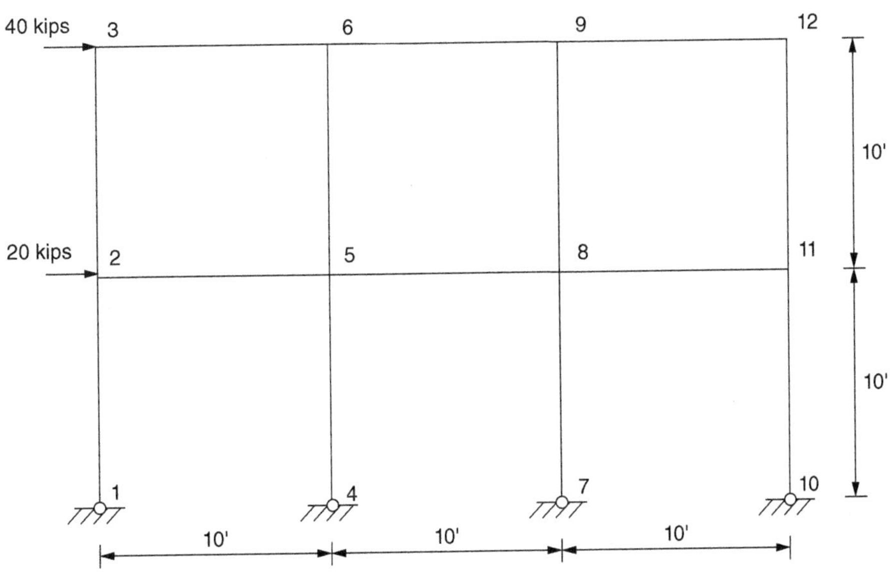


FIGURE 3.39 Details for Examples 3.10 and 3.11.

Resolving horizontally gives

$$H_1 + H_4 + H_7 + H_{10} = 40 + 20$$
$$= 60 \text{ kips}$$

Since

$$H_4 = H_7$$
$$= 2H_1 = 2H_{10}$$
$$H_1 = 60/6$$
$$= 10 \text{ kips}$$
$$= H_{10}$$

and

$$H_4 = 20 \text{ kips}$$
$$= H_7$$

Imposing unit virtual sway displacement on the bottom story gives

$$M_{21} + M_{54} + M_{87} + M_{1110} = (40 + 20) \times 10$$
$$= 600 \text{ kip-ft}$$

Since

$$M_{54} = M_{87}$$
$$= 2M_{21} = 2M_{1110}$$
$$M_{21} = 600/6$$
$$= 100 \text{ kip-ft}$$
$$= M_{1110}$$

and

$$M_{54} = 200 \text{ kip-ft}$$
$$= M_{87}$$

Imposing unit virtual sway displacement on the top story gives

$$M_{23} + M_{56} + M_{89} + M_{1112} + M_{32} + M_{65} + M_{98} + M_{1211} = 40 \times 10$$
$$= 400 \text{ kip-ft}$$

Since

$$M_{56} = M_{65} = M_{89} = M_{98}$$
$$= 2M_{23} = 2M_{1112} = 2M_{32} = 2M_{1211}$$
$$M_{23} = 400/12$$
$$= 33 \text{ kip-ft}$$
$$= M_{1112} = M_{32} = M_{1211}$$

and

$$M_{56} = 66 \text{ kip-ft}$$
$$= M_{65} = M_{89} = M_{98}$$

Equating moments at node 2 gives

$$M_{25} = M_{21} + M_{23}$$
$$= 100 + 33$$
$$= 133 \text{ kip-ft}$$
$$= M_{52} = M_{58} = M_{85} = M_{811} = M_{118}$$

Equating moments at node 3 gives

$$M_{36} = M_{32}$$
$$= 33 \text{ kip-ft}$$
$$= M_{63} = M_{69} = M_{96} = M_{912} = M_{129}$$

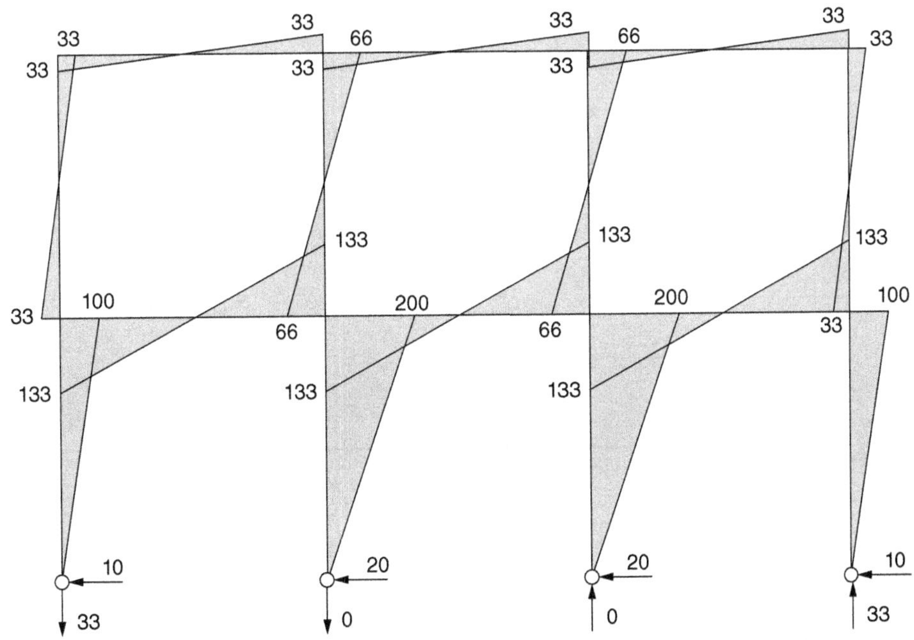


**FIGURE 3.40** Member forces for Example 3.10.

The member forces are shown in Fig. 3.40 with moments drawn on the tension side of the members and units of kips and kip-ft.

## Cantilever Method

The cantilever method assumes that in each story, the axial *stress* in a column is proportional to its distance from the centroid of all the columns in that story. As shown in Fig. 3.41, the axial stress in a column in a given story is given by

$$f_i = Mr_i/I$$

where $r_i$ = distance of column $i$ from the column centroid
$= x - x_i$
$x$ = distance of the column centroid from a baseline
$= \Sigma(A_i x_i)/\Sigma A_i$
$x_i$ = distance of a column from the baseline
$I = \Sigma A_i(r_i)^2$
$A_i$ = area of column $i$
$M$ = moment of applied loads about inflection point of columns in the story or about the hinges of hinged base columns
$V_i$ = axial force in column
$= f_i A_i$

When all columns have equal areas, the axial *load* in a column is proportional to its distance from the centroid of all the columns.

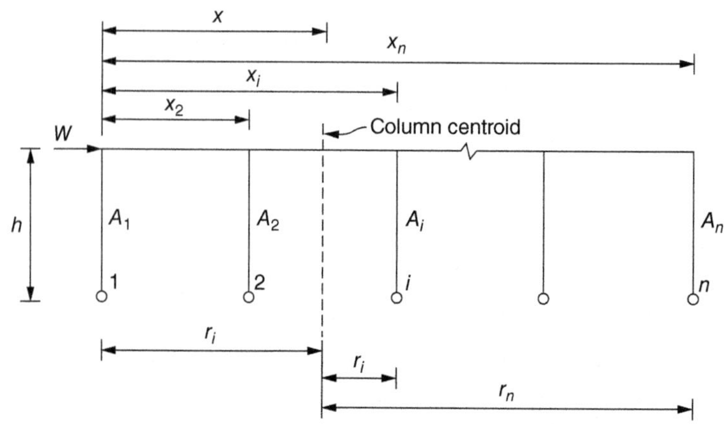


**FIGURE 3.41** Axial force in columns.

The fundamental assumptions of the cantilever method are

- A point of inflection occurs at the midheight of each column.
- A point of inflection occurs at the midpoint of each girder.
- The axial stress in a column is proportional to the distance of the column from the centroid of all the columns.

**Example 3.11.** Analysis by the Cantilever Method
Use the cantilever method to determine the forces in the members of the frame shown in Fig. 3.39. All columns have the same cross-sectional area.

The centroid of this symmetrical bent is at the middle of the center panel. Since all the columns have equal areas, the axial *load* in a column is proportional to its distance from the centroid of all the columns. The area of the columns may be taken as 1.0 and the moment of inertia of the columns about their centroid is

$$I = \Sigma A_i (r_i)^2$$
$$= 2 \times 15^2 + 2 \times 5^2$$
$$= 500$$

The moment of applied loads about the column bases is

$$M_0 = \Sigma Wh$$
$$= 40 \times 20 + 20 \times 10$$
$$= 1000 \text{ kip-ft}$$

The axial force in a column at its base is

$$V_1 = M_0 r_1 / I$$
$$= 1000 \times 15/500$$
$$= 30 \text{ kips}$$
$$= V_{10}$$
$$V_4 = 1000 \times 5/500$$
$$= 10 \text{ kips}$$
$$= V_7$$

The moment of applied loads about the inflection points of the columns in the top story is

$$M_1 = \Sigma Wh$$
$$= 40 \times 5$$
$$= 200 \text{ kip-ft}$$

The axial forces in the columns in the top story are

$$V_{23} = M_1 r_1 / I$$
$$= 200 \times 15/500$$
$$= 6 \text{ kips}$$
$$= V_{1112}$$
$$V_{56} = 200 \times 5/500$$
$$= 2 \text{ kips}$$
$$= V_{89}$$

From Fig. 3.42*a*, resolving forces vertically, the shear force in girder 25 is

$$S_{25} = V_1 - V_{23}$$
$$= 30 - 6$$
$$= 24 \text{ kips}$$

Hence $$M_{25} = S_{25} \times L_{25}/2$$
$$= 24 \times 5$$
$$= 120 \text{ kip-ft}$$
$$= M_{23} + M_{21}$$
$$= 5S_{23} + 10H_1$$

and $$5S_{23} + 10H_1 = 120 \text{ kip-ft}$$

also $H_1 = S_{23} \times 60/40$ ... column shears are in same ratio as total story shears

Hence $$5S_{23} + 15S_{23} = 120 \text{ kip-ft}$$
$$S_{23} = 6 \text{ kips}$$
$$H_1 = 1.5S_{23}$$
$$= 9 \text{ kips}$$

From symmetry

$$H_4 = (40 + 20 - 2 \times 9)/2$$
$$= 21$$
$$M_{23} = 5S_{23}$$
$$= 5 \times 6$$
$$= 30 \text{ kip-ft}$$
$$M_{21} = 10H_1$$
$$= 10 \times 9$$
$$= 90 \text{ kip-ft}$$

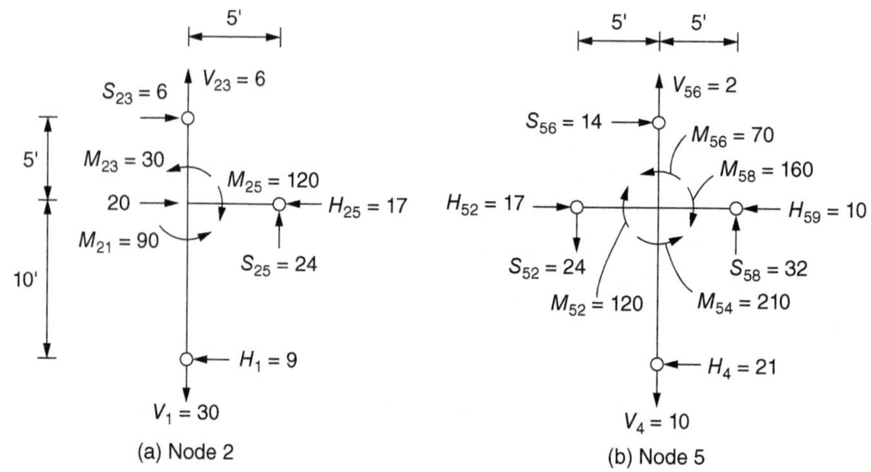


**FIGURE 3.42** Member forces at node 2 and node 5.

Resolving forces horizontally, the axial force in girder 25 is

$$H_{25} = 20 + S_{23} - H_1$$
$$= 20 + 6 - 9$$
$$= 17 \text{ kips}$$

From Fig. 3.42*b*, resolving forces vertically, the shear force in girder 58 is

$$S_{58} = V_4 - V_{56} + S_{52}$$
$$= 10 - 2 + 24$$
$$= 32 \text{ kips}$$

Hence

$$M_{58} = S_{58} \times L_{58}/2$$
$$= 32 \times 5$$
$$= 160 \text{ kip-ft}$$
$$M_{52} = S_{52} \times L_{52}/2$$
$$= 24 \times 5$$
$$= 120 \text{ kip-ft}$$
$$M_{54} = H_4 \times L_{54}$$
$$= 21 \times 10$$
$$= 210 \text{ kip-ft}$$
$$M_{56} = M_{52} + M_{58} - M_{54}$$
$$= 120 + 160 - 210$$
$$= 70 \text{ kip-ft}$$

The member forces are shown in Fig. 3.43 with moments drawn on the tension side of the members and units of kips and kip-ft.

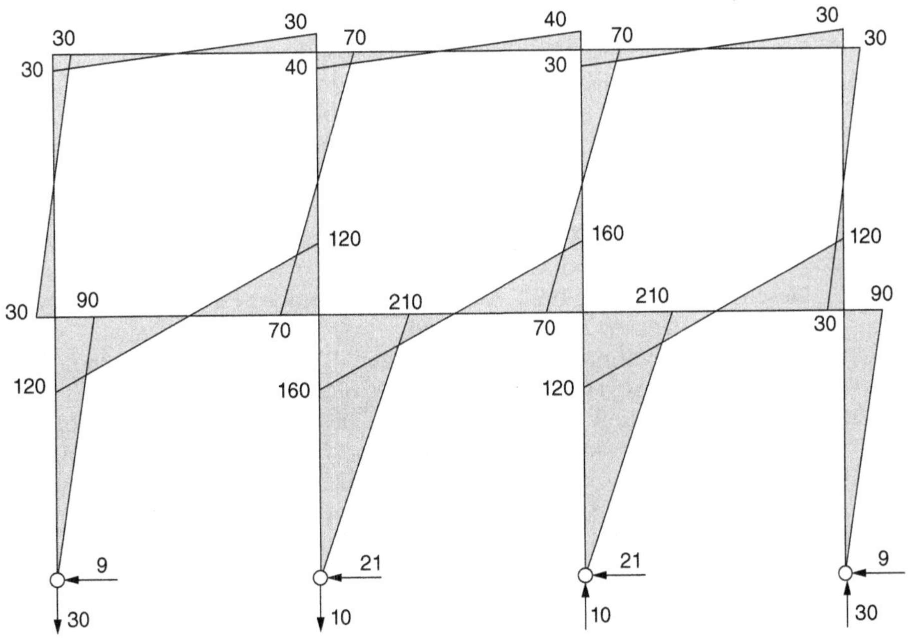


**Figure 3.43** Member forces for Example 3.11.

## References


1. International Code Council (ICC). 2015. *International Building Code*, 2015 edition. ICC, Falls Church, VA.
2. American Institute of Steel Construction (AISC). 2010. *Specification for Structural Steel Buildings* (AISC 360-10), AISC, Chicago, IL.
3. Kennedy, D. J. L. 1969. "Moment-Rotation Characteristics of Shear Connections," *Engineering Journal*, October 1969.
4. American Institute of Steel Construction (AISC). 2011. *Steel Construction Manual.* 14th edition. AISC, Chicago, IL.
5. Astaneh, A., Call, S. M., and McMullin, K. M. 1989. "Design of Single Plate Shear Connections," *Engineering Journal*, 1st Quarter 1989.
6. Astaneh, A. and Nader, M. N. 1989. "Design of Tee Framing Shear Connections," *Engineering Journal*, 1st Quarter 1989.
7. Garrett, J. R., and Brockenbrough, R. L. 1986. "Design Loads for Seated-beam in LRFD," *Engineering Journal*, 2nd Quarter 1986.
8. Yang, W. H., Chen, W. F., and Bowman, M. D. 1987. "The Behavior and Load Carrying Capacity of Unstiffened Seated-beam Connections," *Engineering Journal*, 3rd Quarter 1997.
9. Ellifritt, D. S. and Sputo, T. 1999. "Design Criteria for Stiffened Seated Connections to Column Webs," *Engineering Journal*, 4th Quarter 1999.
10. Roeder, C. W. and Dailey, R. H. 1989 "The Results of Experiments on Seated Beam Connections," *Engineering Journal*, 3rd Quarter 1989.

11. American Institute of Steel Construction (AISC). 2010. *Seismic Provisions for Structural Steel Buildings* (AISC 341-10), AISC, Chicago, IL.
12. Astaneh, H. 2003. *Field Bolted Moment Connections*, AISC, Chicago, IL.
13. Murray, T. M. and Sumner, E. A. 2003. *Extended End-Plate Moment Connections*, Design Guide No. 4, 2nd Edition. AISC, Chicago, IL.
14. Williams, A. 2009. *Structural Analysis in Theory and Practice.* Elsevier/International Code Council, New York, NY.
15. Blodgett, O. W. 1966. *Design of Welded Structures.* The James F. Lincoln Arc Welding Foundation, Cleveland, OH.
16. Geschwindner, L. R. 1991. "A Simplified Look at Partially Restrained Beams," *Engineering Journal*, 2nd Quarter 1991.
17. Leon, R. T., Hoffman, J. J., and Staeger, T. 2003. *Partially Restrained Composite Connections*, Design Guide No. 8. AISC, Chicago, IL.
18. Geschwindner, L. R. and Disque, R. O. 2005. "Flexible Moment Connections for Unbraced Frames—a Return to Simplicity," *Engineering Journal*, 2nd Quarter 2005.
19. American Society of Civil Engineers (ASCE). 2010. *Minimum Design Loads for Buildings and Other Structures* (ASCE 7-10), ASCE, Reston, VA.
20. Prasad, B. K., Thompson, D. S., and Sabelli, R. 2009. *Guide to the Design of Diaphragms, Chords, and Collectors.* ICC, Falls Church, VA.
21. Luttrell, L. D. 1996. "Designing Roof and Floor Diaphragms," *Modern Steel Construction*, March 1996.
22. Steel Deck Institute (SDI). 2010. *Standard for Steel Roof Deck.* ANSI/SDI RD1.0, SDI, Fox River Grove, IL.
23. Steel Deck Institute (SDI). 2004. *Diaphragm Design Manual.* SDI, DDM03, Fox River Grove, IL.
24. Grinter, L. E. 1949. *Theory of Modern Steel Structures*, Vol. 1, *Statically Determinate Structures.* Macmillan, New York, NY.

## Problems

**3.1** *Given:* A framed building to be designed to resist lateral loads

*Find:* *a.* Two possible approximate methods that may be used
*b.* The advantages of each method
*c.* The disadvantages of each method

**3.2** *Given:* The two-story frame shown in Fig. 3.34

*Find:* Using the portal method, determine the forces in the members.

**3.3** *Given:* The two-story frame shown in Fig. 3.34

*Find:* Using the cantilever method, determine the forces in the members.

**3.4** *Given:* A girder/column connection to be designed as a simple connection

*Find:* *a.* Two possible methods that may be used
*b.* The advantages of each method
*c.* The disadvantages of each method

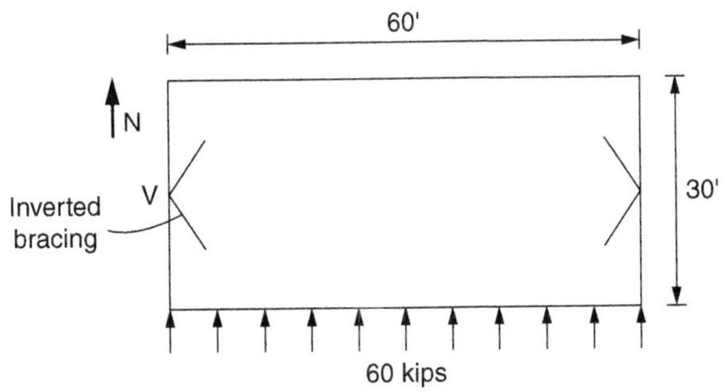


**FIGURE 3.44** Details for Problems 3.7 and 3.8.

**3.5** *Given:* A girder/column connection to be designed as a fully restrained (FR) connection

*Find:* *a.* Two possible methods that may be used
*b.* The advantages of each method
*c.* The disadvantages of each method

**3.6** *Given:* A topped steel deck to be used as a floor diaphragm

*Find:* The function of the boundary elements.

**3.7** *Given:* The single-story building in Fig. 3.44. The bracing in the east and west wall is inverted V, or chevron, type bracing located at the midpoint of the wall and is 10 ft long. The seismic base shear in the north-south direction is 60 kips. The roof diaphragm may be considered flexible.

*Find:* The maximum collector force in the east wall.

**3.8** *Given:* The single-story building in Fig. 3.44. The bracing in the east and west wall is inverted V, or chevron, type bracing located at the midpoint of the wall and is 10 ft long. The seismic base shear in the north-south direction is 60 kips. The roof diaphragm may be considered rigid

*Find:* The maximum collector force in the east wall. Ignore torsional amplification.

# CHAPTER 4

# Design of Steel Beams in Flexure

## 4.1 Introduction

As shown in Fig. 4.1, a beam is a structural element that supports loads transverse to its longitudinal axis. The beam is usually aligned horizontally and the loads are usually gravity loads. The applied loading results in the bending moment $M_x$ and the shear force $V_x$ at any section $x$. Provided that the stresses produced in the beam are within the elastic limit, the stresses produced in a doubly symmetric section are

$f_c = M_x c/I$ ... compressive stress in the extreme top fiber

$= f_t$ ... tensile stress in the extreme bottom fiber

$c =$ distance from extreme fiber to neutral axis

$= d/2$ ... for a symmetrical section

$d =$ depth of beam

$I =$ moment of inertia of beam

$f_v =$ shear stress in web

$= V_x/dt_w$

$t_w =$ web thickness

### Flexural Limit States

Increasing the loading on the beam will eventually cause failure of the beam by either

- Yielding of the material and formation of a plastic hinge and a collapse mechanism
- Lateral-torsional buckling of the beam
- Local buckling of the flange or web of a beam that is not compact

Adequate lateral bracing of a beam will ensure that full plasticity is achieved prior to lateral-torsional buckling occurring. Using a shape that is compact will ensure that full plasticity is achieved prior to flange or web local buckling.

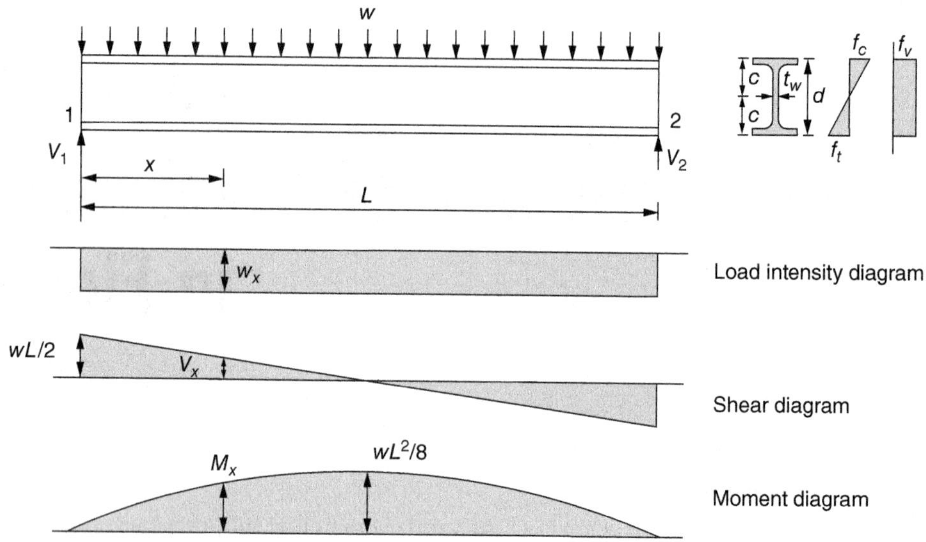


**FIGURE 4.1** Beam loading.

## Lateral Bracing of Beams

If the loading on a compact beam is increased, collapse will eventually occur in one of three possible modes:

- The plastic mode occurs when the beam is adequately braced to prevent lateral-torsional buckling. An example of a braced beam is shown in Fig. 4.2, where the distance between braces is $L_b$. The limiting laterally unbraced length for the limit state of yielding is denoted by $L_p$. When the distance between braces does not exceed $L_p$, yielding of the steel occurs over the full depth of the beam at the point of maximum moment. The full plastic moment $M_p$ is developed and the nominal flexural strength of the beam is

$$M_n = M_p$$

- When the distance between braces exceeds $L_p$ and is less than $L_r$, the limiting laterally unbraced length for the limit state of inelastic lateral-torsional buckling,

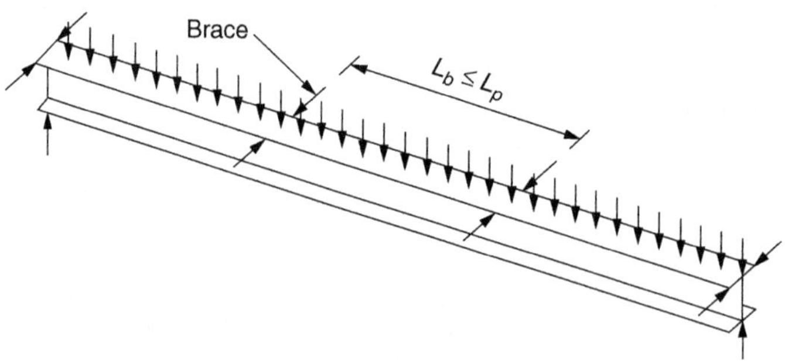


**FIGURE 4.2** Beam bracing.

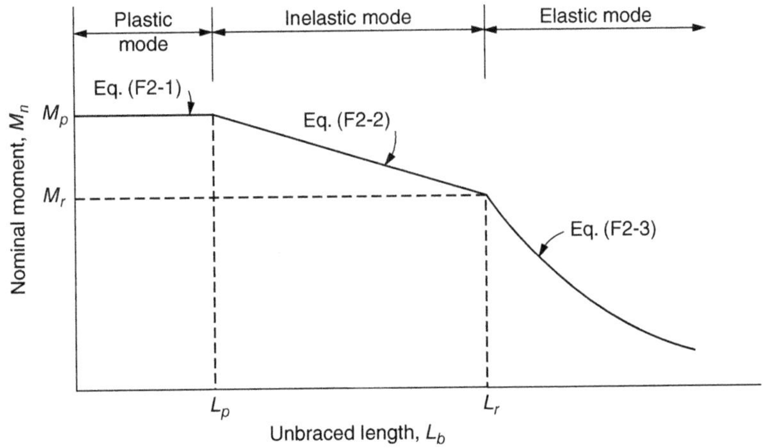


**FIGURE 4.3** Nominal moment and unbraced length, beams with $C_b = 1.0$.

collapse occurs prior to the development of the full plastic moment. When $L_b = L_r$ the nominal flexural strength of the beam is

$$M_n = M_r \text{ ... limiting buckling moment}$$

- When the distance between braces exceeds $L_r$, collapse occurs by elastic lateral-torsional buckling. The nominal flexural strength of the beam is

$$M_n = M_{cr} \text{ ... elastic buckling moment}$$

These effects are considered by White and Chang[1] and the three modes of collapse are plotted in Fig. 4.3.

The required brace strength, for nodal bracing, is given by American Institute of Steel Construction, *Specification for Structural Steel Buildings* (AISC 360)[2] App. 6 Eq. (A-6-7) as

$$P_{br} = 0.02M_rC_d/h_o$$

where $M_r$ = required flexural strength of beam using ASD or LRFD load combinations as appropriate
$h_o$ = distance between beam flange centroids
$C_d$ = 1.0 except in the following case;
= 2.0 for bending in double curvature for brace closest to the inflection point

## Design Flexural Strength and Allowable Flexural Strength

After the nominal flexural strength is determined, the design flexural strength and the allowable flexural strength may be obtained from AISC 360, Sec. F1 as

$$\varphi_bM_n = \text{design flexural strength}$$
$$\geq M_u \text{ ... required flexural strength using LRFD load combinations}$$
$$M_n/\Omega_b = \text{allowable flexural strength}$$
$$\geq M_a \text{ ... required flexural strength using ASD load combinations}$$

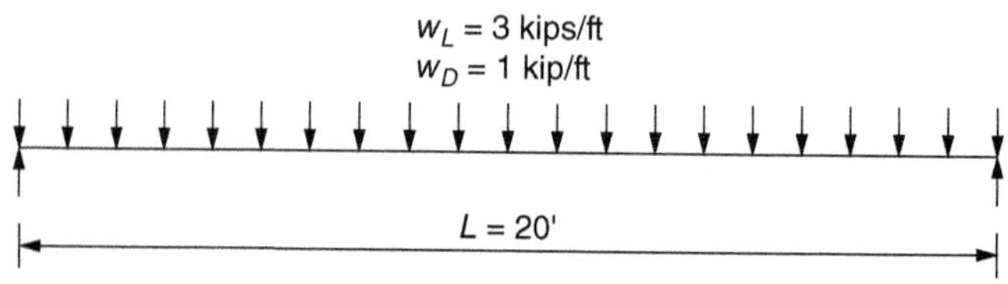


**FIGURE 4.4** Details for Example 4.1.

where $\varphi_b$ = resistance factor for flexure
= 0.9
$\Omega_b$ = safety factor for flexure
= 1.67

American Institute of Steel Construction, *Steel Construction Manual* (AISC Manual)[3] Table 3-2, for a value of $C_b = 1.0$, provides values of $\varphi_b M_p$ and $M_p/\Omega_b$.

**Example 4.1.** Beam Selection
Determine the lightest W10 section suitable for the beam shown in Fig. 4.4. The loading consists of a uniformly distributed dead load of $w_D = 1.0$ kip/ft, which includes the weight of the beam, and a uniformly distributed live load of $w_L = 3.0$ kips/ft. The beam is continuously braced on its compression flange and has a yield stress of $F_y = 50$ ksi.

Since the beam is continuously braced, lateral-torsional buckling does not occur. Hence, full plasticity of the section is possible and the nominal moment is

$$M_n = M_p$$

Applying American Society of Civil Engineers, *Minimum Design Loads for Buildings and Other Structures* (ASCE 7)[4] Secs. 2.3 and 2.4 gives

| LRFD | ASD |
|---|---|
| From ASCE 7 Sec. 2.3.2 combination 2: | From ASCE 7 Sec. 2.4.1 combination 2: |
| $w_u$ = factored load | $w_a$ = factored load |
| $= 1.2w_D + 1.6w_L$ | $= w_D + w_L$ |
| $= 1.2 \times 1.0 + 1.6 \times 3.0$ | $= 1.0 + 3.0$ |
| = 6 kips/ft | = 4 kips/ft |
| $M_u$ = factored moment | $M_a$ = factored moment |
| $= w_u L^2/8$ | $= w_a L^2/8$ |
| $= 6 \times 20^2/8$ | $= 4 \times 20^2/8$ |
| = 300 kip-ft | = 200 kip-ft |
| = required strength | = required strength |
| From AISC Manual Table 3-2, for a value of $C_b = 1.0$, a W10 × 68 provides a design flexural strength of | From AISC Manual Table 3-2, for a value of $C_b = 1.0$, a W10 × 68 provides an allowable flexural strength of |
| $\varphi_b M_p$ = 320 kip-ft ... satisfactory | $M_p/\Omega_b$ = 213 kip-ft ... satisfactory |
| > 300 kip-ft | > 200 kip-ft |

In addition to designing a beam for flexural capacity, the serviceability limit states must also be checked. These include deflection and vibration limitations.

## 4.2 Plastic Moment of Resistance

The beam shown in Fig. 4.5 is braced to prevent lateral torsional buckling, and the flanges and web are compact to prevent local buckling. The bending moment applied to the central portion of the beam is

$$M = WL/3$$
$$= f_b S_x$$

where $f_b$ = stress in the extreme top and bottom fibers shown at (*a*) in Fig. 4.5
$S_x$ = elastic section modulus about the *x*-axis
$= I/c$
$c$ = distance from extreme fiber to neutral axis
$= d/2$ ... for a symmetrical section
$d$ = depth of beam
$I$ = moment of inertia of beam

By increasing the load on the beam, the stress in the extreme fibers reaches the yield stress $F_y$ as shown at (*b*) in Fig. 4.5. Allowing for the residual fabrication stresses in the beam, the applied moment is now

$$M_r = 0.7F_y S_x$$

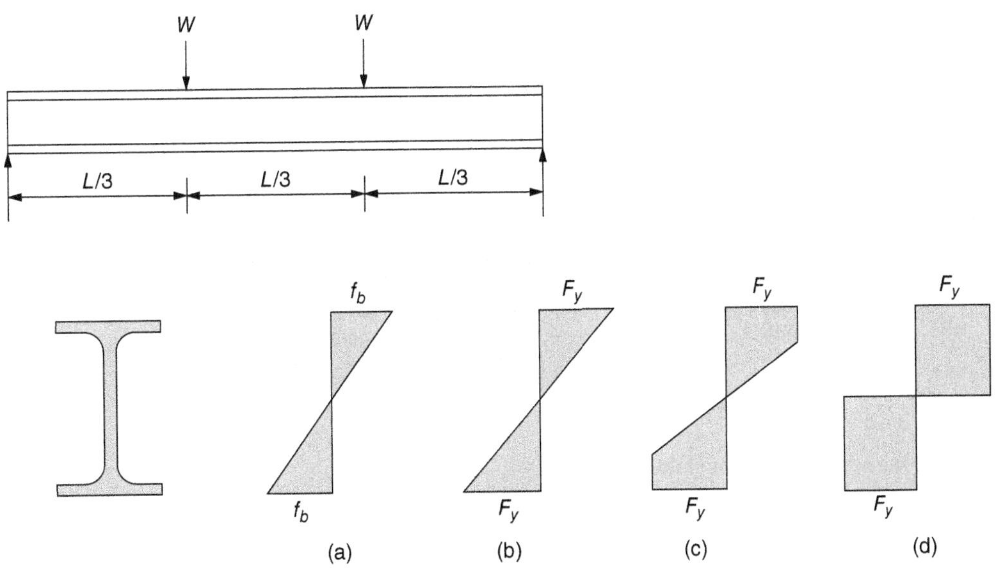


**Figure 4.5** Plastic hinge formation.

Further increase in the load causes the plasticity to spread toward the center of the beam as shown at (*c*). Eventually, as shown at (*d*), all fibers in the cross section have yielded, a plastic hinge has formed, and collapse occurs. The nominal flexural strength of the section is

$$M_n = M_p$$
$$= F_y Z_x$$

where $Z_x$ is plastic section modulus and $M_p$ is plastic moment of resistance.

## Shape Factor and ASD

The shape factor is defined as

$$M_p/M_y = Z_x/S_x$$
$$\approx 1.1 \text{ to } 1.3 \text{ for a W-shape}$$

Using a conservative value for the shape factor of 1.1, the traditional allowable stress design expressions may be derived. The nominal flexural strength is

$$M_n = F_y Z_x$$
$$= 1.1 F_y S_x$$

The allowable flexural *strength* is

$$M_n/\Omega_b = 1.1 F_y S_x/\Omega_b$$
$$= 1.1 F_y S_x/1.67$$
$$= 0.66 F_y S_x$$

Hence, the allowable flexural *stress* is

$$F_b = M_n/\Omega_b S_x$$
$$= 0.66 F_y$$

**Example 4.2.** Allowable Stress Design

Using the traditional allowable stress design method, determine the lightest W10 section suitable for the beam shown in Fig. 4.4. The loading consists of a uniformly distributed dead load of $w_D = 1.0$ kip/ft, which includes the weight of the beam, and a uniformly distributed live load of $w_L = 3.0$ kips/ft. The beam is continuously braced on its compression flange and has a yield stress of $F_y = 50$ ksi.

From Example 4.1, the required strength using ASD load combinations is

$$M_a = 200 \text{ kip-ft}$$
$$= 0.66 F_y S_x$$

The required elastic modulus is

$$S_x = M_a/0.66 F_y$$
$$= 200 \times 12/(0.66 \times 50)$$
$$= 72.7 \text{ in}^3$$

From AISC Manual Table 1-1, a W10 × 68 provides an elastic section modulus of

$$S_x = 75.7 \text{ in}^3 \text{ ... satisfactory}$$
$$> 72.7 \text{ in}^3$$

## Built-Up Sections

Built-up sections are used to reinforce existing structures and, in the case of W-shapes with cap channels, to provide runway beams for overhead cranes. AISC Manual Tables 1-19 and 1-20 provide the section properties of W- and S-shapes with cap channels. The properties of other built-up sections must be determined manually.

**Example 4.3.** Built-Up Section
Determine the plastic moment of resistance for the cover plated W10 × 88 section shown in Fig. 4.6. Ignore the effect of the web fillets.

The relevant properties of the W10 × 88 are

$$A_W = \text{area of W10} \times 88$$
$$= 26.0 \text{ in}^2$$
$$t_f = \text{flange thickness}$$
$$= 0.99 \text{ in}$$
$$t_w = \text{web thickness}$$
$$= 0.605 \text{ in}$$
$$d = \text{depth} = 10.8 \text{ in}$$
$$A_f = \text{flange area}$$
$$= 10.3 \times 0.99$$
$$= 10.20 \text{ in}^2$$
$$A_w = \text{web area}$$
$$= 8.82 \times 0.605$$
$$= 5.34 \text{ in}^2$$

The area of the 12 × 1/4 in plate is

$$A_p = 12 \times 1/4$$
$$= 3 \text{ in}^2$$

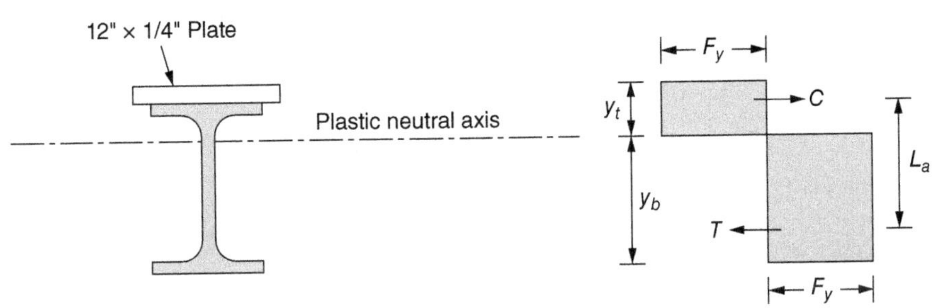


**FIGURE 4.6** Details for Example 4.3.

The total area of the built-up section is

$$A_{comp} = A_W + A_P$$
$$= 26.0 + 3.0$$
$$= 29.0 \text{ in}^2$$

The areas in tension and compression after the formation of a plastic hinge are equal and are given by

$$A_T = A_C = A_{comp}/2$$
$$= 29.0/2$$
$$= 14.5 \text{ in}^2$$

The plastic neutral axis is at a height $y_b$ above the base given by

$$A_T = A_f + t_w(y_b - t_f)$$
$$14.5 = 10.20 + 0.605(y_b - 0.99)$$

and

$$y_b = 8.10 \text{ in}$$
$$y_t = 11.05 - 8.10$$
$$= 2.95 \text{ in}$$

The centroid of $A_T$ is at a height above the base of

$$[10.20 \times 0.99/2 + 0.605(8.10 - 0.99)(0.99 + 7.11/2)]/14.5$$
$$= 1.70 \text{ in}$$

The centroid of $A_C$ is at a height above the base of

$$[3.0 \times 10.93 + 10.20 \times 10.31 + 0.605 \times 1.71(8.10 + 1.71/2)]/14.45$$
$$= 10.19 \text{ in}$$

The lever arm is

$$L_a = 10.19 - 1.70$$
$$= 8.49 \text{ in}$$

The plastic section modulus is

$$Z_x = A_T L_a$$
$$= 14.50 \times 8.49$$
$$= 123.11 \text{ in}^3$$

The plastic moment of resistance is

$$M_p = F_y Z_x$$
$$= 50 \times 123.11/12$$
$$= 513 \text{ kip-ft}$$

## 4.3 Compact, Noncompact, and Slender Sections

The flexural capacity of an adequately braced beam depends on the slenderness ratio of the compression flange and the web. When the slenderness ratios are sufficiently small, the beam can attain its full plastic moment and the cross section is classified as compact. When the slenderness ratios are larger, the compression flange or the web may buckle locally before a full plastic moment is attained and the cross section is classified as noncompact. When the slenderness ratios are sufficiently large, local buckling will occur before the yield stress of the material is reached and the cross section is classified as slender. The flexural response of the three classifications is shown in Fig. 4.7.

### Compact Section

A compact section is one that can develop a plastic hinge prior to local buckling of the flange or web, provided that adequate lateral bracing is provided. The criteria for determining compactness in the flange of rolled beams are defined in AISC 360 Table B4.1b as

$$b_f/2t_f < \lambda_{pf} = 0.38(E/F_y)^{0.5}$$

where $b_f$ = flange width
$t_f$ = flange thickness
$\lambda_{pf}$ = limiting slenderness parameter for compact flange
$b_f/2t_f$ = beam flange slenderness parameter
$= \lambda$

The criteria for determining compactness in the web of rolled beams is defined in AISC 360 Table B4.1b as

$$h/t_w < \lambda_{pw} = 3.76(E/F_y)^{0.5}$$

where $h$ = clear distance between flanges less the corner radius at each flange
$t_w$ = web thickness
$\lambda_{pw}$ = limiting slenderness parameter for compact web
$h/t_w$ = beam web slenderness parameter
$= \lambda$

Values of $b_f/2t_f$ and $h/t_w$ are listed in AISC Manual Tables 1-1 to 1-4.

All rolled I-shapes have compact webs for a yield stress of $F_y$ = 50 ksi. All rolled I-shapes have compact flanges for a yield stress of $F_y$ = 50 ksi with the exception of

W21 × 48, W14 × 99, W14 × 90, W12 × 65, W10 × 12, W8 × 31, W8 × 10, W6 × 15, W6 × 9, W6 × 8.5, and M4 × 6.

The nominal flexural strength of a compact section is given by AISC 360 Eq. (F2-1) as

$$M_n = M_p$$
$$= F_y Z_x$$

where $M_p$ is plastic moment of resistance and $Z_x$ is plastic section modulus about the $x$-axis.

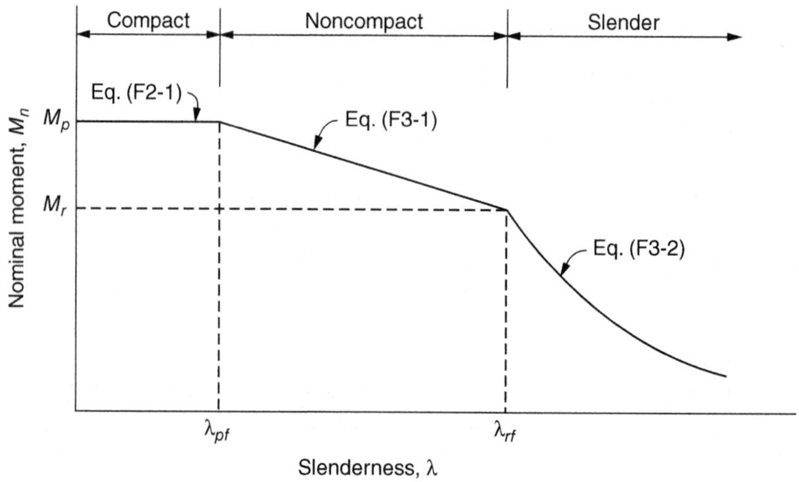


**FIGURE 4.7** Nominal moment and slenderness ratio.

Tabulated values of $\varphi_b M_p$ and $M_p/\Omega_b$, in AISC Manual Table 3-2 allow for any reduction because of noncompactness.

**Example 4.4.** Compact Section Criteria

Determine if a W10 × 12, with a yield stress of 50 ksi, satisfies the compact section criteria.

The limiting slenderness parameter for the web of a rolled I-shape is given by AISC 360 Table B4.1b as

$$\begin{aligned}\lambda_{pw} &= 3.76(E/F_y)^{0.5}\\ &= 3.76(29{,}000/50)^{0.5}\\ &= 90.6\end{aligned}$$

From AISC Manual Table 1-1, a W10 × 12 has a value of

$$\begin{aligned}h/t_w &= 46.6\\ &< 90.6 \ \ldots\ \text{web is compact}\end{aligned}$$

The limiting slenderness parameter for the flange of a rolled I-shape is given by AISC 360 Table B4.1b as

$$\begin{aligned}\lambda_{pf} &= 0.38(E/F_y)^{0.5}\\ &= 0.38(29{,}000/50)^{0.5}\\ &= 9.15\end{aligned}$$

From AISC Manual Table 1-1, a W10 × 12 has a value of

$$\begin{aligned}b_f/2t_f &= 9.43\\ &> 9.15 \ \ldots\ \text{flange is not compact}\end{aligned}$$

Hence a W10 × 12 with a yield stress of 50 ksi is not a compact section.

## Noncompact Section

A noncompact section is one that can, prior to local buckling of the flange or web, develop a nominal flexural strength $M_n$ given by

$$M_r \le M_n < M_p$$

The criteria for determining noncompactness in the flange of rolled beams is defined in AISC 360 Table B4.1b as

$$0.38(E/F_y)^{0.5} < b_f/2t_f < \lambda_{rf} = 1.0(E/F_y)^{0.5}$$

where $\lambda_{rf}$ is limiting slenderness parameter for a noncompact flange.

The criteria for determining noncompactness in the web of rolled beams is defined in AISC 360 Table B4.1b as

$$3.76(E/F_y)^{0.5} < h/t_w < \lambda_{rw} = 5.70(E/F_y)^{0.5}$$

where $\lambda_{rw}$ is limiting slenderness parameter for a noncompact web.

The nominal flexural strength of a section with compact web and noncompact flange with adequate lateral bracing, is given by AISC 360 Eq. (F3-1) as

$$M_n = M_p - (M_p - 0.7F_yS_x)(\lambda - \lambda_{pf})/(\lambda_{rf} - \lambda_{pf})$$

where $\lambda$ is $b_f/2t_f$ = beam flange slenderness parameter.

Tabulated values of $\varphi_b M_p$ and $M_p/\Omega_b$, in AISC Manual Table 3-2 allow for any reduction because of noncompactness.

**Example 4.5.** Noncompact Section
Determine the design flexural strength and allowable flexural strength of a W10 × 12, with a yield stress of 50 ksi.

From Example 4.4, the relevant slenderness parameters for a W10 × 12 are

$$b_f/2t_f = 9.43$$
$$= \lambda$$
$$\lambda_{pf} = 9.15 \text{ ... flange is noncompact}$$
$$h/t_w = 46.6$$
$$\lambda_{pw} = 90.6 \text{ ... web is compact}$$

From AISC 360 Table 1-1

$$S_x = \text{elastic section modulus about the } x\text{-axis}$$
$$= 10.9$$
$$0.7F_yS_x = 0.7 \times 50 \times 10.9/12$$
$$= 31.8 \text{ kip-ft}$$

The plastic moment of resistance is given by

$$M_p = Z_xF_y$$
$$= 12.6 \times 50/12$$
$$= 52.5 \text{ kip-ft}$$

From AISC 360 Table B4.1b, the limiting slenderness parameter for a noncompact flange is

$$\lambda_{rf} = 1.0(E/F_y)^{0.5}$$
$$= 1 \times (29{,}000/50)^{0.5}$$
$$= 24.1$$

From AISC 360 Eq. (F3-1), the nominal flexural strength is given by

$$M_n = M_p - (M_p - 0.7F_yS_x)(\lambda - \lambda_{pf})/(\lambda_{rf} - \lambda_{pf})$$
$$= 52.5 - (52.5 - 31.8)(9.43 - 9.15)/(24.1 - 9.15)$$
$$= 52.1 \text{ kip-ft}$$

| LRFD | ASD |
|---|---|
| The design flexural strength is | The allowable flexural strength is |
| $\varphi_b M_n = 0.9 \times 52.1$ | $M_n/\Omega_b = 52.1/1.67$ |
| $= 46.9$ kip-ft | $= 31.2$ kip-ft |

These values are given in AISC Manual Table 3-2 as $\varphi_b M_p$ and $M_p/\Omega_b$.

### Slender Section

A slender section is one that cannot develop the yield stress prior to web or flange local buckling. In accordance with AISC 360 Sec. B4, a section is classified as slender when the slenderness ratio of the flange or web exceeds the limiting slenderness parameters for a noncompact section. For flange local buckling, a slender section is defined as

$$b_f/2t_f > \lambda_{rf} = 1.0(E/F_y)^{0.5}$$

For web local buckling, a slender section is defined as

$$h/t_w > \lambda_{rw} = 5.70(E/F_y)^{0.5}$$

The nominal flexural strength of a section with slender flanges, with adequate lateral bracing, is given by AISC 360 Eq. (F3-2) as

$$M_n = 0.9Ek_cS_x/\lambda^2$$

where $k_c = 4/(h/t_w)^{0.5}$
$\geq 0.35$
$\leq 0.76$
$\lambda = b_f/2t_f$

There are no rolled I-shapes with slender flanges or webs.

## 4.4 Lateral-Torsional Buckling Modification Factor

The lateral-torsional buckling modification factor accounts for the effect that a variation in bending moment has on the lateral-torsional buckling of a beam. Beams in which the applied moments cause reversed curvature have a greater resistance to lateral-torsional buckling than beams subjected to a uniform bending moment. The derivation of this factor is covered in Zoruba and Dekker.[5]

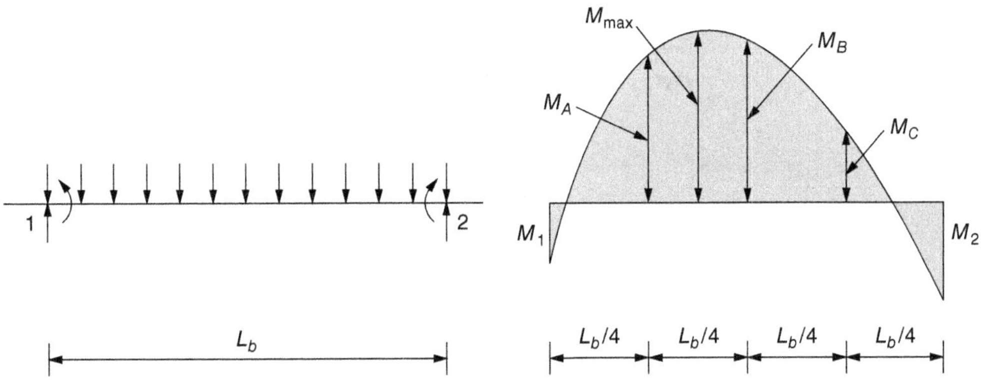


**FIGURE 4.8** Determination of $C_b$.

The modification factor $C_b$ is given by AISC 360 Eq. (F1-1) as

$$C_b = 12.5M_{max}/(2.5M_{max} + 3M_A + 4M_B + 3M_C)$$

where $M_A$ = absolute value of the bending moment at the quarter point of an unbraced segment
$M_B$ = absolute value of the bending moment at the centerline of an unbraced segment
$M_C$ = absolute value of the bending moment at the three-quarter point of an unbraced segment
$M_{max}$ = absolute value of the maximum bending moment in the unbraced segment

The terms used in determining $C_b$ are illustrated in Fig. 4.8.

A beam segment bent in single curvature and subjected to a uniform bending moment with equal values of $M_A$, $M_B$, $M_C$, and $M_{max}$ has a value of $C_b = 1.0$. Other moment diagrams increase the resistance of the segment to lateral torsional buckling and result in a corresponding increase in the value of $C_b$. For any loading condition, the value of $C_b$ may conservatively be taken as 1.0 and this is the value adopted in AISC Manual Table 3-10. For cantilevers, without bracing at the free end, the bending coefficient is taken as unity. Values of $C_b$ for various loading and restraint conditions are illustrated in Table 4.1. Examples of the calculation of $C_b$ are given by Aminmansour.[6]

**Example 4.6.** Calculation of $C_b$

Determine the value of $C_b$ for each span of the two-span continuous beam shown in Fig. 4.9. The loading consists of a uniformly distributed dead load of $w_D = 1.0$ kip/ft, which includes the weight of the beam, and a uniformly distributed live load of $w_L$ = of 3.0 kips/ft. The beam is braced only at the supports.

The total load acting on the continuous beam is

$$\begin{aligned} w &= w_D + w_L \\ &= 1 + 3 \\ &= 4 \text{ kips/ft} \end{aligned}$$

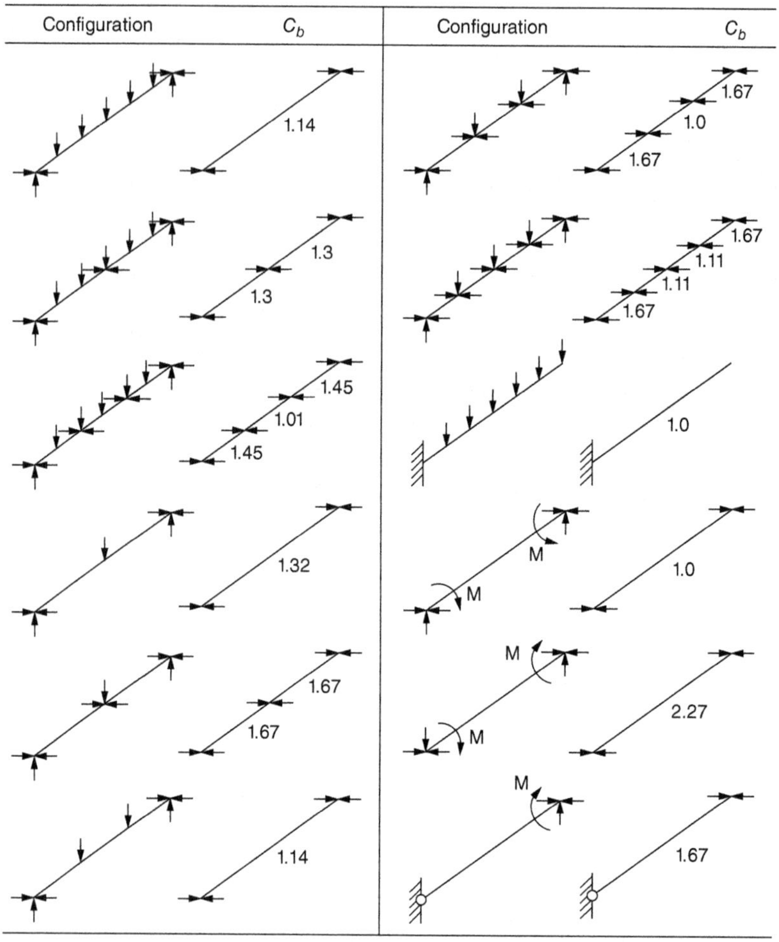


**TABLE 4.1** Typical Values of $C_b$

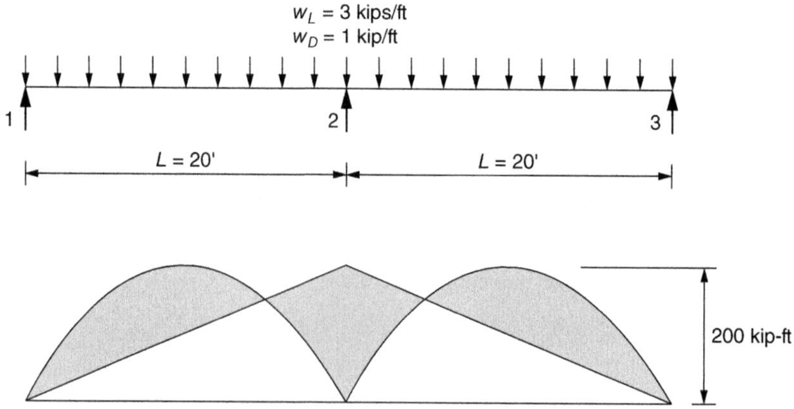


**FIGURE 4.9** Details for Example 4.6.

The free moment in span 12 is

$$M = wL^2/8$$
$$= 4 \times 20^2/8$$
$$= 200 \text{ kip-ft}$$

The fixing moment at support 2 is

$$M_2 = wL^2/8$$
$$= 4 \times 20^2/8$$
$$= 200 \text{ kip-ft}$$

The moment diagram is shown in Fig. 4.9 and for span 12

$$M_A = 0.75M - 0.25M_2$$
$$= 0.75 \times 200 - 0.25 \times 200$$
$$= 100 \text{ kip-ft}$$
$$M_B = M - 0.5M_2$$
$$= 200 - 0.5 \times 200$$
$$= 100 \text{ kip-ft}$$
$$M_C = 0.75M - 0.75M_2$$
$$= 0.75 \times 200 - 0.75 \times 200$$
$$= 0 \text{ kip-ft}$$
$$M_{max} = 0.07wL^2$$
$$= 0.07 \times 4 \times 20^2$$
$$= 112 \text{ kip-ft}$$

The modification factor $C_b$ is given by AISC 360 Eq. (F1-1) as

$$C_b = 12.5M_{max}/(2.5M_{max} + 3M_A + 4M_B + 3M_C)$$
$$= 12.5 \times 112/(2.5 \times 112 + 3 \times 100 + 4 \times 100 + 3 \times 0)$$
$$= 1.43$$

## 4.5 Lateral-Torsional Buckling

### Plastic Mode: $L_b \leq L_p$

The maximum nominal moment capacity of a compact rolled I-shape is $M_n = M_p$. As shown in Fig. 4.3, for a compact beam with $C_b = 1.0$, as the unbraced length increases beyond $L_p$, the nominal moment capacity decreases. When $L_b$ does not exceed the length $L_p$, full plasticity is possible and $L_p$ is given by AISC 360 Eq. (F2-5) as

$$L_p = 1.76r_y(E/F_y)^{0.5}$$

where $r_y$ is radius of gyration about the $y$-axis. Values of $L_p$ are given in AISC Manual Table 3-2 for beams with a yield stress of 50 ksi.

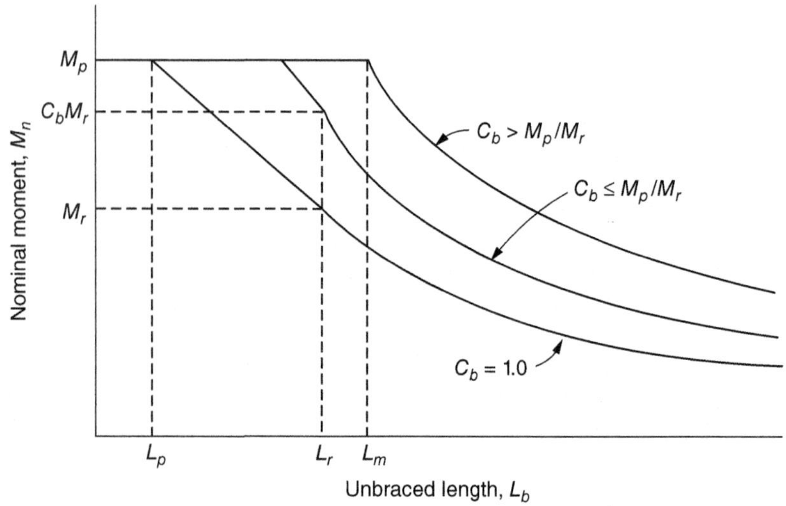


**FIGURE 4.10** Lateral-torsional buckling.

As shown in Fig. 4.10, for a beam with a value of $C_b$ greater than 1.0, the unbraced length for full plastic flexural strength is extended beyond $L_p$ to $L_m$. The increased length is

$$L_m = L_p + M_p(C_b - 1.0)(L_r - L_p)/C_b(M_p - M_r) \text{ ... for } L_m \le L_r$$
$$= L_p + \varphi_b M_p(C_b - 1.0)/C_b(\varphi_b BF) \text{ ... LRFD}$$
$$= L_p + M_p(C_b - 1.0)/C_b\Omega_b(BF/\Omega_b) \text{ ... ASD}$$

where $M_p = F_y Z_x$
$M_r = 0.7F_y S_x$
$(BF)$ = flexural strength factor
$= (M_p - M_r)/(L_r - L_p)$

$BF/\Omega_b$ and $\varphi_b BF$ are tabulated in AISC Manual Table 3-2.

**Example 4.7.** Beam Capacity with $L_b < L_p$

A W12 × 58 beam with a yield stress of 50 ksi is simply supported over a span of 8 ft and loaded with a uniformly distributed load. If the beam is laterally braced at the ends only, determine the flexural capacity.

From Table 4.1, the modification factor is

$$C_b = 1.14$$

The unbraced segment length is

$$L_b = 8 \text{ ft}$$

AISC Manual Table 3-2 indicates that the section is compact with

$$L_p = 8.87 \text{ ft}$$
$$> L_b$$

Hence, full plastic hinging occurs with $M_n = M_p$ and it is unnecessary to account for $C_b > 1.0$. From AISC Manual Table 3-2

| LRFD | ASD |
|---|---|
| The design flexural strength is | The allowable flexural strength is |
| $\varphi_b M_p = \varphi_b M_n$ | $M_n/\Omega_b = M_p/\Omega_b$ |
| $= 324$ kip-ft | $= 216$ kip-ft |

## Plastic Mode Extended: $L_p < L_b \leq L_m$

The limiting laterally unbraced length for the limit state of inelastic lateral-torsional buckling is given by AISC 360 Eq. (F2-6) as

$$L_r = 1.95 r_{ts} E\{Jc/S_x h_o + [(Jc/S_x h_o)^2 + 6.76(0.7F_y/E)^2]^{0.5}\}^{0.5}/0.7F_y$$

where $(r_{ts})^2 = (I_y C_w)^{0.5}/S_x$
$= I_y h_o/2S_x$ ... for doubly symmetric I-shapes with rectangular flanges
$C_w$ = warping constant
$= I_y h_o^2/4$ ... for doubly symmetric I-shapes with rectangular flanges
$S_x$ = section modulus about the major axis
$E$ = modulus of elasticity of steel
= 29,000 ksi
$J$ = torsional constant
$I_y$ = moment of inertia about the minor axis
$c = 1.0$ ... for doubly symmetric I-shapes
$h_o$ = distance between flange centroids

Or conservatively,

$$L_r = \pi r_{ts}(E/0.7F_y)^{0.5}$$

Values of $L_r$ are given in AISC Manual Table 3-2 for beams with a yield stress of 50 ksi.

The value of $C_b$ for which $L_m = L_r$ is given by

$$C_b = M_p/M_r$$
$$= Z_x/0.7S_x$$
$$= 1.57 \text{ ... for a shape factor of 1.1}$$

For values of $L_b \leq L_m$ the nominal flexural strength of the section is

$$M_n = M_p$$

**Example 4.8.** Beam Capacity with $L_p < L_b \leq L_m$
A W12 × 58 beam with a yield stress of 50 ksi is simply supported over a span of 12 ft and loaded with a uniformly distributed load. If the beam is laterally braced at the ends only, determine the flexural capacity.

From Table 4.1, the modification factor is

$$C_b = 1.14$$

The unbraced segment length is

$$L_b = 12 \text{ ft}$$

AISC Manual Table 3-2 indicates that the section is compact with

$$L_p = 8.87 \text{ ft}$$
$$< L_b$$

and

$$L_r = 29.8 \text{ ft}$$
$$> L_b$$

Allowing for the modification factor, the extended limiting laterally unbraced length for full plastic flexural strength is obtained as follows

| LRFD | ASD |
|---|---|
| From AISC Manual Table 3-2: | From AISC Manual Table 3-2: |
| $\varphi_b M_p = 324$ kip-ft | $M_p/\Omega_b = 216$ kip-ft |
| $\varphi_b M_r = 205$ kip-ft | $M_r/\Omega_b = 136$ kip-ft |
| $(\varphi_b BF) = 5.69$ kips | $(BF/\Omega_b) = 3.82$ kips |
| The unbraced length for full plastic yielding is | The unbraced length for full plastic yielding is |
| $L_m = L_p + \varphi_b M_p(C_b - 1.0)/C_b(\varphi_b BF)$ | $L_m = L_p + M_p(C_b - 1.0)/C_b\Omega_b(BF/\Omega_b)$ |
| $= 8.87 + 324(1.14 - 1.0)/1.14(5.69)$ | $= 8.87 + 216(1.14 - 1.0)/1.14(3.82)$ |
| $= 15.9$ ft | $= 15.8$ ft |
| $> 12$ ft | $> 12$ ft |
| Hence, full plastic yielding occurs and the design flexural strength is | Hence, full plastic hinging occurs and the allowable flexural strength is |
| $\varphi_b M_n = \varphi_b M_p$ | $M_n/\Omega_b = M_p/\Omega_b$ |
| $= 324$ kip-ft | $= 216$ kip-ft |

## Inelastic Mode: $L_p < L_b \leq L_r$

As shown in Fig. 4.3, for a compact beam with $C_b = 1.0$, as the unbraced length increases beyond $L_p$, the nominal moment capacity decreases linearly from a maximum value of $M_p$ to a value of $M_r$ at $L_b = L_r$. Values of $L_p$, $L_r$, $\varphi_b M_p$, $\varphi_b M_r$, $M_p/\Omega_b$, and $M_r/\Omega_b$ are given in AISC Manual Table 3-2 for beams with a yield stress of 50 ksi.

As shown in Fig. 4.10, for a beam with a value of $C_b$ greater than 1.0, the unbraced length for full plastic flexural strength is extended beyond $L_p$ to $L_m$. For a value of $C_b$ not exceeding $M_p/M_r$, the nominal moment in the inelastic region is obtained by multiplying the basic strength values by $C_b$. The maximum permitted value of the nominal moment is $M_n = M_p$. For a value of $C_b = M_p/M_r$, $M_n = M_p$. For a value of $C_b \geq M_p/M_r$, no inelastic region exists and the beam transitions directly from the yield failure mode to elastic lateral-torsional buckling. The nominal flexural strength of a beam in the inelastic range is determined by linear interpolation using AISC 360 Eq. (F2-2) which gives

$$M_n = C_b[M_p - (M_p - M_r)(L_b - L_p)/(L_r - L_p)]$$

The design flexural strength is

$$\varphi_b M_n = C_b[\varphi_b M_p - (\varphi_b BF)(L_b - L_p)]$$
$$\leq \varphi_b M_p$$

The allowable strength is

$$M_n/\Omega_b = C_b[M_p/\Omega_b - (BF/\Omega_b)(L_b - L_p)]$$
$$\leq M_p/\Omega_b$$

where $M_p = F_y Z_x$
$M_r = 0.7F_y S_x$
$(BF)$ = flexural strength factor
$= (M_p - M_r)/(L_r - L_p)$

$BF/\Omega_b$ and $\varphi_b BF$ are tabulated in AISC Manual Table 3-2.

**Example 4.9.** Beam Capacity with $L_p < L_b \leq L_r$
A W12 × 58 beam with a yield stress of 50 ksi is simply supported over a span of 24 ft and loaded with a uniformly distributed load. If the beam is laterally braced at the ends only, determine the flexural capacity.
From Table 4.1, the modification factor is

$$C_b = 1.14$$

The unbraced segment length is

$$L_b = 24 \text{ ft}$$

AISC Manual Table 3-2 indicates that the section is compact with

$$L_p = 8.87 \text{ ft}$$
$$< L_b$$

and

$$L_r = 29.8 \text{ ft}$$
$$> L_b$$

Allowing for the modification factor, the flexural capacity is obtained as follows

| LRFD | ASD |
|---|---|
| From AISC Manual Table 3-2: | From AISC Manual Table 3-2: |
| $\varphi_b M_p = 324$ kip-ft | $M_p/\Omega_b = 216$ kip-ft |
| $\varphi_b M_r = 205$ kip-ft | $M_r/\Omega_b = 136$ kip-ft |
| $(\varphi_b BF) = 5.69$ kips | $(BF/\Omega_b) = 3.82$ kips |
| From AISC 360 Eq. (F2-2), the design flexural strength is | From AISC 360 Eq. (F2-2), the allowable flexural strength is |
| $\varphi_b M_n = C_b[\varphi_b M_p - (\varphi_b BF)(L_b - L_p)]$ | $M_n/\Omega_b = C_b[M_p/\Omega_b - (BF/\Omega_b)(L_b - L_p)]$ |
| $= 1.14[324 - (5.69)(24 - 8.87)]$ | $= 1.14[216 - (3.82)(24 - 8.87)]$ |
| = 271 kip-ft ... satisfactory | = 180 kip-ft ... satisfactory |
| $< \varphi_b M_p$ | $< M_p/\Omega_b$ |
| Hence, full plasticity can not be achieved before inelastic buckling occurs and the design flexural strength is | Hence, full plasticity can not be achieved before inelastic buckling occurs and the allowable flexural strength is |
| $\varphi_b M_n = 271$ kip-ft | $M_n/\Omega_b = 180$ kip-ft |

### Elastic Mode: $L_b > L_r$

As shown in Fig. 4.10, for a compact beam with $C_b = 1.0$, when the unbraced length exceeds $L_r$, the beam failure mode transitions from inelastic lateral-torsional buckling to elastic lateral-torsional buckling. The nominal moment strength is then equal to the critical elastic moment given by AISC 360 Eq. (F2-3) as

$$M_n = F_{cr}S_x$$
$$\leq M_p$$

The critical stress is given by LRFD Eq. (F2-4) as

$$F_{cr} = C_b\pi^2E[1 + 0.078Jc(L_b/r_{ts})^2/S_xh_o]^{0.5}/(L_b/r_{ts})^2$$

where $C_b$ = modification factor
$r_{ts}^2 = (I_yC_w)^{0.5}/S_x$
$= I_yh_o/2S_x$ ... for doubly symmetric I-shapes with rectangular flanges
$C_w$ = warping constant
$= I_yh_o^2/4$ ... for doubly symmetric I-shapes with rectangular flanges
$S_x$ = section modulus about the major axis
$E$ = modulus of elasticity of steel
= 29,000 ksi
$J$ = torsional constant
$I_y$ = moment of inertia about the minor axis
$c$ = 1.0 ... for doubly symmetric I-shapes
$= h_o(I_y/4C_w)^{0.5}$ ... for a channel
$h_o$ = distance between flange centroids

Values of $\varphi_bM_n$, and $M_n/\Omega_b$ are graphed in AISC Manual Table 3-10 for W-shapes with a yield stress of 50 ksi and a value of $C_b = 1.0$.

As shown in Fig. 4.10, for a beam with a value of $C_b$ greater than 1.0, the unbraced length for full plastic flexural strength is extended beyond $L_p$ to $L_m$. For a value of $C_b = M_p/M_r$, the extended unbraced length for full plastic yielding is $L_m = L_r$, no inelastic region exists and the beam transitions directly from the yield failure mode to elastic lateral-torsional buckling. For a beam with $C_b > 1.0$, the values of $\varphi_bM_n$, and $M_n/\Omega_b$ obtained from AISC Manual Table 3-10 are multiplied by $C_b$ to provide the adjusted values, with a value not exceeding $\varphi_bM_p$, or $M_p/\Omega_b$ as appropriate.

**Example 4.10.** Beam Capacity with $L_b > L_r$

A W12 × 58 beam with a yield stress of 50 ksi is simply supported over a span of 32 ft and loaded with a uniformly distributed load. If the beam is laterally braced at the ends only, determine the flexural capacity.

From Table 4.1, the modification factor is

$$C_b = 1.14$$

The unbraced segment length is

$$L_b = 32 \text{ ft}$$

AISC Manual Table 3-2 indicates that the section is compact with

$$L_p = 8.87 \text{ ft}$$

and

$$L_r = 29.8 \text{ ft}$$
$$< L_b$$

Allowing for the modification factor, the flexural capacity is obtained as follows

| LRFD | ASD |
|---|---|
| From AISC Manual Table 3-10 for $C_b = 1.0$ and $L_b = 32$ ft: | From AISC Manual Table 3-10 for $C_b = 1.0$ and $L_b = 32$ ft: |
| $\varphi_b M_n = 188$ kip-ft ... elastic lateral-torsional buckling governs | $M_n/\Omega_b = 125$ kip-ft ... elastic lateral-torsional buckling governs |
| The adjusted value for $C_b = 1.14$ is | The adjusted value for $C_b = 1.14$ is |
| $\varphi_b M_n = 1.14 \times 188$ | $M_n/\Omega_b = 1.14 \times 125$ |
| $= 214$ kip-ft | $= 143$ kip-ft |
| The design flexural strength is | The allowable flexural strength is |
| $\varphi_b M_n = 214$ kip-ft ... satisfactory | $M_n/\Omega_b = 143$ kip-ft ... satisfactory |
| $< \varphi_b M_p$ | $< M_p/\Omega_b$ |

## 4.6 Weak Axis Bending

In accordance with AISC 360 Commentary Sec. F6, I-shaped members and channels bent about the minor axis are not subject to lateral torsional instability or local web buckling. The only limit states to consider are yielding and flange local buckling. Lateral bracing is not required and consideration of $C_b$ is unnecessary.

### Compact Flanges

All rolled I-shapes have compact flanges for a yield stress of $F_y = 50$ ksi with the exception of

W21 × 48, W14 × 99, W14 × 90, W12 × 65, W10 × 12, W8 × 31, W8 × 10, W6 × 15, W6 × 9, W6 × 8.5, and M4 × 6.

The nominal flexural strength of a compact section bent about the minor axis is given by AISC 360 Eq. (F6-1) as

$$M_n = M_{py}$$
$$= F_y Z_y$$
$$\leq 1.6 F_y S_y$$

where $Z_y$ = plastic section modulus referred to the minor axis
$S_y$ = elastic section modulus referred to the minor axis
$M_{py}$ = plastic moment of resistance referred to the minor axis
$Z_y/S_y$ = shape factor for minor axis bending
= 1.6 ... average

Tabulated values of $\varphi_b M_{ny}$ and $M_{ny}/\Omega_b$, are given in AISC Manual Table 3-4.

**Example 4.11.** Weak Axis Bending
A W12 × 58 beam with a yield stress of 50 ksi is simply supported over a span of 10 ft and is loaded in its minor axis with a uniformly distributed load. Determine the flexural capacity.
AISC Manual Table 3-4 indicates that the section is compact with

$$Z_y = 32.5 \text{ in}^3$$

AISC Manual Table 1-1 indicates that

$$S_y = 21.4 \text{ in}^3$$

The nominal flexural strength of a compact section bent about the minor axis is given by AISC 360 Eq. (F6-1) as

$$\begin{aligned} M_n &= M_{py} \\ &= F_y Z_y \\ &= 50 \times 32.5/12 \\ &= 135 \text{ kip-ft ... governs} \\ 1.6F_y S_y &= 1.6 \times 21.4 \times 50/12 \\ &= 143 \text{ kip-ft} \\ &> 135 \text{ kip-ft} \end{aligned}$$

| LRFD | ASD |
|---|---|
| From AISC 360 Sec. F1, the design flexural strength is | From AISC 360 Sec. F1, the allowable flexural strength is |
| $\varphi_b M_n = 0.9 \times 135$ | $M_n/\Omega_b = 135/1.67$ |
| $= 122$ kip-ft | $= 81$ kip-ft |
| The design flexural strength is also given directly in AISC 360 Table 3-4 as | The allowable flexural strength is also given directly in AISC 360 Table 3-4 as |
| $\varphi_b M_{ny} = 122$ kip-ft | $M_{ny}/\Omega_b = 81.1$ kip-ft |

## Noncompact Flanges

A section with noncompact flanges is one that can develop a nominal flexural strength $M_n$ given by

$$0.7F_y S_y \leq M_n < M_{py}$$

The criteria for determining noncompactness in the flange of rolled beams is defined in AISC 360 Table B4.1b as

$$0.38(E/F_y)^{0.5} < b_f/2t_f < \lambda_{rf} = 1.0(E/F_y)^{0.5}$$

where $\lambda_{rf}$ is limiting slenderness parameter for a noncompact flange. The nominal flexural strength is given by AISC 360 Eq. (F6-2) as

$$M_n = M_{py} - (M_{py} - 0.7F_y S_y)(\lambda - \lambda_{pf})/(\lambda_{rf} - \lambda_{pf})$$

where $M_{py} = F_y Z_y$
$Z_y$ = plastic section modulus referred to the minor axis
$S_y$ = elastic section modulus referred to the minor axis
$\lambda_{pf} = 0.38(E/F_y)^{0.5}$
$\lambda_{rf} = 1.0(E/F_y)^{0.5}$
$\lambda = b_f/2t_f$

Tabulated values of $\varphi_b M_p$ and $M_p/\Omega_b$, in AISC Manual Table 3-4 allow for any reduction because of noncompactness.

**Example 4.12.** Weak Axis Bending: Noncompact Flange
A W10 × 12 beam with a yield stress of 50 ksi is simply supported over a span of 10 ft and is loaded in its minor axis with a uniformly distributed load. Determine the flexural capacity.
From Example 4.4, the relevant slenderness parameters for a W10 × 12 are

$$b_f/2t_f = 9.43$$
$$= \lambda$$
$$\lambda_{pf} = 9.15 \text{ ... flange is noncompact}$$

From AISC 360 Table 1-1

$$S_y = \text{elastic section modulus referred to the minor axis}$$
$$= 1.1$$
$$0.7F_yS_y = 0.7 \times 50 \times 1.1/12$$
$$= 3.21 \text{ kip-ft}$$

The plastic moment of resistance referred to the minor axis is given by

$$M_{py} = F_yZ_y$$
$$= 50 \times 1.74/12$$
$$= 7.25 \text{ kip-ft}$$

From AISC 360 Table B4.1b, the limiting slenderness parameter for a noncompact flange is

$$\lambda_{rf} = 1.0(E/F_y)^{0.5}$$
$$= 1 \times (29{,}000/50)^{0.5}$$
$$= 24.1$$

From AISC 360 Eq. (F6-2), the nominal flexural strength is given by

$$M_n = M_{py} - (M_{py} - 0.7F_yS_y)(\lambda - \lambda_{pf})/(\lambda_{rf} - \lambda_{pf})$$
$$= 7.25 - (7.25 - 3.21)(9.43 - 9.15)/(24.1 - 9.15)$$
$$= 7.17 \text{ kip-ft ... governs}$$
$$1.6F_yS_y = 1.6 \times 1.1 \times 50/12$$
$$= 7.33 \text{ kip-ft}$$
$$> 7.17 \text{ kip-ft}$$

| LRFD | ASD |
|---|---|
| The design flexural strength is | The allowable flexural strength is |
| $\varphi_b M_n = 0.9 \times 7.17$ | $M_n/\Omega_b = 7.17/1.67$ |
| $= 6.46$ kip-ft | $= 4.29$ kip-ft |
| The design flexural strength is also given directly in AISC 360 Table 3-4 as | The allowable flexural strength is also given directly in AISC 360 Table 3-4 as |
| $\varphi_b M_{ny} = 6.46$ kip-ft | $M_{ny}/\Omega_b = 4.30$ kip-ft |

## 4.7 Biaxial Bending

Biaxial bending is produced in a member when bending moments are applied simultaneously about both principal axes. In practice, biaxial bending occurs in overhead crane runway girders, roof purlins, and side sheeting rails. When it can be assumed that the applied loads about both principal axes act through the shear center, twisting effects do not have to be considered. For this situation, AISC 360 Commentary. Sec. H1.1 states that biaxial bending may be considered a special case of AISC 360 Eq. (H1-1b) with the axial load term equated to zero. The interaction expression then reduces to

$$M_{rx}/M_{cx} + M_{ry}/M_{cy} \leq 1.00$$

where $M_{rx}$ = required flexural strength about the $x$-axis
$M_{cx}$ = available flexural strength about the $x$-axis
$M_{ry}$ = required flexural strength about the $y$-axis
$M_{cy}$ = available flexural strength about the $y$-axis

For the LRFD method

$$M_c = \varphi_b M_n$$

For the ASD method

$$M_c = M_n/\Omega_b$$

When the load applied to the weak axis is significantly smaller than that applied to the strong axis, an I-shape is a suitable choice for the loaded member. When both loads are comparable, a box section is more suitable.

In most cases, the lateral load is applied to the compression flange of the member as shown in Fig. 4.11*a*. In this situation, the design method proposed by Fisher,[7] is appropriate. It is assumed that the lateral load is resisted solely by the compression flange and torsional effects are neglected. For I-shapes, the plastic section modulus of one flange about the $y$-axis, is given by

$$Z_t = Z_y/2$$

where $Z_y$ is plastic section modulus of the full section about the $y$-axis. The nominal moment capacity of one flange about the $y$-axis is

$$M_{nt} = F_y Z_t$$

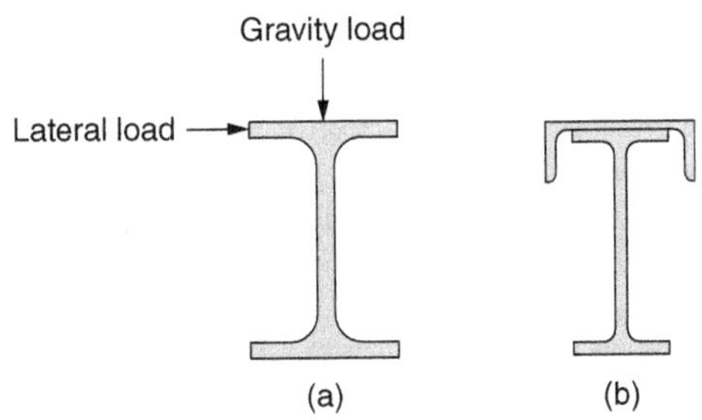


**Figure 4.11** Laterally loaded members.

The design procedure for this technique then consists of applying the modified interaction equation

$$M_{rx}/M_{cx} + M_{ry}/M_{ct} \le 1.00$$

where $M_{ct} = \varphi_b M_{nt}$ ... for the LRFD method
$M_{ct} = M_{nt}/\Omega_b$ ... for the ASD method

To increase the lateral capacity of the compression flange, a built-up section may be used as shown in Fig. 4.11*b* which consists of a W-shape with cap channel. However, using a larger W section without the cap is usually more economical.

## Overhead Traveling Bridge Crane

An overhead bridge crane is shown in Fig. 4.12, reproduced by permission of J. Herbert Corporation. Typically, the crane consists of a bridge girder mounted on end trucks

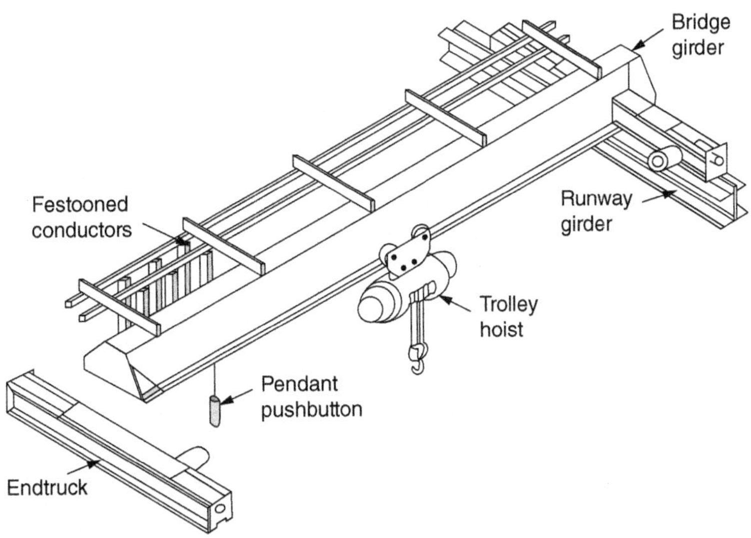


**Figure 4.12** Overhead traveling bridge crane.

with the bridge girder spanning between parallel runway girders. Double bridge girders may also be used and this has the advantage of bigger capacity and better hook height. The bridge girder traverses longitudinally along the runway on the end trucks, each of which runs on two wheels. The bridge girder supports a trolley that carries the hoist. The trolley runs along the bridge girder in a direction perpendicular to the runway girders and the hoist lifts the load vertically. Hence, the load may be moved in any direction.

Figure 4.12 shows a bridge crane consisting of a single bridge girder with end trucks running on a rail mounted on top of the runway girders. Alternatively, the wheels of the trucks may also run directly on the top flange of the runway girders. The trolley runs on the bottom flange of the bridge girder and the hoist and trolley are electric powered and pendant operated from the ground. Alternatively, the hoist may be controlled from a cab mounted on the bridge girder.

Because of the acceleration and deceleration of the trolley on the bridge girder, lateral forces are applied to the runway girders. For electrically powered trolleys, ASCE 7 Sec. 4.9.4 specifies that the lateral force is 20 percent of the sum of the rated capacity of the crane, the weight of the hoist, and weight of the trolley.

The maximum truck wheel loads are obtained as the sum of the applicable weights of the bridge and trucks, the rated capacity of the crane, and the weight of the hoist and trolley. ASCE 7 Sec. 4.9.3 specifies that these wheel loads must be increased by the following percentages to allow for impact and vibration:

- 25 percent for cab-operated powered bridge cranes
- 10 percent for pendant-operated powered bridge cranes
- 0 percent for hand geared bridge cranes

Because of the acceleration and deceleration of the trucks on the runway girders, longitudinal forces are applied to the runway girders. For electrically powered bridges, ASCE 7 Sec. 4.9.5 specifies that the longitudinal force is 10 percent of the maximum wheel loads.

**Example 4.13.** Crane Runway Girder Design
A bridge crane with a rated capacity of 20 kips is electrically powered and pendant operated. The relevant details are

Weight of hoist and trolley = 9 kips
Weight of bridge girder and trucks = 30 kips
Maximum wheel load due to lifted load, hoist, and trolley = 14 kips

The truck wheels are at 6 ft on centers and run directly on the top flange of W24 × 104 runway girders that span 30 ft. Determine if the W24 × 104 girders are satisfactory. Ignore the longitudinal force on the girders.

**Vertical load**
Allowing for impact, the wheel load due to the bridge beam and end trucks is

$$
\begin{aligned}
W_D &= 1.1 \times 30/4 \\
&= 8.25 \text{ kips}
\end{aligned}
$$

Allowing for impact, the wheel load due to the lifted load, hoist, and trolley is

$$W_L = 1.1 \times 14$$
$$= 15.4 \text{ kips}$$

| LRFD | ASD |
|---|---|
| From ASCE 7 Sec. 2.3.2 combination 2: | From ASCE 7 Sec. 2.4.1 combination 2: |
| $W_u$ = factored vertical load | $W_a$ = factored vertical load |
| $= 1.2W_D + 1.6W_L$ | $= W_D + W_L$ |
| $= 1.2 \times 8.25 + 1.6 \times 15.4$ | $= 8.25 + 15.4$ |
| $= 34.54$ kips/wheel | $= 23.65$ kips/wheel |
| From AISC 360 Table 3-23, Item 44, the maximum factored moment is | From AISC 360 Table 3-23, Item 44, the maximum factored moment is |
| $M_u = W_u(L - a/2)^2/2L + 1.2w_{girder}L^2/8$ | $M_a = W_a(L - a/2)^2/2L + w_{girder}L^2/8$ |
| $= 34.54(30 - 6/2)^2/(2 \times 30)$ | $= 23.65(30 - 6/2)^2/(2 \times 30)$ |
| $+ 1.2 \times 0.104 \times 30^2/8$ | $+ 0.104 \times 30^2/8$ |
| $= 434$ kip-ft | $= 299$ kip-ft |

**Strong axis bending**

The unbraced segment length is

$$L_b = 30 \text{ ft}$$

AISC Manual Table 3-2 indicates that for a value of $C_b = 1.0$, the section is compact with

$$L_p = 10.3 \text{ ft}$$
$$< L_b$$
$$L_r = 29.2 \text{ ft}$$
$$< L_b \text{ ... elastic lateral-torsional buckling governs}$$

| LRFD | ASD |
|---|---|
| From AISC Manual Table 3-2: | From AISC Manual Table 3-2: |
| $\varphi_b M_p = 1080$ kip-ft | $M_p/\Omega_b = 721$ kip-ft |
| $\varphi_b M_r = 677$ kip-ft | $M_r/\Omega_b = 451$ kip-ft |
| From AISC Manual Table 3-10 for $C_b = 1.0$ and $L_b = 30$ ft: | From AISC Manual Table 3-10 for $C_b = 1.0$ and $L_b = 30$ ft: |
| $\varphi_b M_n = 649$ kip-ft ... satisfactory | $M_n/\Omega_b = 433$ kip-ft ... satisfactory |
| $> 433$ kip-ft | $> 299$ kip-ft |

**Lateral load**

The lateral load per wheel due to the lifted load, hoist, and trolley is

$$W_h = 0.2(20 + 9)/4$$
$$= 1.45 \text{ kips}$$

| LRFD | ASD |
|---|---|
| From ASCE 7 Sec. 2.3.2 combination 2: $W_{hu}$ = factored lateral load $= 1.6W_h$ $= 1.6 \times 1.45$ $= 2.32$ kips/wheel | From ASCE 7 Sec. 2.4.1 combination 2: $W_{ha}$ = factored lateral load $= W_h$ $= 1.45$ kips/wheel |
| From AISC 360 Table 3-23, Item 44, the maximum factored moment is $M_{hu} = W_{hu}(L - a/2)^2/2L$ $= 2.32(30 - 6/2)^2/(2 \times 30)$ $= 28$ kip-ft | From AISC 360 Table 3-23, Item 44, the maximum factored moment is $M_{ha} = W_{ha}(L - a/2)^2/2L$ $= 1.45(30 - 6/2)^2/(2 \times 30)$ $= 18$ kip-ft |

**Weak axis bending**

| LRFD | ASD |
|---|---|
| From AISC Manual Table 3-4: $\varphi_b M_{py} = 234$ kip-ft | From AISC Manual Table 3-4: $M_{py}/\Omega_b = 156$ kip-ft |
| For the compression flange, the design flexural capacity is $\varphi_b M_{nt} = \varphi_b M_{py}/2$ $= 234/2$ $= 117$ kip-ft ... satisfactory $> 28$ kip-ft | For the compression flange, the allowable flexural capacity is $M_{nt}/\Omega_b = M_{py}/2\Omega_b$ $= 156/2$ $= 78$ kip-ft ... satisfactory $> 18$ kip-ft |

**Biaxial bending**

| LRFD | ASD |
|---|---|
| Applying AISC 360 Eq. (H1-1b) with the axial load term equated to zero, the interaction expression is $M_u/\varphi_b M_n + M_{hu}/\varphi_b M_{nt} \leq 1.00$ $433/649 + 28/117 = 0.91$ $< 1.0$ ... satisfactory | Applying AISC 360 Eq. (H1-1b) with the axial load term equated to zero, the interaction expression is $M_a/(M_n/\Omega_b) + M_{ha}/(M_{nt}/\Omega_b) \leq 1.00$ $299/433 + 18/78 = 0.92$ $< 1.0$ ... satisfactory |

The W24 × 104 section is adequate.

## 4.8 Singly Symmetric Sections in Bending

Examples of singly symmetric sections loaded in the plane of symmetry, such as tees and double angles, are shown in Fig. 4.13.

**FIGURE 4.13** Tees and double angles.

The three modes of collapse of these sections are plastic mode, lateral-torsional buckling, and flange local buckling. In accordance with AISC 360 Commentary Sec. F9, the $C_b$ factor is conservatively taken as 1.0.

## Plastic Mode

When the section is adequately braced, failure occurs by yielding over the full section and the nominal flexural strength is given by AISC 360 Eq. (F9-1) as

$$M_n = M_p$$

The plastic moment of resistance, for stems in tension, is given by AISC 360 Eq. (F9-2) as

$$M_p = F_y Z_x$$
$$\leq 1.6M_y$$

where $Z_x$ = plastic section modulus about the $x$-axis
$M_y$ = yield moment
$= F_y S_x$
$S_x$ = elastic section modulus about the $x$-axis

The plastic moment of resistance, for stems in compression, is given by AISC 360 Eq. (F9-3) as

$$M_p = F_y Z_x$$
$$\leq M_y$$

## Lateral-Torsional Buckling

When lateral-torsional buckling governs, the nominal flexural strength is given by AISC 360 Eq. (F9-4) as

$$M_n = M_{cr}$$
$$= \pi(EI_y GJ)^{0.5}[B + (1 + B^2)^{0.5}]/L_b$$

where $B = 2.3(d/L_b)(I_y/J)^{0.5}$ ... stem in tension
$B = -2.3(d/L_b)(I_y/J)^{0.5}$ ... stem in compression

## Flange Local Buckling

For sections with a compact flange in flexural compression, the limit state of flange local buckling does not apply.

For sections with a noncompact flange in flexural compression, the nominal strength is given by AISC 360 Eq. (F9-6) as

$$M_n = M_p - (M_p - 0.7F_y S_{xc})(\lambda - \lambda_{pf})/(\lambda_{rf} - \lambda_{pf})$$
$$\leq 1.6M_y$$

where $M_p = F_y Z_x$
$S_{xc}$ = elastic section modulus referred to the compression flange
$\lambda_{pf} = 0.38(E/F_y)^{0.5}$
$\lambda_{rf} = 1.0(E/F_y)^{0.5}$
$\lambda = b_f/2t_f$

For sections with a slender flange in flexural compression, the nominal strength is given by AISC 360 Eq. (F9-7) as

$$M_n = 0.7ES_{xc}/(b_f/2t_f)^2$$

## Stem Local Buckling

For sections with a stem in flexural compression, the nominal strength is given by AISC 360 Eq. (F9-8) as

$$M_n = F_{cr} S_x$$

where $F_{cr}$ = critical stress
$= F_y$ ... AISC 360 Eq. (F9-9) when $d/t_w \leq 0.84(E/F_y)^{0.5}$
$= [2.55 - 1.84(d/t_w)(F_y/E)^{0.5}]F_y$ ... AISC 360 Eq. (F9-10) when $0.84(E/F_y)^{0.5} < d/t_w \leq 1.03(E/F_y)^{0.5}$
$= 0.69E/(d/t_w)^2$ ... AISC 360 Eq. (F9-11) when $d/t_w > 1.03(E/F_y)^{0.5}$

**Example 4.14.** WT Section in Flexure
Determine the lightest WT8 section suitable for the beam shown in Fig. 4.14. The loading consists of a uniformly distributed dead load of $w_D$ = of 0.5 kip/ft, which includes the weight of the tee, and a uniformly distributed live load of $w_L$ = of 1.0 kip/ft. The stem of the tee is in tension. The beam is continuously braced on its compression flange and has a yield stress of $F_y$ = 50 ksi.

Since the beam is continuously braced, lateral-torsional buckling does not govern. The applied loads are

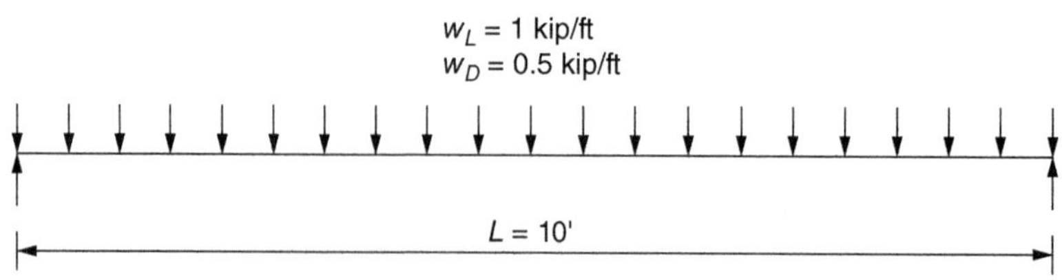


**FIGURE 4.14** Details for Example 4.14.

| LRFD | ASD |
|---|---|
| From ASCE 7 Sec. 2.3.2 combination 2: | From ASCE 7 Sec. 2.4.1 combination 2: |
| $w_u$ = factored load | $w_a$ = factored load |
| $= 1.2w_D + 1.6w_L$ | $= w_D + w_L$ |
| $= 1.2 \times 0.5 + 1.6 \times 1.0$ | $= 0.5 + 1.0$ |
| = 2.2 kips/ft | = 1.5 kips/ft |
| $M_u$ = factored moment | $M_a$ = factored moment |
| $= w_u L^2/8$ | $= w_a L^2/8$ |
| $= 2.2 \times 10^2/8$ | $= 1.5 \times 10^2/8$ |
| = 27.5 kip-ft | = 18.75 kip-ft |
| = required strength | = required strength |

The required plastic section modulus is

| LRFD | ASD |
|---|---|
| $Z_x = M_u/\varphi_b F_y$ | $Z_x = \Omega_b M_a/F_y$ |
| $= 27.5 \times 12/(0.9 \times 50)$ | $= 1.67 \times 18.75 \times 12/50$ |
| $= 7.33\ \text{in}^3$ | $= 7.52\ \text{in}^3$ |
| From AISC Manual Table 1-8, a WT8 × 13 provides a plastic section modulus of | From AISC Manual Table 1-8, a WT8 × 15.5 provides a plastic section modulus of |
| $Z_x = 7.36\ \text{in}^3$ | $Z_x = 8.27\ \text{in}^3$ |
| $> 7.33\ \text{in}^3$ | $> 7.52\ \text{in}^3$ |
| The flange is compact for flexure and flange local buckling does not govern. Hence, the WT8 × 13 section is satisfactory. | The flange is compact for flexure and flange local buckling does not govern. Hence, the WT8 × 15.5 section is satisfactory. |

## 4.9 Redistribution of Bending Moments in Continuous Beams

Redistribution of bending moments occurs in continuous beams and frames after plastic hinges have formed. In an indeterminate structure, the formation of a single hinge in the structure does not cause collapse of the structure. The structure can continue to support increasing load while the moment at the hinge remains constant at a value of $M_p$ and the moments at other locations in the structure continue to increase. This process is utilized in AISC 360 Sec. B3.7.

In accordance with AISC 360 Sec. B3.7, negative moments at supports, produced by gravity loads computed by an elastic analysis, may be reduced by 10 percent provided that span moments are increased by 10 percent of the average adjacent support moments. Moments at cantilevers may not be reduced. The same principle may be applied to beam-columns, provided that the axial force in the column does not exceed $0.15\varphi_c F_y A_g$ for LRFD or $0.15F_y A_g/\Omega_c$ for ASD.

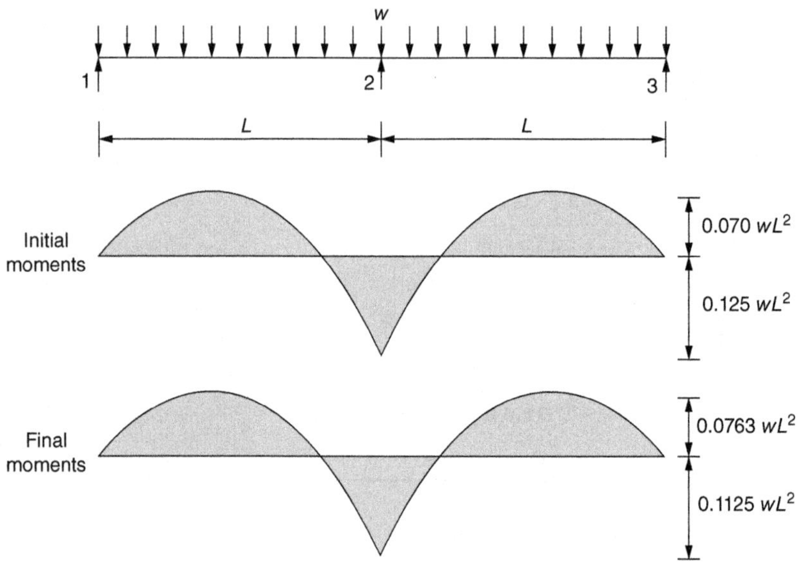


**Figure 4.15** Redistribution of moments.

The technique is illustrated in Fig. 4.15. The bending moment at the central support of the two span beam is

$$M_2 = 0.125wL^2$$

The bending moment in the adjacent spans is

$$M_{12} = 0.070wL^2$$
$$= M_{23}$$

Reducing the support moment by 10 percent gives a revised moment of

$$M_{R2} = 0.9 \times 0.125wL^2$$
$$= 0.1125wL^2$$

Increasing the span moments by 10 percent of the average adjacent support moments gives

$$M_{12} = (0.070 + 0.0125/2)wL^2$$
$$= 0.0763wL^2$$
$$= M_{23}$$

The technique may be applied only to compact sections, as noncompact sections have inadequate plastic hinge rotation capacity to permit redistribution of moments.

**Example 4.15.** Moment Redistribution
The uniform distributed loading, including the beam self-weight, acting on a two span continuous beam is shown in Fig. 4.16. Continuous lateral support is provided to the beam. Determine the lightest adequate W10 section, using steel with a yield stress of 50 ksi.

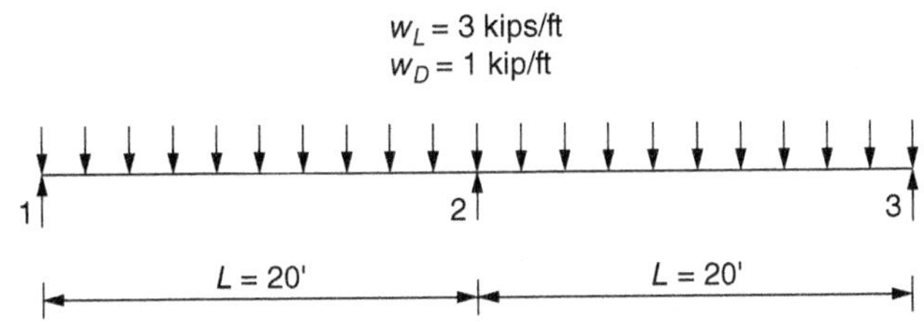


**FIGURE 4.16** Details for Example 4.15.

The beam is continuously braced and $C_b = 1.0$. The applied loads are

| LRFD | ASD |
|---|---|
| From ASCE 7 Sec. 2.3.2 combination 2: | From ASCE 7 Sec. 2.4.1 combination 2: |
| $w_u$ = factored load | $w_a$ = factored load |
| $= 1.2w_D + 1.6w_L$ | $= w_D + w_L$ |
| $= 1.2 \times 1.0 + 1.6 \times 3.0$ | $= 1.0 + 3.0$ |
| $= 6$ kips/ft | $= 4$ kips/ft |
| The corresponding moments are | The corresponding moments are |
| $M_{u2}$ = factored moment | $M_{a2}$ = factored moment |
| $= 0.125w_uL^2$ | $= 0.125w_aL^2$ |
| $= 0.125 \times 6 \times 20^2$ | $= 0.125 \times 4 \times 20^2$ |
| $= 300$ kip-ft | $= 200$ kip-ft |
| $M_{u12} = 0.070w_uL^2$ | $M_{a12} = 0.070w_uL^2$ |
| $= 0.070 \times 6 \times 20^2$ | $= 0.070 \times 4 \times 20^2$ |
| $= 168$ kip-ft | $= 112$ kip-ft |
| $= M_{u23}$ | $= M_{a23}$ |

After redistribution, the revised moments are

| LRFD | ASD |
|---|---|
| $M_{Ru2} = 0.9 \times 300$ | $M_{Ra2} = 0.9 \times 200$ |
| $= 270$ kip-ft ... governs | $= 180$ kip-ft ... governs |
| $M_{Ru12} = 168 + 30/2$ | $M_{Ra12} = 112 + 20/2$ |
| $= 183$ kip-ft | $= 122$ kip-ft |
| $= M_{Ru23}$ | $= M_{Ra23}$ |
| $< 270$ kip-ft | $< 180$ kip-ft |
| = required strength | = required strength |
| From AISC Manual Table 3-2, for a value of $C_b = 1.0$, a W10 × 60 is compact and provides a design flexural strength of | From AISC Manual Table 3-2, for a value of $C_b = 1.0$, a W10 × 60 is compact and provides an allowable flexural strength of |
| $\varphi_b M_p = 280$ kip-ft ... satisfactory | $M_p/\Omega_b = 186$ kip-ft ... satisfactory |
| $> 270$ kip-ft | $> 180$ kip-ft |

## 4.10 Deflection Limits

As stated in Sec. 2.11, the customary deflection limits recommended in AISC 360 Commentary Sec. L3 are 1/360 of the span for floors subjected to reduced live load and 1/240 of the span for roof members.

**Example 4.16.** Beam Deflection

The live load deflection limit imposed on the beam designed in Example 4.1 is 1/240 of the span. A W10 × 68 beam was selected for moment capacity. Determine if this beam satisfies the deflection limitation.

Deflection is calculated at the service load level. For a live load of $w_L = 3$ kips/ft and a span of $L = 20$ ft, the required moment of inertia is

$$I = 240 \times 5\, w_L L^3/384E$$
$$= 240 \times 5(3/12)(20 \times 12)^3/(384 \times 29{,}000)$$
$$= 372 \text{ in}^4$$

The W10 × 68 beam has a moment of inertia of 394 $\text{in}^4$ and is adequate.

## References

1. White, D. W. and Chang C. J. 2007. "Improved Flexural Stability Design of I-Section Members in AISC (2005)—a Case Study Comparison to AISC (1989) ASD," *Engineering Journal*, 4th Quarter 2007.
2. American Institute of Steel Construction (AISC). 2010. *Specification for Structural Steel Buildings* (AISC 360-10), AISC, Chicago, IL.
3. American Institute of Steel Construction (AISC). 2011. *Steel Construction Manual*, 14th edition. AISC, Chicago, IL.
4. American Society of Civil Engineers (ASCE). 2010. *Minimum Design Loads for Buildings and Other Structures* (ASCE 7-10), ASCE, Reston, VA.
5. Zoruba, S. and Dekker, B. 2005. "A Historical and Technical Overview of the $C_b$ Coefficient in the AISC Specification," *Engineering Journal*, 3rd Quarter 2005.
6. Aminmansour, A. 2009. "Technical Note: Optimum Flexural Design of Steel Members Utilizing Moment Gradient and $C_b$," *Engineering Journal*, 1st Quarter 2009.
7. Fisher, J. M. 2005. *Industrial Buildings, Roofs to Anchor Rods.* Design Guide No. 7, 2nd Edition. AISC, Chicago, IL.

## Problems

**4.1** *Given:* Figure 4.17 shows a simply supported beam spanning 30 ft with concentrated loads at the third points of the span. Each load consists of a dead load component of $W_D = 5$ kips, which includes an allowance for the weight of the beam, and a live load component of $W_L = 15$ kips. The beam is continuously braced on its compression flange and has a yield stress of $F_y = 50$ ksi.

*Find:* Using allowable stress level (ASD) load combinations, the lightest W18 section with adequate flexural capacity.

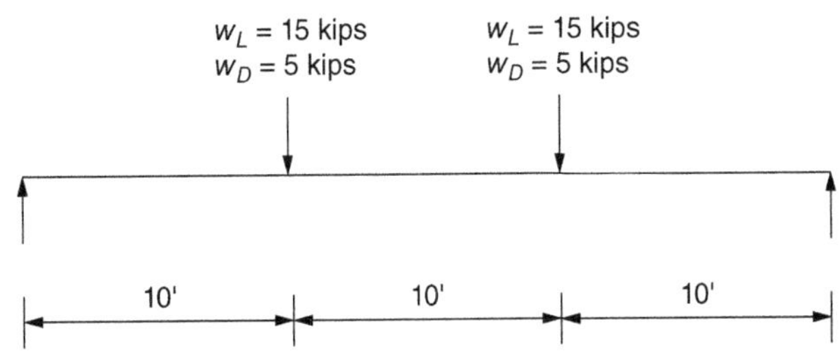


**FIGURE 4.17** Details for Problems 4.1 to 4.7.

**4.2** *Given:* Figure 4.17 shows a simply supported beam spanning 30 ft with concentrated loads at the third points of the span. Each load consists of a dead load component of $W_D = 5$ kips, which includes an allowance for the weight of the beam, and a live load component of $W_L = 15$ kips. The beam is continuously braced on its compression flange and has a yield stress of $F_y = 50$ ksi.

*Find:* The lightest W18 section that will limit the deflection to 1/240 of the span.

**4.3** *Given:* Figure 4.17 shows a simply supported beam spanning 30 ft with concentrated loads at the third points of the span. Each load consists of a dead load component of $W_D = 5$ kips, which includes an allowance for the weight of the beam, and a live load component of $W_L = 15$ kips. The beam is continuously braced on its compression flange and has a yield stress of $F_y = 50$ ksi.

*Find:* Using strength level (LRFD) load combinations, the lightest W18 section with adequate flexural capacity.

**4.4** *Given:* Figure 4.17 shows a simply supported beam spanning 30 ft with concentrated loads at the third points of the span. Each load consists of a dead load component of $W_D = 5$ kips, which includes an allowance for the weight of the beam, and a live load component of $W_L = 15$ kips. The beam is braced at the ends and at the location of the concentrated loads and has a yield stress of $F_y = 50$ ksi.

*Find:* Using allowable stress level (ASD) load combinations, the lightest W18 section with adequate flexural capacity.

**4.5** *Given:* Figure 4.17 shows a simply supported beam spanning 30 ft with concentrated loads at the third points of the span. Each load consists of a dead load component of $W_D = 5$ kips, which includes an allowance for the weight of the beam, and a live load component of $W_L = 15$ kips. The beam is braced at the ends and at the location of the concentrated loads and has a yield stress of $F_y = 50$ ksi.

*Find:* Using strength level (LRFD) load combinations the lightest W18 section with adequate flexural capacity.

**4.6** *Given:* Figure 4.17 shows a simply supported beam spanning 30 ft with concentrated loads at the third points of the span. Each load consists of a dead load component of $W_D = 5$ kips, which includes an allowance for the weight of the beam, and a live load component of $W_L = 15$ kips. The beam is braced at the ends only and has a yield stress of $F_y = 50$ ksi.

*Find:* Using allowable stress level (ASD) load combinations, the lightest W18 section with adequate flexural capacity.

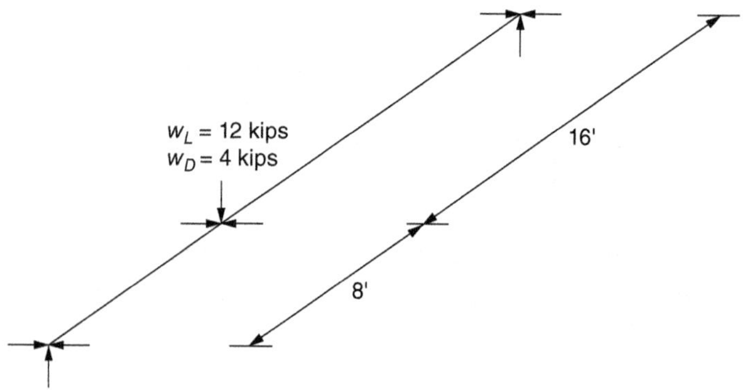


**FIGURE 4.18** Details for Problem 4.8.

**4.7** *Given:* Figure 4.17 shows a simply supported beam spanning 30 ft with concentrated loads at the third points of the span. Each load consists of a dead load component of $W_D = 5$ kips, which includes an allowance for the weight of the beam, and a live load component of $W_L = 15$ kips. The beam is braced at the ends only and has a yield stress of $F_y = 50$ ksi.

*Find:* Using strength level (LRFD) load combinations, the lightest W18 section with adequate flexural capacity.

**4.8** *Given:* Figure 4.18 shows a simply supported beam spanning 24 ft with a single concentrated load as indicated. The beam is braced at the ends and at the location of the concentrated load.

*Find:* The values of $C_b$ for the beam.

**4.9** *Given:* A W14 × 82 beam with a yield stress of 50 ksi subjected to biaxial bending with the lateral load acting on the top flange. The beam is simply supported over a span of 20 ft with lateral bracing at the ends only. The bending moments about the major axis are

$$M_D = 40 \text{ kip-ft ... includes weight of the beam}$$

$$M_L = 120 \text{ kip-ft}$$

The bending moment acting laterally on the top flange is

$$M_{hL} = 15 \text{ kip-ft}$$

*Find:* Using allowable stress level (ASD) load combinations, determine if the W14 × 82 beam is satisfactory.

**4.10** *Given:* A W14 × 82 beam with a yield stress of 50 ksi subjected to biaxial bending with the lateral load acting on the top flange. The beam is simply supported over a span of 20 ft with lateral bracing at the ends only. The bending moments about the major axis are

$$M_D = 40 \text{ kip-ft ... includes weight of the beam}$$

$$M_L = 120 \text{ kip-ft}$$

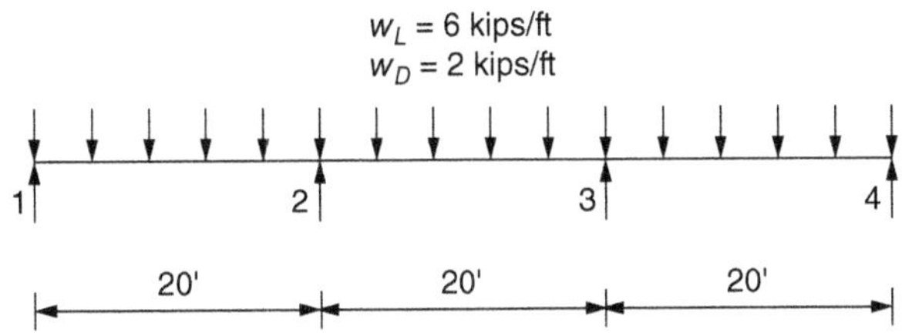


**FIGURE 4.19** Details for Problems 4.11 and 4.12.

The bending moment acting laterally on the top flange is

$$M_{hL} = 15 \text{ kip-ft}$$

*Find:* Using strength level (LRFD) load combinations, determine if the W14 × 82 beam is satisfactory.

**4.11** *Given:* The uniform distributed loading, including the beam self-weight, acting on a three-span continuous beam is shown in Fig. 4.19. Continuous lateral support is provided to the beam which has a yield stress of 50 ksi.

*Find:* Using allowable stress level (ASD) load combinations, and allowing for moment redistribution, the lightest W21 section with adequate flexural capacity.

**4.12** *Given:* The uniform distributed loading, including the beam self-weight, acting on a three-span continuous beam is shown in Fig. 4.19. Continuous lateral support is provided to the beam which has a yield stress of 50 ksi.

*Find:* Using strength level (LRFD) load combinations, and allowing for moment redistribution, the lightest W21 section with adequate flexural capacity.

# CHAPTER 5

# Design of Steel Beams for Shear and Torsion

## 5.1 Introduction

The applied loading on a beam results in a shear force $V$ on the beam. Provided that the stresses produced in the beam are within the elastic limit, the shear stress produced at a specific level in the cross section of a member is given by the expression

$$f_v = VQ/It$$

where $V$ = applied shear force on the section
$Q$ = statical moment of the area, above the level considered, about the neutral axis of the section
$I$ = moment of inertia of the section
$t$ = width of section at the level considered

The plot of this expression over the height of a W-section is shown in Fig. 5.1.

As shown in Fig. 5.1, the maximum shear stress occurs at the neutral axis of the section. Most of the shear capacity of the section is provided by the web of the W-section with only a small portion provided by the flanges. It is customary in design to assume that the applied shear force is resisted by an area equal to the product of the depth of the beam and the thickness of the web. This gives a uniform shear stress over the depth of the beam of

$$f_v = V/A_w$$

where $A_w = dt_w$
$d$ = overall depth of the beam
$t_w$ = web thickness

This uniform shear stress is approximately 88 percent of the maximum shear stress.

**Example 5.1.** Shear Stress
A W16 × 89 beam, with a yield stress of 50 ksi, is simply supported over a span of 21 ft. The beam supports a uniformly distributed load of 6 kips/ft that includes the self-weight of the beam. The beam is laterally braced at the supports and at the third points of the span. Determine the maximum shear stress and the average shear stress in the beam.

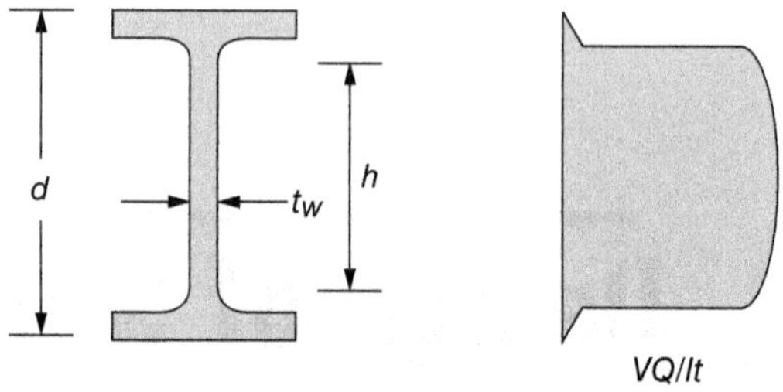


FIGURE 5.1 Shear distribution in a W-section.

The maximum shear stress occurs at the neutral axis of the beam, at the support, where the reaction is

$$V = 6 \times 21/2$$
$$= 63 \text{ kips}$$

The properties of the top half of a W16 × 89 are identical with the properties of a WT8 × 44.5 cut from the W-shape. From American Institute of Steel Construction, *Steel Construction Manual* (AISC Manual)[1] Table 1-8, the properties of a WT8 × 44.5 are

$$A = \text{area of Tee}$$
$$= 13.1 \text{ in}^2$$
$$d = \text{depth of Tee}$$
$$= 8.38 \text{ in}$$
$$y = \text{depth to centroid of the Tee}$$
$$= 1.70 \text{ in}$$

The statical moment of the area of the W16 × 89 above the neutral axis, about the neutral axis of the section is

$$Q = A(d - y)$$
$$= 13.1(8.38 - 1.70)$$
$$= 87.51 \text{ in}^3$$

From AISC Manual Table 1-1, the properties of a W16 × 89 are

$$I = 1300 \text{ in}^4$$
$$t_w = 0.525 \text{ in}$$

The shear stress at the neutral axis of the W16 × 89 is given by

$$f_v = VQ/It_w$$
$$= 63 \times 87.51/(1300 \times 0.525)$$
$$= 8.08 \text{ ksi}$$

The average shear stress over the depth of the beam is

$$f_v = V/dt_w$$
$$= 63/(16.8 \times 0.525)$$
$$= 7.14 \text{ ksi}$$

## 5.2 Shear in Beam Webs

The nominal shear capacity of a W-shape with unstiffened web depends on the slenderness of the web, and the web slenderness parameter is defined as

$$\lambda = h/t_w$$

where $h$ is clear distance between flanges less the corner radius at each flange, for rolled shapes or clear distance between flanges, for built-up welded sections.

As the web slenderness parameter of a beam increases, web shear failure occurs by either

- Plastic yielding of the web in beams with a compact web
- Inelastic web buckling of the web in beams with a noncompact web
- Elastic web buckling of the web in beams with a slender web

The relationship between nominal shear strength and web slenderness is shown in Fig. 5.2.

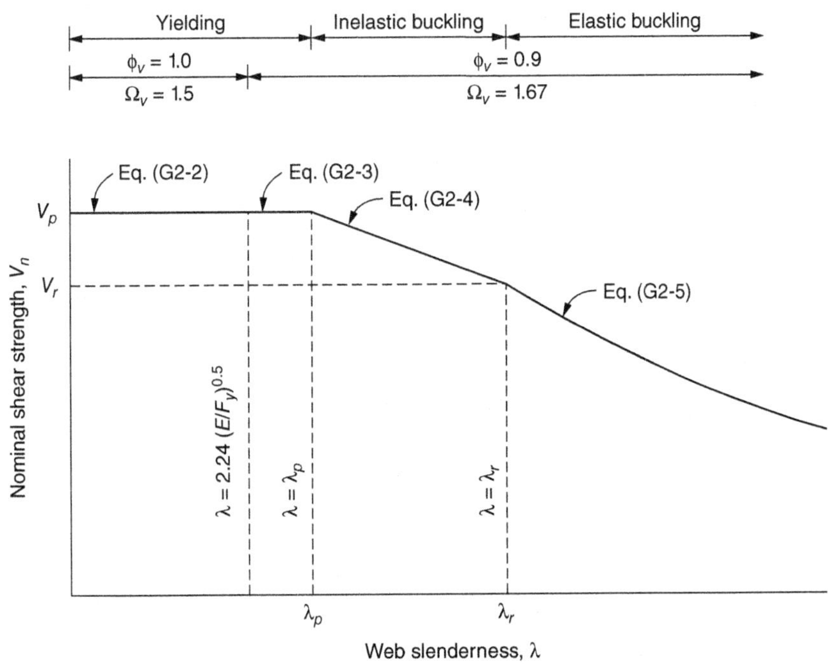


**FIGURE 5.2** Nominal shear strength and web slenderness, beams with unstiffened webs.

## Web Yielding

### (i) $\lambda \leq \lambda_p$

The maximum nominal shear capacity of a rolled I-shape is $V_n = V_p$. As shown in Fig. 5.2, as the web slenderness parameter increases beyond $\lambda_p$, the nominal shear capacity decreases. When $\lambda$ does not exceed $\lambda_p$, full plasticity of the web is possible and the limiting slenderness parameter for the limit state of web yielding is given by American Institute of Steel Construction, *Specification for Structural Steel Buildings* (AISC 360)[2] Sec. G2.1(b)(i) as

$$\lambda_p = 1.10(k_v E/F_y)^{0.5}$$

where $k_v = 5$ ... for unstiffened webs with $\lambda < 260$
hence, $\lambda_p = 2.46(E/F_y)^{0.5}$ ... for unstiffened webs with $\lambda < 260$

The nominal shear strength is given by AISC 360 Eq. (G2-1) as

$$V_n = V_p$$
$$= 0.6F_y A_w C_v$$

where $0.6F_y$ = shear yield strength of the steel
$A_w = dt_w$
$d$ = overall depth of the beam
$t_w$ = web thickness
$C_v$ = web shear coefficient that accounts for the influence of buckling on shear strength
$= 1.0$ ... for $\lambda \leq 2.46(E/F_y)^{0.5}$

With the exception of M12.5 × 12.4, M12.5 × 11.6, M12 × 11.8, M12 × 10.8, M12 × 10, M10 × 8, and M10 × 7.5, all W-, S-, M-, and HP-shapes with a yield stress of 50 ksi meet these criteria.

### (ii) $\lambda \leq 2.24(E/F_y)^{0.5}$

As specified in AISC 360 Sec. G2.1(a), the following criteria apply for unstiffened webs with a limiting slenderness parameter of

$$h/t_w \leq 2.24(E/F_y)^{0.5}$$

| LRFD | ASD |
|---|---|
| The resistance factor for shear is<br>$\varphi_v = 1.0$<br>From AISC 360 Eq. (G2-1), the nominal shear strength is<br>$V_n = 0.6F_y A_w$<br>The design shear strength is<br>$\varphi_v V_n = 1.0 \times 0.6F_y A_w$<br>$= 0.6F_y A_w$<br>AISC Manual Table 3-2 provides values of $\varphi_v V_n$ for W-shapes with a yield stress of 50 ksi. | The safety factor for shear is<br>$\Omega_v = 1.5$<br>From AISC 360 Eq. (G2-1), the nominal shear strength is<br>$V_n = 0.6F_y A_w$<br>The allowable shear strength is<br>$V_n/\Omega_v = 0.6F_y A_w/1.5$<br>$= 0.4F_y A_w$<br>AISC Manual Table 3-2 provides values of $V_n/\Omega_v$ for W-shapes with a yield stress of 50 ksi. |

With the exception of W44 × 230, W40 × 149, W36 × 135, W33 × 118, W30 × 90, W24 × 55, W16 × 26, and W12 × 14, all W-, S-, and HP-shapes with a yield stress of 50 ksi meet these criteria.

### (iii) $2.24(E/F_y)^{0.5} \leq \lambda \leq 2.46(E/F_y)^{0.5}$

As specified in AISC 360 Sec. G1. and Sec. G2.1(b), the following criteria apply for the unstiffened webs of all other doubly symmetric shapes and singly symmetric shapes and channels with a slenderness parameter of

$$2.24(E/F_y)^{0.5} \leq \lambda \leq 2.46(E/F_y)^{0.5}$$

| LRFD | ASD |
|---|---|
| The resistance factor for shear is<br>$\varphi_v = 0.9$<br>From AISC 360 Eq. (G2-1), the nominal shear strength is<br>$V_n = 0.6F_yA_w$<br>The design shear strength is<br>$\varphi_v V_n = 0.9 \times 0.6F_yA_w$<br>$= 0.54F_yA_w$<br>AISC Manual Table 3-2 provides values of $\varphi_v V_n$ for W-shapes with a yield stress of 50 ksi. | The safety factor for shear is<br>$\Omega_v = 1.67$<br>From AISC 360 Eq. (G2-1), the nominal shear strength is<br>$V_n = 0.6F_yA_w$<br>The allowable shear strength is<br>$V_n/\Omega_v = 0.6F_yA_w/1.67$<br>$= 0.36F_yA_w$<br>AISC Manual Table 3-2 provides values of $V_n/\Omega_v$ for W-shapes with a yield stress of 50 ksi. |

**Example 5.2.** Shear Capacity

A W16 × 89 beam, with a yield stress of 50 ksi, is simply supported over a span of 20 ft. The beam supports a uniformly distributed dead load of $w_D = 2$ kips/ft, that includes the self-weight of the beam, and a uniformly distributed live load of $w_L = 6$ kips/ft. The beam is laterally braced at the supports and at the third points of the span. Determine if the beam is adequate for shear.

From AISC Manual Table 1-1, the properties of a W16 × 89 are

$$d = 16.8 \text{ in}$$

$$t_w = 0.525 \text{ in}$$

$$h/t_w = 27.0$$

$$\begin{aligned} 2.24(E/F_y)^{0.5} &= 2.24(29{,}000/50)^{0.5} \\ &= 53.95 \\ &> 27.0 \ldots \text{ hence, } \varphi = 1.0 \text{ and } \Omega = 1.5 \end{aligned}$$

$$\begin{aligned} V_n &= \text{nominal shear} \\ &= 0.6F_yA_w \\ &= 0.6 \times 50 \times 16.8 \times 0.525 \\ &= 265 \text{ kips} \end{aligned}$$

Applying American Society of Civil Engineers, *Minimum Design Loads for Buildings and Other Structures* (ASCE 7)[3] Secs. 2.3 and 2.4 gives

| LRFD | ASD |
|---|---|
| From ASCE 7 Sec. 2.3.2 combination 2: | From ASCE 7 Sec. 2.4.1 combination 2: |
| $w_u$ = factored load | $w_a$ = factored load |
| $= 1.2w_D + 1.6w_L$ | $= w_D + w_L$ |
| $= 1.2 \times 2.0 + 1.6 \times 6.0$ | $= 2.0 + 6.0$ |
| $= 12$ kips/ft | $= 8$ kips/ft |
| $V_u$ = factored shear | $V_a$ = factored shear |
| $= w_u L/2$ | $= w_a L/2$ |
| $= 12 \times 20/2$ | $= 8 \times 20/2$ |
| $= 120$ kips | $= 80$ kips |
| = required strength | = required strength |
| $\varphi_v V_n$ = design shear | $V_n/\Omega_v$ = allowable shear |
| $= 1.0 \times 265$ | $= 265/1.5$ |
| $= 265$ kips | $= 177$ kips |
| $> V_u$ ... satisfactory | $> V_a$ ... satisfactory |
| From AISC Manual Table 3-2, a W16 × 89 provides a design shear strength of $\varphi_v V_n = 265$ kips | From AISC Manual Table 3-2, a W16 × 89 provides an allowable shear strength of $V_n/\Omega_b = 177$ kips |

## Inelastic Buckling

In accordance with AISC 360 Sec. G2.1(b)(ii), the nominal shear capacity is governed by inelastic buckling of the web when the web slenderness parameter is

$$1.10(k_v E/F_y)^{0.5} < \lambda \le 1.37(k_v E/F_y)^{0.5}$$

As shown in Fig. 5.2, as the web slenderness increases beyond $\lambda_p$, the nominal shear capacity decreases linearly from a maximum value of $V_p$ to a value of $V_r$ at $\lambda = \lambda_r$. As specified in AISC 360 Sec. G2.1(i) for unstiffened webs, when $\lambda < 260$ then $k_v = 5$ and the limiting web slenderness parameter for the limit state of web inelastic buckling is

$$\lambda_r = 1.37(k_v E/F_y)^{0.5}$$
$$= 3.06(E/F_y)^{0.5}$$

When the limit state of web inelastic buckling is applicable, the nominal shear strength is given by AISC 360 Eq. (G2-1) as

$$V_n = 0.6F_y A_w C_v$$
$$= 0.6F_y A_w [2.46(E/F_y)^{0.5}]/(h/t_w)$$

where $C_v = 1.10(k_v E/F_y)^{0.5}/(h/t_w)$
$= 2.46(E/F_y)^{0.5}/(h/t_w)$ ... for unstiffened webs with $\lambda < 260$ and $k_v = 5$

For a value of $\lambda = \lambda_r$, the web shear coefficient is

$$C_v = 1.10(k_v E/F_y)^{0.5}/1.37(k_v E/F_y)^{0.5}$$
$$= 0.803$$

and,

$$V_n = V_r$$
$$= 0.6F_y A_w \times 0.803$$
$$= 0.48F_y A_w$$

**Example 5-3.** Inelastic Shear Strength
An M12 × 10.8 beam, with a yield stress of 50 ksi, is simply supported over a span of 10 ft. The beam supports a uniformly distributed dead load of $w_D = 0.6$ kip/ft, that includes the self-weight of the beam, and a uniformly distributed live load of $w_L = 1.8$ kips/ft. Adequate lateral support is provided to the beam. Determine if the beam is satisfactory for shear.

From AISC Manual Table 1-2, the properties of a M12 × 10.8 section are

$$d = \text{overall depth of section}$$
$$= 12.0 \text{ in}$$
$$t_w = \text{web thickness}$$
$$= 0.160 \text{ in}$$
$$A_w = \text{area of the web}$$
$$= dt_w$$
$$= 12.0 \times 0.160$$
$$= 1.92 \text{ in}^2$$
$$h/t_w = 69.2$$
$$\lambda_p = 2.46(E/F_y)^{0.5}$$
$$= 2.46(29{,}000/50)^{0.5}$$
$$= 59.2$$
$$< 69.2$$
$$\lambda_r = 3.07(E/F_y)^{0.5}$$
$$= 3.07(29{,}000/50)^{0.5}$$
$$= 73.9$$
$$> 69.2$$

Hence, inelastic buckling of the web governs and the nominal shear strength is given by AISC 360 Eq. (G2-1) as

$$V_n = 0.6F_y A_w C_v$$
$$= 0.6F_y A_w [2.46(E/F_y)^{0.5}]/(h/t_w)$$
$$= 0.6 \times 50 \times 1.92 \times 2.46(29{,}000/50)^{0.5}/69.2$$
$$= 49.3 \text{ kips}$$

| LRFD | ASD |
|---|---|
| From ASCE 7 Sec. 2.3.2 combination 2: | From ASCE 7 Sec. 2.4.1 combination 2: |
| $w_u$ = factored load | $w_a$ = factored load |
| $= 1.2w_D + 1.6w_L$ | $= w_D + w_L$ |
| $= 1.2 \times 0.6 + 1.6 \times 1.8$ | $= 0.6 + 1.8$ |
| = 3.6 kips/ft | = 2.4 kips/ft |
| $V_u$ = factored shear | $V_a$ = factored shear |
| $= w_u L/2$ | $= w_a L/2$ |
| $= 3.6 \times 10/2$ | $= 2.4 \times 10/2$ |
| = 18 kips | = 12 kips |
| = required strength | = required strength |
| $\varphi_v V_n$ = design shear | $V_n/\Omega_v$ = allowable shear |
| $= 0.9 \times 49.3$ | $= 49.3/1.67$ |
| = 44.4 kips | = 29.5 kips |
| $> V_u$ ... satisfactory | $> V_a$ ... satisfactory |

## Elastic Buckling

In accordance with AISC 360 Sec. G2.1(b)(iii), the nominal shear capacity is governed by elastic buckling of the web when the web slenderness parameter exceeds

$$\lambda_r = 1.37(k_v E/F_y)^{0.5}$$
$$= 3.06(E/F_y)^{0.5} \text{ ... for unstiffened webs with } \lambda < 260 \text{ and } k_v = 5$$

When this limiting slenderness parameter is applicable, the nominal shear strength is given by AISC 360 Eq. (G2-1) as

$$V_n = 0.6F_y A_w C_v$$
$$= 0.6F_y A_w[7.55E/F_y(h/t_w)^2]$$
$$= 4.53A_w E/(h/t_w)^2 \text{ ... for unstiffened webs with } \lambda < 260 \text{ and } k_v = 5$$

where $C_v = 1.51k_v E/F_y(h/t_w)^2$
$= 7.55E/F_y(h/t_w)^2$ ... for unstiffened webs with $\lambda < 260$ and $k_v = 5$

**Example 5.4.** Elastic Web Buckling
An M12.5 × 12.4 beam, with a yield stress of 50 ksi, is simply supported over a span of 10 ft. The beam supports a uniformly distributed dead load of $w_D = 0.8$ kip/ft, that includes the self-weight of the beam, and a uniformly distributed live load of $w_L = 2.4$ kips/ft. Adequate lateral support is provided to the beam. Determine if the beam is satisfactory for shear.

From AISC Manual Table 1-2, the properties of a M12.5 × 12.4 section are

$$d = \text{overall depth of section}$$
$$= 12.5 \text{ in}$$
$$t_w = \text{web thickness}$$
$$= 0.155 \text{ in}$$
$$A_w = \text{area of the web}$$
$$= dt_w$$
$$= 12.5 \times 0.155$$
$$= 1.94 \text{ in}^2$$
$$h/t_w = 74.8$$
$$\lambda_r = 3.06(E/F_y)^{0.5}$$
$$= 3.06(29{,}000/50)^{0.5}$$
$$= 73.7$$
$$< 74.8$$

Hence, elastic buckling of the web governs and the nominal shear strength is given by AISC 360 Eq. (G2-1) as

$$V_n = 0.6F_yA_wC_v$$
$$= 4.53A_wE/(h/t_w)^2 \text{ ... for unstiffened webs with } \lambda < 260 \text{ and } k_v = 5$$
$$= 4.53 \times 1.94 \times 29{,}000/(74.8)^2$$
$$= 45.6 \text{ kips}$$

| LRFD | ASD |
|---|---|
| From ASCE 7 Sec. 2.3.2 combination 2: | From ASCE 7 Sec. 2.4.1 combination 2: |
| $w_u$ = factored load | $w_a$ = factored load |
| $= 1.2w_D + 1.6w_L$ | $= w_D + w_L$ |
| $= 1.2 \times 0.8 + 1.6 \times 2.4$ | $= 0.8 + 2.4$ |
| = 4.8 kips/ft | = 3.2 kips/ft |
| $V_u$ = factored shear | $V_a$ = factored shear |
| $= w_uL/2$ | $= w_aL/2$ |
| $= 4.8 \times 10/2$ | $= 3.2 \times 10/2$ |
| = 24 kips | = 16 kips |
| = required strength | = required strength |
| $\varphi_vV_n$ = design shear | $V_n/\Omega_v$ = allowable shear |
| $= 0.9 \times 45.6$ | $= 45.6/1.67$ |
| = 41 kips | = 27.3 kips |
| $> V_u$ ... satisfactory | $> V_a$ ... satisfactory |

## 5.3 Weak Axis Shear

For singly and doubly symmetric shapes loaded in the weak axis, the flanges of the member act as rectangular sections in resisting shear force. The distribution of shear stress across a rectangular section is shown in Fig. 5.3.

The maximum shear stress occurs at the neutral axis of the section and may be determined by applying the expression

$$f_v = VQ/It$$

where $Q$ = statical moment of the area of the rectangle above the neutral axis about the neutral axis

$= (td/2) \times d/4$

$= td^2/8$

$I$ = moment of inertia of the rectangle

$= td^3/12$

$t$ = width of rectangle at the neutral axis

$f_v = V(td^2/8)/(td^3/12)t$

$= 1.5V/td$

The average shear stress over the depth of the rectangle is

$$f_v = V/td$$

Hence, the maximum shear stress in the rectangle is 1.5 times the average shear stress.

In accordance with AISC 360 Sec. G7, the nominal shear strength of singly and doubly symmetric shapes loaded in the weak axis may be determined using AISC 360 Eq. (G2-1) and AISC 360 Sec. G2.1(b) with $k_v = 1.2$ and $A_w = b_f t_f$ for each flange. All W-, S-, M-, and HP-shapes with a yield stress of 50 ksi fail by the limit state of yielding with $C_v = 1.0$. The resistance factor is $\varphi_v = 0.9$ and the safety factor is $\Omega_v = 1.67$.

**Example 5.5.** Weak Axis Shear Capacity

A W16 × 89 beam, with a yield stress of 50 ksi, is loaded about the weak axis and simply supported over a span of 10 ft. The beam supports a uniformly distributed dead load of $w_D = 2$ kips/ft, that includes the self-weight of the beam, and a uniformly distributed live load of $w_L = 6$ kips/ft. Determine if the beam is adequate for shear.

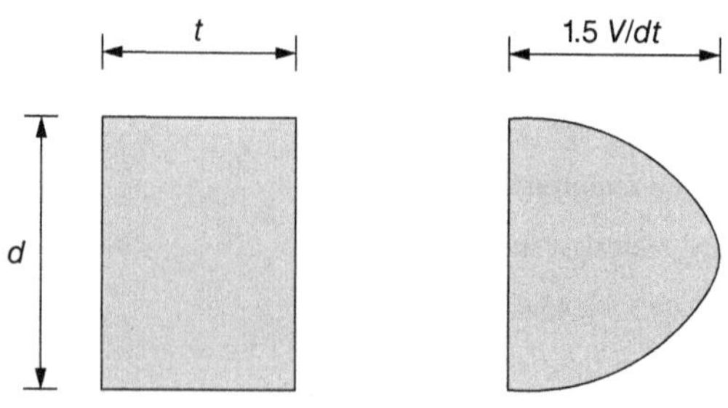


**FIGURE 5.3** Shear distribution in a rectangular section.

From AISC Manual Table 1-1, the properties of a W16 × 89 are

$$b_f = 10.4 \text{ in}$$

$$t_f = 0.875 \text{ in}$$

$$\begin{aligned} A_w &= 2b_f t_f \\ &= 2 \times 10.4 \times 0.875 \\ &= 18.2 \text{ in}^2 \end{aligned}$$

$$\begin{aligned} V_n &= \text{nominal shear from AISC 360 Eq. (G2-1)} \\ &= 0.6F_y A_w C_v \\ &= 0.6 \times 50 \times 18.2 \times 1.0 \\ &= 546 \text{ kips} \end{aligned}$$

| LRFD | ASD |
|---|---|
| From ASCE 7 Sec. 2.3.2 combination 2: | From ASCE 7 Sec. 2.4.1 combination 2: |
| $w_u$ = factored load | $w_a$ = factored load |
| $= 1.2w_D + 1.6w_L$ | $= w_D + w_L$ |
| $= 1.2 \times 2.0 + 1.6 \times 6.0$ | $= 2.0 + 6.0$ |
| = 12 kips/ft | = 8 kips/ft |
| $V_u$ = factored shear | $V_a$ = factored shear |
| $= w_u L/2$ | $= w_a L/2$ |
| $= 12 \times 10/2$ | $= 8 \times 10/2$ |
| = 60 kips | = 40 kips |
| = required strength | = required strength |
| $\varphi_v V_n$ = design shear | $V_n/\Omega_v$ = allowable shear |
| $= 0.9 \times 546$ | $= 546/1.67$ |
| = 491 kips | = 327 kips |
| $> V_u$ ... satisfactory | $> V_a$ ... satisfactory |

## 5.4 Longitudinal Shear in Built-Up Sections

As shown in Fig. 5.4*a*, plates may be added to rolled members to reinforce the member. Also, as shown in Fig. 5.4*b*, overhead crane runway girders are frequently composed of a built-up section consisting of a W-shape and a channel as detailed by Fisher.[4] In addition, as shown in Fig. 5.4*c*, plate web girders are built up from individual plates to provide a member with a flexural strength larger than available rolled sections. In order to determine the strength of weld required to connect together the individual parts of a built-up section, it is necessary to calculate the longitudinal shear at the interface of the parts.

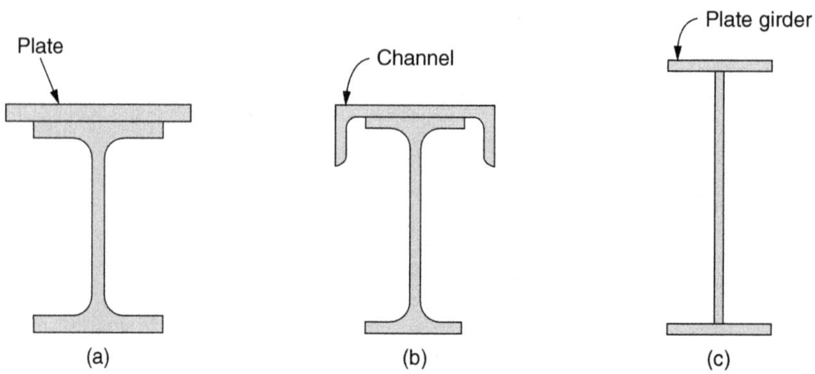


**FIGURE 5.4** Built-up sections.

Shear at the interface is given by

$$v = VQ/I$$

where $V$ = applied shear force at the section
$Q$ = statical moment of the area of the part above the interface about the neutral axis of the built-up section
$I$ = moment of inertia of the built-up section

**Example 5.6.** Longitudinal Shear
The built-up girder shown in Fig. 5.4*a* consists of a W10 × 88 with a 12- × 1/4-in plate welded to the top flange. Shear at the end supports consists of a shear due to dead load of $V_D = 10$ kips, which includes the weight of the beam, and a shear due to live load of $V_L = 30$ kips. Determine the longitudinal shear at the interface.

The relevant properties of the W10 × 88 are

$$A_W = \text{area}$$
$$= 26.0 \text{ in}^2$$
$$d = \text{depth}$$
$$= 10.8 \text{ in}$$
$$I_W = \text{moment of inertia}$$
$$= 534 \text{ in}^4$$

The properties of the 12- × 1/4-in plate are

$$A_p = \text{area}$$
$$= 3 \text{ in}^2$$
$$I_p = \text{moment of inertia}$$
$$= 0.016 \text{ in}^4$$

The properties of the built-up section are obtained as shown in Table 5.1.

| Part | A | y | I | Ay | Ay² |
|---|---|---|---|---|---|
| Plate | 3.0 | 10.925 | 0 | 32.8 | 358 |
| Beam | 26.0 | 5.400 | 534 | 140.4 | 758 |
| Total | 29.0 | | 534 | 173.2 | 1116 |

**Table 5.1** Details for Example 5.6

The height of the centroid of the built-up section is

$$y' = \Sigma Ay/\Sigma A$$
$$= 173.2/290$$
$$= 5.97 \text{ in}$$

The moment of inertia of the built-up section is

$$I = \Sigma I + \Sigma A(y - y')^2$$
$$= 534 + 3 \times 4.955^2 + 26.0 \times 0.57^2$$
$$= 616 \text{ in}^4$$

$Q$ = statical moment of the area of the plate about the neutral axis of the built-up section

$$= A_p(y - y')$$
$$= 3(10.925 - 5.97)$$
$$= 14.87 \text{ in}^3$$

| LRFD | ASD |
|---|---|
| From ASCE 7 Sec. 2.3.2 combination 2: | From ASCE 7 Sec. 2.4.1 combination 2: |
| $V_u$ = factored shear | $V_a$ = factored shear |
| $= 1.2V_D + 1.6V_L$ | $= V_D + V_L$ |
| $= 1.2 \times 10 + 1.6 \times 30$ | $= 10 + 30$ |
| $= 60$ kips | $= 40$ kips |
| The longitudinal shear at the interface is | The longitudinal shear at the interface is |
| $v_u = V_uQ/I$ | $v_a = V_aQ/I$ |
| $= 60 \times 14.87/616$ | $= 40 \times 14.87/616$ |
| $= 1.45$ kips/in | $= 0.97$ kip/in |

## 5.5 Block Shear

Block shear failure in a connection involves both shear and tension components, and is specified in AISC 360 Sec. J4.3. The application of the AISC provisions is covered by Geschwindner[5] and Epstein and Aleksiewicz.[6] In any connection, several different modes of block shear failure may be possible. The bolted connection of a tension

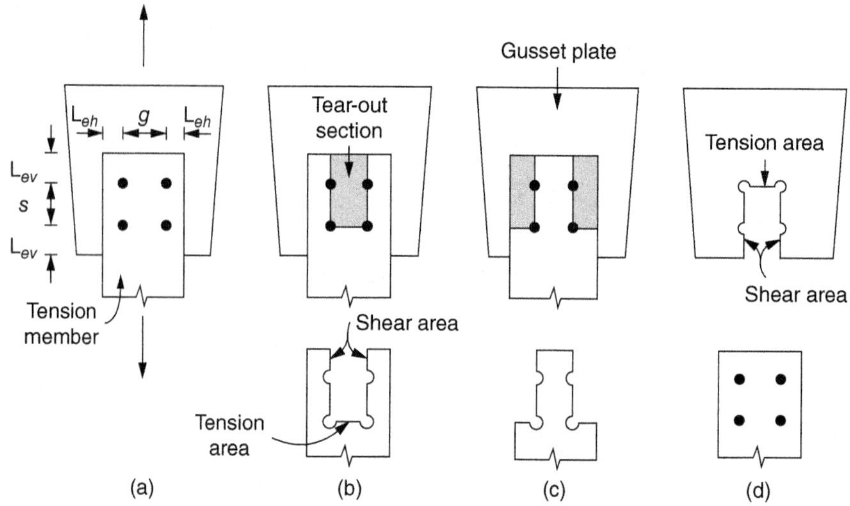


**FIGURE 5.5** Block shear failure.

member to a gusset plate is shown in Fig. 5.5*a* and three possible modes of block shear failure are indicated in Fig. 5.5*b*, *c*, and *d*. Each mode involves shear failure along a plane parallel to the applied force and tension failure along a plane perpendicular to the applied force. Failure may occur in either the tension member, as in modes (*b*) and (*c*), or the gusset plate, as in mode (*d*). The failure path is defined by the centerlines of the bolt holes and shear of the bolts does not occur.

## Block Shear Strength for Bolted Connections

The total block shear nominal strength is the sum of the strengths of the tensile area and the shear area. Tension failure occurs by rupture in the net tension area. Shear failure then occurs either by rupture in the net shear area or by shear yielding in the gross shear area and the minimum value governs.

The nominal rupture strength in tension is given by

$$P_n = U_{bs}F_uA_{nt}$$

The nominal rupture strength in shear is given by

$$V_n = 0.6F_uA_{nv}$$

The nominal yield strength in shear is given by

$$V_n = 0.6F_yA_{gv}$$

where $A_{nv}$ = net shear area
$A_{nt}$ = net tension area
$A_{gv}$ = gross shear area
$F_y$ = specified minimum yield stress
$F_u$ = specified minimum tensile strength
$U_{bs}$ = reduction coefficient for tensile stress
= 1.0 for uniform tensile stress
= 0.5 for nonuniform tensile stress

When $0.6F_yA_{gv}$ exceeds $0.6F_uA_{nv}$, tension rupture occurs in combination with shear rupture and the nominal strength is given by AISC 360 Eq. (J4-5) as

$$R_n = 0.6F_uA_{nv} + U_{bs}F_uA_{nt}$$

When $0.6F_uA_{nv}$ exceeds $0.6F_yA_{gv}$, tension rupture occurs in combination with shear yielding and the nominal strength is given by AISC 360 Eq. (J4-5) as

$$R_n = 0.6F_yA_{gv} + U_{bs}F_uA_{nt}$$

For block shear, AISC 360 Sec. J4.3 gives the resistance factor and the safety factor as

$$\begin{aligned} \varphi &= \text{resistance factor ... LRFD} \\ &= 0.75 \\ \Omega &= \text{safety factor ... ASD} \\ &= 2.00 \end{aligned}$$

## Effective Bolt Hole Diameter and Net Area

In order to determine the net area of a failure path, it is necessary to deduct the effective hole areas along the path from the gross area of the failure path. The nominal diameter of a standard hole is detailed in AISC 360 Table J3.3 as 1/16 in larger than the bolt diameter. As the hole is formed, some deterioration occurs in the surrounding material and AISC 360 Sec. B4.3b specifies that the effective hole diameter shall be taken as 1/16 in larger than the nominal hole diameter. Hence, the effective hole diameter is

$$\begin{aligned} d_h &= d_b + 1/16 \text{ in} + 1/16 \text{ in} \\ &= d_b + 1/8 \text{ in} \end{aligned}$$

where $d_b$ is diameter of fastener.

Referring to Fig. 5.5*d*, the net tension area is

$$A_{nt} = t_g(g - d_h)$$

The gross shear area is

$$A_{gv} = 2t_g(L_{ev} + s)$$

The net shear area is

$$A_{nv} = 2t_g(L_{ev} + s - 1.5d_h)$$

where $t_g$ is gusset plate thickness.

**Example 5.7.** Block Shear, Bolted Connection

Determine the block shear strength of the connection shown in Fig. 5.5*a*. Both members have a yield stress of 36 ksi and a tensile strength of 58 ksi. The tension member thickness is $t = 1/2$ in, the gusset plate thickness is $t_g = 1.0$ in, and the bolts are 7/8-in diameter. The relevant dimensions are $s = 3$ in, $g = 2$ in, $L_{eh} = 2$ in, and $L_{ev} = 2$ in.

The effective hole diameter for a 7/8-in diameter bolt is defined in AISC 360 Sec. B4.3b as

$$\begin{aligned} d_h &= d_b + 1/8 \text{ in} \\ &= 0.875 + 0.125 \\ &= 1.0 \text{ in} \end{aligned}$$

The tension stress is uniform and

$$U_{bs} = 1.0$$

Because of the large gusset plate thickness, block shear will not occur in the gusset plate and it is necessary to check failure modes (*b*) and (*c*) only.

*Mode (b)*

From Fig. 5.5*b*, the gross shear area is

$$\begin{aligned} A_{gv} &= 2t(L_{ev} + s) \\ &= 2 \times 0.5(2 + 3) \\ &= 5 \text{ in}^2 \end{aligned}$$

From Fig. 5.5*b*, the net shear area is

$$\begin{aligned} A_{nv} &= 2t(L_{ev} + s - 1.5d_h) \\ &= 2 \times 0.5(2 + 3 - 1.5 \times 1.0) \\ &= 3.5 \text{ in}^2 \end{aligned}$$

From Fig. 5.5*b*, the net tension area is

$$\begin{aligned} A_{nt} &= t_g(g - d_h) \\ &= 0.5(2 - 1) \\ &= 0.5 \text{ in}^2 \end{aligned}$$

The rupture strength in shear is given by

$$\begin{aligned} 0.6F_uA_{nv} &= 0.6 \times 58 \times 3.5 \\ &= 122 \text{ kips} \end{aligned}$$

The yield strength in shear is given by

$$\begin{aligned} 0.6F_yA_{gv} &= 0.6 \times 36 \times 5 \\ &= 108 \text{ kips ... shear yielding governs} \\ &< 0.6F_uA_{nv} \end{aligned}$$

The rupture strength in tension is given by

$$\begin{aligned} U_{bs}F_uA_{nt} &= 1.0 \times 58 \times 0.5 \\ &= 29 \text{ kips} \end{aligned}$$

Hence, the block shear nominal strength is given by AISC 360 Eq. (J4-5) as

$$\begin{aligned} R_n &= 0.6F_yA_{gv} + U_{bs}F_uA_{nt} \\ &= 108 + 29 \\ &= 137 \text{ kips} \end{aligned}$$

*Mode (c)*

The shear areas for mode (*c*) are identical with those in mode (*b*) and the net tension area is larger. Hence, mode (*b*) is the critical failure mode.

| LRFD | ASD |
|---|---|
| The resistance factor for block shear is<br>$\varphi = 0.75$<br>The design shear strength is<br>$\varphi R_n = 0.75 \times 137$<br>$= 103$ kips | The safety factor for block shear is<br>$\Omega = 2.0$<br>The allowable block shear strength is<br>$R_n/\Omega = 137/2.0$<br>$= 69$ kips |

## Block Shear Strength for Welded Connections

As shown in Fig. 5.6, block shear may also occur in welded connections. For this situation, net areas are not applicable and AISC 360 Eq. (J4-5) reduces to

$$R_n = 0.6F_yA_{gv} + U_{bs}F_uA_{gt}$$

**Example 5.8.** Block Shear, Welded Connection
Determine the block shear strength of the 1/2-in plate, shown in Fig. 5.6, which has a yield stress of 36 ksi and a tensile strength of 58 ksi. The angle is $2 \times 2 \times 1/2$ in and dimension $L = 4$ in.

The tension stress is uniform and

$$U_{bs} = 1.0$$

From Fig. 5.6, the gross shear area is

$$A_{gv} = 2 \times 4 \times 0.5$$
$$= 4 \text{ in}^2$$

From Fig. 5.6, the gross tension area is

$$A_{gt} = 2 \times 0.5$$
$$= 1.0 \text{ in}^2$$

The yield strength in shear is given by

$$0.6F_yA_{gv} = 0.6 \times 36 \times 4$$
$$= 86 \text{ kips}$$

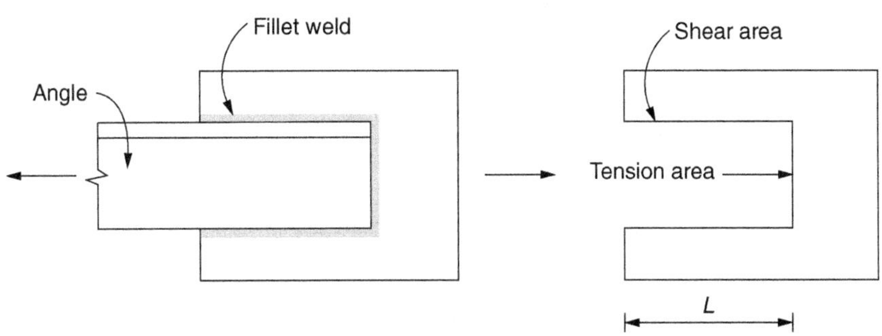


**FIGURE 5.6** Block shear failure, welded connection.

The rupture strength in tension is given by

$$U_{bs}F_uA_{gt} = 1.0 \times 58 \times 1.0$$
$$= 58 \text{ kips}$$

The block shear nominal strength is

$$R_n = 0.6F_yA_{gv} + U_{bs}F_uA_{gt}$$
$$= 86 + 58$$
$$= 144 \text{ kips}$$

| LRFD | ASD |
|---|---|
| The resistance factor for block shear is | The safety factor for block shear is |
| $\varphi = 0.75$ | $\Omega = 2.0$ |
| The design shear strength is | The allowable block shear strength is |
| $\varphi R_n = 0.75 \times 144$ | $R_n/\Omega = 144/2.0$ |
| $= 108$ kips | $= 72$ kips |

## Block Shear Strength for Coped Beams

The connection of beams to a girder is usually made with the top flanges held at the same elevation, as shown in Fig. 5.7. This necessitates coping the top flanges of the beams as illustrated. The webs of the coped beams are then subjected to block shear failure.

Figure 5.8*a* shows a typical coped beam with a bolted shear tab. Two configurations are shown at (*b*) and (*c*). In case (*b*), one line of bolts is used to connect the shear tab to the beam web. This results in a uniform stress distribution on the tensile plane and the reduction coefficient is $U_{bs} = 1.0$. In case (*c*), two lines of bolts are used and the line of bolts nearest to the beam end picks up most of the load and the tensile stress distribution is non-uniform resulting in a reduction coefficient of $U_{bs} = 0.5$.

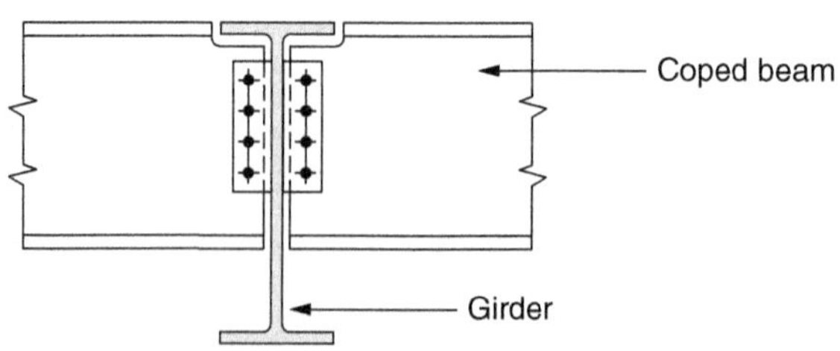


**FIGURE 5.7** Coped beams.

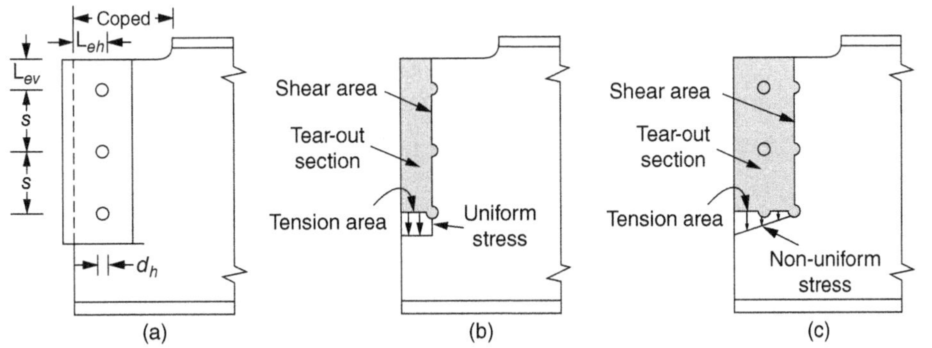


**FIGURE 5.8** Web tear out in coped beams.

The AISC Manual provides tables to facilitate the determination of block shear strength in coped beams. AISC Manual Table 9-3a provides values of $\varphi(F_uA_{nt})/t$ and $(F_uA_{nt})/t\Omega$, AISC Manual Table 9-3b provides values of $\varphi(0.6F_yA_{gv})/t$ and $(0.6F_yA_{gv})/t\Omega$, and AISC Manual Table 9-3c provides values of $\varphi(0.6F_uA_{nv})/t$ and $(0.6F_uA_{nv})/t\Omega$.

**Example 5.9.** Block Shear, Coped Beam

Determine the block shear design strength of the coped W14 × 74 beam, shown in Fig. 5.8*a*, which has a yield stress of 50 ksi and a tensile strength of 65 ksi. The relevant dimensions are $L_{eh} = L_{ev} = 1.5$ in and $s = 3$ in. The bolt diameter is 3/4 in.

The effective hole diameter for a 3/4-in-diameter bolt is defined in AISC 360 Sec. B4.3b as

$$\begin{aligned} d_h &= d_b + 1/8 \text{ in} \\ &= 0.75 + 0.125 \\ &= 0.875 \text{ in} \end{aligned}$$

For a single line of bolts, the tension stress is uniform and

$$U_{bs} = 1.0$$

From Fig. 5.8*a*, the gross shear area is

$$\begin{aligned} A_{gv} &= t_w(L_{ev} + 2s) \\ &= 0.45(1.5 + 2 \times 3) \\ &= 3.38 \text{ in}^2 \end{aligned}$$

From Fig. 5.8*a*, the net shear area is

$$\begin{aligned} A_{nv} &= t_w(L_{ev} + 2s - 2.5d_h) \\ &= 0.45(1.5 + 2 \times 3 - 2.5 \times 0.875) \\ &= 2.39 \text{ in}^2 \end{aligned}$$

From Fig. 5.8*a*, the net tension area is

$$\begin{aligned} A_{nt} &= t_w(L_{eh} - 0.5d_h) \\ &= 0.45(1.5 - 0.5 \times 0.875) \\ &= 0.48 \text{ in}^2 \end{aligned}$$

The rupture strength in tension is given by

$$\begin{aligned} U_{bs}F_uA_{nt} &= 1.0 \times 65 \times 0.48 \\ &= 31.20 \text{ kips} \end{aligned}$$

The yield strength in shear is given by

$$\begin{aligned} 0.6F_yA_{gv} &= 0.6 \times 50 \times 3.38 \\ &= 101.40 \text{ kips} \end{aligned}$$

The rupture strength in shear is given by

$$\begin{aligned} 0.6F_uA_{nv} &= 0.6 \times 65 \times 2.39 \\ &= 93.21 \text{ kips ... shear rupture governs} \\ &< 0.6F_yA_{gv} \end{aligned}$$

Hence, the block shear nominal strength is given by AISC 360 Eq. (J4-5) as

$$\begin{aligned} R_n &= 0.6F_uA_{nv} + U_{bs}F_uA_{nt} \\ &= 93.21 + 31.20 \\ &= 124.41 \text{ kips} \end{aligned}$$

| LRFD | ASD |
|---|---|
| The resistance factor for block shear is<br>$\varphi = 0.75$<br>The design block shear strength is<br>$\varphi R_n = 0.75 \times 124.41$<br>$= 93$ kips<br>Using AISC Manual Tables 9-3a and 9-3c:<br>$\varphi R_n = t[\varphi(F_uA_{nt})/t + \varphi(0.6F_uA_{nv})/t]$<br>$= 0.45(51.8 + 155)$<br>$= 93$ kips | The safety factor for block shear is<br>$\Omega = 2.0$<br>The allowable block shear strength is<br>$R_n/\Omega = 124.41/2.0$<br>$= 62$ kips<br>Using AISC Manual Tables 9-3a and 9-3c:<br>$R_n/\Omega = t[(F_uA_{nt})/\Omega t + (0.6F_uA_{nv})/\Omega t]$<br>$= 0.45(34.5 + 104)$<br>$= 62$ kips |

## 5.6 Web Local Yielding

Girders that frame into columns are supported by a steel-to-steel connection. In other situations, girders may be supported on concrete or masonry walls and, in these instances, the beam is provided with a bearing plate as shown in Fig. 5.9. The function of the bearing plate is twofold:

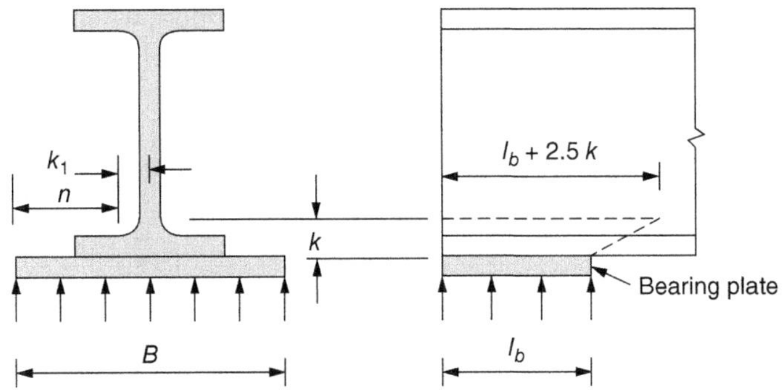


**FIGURE 5.9** Web local yielding at support.

- To distribute the support reaction to the beam web so that yielding of the web does not occur
- To distribute the support reaction to the concrete or masonry wall so as to prevent compression failure in the concrete or masonry

## Bearing on Concrete

The distribution of bearing pressure on a bearing plate is highly complex. The deflection of the girder produces larger pressures on the inner edge of the plate. The bending of the plate in a direction perpendicular to the girder relieves pressure at the sides of the plate. However, it is customary to assume that the pressure is uniformly distributed over the plate. It is also assumed that the plate cantilevers from the flange toe of fillet, shown as dimension $n$ in Fig. 5.9, and the stiffness of the beam flange is ignored. Hence, the bending moment on the plate is

$$M = qn^2/2 \text{ ... per unit length}$$

where $q$ = pressure under the plate, assumed uniform
$n = B/2 - k_1$
$B$ = width of plate
$k_1$ = distance from web center line to flange toe of fillet

The nominal flexural strength of the plate bending about the minor axis is given by AISC 360 Eq. (F11-1) as

$$\begin{aligned} M_n &= M_p \\ &= F_y Z \\ &\leq 1.6F_y S \end{aligned}$$

where $Z$ = plastic section modulus referred to the minor axis
$S$ = elastic section modulus referred to the minor axis
$M_p$ = plastic moment of resistance referred to the minor axis

The nominal bearing capacity of the concrete support, when the bearing plate covers the full area of the support, is given by AISC 360 Eq. (J8-1) as

$$P_p = 0.85 f'_c A$$

where $A$ = area of bearing plate
$= B\ell_b$
$f'_c$ = concrete compressive strength

For bearing on concrete, AISC 360 Sec. J8 gives the resistance factor and the safety factor as

$$\varphi = \text{resistance factor ... LRFD}$$
$$= 0.65$$
$$\Omega = \text{safety factor ... ASD}$$
$$= 2.31$$

**Example 5.10.** Bearing Plate Dimensions
The W14 × 30 girder, shown in Fig. 5.9, has a yield stress of 50 ksi. The girder is supported on a bearing plate with a yield stress of 36 ksi that has dimensions $B = 7$ in, $\ell_b = 6$ in, and a thickness $t = 7/8$ in. The bearing plate sits on a concrete wall with a cylinder strength of $f'_c = 3000$ psi. The support reaction consists of a shear due to dead load of $V_D = 10$ kips and a shear due to live load of $V_L = 30$ kips. Check the concrete bearing pressure and the bearing plate thickness.

The relevant properties of the W14 × 30 girder are

$$t_w = 0.27 \text{ in}$$
$$k_1 = 0.75 \text{ in}$$

The plate cantilevers from the flange toe of fillet a distance

$$n = (B/2) - k_1$$
$$= 3.5 - 0.75$$
$$= 2.75 \text{ in}$$

The nominal flexural strength of the plate is

$$M_n = M_p$$
$$= F_y Z$$
$$= 36 \times t^2/4 \text{ ... for a 1-in length}$$
$$= 36 \times 0.875^2/4$$
$$= 6.89 \text{ kip-in ... satisfactory}$$
$$< 1.6 F_y S = 7.35 \text{ kip-in}$$

The nominal bearing capacity of the concrete is

$$P_n = 0.85 f'_c A$$
$$= 0.85 \times 3 \times 7 \times 6$$
$$= 107.1 \text{ kips}$$

| LRFD | ASD |
|---|---|
| From ASCE 7 Sec. 2.3.2 combination 2: | From ASCE 7 Sec. 2.4.1 combination 2: |
| $V_u$ = factored support reaction | $V_a$ = factored support reaction |
| $= 1.2V_D + 1.6V_L$ | $= V_D + V_L$ |
| $= 1.2 \times 10 + 1.6 \times 30$ | $= 10 + 30$ |
| = 60 kips | = 40 kips |
| $\varphi P_n$ = design bearing capacity | $P_n/\Omega$ = allowable bearing capacity |
| $= 0.65 \times 107.1$ | $= 107.1/2.31$ |
| = 69.6 kips | = 46.4 kips |
| $> V_u$ ... satisfactory | $> V_a$ ... satisfactory |
| $q_u$ = pressure under the plate | $q_a$ = pressure under the plate |
| $= 60/(7 \times 6)$ | $= 40/(7 \times 6)$ |
| = 1.43 kips/in$^2$ | = 0.95 kip/in$^2$ |
| $M_u$ = factored moment on the bearing plate | $M_a$ = factored moment on the bearing plate |
| $= q_u n^2/2$ ... for a 1-in length | $= q_a n^2/2$ ... for a 1-in length |
| $= 1.43 \times 2.75^2/2$ | $= 0.95 \times 2.75^2/2$ |
| = 5.41 kip-in | = 3.59 kip-in |
| $\varphi M_n$ = design flexural capacity | $M_n/\Omega$ = design flexural capacity |
| $= 0.9 \times 6.89$ | $= 6.89/1.67$ |
| = 6.20 kip-in | = 4.13 kip-in |
| $> M_u$ ... satisfactory | $> M_a$ ... satisfactory |

## Web Yielding at Support

The support reaction at the end of a girder is transmitted through the flange to the girder web. The failure stress in the web equals the yield stress $F_y$. It has been established empirically that the load from a bearing plate is dispersed at a slope of 2.5 to 1.0 into the web toe of fillet, as shown in Fig. 5.9. The critical section in the web is at the top of the web fillet. In accordance with AISC 360 Sec. J10.2(b), for loads applied at a distance of not more than $d$ from the end of the girder, the nominal web yield strength is given by AISC 360 Eq. (J10-3) as

$$R_n = (2.5k + \ell_b)F_y t_w$$

where $\ell_b$ = length of bearing plate
$k$ = distance from outer face of flange to web toe of fillet
$t_w$ = web thickness

When the required web strength exceeds the available strength, a pair of transverse stiffeners must be provided to distribute the support reaction over the depth of the web. The stiffeners, one on either side of the web, are welded to the loaded flange and extend to the mid height of the girder. Alternatively, a doubler plate may be used. This consists of a plate welded to the web to increase the effective area of the web.

For local web yielding, AISC 360 Sec. J10.2 gives the resistance factor and the safety factor as

$$\varphi = \text{resistance factor ... LRFD}$$
$$= 1.00$$
$$\Omega = \text{safety factor ... ASD}$$
$$= 1.50$$

To facilitate the determination of web local yielding strength, AISC Manual Table 9-4 provides values of $\varphi R_1$, $R_1/\Omega$, $\varphi R_2$, and $R_2/\Omega$, where

$$R_1 = 2.5kF_y t_w$$

and

$$R_2 = F_y t_w$$

**Example 5.11.** Web Local Yielding
The W14 × 30 girder, shown in Fig. 5.9, has a yield stress of 50 ksi. The girder is supported on a bearing plate with a yield stress of 36 ksi that has dimensions $B = 7$ in, $\ell_b = 6$ in, and a thickness $t = 7/8$ in. The bearing plate sits on a concrete wall with a cylinder strength of $f'_c = 3000$ psi. The support reaction consists of a shear due to dead load of $V_D = 10$ kips and a shear due to live load of $V_L =$ of 30 kips. Check web local yielding.

The relevant properties of the W14 × 30 girder are

$$t_w = 0.27 \text{ in}$$
$$k = 0.785 \text{ in}$$

The nominal web yield strength is

$$R_n = (2.5k + \ell_b)F_y t_w$$
$$= (2.5 \times 0.785 + 6) \times 50 \times 0.27$$
$$= 107 \text{ kips}$$

| LRFD | ASD |
|---|---|
| From ASCE 7 Sec. 2.3.2 combination 2: | From ASCE 7 Sec. 2.4.1 combination 2: |
| $V_u$ = factored support reaction | $V_a$ = factored support reaction |
| $= 1.2V_D + 1.6V_L$ | $= V_D + V_L$ |
| $= 1.2 \times 10 + 1.6 \times 30$ | $= 10 + 30$ |
| $= 60$ kips | $= 40$ kips |
| $\varphi R_n$ = design web yield strength | $R_n/\Omega$ = allowable web shear strength |
| $= 1.0 \times 107$ | $= 107/1.5$ |
| $= 107$ kips | $= 71$ kips |
| $> V_u$ ... satisfactory | $> V_a$ ... satisfactory |
| Using AISC Manual Table 9-4: | Using AISC Manual Table 9-4: |
| $\varphi R_n = \varphi(R_1 + \ell_b R_2)$ | $R_n/\Omega = (R_1 + \ell_b R_2)/\Omega$ |
| $= 26.5 + 6 \times 13.5$ | $= 17.7 + 6 \times 9$ |
| $= 108$ kips | $= 72$ kips |

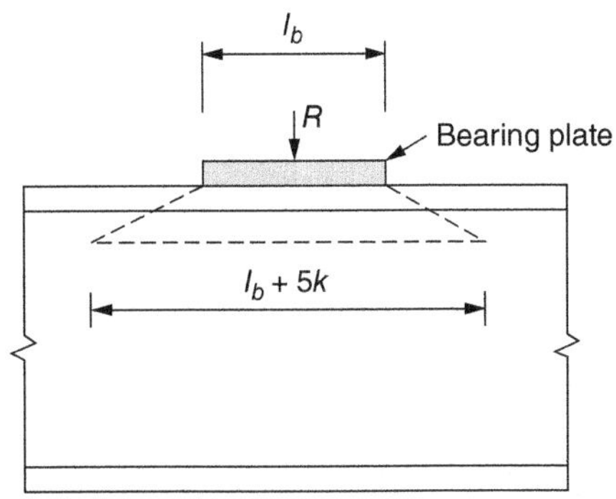


**FIGURE 5.10** Web local yielding at interior of girder.

## Web Yielding at Girder Interior

When a concentrated load is applied to a girder at a distance of more than $d$ from the end of the girder, the load can disperse into the web on both sides of the bearing plate as shown in Fig. 5.10. In accordance with AISC 360 Sec. J10.2(a), the nominal web yield strength is given by AISC 360 Eq. (J10-2) as

$$R_n = (5k + \ell_b)F_y t_w$$

where $\ell_b$ = length of bearing plate
$k$ = distance from outer face of flange to web toe of fillet
$t_w$ = web thickness

**Example 5.12.** Web Local Yielding at Interior of Girder
The W14 × 30 girder, shown in Fig. 5.10, has a yield stress of 50 ksi. The girder supports a column at a distance from the girder end that exceeds the depth of the girder. The column sits on a bearing plate as shown. The load applied by the column consists of a dead load of $V_D = 20$ kips and a live load of $V_L = 60$ kips. Determine the bearing plate length to prevent local web yielding failure.
The relevant properties of the W14 × 30 girder are

$$t_w = 0.27 \text{ in}$$
$$k = 0.785 \text{ in}$$

| LRFD | ASD |
|---|---|
| From ASCE 7 Sec. 2.3.2 combination 2: | From ASCE 7 Sec. 2.4.1 combination 2: |
| $V_u$ = factored load | $V_a$ = factored load |
| $= 1.2V_D + 1.6V_L$ | $= V_D + V_L$ |
| $= 1.2 \times 20 + 1.6 \times 60$ | $= 20 + 60$ |
| = 120 kips | = 80 kips |
| $R_n$ = required nominal web yield strength | $R_n$ = required nominal web shear strength |
| $= V_u/\varphi$ | $= \Omega V_a$ |
| $= 120/1.0$ | $= 1.5 \times 80$ |
| = 120 kips | = 120 kips |
| $\ell_b$ = required length of base plate | $\ell_b$ = required length of base plate |
| $= R_n/F_y t_w - 5k$ | $= R_n/F_y t_w - 5k$ |
| $= 120/(50 \times 0.27) - 5 \times 0.785$ | $= 120/(50 \times 0.27) - 5 \times 0.785$ |
| = 4.96 in ... use 5 in | = 4.96 in ... use 5 in |

## 5.7 Web Crippling

A concentrated load applied to a girder flange produces a compressive stress in the web. If the compressive stress is excessive, local buckling of the web may occur near the junction of the flange and the web. This is known as web crippling and is more critical at the ends of a girder than in the interior.

The nominal web crippling strength of a beam, with a concentrated load applied at a distance of not less than $d/2$ from the end of the beam, is given by AISC 360 Eq. (J10-4) as

$$R_n = 0.80t_w^2[1 + 3(\ell_b/d)(t_w/t_f)^{1.5}](EF_yt_f/t_w)^{0.5}$$

For loads applied at a distance of less than $d/2$ from the end of the beam, and for $\ell_b/d \leq 0.2$, the value of the nominal web crippling strength is given by AISC 360 Eq. (J10-5a) as

$$R_n = 0.40t_w^2[1 + 3(\ell_b/d)(t_w/t_f)^{1.5}](EF_yt_f/t_w)^{0.5}$$

For loads applied at a distance of less than $d/2$ from the end of the beam, and for $\ell_b/d > 0.2$, the value of the nominal web crippling strength is given by AISC 360 Eq. (J10-5b) as

$$R_n = 0.40t_w^2[1 + (4\ell_b/d - 0.2)(t_w/t_f)^{1.5}](EF_yt_f/t_w)^{0.5}$$

where $d$ = overall depth of the beam
$t_f$ = flange thickness
$t_w$ = web thickness
$\ell_b$ = length of bearing plate

For web crippling, AISC 360 Sec. J10.3 gives the resistance factor and the safety factor as

$$\varphi = \text{resistance factor ... LRFD}$$
$$= 0.75$$
$$\Omega = \text{safety factor ... ASD}$$
$$= 2.00$$

To facilitate the determination of web crippling strength, AISC Manual Table 9-4 provides values of $\varphi R_3$, $R_3/\Omega$, $\varphi R_4$, $R_4/\Omega$, $\varphi R_5$, $R_5/\Omega$, $\varphi R_6$, and $R_6/\Omega$, where

$$R_3 = 0.4t_w^2(EF_yt_f/t_w)^{0.5}$$
$$R_4 = 0.4t_w^2[(3/d)(t_w/t_f)^{1.5}](EF_yt_f/t_w)^{0.5}$$
$$R_5 = 0.4t_w^2[1 - 0.2(t_w/t_f)^{1.5}](EF_yt_f/t_w)^{0.5}$$
$$R_6 = 0.4t_w^2[(4/d)(t_w/t_f)^{1.5}](EF_yt_f/t_w)^{0.5}$$

Either a pair of transverse stiffeners, or a doubler plate is required with a minimum length equal to half the depth of the beam, at the location of the concentrated load, when the required web crippling strength exceeds the available strength.

**Example 5.13.** Web Crippling at Interior of Girder
Determine the web crippling capacity of the W14 × 30 girder, shown in Fig. 5.10. The beam has a yield stress of 50 ksi and the base plate length is 8.5 in.

For loads applied at a distance of more than $d/2$ from the end of the beam, the web crippling nominal strength is given by AISC 360 Eq. (J10-4) as

$$R_n = 0.80t_w^2[1 + 3(\ell_b/d)(t_w/t_f)^{1.5}](EF_yt_f/t_w)^{0.5}$$

| LRFD | ASD |
|---|---|
| The design crippling strength is obtained from AISC Manual Table 9-4 as | The allowable crippling strength is obtained from AISC Manual Table 9-4 as |
| $\varphi R_n = 2(\varphi R_3 + \ell_b \varphi R_4)$ | $R_n/\Omega = 2(R_3/\Omega + \ell_b R_4/\Omega)$ |
| $= 2(31.4 + 8.5 \times 4.01)$ | $= 2(21.0 + 8.5 \times 2.68)$ |
| $= 131$ kips | $= 87$ kips |

## 5.8 Web Sidesway Buckling

As shown in Fig. 5.11, web sidesway buckling may occur at the tension flange of a beam loaded with a concentrated load on the compression flange. It occurs when relative lateral displacement between the loaded compression flange and the tension flange is not restrained at the point of application of the load. Sidesway buckling can be prevented by providing lateral bracing to both the tension and compression flanges at the load point and preventing rotation of the compression flange. The lateral bracing at the flanges should each be designed for 1 percent of the concentrated load applied at that point. Alternatively, a pair of stiffeners may be used, designed to carry the full load, and extending from the load point to at least half the depth of the beam. Rotation of the compression flange must also be prevented.

In accordance with AISC 360 Commentary Eq. (C-J10-1), if the loaded flange *is restrained* against rotation, such as when connected to a slab, sidesway buckling will not occur when

$$(hb_f/L_bt_w) > 2.3$$

where $h$ = the clear distance between flanges less the corner radius at each flange, for rolled shapes
= the clear distance between flanges, for built-up welded sections
$t_w$ = web thickness
$b_f$ = flange width
$L_b$ = largest laterally unbraced length along either flange at the load location as defined in AISC 360 Commentary Fig. C-J10.2

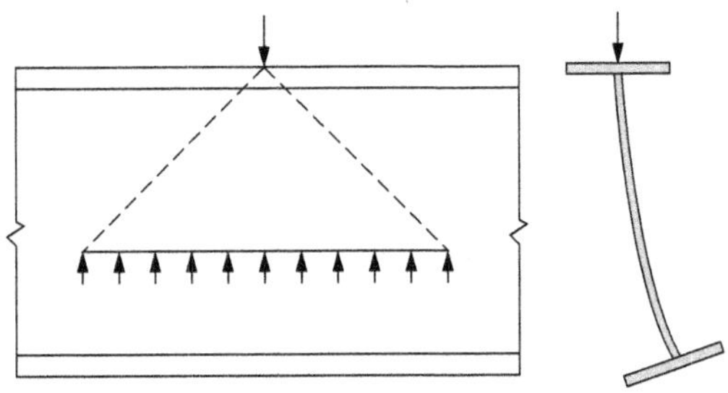

**Figure 5.11** Web sidesway buckling.

In accordance with AISC 360 Commentary Eq. (C-J10-2), if the loaded flange *is not restrained* against rotation sidesway buckling will not occur when

$$(hb_f/L_bt_w) > 1.7$$

The nominal web sidesway buckling strength of a beam, when the loaded flange *is restrained* against rotation and when $(hb_f/L_bt_w) \leq 2.3$, is given by AISC 360 Eq. (J10-6) as

$$R_n = (C_rt_w^3t_f/h^2)[1 + 0.4(hb_f/L_bt_w)^3]$$

where $C_r$ = 960,000 ... for $M_u < M_y$ (LRFD) or $1.5M_a < M_y$ (ASD) at the location of the load

$C_r$ = 480,000 ... for $M_u \geq M_y$ (LRFD) or $1.5M_a \geq M_y$ (ASD) at the location of the the load

$M_y = F_yS$

The nominal web sidesway buckling strength of a beam, when the loaded flange *is not restrained* against rotation and when $(hb_f/L_bt_w) \leq 1.7$, is given by AISC 360 Eq. (J10-7) as

$$R_n = (C_rt_w^3t_f/h^2)[0.4(hb_f/L_bt_w)^3]$$

For web sidesway buckling, AISC 360 Sec. J10.4 gives the resistance factor and the safety factor as

$$\varphi = \text{resistance factor ... LRFD}$$
$$= 0.85$$
$$\Omega = \text{safety factor ... ASD}$$
$$= 1.76$$

**Example 5.14.** Web Sidesway Buckling

A W14 × 30 girder with a yield stress of 50 ksi is simply supported over a span of 20 ft and has a concentrated load applied at midspan. The top flange is restrained against rotation and is laterally braced at the ends and at the location of the load. It may be assumed that $M_u$ and $1.5M_a$ exceed $M_y$ at the location of the load. Determine the nominal web sidesway buckling strength of the W14 × 30 girder.

The properties of a W14 × 30 are

$$t_f = \text{flange thickness}$$
$$= 0.385 \text{ in}$$
$$h/t_w = \text{web slenderness parameter}$$
$$= 45.4 \text{ in}$$
$$t_w = \text{web thickness}$$
$$= 0.27 \text{ in}$$
$$b_f = \text{flange width}$$
$$= 6.73 \text{ in}$$

$h$ = the clear distance between flanges less the corner radius at each flange

$$= t_w(h/t_w)$$

$$= 0.27 \times 45.4$$

$$= 12.26 \text{ in}$$

$$L_b = 240 \text{ in ... from AISC 360 Commentary Fig. C-J10.2}$$

$$hb_f/L_b t_w = 45.4 \times 6.73/240$$

$$= 1.27$$

$$< 2.3 \text{ ... web sidesway buckling may occur}$$

The nominal web sidesway buckling strength is given by AISC 360 Eq. (J10-6) as

$$R_n = (C_r t_w^3 t_f/h^2)[1 + 0.4(hb_f/L_b t_w)^3]$$

$$= (480{,}000 \times 0.27^3 \times 0.385/12.26^2)[1 + 0.4(1.27)^3]$$

$$= 44.03 \text{ kips}$$

## 5.9 Design for Torsion

### Torsion in Closed Sections

A constant torsional moment, or torque, applied to a cylindrical tube produces an angle of twist between the ends of

$$\theta = TL/GJ$$

where $T$ = applied torque
$L$ = length of tube
$G$ = shear modulus of elasticity
= 11,200 ksi for steel
$J$ = torsional constant of the section
= $I_o$ ... for cylindrical members
$I_o$ = polar moment of inertia

The angle of twist per unit length is

$$\theta' = \theta/L$$

$$= T/GJ$$

In a cylindrical tube, plane sections remain plane when a torque is applied. This is an example of pure torsion, also known as St. Venant torsion, and the St. Venant torsional resistance is

$$T_t = GJ\theta'$$

The shear stress produced in a thin cylinder is assumed uniformly distributed over the thickness of the cylinder and is given by

$$\tau_t = T_t R/J$$

and,

$$T_t = \tau_t J/R$$
$$= \tau_t C$$

where $R$ is the outside radius of the tube and $C$ is the torsional constant for hollow structural sections.

In a rectangular tube, a similar approach is adopted since relative longitudinal displacements of points on the cross section are restrained and warping effects are negligible. Values of the torsional constant C are given in AISC Manual Tables 1-11, 1-12, and 1-13 for rectangular, square, and round hollow structural sections.

## Torsion in Open Sections

When torsion is applied to an open section, such as a W-shape, plane sections do not remain plane after deformation. Warping occurs due to relative longitudinal displacements of points on the cross section. The total torsional resistance is then the sum of the St. Venant resistance and the warping resistance giving

$$T = GJ\theta' - EC_w\theta'''$$

where $E$ = modulus of elasticity
$C_w$ = warping constant
$\theta'''$ = third derivative of $\theta$ with respect to the length of the member

Values of $J$ and $C_w$ are given in AISC Manual Tables 1-1 to 1-10.

As shown in Fig. 5.12, shear stress is produced in a W-shape by both St. Venant and warping torsion and normal stress is produced in the flange by warping torsion. The shear stress produced in the flange and web of a W-shape by St. Venant torsion is

$$\tau_t = Gt\theta'$$

where $t$ is member thickness.

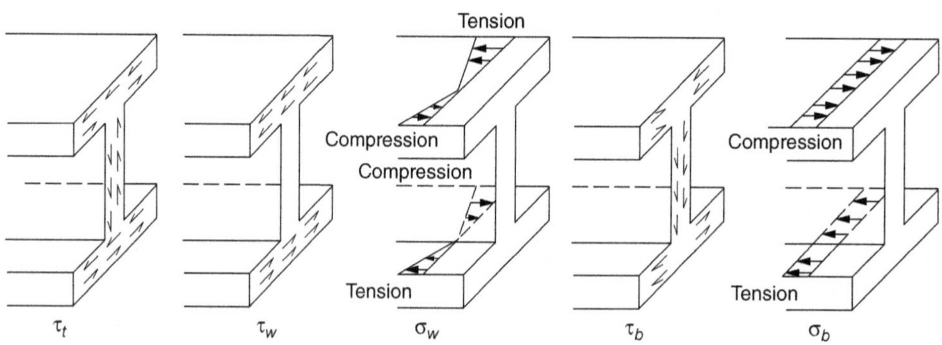


**FIGURE 5.12** Torsion effects in a W-shape.

Warping produces negligible shear in the web of a W-shape. The shear stress produced in the flange varies parabolically and at the center of the flange is given by

$$\tau_w = -ES_{w1}\theta''' / t_f$$

where $S_{w1}$ is warping statical moment for a point in the center of the flange and $t_f$ is flange thickness.

The normal stress at the edge of the flange due to warping is

$$\sigma_w = EW_{no}\theta''$$

where $W_{no}$ is normalized warping function at a point on the flange edge.

Values of $\theta$, $\theta'$, $\theta''$, $\theta'''$, $W_{no}$, and $S_{w1}$ are given by Seaburg[7] for several loading cases and support conditions for W-shapes, channels, T-shapes, and angles.

Support restraints modify values of $\theta$ and its derivatives and influence the warping effects. At a free end, where cross sections can warp freely

$$\theta'' = 0$$

At a torsionally pinned end, such as a web shear tab connection, where cross sections can rotate laterally

$$\theta = \theta''$$
$$= 0$$

At a torsionally fixed end, such as a welded connection with stiffener plates, where cross sections cannot warp

$$\theta = \theta'$$
$$= 0$$

## Specification Provisions

For torsion, AISC 360 Sec. H3.1 gives the resistance factor and the safety factor as

$$\varphi = \text{resistance factor ... LRFD}$$
$$= 0.90$$
$$\Omega = \text{safety factor ... ASD}$$
$$= 1.67$$

When considering the interaction of flexure, compression, and torsion on a member, if the required torsional strength is not more than 20 percent of the available torsional strength, torsional effects may be neglected. Otherwise, it is necessary to check the interaction expression given by AISC 360 Eq. (H3-6) as

$$(P_r/P_c + M_r/M_c) + (V_r/V_c + T_r/T_c)^2 \leq 1.00$$

where $P_r$ = required axial strength
$M_r$ = required flexural strength
$V_r$ = required shear strength
$T_r$ = required torsional strength
$P_c$ = available axial strength
$M_c$ = available flexural strength
$V_c$ = available shear strength
$T_c$ = available torsional strength

For the LRFD method

$$P_c = \varphi P_n$$
$$M_c = \varphi_b M_n$$
$$V_c = \varphi_v V_n$$
$$T_c = \varphi_T T_n$$

For the ASD method

$$P_c = P_n/\Omega$$
$$M_c = M_n/\Omega_b$$
$$V_c = V_n/\Omega_v$$
$$T_c = T_n/\Omega_T$$

## Round HSS Subject to Torsion

The nominal torsional strength is given by AISC 360 Eq. (H3-1) as

$$T_n = F_{cr} C$$

where $C$ = torsional constant for hollow structural sections
$F_{cr}$ = critical stress given by the larger value of

$$F_{cr} = 1.23E/[(L/D)^{0.5}(D/t)^{5/4}] \text{ ... for torsional yielding}$$

or,

$$F_{cr} = 0.60E/(D/t)^{3/2}$$

but shall not exceed,

$$F_{cr} \leq 0.6F_y$$

where $L$ = length of member
$D$ = outside diameter
$t$ = wall thickness

**Example 5.15.** Torsion in a Round HSS
Determine the nominal torsional strength of an HSS10.000 × 0.500 with a yield stress of 42 ksi and a length of 20 ft.

The relevant properties are obtained from AISC Manual Table 1-13 as

$$t = 0.465 \text{ in}$$

$$D/t = 21.5$$

$$D = 10.0 \text{ in}$$

$$C = 63.5 \text{ in}^3$$

$F_{cr}$ = critical stress given by the larger value of

$$F_{cr} = 1.23E/[(L/D)^{0.5}(D/t)^{5/4}] \text{ ... for torsional yielding}$$

$$= 1.23 \times 29{,}000/[(240/10)^{0.5}(21.5)^{5/4}]$$

$$= 157 \text{ ksi}$$

or,

$$F_{cr} = 0.60E/(D/t)^{3/2}$$

$$= 0.60 \times 29{,}000/21.5^{3/2}$$

$$= 175 \text{ ksi}$$

However, the maximum value of $F_{cr}$ is

$$F_{cr} = 0.6F_y$$

$$= 0.6 \times 42$$

$$= 25.2 \text{ ksi ... governs}$$

$T_n$ = nominal torsional strength

$$= F_{cr}C$$

$$= 25.2 \times 63.5$$

$$= 1600 \text{ kip-in}$$

## Rectangular HSS Subject to Torsion

The nominal torsional strength is given by AISC 360 Eq. (H3-1) as

$$T_n = F_{cr}C$$

In accordance with AISC 360 Eq. (H3-3), the nominal torsional strength is governed by torsional yield when the slenderness parameter is

$$h/t \leq 2.45(E/F_y)^{0.5}$$

then,

$$F_{cr} = 0.6F_y$$

where $h$ is clear distance between the flanges less the inside corner radius on each side and $t$ is design wall thickness.

In accordance with AISC 360 Eq. (H3-4), the nominal torsional capacity is governed by inelastic torsional buckling when the slenderness parameter is

$$2.45(E/F_y)^{0.5} < h/t \leq 3.07(E/F_y)^{0.5}$$

then,

$$F_{cr} = 0.6F_y[2.45(E/F_y)^{0.5}]/(h/t)$$

In accordance with AISC 360 Eq. (H3-5), the nominal torsional capacity is governed by elastic torsional buckling when the slenderness parameter is

$$3.07(E/F_y)^{0.5} < h/t \leq 260$$

then,

$$F_{cr} = 0.458\pi^2 E/(h/t)^2$$

**Example 5.16.** Torsion in a Rectangular HSS
An HSS10 × 5 × 3/8 beam with a yield stress of 46 ksi is simply supported over a span of 10 ft and has a concentrated load applied at midspan. The load is applied at an eccentricity of $e = 12$ in with respect to the centroid of the beam and consists of a dead load of $W_D = 4$ kips and a live load of $W_L = 12$ kips. The beam ends are flexurally and torsionally pinned and the self-weight of the beam may be neglected. The beam is oriented with flexure occurring about the major axis. Determine the adequacy of the beam.

The relevant properties obtained from AISC Manual Table 1-11 as

$$t = 0.349 \text{ in}$$

$$b/t = 11.3$$

$$h/t = 25.7$$

$$< 2.46(E/F_y)^{0.5} = 61.8 \text{ ... hence, from AISC 360 Eq. (G2-3), } C_v = 1.0$$

$$< 2.45(E/F_y)^{0.5} \text{ ... hence, from AISC 360 Eq. (H3-3), torsional yield governs}$$

$$Z = 30.4 \text{ in}^3$$

$$C = 31.2 \text{ in}^3$$

$$h = 25.7 \times 0.349$$

$$= 8.97 \text{ in}$$

In accordance with AISC Manual Table 1-12A, the section is compact. Hence, flexural yield governs and

$$M_n = F_y Z$$

$$= 46 \times 30.4/12$$

$$= 117 \text{ kip-ft}$$

$$V_n = 0.6F_y A_w C_v$$

$$= 0.6F_y(2ht)C_v \text{ ... from AISC 360 Sec. G5}$$

$$= 0.6 \times 46 \times 2 \times 8.97 \times 0.349 \times 1.0$$

$$= 173 \text{ kips}$$

$$T_n = F_{cr} C \text{ ... from AISC 360 Eq. (H3-1)}$$

$$= 0.6 \times 46 \times 31.2/12$$

$$= 72 \text{ kip-ft}$$

The factored loads are

| LRFD | ASD |
|---|---|
| From ASCE 7 Sec. 2.3.2 combination 2: | From ASCE 7 Sec. 2.4.1 combination 2: |
| $W_u$ = factored load | $W_a$ = factored load |
| $= 1.2W_D + 1.6W_L$ | $= W_D + W_L$ |
| $= 1.2 \times 4.0 + 1.6 \times 12.0$ | $= 4.0 + 12.0$ |
| = 24 kips | = 16 kips |

The required strengths are

| LRFD | ASD |
|---|---|
| $M_r$ = required flexural strength | $M_r$ = required flexural strength |
| $= W_u L/4$ | $= W_a L/4$ |
| $= 24 \times 10/4$ | $= 16 \times 10/4$ |
| = 60 kip-ft | = 40 kips |
| $V_r$ = required shear strength | $V_r$ = required shear strength |
| $= W_u/2$ | $= W_a/2$ |
| $= 24/2$ | $= 16/2$ |
| = 12 kips | = 8 kips |
| $T_r$ = required torsional strength | $T_r$ = required torsional strength |
| $= W_u e$ | $= W_a e$ |
| $= 24 \times 12/12$ | $= 16 \times 12/12$ |
| = 24 kip-ft | = 16 kip-ft |

The available strengths are

| LRFD | ASD |
|---|---|
| $M_c$ = design flexural strength | $M_c$ = allowable flexural strength |
| $= \varphi_b M_n$ | $= M_n/\Omega_b$ |
| $= 0.9 \times 117$ | $= 117/1.67$ |
| = 105 kip-ft | = 70 kip-ft |
| $V_c$ = design shear strength | $V_c$ = allowable shear strength |
| $= \varphi_v V_n$ | $= V_n/\Omega_v$ |
| $= 0.9 \times 173$ | $= 173/1.67$ |
| = 156 kips | = 104 kips |
| $T_c$ = design torsional strength | $T_c$ = allowable torsional strength |
| $= \varphi_T T_n$ | $= T_n/\Omega_v$ |
| $= 0.9 \times 72$ | $= 72/1.67$ |
| = 65 kip-ft | = 43 kip-ft |

Hence, $T_r > 0.2 \times T_c$ and torsional effects must be considered. Substituting in the interaction expression AISC 360 Eq. (H3-6) gives

| LRFD | ASD |
|---|---|
| $M_r/M_c + (V_r/V_c + T_r/T_c)^2$ | $M_r/M_c + (V_r/V_c + T_r/T_c)^2$ |
| $= 60/105 + (12/156 + 24/65)^2$ | $= 40/70 + (8/104 + 16/43)^2$ |
| $= 0.77$ | $= 0.77$ |
| $< 1.0$ ... satisfactory | $< 1.0$ ... satisfactory |

## W-Shape Subject to Torsion

In accordance with AISC 360 Sec. H3.3, three limit states must be considered. These are

- The limit state of yielding under normal stress given by AISC 360 Eq. (H3-7) as

$$F_n = F_y$$

The required normal stress is

$$f_r = \sigma_b + \sigma_w$$

where $F_n$ = nominal torsional strength
$F_y$ = yield stress
$\sigma_b$ = normal stress due to bending
$\sigma_w$ = normal stress due to warping torsion

- The limit state of shear yielding under shear stress given by AISC 360 Eq. (H3-8) as

$$F_n = 0.6F_y$$

The required shear stress is

$$f_r = \tau_b + \tau_w + \tau_t$$

where $\tau_b$ = shear stress due to bending
$\tau_w$ = shear stress due to warping torsion
$\tau_t$ = shear stress due to St. Venant torsion

- The limit state of buckling given by AISC 360 Eq. (H3-9) as

$$F_n = F_{cr}$$

where $F_{cr}$ is critical buckling stress for the section.

AISC 360 Commentary Sec. H3.3 indicates that the required stresses may be obtained by elastic analysis. Normal stresses and shear stresses may be considered separately as the maximum values rarely occur at the same location.

**Example 5.17.** Torsion in a W-Shape
A W14 × 109 beam, with a yield stress of 50 ksi and a span of $L = 10$ ft, is loaded as shown in Fig. 5.13. The beam supports a uniformly distributed dead load of $w_D = 1$ kip/ft, that includes the self-weight of

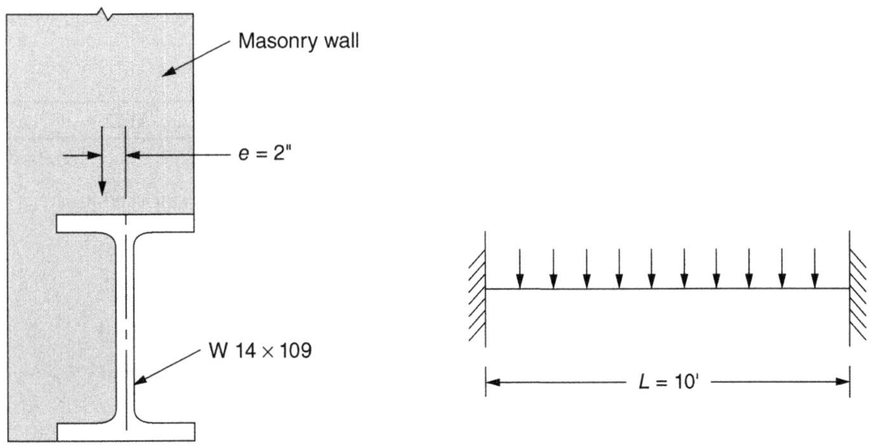


**Figure 5.13** Torsion load on a W-shape.

the beam, and a uniformly distributed live load of $w_L = 3$ kips/ft. The ends of the beam are flexurally and torsionally fixed and the load has an eccentricity of $e = 2$ in. Determine the stresses produced in the beam at the supports by the torsional effects.

The relevant torsional properties of a W14 × 109 are given in Seaburg App. A as

$$J = 7.12 \text{ in}^4$$

$$a = 85.7 \text{ in}$$

$$W_{no} = 49.1 \text{ in}^2$$

$$S_{w1} = 154 \text{ in}^4$$

$$L/a = 10 \times 12/85.7$$

$$= 1.40$$

From AISC Manual Table 1-1, the relevant properties of a W14 × 109 are

$$t_f = 0.860 \text{ in}$$

$$t_w = 0.525 \text{ in}$$

At a fixed ended support, the rotation of the beam and the first derivative of the rotation are

$$\theta = 0$$

$$\theta' = 0$$

Hence, the shear stress in the beam web and flanges at the support, due to St. Venant torsion, is

$$\tau_{tf} = Gt\theta'$$

$$= 0$$

$$= \tau_{tw}$$

Applying American Society of Civil Engineers, *Minimum Design Loads for Buildings and Other Structures* ASCE 7, Secs. 2.3 and 2.4 gives

| LRFD | ASD |
|---|---|
| From ASCE 7 Sec. 2.3.2 combination 2: | From ASCE 7 Sec. 2.4.1 combination 2: |
| $w_u$ = factored load | $w_a$ = factored load |
| $= 1.2w_D + 1.6w_L$ | $= w_D + w_L$ |
| $= 1.2 \times 1.0 + 1.6 \times 3.0$ | $= 1.0 + 3.0$ |
| = 6 kips/ft | = 4 kips/ft |
| $T_r$ = required torsion | $T_r$ = required torsion |
| $= w_u e$ | $= w_a e$ |
| $= 6 \times 2$ | $= 4 \times 2$ |
| = 12 kip-in/ft | = 8 kip-in/ft |
| $T_r L/GJ = 12 \times 10/(11{,}200 \times 7.12)$ | $T_r L/GJ = 8 \times 10/(11{,}200 \times 7.12)$ |
| $= 1.505 \times 10^{-3}$ rad/in | $= 1.003 \times 10^{-3}$ rad/in |
| $= \eta$ | $= \eta$ |

The torsional functions at the support, for a fixed ended member with a distributed torsional load are obtained from Seaburg App. B, Case 7 as

| LRFD | ASD |
|---|---|
| $\theta'' = 0.22\eta/2a$ | $\theta'' = 0.22\eta/2a$ |
| $= 0.22 \times 1.505 \times 10^{-3}/(2 \times 85.7)$ | $= 0.22 \times 1.003 \times 10^{-3}/(2 \times 85.7)$ |
| $= 1.931 \times 10^{-6}$ rad/in$^2$ | $= 1.287 \times 10^{-6}$ rad/in$^2$ |
| $\theta''' = -1.0\eta/2a^2$ | $\theta''' = -1.0\eta/2a^2$ |
| $= -1.0 \times 1.505 \times 10^{-3}/(2 \times 85.7^2)$ | $= -1.0 \times 1.003 \times 10^{-3}/(2 \times 85.7^2)$ |
| $= -1.025 \times 10^{-7}$ rad/in$^3$ | $= -6.828 \times 10^{-8}$ rad/in$^3$ |

The maximum normal stress in the beam flange at the support, due to warping torsion, is obtained from Seaburg Eq. (4.3a) as

| LRFD | ASD |
|---|---|
| $\sigma_{wf} = EW_{no}\theta''$ | $\sigma_{wf} = EW_{no}\theta''$ |
| $= 29{,}000 \times 49.1 \times 1.931 \times 10^{-6}$ | $= 29{,}000 \times 49.1 \times 1.287 \times 10^{-6}$ |
| = 2.75 ksi | = 1.83 ksi |

The maximum shear stress in the beam flange at the support, due to warping torsion, is obtained from Seaburg Eq. (4.2a) as

| LRFD | ASD |
|---|---|
| $\tau_{wf} = -ES_{w1}\theta'''/t_f$ | $\tau_{wf} = -ES_{w1}\theta'''/t_f$ |
| $= -29{,}000 \times 154 \times -1.025 \times 10^{-7}/0.860$ | $= -29{,}000 \times 154 \times -6.828 \times 10^{-8}/0.860$ |
| $= 0.53$ ksi | $= 0.36$ ksi |

As shown in Fig. 5.12, shear stress and normal stress are produced in a W-shape by flexure. The normal stress produced in the flange is

$$\sigma_{bf} = M/S$$

where $M$ is applied moment and $S$ is elastic section modulus.

The shear stress produced in the web by flexure is

$$\tau_{bw} = VQ_w/It_w$$

where $V$ = applied shear
$Q_w$ = statical moment at mid-depth of the section
$t_w$ = web thickness

The shear stress produced in the flange by flexure is

$$\tau_{bf} = VQ_f/It_f$$

where $Q_f$ is statical moment for a point in the flange directly above the web and $t_f$ is flange thickness.

**Example 5.18.** Torsion and Combined Stress in a W-Shape
For the W14 × 109 beam, analyzed in Example 5.17, determine the combined flexural and torsional stresses at the supports and determine the adequacy of the beam.

The relevant torsional properties of a W14 × 109 are given in Seaburg App. A as

$Q_f$ = statical moment for a point in the flange directly above the web

$= 41.2$ in$^3$

$Q_w$ = statical moment at mid-depth of the section

$= 95.9$ in$^3$

From AISC Manual Table 1-1, the relevant properties of a W14 × 109 are

$$I = 1240 \text{ in}^4$$

$$S = 173 \text{ in}^3$$

$$t_f = 0.860 \text{ in}$$

$$t_w = 0.525 \text{ in}$$

From AISC Manual Table 3-2, the relevant properties of a W14 × 109 are

$$L_p = 13.2 \text{ ft}$$

$$L_r = 48.5 \text{ ft}$$

**Check limit state of yielding under normal stress**

| LRFD | ASD |
|---|---|
| $w_u$ = factored load<br>$= 6$ kips/ft<br>Required bending moment at the support is<br>$M_r = w_u L^2/12$<br>$= 6 \times 10^2/12$<br>$= 50$ kip-ft<br>Corresponding normal stress in the flange due to bending is<br>$\sigma_{bf} = M_r/S$<br>$= 50 \times 12/173$<br>$= 3.47$ ksi<br>Required normal stress in the flange due to combined bending and torsion is<br>$f_{rn} = \sigma_{bf} + \sigma_{wf}$<br>$= 3.47 + 2.75$<br>$= 6.22$ ksi<br>$< 0.9F_y$ ... *limit state of yielding under normal stress is satisfactory* | $w_a$ = factored load<br>$= 4$ kips/ft<br>Required bending moment at the support is<br>$M_r = w_a L^2/12$<br>$= 4 \times 10^2/12$<br>$= 33.3$ kip-ft<br>Corresponding normal stress in the flange due to bending is<br>$\sigma_{bf} = M_r/S$<br>$= 33.3 \times 12/173$<br>$= 2.31$ ksi<br>Required normal stress in the flange due to combined bending and torsion is<br>$f_{rn} = \sigma_{bf} + \sigma_{wf}$<br>$= 2.31 + 1.83$<br>$= 4.14$ ksi<br>$< F_y/1.67$ ... *limit state of yielding under normal stress is satisfactory* |

**Check limit state of flange shear yielding**

| LRFD | ASD |
|---|---|
| Required shear force at the support is<br>$V_r = w_u L/2$<br>$= 6 \times 10/2$<br>$= 30$ kips<br>Corresponding shear stress in the flange due to bending is<br>$\tau_{bf} = V_r Q_f/It_f$<br>$= 30 \times 41.2/(1240 \times 0.860)$<br>$= 1.16$ ksi<br>Required shear stress in the flange due to combined bending and torsion is<br>$f_{rv} = \tau_{bf} + \tau_{wf} + \tau_{tf}$<br>$= 1.16 + 0.53 + 0$<br>$= 1.69$ ksi<br>$< 0.6F_y$ ... *limit state of flange shear yielding is satisfactory* | Required shear force at the support is<br>$V_r = w_u L/2$<br>$= 4 \times 10/2$<br>$= 20$ kips<br>Corresponding shear stress in the flange due to bending is<br>$\tau_{bf} = V_r Q_f/It_f$<br>$= 20 \times 41.2/(1240 \times 0.860)$<br>$= 0.77$ ksi<br>Required shear stress in the flange due to combined bending and torsion is<br>$f_{rv} = \tau_{bf} + \tau_{wf} + \tau_{tf}$<br>$= 0.77 + 0.36 + 0$<br>$= 1.13$ ksi<br>$< 0.4F_y$ ... *limit state of flange shear yielding is satisfactory* |

**Check limit state of web shear yielding**

| LRFD | ASD |
|---|---|
| Shear stress in the web due to flexure is | Shear stress in the web due to flexure is |
| $\tau_{bw} = V_r Q_w / I t_w$ | $\tau_{bw} = V_r Q_w / I t_w$ |
| $= 30 \times 95.9/(1240 \times 0.525)$ | $= 20 \times 95.9/(1240 \times 0.525)$ |
| $= 4.42$ ksi | $= 2.95$ ksi |
| Required shear stress in the web due to combined bending and torsion is | Required shear stress in the web due to combined bending and torsion is |
| $f_{rv} = \tau_{bw} + \tau_{ww} + \tau_{tw}$ | $f_{rv} = \tau_{bw} + \tau_{ww} + \tau_{tw}$ |
| $= 4.42 + 0 + 0$ | $= 2.95 + 0 + 0$ |
| $= 4.42$ ksi | $= 2.95$ ksi |
| $< 0.6F_y$ ... *limit state of web shear yielding is satisfactory* | $< 0.4F_y$ ... *limit state of web shear yielding is satisfactory* |

**Check limit state of lateral-torsional buckling**

| LRFD | ASD |
|---|---|
| The unbraced length is | The unbraced length is |
| $L_b = 10$ ft: | $L_b = 10$ ft: |
| $< L_p$ ... *limit state of lateral-torsional buckling does not govern* | $< L_p$ ... *limit state of lateral-torsional buckling does not govern* |
| The beam is adequate. | The beam is adequate. |

# References


1. American Institute of Steel Construction (AISC). 2011. *Steel Construction Manual*, 14th edition, AISC, Chicago, IL.
2. American Institute of Steel Construction (AISC). 2010. *Specification for Structural Steel Buildings* (AISC 360-10), AISC, Chicago, IL.
3. American Society of Civil Engineers (ASCE). 2010. *Minimum Design Loads for Buildings and Other Structures* (ASCE 7-10), ASCE, Reston, VA.
4. Fisher, J. M. 2005. *Industrial Buildings, Roofs to Anchor Rods*, Design Guide No. 7, 2nd Edition. AISC, Chicago, IL.
5. Geschwindner, L. R. 2006. "Evolution of Shear Lag and Block Shear Provisions in the AISC Specification," *Engineering Journal*, 4th Quarter 2006.
6. Epstein, H. I. and Aleksiewicz, L. J. 2008. "Block Shear Equations Revisited...Again," *Engineering Journal*, 1st Quarter 2008.
7. Seaburg, P. A. and Carter, C. J. 1997. *Torsional Analysis of Structural Steel Members*, Design Guide No. 9. AISC, Chicago, IL.

## Problems

**5.1** *Given:* A C12 × 30 beam, with a yield stress of 36 ksi, is simply supported over a span of 10 ft. The beam supports a uniformly distributed load consisting of a dead load component of $w_D = 1$ kip/ft, which includes an allowance for the weight of the beam, and a live load component of $w_L = 3$ kips/ft. The beam is continuously braced on its compression flange.

*Find:* Using allowable stress level (ASD) load combinations, the maximum shear stress and the average shear stress in the beam.

**5.2** *Given:* A C12 × 30 beam, with a yield stress of 36 ksi, is simply supported over a span of 10 ft. The beam supports a uniformly distributed load consisting of a dead load component of $w_D = 1$ kip/ft, which includes an allowance for the weight of the beam, and a live load component of $w_L = 3$ kips/ft. The beam is continuously braced on its compression flange.

*Find:* Using strength level (LRFD) load combinations, the maximum shear stress and the average shear stress in the beam.

**5.3** *Given:* A W30 × 90 beam, with a yield stress of 50 ksi, is simply supported over a span of 20 ft. The beam supports a uniformly distributed dead load of $w_D = 3$ kips/ft, which includes the self weight of the beam, and a uniformly distributed live load of $w_L = 9$ kips/ft. The beam is continuously braced on its compression flange.

*Find:* Using allowable stress level (ASD) load combinations, whether the beam is adequate in shear.

**5.4** *Given:* A W30 × 90 beam, with a yield stress of 50 ksi, is simply supported over a span of 20 ft. The beam supports a uniformly distributed dead load of $w_D = 3$ kips/ft, which includes the self-weight of the beam, and a uniformly distributed live load of $w_L = 9$ kips/ft. The beam is continuously braced on its compression flange.

*Find:* Using strength level (LRFD) load combinations, whether the beam is adequate in shear.

**5.5** *Given:* A W30 × 99 beam, with a yield stress of 50 ksi, is simply supported over a span of 10 ft. The beam supports a uniformly distributed dead load of $w_D = 15$ kips/ft, which includes the self-weight of the beam, and a uniformly distributed live load of $w_L = 45$ kips/ft. The beam is continuously braced on its compression flange.

*Find:* Using allowable stress level (ASD) load combinations, whether the beam is adequate in shear.

**5.6** *Given:* A W30 × 99 beam, with a yield stress of 50 ksi, is simply supported over a span of 10 ft. The beam supports a uniformly distributed dead load of $w_D = 15$ kips/ft, which includes the self-weight of the beam, and a uniformly distributed live load of $w_L = 45$ kips/ft. The beam is continuously braced on its compression flange.

*Find:* Using strength level (LRFD) load combinations, whether the beam is adequate in shear.

**5.7** *Given:* An M12 × 10 beam, with a yield stress of 50 ksi, is simply supported over a span of 10 ft. The beam supports a uniformly distributed dead load of $w_D = 0.6$ kip/ft, which includes the self-weight of the beam, and a uniformly distributed live load of $w_L = 1.8$ kips/ft. The beam is continuously braced on its compression flange.

*Find:* Using allowable stress level (ASD) load combinations, whether the beam is satisfactory for shear.

**5.8** *Given:* An M12 × 10 beam, with a yield stress of 50 ksi, is simply supported over a span of 10 ft. The beam supports a uniformly distributed dead load of $w_D = 0.6$ kip/ft, which includes the self-weight of the beam, and a uniformly distributed live load of $w_L = 1.8$ kips/ft. The beam is continuously braced on its compression flange.

*Find:* Using strength level (LRFD) load combinations, whether the beam is satisfactory for shear.

**5.9** *Given:* An HP12 × 84 beam, with a yield stress of 50 ksi, is loaded about the weak axis.

*Find:* Using the allowable stress level (ASD) method, the allowable shear capacity of the beam.

**5.10** *Given:* An HP12 × 84 beam, with a yield stress of 50 ksi, is loaded about the weak axis.

*Find:* Using the strength level (LRFD) method, the design shear capacity of the beam.

**5.11** *Given:* The built-up girder shown in Fig. 5.14 consists of a W30 × 99 with a C15 × 33.9 welded to the top flange. Shear at the end supports consists of a shear due to dead load of $V_D = 30$ kips, which includes the weight of the beam, and a shear due to live load of $V_L =$ of 90 kips.

*Find:* Using allowable stress level (ASD) load combinations, the longitudinal shear at the interface.

**5.12** *Given:* The built-up girder shown in Fig. 5.14 consists of a W30 × 99 with a C15 × 33.9 welded to the top flange. Shear at the end supports consists of a shear due to dead load of $V_D = 30$ kips, which includes the weight of the beam, and a shear due to live load of $V_L = 90$ kips.

*Find:* Using strength level (LRFD) load combinations, the longitudinal shear at the interface.

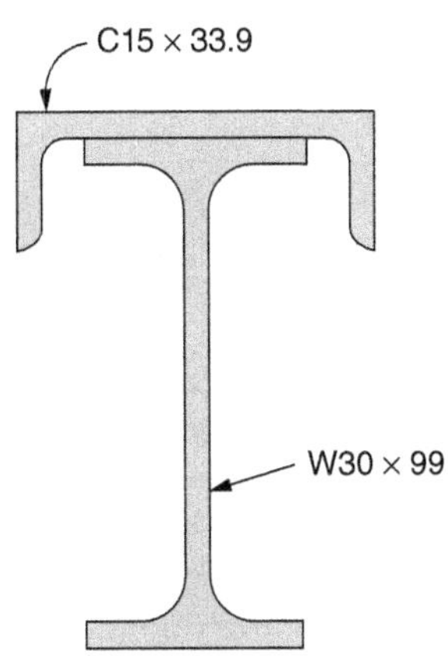


**FIGURE 5.14** Details for Problems 5.11 and 5.12.

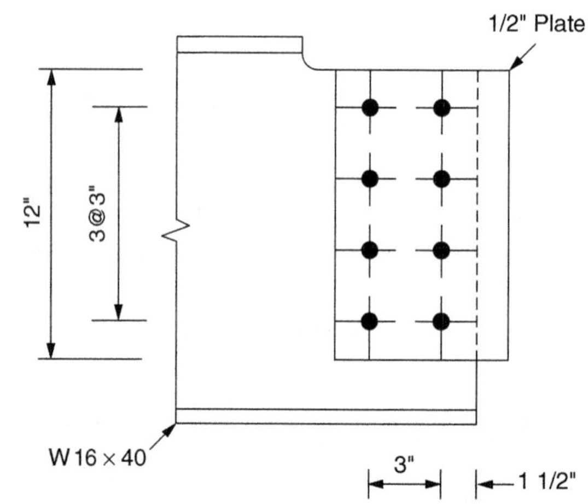


**FIGURE 5.15** Details for Problems 5.13 and 5.14.

**5.13** *Given:* The coped W16 × 40 beam, shown in Fig. 5.15, which has a yield stress of 50 ksi and a tensile strength of 65 ksi, is connected to a 1/2-in shear tab. The bolt diameter is 3/4 in.

*Find:* Using allowable stress level (ASD) load combinations, the block shear strength of the beam web.

**5.14** *Given:* The coped W16 × 40 beam, shown in Fig. 5.15, which has a yield stress of 50 ksi and a tensile strength of 65 ksi, is connected to a 1/2-in shear tab. The bolt diameter is 3/4 in.

*Find:* Using strength level (LRFD) load combinations, the block shear strength of the beam web.

**5.15** *Given:* The two 3½- × 3½- × 3/8-in angles, shown in Fig. 5.16, are bolted to a 3/4-in plate with 3/4-in diameter bolts. All components are grade 36 steel.

*Find:* Using allowable stress level (ASD) load combinations, the block shear strength of the angles.

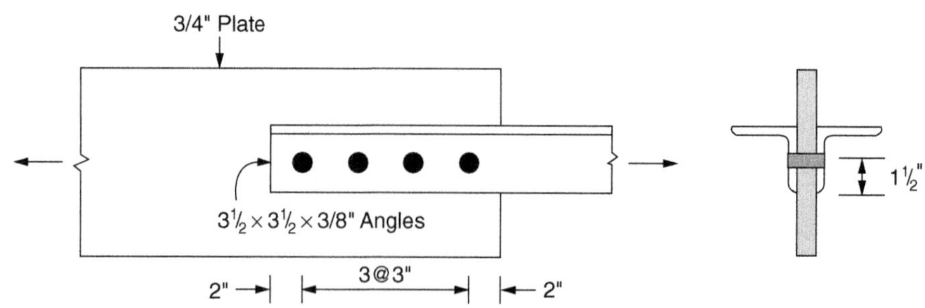


**FIGURE 5.16** Details for Problems 5.15 and 5.16.

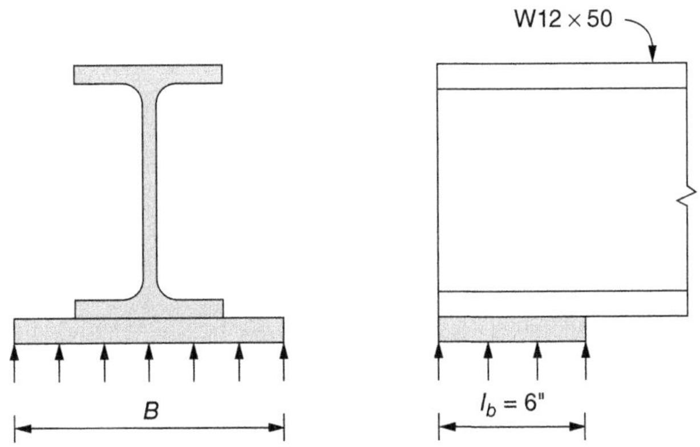


**FIGURE 5.17** Details for Problems 5.17 and 5.18.

**5.16** *Given:* The two 3½- × 3½- × 3/8-in angles, shown in Fig. 5.16, are bolted to a 3/4-in plate with 3/4-in diameter bolts. All components are grade 36 steel.

*Find:* Using strength level (LRFD) load combinations, the block shear strength of the angles.

**5.17** *Given:* The W12 × 50 girder, shown in Fig. 5.17, has a yield stress of 50 ksi. The girder is supported on a bearing plate with a yield stress of 36 ksi and a length $l_b$ = 6 in. The bearing plate sits on a concrete wall with a cylinder strength of $f'_c$ = 3000 psi. The support reaction consists of a shear due to dead load of $V_D$ = 15 kips and a shear due to live load of $V_L$ = 45 kips.

*Find:* Using allowable stress level (ASD) load combinations, the bearing plate width and thickness.

**5.18** *Given:* The W12 × 50 girder, shown in Fig. 5.17, has a yield stress of 50 ksi. The girder is supported on a bearing plate with a yield stress of 36 ksi and a length $l_b$ = 6 in. The bearing plate sits on a concrete wall with a cylinder strength of $f'_c$ = 3000 psi. The support reaction consists of a shear due to dead load of $V_D$ = 15 kips and a shear due to live load of $V_L$ = 45 kips.

*Find:* Using strength level (LRFD) load combinations, the bearing plate width and thickness.

**5.19** *Given:* The W12 × 50 girder, shown in Fig. 5.18, has a yield stress of 50 ksi. The girder supports a concentrated load at midspan on a bearing plate with a length $l_b$ = 5 in.

*Find:* Using the allowable stress level (ASD) method, the beam web crippling strength.

**5.20** *Given:* The W12 × 50 girder, shown in Fig. 5.18, has a yield stress of 50 ksi. The girder supports a concentrated load at midspan on a bearing plate with a length $l_b$ = 5 in.

*Find:* Using the strength level (LRFD) method, the beam web crippling strength.

**5.21** *Given:* An HSS10.000 × 0.500 with a yield stress of 42 ksi.

*Find:* The nominal torsional strength.

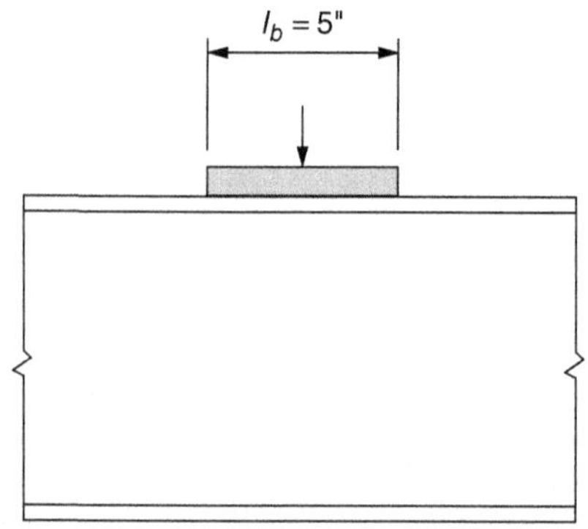


**FIGURE 5.18** Details for Problems 5.19 and 5.20.

**5.22** *Given:* An HSS10 × 4 × 5/8 beam with a yield stress of 46 ksi is simply supported over a span of 10 ft and has a concentrated load applied at midspan. The load is applied at an eccentricity of $e$ = 12 in with respect to the centroid of the beam and consists of a dead load of $W_D = 5$ kips and a live load of $W_L = 15$ kips. The beam ends are flexurally and torsionally pinned and the self-weight of the beam may be neglected. The beam is oriented with flexure occurring about the major axis.

*Find:* Using allowable stress level (ASD) load combinations, whether the beam is adequate.

**5.23** *Given:* An HSS10 × 4 × 5/8 beam with a yield stress of 46 ksi is simply supported over a span of 10 ft and has a concentrated load applied at midspan. The load is applied at an eccentricity of $e$ = 12 in with respect to the centroid of the beam and consists of a dead load of $W_D = 5$ kips and a live load of $W_L = 15$ kips. The beam ends are flexurally and torsionally pinned and the self-weight of the beam may be neglected. The beam is oriented with flexure occurring about the major axis.

*Find:* Using strength level (LRFD) load combinations, whether the beam is adequate.

# CHAPTER 6

# Design of Compression Members

## 6.1 Introduction

As shown in Fig. 6.1, a compression member is a structural element that supports loads applied along its longitudinal axis. Axially loaded members are compression members that are nominally free from applied bending moments and consist of the several types illustrated. The column in a building frame, as shown in Fig. 6.1*a*, supports the gravity loads applied to the frame. Failure of a column may cause complete collapse of the structure above the failed column. A brace in a braced frame, as shown in Fig. 6.1*b*, provides the lateral restraint to resist the horizontal forces caused by wind or earthquake. A strut in a roof truss, as shown in Fig. 6.1*c*, is a web member that provides the required compression force. Similarly, as shown in Fig. 6.1*d*, the top chord provides the compression members in a truss.

### Compression Limit State

For doubly symmetric compact and noncompact compression members, flexural buckling is normally the governing limit state. An ideal pin-ended column with an applied axial load causing flexural buckling in the elastic range, fails at a critical stress given by the Euler expression as

$$\begin{aligned} F_e &= \text{elastic critical buckling stress} \\ &= \pi^2 E/(L/r)^2 \end{aligned}$$

where $L$ is length of the column and $r$ is governing radius of gyration.

The Euler expression may be modified to account for alternative support conditions by using the factor $K$ to give

$$F_e = \pi^2 E/(KL/r)^2$$

where $KL/r$ = slenderness ratio
$K$ = effective length factor
= factor that modifies actual column length and support conditions to an equivalent pin-ended column

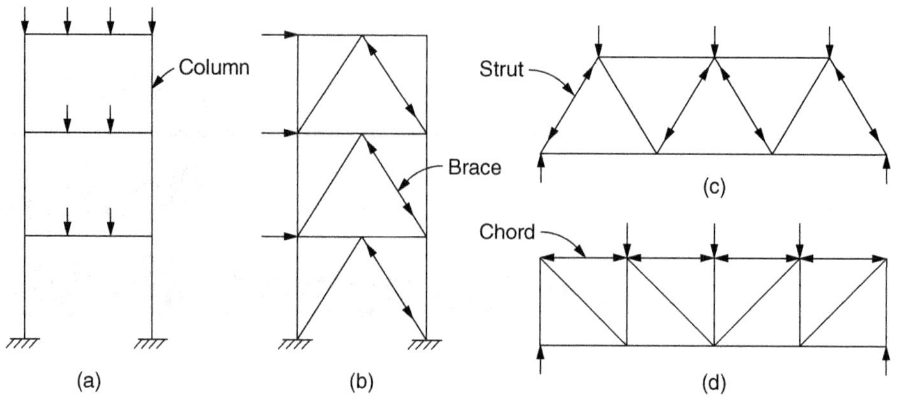


**FIGURE 6.1** Types of compression members.

This expression indicates that the buckling stress is directly proportional to the modulus of elasticity of the material and is independent of the yield stress. The slenderness ratio has a pronounced effect on the critical stress and should preferably be limited to a maximum recommended value of 200. Members with a slenderness ratio exceeding 200 are excessively flexible and are liable to damage during fabrication, transport, and erection.

In practice, several factors tend to cause buckling at a stress less than the Euler critical stress and these include

- Initial out-of-straightness of the column causes bending stresses.
- Residual stresses in the column produce higher than anticipated stresses on the section.
- Eccentricity of the applied load causes flexural stresses in the column.
- Actual column end restraints differ from the assumed criteria.
- *P*-delta effects produce additional stress in the column.

Hence, the American Institute of Steel Construction, *Specification for Structural Steel Buildings* (AISC 360)[1] Eq. (E3-3) provides an empirical expression to model the behavior of actual columns in the elastic range. As shown in Fig. 6.2, this expression is valid for a slenderness ratio of

$$KL/r > 4.71(E/F_y)^{0.5}$$

For smaller values of the slenderness ratio, the outer fibers of the column cross-section yield before the critical load is reached and the Euler expression is no longer valid. Hence, AISC 360 Eq. (E3-2) provides an empirical expression to model the behavior of the actual column. As shown in Fig. 6.2, this diverges rapidly from the Euler expression. As the slenderness ratio becomes smaller, the failure load approaches the squash load and the critical stress approaches the yield stress.

For singly symmetric, unsymmetric, and cruciform sections, additional limit states to consider are torsional and flexural-torsional buckling. Local buckling may also be a factor to consider. This occurs when slender elements of the cross-section buckle before the overall strength of the section is reached. The limiting width-to-thickness ratio for

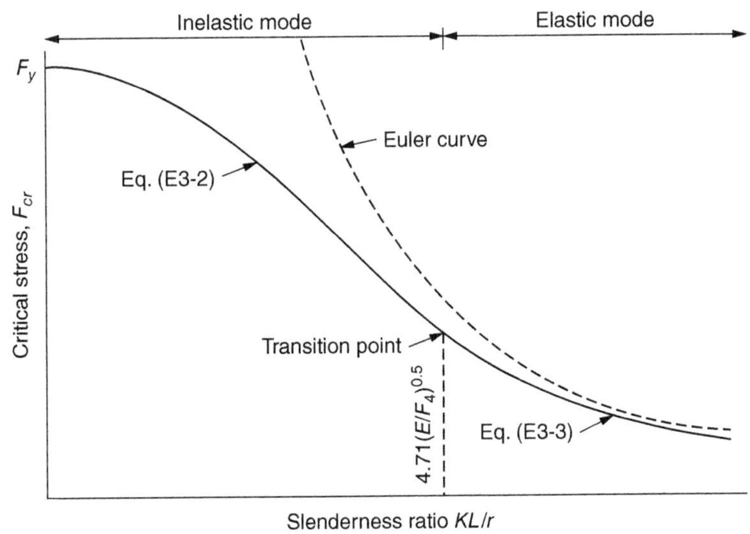


**FIGURE 6.2** Standard column curve.

slender elements in compression are given in AISC 360 Table B4.1a. All W-shapes with a yield stress of $F_y = 50$ ksi have nonslender flanges. All W-shapes in the range W8 to W14, with a yield stress of $F_y = 50$ ksi, have nonslender webs with the exception of W14 × 43.

## 6.2 Effective Length

The effective length factor $K$ converts the actual column length $L$ to an equivalent pin-ended column of length $KL$. The factor accounts for the influence of restraint conditions on the behavior of the column and $KL$ represents the length over which the column actually buckles. The effective length factor is determined in accordance with AISC 360 Commentary App. 7 and two methods are presented:

- Tabulated factors for stand-alone columns with well-defined support conditions.
- Alignment charts for columns in a rigid framed structure.

### Tabulated Factors

AISC 360 Table C-A-7.1 specifies effective length factors for well-defined, standard conditions of restraint and these are illustrated in Fig. 6.3 for sway and nonsway columns. The values for ideal end conditions are indicated and also recommended values that allow for practical site conditions. These values may only be used in simple cases when the tabulated end conditions are approximated in practice.

#### *K* Values for Braced Frames

In accordance with AISC 360 Commentary Sec. A-7.2, braced frames may be analyzed and designed as vertical cantilevered pin-connected truss systems, ignoring any secondary moments. The effective length factor for components of the frame is then taken as 1.0.

| Braced frame | | | | Sway frame | | | |
|---|---|---|---|---|---|---|---|
| End restraints | Ideal K | Practical K | Shape | End restraints | Ideal K | Practical K | Shape |
| Fixed at both ends | 0.5 | 0.65 | 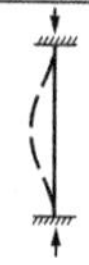 | Fixed at one end with the other end fixed in direction but not held in position | 1.0 | 1.2 |  |
| Fixed at one end, pinned at the other end | 0.7 | 0.8 |  | Pinned at one end with the other end fixed in direction but not held in position | 2.0 | 2.0 |  |
| Pinned at both ends | 1.0 | 1.0 |  | Fixed at one end with the other end free | 2.0 | 2.1 |  |

**FIGURE 6.3** Effective length factors for sway and nonsway columns.

**Example 6.1.** Braced Frame Effective Length Factors
For the braced frame shown in Fig. 6.4, determine the effective length factors of the columns. The girder may be considered infinitely rigid and the columns are adequately braced in the transverse direction.

The effective length factors may be obtained from Fig. 6.3. For column 12 which is fixed at one end and pinned at the other

$$K = 0.8$$

For column 34 which is fixed at both ends

$$K = 0.65$$

## *K* Values for Sway Frames

Effective length factors of sway columns are similarly obtained. In general, the effective length factor exceeds 1.0 except for frames with high structural stiffness. For these frames, the sidesway amplification factor is

$$B_2 = \Delta_{2nd}/\Delta_{1st}$$

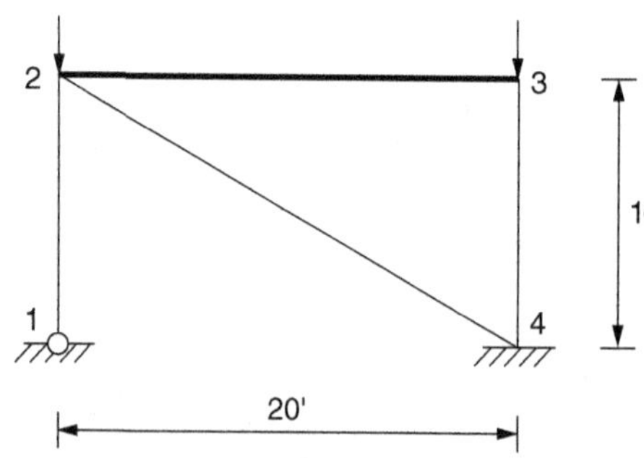


**FIGURE 6.4** Details for Example 6.1.

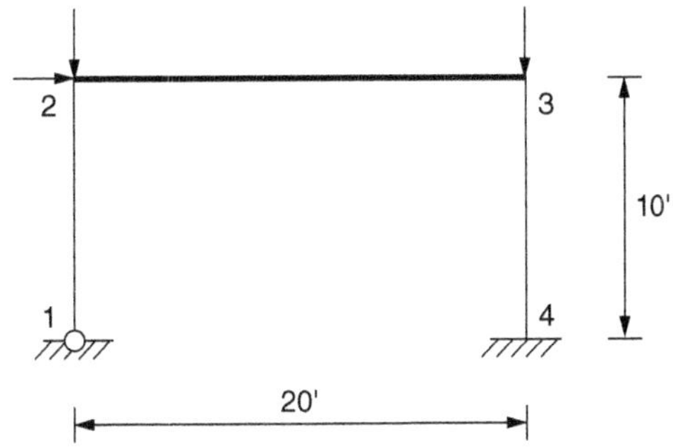


**FIGURE 6.5** Details for Example 6.2.

where $\Delta_{2nd}$ is second-order drift and $\Delta_{1st}$ is first-order drift. When $B_2 \leq 1.1$, AISC 360 Sec. A-7.2 permits the use of an effective length factor of $K = 1.0$.

Leaning columns are columns that contribute little or nothing to the sway stiffness of a story or the resistance to lateral loads. These columns may be designed as pin-ended, with an effective length factor of $K = 1.0$, in accordance with AISC 360 Sec. A-7.2. However, all other columns in the story must be designed to support the destabilizing $P$-$\Delta$ moments developed from the loads on the leaning columns.

**Example 6.2.** Sway Frame Effective Length Factors

For the sway frame shown in Fig. 6.5, determine the effective length factors of the columns. The girder may be considered infinitely rigid and the columns are adequately braced in the transverse direction.

For column 12 which is fixed at one end and pinned at the other

$$K = 2.0$$

For column 34 which is fixed at both ends

$$K = 1.2$$

## 6.3 Alignment Charts

For columns in a frame with rigid joints, the effective length factor may be determined based on the restraint provided at each end of the column. The alignment charts are given in AISC 360 Figs. C-A-7.1 and C-A-7.2 and are shown combined in Fig. 6.6. To utilize the alignment chart, the stiffness ratio at the two ends of the column under consideration must be determined and this is defined by

$$G = \Sigma(E_cI_c/L_c)/\Sigma(E_gI_g/L_g)$$

where $\Sigma(E_cI_c/L_c)$ is the sum of the $EI/L$ values for all columns meeting at the joint and $\Sigma(E_gI_g/L_g)$ is the sum of the $EI/L$ values for all girders meeting at the joint. Using the stiffness ratio $G_A$ at one end of a column and the stiffness ratio $G_B$ at the other end of the column, the alignment chart is entered and the effective length factor is obtained.

For a column with a pinned base, the stiffness ratio is theoretically infinity and AISC 360 Commentary Sec. A-7.2 recommends a practical value of $G = 10$. For a column with a fixed base, the stiffness ratio is theoretically zero and AISC 360 Commentary Sec. A-7.2 recommends a practical value of $G = 1.0$.

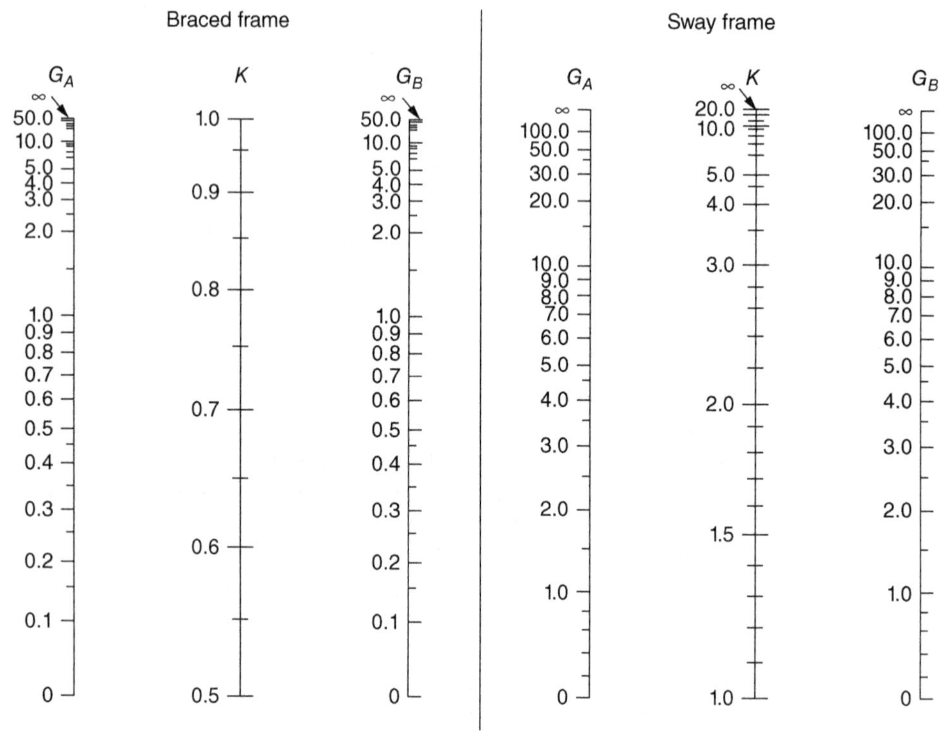


**FIGURE 6.6** Alignment chart for effective length.

## Alignment Chart for Braced Frame

For a braced frame, with rigid joints, the girders are assumed bent in single curvature and the alignment chart is based on a stiffness value for the girders of $2EI/L$. If the far end of a girder is pinned, its stiffness is $3EI/L$, as determined in Williams[2] Part 2, Sec. 7.4. Hence, the calculated $(E_gI_g/L_g)$ value is multiplied by 1.5 before determining the value of $\Sigma(E_gI_g/L_g)$ and entering the chart. If the far end of a girder is fixed, its stiffness is $4EI/L$, as determined in Williams Part 2, Sec. 7.4. Hence, the calculated $(E_gI_g/L_g)$ value is multiplied by 2.0 before determining the value of $\Sigma(E_gI_g/L_g)$ and entering the chart.

**Example 6.3.** Braced Frame Effective Length Factors by Alignment Chart
For the braced frame shown in Fig. 6.7, determine the effective length factors of columns 45 and 56. The girders have a moment of inertia of twice that of the columns. The braces may be considered pinned at each end and the columns are adequately braced in the transverse direction.

For the fixed connection at joint 6, AISC 360 Commentary Sec. A-7.2 recommends a practical value of

$$G_6 = 1.0$$

At joint 5

$$\begin{aligned} G_5 &= \Sigma(E_cI_c/L_c)/\Sigma(E_gI_g/L_g) \\ &= (2 \times 1/10)/(2/20) \\ &= 2.0 \end{aligned}$$

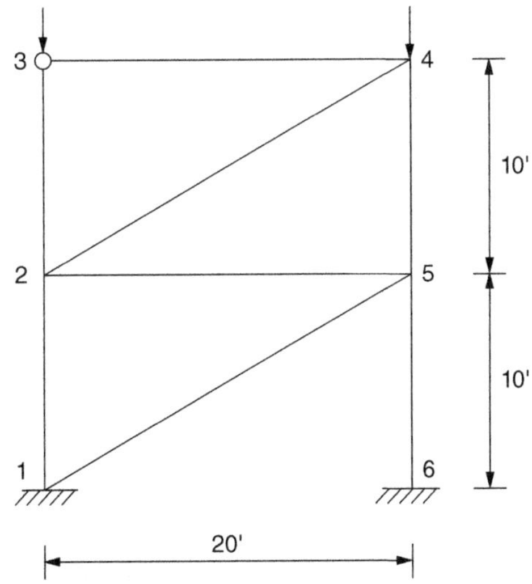


**Figure 6.7** Details for Example 6.3.

From the alignment chart for braced frames, the effective length factor is

$$K_{56} = 0.82$$

At joint 4, allowing for the pinned end at joint 3

$$\begin{aligned} G_4 &= \Sigma(E_cI_c/L_c)/\Sigma 1.5(E_gI_g/L_g) \\ &= (1/10)/1.5(2/20) \\ &= 0.67 \end{aligned}$$

From the alignment chart for braced frames, the effective length factor is

$$K_{45} = 0.78$$

## Alignment Chart for Sway Frame

The alignment chart for sway frames is illustrated in Fig. 6.6. For a sway frame with rigid joints, the girders are assumed bent in double curvature and the alignment chart is based on a stiffness value for the girders of $6EI/L$. If the far end of a girder is pinned, its stiffness is $3EI/L$ and the calculated $(E_gI_g/L_g)$ value is multiplied by 0.5 before determining the value of $\Sigma(E_gI_g/L_g)$ and entering the chart. If the far end of a girder is fixed, its stiffness is $4EI/L$ and the calculated $(E_gI_g/L_g)$ value is multiplied by 0.67 before determining the value of $\Sigma(E_gI_g/L_g)$ and entering the chart. In addition, the girder length is modified by AISC 360 Eq. (C-A-7-4) to give a revised girder length of

$$L_g' = L_g(2 - M_F/M_N)$$

where $M_F$ = far end girder moment

$M_N$ = near end girder moment

$M_F/M_N$ = +ve when the girder is in reverse curvature

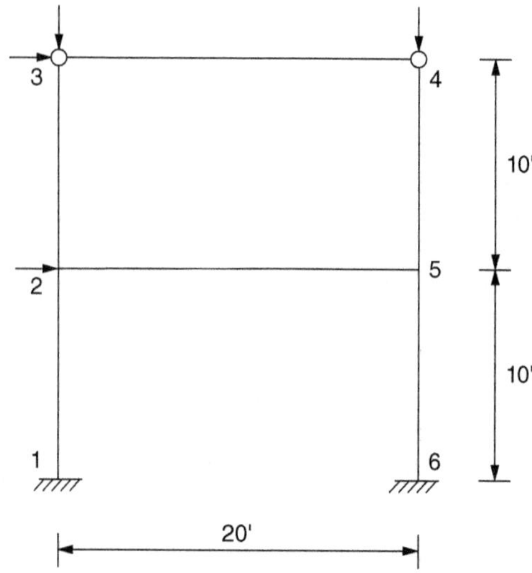


**Figure 6.8** Details for Example 6.4.

**Example 6.4.** Sway Frame Effective Length Factors by Alignment Chart

For the sway frame shown in Fig. 6.8, determine the effective length factors of columns 45 and 56. The girders have a moment of inertia of twice that of the columns. The columns are adequately braced in the transverse direction.

For the fixed connection at joint 6, AISC 360 Commentary Sec. A-7.2 recommends a practical value of

$$G_6 = 1.0$$

At joint 5, because of the skew symmetrical loading and symmetrical structure, the girder is in reverse curvature with $M_{25} = M_{52}$ and

$$\begin{aligned} L_g' &= L_g(2 - M_{25}/M_{52}) \\ &= L_g(2 - 1) \\ &= L_g \end{aligned}$$

$$\begin{aligned} G_5 &= \Sigma(E_cI_c/L_c)/\Sigma(E_gI_g/L_g) \\ &= (2 \times 1/10)/(2/20) \\ &= 2.0 \end{aligned}$$

From the alignment chart for sway frames, the effective length factor is

$$K_{56} = 1.45$$

For the pinned connection at joint 4, AISC 360 Commentary Sec. A-7.2 recommends a practical value of

$$G_4 = 10$$

From the alignment chart for sway frames, the effective length factor is

$$K_{45} = 2.1$$

## Stiffness Reduction Factors

The alignment charts given in Fig. 6.6 are applicable to columns in the elastic range of stress. When the axial load on a column is increased and portions of the section yield, the effective modulus of elasticity of the yielded areas reduces to zero. Hence, in the inelastic range, this reduction in the modulus of elasticity has the effect of reducing the overall stiffness of the column. This may be compensated for by multiplying the stiffness ratio $G$, by the stiffness reduction factor given by

$$\tau_b = E_T/E$$

where $E$ is the elastic modulus of elasticity and $E_T$ is the tangent modulus.

Hence, the reduced stiffness ratio is

$$G_a = \tau_b G$$

Values of the stiffness reduction factor are tabulated in American Institute of Steel Construction, *Steel Construction Manual* (AISC Manual)[3] Table 4-21 for steel members with a yield stress of 35, 36, 42, 46, or 50 ksi for varying values of $P_r/A_g$. Where $P_r$ is the required axial compressive strength using ASD or LRFD load combinations. It is conservative to ignore the stiffness reduction factor.

**Example 6.5.** Effective Length Factors Allowing for Stiffness Reduction

For the sway frame shown in Fig. 6.8, determine the effective length factor of column 56. The girders have a moment of inertia of twice that of the columns and the columns are adequately braced in the transverse direction. The ratio $P_r/A_g = 20$ ksi for ASD load combinations and 30 ksi for LRFD load combinations. The yield stress of all members is $F_y = 50$ ksi.

For the fixed connection at joint 6, AISC 360 Commentary Sec. A-7.2 recommends a practical value of

$$G_6 = 1.0$$

At joint 5, because of the skew symmetrical loading and symmetrical structure, the girder is in reverse curvature with $M_{25} = M_{52}$ and

$$\begin{aligned} L_g' &= L_g(2 - M_{25}/M_{52}) \\ &= L_g(2 - 1) \\ &= L_g \end{aligned}$$

Ignoring stiffness reduction, the stiffness ratio at joint 5 is

$$\begin{aligned} G_5 &= \Sigma(E_c I_c/L_c)/\Sigma(E_g I_g/L_g) \\ &= (2 \times 1/10)/(2/20) \\ &= 2.0 \end{aligned}$$

Allowing for stiffness reduction, the amended stiffness ratio at joint 5 is

| LRFD | ASD |
|---|---|
| From AISC Manual Table 4-21, the stiffness reduction factor for $P_u/A_g = 30$ ksi is<br>$\tau_b = 0.960$<br>The amended stiffness ratio at joint 5 is<br>$G_{a5} = \tau_b G_5$<br>$= 0.960 \times 2.0$<br>$= 1.92$<br>From the alignment chart for sway frames, the effective length factor is<br>$K_{56} = 1.42$ | From AISC Manual Table 4-21, the stiffness reduction factor for $P_a/A_g = 20$ ksi is<br>$\tau_b = 0.922$<br>The amended stiffness ratio at joint 5 is<br>$G_{a5} = \tau_b G_5$<br>$= 0.922 \times 2.0$<br>$= 1.844$<br>From the alignment chart for sway frames, the effective length factor is<br>$K_{56} = 1.41$ |

## 6.4 Axially Loaded Compression Members

### Flexural Buckling of Members without Slender Elements

Inelastic buckling governs when $KL/r \leq 4.71(E/F_y)^{0.5}$ and $F_y/F_e \leq 2.25$. Then AISC 360 Eq. (E3-2) governs and the critical stress is

$$F_{cr} = (0.658^{\kappa})F_y$$

where $\kappa = F_y/F_e$
$F_e$ = elastic critical buckling stress
$= \pi^2 E/(KL/r)^2$ ... from AISC 360 Eq. (E3-4)
$\geq F_y/2.25$

Elastic buckling governs when $KL/r > 4.71(E/F_y)^{0.5}$ and $F_y/F_e > 2.25$. Then AISC 360 Eq. (E3-3) governs and the critical stress is

$$F_{cr} = 0.877F_e$$

The transition point between AISC 360 Eqs. (E3-2) and (E3-3) is shown in Table 6.1.

The nominal compressive strength is given by AISC 360 Eq. (E3-1) as

$$P_n = F_{cr}A_g$$

where $A_g$ is gross area of member.

| $F_y$, ksi | Limiting $KL/r$ | $F_e$, ksi |
|---|---|---|
| 36 | 134 | 16.0 |
| 50 | 113 | 22.2 |

**Table 6.1** Transition Point on Standard Column Curve

The design compressive strength and the allowable compressive strength may be obtained from AISC 360, Sec. E1 as

$$\varphi_c P_n = \text{design compressive strength}$$
$$\geq P_u \text{ ... required compressive strength using LRFD load combinations}$$
$$P_n/\Omega_c = \text{allowable compressive strength}$$
$$\geq P_a \text{ ... required compressive strength using ASD load combinations}$$

where $\varphi_c$ = resistance factor for compression
= 0.9
$\Omega_c$ = safety factor for compression
= 1.67

Once the governing slenderness ratio of a column is established, the available critical stress may be obtained directly from AISC Manual Table 4-22 for steel members with a yield stress of 35, 36, 42, 46, or 50 ksi.

**Example 6.6.** Analysis of a W-Shape Column
A W12 × 50 column in a braced frame has a yield stress of 50 ksi and a height of 15 ft. The column is pinned at the top and fixed at the bottom and has no intermediate bracing. Determine the available strength of the column.

The relevant properties of the W12 × 50 section are obtained from AISC Manual Table 1-1 as

$$A_g = 14.6 \text{ in}^2 \qquad r_x = 5.18 \text{ in} \qquad r_y = 1.96 \text{ in}$$

From Fig. 6.3, for a braced frame, the effective length factor about both axes is

$$K = 0.8$$

Hence, the slenderness ratio about each axis is

$$KL/r_x = 0.8 \times 15 \times 12/5.18$$
$$= 27.80$$
$$KL/r_y = 0.8 \times 15 \times 12/1.96$$
$$= 73.47 \text{ ... governs}$$
$$> KL/r_x$$
$$KL/r_y < 4.71(E/F_y)^{0.5} = 113 \text{ ... as given in Table 6.1}$$

Hence, inelastic buckling governs and

$$F_e = \text{elastic critical buckling stress}$$
$$= \pi^2 E/(KL/r)^2 \text{ ... from AISC 360 Eq. (E3-4)}$$
$$= 3.14^2 \times 29{,}000/73.47^2$$
$$= 52.97 \text{ ksi}$$
$$\kappa = F_y/F_e$$
$$= 50/52.97$$
$$= 0.94$$

AISC 360 Eq. (E3-2) governs and the critical stress is

$$\begin{aligned} F_{cr} &= (0.658^{\kappa})F_y \\ &= 0.658^{0.94} \times 50 \\ &= 33.74 \text{ ksi} \end{aligned}$$

The nominal compressive strength is given by AISC 360 Eq. (E3-1) as

$$\begin{aligned} P_n &= F_{cr}A_g \\ &= 33.74 \times 14.6 \\ &= 493 \text{ kips} \end{aligned}$$

| LRFD | ASD |
|---|---|
| The design compressive strength is | The allowable compressive strength is |
| $\varphi_c P_n = 0.9 \times 493$ | $P_n/\Omega_c = 493/1.67$ |
| $= 444$ kips | $= 295$ kips |

Alternatively, from AISC Manual Table 4-22 for a value of $KL/r = 73.47$ and $F_y = 50$ ksi, the available critical stress is obtained as

| LRFD | ASD |
|---|---|
| $\varphi_c F_{cr} = 30.35$ ksi | $F_{cr}/\Omega_c = 20.20$ ksi |
| The design compressive strength is | The allowable compressive strength is |
| $\varphi_c P_n = \varphi_c F_{cr} A_g$ | $P_n/\Omega_c = (F_{cr}/\Omega_c)A_g$ |
| $= 30.35 \times 14.6$ | $= 20.20 \times 14.6$ |
| $= 443$ kips | $= 295$ kips |

When the slenderness ratio about the $y$-axis of a member governs, values of the available axial load may be obtained from AISC Manual Tables 4-1 to 4-12 for varying effective lengths and for different column sections. The available loads are tabulated with respect to the minor axis and these tabulated values may be used directly when the slenderness ratio about the minor axis exceeds the slenderness ratio about the major axis.

**Example 6.7.** Design of a W-Shape Column

A column in a braced frame with a height of 15 ft is pinned at the top and fixed at the bottom and has no intermediate bracing. The loading consists of an axial dead load of $P_D = 70$ kips, which includes the weight of the column, and an axial live load of $P_L = 220$ kips. Determine the lightest W12 column, with a yield stress of 50 ksi, which can support the load.

From Example 6.6, the slenderness ratio about the $y$-axis governs. Hence, the tabulated available loads in AISC Manual Table 4-1 may be used directly.

From Fig. 6.3, the effective length about each axis is

$$\begin{aligned} KL &= 0.8 \times 15 \\ &= 12 \text{ ft} \end{aligned}$$

Applying American Society of Civil Engineers, *Minimum Design Loads for Buildings and Other Structures* (ASCE 7, Ref. 4) Secs. 2.3 and 2.4 gives

| LRFD | ASD |
|---|---|
| From ASCE 7 Sec. 2.3.2 combination 2: | From ASCE 7 Sec. 2.4.1 combination 2: |
| $P_u$ = factored axial load | $P_a$ = factored axial load |
| $= 1.2P_D + 1.6P_L$ | $= P_D + P_L$ |
| $= 1.2 \times 70 + 1.6 \times 220$ | $= 70 + 220$ |
| = 436 kips | = 290 kips |
| From AISC Manual Table 4-1 for a value of $KL$ = 12 ft, a W12 × 50 provides a design strength of | From AISC Manual Table 4-1 for a value of $KL$ = 12 ft, a W12 × 50 provides an allowable strength of |
| $\varphi_c P_n$ = 443 kips | $P_n/\Omega_c$ = 295 kips |
| > 436 kips ... satisfactory | > 290 kips ... satisfactory |

When the minor axis of a W-section is braced at closer intervals than the major axis, the slenderness ratio about both axes must be investigated to determine which governs. The larger of the two values will control the design. To facilitate the procedure, the ratio $r_x/r_y$ is included in AISC Manual Tables 4-1 and 4-2. The effective length about the major axis is divided by $r_x/r_y$ to give an equivalent effective length, with respect to the minor axis, which has the same load carrying capacity as the actual effective length about the major axis. This equivalent effective length is compared with the actual effective length about the minor axis and the design load is obtained directly from the load tables using the larger of the two values.

**Example 6.8.** Analysis of a W-Shape Column with Intermediate Bracing
Determine the available axial strength of a W12 × 50 column, with a yield stress of 50 ksi, which is 15 ft high, is pinned at each end, and is braced at 5 ft intervals about the $y$-axis.

The effective length about the $y$-axis is

$$KL_y = 5 \text{ ft}$$

The effective length about the $x$-axis is

$$KL_x = 15 \text{ ft}$$

From AISC Manual Table 4-1, a W12 × 50 column has a value of

$$r_x/r_y = 2.64$$

The equivalent effective length about the major axis with respect to the $y$-axis is

$$\begin{aligned}(KL_y)_{\text{equiv}} &= (KL_x)/(r_x/r_y)\\ &= 15/2.64\\ &= 5.68 \text{ ft ... governs}\\ &> KL_y\end{aligned}$$

| LRFD | ASD |
|---|---|
| From AISC Manual Table 4-1 for a value of $KL$ = 5.68 ft, a W12 × 50 provides a design strength of<br>$\varphi_c P_n = 598$ kips | From AISC Manual Table 4-1 for a value of $KL$ = 5.68 ft, a W12 × 50 provides an allowable strength of<br>$P_n/\Omega_c = 398$ kips |

## Torsional and Flexural-Torsional Buckling of Members without Slender Elements

For singly symmetric and unsymmetric members, the shear center does not coincide with the centroid of the section and it is necessary to consider torsional and flexural-torsional buckling. Similarly, in doubly symmetric cruciform or built-up columns and single angles with $b/t > 20$, the torsional unbraced length may exceed the flexural unbraced length of the member.

The nominal strength in compression for these members is given by AISC 360 Eq. (E4-1) as

$$P_n = A_g F_{cr}$$

where $A_g$ is gross area of member and $F_{cr}$ is critical stress.

### Double-Angle and T-Shaped Members

For double-angle and T-shaped members, AISC 360 Eq. (E4-2) defines the critical stress as

$$F_{cr} = [(F_{cry} + F_{crz})/2H]\{1 - [1 - 4F_{cry}F_{crz}H/(F_{cry} + F_{crz})^2]^{0.5}\}$$

where $F_{crz}$ = critical torsional buckling stress given by AISC 360 Eq. (E4-3) as
$= GJ/A_g\overline{r}_o^2$
$G$ = shear modulus of elasticity of steel
= 11,200 ksi
$J$ = torsional constant
$\overline{r}_o$ = polar radius of gyration about the shear center
$\overline{r}_o^2 = x_o^2 + y_o^2 + (I_x + I_y)/A_g$
$I_x$ = moment of inertia about the $x$-axis
$I_y$ = moment of inertia about the $y$-axis
$x_o$ = $x$-ordinate of the shear center with respect to the centroid
$y_o$ = $y$-ordinate of the shear center with respect to the centroid
$H = 1 - (x_o^2 + y_o^2)/\overline{r}_o^2$

The critical stress for flexural buckling about the axis of symmetry $F_{cry}$ is obtained from AISC 360 Eq. (E3-2) or (E3-3) as appropriate. Values of the available axial load are tabulated in AISC Manual Tables 4-8 to 4-10 for double angle members, for varying effective lengths, and in AISC Manual Table 4-7 for WT-shapes.

### Doubly Symmetric Members

For doubly symmetric members, the elastic buckling stress is determined from AISC 360 Eq. (E4-4) as

$$F_e = [\pi^2 EC_w/(K_z L)^2 + GJ]/(I_x + I_y)$$

where $C_w$ is warping constant and $K_z$ is effective length factor for torsional buckling. The critical stress for torsional or flexural-torsional buckling $F_{cr}$ is then obtained from AISC 360 Eq. (E3-2) or (E3-3) as appropriate.

### Singly Symmetric Members Where y Is the Axis of Symmetry

For singly symmetric members where $y$ is the axis of symmetry, the elastic buckling stress is determined from AISC 360 Eq. (E4-5) as

$$F_e = (F_{ey} + F_{ez})\{1 - [1 - 4F_{ey}F_{ez}H/(F_{ey} + F_{ez})^2]^{0.5}\}/2H$$
$$F_{ey} = \pi^2 E/(K_y L/r_y)^2$$
$$F_{ez} = [\pi^2 EC_w/(K_z L)^2 + GJ]/A_g\overline{r}_o^2$$

### Unsymmetric Members

For unsymmetric members, the elastic buckling stress $F_e$ is determined as the lowest root of the cubic equation given by AISC 360 Eq. (E4-6) as

$$(F_e - F_{ex})(F_e - F_{ey})(F_e - F_{ez}) - F_e^2(F_e - F_{ey})(x_o/\overline{r}_o)^2 - F_e^2(F_e - F_{ex})(y_o/\overline{r}_o)^2 = 0$$

where $F_{ex}$ is equal to $\pi^2 E/(K_x L/r_x)^2$.

**Example 6.9.** Analysis of a WT Column

A WT6 × 25 column in a braced frame has a yield stress of 50 ksi and a height of 12 ft. The column is pinned at the top and bottom and has no intermediate bracing. Determine the available strength of the column.

The relevant properties of the WT6 × 25 section are obtained from AISC Manual Table 1-8 as

$$A_g = 7.3 \text{ in}^2 \qquad r_x = 1.60 \text{ in} \qquad r_y = 1.96 \text{ in} \qquad \overline{y} = 1.17 \text{ in}$$
$$I_x = 18.7 \text{ in}^4 \qquad I_y = 28.2 \text{ in}^4 \qquad t_f = 0.64 \text{ in} \qquad J = 0.855 \text{ in}^4$$

The shape contains no slender elements.

The shear center lies on the intersection of the longitudinal axis of the flange and the longitudinal axis of the stem. Hence,

$$x_o = 0 \text{ in}$$
$$y_o = \overline{y} - t_f/2$$
$$= 1.17 - 0.64/2$$
$$= 0.85 \text{ in}$$
$$\overline{r}_o^2 = x_o^2 + y_o^2 + (I_x + I_y)/A_g$$
$$= 0 + 0.85^2 + (18.7 + 28.2)/7.3$$
$$= 7.15$$
$$H = 1 - (x_o^2 + y_o^2)/\overline{r}_o^2$$
$$= 1 - (0 + 0.85^2)/7.15$$
$$= 0.90$$
$$F_{crz} = GJ/A_g\overline{r}_o^2$$
$$= 11{,}200 \times 0.855/(7.30 \times 7.15)$$
$$= 183 \text{ ksi}$$

**Calculate critical stress for flexural buckling about the *x*-axis**

From Fig. 6.3, for a braced frame, the effective length factor about both axes is

$$K = 1.0$$

Hence, the slenderness ratio about the $x$-axis governs and the elastic buckling stress about the $x$-axis is given by AISC 360 Eq. (E3-4) as

$$\begin{aligned} F_e &= \pi^2 F(KL/r_x)^2 \\ &= 3.14^2 \times 29{,}000/(1.0 \times 12 \times 12/1.60)^2 \\ &= 35.30 \text{ ksi} \\ F_y/F_e &= \kappa \\ &= 50/35.30 \\ &= 1.42 \\ &< 2.25 \end{aligned}$$

Hence, inelastic buckling governs and the critical stress for flexural buckling about the $x$-axis is given by AISC 360 Eq. (E3-2) as

$$\begin{aligned} F_{cr} &= (0.658^{\kappa})F_y \\ &= 0.658^{1.42} \times 50 \\ &= 27.60 \text{ ksi} \end{aligned}$$

**Calculate critical stress for torsional and flexural-torsional buckling**

The elastic buckling stress about the $y$-axis is given by AISC 360 Eq. (E3-4) as

$$\begin{aligned} F_e &= \pi^2 E/(KL/r_y)^2 \\ &= 3.14^2 \times 29{,}000/(1.0 \times 12 \times 12/1.96)^2 \\ &= 52.97 \text{ ksi} \\ F_y/F_e &= \kappa \\ &= 50/52.97 \\ &= 0.94 \\ &< 2.25 \end{aligned}$$

The critical stress for flexural buckling about the axis of symmetry $F_{cry}$ is obtained from AISC 360 Eq. (E3-2) as

$$\begin{aligned} F_{cry} &= F_{cr} \\ &= (0.658^{\kappa})F_y \\ &= 0.658^{0.94} \times 50 \\ &= 33.74 \text{ ksi} \end{aligned}$$

The critical stress for T-shaped members is given by AISC 360 Eq. (E4-2) as

$$\begin{aligned} F_{cr} &= [(F_{cry} + F_{crz})/2H]\{1 - [1 - 4F_{cry}F_{crz}H/(F_{cry} + F_{crz})^2]^{0.5}\} \\ &= [(33.74 + 183)/(2 \times 0.90]\{1 - [1 - 4 \times 33.74 \times 183 \times 0.90/(33.74 + 183)^2]^{0.5}\} \\ &= 33.01 \text{ ksi ... does not govern} \\ &> 27.60 \text{ ksi} \end{aligned}$$

Hence, the nominal compressive strength for flexural buckling about the $x$-axis governs and is given by AISC 360 Eq. (E4-1) as

$$P_n = F_{cr}A_g$$
$$= 27.60 \times 7.3$$
$$= 201 \text{ kips}$$

| LRFD | ASD |
|---|---|
| The design compressive strength is<br>$\varphi_c P_n = 0.9 \times 201$<br>$= 181$ kips | The allowable compressive strength is<br>$P_n/\Omega_c = 201/1.67$<br>$= 120$ kips |

Alternatively, using AISC Manual Table 4-7

| LRFD | ASD |
|---|---|
| From AISC Manual Table 4-7 for a value of $KL = 12$ ft, a WT6 × 25 provides a design strength of<br>$\varphi_c P_n = 182$ kips | From AISC Manual Table 4-7 for a value of $KL = 12$ ft, a WT6 × 25 provides an allowable strength of<br>$P_n/\Omega_c = 121$ kips |

## Single Angle Compression Members without Slender Elements

Single angles may be assumed loaded through the centroid of the cross section when, as specified in AISC 360 Sec. E5:

- The members are loaded at the ends through the same leg.
- The members are attached by welding or by a minimum of two bolts.
- There are no intermediate transverse loads.

The eccentricity is accounted for by an assumed increase in the slenderness ratio. For equal-leg angles or unequal-leg angles connected through the longer leg that are *individual* members or web members of a *planar* truss, with adjacent web members attached to the same side of the gusset plate or chord.

When $$L/r_x \le 80$$

the effective slenderness ratio is given by AISC 360 Eq. (E5-1) as

$$KL/r = 72 + 0.75L/r_x$$

When $$L/r_x > 80$$

the effective slenderness ratio is given by AISC 360 Eq. (E5-2) as

$$KL/r = 32 + 1.25L/r_x$$
$$\le 200$$

For equal-leg angles or unequal-leg angles connected through the longer leg that are web members of *box* or *space* trusses, with adjacent web members attached to the same side of the gusset plate or chord.

When $$L/r_x \le 75$$

the effective slenderness ratio is given by AISC 360 Eq. (E5-3) as

$$KL/r = 60 + 0.8L/r_x$$

When $$L/r_x > 75$$

the effective slenderness ratio is given by AISC 360 Eq. (E5-4) as

$$KL/r = 45 + L/r_x$$
$$\le 200$$

Using these effective slenderness ratios, the critical stress is determined from AISC 360 Eq. (E3-2) or (E3-3) as appropriate.

**Example 6.10.** Analysis of a Single Angle Strut
Determine the design axial load for a single angle $5 \times 5 \times 1/2$ strut with a length of 12 ft and a yield stress of $F_y = 36$ ksi. The strut is attached at each end by welding through the same leg. The strut has no intermediate bracing.

From AISC Manual Table 1-7, the relevant values are

$$A_g = 4.79 \text{ in}^2 \qquad r_x = 1.53 \text{ in} \qquad r_z = 0.98 \text{ in}$$

The $L/r_x$ ratio is given by

$$L/r_x \le 12 \times 12/1.53$$
$$= 94.12$$
$$> 80$$

Hence, the effective slenderness ratio is given by AISC 360 Eq. (E5-2) as

$$KL/r = 32 + 1.25L/r_x$$
$$= 32 + 1.25 \times 94.12$$
$$= 150$$
$$< 200 \text{ ... satisfactory}$$
$$> 4.71(E/F_y)^{0.5} = 134$$

Hence, elastic buckling governs and the elastic critical buckling stress is given by AISC 360 Eq. (E3-4) as

$$F_e = \pi^2 E/(KL/r)^2$$
$$= (3.14/150)^2 \times 29{,}000$$
$$= 12.71 \text{ ksi}$$

The critical stress is given by AISC 360 Eq. (E3-3) as

$$F_{cr} = 0.877F_e$$
$$= 0.877 \times 12.71$$
$$= 11.15 \text{ ksi}$$

The nominal compressive strength is given by AISC 360 Eq. (E3-1) as

$$P_n = F_{cr}A_g$$
$$= 11.15 \times 4.79$$
$$= 53.4 \text{ kips}$$

| LRFD | ASD |
|---|---|
| The design compressive strength is<br>$\varphi_c P_n = 0.9 \times 53.4$<br>$= 48.1$ kips | The allowable compressive strength is<br>$P_n/\Omega_c = 53.4/1.67$<br>$= 32.0$ kips |

Alternatively, the design strength may be obtained from AISC Manual Table 4-11 for concentrically loaded single angles. The effective value of *KL* referred to the *z*-axis is

$$KL_{eff} = (KL/r)r_z$$
$$= 150 \times 0.98/12$$
$$= 12.25 \text{ ft}$$

| LRFD | ASD |
|---|---|
| From AISC Manual Table 4-11 for a value of $KL = 12.25$ ft, an L5 × 5 × 1/2 provides a design strength of<br>$\varphi_c P_n = 48.3$ kips | From AISC Manual Table 4-11 for a value of $KL = 12.25$ ft, an L5 × 5 × 1/2 provides an allowable strength of<br>$P_n/\Omega_c = 32.1$ kips |

## Members with Slender Elements

A nonslender element in a column can support the full yield stress without local buckling. A slender element in a column may fail by local buckling prior to overall member instability. When a column cross-section contains slender elements, the slender element reduction factor $Q$ defines the ratio of stress at local buckling to the yield stress $F_y$. Section E7 of the AISC Specification addresses the design of columns with slender elements. Expressions presented in Sec. E7 for the determination of the critical stress are similar to those given in Sec. E3 with the yield stress $F_y$ replaced with the value $QF_y$. The slender element reduction factor is given in AISC 360 Sec. E7 as

$Q = 1.0$ ... for a shape without slender elements

$= Q_s$ ... for a shape with only unstiffened slender elements

$= Q_a$ ... for a shape with only stiffened slender elements

$= Q_sQ_a$ ... for a shape with both stiffened and unstiffened slender elements

The limiting width-to-thickness ratio for slender elements in compression are given in AISC 360 Table B4.1a. All W-shapes with a yield stress of $F_y = 50$ ksi have nonslender flanges. All W-shapes in the range W8 to W14 with a yield stress of $F_y = 50$ ksi, have nonslender webs with the exception of W14 × 43. Angles with thin legs and T-shapes

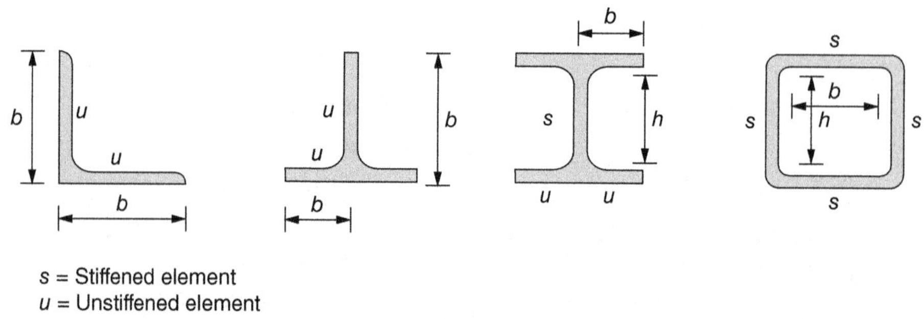


**FIGURE 6.9** Stiffened and unstiffened elements.

with thin stems may be slender elements. Shapes that are slender are indicated in the appropriate table in the AISC Manual. Values of $Q_s$ for angles are given in AISC Manual Table 1-7 and for MT-shapes in AISC Manual Table 1-9. Compactness criteria for rectangular and square HSS-shapes are given in AISC Manual Table 1-12A.

A compression element is considered to be slender when its width-to-thickness ratio exceeds the limiting value $\lambda_r$ given in AISC 360 Table B4.1a. A stiffened element is an element that is supported along both of its edges. An unstiffened element is an element that is supported along only one edge. Figure 6.9 gives examples of stiffened and unstiffened elements.

### Slender Unstiffened Elements, $Q_s$

The determination of the reduction factor $Q_s$ is covered in AISC 360 Sec. E7.1.

1. For flanges, angles, and plates projecting from rolled columns or other compression members:

   For $b/t \le 0.56(E/F_y)^{0.5}$, the reduction factor is given by AISC 360 Eq. (E7-4) as

   $$Q_s = 1.0$$

   For $0.56(E/F_y)^{0.5}, < b/t < 1.03(E/F_y)^{0.5}$, the reduction factor is given by AISC 360 Eq. (E7-5) as

   $$Q_s = 1.415 - 0.74(b/t)(F_y/E)^{0.5}$$

   For $b/t \ge 1.03(E/F_y)^{0.5}$, the reduction factor is given by AISC 360 Eq. (E7-6) as

   $$Q_s = 0.69E/F_y(b/t)^2$$

   where $b$ = width of unstiffened compression element
   = $b_f/2$ for flange of an I-shape or T-shape
   = full nominal dimension of the leg of an angle
   = full nominal dimension of the flange of a channel
   $t$ = thickness of element

   The appropriate widths are dimensioned in Fig. 6.9.

2. For flanges, angle legs, and plates projecting from built-up I-shaped columns or other compression members:

   For $b/t \le 0.64(Ek_c/F_y)^{0.5}$, the reduction factor is given by AISC 360 Eq. (E7-7) as

$$Q_s = 1.0$$

   For $0.64(Ek_c/F_y)^{0.5} < b/t \le 1.17(Ek_c/F_y)^{0.5}$, the reduction factor is given by AISC 360 Eq. (E7-8) as

$$Q_s = 1.415 - 0.65(b/t)(F_y/Ek_c)^{0.5}$$

   For $b/t > 1.17(Ek_c/F_y)^{0.5}$, the reduction factor is given by AISC 360 Eq. (E7-9) as

$$Q_s = 0.90Ek_c/F_y(b/t)^2$$

   where $k_c = 4/(h/t)^{0.5}$
   $\ge 0.35$
   $\le 0.76$
   $h$ = clear distance between flanges

3. For single angles:

   For $b/t \le 0.45(E/F_y)^{0.5}$, the reduction factor is given by AISC 360 Eq. (E7-10) as

$$Q_s = 1.0$$

   For $0.45(E/F_y)^{0.5} < b/t \le 0.91(E/F_y)^{0.5}$, the reduction factor is given by AISC 360 Eq. (E7-11) as

$$Q_s = 1.34 - 0.76(b/t)(F_y/E)^{0.5}$$

   For $b/t > 0.91(E/F_y)^{0.5}$, the reduction factor is given by AISC 360 Eq. (E7-12) as

$$Q_s = 0.53E/F_y(b/t)^2$$

   where $b$ is full width of longest leg.

4. For stems of tees:

   For $d/t \le 0.75(E/F_y)^{0.5}$, the reduction factor is given by AISC 360 Eq. (E7-13) as

$$Q_s = 1.0$$

   For $0.75(E/F_y)^{0.5} < d/t \le 1.03(E/F_y)^{0.5}$, the reduction factor is given by AISC 360 Eq. (E7-14) as

$$Q_s = 1.908 - 1.22(d/t)(F_y/E)^{0.5}$$

   For $d/t > 1.03(E/F_y)^{0.5}$, the reduction factor is given by AISC 360 Eq. (E7-15) as

$$Q_s = 0.69E/F_y(d/t)^2$$

   where $b$ = width of unstiffened compression element
   $= b_f/2$
   $d$ = full nominal depth of tee

### Slender Stiffened Elements, $Q_a$

The determination of the reduction factor $Q_a$ is covered in AISC 360 Sec. E7.2. For slender stiffened elements, the reduction factor $Q_a$ is given by AISC 360 Eq. (E7-16) as

$$Q_a = A_e/A_g$$

where $A_g$ = gross cross-sectional area of member
$A_e$ = summation of the effective areas based on the reduced effective width $b_e$
$= A_g - \Sigma(b - b_e)t$

For uniformly compressed slender elements:

1. For $b/t \geq 1.49(E/f)^{0.5}$, the reduced effective width is given by AISC 360 Eq. (E7-17) as

$$b_e = 1.92t(E/f)^{0.5}[1 - 0.34(E/f)^{0.5}/(b/t)]$$
$$\leq b$$

where $f = F_{cr}$ calculated with $Q = 1.0$
$t$ = web thickness for I-shapes
$= t_w$
$b$ = web depth for I-shapes
= clear distance between flanges less the corner radius for rolled shapes
= clear distance between flanges for built-up sections
$= h$

2. For flanges of square and rectangular hollow sections:

For $b/t \geq 1.40(E/f)^{0.5}$, the reduced effective width is given by AISC 360 Eq. (E7-18) as

$$b_e = 1.92t(E/f)^{0.5}[1 - 0.38(E/f)^{0.5}/(b/t)]$$
$$\leq b$$

where $f = P_n/A_e$
$\approx F_y$

3. For axially loaded circular hollow sections:

For $0.11E/F_y < D/t < 0.45E/F_y$, the slender element reduction factor is given by AISC 360 Eq. (E7-19) as

$$Q = Q_a$$
$$= 0.038E/F_y(D/t) + 2/3$$

where $D$ is outside diameter of round HSS and $t$ is wall thickness.

The critical stress is determined using the expressions in AISC 360 Sec. E7(a) and (b).
When the slenderness ratio is given by $KL/r \leq 4.71(E/QF_y)^{0.5}$, AISC 360 Eq. (E7-2) defines the critical stress as

$$F_{cr} = Q(0.658^{Q\kappa})F_y$$

where $\kappa = F_y/F_e$
$F_e$ = elastic critical buckling stress
$= \pi^2E/(KL/r)^2$ ... for doubly symmetric members from AISC 360 Eq. (E3-4)

When the slenderness ratio is given by $KL/r > 4.71(E/QF_y)^{0.5}$, AISC 360 Eq. (E7-3) defines the critical stress as

$$F_{cr} = 0.877F_e$$

**Example 6.11.** Analysis of a W-Shape Column with Slender Web
A W14 × 43 column in a braced frame has a yield stress of 50 ksi and a height of 10 ft. The column is pinned at the top and bottom and has no intermediate bracing. Determine the available strength of the column.

From AISC Manual Table 1-1, the relevant values of a W14 × 43 are

$$A = 12.6 \text{ in}^2 \quad t_w = 0.305 \text{ in} \quad b_f/2t_f = 7.54 \quad h/t_w = 37.4$$

$$r_x = 5.82 \text{ in} \quad r_y = 1.89 \text{ in} \quad I_x = 428 \text{ in}^4 \quad I_y = 45.2 \text{ in}^4$$

$$G = 11{,}200 \text{ ksi} \quad J = 1.05 \text{ in}^4 \quad C_w = 1950 \text{ in}^6$$

The shape is slender for compression.

The slenderness parameter of the flange is

$$\begin{aligned}\lambda_{rf} &= b_f/2t_f \\ &= 7.54 \\ &< 0.56(E/F_y)^{0.5} \\ &= 13.5\end{aligned}$$

Hence, the flange is not slender and the reduction factor for unstiffened slender elements is

$$Q_s = 1.0$$

The slenderness parameter of the web is

$$\begin{aligned}\lambda_{rw} &\leq h/t_w \\ &= 37.4 \\ &> 1.49(E/F_y)^{0.5} \\ &= 35.9\end{aligned}$$

Hence, the web is slender and $Q = Q_sQ_a = Q_a$.

**Calculate critical stress for flexural buckling about the y-axis**

From Fig. 6.3, for a braced frame, the effective length factor about the $x$- and $y$-axes is

$$K = 1.0$$

Hence, the slenderness ratio about the $y$-axis governs and the elastic buckling stress about the $y$-axis, for a member without slender elements, is given by AISC 360 Eq. (E3-4) as

$$\begin{aligned}F_e &= \pi^2E/(KL/r_y)^2 \\ &= 3.14^2 \times 29{,}000/(1.0 \times 10 \times 12/1.89)^2 \\ &= 70.93 \text{ ksi}\end{aligned}$$

$$\begin{aligned}F_y/F_e &= \kappa \\ &= 50/70.93 \\ &= 0.71 \\ &< 2.25\end{aligned}$$

Hence, inelastic buckling governs and the critical stress for flexural buckling about the $y$-axis is given by AISC 360 Eq. (E3-2) as

$$F_{cr} = (0.658^{\kappa})F_y$$
$$= 0.658^{0.71} \times 50$$
$$= 37.15 \text{ ksi}$$

**Calculate critical stress for torsional and flexural-torsional buckling**

For a doubly symmetric section and considering flexural-torsional buckling of a member without slender elements, the elastic critical stress is given by AISC 360 Eq. (E4-4) as

$$F_e = [\pi^2 EC_w/(K_z L)^2 + GJ]/(I_x + I_y)$$
$$= [3.14^2 \times 29{,}000 \times 1950/120^2 + 11{,}200 \times 1.05]/(428 + 45.2)$$
$$= (38{,}719 + 11{,}760)/473.2$$
$$= 106.7 \text{ ksi}$$
$$> 0.44F_y$$

Hence, the critical stress is given by AISC 360 Eq. (E3-2) as

$$F_{cr} = (0.658^{50/106.7})50$$
$$= 41.10 \text{ ksi ... does not govern}$$
$$> 37.15 \text{ ksi}$$

**Calculate critical stress for local buckling**

The value of the stress $f$ in AISC 360 Sec. E7.2(a) is

$$f = F_{cr} \text{ ... with } Q = 1.0$$
$$= 37.15 \text{ ksi}$$

The web depth is

$$b = h$$
$$= t_w(h/t_w)$$
$$= 0.305 \times 37.4$$
$$= 11.41 \text{ in}$$

and

$$b/t = 11.41/0.305$$
$$= 37.4$$
$$1.49(E/f)^{0.5} = 1.49(29{,}000/37.15)^{0.5}$$
$$= 41.63$$
$$> 37.4 \text{ ... AISC 360 Eq. (E7-17) is not applicable}$$

Hence, there is no reduction in the web width and the reduction factor for stiffened elements is given by AISC 360 Eq. (E7-16) as

$$Q_a = A_e/A_g$$
$$= 1.0$$

and

$$Q = Q_s Q_a$$
$$= 1.0$$

The critical stress for flexural buckling governs and the nominal axial load is given by AISC 360 Eq. (E7-1) as

$$P_n = A_g F_{cr}$$
$$= 12.6 \times 37.15$$
$$= 468 \text{ kips}$$

| LRFD | ASD |
|---|---|
| The design compressive strength is<br>$\varphi_c P_n = 0.9 \times 468$<br>$= 421$ kips | The allowable compressive strength is<br>$P_n/\Omega_c = 468/1.67$<br>$= 280$ kips |

Alternatively, using AISC Manual Table 4-1

| LRFD | ASD |
|---|---|
| From AISC Manual Table 4-1 for a value of $KL = 10$ ft, a W14 × 43 provides a design strength of<br>$\varphi_c P_n = 422$ kips | From AISC Manual Table 4-1 for a value of $KL = 10$ ft, a W14 × 43 provides an allowable strength of<br>$P_n/\Omega_c = 281$ kips |

## 6.5 Built-Up Sections

Built-up columns are used when the height of the column is such that a rolled section cannot provide a sufficiently large radius of gyration. Built-up columns consist generally of two or four shapes connected together by cover plates perforated at intervals with access holes. Alternatively, as shown in Fig. 6.10, lacing bars may be used to hold the component parts together. The lacing bars do not provide assistance in supporting the axial load but they do provide resistance to shear caused by accidental eccentricity in the applied load and by the buckling effect. The shearing force is specified in AISC 360 Sec. E6.2 as 2 percent of the available compressive strength of the column, perpendicular to the axis of the column. The $L/r$ ratio for lacing bars arranged in a single system, as shown, shall not exceed 140. For double lacing, this ratio shall not exceed 200. Lacing bars are spaced so that the controlling ratio of the chord members, between lacing bars, shall not exceed 75 percent of the governing slenderness ratio of the column. The inclination of lacing bars to the axis of the column shall preferably be not less than 60° for single lacing and 45° for double lacing. When the distance between the fasteners in the chords exceeds 15 in, the lacing shall preferably be double or be made of angles. Tie plates are required at the ends of the column with a thickness not less than 1/50 times the distance between lines of welds or fasteners connecting them to the segments of the members.

**Example 6.12.** Latticed Column
The latticed column, shown in Fig. 6.11, consists of four 3 × 3 × 1/2 in angles with a yield stress of 36 ksi. The column is 30 ft high with pinned ends and the 1- × 3/8-in lacing bars are arranged in a single system at an angle of 60° to the axis of the column. Determine the available strength of the column and determine if the lacing members are adequate.

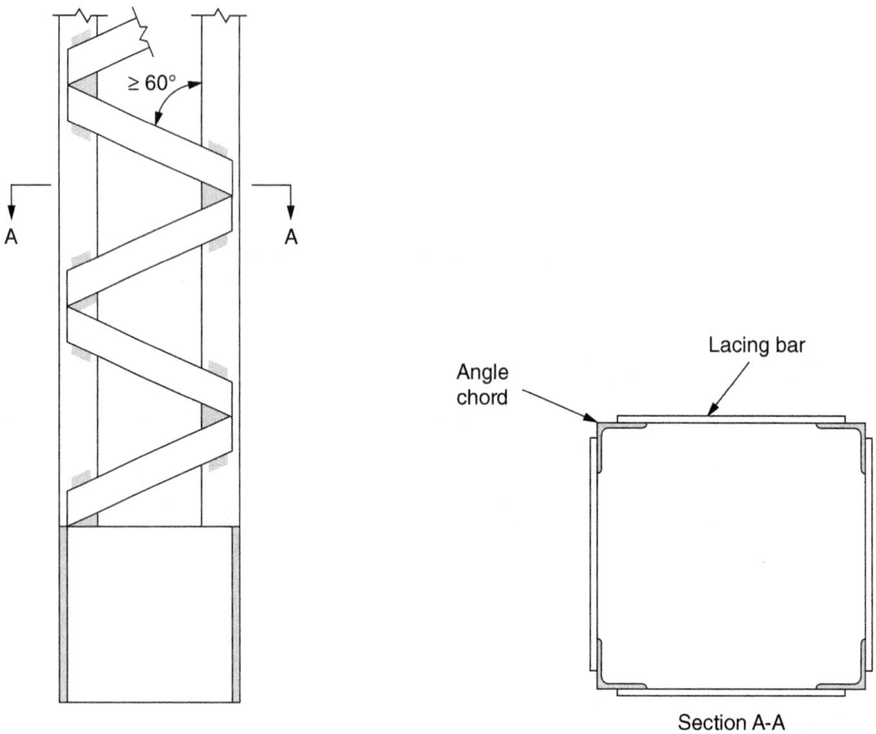


**Figure 6.10** Latticed column details.

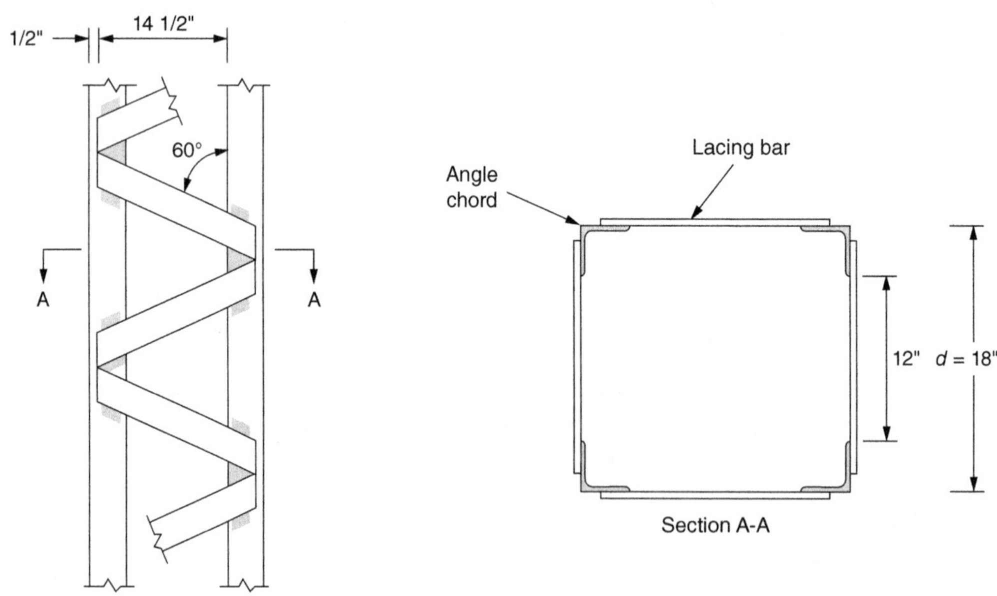


**Figure 6.11** Details for Example 6.12.

The relevant properties of a 3- × 3- × 1/2-in angle are obtained from AISC Manual Table 1-7 as

$$A_a = 2.76 \text{ in}^2 \qquad I_a = 2.20 \text{ in}^4 \qquad r_z = 0.580 \text{ in} \qquad \bar{y}_a = 0.929 \text{ in}$$

The relevant properties of the latticed column are

$$\Sigma A = 4A_a$$
$$= 4 \times 2.76$$
$$= 11 \text{ in}^2$$
$$\Sigma I = 4I_a + \Sigma A(d/2 - \bar{y}_a)^2$$
$$= 4 \times 2.20 + 11(9 - 0.929)^2$$
$$= 725 \text{ in}^4$$

The radius of gyration of the latticed column is

$$r = (\Sigma I/\Sigma A)^{0.5}$$
$$= (725/11)^{0.5}$$
$$= 8.12 \text{ in}$$

The slenderness ratio of the latticed column is

$$KL/r = 1.0 \times 30 \times 12/8.12$$
$$= 44.34$$
$$< 200 \text{ ... satisfactory}$$

From AISC Manual Table 4-22 for a value of $KL/r = 44.34$ and $F_y = 36$ ksi, the available critical stress is obtained as

| LRFD | ASD |
|---|---|
| $\varphi_c F_{cr} = 29.23$ ksi | $F_{cr}/\Omega_c = 19.47$ ksi |
| The design compressive strength is | The allowable compressive strength is |
| $\varphi_c P_n = \varphi_c F_{cr} A_g$ | $P_n/\Omega_c = (F_{cr}/\Omega_c)A_g$ |
| $= 29.23 \times 11$ | $= 19.47 \times 11$ |
| $= 322$ kips | $= 214$ kips |

The length of angle chord between lacing bars is

$$L_a = 2 \times 14.5 \times \tan 30°$$
$$= 16.74 \text{ in}$$

and

$$L_a/r_z = 16.74/0.580$$
$$= 28.86$$
$$< 0.75 \times 44.34 \text{ ... satisfactory}$$

The length of a lacing bar between angle chords is

$$L_b = 12/\cos 30°$$
$$= 13.86 \text{ in}$$

The radius of gyration of a lacing bar is

$$r_b = t_b/(12)^{0.5}$$
$$= 0.375/3.46$$
$$= 0.108 \text{ in}$$

and

$$L_b/r_b = 13.86/0.108$$
$$= 128$$
$$< 140 \text{ ... satisfactory}$$

From AISC Manual Table 4-22 for a value of $KL/r = 128$ and $F_y = 36$ ksi, the available critical stress is obtained as

| LRFD | ASD |
|---|---|
| $(\varphi_c F_{cr})_{bar} = 13.7$ ksi | $(F_{cr}/\Omega_c)_{bar} = 9.10$ ksi |
| The design compressive strength is | The allowable compressive strength is |
| $(\varphi_c P_n)_{bar} = (\varphi_c F_{cr})_{bar} A_b$ | $(P_n/\Omega_c)_{bar} = (F_{cr}/\Omega_c)_{bar} A_b$ |
| $= 13.7 \times 1.0 \times 0.375$ | $= 9.10 \times 1.0 \times 0.375$ |
| $= 5.14$ kips | $= 3.41$ kips |
| The required shearing strength on each face of the latticed column is | The required shearing strength on each face of the latticed column is |
| $V_u = 0.02 \times \varphi_c P_n/2$ | $V_a = 0.02 \times P_n/2\Omega_c$ |
| $= 0.02 \times 322/2$ | $= 0.02 \times 214/2$ |
| $= 3.22$ kips | $= 2.14$ kips |
| The axial force in one lacing bar is | The axial force in one lacing bar is |
| $P_u = V_u/\cos 30°$ | $P_a = V_a/\cos 30°$ |
| $= 3.22/\cos 30°$ | $= 2.14/\cos 30°$ |
| $= 3.72$ kips | $= 2.47$ kips |
| $< (\varphi_c P_n)_{bar}$ ... satisfactory | $< (P_n/\Omega_c)_{bar}$ ... satisfactory |

## 6.6 Column Base Plates

### Concrete Footing Capacity

Base plates are provided to columns to ensure that the column load is distributed to the concrete footing without exceeding the capacity of the concrete. As shown in Fig. 6.12, the column load is assumed dispersed in the footing at a slope of 2 in 1. The nominal bearing strength of the concrete is given by AISC 360 Eq. (J8-2) as

$$P_p = 0.85 f'_c A_1 (A_2/A_1)^{0.5}$$
$$\leq 1.7 f'_c A_1$$

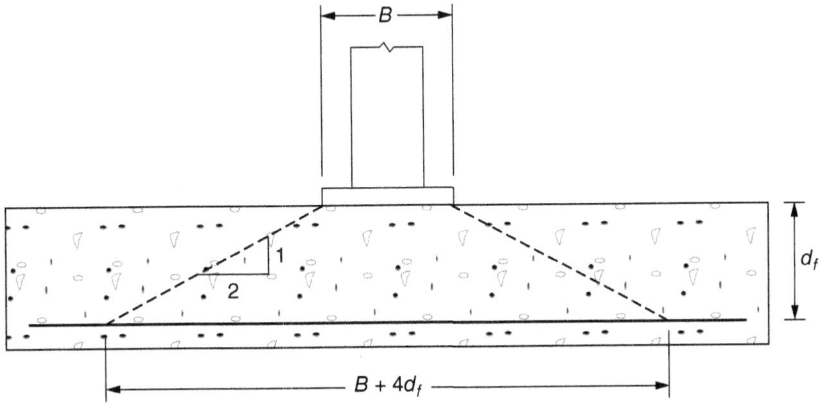


**FIGURE 6.12** Concrete footing.

where $f'_c$ = compressive strength of footing concrete
$A_1$ = area of the base plate
$= N \times B$
$A_2$ = area of the base of the pyramid, with side slopes of 1:2, formed within the footing by the base plate, as shown in Fig. 6.12
$= (B + 4d_f)(N + 4d_f)$
$d_f$ = effective depth of the concrete footing
$N$ = length of base plate
$B$ = width of base plate

When the base plate is the same size as the concrete footing, $A_2 = A_1$ and the nominal bearing strength of the concrete is given by AISC 360 Eq. (J8-1) as

$$P_p = 0.85f'_cA_1$$

The available bearing stress is obtained from AISC 360, Sec. J8 as

| LRFD | ASD |
|---|---|
| $\varphi_c P_p$ = design compressive strength $\geq P_u$<br>where $\varphi_c$ = resistance factor $= 0.65$<br>$P_u$ = required compressive strength using LRFD load combinations | $P_n/\Omega_c$ = allowable compressive strength $\geq P_a$<br>where $\Omega_c$ = safety factor $= 2.31$<br>$P_a$ = required compressive strength using ASD load combinations |

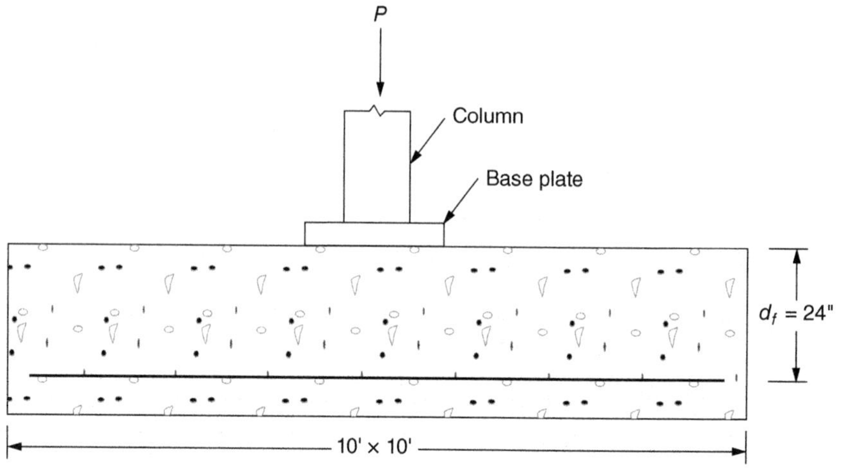


**Figure 6.13** Details for Example 6.13.

**Example 6.13.** Concrete Footing

A W14 × 109 column is seated on an 18-in square base plate that is supported on a reinforced concrete footing as shown in Fig. 6.13. The loading consists of an axial dead load of $P_D = 100$ kips, which includes the weight of the column, and an axial live load of $P_L = 300$ kips. The concrete footing is 10 ft square with an effective depth to reinforcement of $d_f = 24$ in and a compressive strength of $f'_c = 3000$ psi. Determine the adequacy of the footing concrete.

The area of the base of the pyramid, with side slopes of 1:2, formed within the footing by the base plate area is

$$A_2 = (B + 4d_f)^2$$
$$= (18 + 4 \times 24)^2$$
$$= 12{,}996 \text{ in}^2$$
$$(A_2)^{0.5} = (12{,}996)^{0.5}/12$$
$$= 9.5 \text{ ft}$$
$$< 10 \text{ ft ... the distributed load is contained within the concrete footing}$$
$$(A_2/A_1)^{0.5} = (12{,}996/18^2)^{0.5}$$
$$= 6.33$$
$$> 1.7 \text{ ... use 1.7 maximum}$$

The nominal bearing strength of the concrete is given by AISC 360 Eq. (J8-2) as

$$P_p = 0.85f'_cA_1(A_2/A_1)^{0.5}$$
$$= 1.7f'_cA_1$$
$$= 1.7 \times 3000 \times 18^2/1000$$
$$= 1652 \text{ kips}$$

| LRFD | ASD |
|---|---|
| The design bearing strength is | The allowable bearing strength is |
| $\varphi_c P_p = 0.65 \times 1652$ | $P_p/\Omega_c = 1652/2.31$ |
| $= 1074$ kips | $= 715$ kips |
| From ASCE 7 Sec. 2.3.2 combination 2: | From ASCE 7 Sec. 2.4.1 combination 2: |
| $P_u$ = factored axial load | $P_a$ = factored axial |
| $= 1.2P_D + 1.6P_L$ | $= P_D + P_L$ |
| $= 1.2 \times 100 + 1.6 \times 300$ | $= 100 + 300$ |
| $= 600$ kips | $= 400$ kips |
| $< \varphi_c P_p$ ... satisfactory | $< P_p/\Omega_c$ ... satisfactory |

## Base Plate Thickness

Design of the base plate is based on an assumed uniform bearing pressure over the entire base plate surface area. As shown in Fig. 6.14, the plate is assumed to cantilever beyond the column shaft with the maximum moments occurring at sections $0.95d$ apart in one direction and $0.8b_f$ apart in the other. The cantilever lengths are, then

$$m = (N - 0.95d)/2$$

$$n = (B - 0.8b_f)/2$$

In addition, a yield line solution proposed by Thornton[5] assumes the plate is supported and completely fixed on three edges and unsupported on the fourth. From this, an equivalent cantilever length is derived given by AISC Manual Part 14 Eqs. (14-4) and (14-5) as

$$\lambda n' = \lambda(db_f)^{0.5}/4$$

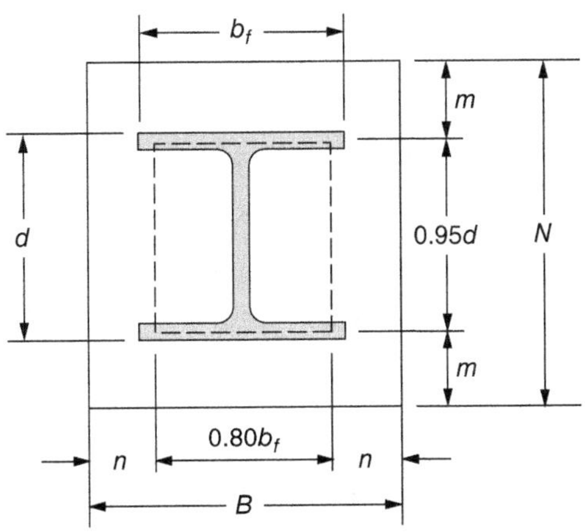


**Figure 6.14** Column base plate.

where $\lambda = 2(X)^{0.5}/[1 + (1 - X)^{0.5}]$

$\leq 1.0$

and, from Eqs. (14-6a) and (14-6b)

| LRFD | ASD |
|---|---|
| $X = [4db_f/(d + b_f)^2](P_u/\varphi_c P_p)$ | $X = [4db_f/(d + b_f)^2](\Omega_c P_a/P_p)$ |

The term $\lambda$ may conservatively be taken as 1.0.

The critical base plate cantilever dimension is then determined as

$$\ell = \max(m, n, \lambda n')$$

The minimum required base plate thickness may now be calculated from Eqs. (14-7a) and (14-7b) as

| LRFD | ASD |
|---|---|
| $t_{min} = \ell[2P_u/(0.9F_y BN)]^{0.5}$ | $t_{min} = \ell[3.33P_a/(F_y BN)]^{0.5}$ |

**Example 6.14** Column Base Plate

Determine the required minimum base plate thickness for the W14 × 109 column of Example 6.13. The plate has a yield stress of $F_y = 36$ ksi

The relevant properties of a W14 × 109 are

$$d = 14.3 \text{ in} \qquad b_f = 14.6 \text{ in}$$

The relevant design parameters are

$$m = (N - 0.95d)/2$$
$$= (18 - 0.95 \times 14.3)/2$$
$$= 2.21 \text{ in}$$
$$n = (B - 0.80b_f)/2$$
$$= (18 - 0.80 \times 14.6)/2$$
$$= 3.16 \text{ in}$$
$$\lambda n' = \lambda(db_f)^{0.5}/4$$
$$= 1.0(14.3 \times 14.6)^{0.5}/4 \text{ ... for } \lambda = 1.0$$
$$= 3.61 \text{ ... governs}$$
$$= \ell$$

Hence, the required base plate thickness is given by

| LRFD | ASD |
|---|---|
| $t_{min} = \ell[2P_u/(0.9F_y BN)]^{0.5}$ | $t_{min} = \ell[3.33P_a/(F_y BN)]^{0.5}$ |
| $= 3.61[2 \times 600/(0.9 \times 36 \times 18 \times 18)]^{0.5}$ | $= 3.61[3.33 \times 400/(36 \times 18 \times 18)]^{0.5}$ |
| = 1.22 in … use 1¼ in | = 1.22 in … use 1¼ in |

## 6.7 Column Flanges with Concentrated Forces

### Introduction

In frames with moment connections, girders framing into a column flange apply both tensile and compressive concentrated forces to the column flange, forming a couple on the same side of the column. As shown in Fig. 6.15, these double-concentrated forces may be applied on either one side or both sides of the column. For gravity loaded frames with girders on both sides of a column, the applied forces are both tensile at the top of the girder and both compressive at the bottom of the girder. When lateral forces are applied to the frame, the forces on opposite sides of the column act in the same direction.

Tensile forces applied to a column flange may cause failure by flange local bending. Compressive forces applied to a column flange may cause web failure by compression buckling or panel zone shear. When the required strength exceeds the available strength, transverse stiffeners or web doubler plates or both may be provided to resist the difference between the required strength and the available strength. Alternatively, a heaver column may be selected to eliminate the need for stiffeners or doubler plates and this is often the most economical solution as indicated by Carter.[6] Columns are also subject to failure by web local yielding and web crippling and these issues have been covered in Secs. 5.6 and 5.7 of Chap. 5.

### Flange Local Bending

Flange local bending is produced when a tensile force is applied to a column flange. This occurs at a moment connection with girders framing into the column on one or two sides. As shown in Fig. 6.16, stiffener plates, also known as continuity plates, may be used to reinforce slender column flanges. Stiffeners are placed in pairs and welded to the loaded column flange. For a girder framing into only one side of a column, the stiffeners need not extend more than one-half the depth of the column web. The force delivered by the girder flange to the column is

$$P_{rf} = M_r/d_m + P_r/2$$

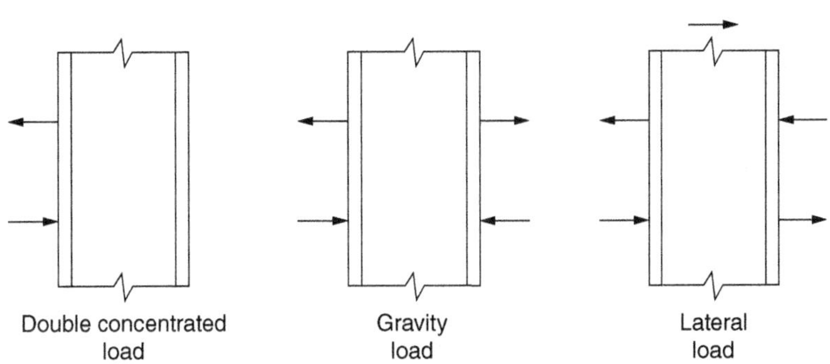


**FIGURE 6.15** Column flange forces.

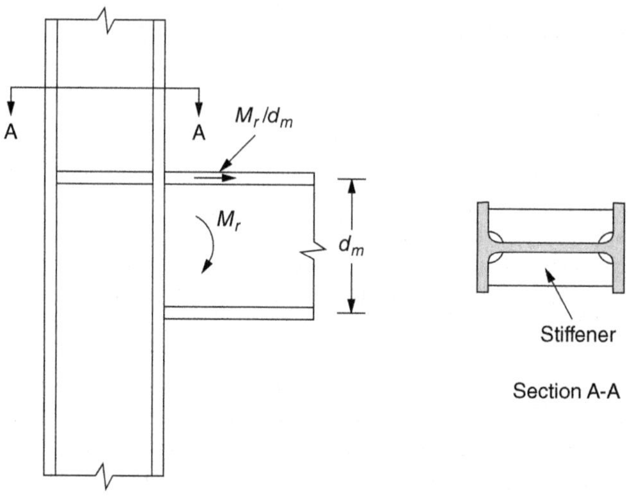


**FIGURE 6.16** Stiffener plates.

where $M_r$ = girder moment
$d_m$ = distance between flange forces
$= d - t_f$
$d$ = girder depth
$t_f$ = girder flange thickness
$P_r$ = axial force in girder

The axial force in the girder is normally small and may be neglected.

AISC 360 Sec. J10.1 requires the provision of stiffeners opposite the tension girder flange, in a girder-column connection, if the tensile force exceeds the available capacity of the column flange. When the loading is applied to the column flange a distance greater than $10t_f$ from the end of the column, the nominal strength of the column flange is given by AISC 360 Eq. (J10-1) as

$$R_n = 6.25t_f^2F_{yf}$$

where $F_{yf}$ is column flange yield stress and $t_f$ is column flange thickness.

When the length of the applied loading measured across the column flange is less than $0.15b_f$, where $b_f$ is the column flange width, the design capacity of the column flange need not be checked. When the loading is applied to the column flange a distance less than $10t_f$ from the end of the column, the value of $R_n$ must be reduced by 50 percent.

The design flange strength and the allowable flange strength may be obtained from AISC 360, Sec. J10.1 as

$\varphi R_n$ = design flange strength

$\geq P_{rf}$ ... force delivered by the girder flange using LRFD load combinations

$$R_n/\Omega = \text{allowable flange strength}$$
$$\geq P_{rf} \ldots \text{force delivered by the girder flange using ASD load combinations}$$

where $\varphi$ = resistance factor
$= 0.9$
$\Omega$ = safety factor
$= 1.67$

AISC Manual Tables 4-1 and 4-2 tabulate the values of

$$P_{fb} = \varphi R_n \text{ for LRFD design}$$
$$= R_n/\Omega \text{ for ASD design}$$

**Example 6.15.** Flange Local Bending
For the girder-column connection shown in Fig. 6.16, determine if horizontal stiffener plates are required opposite the tension flange of the girder. The moment indicated is caused by gravity loads and consists of a dead load moment of $M_D = 40$ kip-ft, and a live load moment of $M_L = 120$ kip-ft. The column section is a W14 × 109 and the girder is a W16 × 100. The steel sections have a yield stress of 50 ksi.

The relevant properties of a W16 × 100 girder are

$$d = 17.0 \text{ in} \qquad t_f = 0.985 \text{ in}$$

The relevant properties of a W14 × 109 column are

$$t_f = 0.860 \text{ in}$$

The nominal capacity of the column flange is given by AISC 360 Eq. (J10-1) as

$$R_n = 6.25t_f^2F_{yf}$$
$$= 6.25 \times (0.86)^2 \times 50$$
$$= 231 \text{ kips}$$

| LRFD | ASD |
|---|---|
| $\varphi R_n$ = design capacity of the column flange | $R_n/\Omega$ = allowable capacity of the column flange |
| $= 0.9 \times 231$ | $= 231/1.67$ |
| $= 208$ kips | $= 138$ kips |
| Alternatively, from AISC Manual Table 4-1: | Alternatively, from AISC Manual Table 4-1: |
| $P_{fb} = \varphi R_n$ | $P_{fb} = R_n/\Omega$ |
| $= 208$ kips | $= 138$ kips |

Applying American Society of Civil Engineers, *Minimum Design Loads for Buildings and Other Structures* (ASCE 7)[4] Secs. 2.3 and 2.4 gives

| LRFD | ASD |
|---|---|
| From ASCE 7 Sec. 2.3.2 combination 2: | From ASCE 7 Sec. 2.4.1 combination 2: |
| $M_u$ = factored moment | $M_a$ = factored moment |
| $= 1.2M_D + 1.6M_L$ | $= M_D + M_L$ |
| $= 1.2 \times 40 + 1.6 \times 120$ | $= 40 + 120$ |
| = 240 kip-ft | = 160 kips |
| $P_{rf}$ = force delivered by the girder flange | $P_{rf}$ = force delivered by the girder flange |
| $= M_u/(d - t_f)$ | $= M_a/(d - t_f)$ |
| $= 12 \times 240/(17.0 - 0.985)$ | $= 12 \times 160/(17.0 - 0.985)$ |
| = 180 kips | = 120 kips |
| $< \varphi R_n$ | $< R_n/\Omega$ |
| Hence, stiffeners are not required. | Hence, stiffeners are not required. |

## Web Compression Buckling

Web compression buckling may occur when a pair of compressive forces is applied at opposite flanges of a column at the same location. As shown in Fig. 6.17, this occurs at the bottom flange of two back-to-back moment connections under gravity load. A single stiffener plate, a pair of stiffener plates, or a doubler plate extending the full depth of the web may be used to reinforce a slender column web.

The force delivered by the two girder flanges to the column is

$$P_{rf} = (M_{r1} + M_{r2})/d_m$$

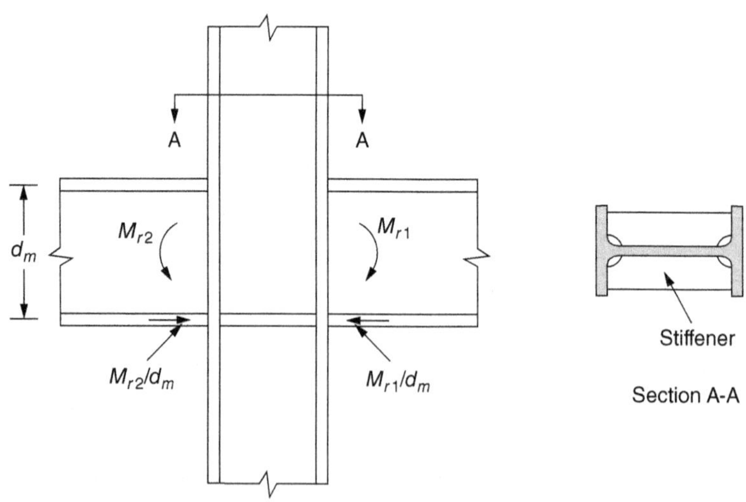


**FIGURE 6.17** Compression buckling.

where $M_{r1}$ = girder moment acting on the right-hand girder
$M_{r2}$ = girder moment acting on the left-hand girder
$d_m$ = distance between flange forces
$= d - t_f$
$d$ = girder depths
$t_f$ = girder flange thickness

The nominal capacity of a column web is given by AISC 360 Eq. (J10-8) as

$$R_n = 24t_w^3(EF_{yw})^{0.5}/h$$

where $t_w$ = thickness of column web
$F_{yw}$ = yield stress of column web
$h$ = clear distance between column flanges less the corner radius at each flange, for rolled shapes
$= t_w \times h/t_w$

When the loading is applied to the column flange a distance less than $d/2$ from the end of the column, the value of $R_n$ is reduced by 50 percent.

The available compression strength is obtained from AISC 360, Sec. J10.5 as

| LRFD | ASD |
|---|---|
| $\varphi R_n$ = design compression strength<br>$\geq P_{rf}$<br>where $\varphi$ = resistance factor<br>$= 0.90$<br>$P_{rf}$ = required compression strength using LRFD load combinations | $R_n/\Omega$ = allowable compression strength<br>$\geq P_{rf}$<br>where $\Omega$ = safety factor<br>$= 1.67$<br>$P_{rf}$ = required compression strength using ASD load combinations |

AISC Manual Tables 4-1 and 4-2 tabulate the values of

$$P_{wb} = \varphi R_n \text{ for LRFD design}$$
$$= R_n/\Omega \text{ for ASD design}$$

**Example 6.16.** Web Compression Buckling
For the girder-column connection shown in Fig. 6.17, determine if horizontal stiffener plates are required opposite the compression flanges of the girders. The moments indicated are caused by gravity loads and consist of dead load moments of $M_D$ = 40 kip-ft, and live load moments of $M_L$ = 120 kip-ft. The column section is a W14 × 109 and the girder is a W16 × 100. The steel sections have a yield stress of 50 ksi.

The relevant properties of a W14 × 109 column are

$$t_w = 0.525 \text{ in} \qquad h/t_w = 21.7 \text{ in}$$

The clear distance between flanges less the corner radius at each flange is

$$h = t_w \times h/t_w$$
$$= 0.525 \times 21.7$$
$$= 11.39 \text{ in}$$

The compression buckling nominal capacity of the web is given by AISC 360 Eq. (J10-8) as

$$R_n = 24t_w^3(EF_{yw})^{0.5}/h$$
$$= 24 \times 0.525^3(29{,}000 \times 50)^{0.5}/11.39$$
$$= 367 \text{ kips}$$

| LRFD | ASD |
|---|---|
| $\varphi R_n$ = design capacity of the column flange | $R_n/\Omega$ = allowable capacity of the column flange |
| $= 0.9 \times 367$ | $= 367/1.67$ |
| $= 330$ kips | $= 220$ kips |
| Alternatively, from AISC Manual Table 4-1: | Alternatively, from AISC Manual Table 4-1: |
| $P_{wb} = \varphi R_n$ | $P_{wb} = R_n/\Omega$ |
| $= 330$ kips | $= 220$ kips |

Applying American Society of Civil Engineers, *Minimum Design Loads for Buildings and Other Structures* (ASCE 7)[4] Secs. 2.3 and 2.4 gives

| LRFD | ASD |
|---|---|
| From ASCE 7 Sec. 2.3.2 combination 2: | From ASCE 7 Sec. 2.4.1 combination 2: |
| $M_u$ = factored moment | $M_a$ = factored moment |
| $= 1.2M_D + 1.6M_L$ | $= M_D + M_L$ |
| $= 1.2 \times 40 + 1.6 \times 120$ | $= 40 + 120$ |
| $= 240$ kip-ft | $= 160$ kips |
| $P_{rf}$ = force delivered by the girder flange | $P_{rf}$ = force delivered by the girder flange |
| $= 2M_u/(d - t_f)$ | $= 2M_a/(d - t_f)$ |
| $= 2 \times 12 \times 240/(17.0 - 0.985)$ | $= 2 \times 12 \times 160/(17.0 - 0.985)$ |
| $= 360$ kips | $= 240$ kips |
| $> \varphi R_n$ | $> R_n/\Omega$ |
| Hence, stiffeners are required. | Hence, stiffeners are required. |

## Web Panel Zone Shear

The column panel zone is the web area bounded by the top and bottom girder flanges as shown in Fig. 6.18. Web panel zone shear occurs when double-concentrated forces are applied on either one side or both sides of the column at the same location. Panel zone shear results from the combined effects of the girder flange forces and the story shear.

For wind load acting from right to left, the required panel zone shear strength is obtained from AISC 360 Eqs. (C-J10-3a) and (C-J10-3b) as

$$F_r = M_1/d_{m1} + M_2/d_{m2} - V$$

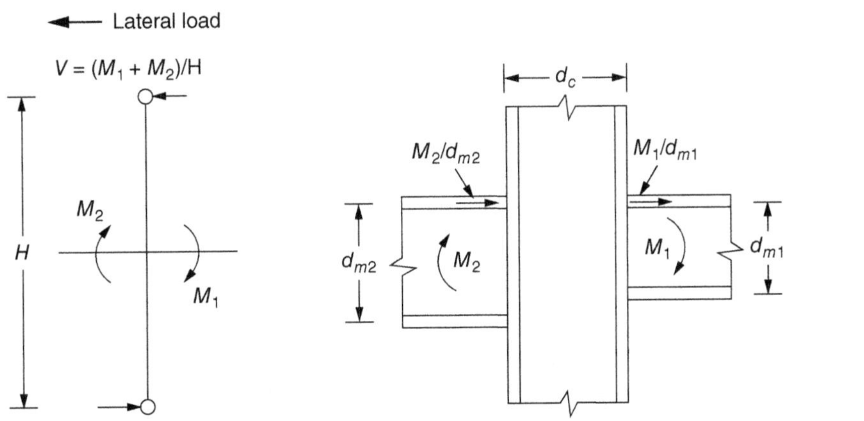


**FIGURE 6.18** Panel zone shear.

where $d_{m1}$ = distance between flange forces of girder framing into column on windward side of connection
$= d_{b1} - t_{f1}$
$d_{m2}$ = distance between flange forces of girder framing into column on leeward side of connection
$= d_{b2} - t_{f2}$
$M_1$ = factored moment due to gravity load plus wind load acting on the windward girder
$= M_{u1}$ for LRFD load combinations
$= M_{a1}$ for ASD load combinations
$M_2$ = factored moment due to gravity load plus wind load acting on the leeward girder
$= M_{u2}$ for LRFD load combinations
$= M_{a2}$ for ASD load combinations
$V$ = story shear
$= (M_1 + M_2)/H$
$= V_u$ for LRFD load combinations
$= V_a$ for ASD load combinations
$H$ = story height

When panel zone behavior is restricted within the elastic range and the effect of panel zone deformation on frame stability is *not* considered, the nominal shear strength is given by AISC 360 Eqs. (J10-9) and (J10-10) as

$$R_n = 0.60F_y d_c t_w \; \ldots \text{ for } P_r \le 0.4P_c$$

$$R_n = 0.60F_y d_c t_w (1.4 - P_r/P_c) \; \ldots \text{ for } P_r > 0.4P_c$$

When adequate ductility is provided to accommodate deformations in the inelastic range, the post-yield panel zone strength may be utilized. Hence, when the effect of panel zone deformation on frame stability *is* considered, the nominal shear strength is given by AISC 360 Eqs. (J10-11) and (J10-12) as

$$R_n = 0.60F_y d_c t_w[1 + 3b_{cf}t_{cf}^2/(d_b d_c t_w)] \ldots \text{ for } P_r \leq 0.75P_c$$

$$R_n = 0.60F_y d_c t_w[1 + 3b_{cf}t_{cf}^2/(d_b d_c t_w)][1.9 - (1.2P_r)/P_c] \ldots \text{ for } P_r > 0.75P_c$$

where $t_w$ = column web thickness
$b_{cf}$ = width of column flange
$t_{cf}$ = thickness of column flange
$d_c$ = column depth
$d_b$ = girder depth
$F_y$ = yield stress of the column web
$P_c = P_y$ for LRFD load combinations
$= 0.6P_y$ for ASD load combinations
$P_r$ = required axial strength using LRFD or ASD load combinations
$P_y$ = axial yield strength of the column
$= F_y A_g$
$A_g$ = column cross-sectional area

The available shear strength is obtained from AISC 360, Sec. J10.6 as

| LRFD | ASD |
|---|---|
| $\varphi R_n$ = design shear strength | $R_n/\Omega$ = allowable shear strength |
| $\geq F_r$ | $\geq F_r$ |
| where $\varphi$ = resistance factor<br>= 0.90<br>$F_r$ = required shear strength using LRFD load combinations | where $\Omega$ = safety factor<br>= 1.67<br>$F_r$ = required shear strength using ASD load combinations |

When the web shear strength is inadequate, a web doubler plate or a pair of diagonal stiffeners may be welded to the column web to carry the additional force. Any required doubler plate must extend over the full area of the panel zone and provide the additional thickness necessary to equal or exceed the strength requirements. The doubler plate must be welded to develop the proportion of the total load transmitted to the doubler plate.

**Example 6.17.** Web Panel Zone Shear
The controlling load cases for a girder-column connection, for ASD and LRFD load combinations, are shown in Fig. 6.19. The forces indicated are caused by wind and gravity loads and the connection is in a moment resisting frame with a story height of 12 ft. The column section is W14 × 109 and the girder is W16 × 100. The effect of panel zone deformation on frame stability is not considered in the frame analysis and the steel sections have a yield stress of 50 ksi. Determine if the web panel zone shear capacity is satisfactory.

The relevant properties of a W14 × 109 column are

$$t_w = 0.525 \text{ in} \qquad d_c = 14.3 \text{ in} \qquad A = 32.0 \text{ in}^2$$

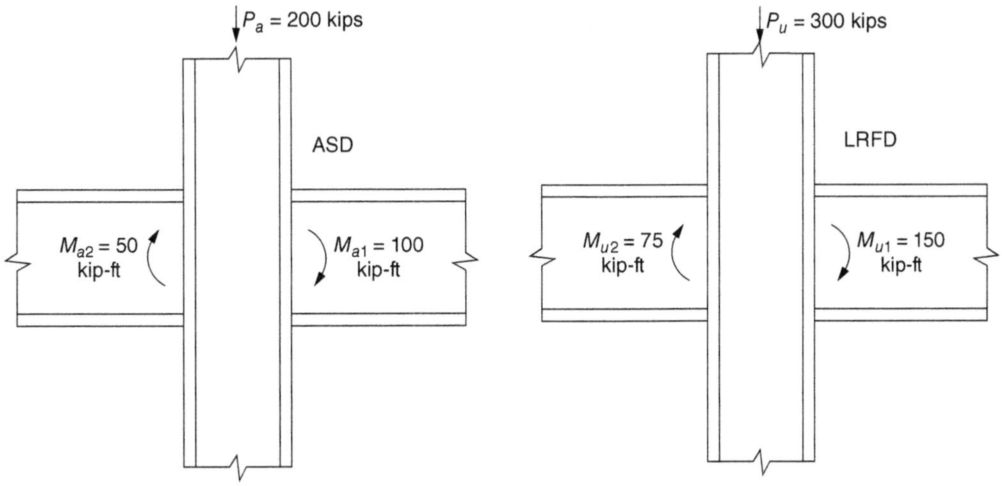


**FIGURE 6.19** Details for Example 6.17.

The relevant properties of a W16 × 100 girder are

$$d_b = 17.0 \text{ in} \qquad t_f = 0.985 \text{ in}$$

$$\begin{aligned} P_y &= F_y A_g \\ &= 50 \times 32 \\ &= 1600 \text{ kips} \end{aligned}$$

The required panel zone shear strength is obtained from AISC 360 Eq. (C-J10-3) as

| LRFD | ASD |
|---|---|
| $F_u = M_{u1}/d_{m1} + M_{u2}/d_{m2} - V_u$ | $F_a = M_{a1}/d_{m1} + M_{a2}/d_{m2} - V_a$ |
| $= (M_{u1} + M_{u2})/(d_b - t_f) - (M_{u1} + M_{u2})/H$ | $= (M_{a1} + M_{a2})/(d_b - t_f) - (M_{a1} + M_{a2})/H$ |
| $= 12(150 + 75)/(17.0 - 0.985) - 225/12$ | $= 12(100 + 50)/(17.0 - 0.985) - 150/12$ |
| $= 150$ kips | $= 100$ kips |
| $P_c = P_y = 1600$ kips | $P_c = 0.6P_y = 0.6 \times 1600 = 960$ kips |
| $P_r/P_c = P_u/P_y = 300/1600 = 0.19$ | $P_r/P_c = P_a/0.6P_y = 200/960 = 0.21$ |
| $< 0.4$ ... AISC 360 Eq. (J10-9) applies | $< 0.4$ ... AISC 360 Eq. (J10-9) applies |
| The design web panel shear strength is | The allowable web panel shear strength is |
| $\varphi R_n = 0.9 \times 0.60 F_y d_c t_w$ | $R_n/\Omega = 0.60 F_y d_c t_w/\Omega$ |
| $= 0.9 \times 0.6 \times 50 \times 14.3 \times 0.525$ | $= 0.6 \times 50 \times 14.3 \times 0.525/1.67$ |
| $= 203$ kips | $= 135$ kips |
| $> F_u$ ... satisfactory | $> F_a$ ... satisfactory |

## Transverse Stiffener Requirements

When the required strength exceeds the available strength of a column for concentrated forces applied to the flange, stiffeners are required to resist the difference between the required strength and available limit state strength. Hence, the force delivered to the stiffener is

$$R_{r,\,st} = F_r - R_{c,\,min}$$

where $R_{r,\,st}$ = required stiffener strength
= $R_{u,\,st}$ for LRFD load combinations
= $R_{a,\,st}$ for ASD load combinations
$F_r$ = girder flange force
= $F_u$ for LRFD load combinations
= $F_a$ for ASD load combinations
$R_{c,\,min}$ = lesser of the available strengths in flange bending and web yielding at the location of tensile concentrated force, or the lesser of the available strengths in web yielding, web crippling, and compression buckling at the location of compressive concentrated force
= $\varphi R_n$ for LRFD load combinations
= $R_n/\Omega$ for ASD load combinations

If $R_{r,\,st}$ is negative, stiffeners are not required. If $R_{r,\,st}$ is positive, the required stiffener area is

$$A_{st} = R_{u,\,st}/\varphi F_{y,\,st} \text{ for LRFD load combinations}$$

$$= R_{a,\,st}\Omega/F_{y,\,st} \text{ for ASD load combinations}$$

where $\varphi$ = resistance factor
= 0.9
$\Omega$ = safety factor
= 1.67

Stiffeners resisting tensile forces are designed in accordance with AISC 360 Sec. J10.8 and are welded to the loaded flange and the web. The welds to the flange are designed for the difference between the required strength and available limit state strength. Stiffeners to web welds are designed to transfer to the web the algebraic difference in tensile force at the ends of the stiffener.

Stiffeners resisting compressive forces either bear on or are welded to the loaded flange and welded to the web. The welds to the flange are designed for the difference between the required strength and the available limit state strength. Stiffeners to web welds are designed to transfer to the web the algebraic difference in compression force at the ends of the stiffener.

The stiffener dimensions are defined by AISC 360 Sec. J10.8 as

$$t_{st} \geq t_f/2$$

$$b_{st} \geq b_f/3 - t_w/2$$

$$b_{st}/t_{st} \leq 16$$

where $t_{st}$ = thickness of stiffener
$b_{st}$ = width of stiffener
$t_f$ = flange thickness of girder framing into column
$b_f$ = flange width of girder framing into column
$t_w$ = thickness of column web

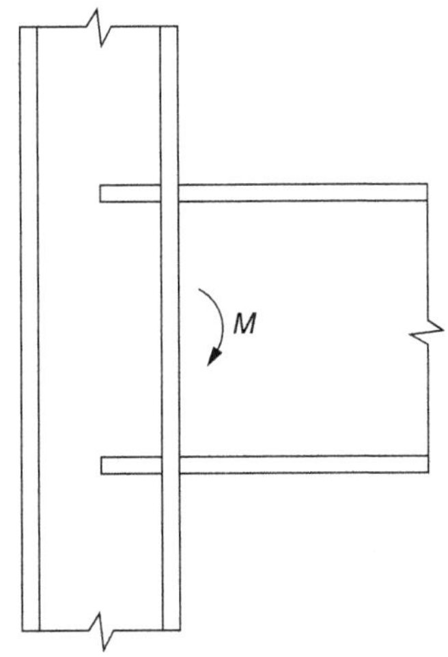


**Figure 6.20** Details for Example 6.18.

The stiffener depth need not exceed half the depth of the column when the available strength in flange local bending, web local yielding, and web crippling is exceeded. A full depth stiffener is required when the available strength in web compression buckling is exceeded. Diagonal stiffeners are required when the available strength in web panel zone shear is exceeded.

In welding stiffener plates, problems have been experienced with cracking in the *k*-area of the column, where the *k*-area is defined as the region extending from the midpoint of the fillet into the web approximately 1 to 1½ in. To avoid this, AISC 360 Commentary Sec. J10.8 recommends that the stiffener corner is clipped 1½ in and the fillet welds are stopped short by a weld leg length from the edges of the cutout.

**Example 6.18.** Transverse Stiffeners
For the girder-column connection, shown in Fig. 6.20, determine the size of stiffeners required. The moment indicated is caused by gravity loads and consists of a dead load moment of $M_D = 60$ kip-ft, and a live load moment of $M_L = 180$ kip-ft. The column section is W14 × 109 and the girder is W16 × 100. The steel sections have a yield stress of 50 ksi and the stiffeners are grade A36 steel.

The capacity of the column must be checked for flange local bending, web local yielding, and web crippling. The factored force delivered by the girder flange to the column is given by

| LRFD | ASD |
|---|---|
| From ASCE 7 Sec. 2.3.2 combination 2: | From ASCE 7 Sec. 2.4.1 combination 2: |
| $M_u$ = factored moment | $M_a$ = factored moment |
| $= 1.2M_D + 1.6M_L$ | $= M_D + M_L$ |
| $= 1.2 \times 60 + 1.6 \times 180$ | $= 60 + 180$ |
| $= 360$ kip-ft | $= 240$ kips |
| $P_{rf}$ = force delivered by the girder flange | $P_{rf}$ = force delivered by the girder flange |
| $= M_u/(d - t_f)$ | $= M_a/(d - t_f)$ |
| $= 12 \times 360/(17.0 - 0.985)$ | $= 12 \times 240/(17.0 - 0.985)$ |
| $= 270$ kips | $= 180$ kips |

The web crippling capacity is determined as

| LRFD | ASD |
|---|---|
| The design crippling strength is obtained from AISC Manual Table 9-4 as<br>$\varphi R_n = 2(\varphi R_3 + t_f \varphi R_4)$<br>$= 2(127 + 0.985 \times 12.8)$<br>$= 279$ kips | The allowable crippling strength is obtained from AISC Manual Table 9-4 as<br>$R_n/\Omega = 2(R_3/\Omega + t_f R_4/\Omega)$<br>$= 2(85.0 + 0.985 \times 8.50)$<br>$= 187$ kips |

The web local yielding capacity is determined as

| LRFD | ASD |
|---|---|
| The design local yielding strength is obtained from AISC Manual Table 9-4 as<br>$\varphi R_n = 2\varphi R_1 + t_f \varphi R_2)$<br>$= 2 \times 95.8 + 0.985 \times 26.3$<br>$= 218$ kips | The allowable local yielding strength is obtained from AISC Manual Table 9-4 as<br>$R_n/\Omega = 2R_1/\Omega + t_f R_2/\Omega$<br>$= 2 \times 63.9 + 0.985 \times 17.5$<br>$= 145$ kips |

The flange local bending capacity is obtained from Example 6.15 as

| LRFD | ASD |
|---|---|
| $\varphi R_n =$ design capacity of the column flange<br>$= 0.9 \times 231$<br>$= 208$ kips | $R_n/\Omega =$ allowable capacity of the column flange<br>$= 231/1.67$<br>$= 138$ kips |

Hence, web local yielding governs and a pair of transverse stiffeners, extending at least one-half the depth of the web, must be provided adjacent to the bottom flange of the girder. However, the available capacity for flange local bending is also exceeded and transverse stiffeners are also required adjacent to the top flange of the girder. The force delivered to the stiffener is

| LRFD | ASD |
|---|---|
| $R_{u,st} = P_{rf} - \varphi R_{n,min}$<br>$= 270 - 218$<br>$= 52$ kips | $R_{a,st} = P_{rf} - R_{n,min}/\Omega$<br>$= 180 - 145$<br>$= 35$ kips |

The required area of a pair of stiffeners adjacent to the top flange is

| LRFD | ASD |
|---|---|
| $A_{st} = R_{u,st}/\varphi F_{yst}$<br>$= 52/(0.9 \times 36)$<br>$= 1.61 \text{ in}^2$ | $A_{st} = \Omega R_{a,st}/F_{yst}$<br>$= 1.67 \times 35/36$<br>$= 1.62 \text{ in}^2$ |

Using a pair of stiffener plates 4¼ × 1/2 in, with a 1½ in corner clip, provides a stiffener area at the column flange of

$$A_{st} = 2 \times (4.25 - 1.50) \times 0.5$$
$$= 2.75 \text{ in}^2$$
$$> 1.62 \text{ in}^2 \text{ ... satisfactory}$$

The minimum stiffener thickness required is

$$\begin{aligned} t_{st} &= t_f/2 \\ &= 0.985/2 \\ &= 0.493 \text{ in} \\ &< 0.5 \text{ in ... satisfactory} \end{aligned}$$

The minimum stiffener width required is

$$\begin{aligned} b_{st} &= b_f/3 - t_w/2 \\ &= 10.4/3 - 0.525/2 \\ &= 3.20 \text{ in} \\ &< 4.25 \text{ in ... satisfactory} \end{aligned}$$

The width-thickness ratio provided is

$$\begin{aligned} b_{st}/t_{st} &= 4.25/0.5 \\ &= 8.5 \\ &< 16 \text{ ... satisfactory} \end{aligned}$$

Hence, the stiffeners proposed are adequate.

The stiffeners are welded to the loaded flange to resist the unbalanced flange force. The length of weld on each side of the stiffeners is

$$\begin{aligned} \ell &= b_{st} - \text{corner clip} \\ &= 4.25 - 1.5 \\ &= 2.75 \text{ in} \end{aligned}$$

Using E70XX fillet welds, on both sides of the pair of stiffeners, the fillet weld size per 1/16 in required to develop the pair of stiffeners is given by AISC 360 Eq. (J2-5) as

| LRFD | ASD |
|---|---|
| $D = R_{u,st}/(\ell q_u \times 1.5)$<br>$= 52/(2 \times 2 \times 2.75 \times 1.39 \times 1.5)$<br>$= 2.27$ sixteenths | $D = R_{a,st}/(\ell q_a \times 1.5)$<br>$= 35/(2 \times 2 \times 2.75 \times 0.928 \times 1.5)$<br>$= 2.29$ sixteenths |

To develop the capacity of the 2.31/16 in welds on each side of the stiffener, the minimum thickness of the stiffener is given by AISC Manual Part 9 Eq. (9-3) as

$$\begin{aligned} t_{st} &= 6.19D/F_u \\ &= 6.19 \times 2.31/58 \\ &= 0.25 \text{ in} \\ &< 0.5 \text{ in ... satisfactory} \end{aligned}$$

From LRFD Table J2.4, the minimum size of fillet weld required for the 0.5-in-thick stiffener is

$$w = 3/16 \text{ in}$$

Use a 3/16-in weld.

The stiffeners are welded to the column web to resist the unbalanced flange force. Using 3/16 in E70XX fillet welds, on both sides of the pair of stiffeners, and allowing for the 1½-in corner clip, the length of stiffener required is

| LRFD | ASD |
|---|---|
| $\ell_{st} = R_{u,st}/(2 \times 2 \times 3q_u) + 1.5$<br>$= 52/(2 \times 2 \times 3 \times 1.39) + 1.5$<br>$= 4.62$ in | $\ell_{st} = R_{a,st}/(2 \times 2 \times 3q_a) + 1.5$<br>$= 35/(2 \times 2 \times 3 \times 0.928) + 1.5$<br>$= 4.64$ in |

The minimum required stiffener length, in accordance with AISC 360 Sec. J10.8, is

$$\ell_{st} = d_c/2 - t_{cf}$$
$$= 14.3/2 - 0.86$$
$$= 6.29 \text{ in}$$

Use a pair of stiffeners 4¼ × 1/2 × 6½ in long.

The same size stiffener may be adopted adjacent to the top flange.

## Doubler Plate Requirements

When the column web thickness is inadequate to resist the applied forces, doubler plates may be used in place of stiffeners in the case of web local yielding, web crippling, web compression buckling, and web panel zone shear. Doubler plates are required to resist the difference between the required strength and available limit state strength. Hence, the force delivered to the doubler plate is

$$R_{r,\,dp} = F_r - R_{c,\,min}$$

where $R_{r,\,dp}$ = required doubler plate strength
= $R_{u,\,dp}$ for LRFD load combinations
= $R_{a,\,dp}$ for ASD load combinations
$F_r$ = girder flange force
= $F_u$ for LRFD load combinations
= $F_a$ for ASD load combinations
$R_{c,\,min}$ = lesser of the available strength in web yielding at the location of tensile concentrated force, or the lesser of the available strengths in web yielding, web crippling, compression buckling, and panel zone shear at the location of compressive concentrated force
= $\varphi R_n$ for LRFD load combinations
= $R_n/\Omega$ for ASD load combinations
$\varphi$ = resistance factor
= 0.9
$\Omega$ = safety factor
= 1.67

The web doubler plate thickness is selected to provide the additional strength needed to resist the applied forces. The required doubler plate thickness is

$$t_{p,\,req} = t_e - t_{wc}$$

where $t_e$ is total required effective thickness and $t_{wc}$ is actual column web thickness.

As shown in Fig. 6.21, the doubler plate is placed against the column web with its ends extending beyond the girder flanges to clear the area of the column web subject to crippling and buckling. This distance is not less than 2.5 times the column *k*-distance. The top and bottom of the plate are fillet welded to the column web and the plate is connected to the column web with plug welds to prevent local buckling. When the effect of panel zone deformation on frame stability is considered in the design and the plate thickness is not less than 1/90 of the sum of its length and width, plug welds are not required. The plate sides may be fillet welded to the column flanges and a beveled edge is necessary to clear the flange-to-web fillet radius.

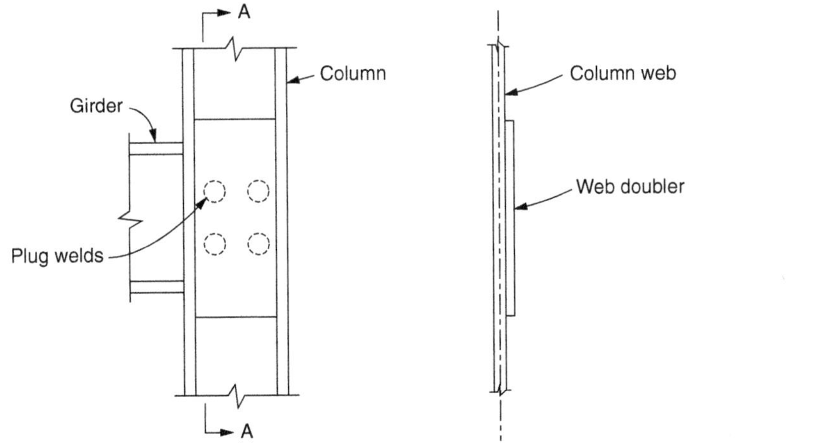


FIGURE 6.21 Detail of web doubler plate.

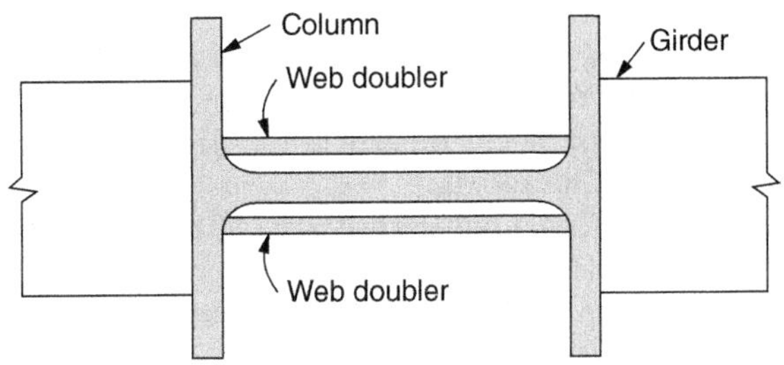


FIGURE 6.22 Alternative web doubler plate detail.

An alternative detail is shown in Fig. 6.22, where doubler plates are placed symmetrically in pairs spaced away from the column web.

**Example 6.19.** Doubler Plate
For the girder-column connection analyzed in Example 6.18, determine the thickness of doubler plate required to replace the stiffener plates.

The factored force delivered by the girder flange to the column is obtained from Example 6.18 as

| LRFD | ASD |
|---|---|
| $F_r$ = force delivered by the girder flange<br>$= 270$ kips | $F_r$ = force delivered by the girder flange<br>$= 180$ kips |

Web local yielding governs and the local yielding capacity for a web thickness of 0.525 in is obtained from Example 6.18 as

| LRFD | ASD |
|---|---|
| The design local yielding strength is<br>$\varphi R_n = 218$ kips | The allowable local yielding strength is<br>$R_n/\Omega = 145$ kips |

The force delivered to the doubler plate is

$$R_{r,\,dp} = F_r - R_{c,\,min}$$

| LRFD | ASD |
|---|---|
| Required doubler plate capacity is | Required doubler plate capacity is |
| $R_{u,\,dp} = 270 - 218$ | $R_{a,\,dp} = 180 - 145$ |
| $= 52$ kips | $= 35$ kips |
| Required doubler plate thickness is | Required doubler plate thickness is |
| $t_{dp} = 0.525 \times 52/218$ | $t_{dp} = 0.525 \times 35/145$ |
| $= 0.125$ in | $= 0.127$ in |

## References


1. American Institute of Steel Construction (AISC). 2010. *Specification for Structural Steel Buildings* (AISC 360-10), AISC, Chicago, IL.
2. Williams, A. 2009. *Structural Analysis in Theory and Practice.* Elsevier/International Code Council, New York, NY.
3. American Institute of Steel Construction (AISC). 2011. *Steel Construction Manual,* 14th edition. AISC, Chicago, IL.
4. American Society of Civil Engineers (ASCE). 2010. *Minimum Design Loads for Buildings and Other Structures* (ASCE 7-10), ASCE, Reston, VA.
5. Thornton, W. A. 1990. "Design of Small Base Plates for Wide Flange Columns," *Engineering Journal*, 3rd Quarter 1990.
6. Carter, C. J. 2003. *Stiffening of Wide-flange Columns at Moment Connections: Wind and Seismic Applications.* Design Guide No. 13. AISC, Chicago, IL.


## Problems

**6.1** *Given:* The braced frame shown in Fig. 6.23. The girder may be considered infinitely rigid and the brace is pinned at each end. The columns are adequately braced in the transverse direction.

*Find:* The effective length factors of the columns using AISC 360 Table C-A-7.1.

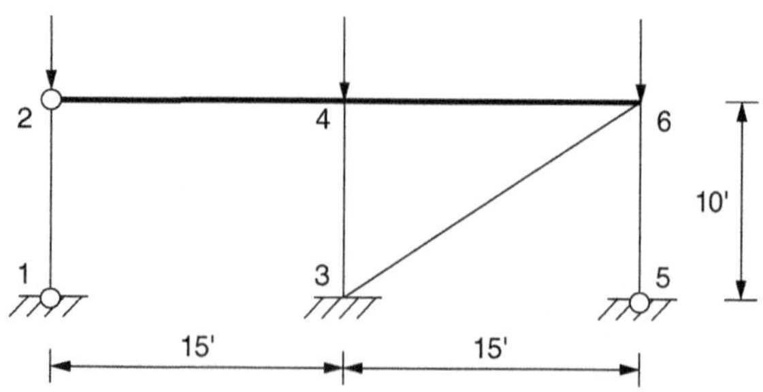


**FIGURE 6.23** Details for Problems 6.1 and 6.3.

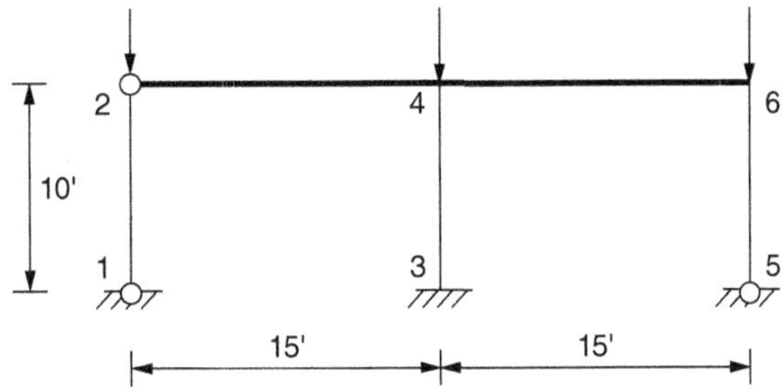


**FIGURE 6.24** Details for Problems 6.2 and 6.4.

**6.2** *Given:* The sway frame shown in Fig. 6.24. The girder may be considered infinitely rigid and the columns are adequately braced in the transverse direction.

*Find:* The effective length factors of the columns using AISC 360 Table C-A-7.1.

**6.3** *Given:* The braced frame shown in Fig. 6.23. The girders have a moment of inertia of five times that of the columns and the brace is pinned at each end. The columns are adequately braced in the transverse direction.

*Find:* The effective length factors of the columns using AISC 360 Fig. C-A-7.1.

**6.4** *Given:* The sway frame shown in Fig. 6.24. The girders have a moment of inertia of five times that of the columns. The columns are adequately braced in the transverse direction.

*Find:* The effective length factors of the columns using AISC 360 Fig. C-A-7.2.

**6.5** *Given:* The sway frame shown in Fig. 6.25. The girder is W16 × 57 and both columns are W12 × 50. The columns are adequately braced in the transverse direction. The value of W for the governing ASD load combination is 200 kips.

*Find:* Allowing for the stiffness reduction in the columns, determine if the frame is adequate.

**6.6** *Given:* The sway frame shown in Fig. 6.25. The girder is a W16 × 57 and both columns are W12 × 50. The columns are adequately braced in the transverse direction. The value of W for the governing LRFD load combination is 300 kips.

*Find:* Allowing for the stiffness reduction in the columns, determine if the frame is adequate.

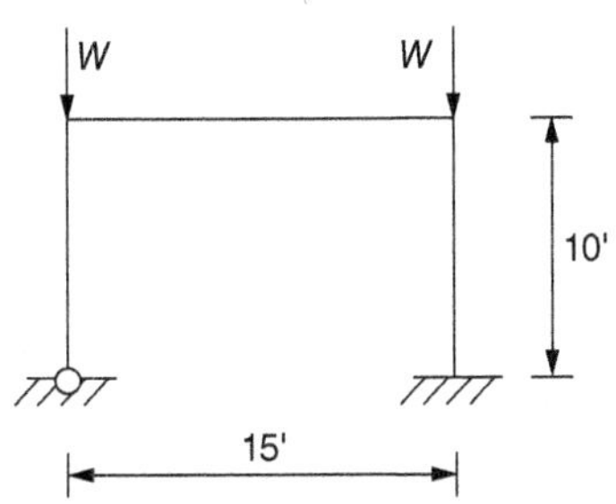


**FIGURE 6.25** Details for Problems 6.5 and 6.6.

**6.7** *Given:* A W12 × 58 column in a sway frame with a yield stress of 50 ksi and a height of 18 ft. The column may be considered fixed at each end and has no intermediate bracing.

*Find:* The available strength of the column for ASD loads using AISC 360 Sec. E3.

**6.8** *Given:* A W12 × 58 column in a sway frame with a yield stress of 50 ksi and a height of 18 ft. The column may be considered fixed at each end and has no intermediate bracing.

*Find:* The available strength of the column for LRFD loads using AISC 360 Sec. E3.

**6.9** *Given:* A column in a sway frame with a height of 18 ft is fixed at both ends and has no intermediate bracing. The loading consists of an axial dead load of $P_D = 100$ kips, which includes the weight of the column, and an axial live load of $P_L = 300$ kips.

*Find:* Using ASD load combinations, the lightest W12 column, with a yield stress of 50 ksi, which can support the load.

**6.10** *Given:* A column in a sway frame with a height of 18 ft is fixed at both ends and has no intermediate bracing. The loading consists of an axial dead load of $P_D = 100$ kips, which includes the weight of the column, and an axial live load of $P_L = 300$ kips.

*Find:* Using LRFD load combinations, the lightest W12 column, with a yield stress of 50 ksi, which can support the load.

**6.11** *Given:* A column in a braced frame with a height of 18 ft is pinned at each end and is braced at 6 ft intervals about the *y*-axis.

*Find:* Using ASD load combinations, the lightest W12 column, with a yield stress of 50 ksi, which can support the load.

**6.12** *Given:* A column in a braced frame with a height of 18 ft is pinned at each end and is braced at 6 ft intervals about the *y*-axis.

*Find:* Using LRFD load combinations, the lightest W12 column, with a yield stress of 50 ksi, which can support the load.

**6.13** *Given:* A single angle 4 × 4 × 1/2 strut with a length of 10 ft and a yield stress of $F_y = 36$ ksi. The strut is attached at each end by welding through the same leg. The strut has no intermediate bracing.

*Find:* The available strength of the strut for ASD loads using AISC 360 Sec. E5.

**6.14** *Given:* A single angle 4 × 4 × 1/2 strut with a length of 10 ft and a yield stress of $F_y = 36$ ksi. The strut is attached at each end by welding through the same leg. The strut has no intermediate bracing.

*Find:* The available strength of the strut for LRFD loads using AISC 360 Sec. E5.

**6.15** *Given:* A latticed column, shown in Fig. 6.26, consisting of four 4 × 4 × 1/2 in angles with a yield stress of 36 ksi. The column is 40 ft high with pinned ends and the 1- × 3/8-in lacing bars are arranged in a single system at an angle of 60° to the axis of the column.

*Find:* The available strength of the column for ASD loads.

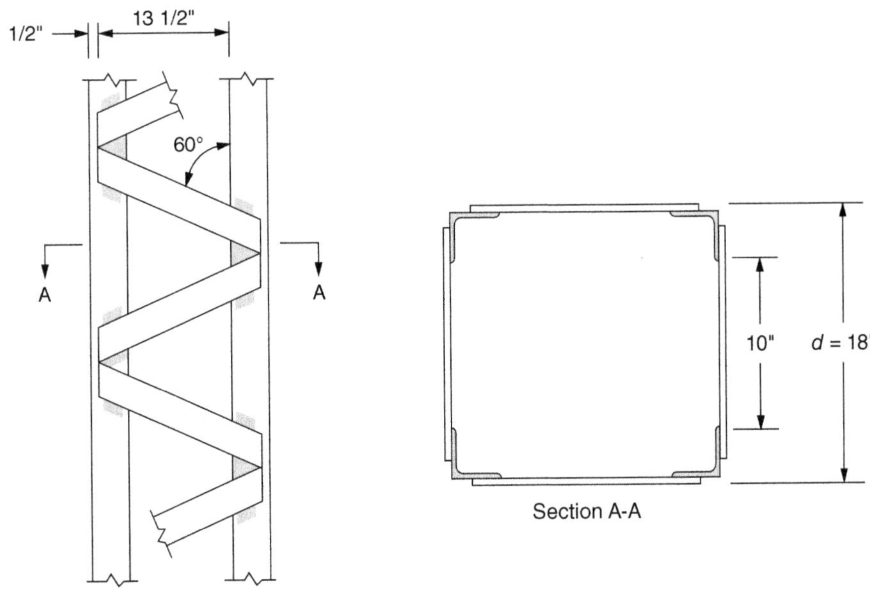


**FIGURE 6.26** Details for Problems 6.15 and 6.16.

**6.16** *Given:* A latticed column, shown in Fig. 6.26, consisting of four $4 \times 4 \times 1/2$ in angles with a yield stress of 36 ksi. The column is 40 ft high with pinned ends and the 1- $\times$ 3/8-in lacing bars are arranged in a single system at an angle of 60° to the axis of the column.

*Find:* The available strength of the column for LRFD loads.

**6.17** *Given:* A W12 $\times$ 106 column seated on a 16-in square base plate and supported on a reinforced concrete footing. The loading consists of an axial dead load of $P_D = 80$ kips, which includes the weight of the column, and an axial live load of $P_L = 240$ kips. The concrete footing is 8 ft square with an effective depth to reinforcement of 12 in and a compressive strength of 3000 psi.

*Find:* Using ASD load combinations, determine if the concrete footing is adequate.

**6.18** *Given:* A W12 $\times$ 106 column seated on a 16-in square base plate and supported on a reinforced concrete footing. The loading consists of an axial dead load of $P_D = 80$ kips, which includes the weight of the column, and an axial live load of $P_L = 240$ kips. The concrete footing is 8 ft square with an effective depth to reinforcement of 12 in and a compressive strength of 3000 psi.

*Find:* Using LRFD load combinations, determine if the concrete footing is adequate.

**6.19** *Given:* The W12 $\times$ 106 column with a 16-in square base plate of Problem 6.17. The plate has a yield stress of $F_y = 36$ ksi.

*Find:* Using ASD load combinations, determine the required minimum base plate thickness.

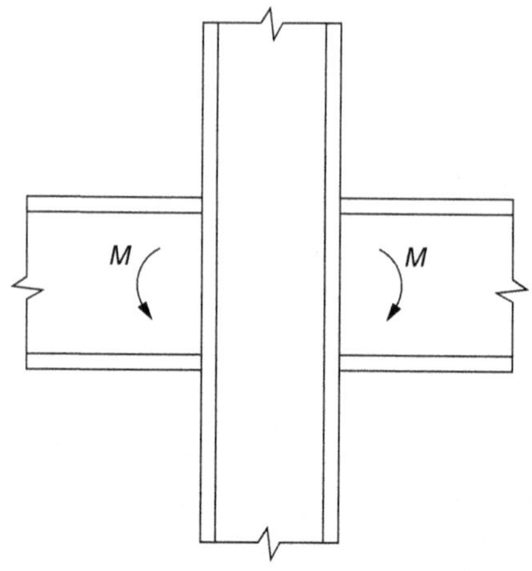


**FIGURE 6.27** Details for Problems 6.21 and 6.22.

**6.20** *Given:* The W12 × 106 column with a 16-in square base plate of Problem 6.18. The plate has a yield stress of $F_y = 36$ ksi.

*Find:* Using LRFD load combinations, determine the required minimum base plate thickness.

**6.21** *Given:* The girder-column connection shown in Fig. 6.27. The moments indicated are caused by gravity loads and consists of a dead load moment of $M_D = 50$ kip-ft, and a live load moment of $M_L = 150$ kip-ft. The column section is W14 × 82, the girder is W16 × 57, and both sections have a yield stress of 50 ksi.

*Find:* Using ASD load combinations, determine the size of stiffeners required using grade A36 steel.

**6.22** *Given:* The girder-column connection shown in Fig. 6.27. The moments indicated are caused by gravity loads and consists of a dead load moment of $M_D = 50$ kip-ft, and a live load moment of $M_L = 150$ kip-ft. The column section is W14 × 82, the girder is a W16 × 57, and both sections have a yield stress of 50 ksi.

*Find:* Using LRFD load combinations, determine the size of stiffeners required using grade A36 steel.

# CHAPTER 7
# Stability of Frames

## 7.1 Introduction

### Beam-Columns

In an actual structure, a column is seldom subjected to a purely concentric axial load. Because of accidental eccentricity and moments induced by frame action and lateral loads, a column is normally subjected to combined flexure and axial force. For this situation, a member is referred to as a beam-column.

The axial load and the bending moments about the $x$- and $y$-axes each produce a normal stress on the column section at a specific location. If elastic behavior is assumed and the principle of superposition is applied, the total stress at the point is given by the sum of the individual stresses. Thus,

$$f_{combined} = f_a + f_{bx} + f_{by}$$

A difficulty arises in determining the value of the allowable combined stress since the allowable stress for axial load is not the same as the allowable stress for flexure. Hence, an interaction expression is adopted in which the ratio of actual stress to allowable stress for the individual stress types is summed. If this sum is not greater than 1.0, the combination is considered acceptable.

This approach is used by American Institute of Steel Construction, *Specification for Structural Steel Buildings* (AISC 360)[1] Eq. (H2-1) which is

$$|f_{ra}/F_{ca} + f_{rbw}/F_{cbw} + f_{rbz}/F_{cbz}| \leq 1.0$$

where $w$ is a subscript relating symbol to major principal axis bending, $z$ is a subscript relating symbol to minor principal axis bending, and

| LRFD | ASD |
|---|---|
| $f_{ra}$ = required axial stress at the point of consideration using LRFD load combinations | $f_{ra}$ = required axial stress at the point of consideration using ASD load combinations |
| $F_{ca} = \varphi_c F_{cr}$ = design axial stress at the point of consideration | $F_{ca} = F_{cr}/\Omega_c$ = allowable axial stress at point of consideration |
| $f_{rbw}, f_{rbz}$ = required flexural stress at the point of consideration using LRFD load combinations | $f_{rbw}, f_{rbz}$ = required flexural stress at the point of consideration using ASD load combinations |
| $F_{cbw}, F_{cbz} = \varphi_b M_n/S$ = design flexural stress at point of consideration | $F_{cbw}, F_{cbz} = M_n/\Omega_b S$ = allowable flexural stress at point of consideration |
| $\varphi_c$ = resistance factor for compression = 0.90 | $\Omega_c$ = safety factor for compression = 1.67 |
| $\varphi_b$ = resistance factor for flexure = 0.90 | $\Omega_b$ = safety factor for flexure = 1.67 |

The interaction expression is applicable to unsymmetric as well as symmetric members and is valid when the axial force is tensile as well as compressive. The expression is evaluated using the principal bending axes by considering the sense of the flexural stresses at the critical points of the cross section. The flexural terms are either added to or subtracted from the axial term as appropriate. When the axial force is compression, second-order effects must be included in the analysis.

By multiplying each stress by its appropriate section property, AISC 360 Eq. (H2-1) can be transformed into an interaction expression based on strength to give

$$|P_r/P_c + M_{rw}/M_{cw} + M_{rz}/M_{cz}| \leq 1.0$$

where
$w$ = subscript relating symbol to major principal axis bending
$z$ = subscript relating symbol to minor principal axis bending
$P_r$ = required axial strength using LRFD or ASD load combinations
$M_{rw}$, $M_{rz}$ = required flexural strength using LRFD or ASD load combinations
$P_c$ = available axial strength
$M_{cw}$, $M_{cz}$ = available flexural strength

## Second-Order Effects

A compression force in a column introduces secondary stresses into the column due to the additional moment produced by the column displacements. The influence of second-order effects on frame stability is considered by Hewitt[2] and Geschwindner.[3] The primary forces in a structure, as calculated by a first-order analysis, are smaller than the actual values due to secondary moments caused by the *P*-delta effects. The *P*-delta effects are produced by the applied loads acting on the deformed shape of the structure caused by the curvature of the members and the sidesway of the structure. As shown in Fig. 7.1, secondary moments are produced by the two effects *P*-δ and *P*-Δ. The *P*-δ effect produces an amplified moment due to the curvature of the member as it deforms with the axial load acting on the displacement between the ends of the member. The *P*-Δ effect produces an amplified moment due to the drift in a sway frame with the applied load acting on the story drift.

The moment magnification factor that, when applied to the primary moments accounts for the *P*-δ effect, is termed $B_1$. The moment magnification factor that, when applied to the primary moments accounts for the *P*-Δ effect, is termed $B_2$. An approximate method of accounting for second-order effects is given in AISC 360 App. 8 Sec. 8.2.

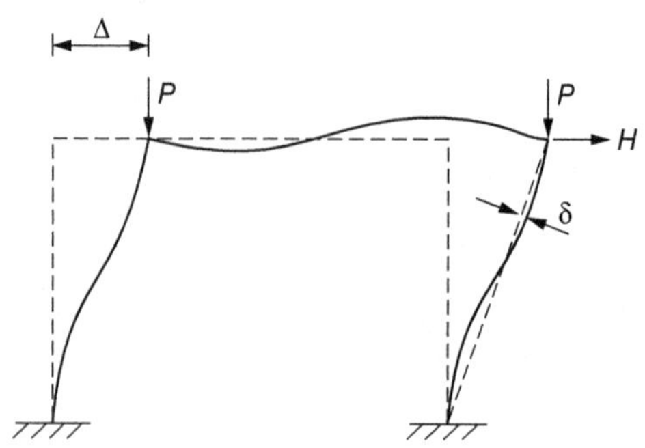


**FIGURE 7.1** *P*-delta effects.

The forces determined by a first-order analysis may be amplified as indicated by AISC 360 App. 8 Eqs. (A-8-1) and (A-8-2) to give the required final second-order forces

$$M_r = B_1 M_{nt} + B_2 M_{lt}$$

$$P_r = P_{nt} + B_2 P_{lt}$$

where $M_r = M_u$ or $M_a$
= required second-order flexural strength using LRFD or ASD load combinations
$P_r = P_u$ or $P_a$
= required second-order axial strength using LRFD or ASD load combinations
$M_{nt}$ = calculated first-order moment using LRFD or ASD load combinations assuming no lateral translation of the frame
$P_{nt}$ = calculated first-order axial force using LRFD or ASD load combinations assuming no lateral translation of the frame
$M_{lt}$ = calculated first-order moment using LRFD or ASD load combinations due to lateral translation of the frame only
$P_{lt}$ = calculated first-order axial force using LRFD or ASD load combinations in a member, due to lateral translation of the frame only
$B_1$ = multiplier to account for $P$-$\delta$ effects on the nonsway moments of compression members
$B_2$ = multiplier determined for each story of the structure to account for the $P$-$\Delta$ effects on the axial force and moment in all members of a sway frame

The multiplier $B_1$ is taken as 1.0 for members not subject to compression. The multiplier $B_2$ applies to both moments and axial forces in all components of the lateral load-resisting system. In accordance with AISC 360 Sec. C2.1(2), it is permissible to neglect the effects of $B_1$ when all of the following conditions are satisfied:

- gravity loads are supported primarily by vertical columns, walls, or frames
- the ratio of second-order drift to first-order drift is less than 1.7 where drift is calculated using LRFD load combinations or 1.6 times ASD load combinations and with stiffness adjusted to compensate for residual stresses
- no more than one-third of the total gravity load is supported on moment-resisting frames

A symmetrical structure subjected to symmetrical gravity loads experiences no lateral translation and secondary moments are produced only by the nonsway effects. However, in general, an unbraced frame must be analyzed for both effects. Two analyses are required in order to determine both $M_{nt}$ and $M_{lt}$ as shown in Fig. 7.2. In the first analysis, imaginary horizontal restraints are introduced at each floor level to prevent lateral translation. The factored loads are then applied, the primary moments $M_{nt}$ calculated, and the magnitudes of the imaginary restraints $R$ are determined. In the second analysis, the frame is analyzed for the reverse of the imaginary restraints in order to determine the primary moments $M_{lt}$. Alternatively, the frame may be analyzed with all factored loads applied, with joint translations prevented and the primary moments $M_{nt}$ calculated. The frame is then again analyzed with all factored loads applied, with joint translations permitted and the primary moments $M_{nt} + M_{lt}$

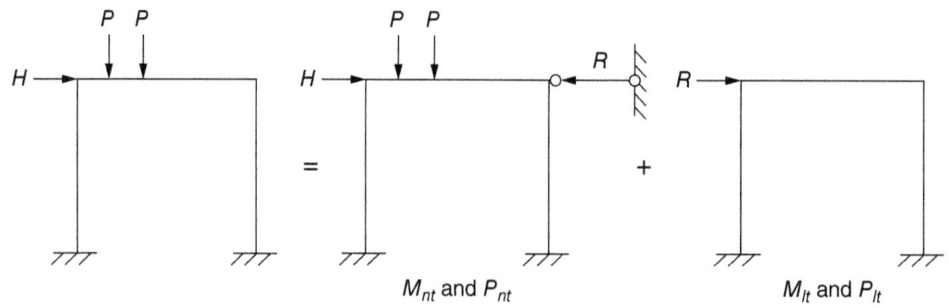


**FIGURE 7.2** Analysis of a sway frame.

calculated. The primary moments $M_{lt}$ are then obtained by subtracting the results of the first analysis from the second. The first-order forces, calculated prior to amplification, may be superimposed to obtain the total first-order forces. The additional moment in a member due to amplification must be distributed to connected members, in proportion to their relative stiffness, in order to maintain equilibrium.

*P*-delta effects are automatically determined in a rigorous, computer generated, second-order frame analysis and the members may then be designed directly for the calculated axial force and bending moment. Second-order analysis is a nonlinear problem and the principle of superposition does not apply. Hence, the results of individual load cases cannot be combined and separate analyses are necessary for each combination of factored loads. Because of this nonlinearity, a computer analysis must be conducted at the strength level of applied loads. For design by ASD, this load level is estimated as 1.6 times the ASD load combinations, and the analysis must be conducted at this elevated load to capture second-order effects at the strength level. The required member forces are then obtained by dividing the computer results by 1.6. This requirement does not apply to the approximate method of accounting for second-order effects. In this method, the second-order effects are captured by introducing the force level adjustment factor $\alpha = 1.6$ in the calculation of $B_1$ and $B_2$.

## 7.2 Design for Combined Forces

The design condition for a column in a typical building frame is that of axial compression plus flexure. This is covered in AISC 360 Sec. H1.1. As shown in Fig. 7.3, two interaction expressions are used to define the lower-bound curve for the nondimensional strengths. The interaction formulae for the design of doubly and singly symmetric members subjected to compression and biaxial bending are presented in AISC 360 Sec. H1.1.

For values of $P_r/P_c \geq 0.2$, AISC 360 Eq. (H1-1a) applies and

$$P_r/P_c + (8/9)(M_{rx}/M_{cx} + M_{ry}/M_{cy}) \leq 1.0$$

For values of $P_r/P_c < 0.2$, AISC 360 Eq. (H1-1b) applies and

$$P_r/2P_c + (M_{rx}/M_{cx} + M_{ry}/M_{cy}) \leq 1.0$$

where *x* is a subscript relating symbol to strong axis bending, *y* is a subscript relating symbol to weak axis bending, and

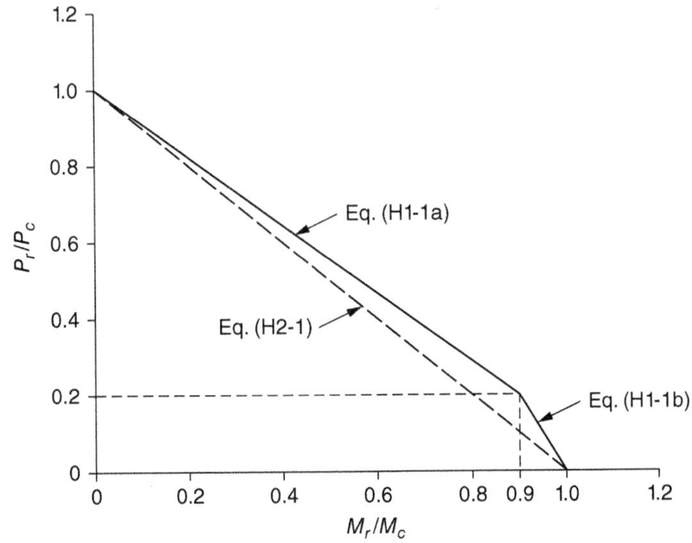


**FIGURE 7.3** Interaction curve.

| LRFD | ASD |
|---|---|
| $P_r$ = required axial strength using LRFD load combinations | $P_r$ = required axial strength using ASD load combinations |
| $P_c = \varphi_c P_n$ = design axial strength | $P_c = P_n/\Omega_c$ = allowable axial strength |
| $M_{rx}$, $M_{ry}$ = required flexural strength using LRFD load combinations | $M_{rx}$, $M_{ry}$ = required flexural strength using ASD load combinations |
| $M_{cx}$, $M_{cy} = \varphi_b M_{nx}$, $\varphi_b M_{ny}$ = design flexural strength | $M_{cx}$, $M_{cy} = M_{nx}/\Omega_b$, $M_{ny}/\Omega_b$ = allowable flexural strength |
| $\varphi_c$ = resistance factor for compression = 0.90 | $\Omega_c$ = safety factor for compression = 1.67 |
| $\varphi_b$ = resistance factor for flexure = 0.90 | $\Omega_b$ = safety factor for flexure = 1.67 |

To facilitate the application of these expressions, American Institute of Steel Construction, *Steel Construction Manual* (AISC Manual)[4] Table 6-1 lists values of $p$, $b_x$, and $b_y$ for W-shapes with a yield stress of $F_y = 50$ ksi and assuming a bending coefficient of $C_b = 1.0$, where

| | LRFD | ASD |
|---|---|---|
| Axial compression: | $p = 1/\varphi_c P_n \times 10^3$ (kips)$^{-1}$ | $p = \Omega_c/P_n \times 10^3$ (kips)$^{-1}$ |
| Strong axis bending: | $b_x = 8/9\varphi_b M_{nx} \times 10^3$ (kip-ft)$^{-1}$ | $b_x = 8\Omega_b/9M_{nx} \times 10^3$ (kip-ft)$^{-1}$ |
| Weak axis bending: | $b_y = 8/9\varphi_b M_{ny} \times 10^3$ (kip-ft)$^{-1}$ | $b_y = 8\Omega_b/9M_{ny} \times 10^3$ (kip-ft)$^{-1}$ |

Using these values, as indicated by Aminmansour,[5] the interaction expressions are reduced to

$$pP_r + b_x M_{rx} + b_y M_{ry} \leq 1.0 \text{ ... for values of } P_r/P_c \geq 0.2$$

$$pP_r/2 + (9/8)(b_x M_{rx} + b_y M_{ry}) \leq 1.0 \text{ ... for values of } P_r/P_c < 0.2$$

**Example 7.1.** Combined Compression and Bending
A pin ended W12 × 72 column has an unbraced length about both axes of 12 ft and a yield stress of $F_y = 50$ ksi. Determine the adequacy of the column to support the second-order forces indicated.

| LRFD | ASD |
|---|---|
| $P_u = 600$ kips | $P_a = 400$ kips |
| $M_{ux} = 80$ kip-ft | $M_{ax} = 60$ kip-ft |
| $M_{uy} = 0$ kip-ft | $M_{ay} = 0$ kip-ft |

The effective length about the $x$- and $y$-axis is 12 ft and the unbraced length for flexure is 12 ft. From AISC Manual Table 6-1

| LRFD | ASD |
|---|---|
| $p = 1.24 \times 10^{-3}$ (kips)$^{-1}$ | $p = 1.87 \times 10^{-3}$ (kips)$^{-1}$ |
| $b_x = 2.23 \times 10^{-3}$ (kip-ft)$^{-1}$ | $b_x = 3.36 \times 10^{-3}$ (kip-ft)$^{-1}$ |
| $P_u/\varphi_c P_n = pP_u = 1.24 \times 10^{-3} \times 600$ | $\Omega_c P_{uc}/P_n = pP_a = 1.87 \times 10^{-3} \times 400$ |
| $= 0.744 > 0.2$ ... Eq. (H1-1a) applies | $= 0.748 > 0.2$ ... Eq. (H1-1a) applies |
| $pP_u + b_x M_{ux} + b_y M_{uy} \le 1.0$ | $pP_a + b_x M_{ax} + b_y M_{ay} \le 1.0$ |
| $0.744 + 2.23 \times 10^{-3} \times 80 = 0.92$ | $0.748 + 3.36 \times 10^{-3} \times 60 = 0.95$ |
| $< 1.0$ ... satisfactory | $< 1.0$ ... satisfactory |

## 7.3 Stability Analysis

Several methods are available for stability analysis and design of steel frames. These include the following techniques:

- Direct analysis method detailed in AISC Sec. C2
- Effective length method, detailed in AISC App. 7 Sec 7.2
- First-order analysis method detailed in AISC App. 7 Sec 7.3
- Simplified method detailed in AISC Manual Part 2

These methods have been extensively reviewed in the literature by White and Griffis,[6] Carter and Geshwindner,[7] and Nair.[8]

The direct analysis method and the effective length method both require a second-order analysis of the structure. This may be accomplished using a rigorous, second-order computer analysis. An alternative approximate approach is presented in AISC 360 App. 8 Sec. 8.2 and this is known as the $B_1$-$B_2$ procedure.

### Approximate Second-Order Analysis

In this procedure, the results of a first-order analysis are amplified using the $B_1$ and $B_2$ multipliers in the expressions

$$M_r = B_1 M_{nt} + B_2 M_{lt} \text{ ... AISC 360 App. 8 Eq. (A-8-1)}$$

$$P_r = P_{nt} + B_2 P_{lt} \text{ ... AISC 360 App. 8 Eq. (A-8-2)}$$

where $B_1$ = multiplier to account of $P$-$\delta$ effects in members subject to compression as defined by AISC 360 App. 8 Eq. (A-8-3)
$= C_m/(1 - \alpha P_r/P_{e1})$
$\geq 1.0$
$P_r$ = required axial strength $\approx P_{nt} + P_{lt}$ ... for AISC 360 App. 8 Eq. (A-8-3)
$\alpha$ = force level adjustment factor = 1.0 ... for LRFD load combinations
= force level adjustment factor = 1.6 ... for ASD load combinations
$P_{e1}$ = Euler buckling strength of the member in the plane of bending as defined by AISC 360 App. 8 Eq. (A-8-5)
$= \pi^2 EI^*/(K_1L)^2$
$L$ = length of member
$K_1$ = effective-length factor in the plane of bending
= 1.0 ... unless analysis justifies a smaller value
$EI^*$ = flexural rigidity used in the analysis
$= 0.8\tau_b EI$ ... for direct analysis method
$= EI$ ... for effective length and first-order methods
$C_m$ = reduction factor given by AISC 360 App. 8 Eq. (A-8-4)
$= 0.6 - 0.4(M_1/M_2)$ ... for members not subjected to transverse loading between supports
= 1.0 ... for a member transversely loaded between supports
= 1.0 ... for a member bent in single curvature under uniform bending moment
$M_1$ = smaller moment at end of member, calculated from a first-order analysis
$M_2$ = larger moment at end of member, calculated from a first-order analysis
$M_1/M_2$ = +ve for a member bent in reverse curvature
= –ve for a member bent in single curvature
$B_2$ = multiplier for each story to account for $P$-$\Delta$ effects as defined by AISC 360 App. 8 Eq. (A-8-6)
$= 1/(1 - \alpha P_{story}/P_{e,\, story})$
$= \Delta_{2nd}/\Delta_{1st}$
$\geq 1.0$
$P_{story}$ = total factored vertical load supported by the story using LRFD or ASD load combinations, as applicable, including loads in columns that are not part of the lateral load-resisting system
$P_{e,story}$ = elastic critical buckling strength for the story in the direction of translation being considered as defined by AISC 360 App. 8 Eq. (A-8-7)
$= R_m HL/\Delta_H$
$R_M$ = coefficient to account for the influence of $P$-$\delta$ on $P$-$\Delta$ as defined by AISC 360 App. 8 Eq. (A-8-8)
$= 1 - 0.15(P_{mf}/P_{story}) = 0.85$ for moment frames
$P_{mf}$ = total vertical load in columns in a story that are part of moment frames, if any, in the direction of translation being considered
= 0 ... for braced frame systems
$L$ = story height
$\Delta_H$ = first-order interstory drift due to lateral forces computed using the appropriate stiffness
$H$ = story shear produced by the lateral forces used to compute $\Delta$H
$\Delta_{1st}$ = first-order interstory drift
$\Delta_{2nd}$ = second-order interstory drift

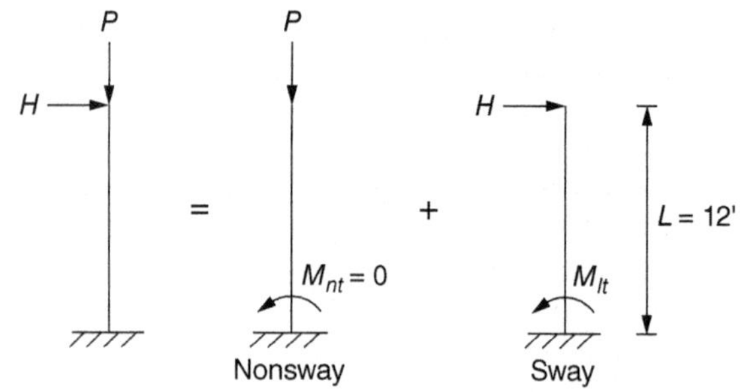


**FIGURE 7.4** Details for Example 7.2.

Note that it is permitted to use the first-order estimate of $P_r = P_{nt} + P_{lt}$ in AISC 360 App. 8 Eq. (A-8-3).

To assist in the determination of the Euler buckling strength of a member, AISC Manual Table 4-1 tabulates values of $P_e(KL)^2/10^4$ for W-shapes with a yield stress of 50 ksi.

**Example 7.2.** Approximate Second-Order Analysis

The W14 × 132 cantilever-column shown in Fig. 7.4 has an unbraced length about both axes of 12 ft and a yield stress of $F_y = 50$ ksi. The nominal applied loads are $D = 200$ kips, $L = 300$ kips, and $W = 40$ kips. Lateral bracing is provided at the top of the column and the column may be considered pinned top and bottom about the $y$-axis. Using the load combinations indicated, from ASCE 7 Sec. 2.3.2.4 and Sec. 2.4.1.6a, determine the adequacy of the column to support the second-order forces.

| LRFD | ASD |
|---|---|
| $1.2D + 0.5L + 1.0W$ | $D + 0.75L + 0.75(0.6W)$ |
| $P = 390$ kips | $P = 425$ kips |
| $H = 40$ kips | $H = 18$ kips |

The relevant properties of a W14 × 132 are

$$r_x = 6.28 \text{ in} \qquad r_y = 3.76 \text{ in} \qquad r_x/r_y = 1.67 \qquad I = 1530 \text{ in}^4$$

Hence, from Fig. 6.3

$$\begin{aligned} KL_y &= 12 \text{ ft} \\ KL_x &= 12 \times 2.1 = 25.2 \text{ ft} \\ (KL_y)_{\text{equiv}} &= (KL_x)/(r_x/r_y) = 25.2/1.67 \\ &= 15.1 \text{ ft ... governs} \\ &> KL_y \end{aligned}$$

As shown in Fig. 7.4, for the nonsway loading case

$$\begin{aligned} P_{nt} &= P \\ M_{nt} &= 0 \text{ ... calculation of } B_1 \text{ is not required} \\ P_{story} &= P_{nt} \\ P_{mf} &= P_{nt} \end{aligned}$$

For the sway loading case

$$\begin{aligned} P_{lt} &= 0 \\ M_{lt} &= 12H \end{aligned}$$

**Calculation of B$_2$**

| LRFD | ASD |
|---|---|
| $\alpha$ = force level adjustment factor = 1.0 | $\alpha$ = force level adjustment factor = 1.6 |
| $\Delta_H/L = HL^2/3EI$ | $\Delta_H/L = HL^2/3EI$ |
| $= 40 \times 144^2/(3 \times 29{,}000 \times 1530)$ | $= 18 \times 144^2/(3 \times 29{,}000 \times 1530)$ |
| $= 0.00623$ | $= 0.002804$ |
| $P_{story} = P_{mf} = P_{nt} = 390$ kips | $P_{story} = P_{mf} = P_{nt} = 425$ kips |
| $R_m = 1 - 0.15(P_{mf}/P_{story}) = 0.85$ | $R_m = 1 - 0.15(P_{mf}/P_{story}) = 0.85$ |
| $P_{e,\,story} = R_m HL/\Delta_H = 0.85 \times 40/0.00623$ | $P_{e,\,story} = R_m HL/\Delta_H = 0.85 \times 18/0.002804$ |
| $= 5457$ kips | $= 5457$ kips |
| $B_2 = 1/(1 - \alpha P_{story}/P_{e,\,story}) = 1/(1 - 1 \times 390/5457)$ | $B_2 = 1/(1 - \alpha P_{story}/P_{e story}) = 1/(1 - 1.6 \times 425/5457)$ |
| $= 1.08$ | $= 1.14$ |

**Calculate second-order forces**

$$M_r = B_1 M_{nt} + B_2 M_{lt}$$
$$= B_2 M_{lt}$$
$$P_r = P_{nt} + B_2 P_{lt}$$
$$= P_{nt}$$

| LRFD | ASD |
|---|---|
| $M_u = B_2 M_{lt} = 1.08 \times 12 \times 40 = 518$ kip-ft | $M_a = B_2 M_{lt} = 1.14 \times 12 \times 18 = 246$ kip-ft |
| $P_u = P_{nt} = 390$ kips | $P_a = P_{nt} = 425$ kips |

**Design for combined compression and bending**

The equivalent effective length about the $y$-axis is 15.1 ft and the unbraced length for flexure is 12 ft. From AISC Manual Table 6-1 for a W14 × 132

| LRFD | ASD |
|---|---|
| $p = 0.68 \times 10^3$ (kips)$^{-1}$ | $p = 1.02 \times 10^3$ (kips)$^{-1}$ |
| $b_x = 1.01 \times 10^3$ (kip-ft)$^{-1}$ | $b_x = 1.52 \times 10^3$ (kip-ft)$^{-1}$ |
| $P_u/\varphi_c P_n = pP_u = 0.68 \times 10^{-3} \times 390$ | $\Omega_c P_a/P_n = pP_a = 1.02 \times 10^{-3} \times 425$ |
| $= 0.265 > 0.2$ ... Eq. (H1-1a) applies | $= 0.434 > 0.2$ ... Eq. (H1-1a) applies |
| $pP_u + b_x M_{ux} + b_y M_{uy} \le 1.0$ | $pP_a + b_x M_{ax} + b_y M_{ay} \le 1.0$ |
| $0.265 + 1.01 \times 10^{-3} \times 518 = 0.79$ | $0.434 + 1.52 \times 10^{-3} \times 246 = 0.81$ |
| $< 1.0$ ... satisfactory | $< 1.0$ ... satisfactory |

## Stability Analysis Procedures

AISC 360 Sec. C1 lists the effects that must be considered in the design of a structure for stability. These are

- Flexural, shear, and axial deformations of members
- All other deformations that contribute to displacements of the structure
- $P$-$\Delta$ second-order effects caused by structure displacements
- $P$-$\delta$ second-order effects caused by member deformations
- Geometric imperfections caused by initial out-of-plumbness
- Reduction in member stiffness due to inelasticity and residual stresses
- Uncertainty in stiffness and strength

The first four of these issues are covered in the analysis of the structure.

Geometric imperfections are caused by the permitted tolerances in the plumbness of columns. As specified by American Institute of Steel Construction, *Code of Standard Practice for Steel Buildings and Bridges* (AISC 303)[9] Sec. 7.13.1.1, the maximum tolerance on out-of-plumbness of a column is $L/500$. As shown in Fig. 7.5 this produces a moment in a column of

$$M = P\Delta = PL/500$$

The same effect may be produced by applying a notional load of $P/500$ at the top of the column. This produces an identical moment at the bottom of the column of

$$M = PL/500 = 0.002PL$$

Residual stresses cause premature yielding in a member with a consequent loss of stiffness and increase in deformations. This may be compensated for by reducing the axial and flexural stiffness of members that contribute to the lateral stability of the structure prior to analysis.

The following stability analysis procedures are presented in the Specification and the Manual.

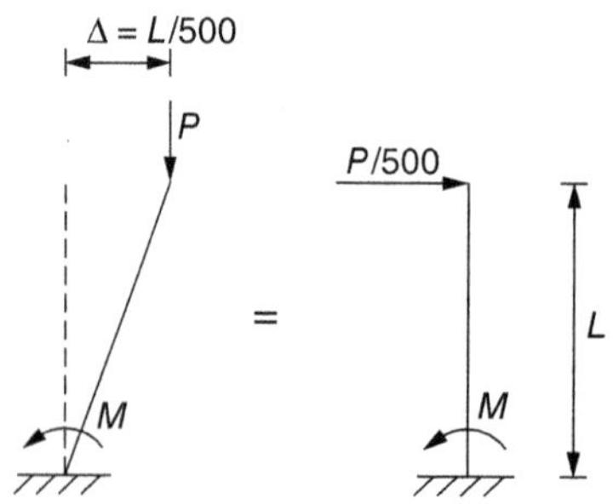


**FIGURE 7.5** Column out-of-plumbness.

### Direct Analysis Method

The direct analysis method is applicable to all types of structures. A second-order analysis is required that considers both $P$-$\Delta$ and $P$-$\delta$ effects. However, in accordance with AISC 360 Sec. C2.1 when $P$-$\delta$ effects are negligible, a $P$-$\Delta$-only analysis is permissible. This typically occurs when the ratio of second-order drift to first-order drift is less than 1.7 and no more than one-third of the total gravity load on the building is on columns that are part of moment-resisting frames. The latter condition is equivalent to an $R_M$ value of 0.95 or greater. It is still necessary to consider $P$-$\delta$ effects in the evaluation of individual beam-columns.

To account for initial imperfections in the members, notional lateral loads are applied at each story, in accordance with AISC 360 Sec. C2.2b(1), as shown in Fig. 7.6 and are given by

$$N_i = 0.002\alpha Y_i$$

where $N_i$ is notional lateral load applied at level $i$, $Y_i$ is gravity load applied at level $i$, and $\alpha = 1.0$ (LRFD), $\alpha = 1.6$ (ASD).

The notional loads are additive to the applied lateral loads when the sidesway amplification ratio, determined with drift calculated for LRFD load combinations or $1.6 \times$ ASD load combinations, is

$$\Delta_{2nd}/\Delta_{1st} = B_2 > 1.7 \text{ ... using the reduced elastic stiffness}$$
$$> 1.5 \text{ ... using the unreduced elastic stiffness}$$

where $\Delta_{2nd}$ is second-order drift and $\Delta_{1st}$ is first-order drift.

Otherwise, as specified in AISC 360 Sec. C2.2b(4), it is permissible to apply the notional load only in gravity-only load combinations.

If the out-of-plumb geometry of the structure is used in the analysis, it is permissible to omit the notional loads. Similarly, when the nominal initial out-of-plumbness differs from $L/500$, it is permissible to adjust the notional load coefficient proportionally.

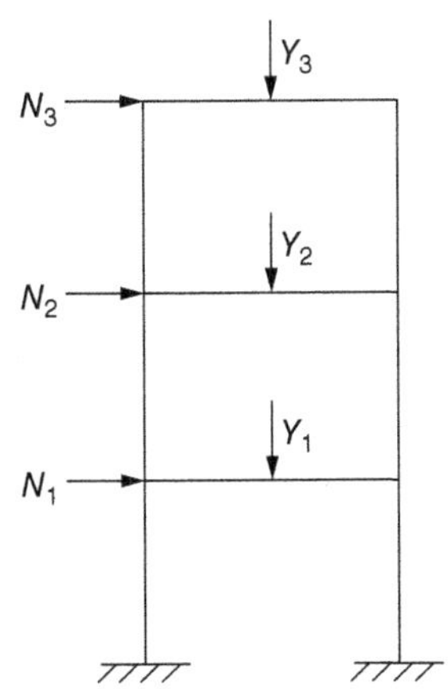


**FIGURE 7.6** Notional lateral loads.

To account for residual stresses and inelastic softening effects, the flexural and axial stiffness of members that contribute to the lateral stability of the structure are reduced as specified in AISC 360 Sec. C2.3 to give

$$EI^* = 0.8\tau_b EI$$

$$EA^* = 0.8EA$$

where $\tau_b$ = stiffness reduction parameter
= 1.0 ... for $\alpha P_r \leq 0.5P_y$
= $4(\alpha P_r/P_y)(1 - \alpha P_r/P_y)$ ... $\alpha P_r > 0.5P_y$
$\alpha$ = force level adjustment factor = 1.0 ... for LRFD load combinations
= force level adjustment factor = 1.6 ... for ASD load combinations
$P_r$ = required second-order axial strength
$P_y$ = member yield strength
= $AF_y$

It is permissible to apply the stiffness reduction to all members, including those that do not contribute to the stability of the structure. When $\alpha P_r > 0.5P_y$ the stiffness reduction factor $\tau_b$ may be taken as 1.0 provided that the actual lateral loads, in all load combinations, are increased by an additional notional lateral load of

$$N_i = 0.001\alpha Y_i$$

Because of the allowance for initial imperfections and inelastic softening effects in the analysis, stability effects are incorporated in the calculated required member strengths. Hence, the available strength of members may be determined using an effective length factor of $K = 1.0$ for all members.

**Example 7.3.** Direct Analysis Method
Analyze the frame shown in Fig. 7.7 using the direct analysis method. All members of the frame consist of W14 × 132 sections with a yield stress of 50 ksi. The self-weight of the frame may be neglected. No intermediate bracing is provided to the columns about either axis. The nominal applied loads are $D$ = 200 kips, $L$ = 300 kips, and $W$ = 40 kips. Using the load combinations indicated, determine the adequacy of column 12 to support the second-order forces. Lateral bracing

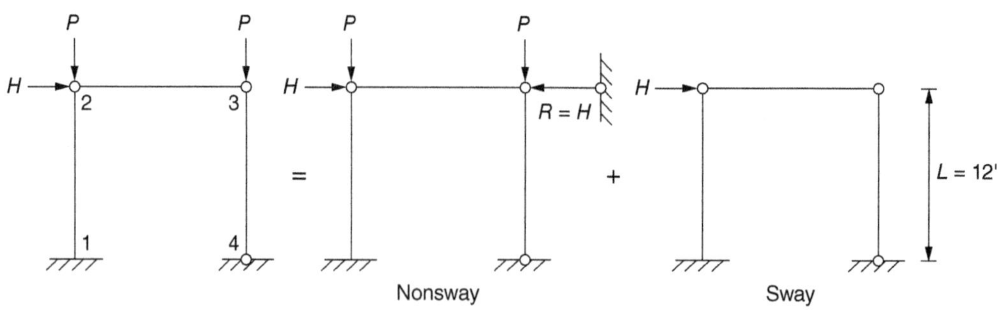


**Figure 7.7** Details for Example 7.3.

is provided at the top of column 12 and the column may be considered pinned top and bottom about the $y$-axis.

| LRFD | ASD |
|---|---|
| $1.2D + 0.5L + 1.0W$ | $D + 0.75L + 0.75(0.6W)$ |
| $P = 390$ kips | $P = 425$ kips |
| $H = 40$ kips | $H = 18$ kips |

The total gravity load applied to the frame is

$$Y_1 = 2P$$

In accordance with AISC 360 Sec. C2.2b(1), the notional lateral load is given by

$$N_i = 0.002 \times 2\alpha P$$

Assuming that $B_2 < 1.7$ using the reduced stiffness, the notional load is not added to the applied lateral load.

For member 12 for the nonsway loading case

$$P_{nt} = P$$

$$M_{nt} = 0 \text{ ... calculation of } B_1 \text{ is not required}$$

For member 12 for the sway loading case

$$P_{lt} = 0$$

$$M_{lt} = 12H$$

Since $P_{lt} = 0$, the second-order axial force in column 12 is

$$P_r = P_{nt}$$

$$= P$$

For the frame,

$P_{story}$ = total factored vertical load supported by the story including loads in columns that are not part of the lateral load-resisting system

$$= P + P = 2P$$

Column 34 is a leaning column which also forms part of the moment frame 1234.

$P_{mf}$ = total vertical load in columns in a story that are part of moment frames in the direction of translation being considered

$$= P + P = 2P$$

The column yield strength is

$$P_y = AF_y$$

$$= 38.8 \times 50$$

$$= 1940 \text{ kips}$$

**Calculation of $EI^*$ and $\Delta H/L$**

| LRFD | ASD |
|---|---|
| $\alpha$ = force level adjustment factor = 1.0 | $\alpha$ = force level adjustment factor = 1.6 |
| $\alpha P_r/P_y = 390/1940 = 0.201 < 0.5$ ... Hence, | $\alpha P_r/P_y = 1.6 \times 425/1940 = 0.351 < 0.5$ |
| $\tau_b = 1.0$ ... Hence, | $\tau_b = 1.0$ ... Hence, |
| $EI^* = 0.8\tau_b EI$ | $EI^* = 0.8\tau_b EI$ |
| $= 0.8 \times 1.0 \times 29{,}000 \times 1530$ | $= 0.8 \times 1.0 \times 29{,}000 \times 1530$ |
| $= 35{,}496{,}000$ kip-in$^2$ | $= 35{,}496{,}000$ kip-in$^2$ |
| $\Delta_H/L = HL^2/3EI^*$ | $\Delta_H/L = HL^2/3EI^*$ |
| $= 40 \times 144^2/(3 \times 35{,}496{,}000)$ | $= 18 \times 144^2/(3 \times 35{,}496{,}000)$ |
| $= 0.00779$ | $= 0.003505$ |

**Calculation of $B_2$**

| LRFD | ASD |
|---|---|
| $P_{story} = 2P = 390 + 390 = 780$ kips | $P_{story} = 2P = 425 + 425 = 850$ kips |
| $P_{mf} = 2P = 390 + 390 = 780$ kips | $P_{mf} = 2P = 425 + 425 = 850$ kips |
| $R_m = 1 - 0.15(P_{mf}/P_{story}) = 1 - 0.15 = 0.85$ | $R_m = 1 - 0.15(P_{mf}/P_{story}) = 1 - 0.15 = 0.85$ |
| $P_{e,story} = R_m HL/\Delta_H = 0.85 \times 40/0.00779$ | $P_{e,story} = R_m HL/\Delta_H = 0.85 \times 18/0.003505$ |
| $= 4365$ kips | $= 4365$ kips |
| $B_2 = 1/(1 - \alpha P_{story}/P_{e,story}) = 1/(1 - 1 \times 780/4365)$ | $B_2 = 1/(1 - \alpha P_{story}/P_{e,story}) = 1/(1 - 1.6 \times 850/4365)$ |
| $= 1.22 < 1.7$ as assumed | $= 1.45 < 1.7$ as assumed. |

**Calculate second-order forces for member 12**

$$M_r = B_1 M_{nt} + B_2 M_{lt}$$
$$= B_2 M_{lt}$$
$$P_r = P_{nt} + B_2 P_{lt}$$
$$= P_{nt}$$

| LRFD | ASD |
|---|---|
| $M_u = B_2 M_{lt} = 1.22 \times 12 \times 40 = 586$ kip-ft | $M_a = B_2 M_{lt} = 1.45 \times 12 \times 18 = 313$ kip-ft |
| $P_u = P_{nt} = 390$ kips | $P_a = P_{nt} = 425$ kips |

**Design of member 12 for combined compression and bending**

The effective length factor is $K = 1.0$. Hence, the effective length of member 12 about the $x$- and $y$-axes is 12 ft and the unbraced length for flexure is 12 ft. From AISC Manual Table 6-1 for a W14 × 132

| LRFD | ASD |
|---|---|
| $p = 0.638 \times 10^3\ (\text{kips})^{-1}$ | $p = 0.958 \times 10^3\ (\text{kips})^{-1}$ |
| $b_x = 1.01 \times 10^3\ (\text{kip-ft})^{-1}$ | $b_x = 1.52 \times 10^3\ (\text{kip-ft})^{-1}$ |
| $P_u/\varphi_c P_n = pP_u = 0.638 \times 10^{-3} \times 390$ | $\Omega_c P_a/P_n = pP_a = 0.958 \times 10^{-3} \times 425$ |
| $= 0.249 > 0.2$ ... Eq. (H1-1a) applies | $= 0.407 > 0.2$ ... Eq. (H1-1a) applies |
| $pP_u + b_x M_{ux} + b_y M_{uy} \leq 1.0$ | $pP_a + b_x M_{ax} + b_y M_{ay} \leq 1.0$ |
| $0.249 + 1.01 \times 10^{-3} \times 586 = 0.84$ | $0.407 + 1.52 \times 10^{-3} \times 313 = 0.88$ |
| $< 1.0$ ... satisfactory | $< 1.0$ ... satisfactory |

## Effective Length Method

Required strengths must be determined by a second-order analysis, either by computer or by the $B_1$-$B_2$ procedure.

In accordance with AISC 360 App. 7 Sec. 7.2, the use of the effective length method is not permitted when

$$\Delta_{2nd}/\Delta_{1st} = B_2 > 1.5 \text{ ... with drift determined for LRFD load combinations or } 1.6 \times \text{ASD load combinations}$$

To account for initial imperfections in the members, notional lateral loads are applied at each story, in accordance with AISC 360 Sec. C2.2b(1), and are given by

$$N_i = 0.002\alpha Y_i$$

where $N_i$ is notional lateral load applied at level $i$, and $\alpha = 1.0$ (LRFD) or $\alpha = 1.6$ (ASD), $Y_i$ is gravity load applied at level $i$. It is permissible to apply the notional load only in gravity-only load combinations.

In the analysis, the nominal stiffness of all members shall be used with no reduction for inelastic softening effects.

The available strength of members may be determined using the appropriate effective length factor $K$ as defined in AISC 360 Commentary App. 7 Table C-A-7.1 or calculated in accordance with AISC 360 Commentary App. 7 Sec. 7.2. The empirical column curve then accounts for geometrical imperfections and inelastic softening effects, as demonstrated by Schwinger.[10] A value of 1.0 may be used for the effective length factor of members in a braced frame. For columns that do not contribute to the lateral resistance of the structure, a value of 1.0 may be used for the effective length factor. A value of 1.0 may also be used, in accordance with AISC 360 App. 7 Sec. 7.2.3, for all columns when

$$\Delta_{2nd}/\Delta_{1st} = B_2 \leq 1.1 \text{ ... with drift determined for LRFD load combinations or } 1.6 \times \text{ASD load combinations}$$

Any increase in column moments due to amplification must be balanced by a compensating increase in beam moments.

**Example 7.4.** Effective Length Method

Analyze the frame shown in Fig. 7.7 using the effective length method. All members of the frame consist of W14 × 132 section with a yield stress of 50 ksi. The self-weight of the frame may be neglected. No intermediate bracing is provided to the columns about either axis. The nominal applied loads are $D = 200$ kips, $L = 300$ kips, and $W = 40$ kips. Using the load combinations indicated, determine the adequacy of column 12 to support the second-order forces. Lateral bracing is provided at the top of column 12 and the column may be considered pinned top and bottom about the $y$-axis.

| LRFD | ASD |
|---|---|
| $1.2D + 0.5L + 1.6W$ | $D + 0.75L + 0.75(0.6W)$ |
| $P = 390$ kips | $P = 425$ kips |
| $H = 40$ kips | $H = 18$ kips |

The load combination is not a gravity-only condition and notional loads are not necessary in the analysis.

For member 12 for the nonsway loading case

$$P_{nt} = P$$

$$M_{nt} = 0 \text{ ... calculation of } B_1 \text{ is not required}$$

For member 12 for the sway loading case

$$P_{lt} = 0$$

$$M_{lt} = 12\text{H}$$

Since $P_{lt} = 0$, the second-order axial force in column 12 is

$$P_r = P_{nt}$$
$$= P$$

For the frame

$P_{story}$ = total factored vertical load supported by the story including loads in columns that are not part of the lateral load-resisting system

$= P + P = 2P$

$P_{mf}$ = total vertical load in columns in a story that are part of moment frames, if any, in the direction of translation being considered

$= P + P = 2P$

**Calculation of $B_2$**

| LRFD | ASD |
|---|---|
| $\alpha$ = force level adjustment factor = 1.0 | $\alpha$ = force level adjustment factor = 1.6 |
| $\Delta_H/L = HL^2/3EI$ | $\Delta_H/L = HL^2/3EI$ |
| $= 40 \times 144^2/(3 \times 29{,}000 \times 1530)$ | $= 18 \times 144^2/(3 \times 29{,}000 \times 1530)$ |
| $= 0.00623$ | $= 0.002804$ |
| $P_{story} = 2P = 390 + 390 = 780$ kips | $P_{story} = 2P = 425 + 425 = 850$ kips |
| $P_{mf} = 2P = 390 + 390 = 780$ kips | $P_{mf} = 2P = 425 + 425 = 850$ kips |
| $R_m = 1 - 0.15(P_{mf}/P_{story}) = 1 - 0.15 = 0.85$ | $R_m = 1 - 0.15(P_{mf}/P_{story}) = 1 - 0.15 = 0.85$ |
| $P_{e,story} = R_m HL/\Delta_H = 0.85 \times 40/0.00623$ | $P_{e,story} = R_m HL/\Delta_H = 0.85 \times 18/0.002804$ |
| $= 5457$ kips | $= 5457$ kips |
| $B_2 = 1/(1 - \alpha P_{story}/P_{e,story}) = 1/(1 - 1 \times 780/5457)$ | $B_2 = 1/(1 - \alpha P_{story}/P_{e,story}) = 1/(1 - 1.6 \times 850/5457)$ |
| $= 1.17$ | $= 1.33$ |
| $< 1.5$ ... AISC 360 App. 7 Sec. 7.2.1 is satisfied $> 1.1$ ... AISC 360 App. 7 Sec. 7.2.3, $K \neq 1.0$ | $< 1.5$ ... AISC 360 App. 7 Sec. 7.2.1 is satisfied $> 1.1$ ... AISC 360 App. 7 Sec. 7.2.3, $K \neq 1.0$ |

**Calculate second-order forces for member 12**

$$M_r = B_1M_{nt} + B_2M_{lt}$$
$$= B_2M_{lt}$$
$$P_r = P_{nt} + B_2P_{lt}$$
$$= P_{nt}$$

| LRFD | ASD |
|---|---|
| $M_u = B_2M_{lt} = 1.17 \times 12 \times 40 = 562$ kip-ft | $M_a = B_2M_{lt} = 1.33 \times 12 \times 18 = 287$ kip-ft |
| $P_u = P_{nt} = 390$ kips | $P_a = P_{nt} = 425$ kips |

**Design of member 12 for combined compression and bending**

For both the ASD and LRFD loading combinations, $B_2 > 1.1$ and effective length factors must be determined. Column 34 is a leaning column and does not contribute to the sway stiffness of the frame and has an effective length factor of $K = 1.0$. As described by Lim and McNamara[11], the columns in a story that are part of the lateral load-resisting system must be designed to resist the second-order effects of the gravity loads on the leaning columns. In effect, the effective length factor for in-plane buckling of the rigid frame columns must be increased to account for the destabilizing effects of the leaning column gravity loads. The modified effective length factor is

$$K_m = K_x(1 + N)^{0.5}$$

where $K_x$ = nominal effective length determined from AISC 360 Commentary App. 7 Sec. 7.2
$N = P_{leaning}/P_{stability}$
$P_{leaning}$ = total vertical load supported by the leaning columns
$P_{stability}$ = total vertical load supported by stability columns

For member 12

$$KL_y = 12 \text{ ft}$$
$$K_x = 2.1$$
$$K_m = K_x(1 + N)^{0.5}$$
$$N = P_{leaning}/P_{stability}$$
$$= P/P$$
$$= 1.0$$
$$K_m = 2.1(1 + 1)^{0.5}$$
$$= 2.97$$
$$K_mL_x = 12 \times 2.97 = 35.64 \text{ ft}$$
$$(K_mL_x)_{\text{equiv}} = (K_mL_x)/(r_x/r_y)$$
$$= 35.64/1.67$$
$$= 21.3 \text{ ft ... governs}$$
$$> KL_y$$

The equivalent effective length about the $y$-axis is 21.3 ft and the unbraced length for flexure is 12 ft. From AISC Manual Table 6-1 for a W14 × 132

| LRFD | ASD |
|---|---|
| $p = 0.805 \times 10^3\ (\text{kips})^{-1}$ | $p = 1.21 \times 10^3\ (\text{kips})^{-1}$ |
| $b_x = 1.01 \times 10^3\ (\text{kip-ft})^{-1}$ | $b_x = 1.52 \times 10^3\ (\text{kip-ft})^{-1}$ |
| $P_u/\varphi_c P_n = pP_u = 0.805 \times 10^{-3} \times 390$ | $\Omega_c P_a/P_n = pP_a = 1.21 \times 10^{-3} \times 425$ |
| $= 0.314 > 0.2$ ... Eq. (H1-1a) applies | $= 0.514 > 0.2$ ... Eq. (H1-1a) applies |
| $pP_u + b_x M_{ux} + b_y M_{uy} \le 1.0$ | $pP_a + b_x M_{ax} + b_y M_{ay} \le 1.0$ |
| $0.314 + 1.01 \times 10^{-3} \times 562 = 0.88$ | $0.514 + 1.52 \times 10^{-3} \times 287 = 0.95$ |
| $< 1.0$ ... satisfactory | $< 1.0$ ... satisfactory |

### First-Order Elastic Analysis Method

The first-order elastic analysis method, as derived by Kuchenbecker et al.,[12] is detailed in AISC 360 App. 7 Sec. 7.3. Required strengths are determined by a first-order analysis only, using the nominal (unreduced) stiffness, with some restrictions imposed.

In accordance with AISC 360 App. 7 Sec. 7.3.1, the use of the first-order elastic analysis method is not permitted when

$$\Delta_{2nd}/\Delta_{1st} = B_2 > 1.5 \text{ ... with drift determined for LRFD load combinations or } 1.6 \times \text{ASD load combinations}$$

In addition, the use of the first-order elastic analysis method is not permitted when

$$\alpha P_r > 0.5P_y$$

where $\alpha$ = force level adjustment factor
= 1.0 ... for LRFD load combinations
= 1.6 ... for ASD load combinations
$P_r$ = required axial strength
$P_y$ = member yield strength
$= AF_y$

All load combinations must include an additional lateral load in combination with other loads at each level of the structure. This presupposes a first-order drift ratio of $\Delta/L = 1/500$ and a sidesway amplification factor of $B_2 = 1.5$. The additional notional lateral load is given by AISC 360 App. 7 Eq. (A-7-2-2) as

$$N_i = 2.1\alpha(\Delta/L)Y_i$$
$$\ge 0.0042Y_i$$

where $N_i$ = notional lateral load applied at level $i$
$Y_i$ = gravity load applied at level $i$
$\Delta/L$ = maximum ratio of $\Delta$ to $L$ for all stories in the structure
$\Delta$ = first-order inter-story drift due to the LRFD or ASD load combination, as applicable
$L$ = story height

The notional loads are applicable to all loading combinations and are additive to the applied lateral loads.

In accordance with AISC 360 App. 7 Sec. 7.3.2(2), the nonsway amplification of column moments is considered by applying the $B_1$ amplifier to the total column moments.

Because of the additional requirements incorporated in the analysis, stability effects are integrated in the calculated required member strengths. Hence, the available strength of members may be determined using an effective length factor of $K = 1.0$ for all members.

**Example 7.5.** First-Order Elastic Analysis Method.
Analyze the frame shown in Fig. 7.7 using the first-order elastic analysis method. All members of the frame consist of W14 × 132 sections with a yield stress of 50 ksi. The self weight of the frame may be neglected. No intermediate bracing is provided to the columns about either axis. The nominal applied loads are $D = 200$ kips, $L = 300$ kips, and $W = 40$ kips. Using the load combinations indicated, determine the adequacy of column 12 to support the factored loads. Lateral bracing is provided at the top of column 12 and the column may be considered pinned top and bottom about the $y$-axis.

| LRFD | ASD |
|---|---|
| $1.2D + 0.5L + 1.6W$ | $D + 0.75L + 0.75(0.6W)$ |
| $P = 390$ kips | $P = 425$ kips |
| $H = 40$ kips | $H = 18$ kips |

For member 12 for the nonsway loading case

$$P_{nt} = P$$

For member 12 for the sway loading case

$$P_{lt} = 0$$
$$M_{lt} = 12H$$

Since $P_{lt} = 0$, the required axial force in column 12 is

$$P_r = P_{nt}$$
$$= P$$

From Example 7.4, the ratio of second-order drift to first-order drift is

$$\Delta_{2nd}/\Delta_{1st} = B_2$$
$$= 1.15 \text{ ... for LRFD load combination}$$
$$< 1.5$$
$$= 1.30 \text{ ... for ASD load combination}$$
$$< 1.5$$

Hence, AISC 360 App. 7 Sec. 7.3.1(2) is satisfied.

For member 12 the column yield strength is

$$P_y = AF_y$$
$$= 38.8 \times 50$$
$$= 1940 \text{ kips}$$
$$> 2P_r \text{ ... LRFD load combinations}$$
$$> 1.6 \times 2P_r \text{ ... ASD load combinations}$$

Hence, AISC 360 App. 7 Sec. 7.3.1(3) is also satisfied and the first-order elastic analysis method may be used.

**Calculation of $N_i$**

| LRFD | ASD |
|---|---|
| $\alpha$ = force level adjustment factor = 1.0 | $\alpha$ = force level adjustment factor = 1.6 |
| $\Delta_H/L = HL^2/3EI$ | $\Delta_H/L = HL^2/3EI$ |
| $= 40 \times 144^2/(3 \times 29{,}000 \times 1530)$ | $= 18 \times 144^2/(3 \times 29{,}000 \times 1530)$ |
| $= 0.00623$ | $= 0.00280$ |
| $Y_i = 2P = 390 + 390 = 780$ kips | $Y_i = 2P = 425 + 425 = 850$ kips |
| $N_i = 0.0042Y_i = 0.0042 \times 780 = 3.28$ | $N_i = 0.0042Y_i = 0.0042 \times 850 = 3.57$ |
| $\geq 2.1\alpha(\Delta/L)Y_i = 2.1 \times 1 \times 0.00623 \times 780$ | $\geq 2.1\alpha(\Delta/L)Y_i = 2.1 \times 1.6 \times 0.00280 \times 850$ |
| $= 10.21$ ... governs | $= 8.00$ ... governs |
| Augmented moment at end 1 of column 12 is $M_{12} = (H + N_i)L = (40 + 10.21)12 = 603$ kip-ft | Augmented moment at end 1 of column 12 is $M_{12} = (H + N_i)L = (18 + 8.00)12 = 312$ kip-ft |

**Calculation of $B_1$**

$K_1$ = effective-length factor in the plane of bending for column 12
= 1.0 ... from AISC 360 App. 8 Sec. 8.2.1
$EI$ = nominal flexural rigidity
$= 29{,}000 \times 1530$
$= 44{,}370 \times 10^3$ kip-in$^2$
$P_{e1}$ = Euler buckling strength of member 12 in the plane of bending
$= \pi^2 EI/(K_1L)^2$
$= 3.14^2 \times 44{,}370{,}000/(1 \times 144)^2$
$= 21{,}097$ kips

| LRFD | ASD |
|---|---|
| $C_m = 0.6 - 0.4(M_1/M_2) = 0.6 - 0.4(0/603) = 0.6$ | $C_m = 0.6 - 0.4(M_1/M_2) = 0.6 - 0.4(0/312) = 0.6$ |
| $B_1 = C_m/(1 - \alpha P_r/P_{e1}) = 0.6/(1 - 1 \times 390/21{,}097)$ | $B_1 = C_m/(1 - \alpha P_r/P_{e1}) = 0.6/(1 - 1.6 \times 425/21{,}097)$ |
| $= 0.61$ ... use 1.0 minimum | $= 0.62$ ... use 1.0 minimum |

**Calculate amplified forces for member 12**

$$M_r = B_1 M_{12}$$
$$P_r = P$$

| LRFD | ASD |
|---|---|
| $M_u = B_1 M_{12} = 1.0 \times 603 = 603$ kip-ft | $M_a = B_1 M_{12} = 1.0 \times 312 = 312$ kip-ft |
| $P_u = P = 390$ kips | $P_a = P = 425$ kips |

**Design of member 12 for combined compression and bending**

The effective length factor is $K = 1.0$. Hence, the effective length of member 12 about the $y$-axis is 12 ft and the unbraced length for flexure is 12 ft. From AISC Manual Table 6-1 for a W14 × 132

| LRFD | ASD |
|---|---|
| $p = 0.638 \times 10^3$ (kips)$^{-1}$ | $p = 0.958 \times 10^3$ (kips)$^{-1}$ |
| $b_x = 1.01 \times 10^3$ (kip-ft)$^{-1}$ | $b_x = 1.52 \times 10^3$ (kip-ft)$^{-1}$ |
| $P_u/\varphi_c P_n = pP_u = 0.638 \times 10^{-3} \times 390$ | $\Omega_c P_a/P_n = pP_a = 0.958 \times 10^{-3} \times 425$ |
| $= 0.249 > 0.2$ ... Eq. (H1-1a) applies | $= 0.407 > 0.2$ ... Eq. (H1-1a) applies |
| $pP_u + b_x M_{ux} + b_y M_{uy} \le 1.0$ | $pP_a + b_x M_{ax} + b_y M_{ay} \le 1.0$ |
| $0.249 + 1.01 \times 10^{-3} \times 603 = 0.86$ | $0.407 + 1.52 \times 10^{-3} \times 312 = 0.88$ |
| $< 1.0$ ... satisfactory | $< 1.0$ ... satisfactory |

**Simplified Method**

The simplified method is detailed in AISC Manual Part 2. Required strengths are determined by multiplying the forces obtained from a first-order analysis by tabulated values of $B_2$. These tabulated values are a function of $\Delta_t$ and $\alpha P_{story}/H_t$ where

$\Delta_t$ = target story drift
$\alpha$ = force level adjustment factor
= 1.0 ... for LRFD load combinations
= 1.6 ... for ASD load combinations
$P_{story}$ = total factored vertical load supported by the story using LRFD or ASD load combinations, as applicable, including loads in columns that are not part of the lateral load-resisting system
$H_t$ = lateral force required to produce $\Delta_t$

The simplified method is not permitted when the sidesway amplification factor is

$$\Delta_{2nd}/\Delta_{1st} = B_2 > 1.5 \text{ ... with drift determined for LRFD load combinations or } \alpha \times \text{ASD load combinations}$$

In addition, the use of the simplified method is not permitted when the ratio of the sway and nonsway amplification factors is

$$B_1/B_2 > 1.0$$

where $\Delta_{2nd}$ is second-order drift and $\Delta_{1st}$ is first-order drift.

For gravity load cases, notional lateral loads are applied at each story and are given by

$$N_i = 0.002\alpha Y_i$$

where $N_i$ is notional lateral applied at level $i$ and $Y_i$ is gravity load applied at level $i$ independent of loads from above.

The procedure consists of performing a first-order elastic analysis of the structure for the applied loads and any necessary notional lateral loads. From this analysis, the lateral load $H_t$ required to produce the target story drift is determined. Hence, the ratio $\alpha P_{story}/H_t$ may be calculated. Using this value and the value of the target story drift, the appropriate amplification factor $B_2$ is obtained from the table in AISC Manual Table 2-1. The amplified forces are then

| LRFD | ASD |
|---|---|
| $M_r = B_1M_{nt} + B_2M_{lt} = B_2M_u$<br>$P_r = P_{nt} + B_2P_{lt} = B_2P_u$ | $M_r = B_1M_{nt} + B_2M_{lt} = B_2M_a$<br>$P_r = P_{nt} + B_2P_{lt} = B_2P_a$ |

For cases where the amplification factor does not exceed 1.1, the members may be designed using a value for the effective length factor of $K = 1.0$. For cases where the amplification factor exceeds 1.1, the effective length factors for the members must be determined by analysis.

**Example 7.6.** Simplified Method

Analyze the frame shown in Fig. 7.7 using the simplified method with a target story drift of $L/160$ for LRFD load combinations and $L/357$ for ASD load combinations. All members of the frame consist of W14 × 132 sections with a yield stress of 50 ksi. The self weight of the frame may be neglected. No intermediate bracing is provided to the columns about either axis. Using the axial loads indicated, determine the adequacy of column 12 to support the factored loads. Lateral bracing is provided at the top of column 12 and the column may be considered pinned top and bottom about the $y$-axis.

| LRFD | ASD |
|---|---|
| $P = 390$ kips | $P = 425$ kips |

From previous examples, $B_2 < 1.5$ and $B_1/B_2 < 1.0$. Hence, the simplified method is admissible.

The load combination is not a gravity-only condition and notional loads are not necessary in the analysis.

The total gravity load applied to the frame is

$$P_{story} = 2P$$

The lateral load applied to the frame is

$$H_{story} = H_t$$

**Calculate $H_t$ and $\alpha P_{story}/H_t$**

| LRFD | ASD |
|---|---|
| $\alpha$ = force level adjustment factor = 1.0 | $\alpha$ = force level adjustment factor = 1.6 |
| Target story drift = $\Delta_t = L/160$, hence, | Target story drift = $\Delta_t = L/357$, hence, |
| $H_t = (3EI/L^2)/160$ | $H_t = (3EI/L^2)/357$ |
| $= (3 \times 29{,}000 \times 1530/144^2)/160 = 40$ kips | $= (3 \times 29{,}000 \times 1530/144^2)/357 = 18$ kips |
| $\alpha P_{story}/H_t = 1.0 \times 780/40 = 20$ | $\alpha P_{story}/H_t = 1.6 \times 850/18 = 76$ |

**Determine amplified forces for column 12 from table in AISC Manual Part 2**

| LRFD | ASD |
|---|---|
| For $\Delta_t = L/160$ and $\alpha P_{story}/H_t = 20$, $B_2 = 1.2$ | For $\Delta_t = L/338$ and $\alpha P_{story}/H_t = 76$, $B_2 = 1.30$ |
| $M_u = B_2H_tL = 1.2 \times 40 \times 12 = 576$ kip-ft | $M_a = B_2H_tL = 1.30 \times 18 \times 12 = 281$ kip-ft |
| $P_u = 1.2P = 1.2 \times 390 = 468$ kips | $P_a = 1.30P = 1.30 \times 425 = 553$ kips |

**Design of member 12 for combined compression and bending**

Since $B_2 > 1.1$, effective length factors must be determined and the effect of the leaning column must be included. From Example 7.4, the equivalent effective length about the $y$-axis is 21.3 ft and the unbraced length for flexure is 12 ft.

| LRFD | ASD |
|---|---|
| $p = 0.805 \times 10^3$ (kips)$^{-1}$ | $p = 1.21 \times 10^3$ (kips)$^{-1}$ |
| $b_x = 1.01 \times 10^3$ (kip-ft)$^{-1}$ | $b_x = 1.52 \times 10^3$ (kip-ft)$^{-1}$ |
| $P_u/\varphi_c P_n = pP_u = 0.805 \times 10^{-3} \times 468$ | $\Omega_c P_a/P_n = pP_a = 1.21 \times 10^{-3} \times 553$ |
| $= 0.377 > 0.2$ ... Eq. (H1-1a) applies | $= 0.669 > 0.2$ ... Eq. (H1-1a) applies |
| $pP_u + b_x M_{ux} + b_y M_{uy} \leq 1.0$ | $pP_a + b_x M_{ax} + b_y M_{ay} \leq 1.0$ |
| $0.377 + 1.01 \times 10^{-3} \times 576 = 0.96$ | $0.669 + 1.52 \times 10^{-3} \times 281 = 1.1$ |
| $< 1.0$ ... satisfactory | $> 1.0$ ... unsatisfactory |

## References


1. American Institute of Steel Construction (AISC). 2010. *Specification for Structural Steel Buildings* (AISC 360-10), AISC, Chicago, IL.
2. Hewitt, C. M. 2008. "Stability Analysis: It's Not As Hard As You Think," *Modern Steel Construction*, September, 2008.
3. Geschwindner, L. F. 2002. "A Practical Look at Frame Analysis, Stability and Leaning Columns," *Engineering Journal*, 4th Quarter 2002.
4. American Institute of Steel Construction (AISC). 2011. *Steel Construction Manual*, 13th edition, AISC, Chicago, IL.
5. Aminmansour, A. 2007. "Design of Structural Steel Members Subject to Combined Loading," *Structure Magazine*, February, 2007.
6. White, D. W. and Griffis, L. G. 2007. "Stability Design of Steel Buildings: Highlights of a New AISC Design Code." *Proceedings of the North American Steel Construction Conference*, New Orleans, LA, April 2007.
7. Carter, C. J. and Geschwindner, L. F. 2008. "A Comparison of Frame Stability Analysis Methods in ANSI/AISC 360-05," *Engineering Journal*, 3rd Quarter 2008.
8. Nair, R. S. 2009. "Simple and Direct," *Modern Steel Construction*, January, 2009.
9. American Institute of Steel Construction (AISC). 2010. *Code of Standard Practice for Steel Buildings and Bridges*. AISC, Chicago, IL.
10. Schwinger, C. 2013. "Stability Analysis and Design," *Modern Steel Construction*, April, 2013.
11. Geschwinder, L. F. 1994. "A Practical Approach to the Leaning Column," *Engineering Journal*, 4th Quarter 1994.
12. Kuchenbecker, G. H., White, D. W., and Surovek-Maleck, A. E. 2004. "Simplified Design of Building Frames Using First-Order Analysis and K = 1." *Proceedings of the Structural Stability Research Council Annual Meeting*, Long Beach, CA, March 2004.


## Problems

**7.1** *Given:* A pin ended column with an unbraced length about both axes of 12 ft and a yield stress of $F_y = 50$ ksi. The applied second-order forces, using ASD load combinations, are $P_a = 200$ kips and $M_{ax} = 75$ kip-ft. The self weight of the column may be neglected.

*Find:* The lightest W12 column to support the second-order forces.

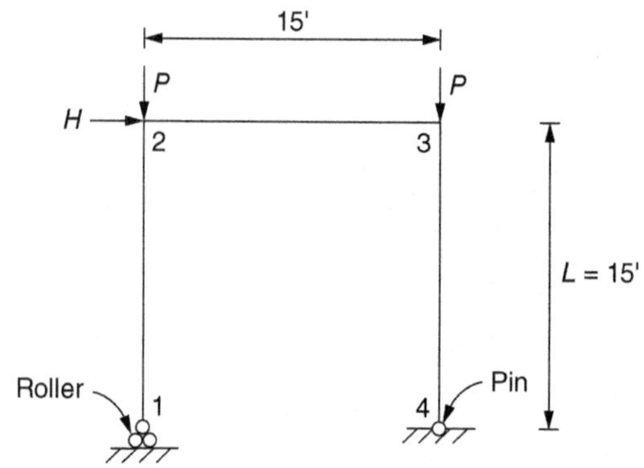


**Figure 7.8** Details for Problem 7.3.

**7.2** *Given:* A pin ended column with an unbraced length about both axes of 12 ft and a yield stress of $F_y = 50$ ksi. The applied second-order forces, using LRFD load combinations, are $P_u = 300$ kips and $M_{ux} = 110$ kip-ft. The self-weight of the column may be neglected.

*Find:* The lightest W12 column to support the second-order forces.

**7.3** *Given:* The frame shown in Fig. 7.8 with all members consisting of W14 × 99 sections with a yield stress of $F_y = 50$ ksi. The first-order ASD load combination is indicated, where $P = 250$ kips and $H = 10$ kips.

*Find:* Using the direct analysis method, determine if column 34 is adequate. The self-weight of the frame may be neglected. No intermediate bracing is provided to the column about either axis.

**7.4** *Given:* The frame shown in Fig. 7.8 with all members consisting of W14 × 99 sections with a yield stress of $F_y = 50$ ksi. The first-order LRFD load combination is indicated, where $P = 300$ kips and $H = 15$ kips.

*Find:* Using the direct analysis method, determine if column 34 is adequate. The self-weight of the frame may be neglected. No intermediate bracing is provided to the column about either axis.

**7.5** *Given:* The frame shown in Fig. 7.8 with all members consisting of W14 × 99 sections with a yield stress of $F_y = 50$ ksi. The first-order ASD load combination is indicated, where $P = 250$ kips and $H = 10$ kips.

*Find:* Using the effective length method, determine if column 34 is adequate. The self-weight of the frame may be neglected. No intermediate bracing is provided to the column about either axis.

**7.6** *Given:* The frame shown in Fig. 7.8 with all members consisting of W14 × 99 sections with a yield stress of $F_y = 50$ ksi. The first-order LRFD load combination is indicated, where $P = 300$ kips and $H = 15$ kips.

*Find:* Using the effective length method, determine if column 34 is adequate. The self-weight of the frame may be neglected. No intermediate bracing is provided to the column about either axis.

**7.7** *Given:* The frame shown in Fig. 7.8 with all members consisting of W14 × 99 sections with a yield stress of $F_y = 50$ ksi. The first-order ASD load combination is indicated, where $P = 250$ kips and $H = 10$ kips.

*Find:* Using the first-order elastic analysis method, determine if column 34 is adequate. The self-weight of the frame may be neglected. No intermediate bracing is provided to the column about either axis.

**7.8** *Given:* The frame shown in Fig. 7.8 with all members consisting of W14 × 99 sections with a yield stress of $F_y = 50$ ksi. The first-order LRFD load combination is indicated, where $P = 300$ kips and $H = 15$ kips.

*Find:* Using the first-order elastic analysis method, determine if column 34 is adequate. The self-weight of the frame may be neglected. No intermediate bracing is provided to the column about either axis.

**7.9** *Given:* The frame shown in Fig. 7.8 with all members consisting of W14 × 99 sections with a yield stress of $F_y = 50$ ksi. The first-order ASD axial load is $P = 250$ kips and the target story drift is $L/400$.

*Find:* Using the simplified method, determine if column 34 is adequate. The self-weight of the frame may be neglected. No intermediate bracing is provided to the column about either axis.

**7.10** *Given:* The frame shown in Fig. 7.8 with all members consisting of W14 × 99 sections with a yield stress of $F_y = 50$ ksi. The first-order LRFD axial load is $P = 300$ kips and the target story drift is $L/300$.

*Find:* Using the simplified method, determine if column 34 is adequate. The self-weight of the frame may be neglected. No intermediate bracing is provided to the column about either axis.

# CHAPTER 8

# Design by Inelastic Analysis

## 8.1 Introduction

### General Principles

The design of a structure by inelastic analysis, also known as plastic design, is based on the determination of the maximum loads that a structure can sustain before collapse. The method has been extensively reviewed in the literature by Martin[1] and ASCE.[2] The collapse load is also known as the ultimate load, the failure load, and the limit load. The ratio of collapse load to applied load is known as the load factor.

The two basic approaches to structural analysis are elastic analysis and inelastic analysis. Structures designed by the elastic method posses a considerable reserve of strength beyond the elastic limit before they reach their ultimate limit. However, the reserve strength and the collapse mechanism are not determined and are unknown. The load factor varies considerably between structures and depends on the type of structure, the degree of redundancy, and the disposition of the loads. On the other hand, the inelastic design method is based on a predetermined load factor and the ultimate strength and collapse mechanism are determined. An additional advantage of inelastic design is that prior to incipient collapse the structure is reduced to a determinate condition, thus simplifying the analysis.

Collapse analysis is based on the assumption of proportional applied loads, that is, a single application of a specific load combination. In practice, loads due to wind and seismic effects are repetitive and may reverse in direction or be applied in a random manner. Hence, some members in a structure may be subject to alternating plasticity with a consequent reduction in the load factor as indicated by Popov and McCarthy.[3] However, in practice, the dead weight of the structure is continuously applied and this significantly raises the load factor against failure due to alternating plasticity above that for proportional loading.

When the loads applied to a structure are random but nonreversible, incremental collapse may occur at a load factor lower than that for proportional loading. This is known as shakedown and is discussed by Xiaofeng.[4] However, to the extent that applied loads in building structures are repetitive, collapse due to shakedown is less probable than collapse due to proportional loading.

In accordance with American Institute of Steel Construction, *Specification for Structural Steel Buildings* (AISC 360)[5] App. 1 Sec. 1.1, allowable strength design methods are not permitted for inelastic analysis and design and the LRFD provisions of AISC 360

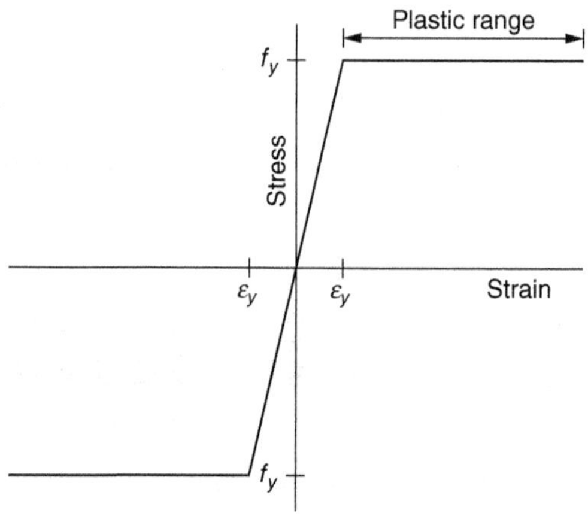


**FIGURE 8.1** Elastic-plastic behavior.

Sec. B3.3 must be used. Hence, this chapter conforms to load and resistance factor design requirements.

### Ductility

Adequate ductility is essential to ensure that the inelastic deformation capacity of members is not less than inelastic deformation demands. This is ensured by the requirement in AISC 360 App. 1 Sec. 1.2.1 that the minimum yield strength of members shall not exceed 65 ksi. In addition, as specified in AISC 360 App. 1 Sec. 1.2.4, the required strength in compression must not exceed $0.75F_yA_g$ for compression members with plastic hinges.

In applying the plastic method of analysis, it is assumed that the structural material used exhibits an ideal linear elastic-plastic behavior and elastic instability effects are ignored. The stress-strain curve shown in Fig. 8.1 is initially linear until the yield point of the material is reached when it is assumed that plastic yielding immediately occurs with the stress remaining constant at the yield stress as strain continues to increase. The additional strength available as strain hardening occurs is ignored and this provides an additional margin of safety to the procedure.

## 8.2 Plastic Moment of Resistance

As shown in Fig. 8.2, the load applied to a simply supported beam produces a bending moment at midspan of

$$M = WL/4$$
$$= f_bS_x$$

where $f_b$ = stress in the extreme top and bottom fibers shown at (*a*) in Fig. 8.2
$S_x$ = elastic section modulus about the *x*-axis
$= I/c$
$c$ = distance from extreme fiber to neutral axis
$= d/2$ ... for a symmetrical section
$d$ = depth of beam
$I$ = moment of inertia of beam

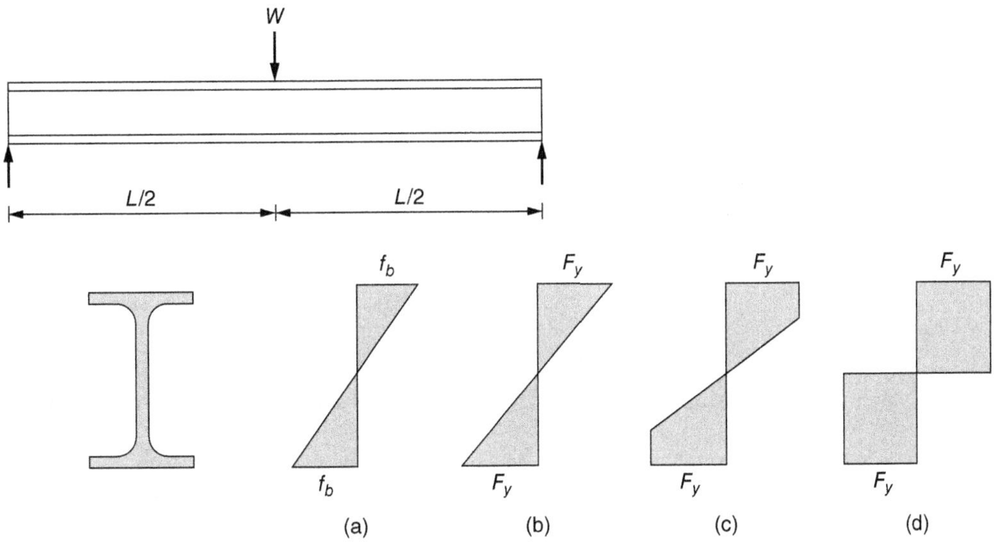


**FIGURE 8.2** Plastic moment of resistance.

By increasing the load on the beam, the stress in the extreme fibers reaches the yield stress $F_y$ as shown at (*b*) in Fig. 8.2. Allowing for the residual fabrication stresses in the beam, the applied moment is now

$$M_r = 0.7F_yS_x$$

Further increase in the load causes the plasticity to spread toward the centroid of the beam as shown at (*c*). Eventually, as shown at (*d*), all fibers in the cross section have yielded, a plastic hinge is assumed to form, the structure becomes a mechanism, and collapse occurs. The nominal flexural strength of the section is

$$\begin{aligned} M_n &= M_p \\ &= F_yZ_x \end{aligned}$$

where $M_p$ = plastic moment of resistance
$Z_x$ = plastic section modulus
= arithmetical sum of the first moments of area about the equal area axis

The ratio of the plastic moment to the yield moment is termed the shape factor and is given by

$$\begin{aligned} \nu &= M_p/M_y \\ &= Z/S \end{aligned}$$

Values of $\varphi_bM_p$ for W-sections with a yield stress of $F_y = 50$ ksi are tabulated in American Institute of Steel Construction, *Steel Construction Manual* (AISC Manual)[6] Tables 3-2 and 3-6, where $\varphi_b = 0.9$.

## 8.3 Plastic Hinge Formation

A fixed-ended beam, supporting a nominal distributed load $w$, is shown in Fig. 8.3*a*. The plastic moment of resistance of the beam section is

$$M_p = wL^2/8$$

Under the working load, the beam remains entirely elastic with the moments at the ends of the beam twice the moment at the center and

$$M_{12} = wL^2/12$$

$$M_3 = wL^2/24$$

By increasing the applied load to $w_1$, the moments shown at (*b*) are produced with

$$M_{12} = M_p$$

$$M_3 = w_1L^2/8 - M_p$$

Hence, plastic hinges have developed simultaneously at both ends of the beam. As shown at (*d*), and ignoring stiffness reduction due to inelastic softening effects, the load-deflection curve is linear up to this point.

Further increase in the applied load causes the two plastic hinges to rotate while the moment at the ends of the beam remains constant at the value $M_p$. Thus, the system is

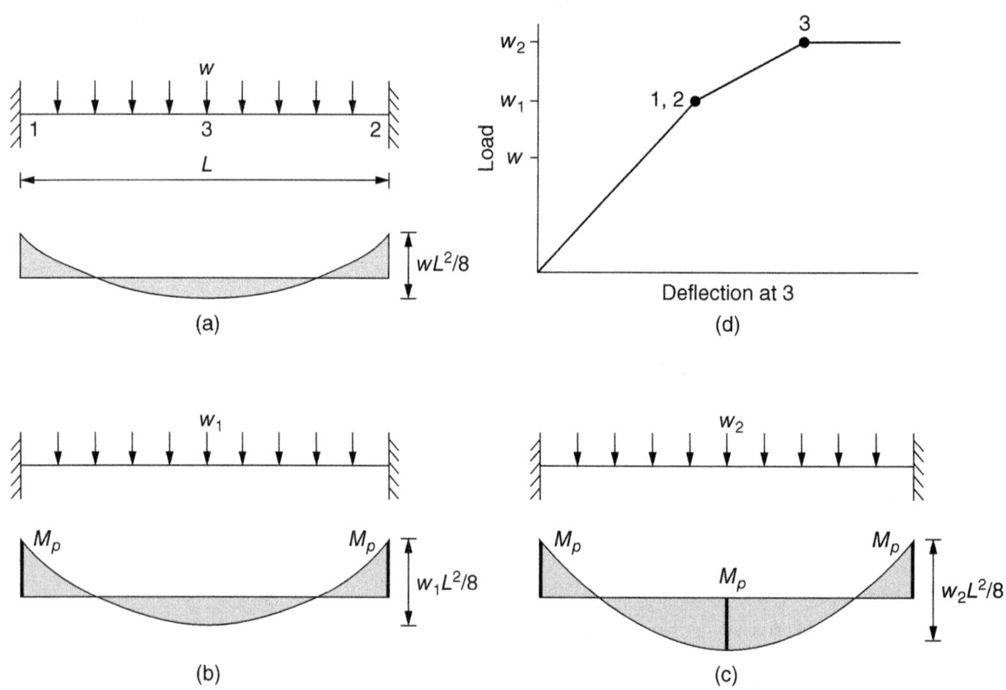


FIGURE 8.3 Plastic hinge formation.

equivalent to a simply supported beam with an applied load and constant end moments $M_p$. As shown at (*d*), the stiffness of the beam is reduced and displacements, for a given increase in applied load, are greater than in the original beam.

Finally, as the load increases to $w_2 = 2w$, the moments shown at (*c*) are produced with

$$M_{12} = M_p$$
$$M_3 = 2wL^2/8 - M_p$$
$$= wL^2/8$$
$$= M_p$$

Hence, a third plastic hinge has formed in the center of the beam producing an unstable collapse mechanism and causing failure. The ultimate load is

$$w_u = w_2 = 2w$$

The load factor is

$$N = w_u/w$$
$$= 2$$

The plastic moment of resistance is

$$M_p = w_u L^2/16$$

The linear elastic-plastic response curve shown at (*d*) is a series of straight lines with the rate of growth of displacements increasing as the stiffness reduces with the formation of each hinge.

## 8.4 Design Requirements

### Local Buckling

The cross-section of members at plastic hinge locations must be doubly symmetric. In addition to satisfying the compact section criteria of AISC 360 Table B4.1b, the webs of W-sections, subject to combined flexure and compression, shall also comply with AISC 360 App. 1 Eqs. (A-1-1) and (A-1-2) which are

$$h/t_w \leq \lambda_{pd}$$
$$= 3.76(E/F_y)^{0.5}(1 - 2.75P_u/\varphi_c P_y) \ldots \text{ for } P_u/\varphi_c P_y \leq 0.125$$
$$h/t_w \leq \lambda_{pd}$$
$$= 1.12(E/F_y)^{0.5}(2.33 - P_u/\varphi_c P_y) \ldots \text{ for } P_u/\varphi_c P_y > 0.125$$
$$\geq 1.49(E/F_y)^{0.5}$$

where $h$ = clear distance between flanges less the corner radius at each flange
$t_w$ = web thickness
$P_u$ = required axial strength in compression
$\varphi_c$ = resistance factor for compression
= 0.90
$P_y$ = axial yield strength
$= F_y A_g$
$A_g$ = area of section

## Unbraced Length

The maximum unbraced length of an I-shaped member bent about the major axis shall not exceed the value given by AISC 360 App. 1 Eq. (A-1-5) as

$$L_{pd} = [0.12 - 0.076(M_1'/M_2)]Er_y/F_y$$

where $r_y$ = radius of gyration of the member about its weak axis, inches
$M_1'/M_2 = +1$ ... when the magnitude of the bending moment at any location within the unbraced length exceeds $M_2$
otherwise $M_1' = 2M_{mid} - M_2$ ... when $M_{mid} > (M_1 + M_2)/2$
$< M_2$
or $M_1' = M_1$ ... when $M_{mid} \leq (M_1 + M_2)/2$
and $L_{pd}$ = limiting laterally unbraced length for plastic analysis, inches
$M_1$ = lesser of the moments at the ends of the unbraced segment, kip-in
$M_2$ = larger of the moments at the ends of the unbraced segment, kip-in
= positive in all cases
$M_{mid}$ = moment at the middle of the unbraced length, kip-in
$M_1'$ = effective moment at the end of the unbraced length opposite from $M_2$

The moments $M_1$ and $M_{mid}$ are taken as positive when they cause compression in the same flange as the moment $M_2$ and negative otherwise.

In the case of the last hinge to form, rotation does not occur, and the bracing requirements of AISC 360 Sec. F2.2 are applicable. Similarly, AISC 360 Sec. F2.2 applies to segments remote from a plastic hinge.

For members subject to axial compression and containing plastic hinges, the maximum permitted laterally unbraced length about both the $x$- and $y$-axis is specified in AISC 360 App. 1 Sec. 1.2.3(b) as

$$L/r = 4.71(E/F_y)^{0.5}$$

$$= 113 \text{ ... for } F_y = 50 \text{ ksi}$$

## Limiting Axial Load

In accordance with AISC 360 App. 1 Sec. 1.2.4, the maximum axial load in a column is limited to

$$P_{max} = 0.75A_g F_y$$

## 8.5 Analysis Requirements

AISC 360 App. 1 Sec. 1.1 requires inelastic analysis to take into account:

- Member deformations
- *P*-delta effects
- Stiffness reduction due to residual stresses and partial yielding
- Geometric imperfections
- Uncertainty of system, member, and connection strength and stiffness

AISC 360 Sec. 1.3 specifies the use of a second-order analysis in order to satisfy these requirements. However, for continuous beams not subject to axial compression, a first-order analysis is permitted. Traditionally, first-order plastic analysis has also been applied to low-rise frames with small axial loads.

### Geometric Imperfections

To account for initial imperfections in the members, AISC 360 App. 1. Sec. 1.3.2 requires the application of notional lateral loads at each story, in accordance with AISC 360 Sec. C2.2b(1). These are given by

$$N_i = 0.002\alpha Y_i$$

where $N_i$ is notional lateral applied at level $i$, $Y_i$ is gravity load applied at level $i$, and $\alpha$ = 1.0 (LRFD) or 1.6 (ASD).

The notional loads are additive to the applied lateral loads when the sidesway amplification ratio is

$$\Delta_{2nd}/\Delta_{1st} = B_2 > 1.7 \text{ ... using the reduced elastic stiffness}$$

where $\Delta_{2nd}$ is second-order drift and $\Delta_{1st}$ is first-order drift.

Otherwise, as specified in AISC 360 Sec. C2.2b(4), it is permissible to apply the notional load only in gravity-only load combinations.

If the out-of-plumb geometry of the structure is used in the analysis, it is permissible to omit the notional loads. Similarly, when the nominal initial out-of-plumbness differs from $L/500$, it is permissible to adjust the notional load coefficient proportionally.

### Residual Stress and Partial Yielding Effects

Residual stresses and inelastic softening may be accounted for by explicitly modeling these effects in the analysis. Alternatively, the flexural and axial stiffness of members that contribute to the lateral stability of the structure may be reduced as specified in AISC 360 Sec. C2.3 to give

$$EI^* = 0.8\tau_b EI$$
$$EA^* = 0.8EA$$

where $\tau_b$ = stiffness reduction parameter
$= 1.0$ ... for $P_r \le 0.5P_y$
$= 4(P_r/P_y)(1 - P_r/P_y)$ ... $P_r > 0.5P_y$
$P_r$ = required second-order axial strength
$P_y$ = member yield strength
$= AF_y$

It is permissible to apply the stiffness reduction to all members, including those that do not contribute to the stability of the structure.

When $P_r > 0.5P_y$ the stiffness reduction factor $\tau_b$ may be taken as 1.0 provided that the actual lateral loads, in all load combinations, are increased by an additional notional lateral load of

$$N_i = 0.001Y_i$$

In applying AISC 360 Eqs. (H1-1a) and (H1-1b) for the design of beam-columns, it is required to use the values $P_c = 0.9P_y$ and $M_c = 0.9M_p$ as specified in AISC 360 App. 1 Sec. 1.3.3(2).

### Material Properties and Yield Criteria

To model these effects, the yield stress $F_y$ and the stiffness of all steel members shall be multiplied by a factor of 0.9. However, the stiffness reduction factor of 0.9 is replaced by the factor 0.8 when the provisions of AISC 360 Sec. C2.3 are used as described in the previous paragraph.

## 8.6 Statical Method of Design

The statical method of design, illustrated in Fig. 8.4, is a simple technique for applying inelastic analysis to continuous beams. The two-span beam has a distributed load $w$ applied as shown at (*a*). The continuous beam is first cut-back to a statically determinate condition and the factored load $w_u$ is applied, as shown at (*b*). As shown at (*c*), the statical, or free, bending moment diagram is drawn for the cut-back structure and, in this case, consists of two moment envelopes each with a maximum value of

$$M_s = w_u L^2/8$$

The fixing moment line due to the redundants in the original structure is superimposed on the statical moment diagram so as to produce the collapse mechanism indicated at

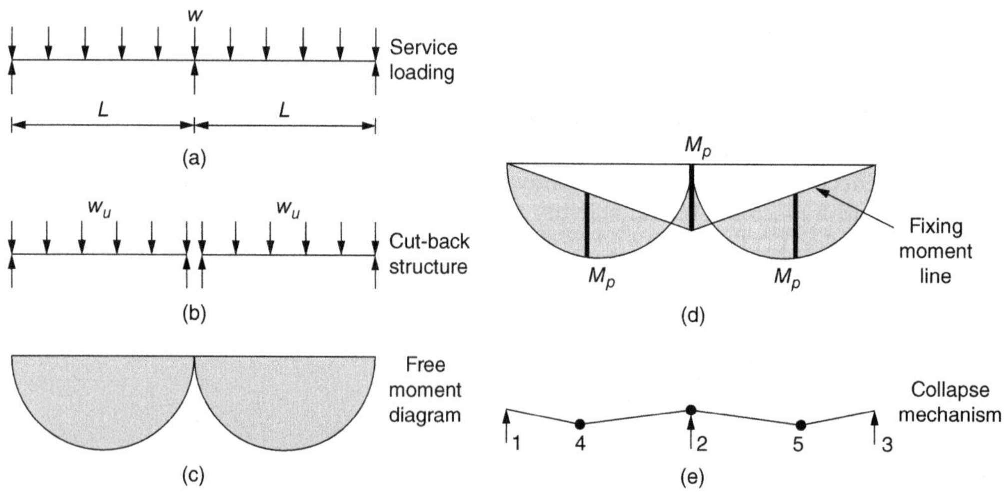


**Figure 8.4** Statical method of design.

(*d*) and (*e*). Three plastic hinges are necessary to produce collapse. For a beam of uniform section, hinges occur simultaneously in the two spans at a plastic moment of

$$M_p = 0.686M_s$$
$$= 0.0858w_uL^2$$

and the plastic hinges occur at a distance of 0.414*L* from the end support.

**Example 8.1.** Continuous Beam Design
Determine the lightest W14 section, with a yield stress of 50 ksi, required for the two-span beam shown in Fig. 8.5. The beam supports a factored distributed load, including its self-weight, of 10 kips/ft and is laterally braced at the supports and at the location of the plastic hinges.

As indicated in the Fig. 8.5, three plastic hinges are necessary to produce collapse and the required plastic moment of resistance is

$$M_p = 0.0858w_uL^2$$
$$= 0.0858 \times 10 \times 20^2$$
$$= 343 \text{ kip-ft}$$

The required plastic section modulus is

$$Z = M_p/\varphi F_y$$
$$= 343 \times 12/(0.9 \times 50)$$
$$= 91.5 \text{ in}^3$$

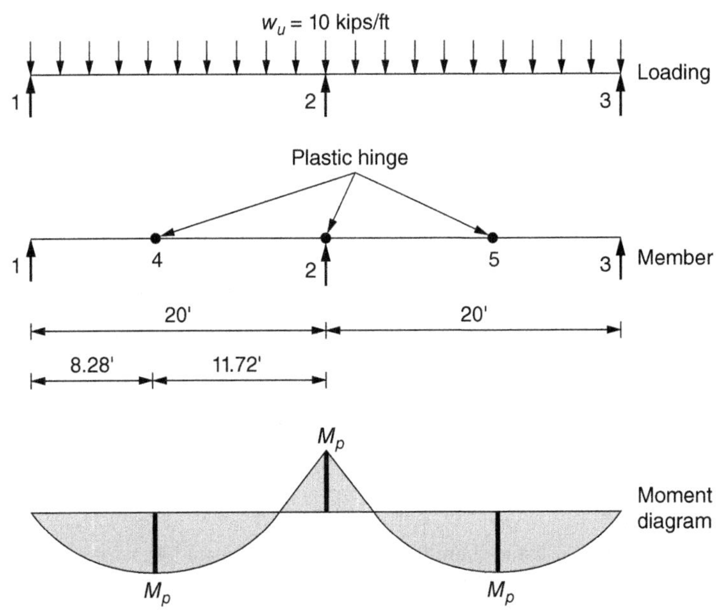


**Figure 8.5** Details for Example 8.1.

From AISC Manual Table 3-6, a W14 × 61 has a plastic section modulus of

$$Z = 102 \text{ in}^3$$
$$> 91.5 \text{ in}^3 \text{ ... satisfactory}$$

**Check compact section requirements**
The W14 × 61 section is compact as indicated in AISC Manual Table 3-6.

**Check shear requirements**
From AISC Manual Table 3-6, the design shear strength of a W14 × 61 is

$$\varphi_v V_n = 0.9 \times 0.6 F_y t_w d$$
$$= 156 \text{ kips}$$

Taking moments about the central support for the collapse mechanism, the end reaction is

$$V_1 = (w_u L^2/2 - M_p)/L$$
$$= (10 \times 20^2/2 - 343)/20$$
$$= 82.85 \text{ kips}$$

The shear force at the central support is

$$V_1 = w_u L - V_1$$
$$= 10 \times 20 - 82.85$$
$$= 117 \text{ kips}$$
$$< \varphi_v V_n \text{ ... satisfactory}$$

**Check lateral bracing requirements**
The distance of hinge 4 from support 1 is

$$L_{14} = 0.414 L_{12}$$
$$= 0.414 \times 20$$
$$= 8.28 \text{ ft}$$

and

$$L_{24} = 20 - 8.28$$
$$= 11.72 \text{ ft}$$

From AISC Manual Tables 3-6 and 1-1, the relevant properties of a W14 × 61 are

$$\varphi_b M_p = 383 \text{ kip-ft}$$
$$L_p = 8.65 \text{ ft}$$
$$L_r = 27.5 \text{ ft}$$
$$r_y = 2.45 \text{ in}$$

Since the hinges at 4 and 5 are the last to form, segment 14 and segment 35 must conform to AISC 360 Sec. F2.2.

$$L_{14} < L_p \text{ ... satisfactory}$$

Hence, segment 14 and segment 35 are adequately braced and full plasticity may be achieved.
Applying AISC 360 App. 1. Sec. 1.2.3(a) to segment 24

$$M_2 = \text{larger moment at end of unbraced length}$$
$$= M_{24} = M_p$$
$$M_1 = \text{smaller moment at end of unbraced length}$$
$$= M_{42} = -M_p$$
$$M_{mid} = \text{moment at midpoint of segment 24}$$
$$= -V_1(L_{14} + L_{42}/2) + w_u(L_{14} + L_{42}/2)^2/2$$
$$= -82.85(8.28 + 11.72/2) + 10(8.28 + 11.72/2)^2/2$$
$$= -172 \text{ kip-ft}$$
$$< (M_1 + M_2)/2 \text{ ... case (ii) applies}$$
$$= 0$$

Hence, $M_1' = M_1$ ... from AISC 360 App. 1. Eq. (A-1-6b)

$$= -M_p$$

Hence, $L_{pd} = [0.12 - 0.076(M_1'/M_2)]Er_y/F_y$ ... from AISC 360 App. 1, Eq. (A-1-5)

$$= (0.12 + 0.076 \times 1.0)(29{,}000 \times 2.45)/50$$
$$= 279 \text{ in}$$
$$= 23.21 \text{ ft}$$
$$> 11.72 \text{ ft ... satisfactory}$$

Hence, segment 24 and segment 25 are adequately braced and full plasticity may be achieved.

**Check web crippling requirements**

The maximum shear force on the beam is

$$V_2 = 117 \text{ kips}$$

For loads applied at a distance of more than $d/2$ from the end of the beam, the web crippling design strength is given by AISC 360 Eq. (J10-4) as

$$\varphi_r R_n = \varphi_r \times 0.80t_w^2[1 + 3(\ell_b/d)(t_w/t_f)^{1.5}](EF_{yw}t_f/t_w)^{0.5}$$
$$= 2\varphi_r R_3 + \ell_b \times 2\varphi_r R_4$$

From AISC Manual Table 9-4, the relevant values are

$$\varphi_r R_3 = 66.6 \text{ kips}$$
$$\varphi_r R_4 = 6.37 \text{ kips}$$

Using a bearing plate length of 6 in, the web crippling design value is

$$\varphi_r R_n = 2 \times 66.6 + 6 \times 2 \times 6.37$$
$$= 210 \text{ kips}$$
$$> V_2 \text{ ... satisfactory}$$

## 8.7 Mechanism Method of Design

The mechanism, or kinematic, design method is a convenient technique to determine the required plastic moments in rigid frames and has been described by Pincus.[7] Plastic hinges are formed at positions of maximum moment which occur at the point of application of a concentrated load, at the ends of members, and at the location of zero shear in a prismatic beam subjected to a distributed load. In the case of two members meeting at a joint, a plastic hinge forms in the weaker member. In the case of three or more members meeting at a joint, a plastic hinge may form at the ends of each of the members. Figure 8.6 illustrates the possible location of plastic hinges in a two-bay frame. As shown, the number of possible hinge locations is

$$p = 10$$

An independent collapse mechanism corresponds to a condition of unstable equilibrium in the structure. An indeterminate structure with a redundancy of $r$ becomes stable and determinate when $r$ plastic hinges have formed since the moment $M_p$ is now known at $r$ locations. Hence, the formation of one more hinge produces a collapse mechanism and the number of hinges required to produce collapse is

$$n = r + 1$$

This expression is generally valid but breaks down when partial collapse occurs leaving part of the structure intact.

Thus, in a structure in which there are $p$ possible hinge positions, the number of independent mechanisms is given by

$$m_i = p - r$$

The frame in Fig. 8.6 is 6 degrees redundant and the number of possible independent mechanisms is

$$\begin{aligned} m_i &= p - r \\ &= 10 - 6 \\ &= 4 \end{aligned}$$

These mechanisms are shown in Fig. 8.7 and are the two beam mechanisms B1 and B2, the sway mechanism S, and the joint mechanism J. The beam mechanisms are each partial collapse mechanisms.

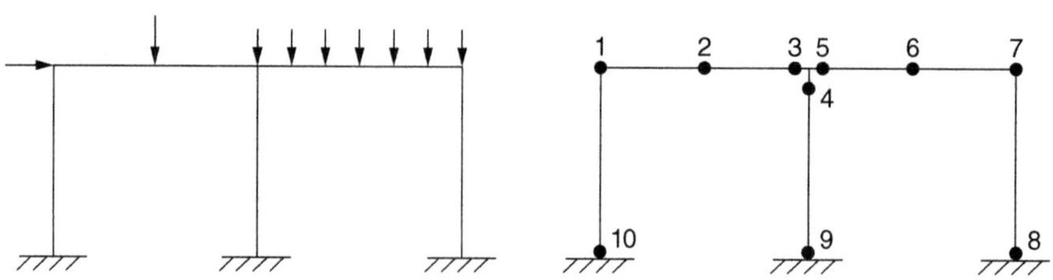


**FIGURE 8.6** Plastic hinges in a two-bay frame.

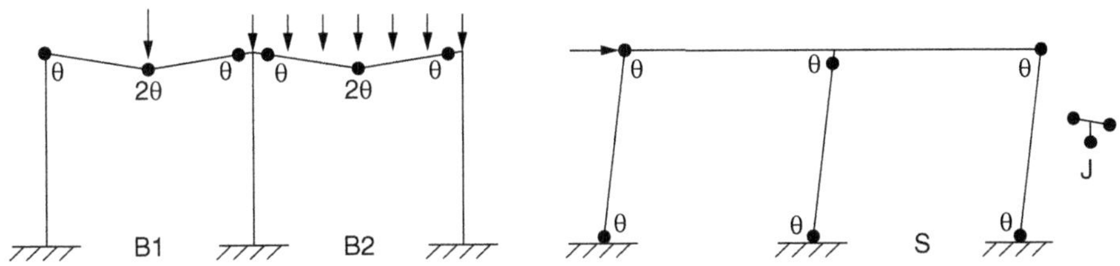


**FIGURE 8.7** Independent collapse mechanisms.

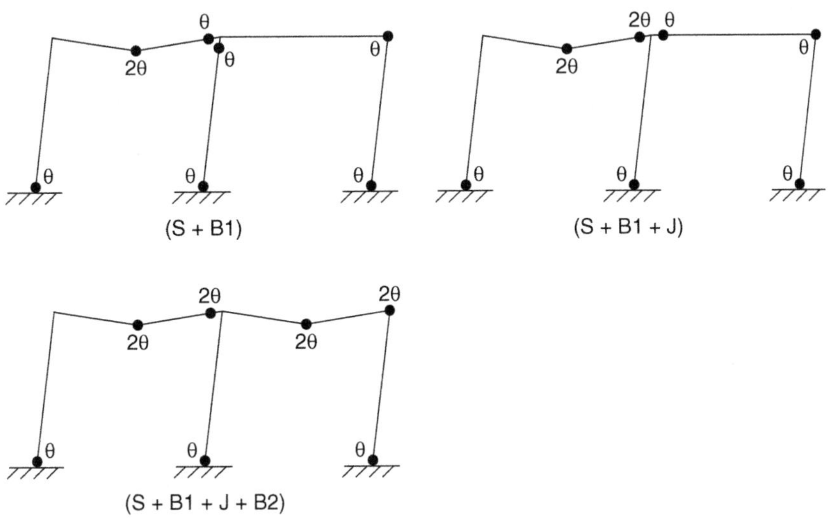


**FIGURE 8.8** Combined mechanisms.

In addition, collapse may occur because of the combination of any of the independent mechanisms. Combinations must be investigated that eliminate hinges as this may produce the most critical collapse mechanism. Three possible combinations are shown in Fig. 8.8.

Combining S and B1 eliminates the hinge at 1. Combining S, B1, and J eliminates the hinge at 4. Combining S, B1, J, and B2 eliminates the hinge at 5.

**Example 8.2.** Single-Bay Frame

The rigid frame shown in Fig. 8.9*a* is fabricated with members of uniform section. The factored working loads are indicated and the frame is designed to collapse at a load factor of $N$. Neglecting the effects of frame deformations and instability, determine the required plastic moment of resistance.

The frame is 3 degrees indeterminate and there are five possible locations of plastic hinges as shown in Fig. 8.9*b*. The number of possible independent mechanisms is, then

$$\begin{aligned} m_i &= p - r \\ &= 5 - 3 \\ &= 2 \end{aligned}$$

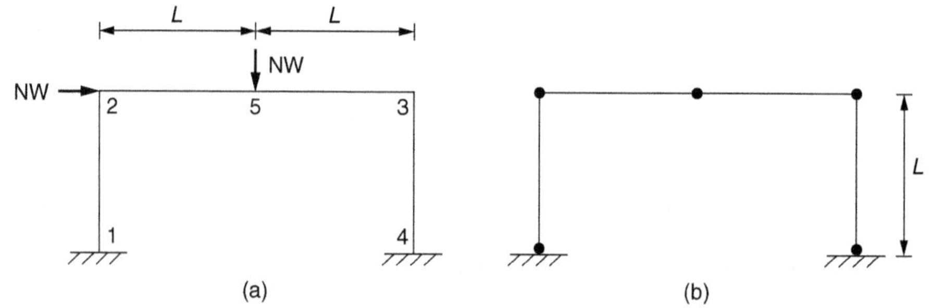


**FIGURE 8.9** Details for Example 8.2.

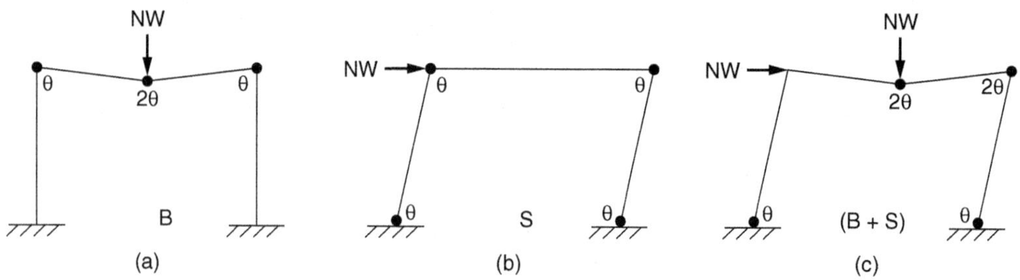


**FIGURE 8.10** Collapse mechanisms.

These mechanisms are the beam mechanism and the sway mechanism. In addition, these may be combined to form the combined mechanism shown in Fig. 8.10.

Applying a virtual displacement to each of these mechanisms in turn and equating internal and external work yields three equations from each of which a value of $M_p$ may be obtained. The largest value of $M_p$ governs.

For the beam mechanism shown at (*a*)

$$4M_p\theta = NWL\theta$$

$$M_p = NWL/4$$

For the sway mechanism shown at (*b*)

$$4M_p\theta = NWL\theta$$

$$M_p = NWL/4$$

For the combined mechanism shown at (*c*)

$$6M_p\theta = NWL\theta + NWL\theta$$

$$M_p = (NWL + NWL)/6$$

$$= NWL/3 \text{ ... governs}$$

Hence, the combined mechanism controls and the required plastic moment of resistance is $M_p = NWL/3$.

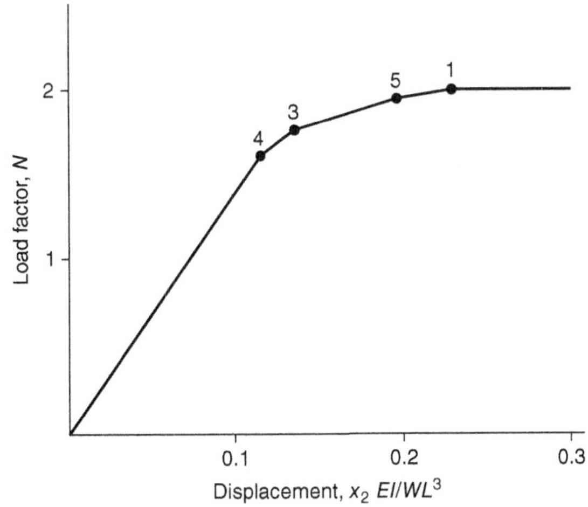


**FIGURE 8.11** Formation of plastic hinges.

## Linear Elastic-Plastic Response Curve

Assigning a value of $N = 2$ to the load factor of the frame in Example 8.2, gives a value for the plastic moment of resistance of the members of

$$M_p = 2WL/3$$

The sequence of formation of the hinges may be traced by an incremental technique, known as pushover analysis, until the ultimate strength is reached and the collapse mechanism has formed. As shown in Fig. 8.11, and ignoring stiffness reduction due to inelastic softening effects, the load-deflection curve is linear until the first hinge forms at joint 4 at a load factor of 1.6. The formation of the hinge may be simulated by the insertion of a frictionless hinge at 4 and the application of a moment of magnitude $M_p$. This reduces the stiffness of the frame and displacements, for a given increase in applied load, are greater than in the original frame. The second hinge forms at joint 3 at a load factor of 1.74 and the third hinge forms at 5 at a load factor of 1.94. The collapse mechanism is produced when the last hinge forms at joint 1 at a load factor of 2.

In practice, elastic instability effects will cause collapse at a load factor of less than 2 unless the frame is adequately braced. Techniques for accounting for elastic instability effects are given by Williams[8] Chap. 12.

The plastic design method has also been used for the design of multistory frames by Daniels,[9] for grid systems by Vukov,[10] and for gable frames and Vierendeel girders by Williams.[8]

**Example 8.3.** Vierendeel Girder
The members of the Vierendeel girder shown in Fig. 8.12 have their relative plastic moments of resistance shown ringed. For the factored load indicated, determine the required plastic modulus.

The number of independent collapse mechanisms is

$$\begin{aligned} m_i &= p - D \\ &= 16 - 9 \\ &= 7 \end{aligned}$$

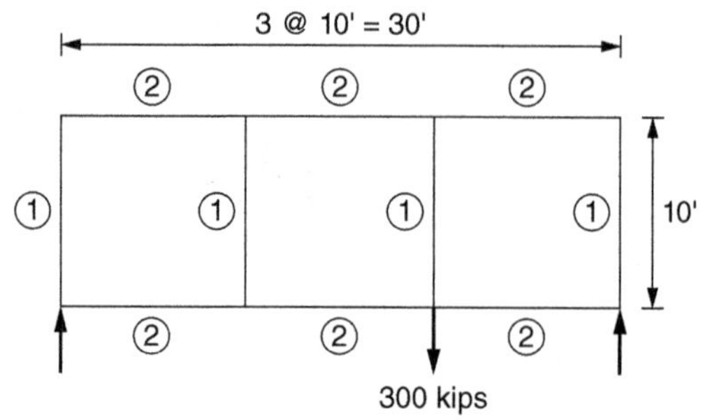


**FIGURE 8.12** Details for Example 8.3.

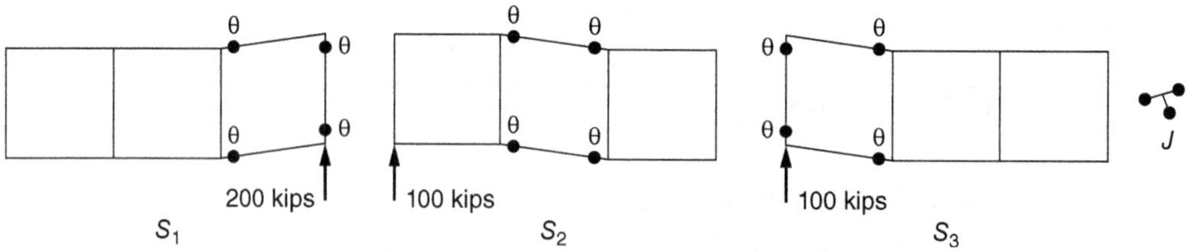


**FIGURE 8.13** Independent mechanisms.

The independent mechanisms consist of three sway mechanisms, and four joint mechanisms at the ends of the interior posts. These are shown in Fig. 8.13.

For mechanism $S_1$

$$6M_p = 200 \times 10$$
$$M_p = 333 \text{ kip-ft}$$

For mechanism $S_2$

$$8M_p = 100 \times 10$$
$$M_p = 125 \text{ kip-ft}$$

For mechanism $S_3$

$$6M_p = 100 \times 10$$
$$M_p = 167 \text{ kip-ft}$$

For mechanism $J$

$$5M_p = 0$$

Two possible combined mechanisms are shown in Fig. 8.14.

For mechanism $(S_1 + 2J)$

$$8M_p = 2000 + 0$$
$$M_p = 250 \text{ kip-ft}$$

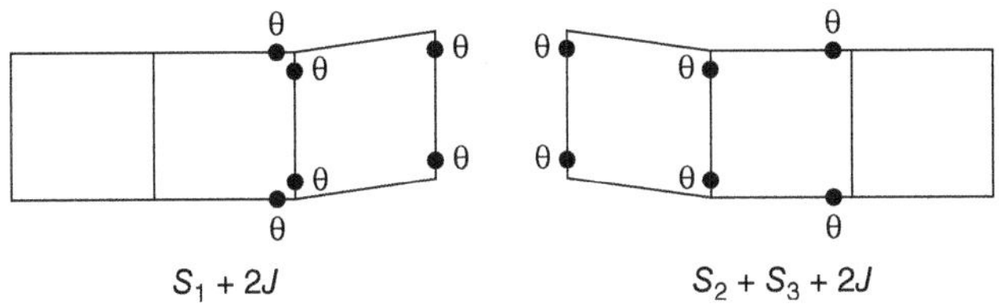


**FIGURE 8.14** Combined mechanisms.

For mechanism $(S_2 + S_3 + 2J)$

$$8M_p = 1000 + 1000 + 0$$
$$M_p = 250 \text{ kip-ft}$$

The $S_1$ mechanism governs and the required plastic modulus is

$$M_p = 333 \text{ kip-ft}$$

## 8.8 Static Equilibrium Check

The mechanism method leads to an upper bound on the collapse load. To confirm that the correct mechanism has been selected, a moment diagram may be constructed using static equilibrium methods to determine that the assumed plastic moment is nowhere exceeded.

**Example 8.4.** Static Equilibrium Check
The assumed collapse mechanism for the frame analyzed in Example 8.2 is shown in Fig. 8.15. Draw the bending moment diagram for the frame to confirm that the assumed mechanism is correct.

The bending moments at locations 1, 3, 4, and 5 are assumed to have a value of $M_p = NWL/3$. Hence, in order to construct the bending moment for the frame, it is necessary to determine

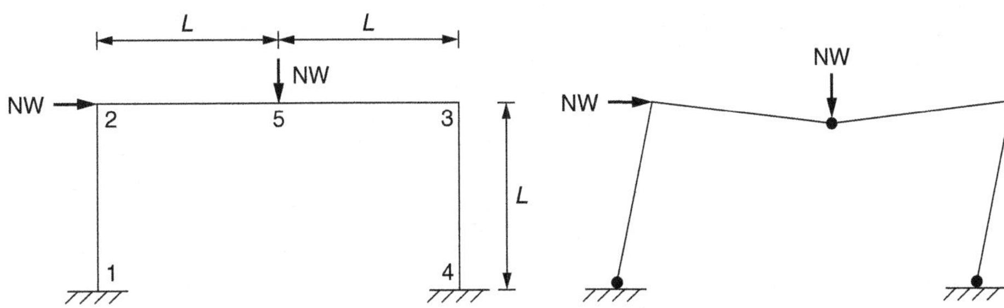


**FIGURE 8.15** Details for Example 8.4.

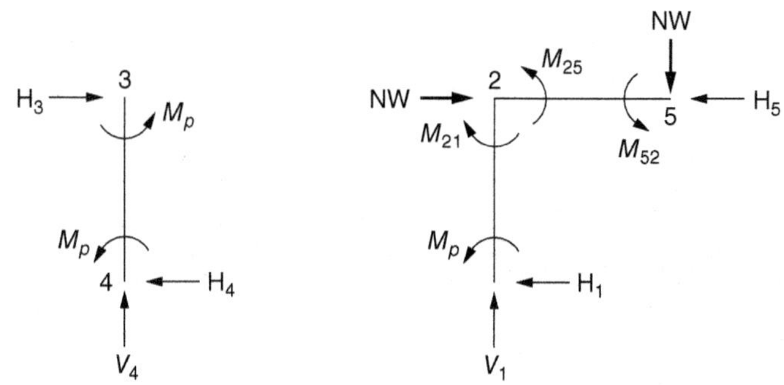


**FIGURE 8.16** Free-body diagrams for Example 8.4.

the moment at joint 2. Taking moments about end 3 of member 34, for the free-body diagram shown in Fig. 8.16

$$\begin{aligned} H_4 &= (M_p + M_p)/L \\ &= 2M_p/L \\ &= 2NW/3 \\ &= H_3 \\ &= H_5 \end{aligned}$$

Resolving horizontal forces for the whole frame

$$\begin{aligned} H_1 &= NW - H_4 \\ &= NW - 2NW/3 \\ &= NW/3 \end{aligned}$$

Taking moments about 4 for the whole frame

$$\begin{aligned} V_1 &= (M_p + M_p - NWL + NWL)/2L \\ &= (2NWL/3)/2L \\ &= NW/3 \end{aligned}$$

Taking moments about end 5 of member 125, for the free-body diagram shown in Fig. 8.16

$$\begin{aligned} M_{52} &= (V_1 \times L + H_1 \times L - M_p) \\ &= NWL/3 + NWL/3 - NWL/3 \\ &= NWL/3 \\ &= M_p \text{ ... correct as assumed} \end{aligned}$$

$$\begin{aligned} M_{21} &= M_p - H_1 \times L \\ &= NWL/3 - NWL/3 \\ &= 0 \end{aligned}$$

The final bending moment diagram is shown in Fig. 8.17.

Since the plastic moment of resistance is not exceeded at any point in the frame, the assumed combined mechanism is the correct failure mode.

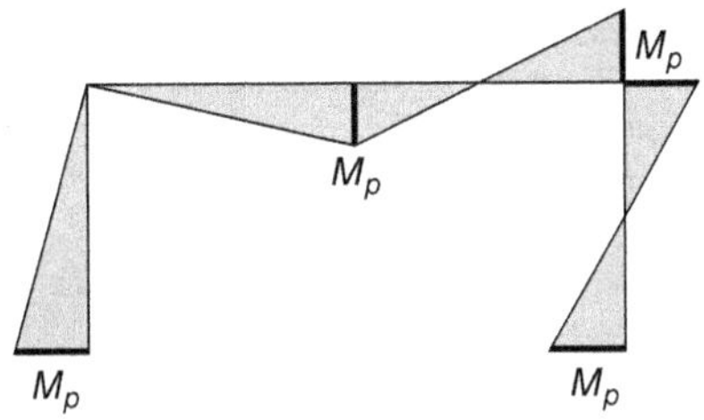


**FIGURE 8.17** Final moment diagram for Example 8.4.

## 8.9 Beam-Column Design

In accordance with AISC 360 App. 1 Sec. 1.1, an inelastic analysis must take into consideration second-order effects. In an elastic analysis, the forces produced in a frame by the no-translation load case (*nt*) and the sway load case (*lt*) may be separately computed by a first-order analysis. The second-order forces are then determined using AISC 360 App. 8 Eqs. (A-8-2-1a) and (A-8-2-1b) which are

$$M_r = B_1 M_{nt} + B_2 M_{lt}$$

$$P_r = P_{nt} + B_2 P_{lt}$$

This approximate approach for obtaining the second-order forces is known as the $B_1$-$B_2$ procedure.

In an inelastic analysis, the principle of superposition is invalid and the two loading cases cannot be separated. The second-order forces may then be determined by

$$M_r = B_2 M_u$$

$$P_r = B_2 P_u$$

These expressions are conservative since in a normal structure $B_2$ is greater than $B_1$.

**Example 8.5.** Beam-Column Design
Determine if column 34 of the frame shown in Fig. 8.18 is adequate. The applicable LRFD loading case is shown and the first-order forces produced in the frame, using the unreduced elastic stiffness,

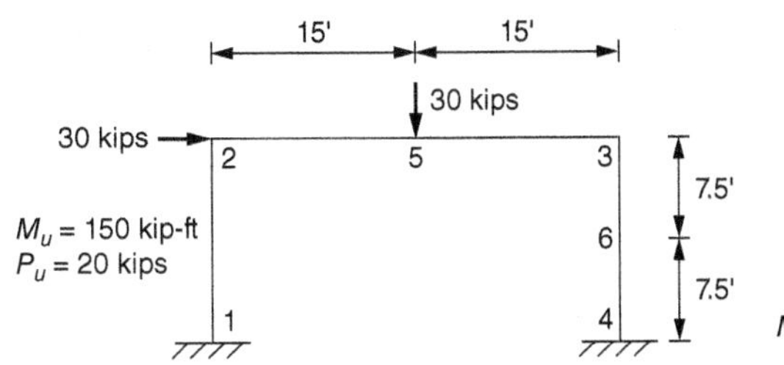


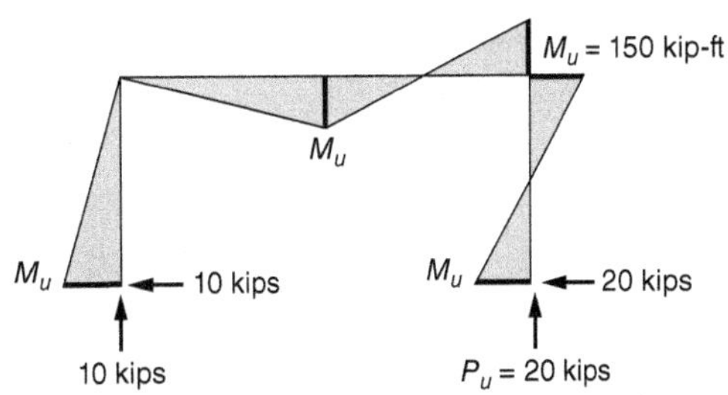


**FIGURE 8.18** Details for Example 8.5.

are indicated. The hinge located at joint 1 is the last to form and the column is laterally braced at its midpoint and at joint 3. About the $y$-axis, column 34 may be considered pinned at top and bottom. All members of the frame are W14 × 53.

The relevant properties of a W14 × 53 section are obtained from AISC Manual Tables 1-1 and 3-2 as

$$A_g = 15.6\ \text{in}^2 \qquad I = 541\ \text{in}^4 \qquad r_x = 5.89\ \text{in} \qquad r_y = 1.92\ \text{in} \quad L_p = 6.78\ \text{ft} \qquad L_r = 22.3\ \text{ft}$$
$$\varphi_b BF = 7.93\ \text{kips} \qquad \varphi_b M_{px} = 327\ \text{kip-ft}$$

**Check compact section requirements**

The W14 × 53 section is compact as indicated in AISC Manual Table 3-6. In addition

$$\begin{aligned} P_y &= AF_y \\ &= 15.6 \times 50 \\ &= 780\ \text{kips} \\ P_u &= 20\ \text{kips} \\ P_u/\varphi_c P_y &= 20/(0.9 \times 780) \\ &= 0.028 \\ &< 0.125 \end{aligned}$$

Hence, AISC 360 App 1 Eq.(A-1-1) applies and

$$\begin{aligned} \lambda_{pd} &= 3.76(E/F_y)^{0.5}(1 - 2.75P_u/\varphi_c P_y) \\ &= 3.76(29{,}000/50)^{0.5}(1 - 2.75 \times 0.028) \\ &= 83.6 \end{aligned}$$

For W14 × 53

$$\begin{aligned} h/t_w &= 30.9 \\ &< \lambda_{pd} \text{ ... the section is compact} \end{aligned}$$

**Geometric imperfections**

Notional lateral loads need not be applied assuming that the sidesway amplification ratio is

$$\Delta_{2nd}/\Delta_{1st} = B_2 < 1.7 \text{ ... using the reduced elastic stiffness}$$

**Adjustments to stiffness**

Apply the provisions of AISC 360 Sec. C2.3

$$\begin{aligned} P_y &= AF_y \\ &= 780\ \text{kips} \\ P_r &\approx P_u \\ &= 20\ \text{kips} \end{aligned}$$

Hence, $\qquad P_r < 0.5P_y$

and $\qquad \tau_b = 1.0$

Hence, residual stresses and inelastic softening may be accounted for by reducing the flexural and axial stiffness of all members to

$$\begin{aligned} EI^* &= 0.8EI \\ &= 0.8 \times 29{,}000 \times 541 \\ &= 12{,}551{,}200 \text{ kip-in}^2 \\ EA^* &= 0.8EA \end{aligned}$$

When axial deformations are ignored, this does not affect the member forces at collapse, but it does affect the deflection of the frame at collapse.

**Material properties and yield criteria**

The stiffness of all members has been multiplied by a factor of 0.8 and, in accordance with AISC 360 Sec. 1.3.3(1), this replaces the reduction factor of 0.9. To account for partial yielding effects when applying AISC 360 Eqs. (H1-1a) and (H1-1b), in accordance with AISC 360 Sec. 1.3.3(2), available strengths are modified to

$$\begin{aligned} P_c &= 0.9P_y \\ M_{cx} &= 0.9M_{px} \\ M_{cy} &= 0.9M_{py} \end{aligned}$$

**Deflection of the frame at incipient collapse**

The horizontal deflection of joint 2 at incipient collapse is given by the virtual work expression

$$\delta = \Sigma \int Mm\, dx/EI + \Sigma m\varphi$$

where $\varphi$ is the total rotation at a hinge during the application of the loads, $m$ is the bending moment at any section due to a unit virtual load applied to the structure at the joint in the direction of the required displacement, and the summations extend over all the hinges and all the members in the structure. The unit virtual load may be applied to any cut-back structure which can support it. Thus, if a cut-back structure is selected which gives a zero value for $m$ at all plastic hinge positions except the last to form, the term $\Sigma m\varphi$ is zero and it is unnecessary to calculate the hinge rotations. A suitable form of the cut-back structure may be obtained by inserting frictionless hinges in the actual structure at all plastic hinge positions except the last to form. The expression then reduces to

$$\delta = \int Mm\, dx/EI$$

The last hinge to form is at joint 1 and the requisite cut-back structure and moment $m$ is shown in Fig. 8.19. The deflection of joint 2 at incipient collapse, using the reduced elastic stiffness, is then

$$\begin{aligned} \Delta_H &= M_u \int_o^\ell x^2\, dx/\ell EI \\ &= M_u \ell^2/3EI^* \\ &= 150 \times 12 \times (15 \times 12)^2/(3 \times 12{,}551{,}200) \\ &= 1.55 \text{ in} \end{aligned}$$

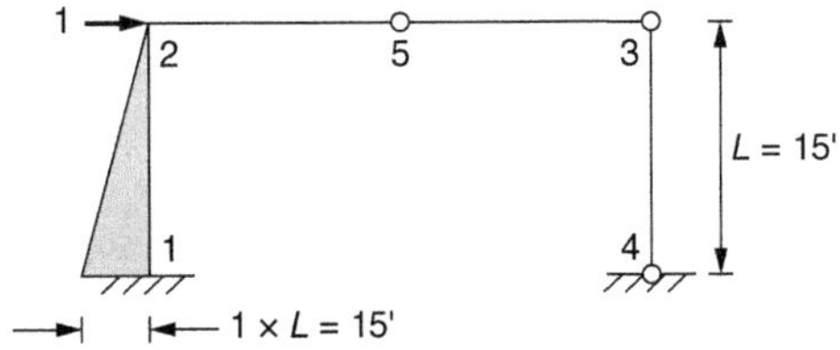


**FIGURE 8.19** Cut-back structure for Example 8.5.

**Calculation of $B_2$**

Apply the provisions of AISC 360 App. 8 Sec. 8.2.

$$\alpha = \text{force level adjustment factor}$$
$$= 1.0$$
$$L/\Delta_H = (12 \times 15)/1.55$$
$$= 116$$
$$P_{story} = P_{mf}$$
$$= 30 \text{ kips}$$
$$R_m = 1 - 0.15(P_{mf}/P_{story})$$
$$= 0.85$$
$$P_{e,\,story} = R_m HL/\Delta_H = 0.85 \times 30 \times 116$$
$$= 2958 \text{ kips}$$
$$B_2 = 1/(1 - \alpha P_{story}/P_{e,\,story})$$
$$= 1/(1 - 1 \times 30/2958)$$
$$= 1.01$$
$$< 1.7$$ ... initial assumption that notional lateral loads need not be applied was correct

**Calculate second-order forces**

$$M_r = B_2 M_u$$
$$= 1.01 \times 150$$
$$= 152 \text{ kip-ft}$$
$$P_r = B_2 P_u$$
$$= 1.01 \times 20$$
$$= 20 \text{ kips}$$

**Check limiting axial load**

The maximum axial load in a column is restricted by AISC 360 App. 1 Sec. 1.2.4 to

$$P_{max} = 0.75 A_g F_y$$
$$= 0.75 \times 15.6 \times 50$$
$$= 585 \text{ kips}$$
$$> P_r$$ ... satisfactory

**Check lateral bracing requirements**

The distance of the hinges at joints 3 and 4 from the lateral support at point 6 is

$$L_{64} = L_{63}$$
$$= 7.5 \text{ ft}$$

Applying AISC 360 Sec. 1.2.3(a) to segment 64

$$M_2 = \text{larger moment at end of unbraced length}$$
$$= M_{46} = M_r$$
$$= 152 \text{ kip-ft}$$
$$M_1 = \text{smaller moment at end of unbraced length}$$
$$= M_{64}$$
$$= 0$$
$$M_{mid} = \text{moment at midpoint of segment 64}$$
$$= (152 + 0)/2$$
$$= 76 \text{ kip-ft}$$
$$= (M_1 + M_2)/2$$

Hence,
$$M_1' = M_1 \text{ ... from AISC 360 App. 1. Eq. (A-1-6b)}$$
$$= 0$$

Hence, the maximum unbraced length is given by AISC 360 App. 1. Eq. (A-1-5) as

$$L_{pd} = [0.12 - 0.076(M_1'/M_2)]Er_y/F_y$$
$$= (0.12 - 0.076 \times 0)(29{,}000 \times 1.92)/50$$
$$= 134 \text{ in}$$
$$= 11 \text{ ft}$$
$$> L_{64} \text{ ... satisfactory}$$

In addition, the maximum permitted laterally unbraced length about both the $x$- and $y$-axis, for members subject to compression, is specified in AISC 360 App. 1 Sec. 1.2.3(b) as

$$L/r = 4.71(E/F_y)^{0.5}$$
$$= 113$$
$$L/r_y = 7.5 \times 12/1.92$$
$$= 47$$
$$< 113 \text{ ... satisfactory}$$
$$L/r_x = 15 \times 12/5.89$$
$$= 31$$
$$< 113 \text{ ... satisfactory}$$

Hence, segment 64 and segment 63 are adequately braced and full plasticity may be achieved.

#### Design for combined compression and bending

*For Compression:*
The effective length about the $y$-axis is obtained from AISC 360 Commentary App. 7 Table C-A-7.1 as

$$KL = 1.0 \times 15/2$$
$$= 7.5 \text{ ft}$$

The slenderness ratio about the $y$-axis is

$$KL/r_y = 7.5 \times 12/1.92$$
$$= 46.88$$

The effective length about the $x$-axis is obtained from AISC 360 Commentary App. 7 Sec. 7.2. For the fixed end at joint 4, the stiffness ratio is

$$G_4 = 1.0$$

At joint 3,

$$G_3 = \Sigma(I_c/L_c)/\Sigma(I_g/L_g)$$
$$= 30/15$$
$$= 2$$

From the alignment chart for sway frames, the effective length factor about the $x$-axis is

$$K_{34} = 1.45$$

The slenderness ratio about the $x$-axis is

$$K_{34}L_x/r_x = 1.45 \times 15 \times 12/5.89$$
$$= 44.31$$
$$< KL/r_y$$

Hence, the slenderness ratio about the $y$-axis governs.

In applying the interaction expressions of AISC 360 Eqs. (H1-1a) and (H1-1b), available strengths are modified by a factor of 0.9 as specified in AISC 360 Sec. 1.3.3(2). Hence, the revised yield stress is

$$0.9F_y = 0.9 \times 50$$
$$= 45 \text{ ksi}$$

For a slenderness ratio about the $y$-axis of $KL/r_y = 46.88$ and a yield stress of $F_y = 45$ ksi, interpolation in AISC Manual Table 4-22 gives the critical stress for member 34 as

$$\varphi_c F_{cr} = 35 \text{ ksi}$$

The available axial strength is

$$P_c = \varphi_c P_n$$
$$= \varphi_c F_{cr} A_g$$
$$= 35 \times 15.6$$
$$= 546 \text{ kips}$$

*For Bending:*

Member 34 is bent in reverse curvature and the modification factor is obtained from Table 4.1 as

$$C_b = 2.27$$

The unbraced segment length is

$$L_b = L_{34} = 15 \text{ ft}$$
$$> L_p = 6.78 \text{ ft}$$
$$< L_r = 22.3 \text{ ft}$$

From AISC 360 Eq. (F2-2), the design flexural strength is

$$\varphi_b M_{nx} = C_b[\varphi_b M_{px} - (\varphi_b BF)(L_b - L_p)]$$
$$= 2.27[327 - (7.93)(15 - 6.78)]$$
$$= 594 \text{ kip-ft}$$

The maximum value permitted by AISC 360 Eq. (F2-2) is

$$\varphi_b M_{nx} = \varphi_b M_{px}$$
$$= 327 \text{ kip-ft}$$

Allowing for the modification factor of 0.9 as specified in AISC 360 Sec. 1.3.3(2), the available flexural strength is

$$M_{cx} = 0.9\varphi_b M_{nx}$$
$$= 0.9 \times 327$$
$$= 294 \text{ kip-ft}$$

*Apply the Interaction Equations:*
The ratio of applied axial force to available axial force is

$$P_r/P_c = 20/546$$
$$= 0.037$$
$$< 0.2$$

Hence, Eq. (H1-1b) applies and

$$P_r/2P_c + (M_{rx}/M_{cx} + M_{ry}/M_{cy}) = 0.037/2 + 152/294 + 0$$
$$= 0.54$$
$$< 1.0 \ldots \text{ satisfactory}$$

## References


1. Martin, J. B. 1963. "Rigid Plastic Analysis of General Structures," *Proceedings American Society of Civil Engineers*, Vol. 89 (EM2), April 1963.
2. American Society of Civil Engineers. 1971. *Plastic Design in Steel: A Guide and Commentary*. Manual No. 41. ASCE, Reston, VA.
3. Popov, E. P. and McCarthy, R. E. 1960. "Deflection Stability of Frames under Repeated Loads," *Proceedings American Society of Civil Engineers*, Vol. 86 (EM1), January 1960.
4. Xiaofeng, Y. 1989. "Shakedown Analysis in Plastic Design of Steel Structures," *Proceedings American Society of Civil Engineers*, Vol. 115 (EM2), February 1989.
5. American Institute of Steel Construction (AISC). 2010. *Specification for Structural Steel Buildings* (AISC 360-10), AISC, Chicago, IL.
6. American Institute of Steel Construction (AISC). 2011. *Steel Construction Manual*, 14th edition, AISC, Chicago, IL.
7. Pincus, G. 1965. "Method of Combining Mechanisms in Plastic Analysis," *Engineering Journal*, 3rd Quarter 1965.

8. Williams, A. 2009. *Structural Analysis in Theory and Practice.* Elsevier/International Code Council, New York, NY.
9. Daniels, J. H. 1966. "A Plastic Method for Unbraced Frame Design," *Engineering Journal*, 4th Quarter 1966.
10. Vukov, A. 1986. "Limit Analysis and Plastic Design of Grid Systems," *Engineering Journal*, 2nd Quarter 1986.

## Problems

**8.1** *Given:* A two-span beam of uniform section with a yield stress of 50 ksi as shown in Fig. 8.20. The beam supports two 100-kip loads as indicated and is laterally braced at the supports and at the location of the plastic hinges.

*Find:* Using the statical method of design, the lightest W14 beam to support the indicated loads.

**8.2** *Given:* A two-span beam of uniform section with a yield stress of 50 ksi as shown in Fig. 8.20. The beam supports two 100-kip loads as indicated and is laterally braced at the supports and at the location of the plastic hinges.

*Find:* Using the mechanism method of design, the lightest W14 beam to support the indicated loads.

**8.3** *Given:* The members of the rigid frame shown in Fig. 8.21 have the relative plastic moments of resistance shown ringed. The LRFD loads are indicated.

*Find:* Neglecting the effects of frame deformations and instability, the required plastic moment of resistance.

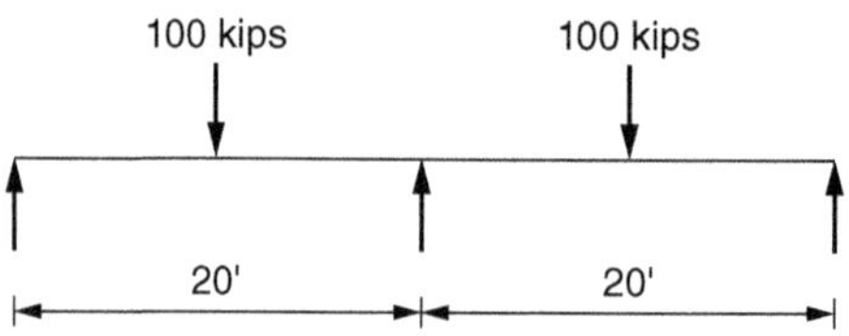


**Figure 8.20** Details for Problem 8.1.

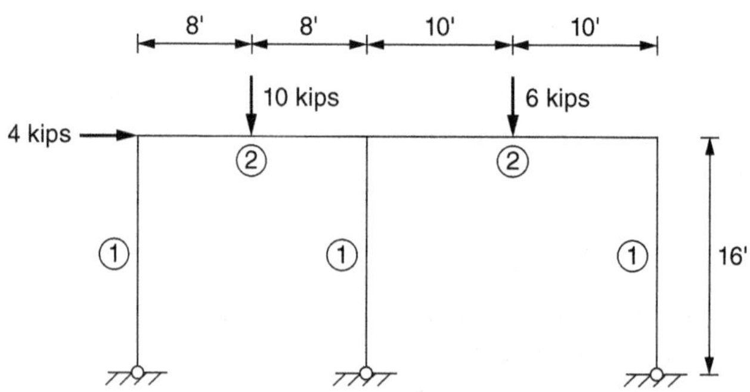


**Figure 8.21** Details for Problem 8.3.

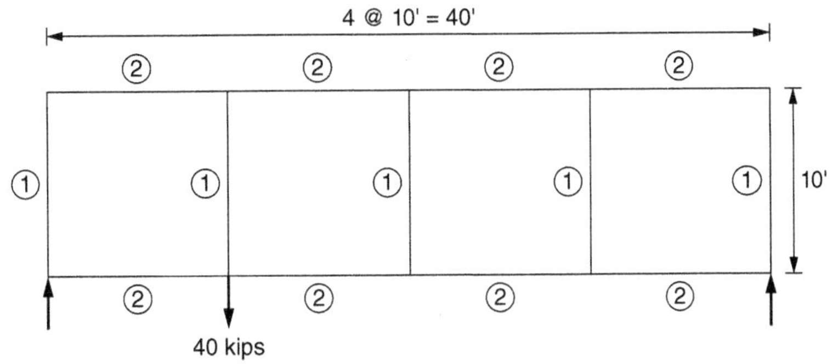


**Figure 8.22** Details for Problem 8.4.

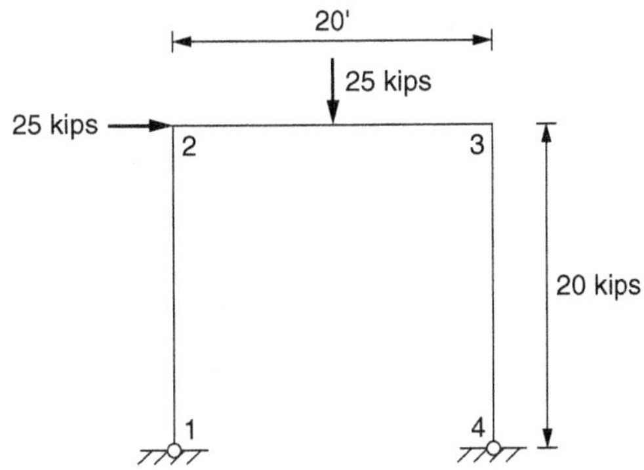


**Figure 8.23** Details for Problem 8.6.

**8.4** *Given:* The members of the Vierendeel girder shown in Fig. 8.22 have relative plastic moments of resistance shown ringed. The LRFD loads are indicated.

*Find:* Neglecting the effects of frame deformations and instability, the required plastic moment of resistance.

**8.5** *Given:* The frame analyzed in Problem 8.3.

*Find:* Draw the bending moment diagram for the frame to confirm that the assumed mechanism is correct.

**8.6** *Given:* The applicable LRFD loading case acting on a frame is shown in Fig. 8.23. Column 34 is laterally braced at its midpoint and at joint 3. About the $y$-axis, column 34 may be considered pinned at top and bottom. All members of the frame are W14 × 61.

*Find:* Using a second-order analysis, determine if column 34 of the frame is adequate.

# CHAPTER 9

# Design of Tension Members

## 9.1 Introduction

Tension members may consist of round bars, wire cables, plates, and rolled sections. End connections may consist of threaded ends, eyebars, pinned, bolted, and welded connections.

Two limit states govern the design of tension members. These are the limit state of yielding and the limit state of rupture. The limit state of yielding applies to failure on the gross cross-sectional area of the member. Yielding of a member may result in excessive elongation and lead to instability of the whole structure. The limit state of rupture applies to failure on the net cross-sectional area of the member. A reduction in area of a tension member occurs at connections where holes are punched in the member to accommodate bolts. Failure by rupture at a connection may be sudden and catastrophic.

A mandatory slenderness ratio is not specified for tension members. However, American Institute of Steel Construction, *Specification for Structural Steel Buildings* (AISC 360)[1] Sec. D1 recommends a maximum slenderness ratio of

$$L/r = 300$$

This limit is based on practical considerations for ease of handling during fabrication, transportation, and erection. It also reduces the possibility of undesirable vibration or slapping in service. This recommendation does not apply to rods and hangers.

## 9.2 Tensile Strength

In accordance with AISC 360 Sec. D2, the available tensile strength of tension members is the lower value obtained according to the limit states of tensile yielding in the gross section and tensile rupture in the net section.

For tensile yielding in the gross section, AISC 360 Eq. (D2-1) gives the nominal strength as

$$P_n = F_y A_g$$

where $F_y$ is yield stress of the member and $A_g$ is gross area of the member.

| LRFD | ASD |
|---|---|
| The resistance factor for tensile yielding is | The safety factor for tensile yielding is |
| $\varphi_t = 0.9$ | $\Omega_t = 1.67$ |
| The design strength in tensile yielding is | The allowable strength in tensile yielding is |
| $\varphi_t P_n = 0.9 F_y A_g$ | $P_n/\Omega_t = F_y A_g/1.67$ |

For tensile rupture in the weakest effective net area, AISC 360 Eq. (D2-2) gives the nominal strength as

$$P_n = F_u A_e$$

where $F_u$ is tensile strength of the member and $A_e$ is effective net area of the member.

| LRFD | ASD |
|---|---|
| The resistance factor for tensile rupture is | The safety factor for tensile rupture is |
| $\varphi_t = 0.75$ | $\Omega_t = 2.00$ |
| The design strength in tensile rupture is | The allowable strength in tensile rupture is |
| $\varphi_t P_n = 0.75 F_u A_e$ | $P_n/\Omega_t = F_u A_e/2.00$ |

The limit state of tensile yielding governs over tensile rupture when

$$A_e/A_g > 0.923 \text{ ... for } F_y = 50 \text{ ksi } F_u = 65 \text{ ksi}$$
$$A_e/A_g > 0.745 \text{ ... for } F_y = 36 \text{ ksi } F_u = 58 \text{ ksi}$$

## 9.3 Effective Net Area

Two plates connected together and subjected to a tensile force are shown in Fig. 9.1. Detail (*a*) indicates a bolted connection and detail (*b*) indicates a welded connection. When only part of the cross-section of a member is connected, the resulting shear lag

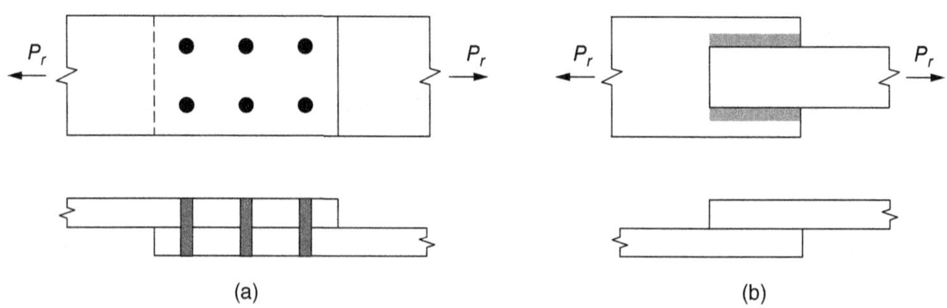


**FIGURE 9.1** Plates in tension.

effect produces a concentration of stress at the connection. To allow for this, a reduction factor is applied to the net area at a connection. Thus, the effective net area is given by AISC 360 Eq. (D3-1) as

$$A_e = A_n U$$

where $A_n$ = net area
= gross area with area of holes deducted
$U$ = shear lag factor given in Table D3.1

## Plates with Bolted Connection

For bolted connections, the net area is shown in Fig. 9.2 and defined in AISC 360 Sec. B4.3b. For the straight perpendicular fracture 1-1, the effective net area of the plate is given by

$$A_n = t(w - 2d_h)$$

For a staggered fracture, the effective net width is obtained by deducting from the gross width the sum of the bolt holes in the failure path and adding, for each gage space traversed by a diagonal portion of the failure path, the quantity $s^2/4g$, where

$g$ = transverse center-to-center spacing between fasteners (gage)

$s$ = bolt spacing in direction of load (pitch) $d_h$ = effective hole diameter

For the staggered fracture 2-2, the effective net area of the plate is given by

$$A_n = t(w - 3d_h + s^2/4g)$$

For the staggered fracture 3-3, the effective net area of the plate is given by

$$A_n = t(w - 4d_h + 3s^2/4g)$$

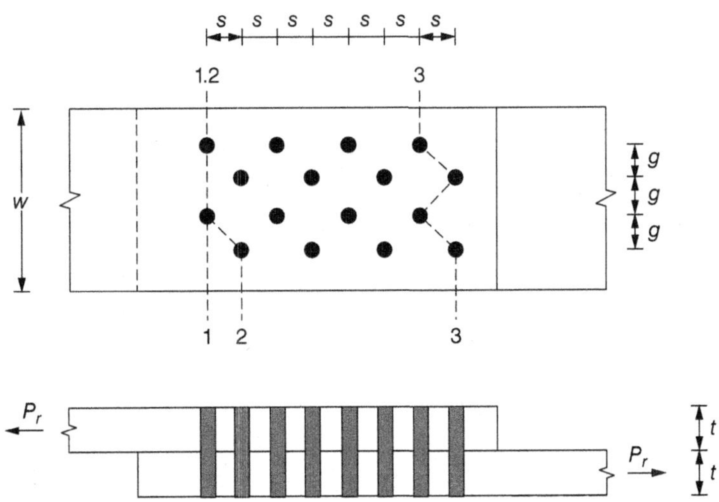


**FIGURE 9.2** Net area of bolted connection.

For flat plates with bolted connections, all of the net area participates in transmitting the load and AISC 360 Table D3-1 specifies a value for the shear lag factor of $U = 1.0$. Hence,

$$A_e = A_n$$

The nominal diameter of a standard hole is detailed in AISC 360 Table J3.3 as 1/16 in larger than the bolt diameter. As the hole is formed, some deterioration occurs in the surrounding material and AISC 360 Sec. B4.3b specifies that the effective hole diameter shall be taken as 1/16 in larger than the nominal hole diameter. Hence, the effective hole diameter is

$$\begin{aligned} d_h &= d + 1/16 \text{ in} + 1/16 \text{ in} \\ &= d + 1/8 \text{ in} \end{aligned}$$

where $d$ is diameter of fastener.

For bolted splice plates the length of the connection is small and inelastic deformation of the gross section is limited. Hence, AISC 360 Sec. J4.1 limits the effective net area of bolted splice plates to

$$\begin{aligned} A_e &= A_n \\ &\leq 0.85 A_g \end{aligned}$$

**Example 9.1.** Bolted Splice

The spliced joint shown in Fig. 9.3 is connected with 3/4-in-diameter bolts in standard holes. The plate material is A36 steel. Assuming that the bolts are satisfactory and that block shear does not govern, determine the available tensile strength of the plates.

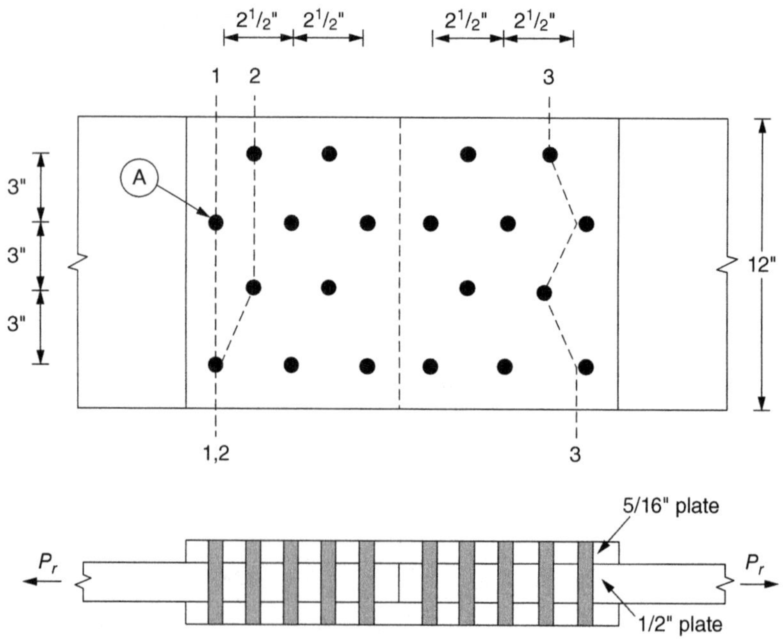


**FIGURE 9.3** Details for Example 9.1.

The 1/2-in plate governs and from Fig. 9.3 the relevant parameters are

$$w = 12 \text{ in} \qquad t = 0.5 \text{ in} \qquad g = 3 \text{ in} \qquad s = 1.25 \text{ in}$$

$$\begin{aligned} d_h &= (d + 1/8 \text{ in}) \\ &= 0.75 + 0.125 \\ &= 0.875 \text{ in} \end{aligned}$$

The gross area of the 1/2-in plate is given by

$$\begin{aligned} A_g &= wt \\ &= 12 \times 0.5 \\ &= 6.0 \text{ in}^2 \end{aligned}$$

For the straight perpendicular fracture 1-1, the net area of the plate is given by

$$\begin{aligned} A_n &= t(w - 2d_h) \\ &= 0.5(12 - 2 \times 0.875) \\ &= 5.125 \text{ in}^2 \end{aligned}$$

For the staggered fracture 2-2, the net area of the plate is given by

$$\begin{aligned} A_n &= t(w - 3d_h + s^2/4g) \\ &= 0.5[12 - 3 \times 0.875 + 1.25^2/(4 \times 3)] \\ &= 4.75 \text{ in}^2 \end{aligned}$$

The connection consists of 10 bolts, each of which may be considered to support 10 percent of the applied tensile force. Since bolt A is in front of fracture plane 2-2, the fracture plane is required to support only 90 percent of the applied force and the equivalent net area is

$$\begin{aligned} A_{n(equiv)} &= A_n/0.9 \\ &= 4.75/0.9 \\ &= 5.28 \text{ in}^2 \end{aligned}$$

For the staggered fracture 3-3, the net area of the plate is given by

$$\begin{aligned} A_n &= t(w - 4d_h + 3s^2/4g) \\ &= 0.5[12 - 4 \times 0.875 + 3 \times 1.25^2/(4 \times 3)] \\ &= 4.45 \text{ in}^2 \text{ ... governs} \end{aligned}$$

For flat plates the shear lag factor is $U = 1.0$. Hence,

$$\begin{aligned} A_e &= A_n U \\ &= 4.45 \text{ in}^2 \end{aligned}$$

The minimum effective area is limited to

$$\begin{aligned} A_e &= 0.85 A_g \\ &= 0.85 \times 6 \\ &= 5.1 \text{ in}^2 \text{ ... does not govern} \end{aligned}$$

| LRFD | ASD |
|---|---|
| The design tensile yield strength is | The allowable tensile yield strength is |
| $\varphi_t P_n = 0.9F_y A_g$ | $P_n/\Omega_t = F_y A_g/1.67$ |
| $= 0.9 \times 36 \times 6.0$ | $= 36 \times 6/1.67$ |
| $= 194.4$ kips | $= 129.3$ kips |
| The design tensile rupture strength is | The allowable tensile rupture strength is |
| $\varphi_t P_n = 0.75F_u A_e$ | $P_n/\Omega_t = F_u A_e/2.00$ |
| $= 0.75 \times 58 \times 4.45$ | $= 58 \times 4.45/2$ |
| $= 193.6$ kips ... governs | $= 129.1$ kips ... governs |

## Plates with Welded Connection

The shear lag factors for flat plates with welded connections are specified in AISC 360 Table D3.1. For the transverse welded connection shown in Fig. 9.4*a*, all of the net area participates in transmitting the load and the shear lag factor is $U = 1.0$. Hence,

$$A_e = A_n = A_g$$

For the longitudinal fillet welded connection shown in Fig. 9.4*b*, shear lag occurs at the end of the plate and the shear lag factor is given by

- $U = 1.0$ when $\ell \geq 2w$
- $U = 0.87$ when $2w > \ell \geq 1.5w$
- $U = 0.75$ when $1.5w > \ell \geq w$

Hence, $A_e = A_n U = A_g U$

**Example 9.2.** Welded Connection
The 1/2-in plate shown in Fig. 9.5 is connected to a gusset plate with longitudinal fillet welds as indicated. The plate material is A36 steel. Assuming that the welds are satisfactory and that block shear does not govern, determine the available tensile capacity of the plate.
The relevant parameters are obtained from Fig. 9.5 as

$$w = 4 \text{ in} \qquad t = 0.5 \text{ in} \qquad \ell = 6 \text{ in} \qquad \ell/w = 1.5$$

From AISC 360 Table D3.1 the shear lag factor is $U = 0.87$.

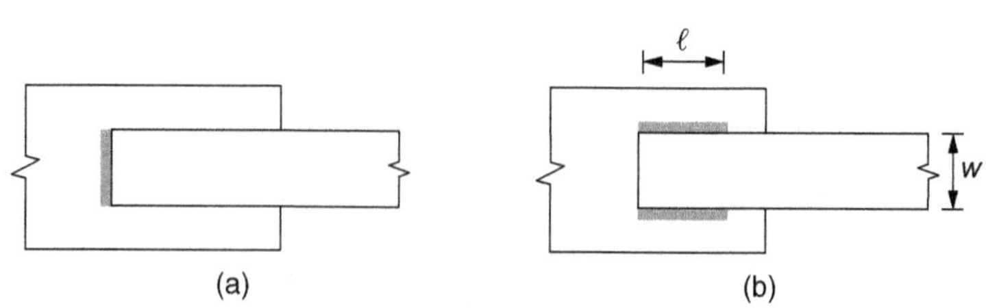


**FIGURE 9.4** Welded connections for plates in tension.

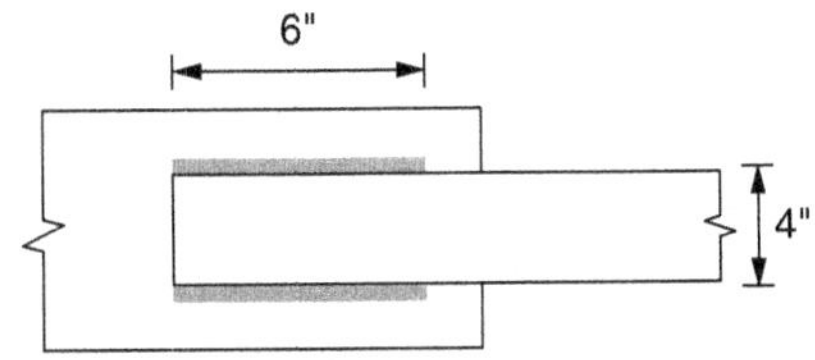


**Figure 9.5** Details for Example 9.2.

The gross area of the 1/2-in plate is given by

$$\begin{aligned} A_g &= wt \\ &= 4 \times 0.5 \\ &= 2 \text{ in}^2 \end{aligned}$$

The effective net area is given by AISC 360 Eq. (D3-1) as

$$\begin{aligned} A_e &= A_n U = A_g U \\ &= 0.87 \times 2 \\ &= 1.74 \text{ in}^2 \end{aligned}$$

| LRFD | ASD |
|---|---|
| The design tensile rupture strength is | The allowable tensile rupture strength is |
| $\varphi_t P_n = 0.75 F_u A_e$<br>$= 0.75 \times 58 \times 1.74$<br>$= 75.7$ kips | $P_n/\Omega_t = F_u A_e/2.00$<br>$= 58 \times 1.74/2$<br>$= 50.5$ kips |
| The design tensile yield strength is | The allowable tensile yield strength is |
| $\varphi_t P_n = 0.9 F_y A_g$<br>$= 0.9 \times 36 \times 2.0$<br>$= 64.8$ kips ... governs | $P_n/\Omega_t = F_y A_g/1.67$<br>$= 36 \times 2/1.67$<br>$= 43.1$ kips ... governs |

## Rolled Sections with Bolted Connection

When the tensile load is transmitted directly to all of the cross-sectional elements of a rolled structural member by connectors, the shear lag factor is $U = 1.0$ and the effective net area is

$$A_e = A_n$$

where $A_n$ is actual net area.

For bolted connections, when rolled structural shapes are connected through only part of their cross-sectional elements, the effective net area is given by AISC 360 Eq. (D3-1) as

$$A_e = A_n U$$

The shear lag factor $U$ allows for the effects of eccentricity and shear lag at the ends of the member and is defined in AISC 360 Table D3.1 Case 2 as

$$U = 1.0 - \bar{x}/\ell$$

where $\bar{x}$ = eccentricity of connection
= distance from the connection plane to the centroid of the member resisting the connection force
$\ell$ = length of the connection
= distance, parallel to the line of force, between the centers of the first and last fasteners in a line

In lieu of applying this expression, it is permitted to adopt the following values for the shear lag factor, as given in AISC 360 Table D3.1 Cases 7 and 8:

- $U = 0.90$ when W-, M-, S-, or HP-shapes and structural tees with $b_f \geq 2d/3$ are connected by the flanges, with not less than three bolts in line in the direction of stress.
- $U = 0.85$ when W-, M-, S-, or HP-shapes and structural tees with $b_f < 2d/3$ are connected by the flanges, with not less than three bolts in line in the direction of stress.
- $U = 0.70$ when W-, M-, S-, or HP-shapes and structural tees are connected by the web with not less than four bolts in line in the direction of stress.
- $U = 0.80$ when single or double angles are connected with not less than four bolts in line in the direction of stress.
- $U = 0.60$ when single or double angles are connected with three bolts in line in the direction of stress.

It is permitted to adopt the larger value for $U$ given by Case 2 and Case 7 or the larger value given by Case 2 and Case 8.

American Institute of Steel Construction, *Steel Construction Manual* (AISC Manual)[2] Tables 5-1 to 5-8 provide values of available strength in tensile rupture and tensile yield for rolled sections. A value of $A_e = 0.75A_g$ is assumed.

**Example 9.3.** Double Angle Bolted Connection
The two 3½- × 3½- × 3/8-in angles, shown in Fig. 9.6, are bolted to a 3/4-in plate with 3/4-in dia bolts. All components are grade 36 steel. Assuming that bolt strength and block shear do not govern, determine the available tensile strength of the double angles.

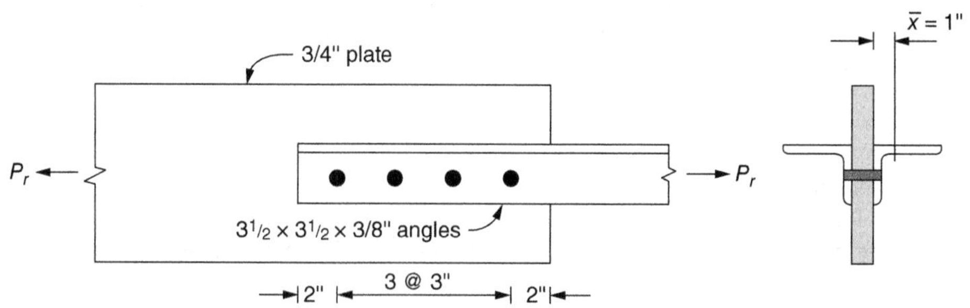


**FIGURE 9.6** Details for Example 9.3.

The hole diameter is defined as

$$d_h = (d + 1/8 \text{ in})$$

where $d$ = diameter of bolt

and

$$d_h = 0.75 + 0.125$$
$$= 0.875 \text{ in}$$

The gross area of the two angles is given by AISC Manual Table 5-8 as

$$A_g = 5.00 \text{ in}^2$$

The net area of the two angles is given by

$$A_n = A_g - 2td_h$$
$$= 5.00 - 2 \times 0.375 \times 0.875$$
$$= 4.34 \text{ in}^2$$

The shear lag factor $U$ is defined in AISC 360 Table D3.1 Case 2 as

$$U = 1.0 - \bar{x}/\ell$$

where

$\bar{x}$ = eccentricity of connection
= 1.0 in ... from AISC Manual Table 1-7
$\ell$ = length of the connection
= 9 in ... from Fig. 9.6

and

$$U = 1.0 - 1.0/9$$
$$= 0.89$$

Alternatively, using AISC 360 Table D3.1 Case 8, the shear lag factor is given as

$$U = 0.80$$
$$< 0.89 \text{ ... governs}$$

Hence,

$$A_e = A_n U$$
$$= 4.34 \times 0.89$$
$$= 3.86 \text{ in}^2$$

| LRFD | ASD |
|---|---|
| The design tensile rupture strength is<br>$\varphi_t P_n = 0.75 F_u A_e$<br>$= 0.75 \times 58 \times 3.86$<br>$= 168$ kips | The allowable tensile rupture strength is<br>$P_n/\Omega_t = F_u A_e/2.00$<br>$= 58 \times 3.86/2$<br>$= 112$ kips |
| The design tensile yield strength is<br>$\varphi_t P_n = 0.9 F_y A_g$<br>$= 0.9 \times 36 \times 5.00$<br>$= 162$ kips ... governs | The allowable tensile yield strength is<br>$P_n/\Omega_t = F_y A_g/1.67$<br>$= 36 \times 5.00/1.67$<br>$= 108$ kips ... governs |

## Rolled Sections with Welded Connection

For welded connections, when the axial force is transmitted only by transverse fillet welds to some but not all of the cross-sectional elements as shown in Fig. 9.7*a*, the shear lag factor is given by AISC 360 Table D3.1 Case 3 as

$$U = 1.0$$

and,

$$A_e = A_n$$
$$= \text{area of directly connected elements}$$

For welded connections, when the axial force is transmitted only by longitudinal fillet welds or by longitudinal fillet welds in combination with transverse fillet welds as shown in Fig. 9.7*b*, the effective net area is given by AISC 360 Eq. (D3-1) as

$$A_e = A_n U$$
$$= A_g U$$

The shear lag factor $U$ allows for the effects of eccentricity and shear lag at the ends of the member and is defined in AISC 360 Table D3.1 Case 2 as

$$U = 1.0 - \bar{x}/\ell$$

where $\bar{x}$ = eccentricity of connection
= distance from the connection plane to the centroid of the member resisting the connection force
$\ell$ = length of the connection defined in AISC 360 Commentary Sec. D3.3
= length of weld, parallel to the line of force

**Example 9.4.** Double Angle Welded Connection
The two 3½- × 3½- × 3/8-in angles, shown in Fig. 9.8, are fillet welded all round to a 3/4-in plate. All components are grade 36 steel. Assuming that weld strength and block shear do not govern, determine the available tensile strength of the double angles.

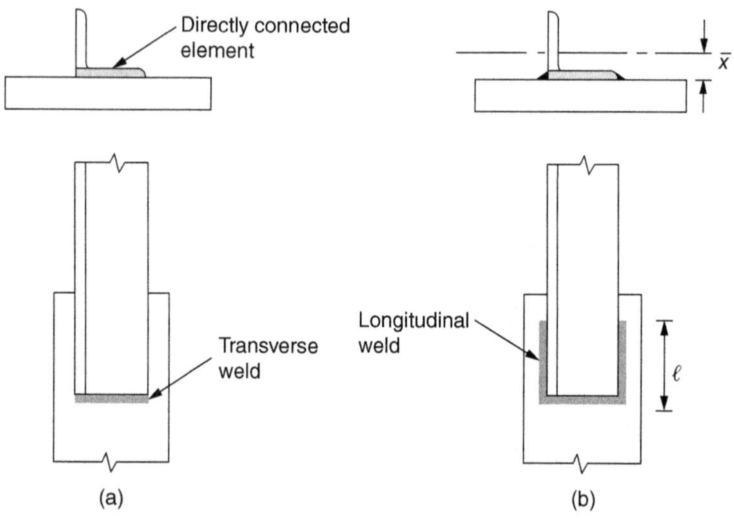


**FIGURE 9.7** Welded connections for rolled sections in tension.

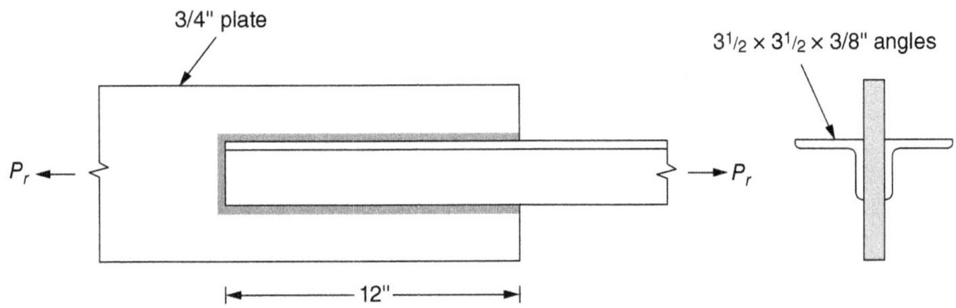


**Figure 9.8** Details for Example 9.4.

The gross area of the two angles is given by AISC Manual Table 5-8 as

$$A_g = 5.00 \text{ in}^2$$
$$= A_n$$

The shear lag factor $U$ is defined in AISC 360 Table D3.1 Case 2 as

$$U = 1.0 - \bar{x}/\ell$$

where $\bar{x}$ = eccentricity of connection
= 1.0 in ... from AISC Manual Table 1-7
$\ell$ = length of the connection
= 12 in ... from Fig. 9.8

and

$$U = 1.0 - 1.0/12$$
$$= 0.917$$

Hence,

$$A_e = A_n U = A_g U$$
$$= 5.00 \times 0.917$$
$$= 4.59 \text{ in}^2$$

| LRFD | ASD |
|---|---|
| The design tensile rupture strength is<br>$\varphi_t P_n = 0.75 F_u A_e$<br>$= 0.75 \times 58 \times 4.59$<br>$= 200$ kips<br>The design tensile yield strength is<br>$\varphi_t P_n = 0.9 F_y A_g$<br>$= 0.9 \times 36 \times 5.00$<br>$= 162$ kips ... governs | The allowable tensile rupture strength is<br>$P_n/\Omega_t = F_u A_e/2.00$<br>$= 58 \times 4.59/2$<br>$= 133$ kips<br>The allowable tensile yield strength is<br>$P_n/\Omega_t = F_y A_g/1.67$<br>$= 36 \times 5.00/1.67$<br>$= 108$ kips ... governs |

## Round Hollow Structural Sections with Welded Connection

For the connection shown in Fig. 9.9 with a gusset plate and a slotted hollow structural section, the effective net area of the hollow structural section is given by AISC 360 Eq. (D3-1) as

$$A_e = A_n U$$

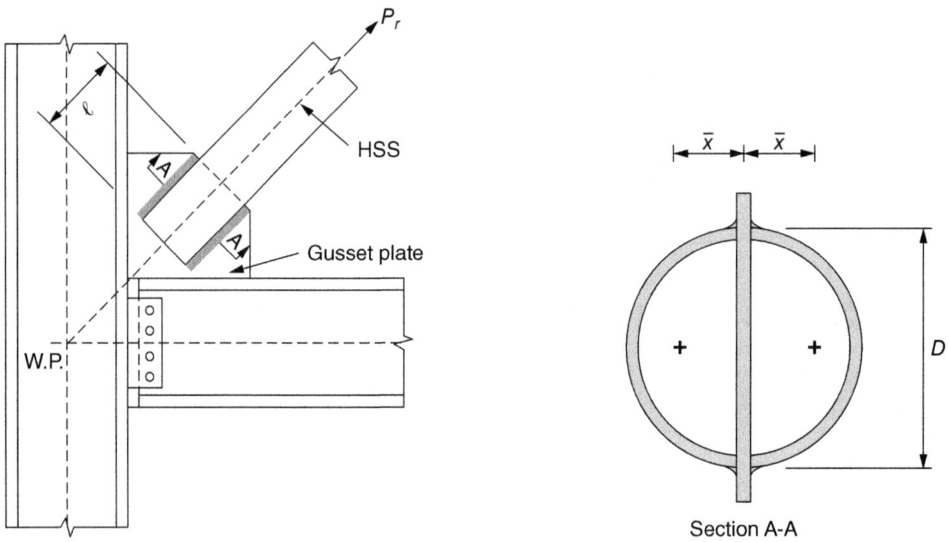


**FIGURE 9.9** Welded connection for hollow structural section.

The shear lag factor is given by AISC 360 Table D3.1 as

- $U = 1.0 - \bar{x}/\ell$ when $D \le \ell < 1.3D$
- $U = 1.0$ when $\ell \ge 1.3D$

where $A_n$ = net area of the hollow structural section at the connection
$D$ = diameter of round hollow structural section
$\ell$ = length of fillet weld
$\bar{x}$ = centroidal height of each segment of the hollow structural section
$= D/\pi$

**Example 9.5.** Hollow Structural Section Welded Connection
Determine the available tensile capacity of the connection shown in Fig. 9.10. The yield stress of the HSS 6.000 × 0.280 is $F_y = 42$ ksi and the tensile strength is $F_u = 58$ ksi. The strength of the 1/2-in gusset plate and the 1/4-in fillet weld is adequate.

The relevant properties of the HSS 6.000 × 0.280 are obtained from AISC Manual Table 1-13 as

$$A_g = 4.69 \text{ in}^2 \qquad t = 0.260 \text{ in}$$

To allow clearance for the gusset plate, the slot is cut 1/8 in oversize, and the net section of the hollow structural section at the gusset plate is

$$\begin{aligned} A_n &= A_g - 2(t_g + 0.125)t \\ &= 4.69 - 2(0.5 + 0.125) \times 0.260 \\ &= 4.37 \text{ in}^2 \end{aligned}$$

The centroidal height of each segment of the hollow structural section, with a diameter of $D = 6.000$ in, is given by AISC 360 Table D3.1 as

$$\begin{aligned} \bar{x} &= D/\pi \\ &= 6.000/3.142 \\ &= 1.91 \text{ in} \end{aligned}$$

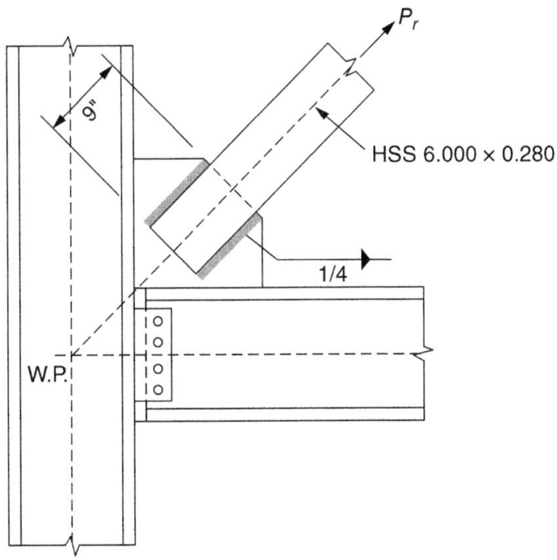


**FIGURE 9.10** Details for Example 9.5.

The ratio of length of weld to diameter of the hollow structural section is

$$\ell/D = 9/6.000$$
$$= 1.50$$
$$> 1.30$$

Hence, from AISC 360 Table D3.1 the shear lag factor is given by

$$U = 1.0$$

The effective net area of the hollow structural section is given by AISC 360 Eq. (D3-1) as

$$A_e = A_n U$$
$$= 4.37 \times 1.0$$
$$= 4.37 \text{ in}^2$$

| LRFD | ASD |
|---|---|
| The design tensile rupture strength is<br>$\varphi_t P_n = 0.75 F_u A_e$<br>$= 0.75 \times 58 \times 4.37$<br>$= 190$ kips | The allowable tensile rupture strength is<br>$P_n/\Omega_t = F_u A_e/2.00$<br>$= 58 \times 4.37/2$<br>$= 127$ kips |
| The design tensile yield strength is<br>$\varphi_t P_n = 0.9 F_y A_g$<br>$= 0.9 \times 42 \times 4.69$<br>$= 177$ kips … governs | The allowable tensile yield strength is<br>$P_n/\Omega_t = F_y A_g/1.67$<br>$= 42 \times 4.69/1.67$<br>$= 118$ kips ... governs |

## 9.4 Pin-Connected Members

Pinned connections are used when free rotation of the connected members is required as in crane booms. In the design of pin-connected members it is necessary to consider the limit states of tensile yield, tensile rupture, shear rupture, and bearing.

### Dimensional Requirements

Dimensions for pin-connected plates are specified in AISC 360 Sec. D5.2 and are shown in Fig. 9.11. The diameter of the pin hole is defined as

$$d_h = d + 1/32 \text{ in}$$

where $d$ is diameter of the pin.

The width of the plate is defined as

$$w \geq 2b_e + d$$

where $b_e$ = effective width on each side of the pin hole
$= 2t + 0.63$
$\leq b$ ... the actual distance from edge of hole to edge of plate
$t$ = thickness of plate

The extension of the plate beyond the bearing end of the pin hole is defined as

$$a \geq 1.33b_e$$

The corners of the plate beyond the pin hole may be cut at an angle of 45° provided that the actual distance from edge of hole to edge of plate, perpendicular to the cut, is

$$c \geq a$$

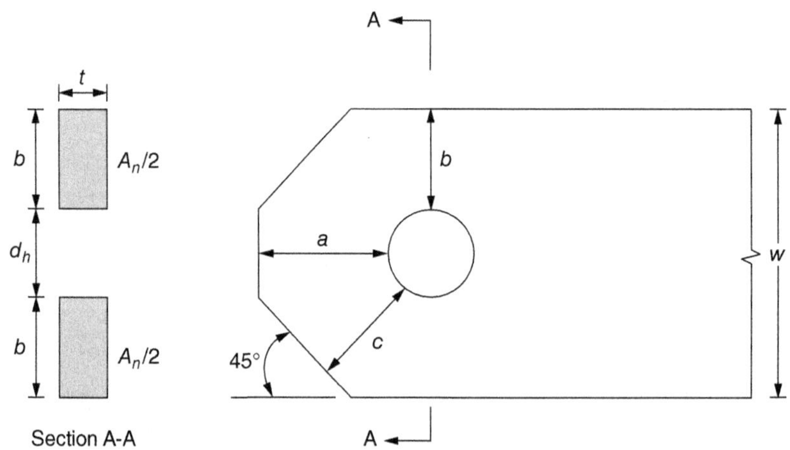


**Figure 9.11** Pin-connected plate.

### Limit States

For tensile yielding in the gross section, AISC 360 Eq. (D2-1) gives the nominal strength as

$$P_n = A_g F_y$$

where $A_g = tw$

$t$ = thickness of plate

and $\varphi_t = 0.9$

$\Omega_t = 1.67$

For tensile rupture in the effective net area, AISC 360 Eq. (D5-1) gives the nominal strength as

$$P_n = 2tb_e F_u$$

where $t$ = thickness of plate

and $\varphi_t = 0.75$

$\Omega_t = 2.00$

For shear rupture on the effective area, AISC 360 Eq. (D5-2) gives the nominal strength as

$$P_n = 0.6F_u A_{sf}$$

where $A_{sf} = 2t(a + d/2)$

and $\varphi_{sf} = 0.75$

$\Omega_{sf} = 2.00$

For bearing on the projected area of the pin, AISC 360 Eq. (J7-1) gives the nominal strength as

$$R_n = 1.8F_y A_{pb}$$

$A_{pb}$ = projected bearing area

$= td$

and $\varphi = 0.75$

$\Omega = 2.00$

In addition, AISC 360 Eq. (J3-1) gives the nominal strength of the pin in double shear as

$$R_n = F_{nv} A_b$$

$A_b$ = area of pin in double shear

$= 2(\pi d^2/4)$

$F_{nv}$ = nominal shear strength from AISC 360 Table J3.2

and $\varphi = 0.75$

$\Omega = 2.00$

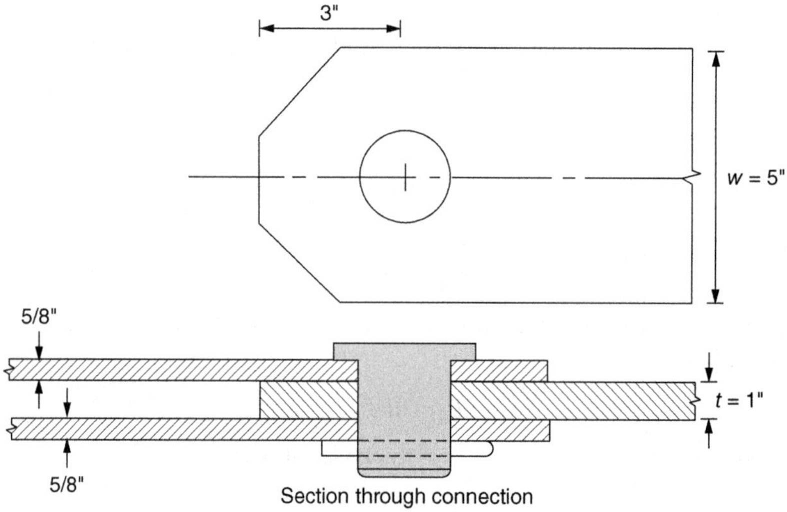


**Figure 9.12** Details for Example 9.6.

**Example 9.6.** Pin-Connected Plate

Determine the available tensile capacity of the pin-connected plate shown in Fig. 9.12. The pin-connected plates are grade A36 steel and the grade A307 pin has a diameter of $d = 2$ in.

The diameter of the pin hole is

$$d_h = d + 1/32 \text{ in}$$
$$= 2.031 \text{ in}$$

The effective width on each side of the pin hole is the lesser value given by

$$b_e = 2t + 0.63$$
$$= 2 \times 1.0 + 0.63$$
$$= 2.63 \text{ in}$$

or

$$b_e = (w - d_h)/2$$
$$= (5 - 2.031)/2$$
$$= 1.49 \text{ in ... governs}$$
$$a_{\text{required}} \geq 1.33b_e$$
$$= 1.33 \times 1.49$$
$$= 1.98 \text{ in}$$
$$a_{\text{provided}} = 3.0 - d_h/2$$
$$= 3.0 - 2.031/2$$
$$= 1.99 \text{ in}$$
$$> 1.98 \text{ in ... satisfactory}$$

The effective area for tensile rupture of the plate is

$$A_n = 2tb_e$$
$$= 2 \times 1.0 \times 1.49$$
$$= 2.98 \text{ in}^2$$

| LRFD | ASD |
|---|---|
| The design shear strength of the pin is | The allowable shear strength of the pin is |
| $\varphi R_n = 0.75 F_{nv} A_b$ | $R_n/\Omega = F_{nv} A_b/2$ |
| $= 0.75 \times 27 \times 2 \times 3.14 \times 2^2/4$ | $= 27 \times 2 \times (3.14 \times 2^2/4)/2$ |
| $= 127$ kips | $= 85$ kips |
| The plate design tensile rupture strength is | The plate allowable tensile rupture strength is |
| $\varphi_t P_n = 0.75 F_u A_n$ | $P_n/\Omega_t = F_u A_n/2.00$ |
| $= 0.75 \times 58 \times 2.98$ | $= 58 \times 2.98/2$ |
| $= 130$ kips | $= 86$ kips |
| The plate design tensile yield strength is | The plate allowable tensile yield strength is |
| $\varphi_t P_n = 0.9 F_y A_g$ | $P_n/\Omega_t = F_y A_g/1.67$ |
| $= 0.9 \times 36 \times 1.0 \times 5.0$ | $= 36 \times 1.0 \times 5.0/1.67$ |
| $= 162$ kips | $= 108$ kips |

The effective area for shear rupture of the plate is

$$\begin{aligned} A_{sf} &= 2t(a + d/2) \\ &= 2 \times 1.0(1.99 + 2/2) \\ &= 5.98 \text{ in}^2 \end{aligned}$$

| LRFD | ASD |
|---|---|
| The plate design shear rupture strength is | The plate allowable shear rupture strength is |
| $\varphi_{sf} P_n = 0.75 \times 0.6 F_u A_{sf}$ | $P_n/\Omega_{sf} = 0.6 F_u A_{sf}/2$ |
| $= 0.75 \times 0.6 \times 58 \times 5.98$ | $= 0.6 \times 58 \times 5.98/2$ |
| $= 156$ kips | $= 104$ kips |
| The pin design bearing strength is | The pin allowable bearing strength is |
| $\varphi R_n = 0.75 \times 1.8 F_y A_{pb}$ | $R_n/\Omega = 1.8 F_y A_{pb}/2$ |
| $= 0.75 \times 1.8 \times 36 \times 1.0 \times 2.0$ | $= 1.8 \times 36 \times 1.0 \times 2.0/2$ |
| $= 97$ kips ... governs | $= 65$ kips ... governs |

## 9.5 Design of Eyebars

Eyebars are used at the end of wire rope hangers as in suspension bridges. In the design of eyebars it is necessary to consider the limit states of tensile yield, tensile rupture, shear rupture, and bearing.

### Dimensional Requirements

The specified dimensions for eyebars are given in AISC 360 Sec. D6.2 and are shown in Fig. 9.13. The diameter of the pin hole is defined as

$$d_h = d + 1/32 \text{ in}$$

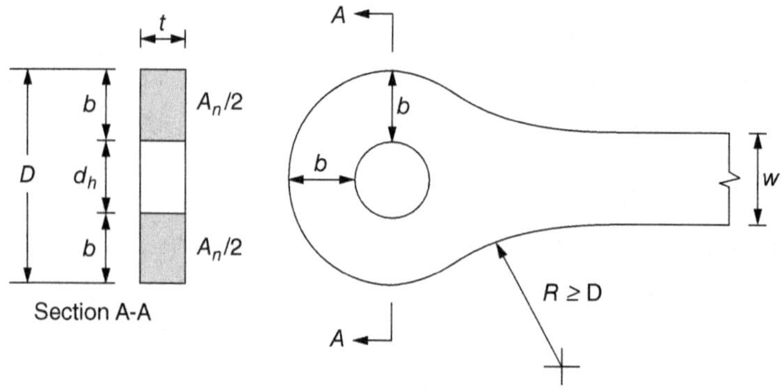


**FIGURE 9-13** Eyebar dimensions.

where $d$ is diameter of the pin. The minimum pin diameter is

$$d = 7w/8$$

where $w$ is eyebar width. The actual width from the hole edge to the plate edge perpendicular to the direction of applied load is defined as

$$b > 2w/3$$

For calculation purposes, the effective width from the hole edge to the plate edge perpendicular to the direction of applied load is defined as

$$b_e \leq 3w/4$$

For calculation purposes, the width of the body of the eyebar is defined as

$$w \leq 8t$$

where $t$ is thickness of plate.

Unless nuts are provided to tighten the pin plates into snug contact, the minimum thickness of the plate is limited to $t = 1/2$ in. The radius of the transition between the circular head and the eyebar body shall not be less than the head diameter. For steels having a yield stress greater than 70 ksi, the hole diameter shall not exceed five times the plate thickness to prevent dishing behind the pin.

**Example 9.7.** Design of Eyebar
Determine the available tensile capacity of the eyebar shown in Fig. 9.14. The eyebar material is grade A36 steel and the grade A307 pin has a diameter of $d = 3$ in.

The width to thickness ratio of the eyebar body is

$$\begin{aligned} w/t &= 3.0/1.5 \\ &= 2 \\ &< 8 \text{ ... satisfactory} \end{aligned}$$

The ratio of pin diameter to width of eyebar shank is

$$\begin{aligned} d/w &= 3.0/3.0 \\ &= 1.0 \\ &> 7/8 \text{ ... satisfactory} \end{aligned}$$

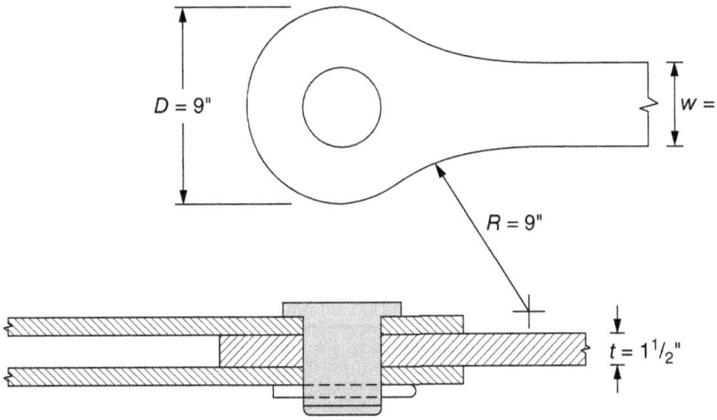


FIGURE 9.14 Details for Example 9.7.

The effective area for shear rupture of the plate is

$$A_{sf} \approx 2t(b + d/2)$$
$$\approx 2 \times 1.5 \times 9.0/2$$
$$= 13.5 \text{ in}^2$$

| LRFD | ASD |
|---|---|
| The plate design shear rupture strength is<br>$\varphi_{sf}P_n = 0.75 \times 0.6F_uA_{sf}$<br>$= 0.75 \times 0.6 \times 58 \times 13.5$<br>$= 352$ kips | The plate allowable shear rupture strength is<br>$P_n/\Omega_{sf} = 0.6F_uA_{sf}/2$<br>$= 0.6 \times 58 \times 13.5/2$<br>$= 235$ kips |
| The pin design bearing strength is<br>$\varphi R_n = 0.75 \times 1.8F_yA_{pb}$<br>$= 0.75 \times 1.8 \times 36 \times 1.5 \times 3.0$<br>$= 219$ kips | The pin allowable bearing strength is<br>$R_n/\Omega = 1.8F_yA_{pb}/2$<br>$= 1.8 \times 36 \times 1.5 \times 3.0/2$<br>$= 146$ kips |

The diameter of the pin hole is

$$d_h = d + 1/32 \text{ in}$$
$$= 3.031 \text{ in}$$

The edge distance provided to the pin hole is

$$b = (D - d_h)/2$$
$$= (9.0 - 3.031)/2$$
$$= 2.99 \text{ in}$$
$$> 2w/3$$
$$= 2 \text{ in} \ldots \text{ satisfactory}$$
$$> 3w/4$$
$$= 2.25 \text{ in} \ldots b \text{ exceeds the limit for calculation of tensile rupture}$$

Hence, for the purpose of calculation, the effective width from the hole edge to the plate edge perpendicular to the direction of applied load is

$$b_e = 3w/4$$

Hence, the effective area for tensile rupture of the plate is

$$\begin{aligned} A_n &= 2tb_e \\ &= 2 \times 1.5 \times 2.25 \\ &= 6.75 \text{ in}^2 \end{aligned}$$

| LRFD | ASD |
|---|---|
| The design shear strength of the pin is | The allowable shear strength of the pin is |
| $\varphi R_n = 0.75F_{nv}A_b$ | $R_n/\Omega = F_{nv}A_b/2$ |
| $= 0.75 \times 27 \times 2 \times 3.14 \times 3^2/4$ | $= 27 \times 2 \times (3.14 \times 3^2/4)/2$ |
| $= 286$ kips | $= 191$ kips |
| The plate design tensile rupture strength is | The plate allowable tensile rupture strength is |
| $\varphi_t P_n = 0.75F_uA_n$ | $P_n/\Omega_t = F_uA_n/2.00$ |
| $= 0.75 \times 58 \times 6.75$ | $= 58 \times 6.75/2$ |
| $= 294$ kips | $= 196$ kips |
| The plate design tensile yield strength is | The plate allowable tensile yield strength is |
| $\varphi_t P_n = 0.9F_yA_g$ | $P_n/\Omega_t = F_yA_g/1.67$ |
| $= 0.9 \times 36 \times 1.5 \times 3.0$ | $= 36 \times 1.5 \times 3.0/1.67$ |
| $= 146$ kips ... governs | $= 97$ kips ... governs |

## 9.6 Design for Fatigue

Fatigue failure may occur when repeated fluctuations in tensile stress cause crack propagation and brittle fracture in a member. Failure is sudden and occurs at a stress lower than the failure stress under static loading. Welded details and details producing stress concentrations are normally associated with fatigue failure. The principal factors influencing fatigue are

- The number of load cycles
- The magnitude of the stress range produced by service live loads
- The severity of the stress concentrations produced by the fabrication details

Issues of fatigue are not normally encountered in conventional building design as the number of load cycles is too small. However, for structures supporting vibrating machinery, crane runway girders, and bridge structures, fatigue is a major consideration. At low levels of cyclic tensile stress, fatigue cracking will not initiate irrespective of the number of load cycles. This stress range is designated the fatigue threshold, $F_{TH}$.

Evaluation of fatigue resistance is not required

- When the number of cycles of application of live load is less than 20,000
- When the live load stress range is less than the threshold allowable stress range

- When the fluctuation in stress does not involve tensile stress
- In members consisting of hollow structural steel sections in buildings subjected to wind loads

## Design Procedure

The design procedure is given in AISC 360 App. 3 and consists of determining the applicable loading condition, from AISC 360 App. 3 Table A-3.1. From the table, the corresponding values of $C_f$ = fatigue constant, $F_{TH}$ = fatigue threshold stress range, and stress category are obtained. The actual stress range is calculated as the change in stress due to the application and removal of the service live load. In the case of a stress reversal, the stress range is calculated as the numerical sum of the maximum tensile and compressive stresses to give

$$f_{SR} = f_{max} - f_{min}$$
$$< F_{SR}$$

where $f_{max}$ = maximum tensile stress in member due to service loads
$f_{min}$ = minimum stress in member due to service loads, negative if compression
$F_{SR}$ = allowable stress range for fatigue loading

When $f_{SR} < F_{TH}$ evaluation of fatigue loading is not necessary.

Eleven stress categories are defined in AISC 360 App. 3 Table A-3.1. For stress categories A, B, B′, C, D, E, and E′ the maximum allowable stress range in the member must not exceed the value given by AISC 360 App. 3 Eq. (A-3-1) as

$$F_{SR} = (C_f/n_{SR})^{0.333}$$
$$\geq F_{TH}$$

where $C_f$ = constant for the fatigue category from AISC 360 App. 3 Table A-3.1
$n_{SR}$ = number of stress range fluctuations in design life
$F_{TH}$ = maximum stress range for an indefinite design life

For stress category F, the maximum allowable stress range in the member shall not exceed the value given by AISC 360 App. 3 Eq. (A-3-2)

$$F_{SR} = (C_f/n_{SR})^{0.167}$$
$$\geq F_{TH}$$

**Example 9.8.** Shear on Throat of Longitudinal Fillet Welds
Two 3½- × 3½- × 3/8-in angles are connected with 1/4-in longitudinal fillet welds to a gusset plate as shown in Fig. 9.15. The additional axial force in the angles, due to live load only, varies from a compression of 20 kips to a tension of 40 kips. During the design life of the structure, the live load may be applied 500,000 times. Determine if the fillet weld provided is adequate to sustain the fatigue load.

From AISC 360 App. 3 Table A-3.1, the loading condition of Sec. 8.2 is applicable and the relevant factors are obtained as

$$F = \text{stress category}$$
$$F_{TH} = \text{threshold stress range}$$
$$= 8 \text{ ksi}$$
$$C_f = \text{fatigue constant}$$
$$= 150 \times 10^{10}$$

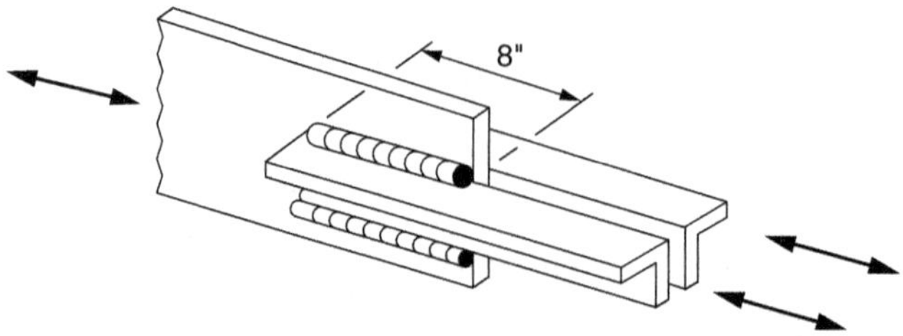


**FIGURE 9.15** Details for Example 9.8.

The applicable expression is AISC 360 App. 3 Eq. (A-3-2) and the allowable stress range is

$$\begin{aligned} F_{SR} &= (C_f/n_{SR})^{0.167} \\ &= [(150 \times 10^{10})/(5 \times 10^5)]^{0.167} \\ &= 12.07 \text{ ksi} \\ &> F_{TH} \text{ ... fatigue effects must be evaluated} \end{aligned}$$

The effective throat thickness of a 1/4-in fillet weld is given by AISC 360 Sec. J2.2a as

$$\begin{aligned} t_e &= 0.25 \times 0.707 \\ &= 0.177 \text{ in} \end{aligned}$$

The effective length of weld on both angles is

$$\begin{aligned} \ell_w &= 4 \times 8 \\ &= 32 \text{ in} \end{aligned}$$

The total effective area of weld is

$$\begin{aligned} A_w &= \ell_w \times t_e \\ &= 32 \times 0.177 \\ &= 5.66 \text{ in}^2 \end{aligned}$$

The maximum force range in the welds is

$$\begin{aligned} T_{FR} &= (T_{max} - T_{min}) \\ &= (40 \text{ kips}) + (20 \text{ kips}) \\ &= 60 \text{ kips} \end{aligned}$$

The maximum stress in the welds is

$$\begin{aligned} f_{SR} &= T_{FR}/A_w \\ &= 60/5.66 \\ &= 10.60 \text{ ksi} \\ &< F_{SR} \text{ ... satisfactory} \end{aligned}$$

## References


1. American Institute of Steel Construction (AISC). 2010. *Specification for Structural Steel Buildings* (AISC 360-10). AISC, Chicago, IL.
2. American Institute of Steel Construction (AISC). 2011. *Steel Construction Manual*, 14th edition. AISC, Chicago, IL.

## Problems

**9.1** *Given:* The bolted connection shown in Fig. 9.16 is connected with 3/4-in-diameter bolts in standard holes. The plate material is A36 steel.

*Find:* The effective net area of the plates.

**9.2** *Given:* The bolted connection shown in Fig. 9.16 is connected with 3/4-in-diameter bolts in standard holes. The plate material is A36 steel.

*Find:* Using allowable stress level design (ASD), the available tensile strength of the plates.

**9.3** *Given:* The bolted connection shown in Fig. 9.16 is connected with 3/4-in-diameter bolts in standard holes. The plate material is A36 steel.

*Find:* Using strength level design (LRFD), the available tensile strength of the plates.

**9.4** *Given:* The bolted connection shown in Fig. 9.17 is connected with 3/4-in-diameter bolts in standard holes. The plate material is A36 steel.

*Find:* The effective net area of the plates.

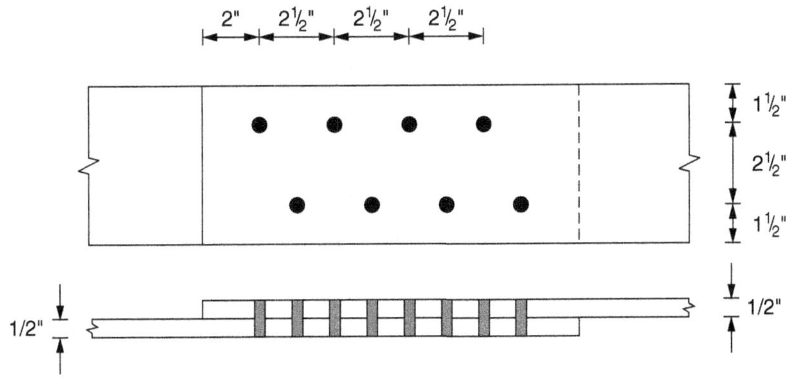


**FIGURE 9.16** Details for Problem 9.1.

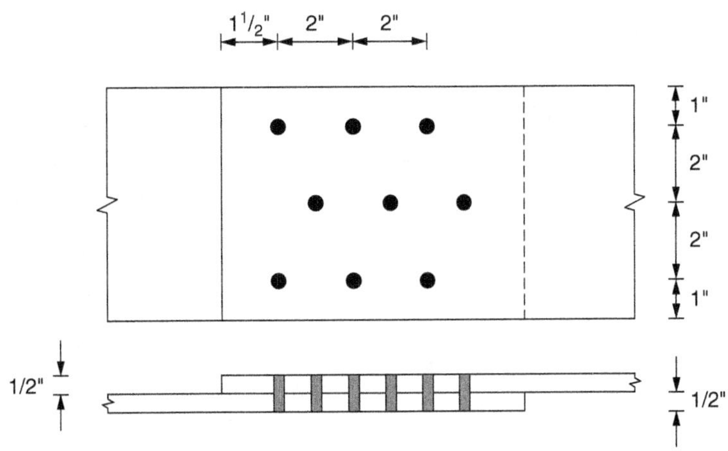


**FIGURE 9.17** Details for Problem 9.4.

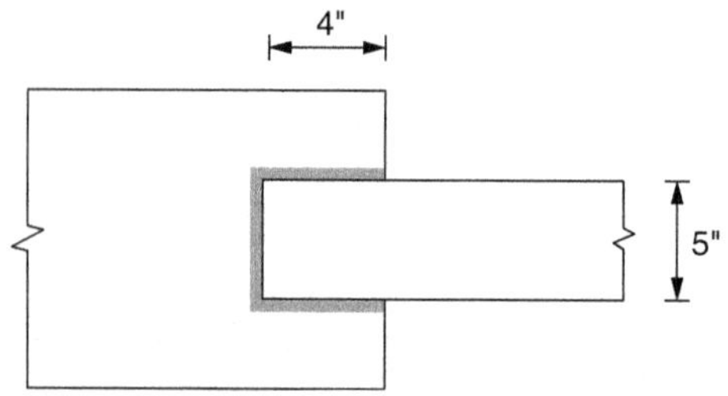


**Figure 9.18** Details for Problem 9.7.

**9.5** *Given:* The bolted connection shown in Fig. 9.17 is connected with 3/4-in-diameter bolts in standard holes. The plate material is A36 steel.

*Find:* Using allowable stress level design (ASD), the available tensile strength of the plates.

**9.6** *Given:* The bolted connection shown in Fig. 9.17 is connected with 3/4-in-diameter bolts in standard holes. The plate material is A36 steel.

*Find:* Using strength level design (LRFD), the available tensile strength of the plates.

**9.7** *Given:* The 1/2-in plate shown in Fig. 9.18 is connected to a gusset plate with fillet welds as indicated. The plate material is A36 steel.

*Find:* The effective net area of the plate.

**9.8** *Given:* The 1/2-in plate shown in Fig. 9.18 is connected to a gusset plate with fillet welds as indicated. The plate material is A36 steel.

*Find:* The available tensile capacity of the plate using allowable stress level design (ASD), assuming that the welds are satisfactory and that block shear does not govern.

**9.9** *Given:* The 1/2-in plate shown in Fig. 9.18 is connected to a gusset plate with fillet welds as indicated. The plate material is A36 steel.

*Find:* The available tensile capacity of the plate using strength level design (LRFD), assuming that the welds are satisfactory and that block shear does not govern.

**9.10** *Given:* The WT5 × 16.5 member shown in Fig. 9.19 is connected to a gusset plate with 3/4-in-diameter bolts in standard holes. The yield stress of the T-section is $F_y = 50$ ksi.

*Find:* The effective net area of the T-section.

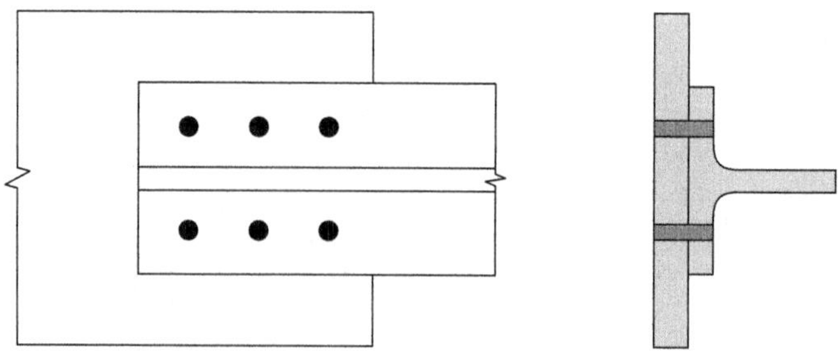

**Figure 9.19** Details for Problem 9.10.

**9.11** *Given:* The WT5 × 16.5 member shown in Fig. 9.19 is connected to a gusset plate with 3/4-in-diameter bolts in standard holes. The yield stress of the T-section is $F_y = 50$ ksi.

*Find:* The available tensile capacity of the T-section using allowable stress level design (ASD), assuming that the bolts are satisfactory and that block shear does not govern.

**9.12** *Given:* The WT5 × 16.5 member shown in Fig. 9.19 is connected to a gusset plate with 3/4-in-diameter bolts in standard holes. The yield stress of the T-section is $F_y = 50$ ksi.

*Find:* The available tensile capacity of the T-section using strength level design (LRFD), assuming that the bolts are satisfactory and that block shear does not govern.

**9.13** *Given:* The WT4 × 15.5 member shown in Fig. 9.20 is connected to a gusset plate with fillet welds as indicated. The yield stress of the T-section is $F_y = 50$ ksi.

*Find:* The effective net area of the T-section.

**9.14** *Given:* The WT4 × 15.5 member shown in Fig. 9.20 is connected to a gusset plate with fillet welds as indicated. The yield stress of the T-section is $F_y = 50$ ksi.

*Find:* The available tensile capacity of the T-section using allowable stress level design (ASD), assuming that the welds are satisfactory and that block shear does not govern.

**9.15** *Given:* The WT4 × 15.5 member shown in Fig. 9.20 is connected to a gusset plate with fillet welds as indicated. The yield stress of the T-section is $F_y = 50$ ksi.

*Find:* The available tensile capacity of the T-section using strength level design (LRFD), assuming that the welds are satisfactory and that block shear does not govern.

**9.16** *Given:* The HSS5 × 5 × 3/8 member shown in Fig. 9.21 is connected to a 3/4-in gusset plate with fillet welds as indicated. The yield stress of the HSS is $F_y = 46$ ksi and the tensile strength is $F_u = 58$ ksi.

*Find:* The effective net area of the HSS.

**9.17** *Given:* The HSS5 × 5 × 3/8 member shown in Fig. 9.21 is connected to a 3/4-in gusset plate with fillet welds as indicated. The yield stress of the HSS is $F_y = 46$ ksi and the tensile strength is $F_u = 58$ ksi.

*Find:* The available tensile capacity of the HSS using allowable stress level design (ASD), assuming that the welds are satisfactory and that block shear does not govern.

**9.18** *Given:* The HSS5 × 5 × 3/8 member shown in Fig. 9.21 is connected to a 3/4-in gusset plate with fillet welds as indicated. The yield stress of the HSS is $F_y = 46$ ksi and the tensile strength is $F_u = 58$ ksi.

*Find:* The available tensile capacity of the HSS using strength level design (LRFD), assuming that the welds are satisfactory and that block shear does not govern.

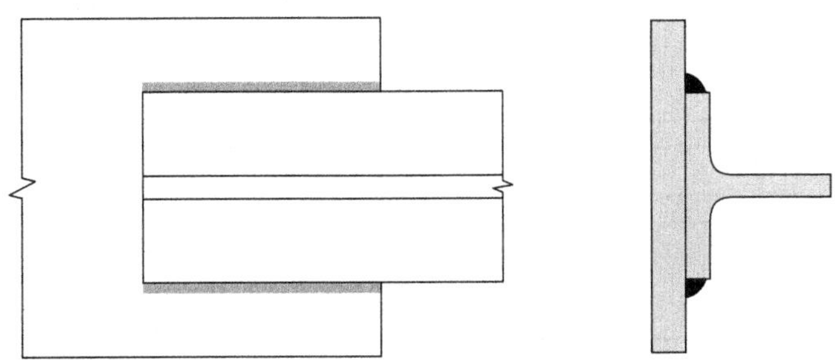

**FIGURE 9.20** Details for Problem 9.13.

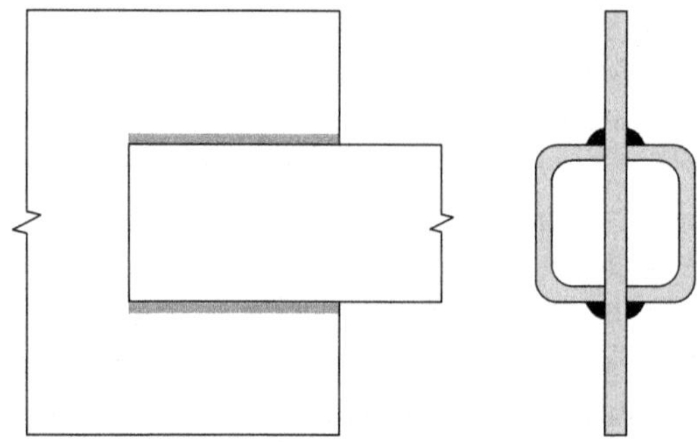

**FIGURE 9.21** Details for Problem 9.16.

**9.19** *Given:* The pin-connected plates shown in Fig. 9.22 are grade A36 steel and the grade A307 pin has a diameter of $d = 3$ in.

*Find:* The available tensile capacity of the connection using allowable stress level design (ASD).

**9.20** *Given:* The pin-connected plates shown in Fig. 9.22 are grade A36 steel and the grade A307 pin has a diameter of $d = 3$ in.

*Find:* The available tensile capacity of the connection using strength level design (LRFD).

**9.21** *Given:* The eyebar shown in Fig. 9.23 is grade A36 steel and the grade A307 pin has a diameter of $d = 3\frac{1}{2}$ in.

*Find:* The available tensile capacity of the connection using allowable stress level design (ASD).

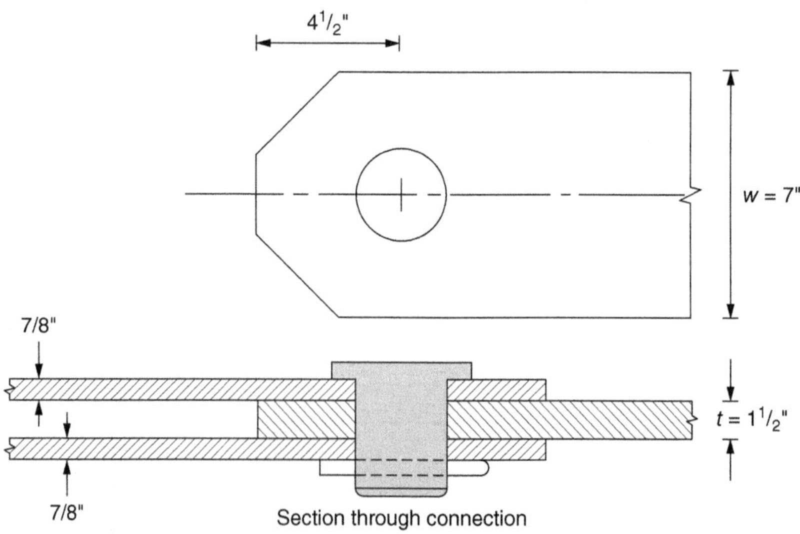


**FIGURE 9.22** Details for Problem 9.19.

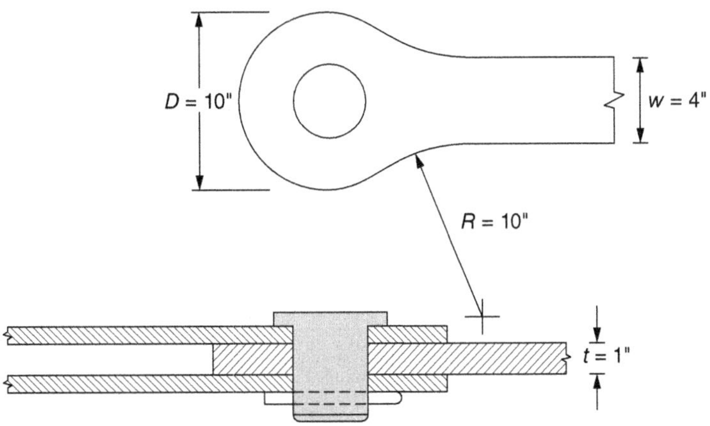


**FIGURE 9.23** Details for Problem 9.21.

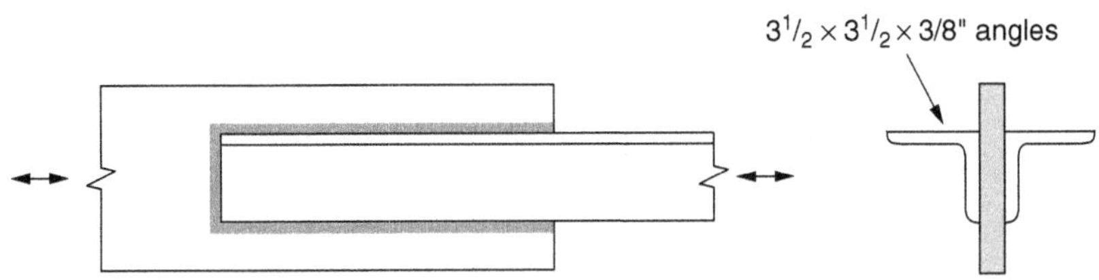


**FIGURE 9.24** Details for Problem 9.23.

**9.22** *Given:* The eyebar shown in Fig. 9.23 is grade A36 steel and the grade A307 pin has a diameter of $d = 3½$ in.

*Find:* The available tensile capacity of the connection using strength level design (LRFD).

**9.23** *Given:* Two 3½- × 3½- × 3/8-in angles are connected by fillet welds to a gusset plate as shown in Fig. 9.24. All components are grade 36 steel. The fluctuating live load varies from a compressive force of 10 kips to a tensile force of 30 kips and, during the design life of the structure, may be applied 500,000 times.

*Find:* Whether fatigue effects in the angles are a concern assuming that the connection is adequate for the maximum applied static load.

# CHAPTER 10

# Design of Bolted Connections

## 10.1 Introduction

### Bolt Types

Bolts are typically used in field connections to connect together members of a structure. Bolts are classified as either common bolts or high-strength bolts. Common bolts are of Grade A307 low-carbon steel with a minimum tensile strength of 60 ksi. High-strength bolts are typically of Grade A325 heat-treated medium-carbon steel with a minimum tensile strength of 120 ksi for bolts up to 1 in diameter and Grade A490 alloy steel with a minimum tensile strength of 150 ksi. Bolts are classified into two categories with similar mechanical properties. These are, Group A bolts–A325, F1852, A354 Grade BC, and A449. Group B bolts–A490, F2280, and A354 Grade BD. The specification for the use of high-strength bolts is provided by the Research Council on Structural Connections, *Specification for Structural Joints Using High-Strength Bolts* (RCSC).[1] Details of the applications of high-strength bolts are covered by Kulak[2] and Shaw.[3]

### Bolt Installation

Common bolts are installed in the snug-tight condition. High-strength bolts may be installed in the snug-tight, pretensioned, or slip-critical condition. In accordance with American Institute of Steel Construction, *Specification for Structural Steel Buildings* (AISC 360)[4] Sec. J3.1, a *snug-tight condition* is defined as the tightness required to bring the plies into firm contact. Transfer of the load from one connected part to another is provided by bearing of the bolts against the side of the holes. A specific level of installed tension is not required. Snug-tight connections may be used where slip of the bolts under service load conditions is acceptable and vibration or load fluctuation are not design considerations.

A *pretensioned condition* requires the bolts to be pretensioned to the level specified in AISC 360 Table J3.1. Transfer of the load from one connected part to another is provided by bearing of the bolts against the side of the holes.

As specified in AISC 360 Sec. J1.10, a pretensioned condition is used for

- Column splices in multistory structures over 125 ft in height.
- Connections of beams and girders to columns or to beams and girders on which the bracing of columns is dependent in structures over 125 ft in height.

- In roof truss splices and connections of trusses to columns, column splices, column bracing, knee braces, and crane supports in all structures carrying cranes of over 5 ton capacity.
- Connections for the support of machinery and other live loads that produce impact or reversal of load.

A *slip-critical condition* requires the bolts to be pretensioned to the level specified in AISC 360 Table J3.1 and, in addition, the faying surfaces are treated to provide a specific slip coefficient between the faying surfaces. Transfer of the load from one connected part to another is provided by the friction between the faying surfaces and no slip occurs at service loads.

As specified in RCSC Sec. 4.3, slip-critical connections are used in situations where

- Stress reversal, impact, fatigue, or vibration may occur.
- Bolts are used in oversize holes or slotted holes parallel to the direction of load as specified in AISC 360 Sec. J3.2.
- Slip at the faying surfaces would be detrimental to the performance of the structure or invalidate the structural analysis.
- Bolts are used in conjunction with welds as specified in AISC 360 Sec. J1.8.

As listed in AISC 360 Table J3.1, all high-strength bolts used in pretensioned or slip-critical connections are tensioned to 70 percent of the minimum tensile strength of the bolts. The installation of tensioned bolts is described in Kulak[2] and may consist of turn-of-nut method, direct-tension indicator, twist-off-type tension-control bolts, or calibrated wrench.

## Connection Types

Figure 10.1 shows a bolted lap joint connection for a tension member. Applying load to the member, in the case of snug-tight bolts, causes the plates to slip until the clearance between the bolt and the edge of the hole is taken up and the bolts bear on the plates.

All bolts are assumed to be equally loaded by the force $P$ applied to the plates. A single bolted connection is shown at Fig. 10.2*a*. The bearing stress produced on the contact area between the bolt and the plate is shown at (*b*) and is given by

$$f_p = P/A_{pb}$$

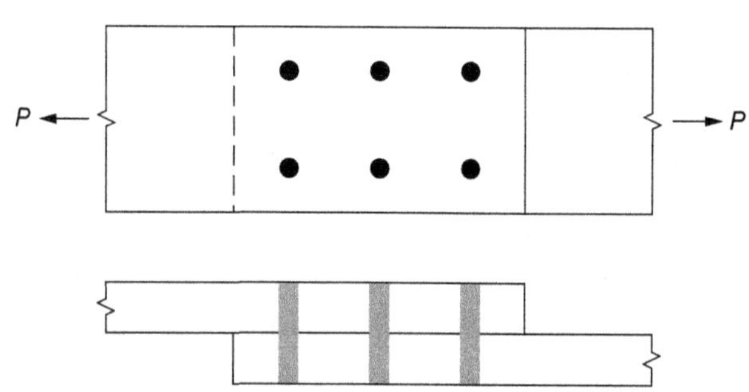


**FIGURE 10.1** Bolted lap joint.

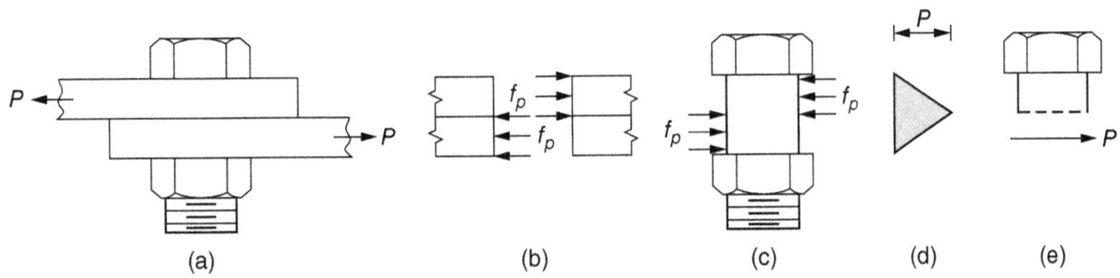


**FIGURE 10.2** Snug-tight bolt stresses.

where $A_{pb}$ = projected area in bearing
= $dt$
$d$ = bolt diameter
$t$ = plate thickness

The bearing stress acting on the bolt is shown at (*c*) and the shear force at (*d*) and (*e*). The maximum shear acting on the bolt, *P*, occurs at the contact surface of the plates. The shear stress on the bolt at the critical shear plane is

$$f_v = P/A_b$$

where $A_b$ is nominal cross-sectional area of the bolt = $\pi d^2/4$.

The slip-critical bolt shown in Fig. 10.3*a*, depends upon the friction induced between the connected parts by the clamping action of the pretensioned bolts to transfer the load from one connected part to another. No slip occurs at service loads and, as shown at (*b*), the force between the connected plates is transferred by friction.

A lap joint is not a preferred connection because of the bending stresses produced in the members by the eccentricity of the loads. A more suitable connection is the butt splice and this is shown in Fig. 10.4. The bolts in a splice joint are in double shear with two critical shear planes. Hence, each bolt has twice the shear capacity of a bolt in a lap joint.

Figure 10.5 shows a hanger connected by bolts to the bottom flange of a girder. Here the bolts are in tension.

An example of a situation where bolts are subjected to combined shear and tension is shown in Fig. 10.6. In this connection, all bolts are equally loaded in shear. However, because of the eccentricity of the applied load, the top row of bolts also resists tensile stress.

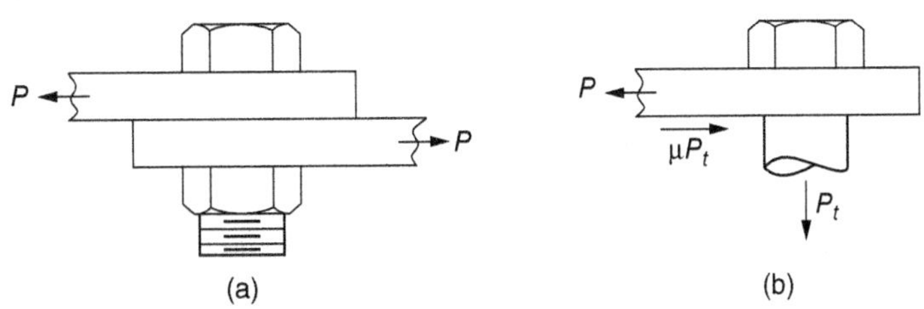


**FIGURE 10.3** Slip-critical connection.

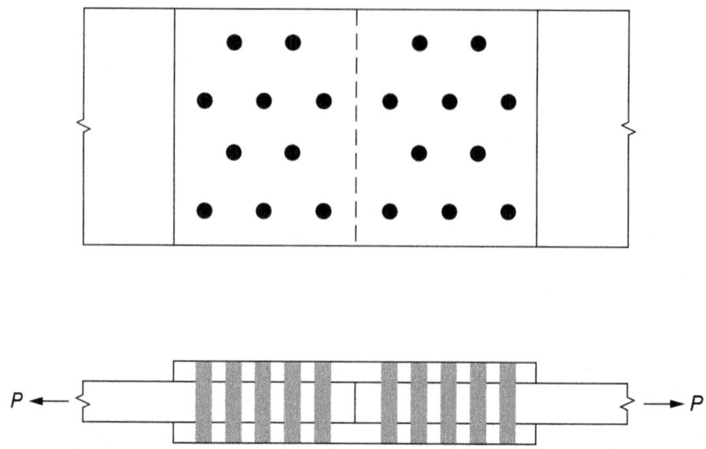


**Figure 10.4** Butt splice joint.

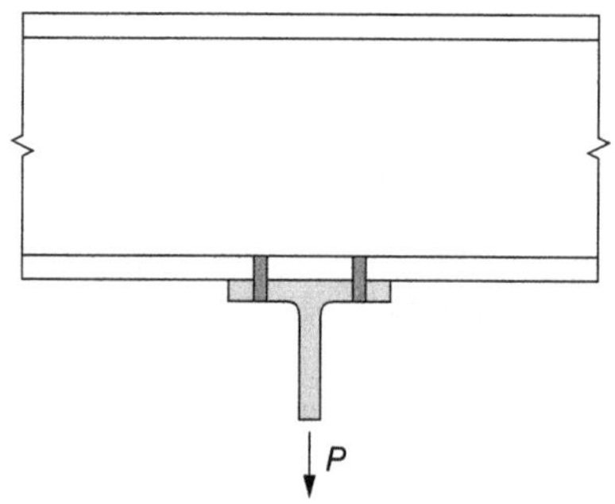


**Figure 10.5** Bolts in tension.

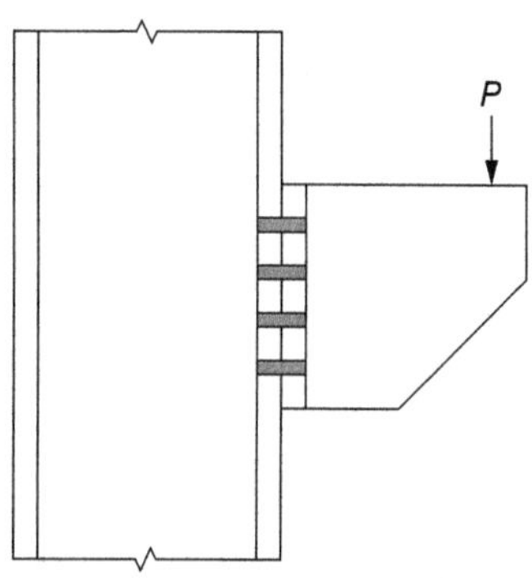


**Figure 10.6** Bolts in combined shear and tension.

## 10.2 Snug-Tight Bolts in Shear and Bearing

### Bolt Spacing

AISC 360 Table J3.4 provides the minimum permissible edge distance from the center of a standard hole, in any direction, toward an edge. These distances are based on fabrication practices and tolerances and may not satisfy the bearing and tearout strength requirements of AISC 360 Sec. J3.10. However, if necessary, lesser edge distances are permitted provided the limitations of AISC 360 Sec. J3.10 are applied. For oversize or slotted holes, an additional increment $C_2$ is added to the distances given in AISC 360 Table J3.4. This increment is given in AISC 360 Table J3.5.

The minimum permissible distance between the centers of standard, oversized, or slotted holes is given by AISC 360 Sec. J3.3 as

$$s_{min} = 2.67d$$

The preferred distance between the centers of standard, oversized, or slotted holes is

$$s_{pref} = 3.0d$$

These distances are to facilitate construction and may not satisfy the bearing and tearout strength requirements of AISC 360 Sec. J3.10.

To prevent the ingress of moisture between elements in contact, the maximum permissible edge distance from the center of a bolt to the nearest edge is given by AISC 360 Sec. J3.5 as

$$\begin{aligned} \ell_{c,\,max} &= 12t \\ &\leq 6 \text{ in} \end{aligned}$$

In addition, the maximum spacing of bolts connecting elements consisting of a plate and a shape or two plates in continuous contact and not subject to corrosion is

$$\begin{aligned} s_{max} &= 24t \\ &\leq 12 \text{ in} \end{aligned}$$

For unpainted elements of weathering steel subject to corrosion, the maximum spacing is

$$\begin{aligned} s_{max} &= 14t \\ &\leq 7 \text{ in} \end{aligned}$$

### Shear Strength

Available shear strength is based on the nominal cross-sectional area of the bolt $A_b$ and the nominal shear strength. Nominal shear strength $F_{nv}$ is given in AISC 360 Table J3.2 and in Table 10.1. For high-strength bolts, a reduced nominal strength is applicable when threads are not excluded from the shear planes. For connection longer than 38 in, the nominal strength is reduced to 83.3 percent of the tabulated values.

| Type of Bolt | A307 | Group A | Group B |
|---|---|---|---|
| Threads included in the shear plane | 27 ksi | 54 ksi | 68 ksi |
| Threads excluded from the shear plane | 27 ksi | 68 ksi | 84 ksi |

**TABLE 10.1** Bolt Nominal Shear Strength, $F_{nv}$

The bolt nominal strength is given by AISC 360 Eq. (J3-1) as

$$R_{nv} = F_{nv}A_b$$

where $A_b$ is nominal unthreaded body area of the bolt.

The available shear strength is obtained from AISC 360, Sec. J3.6 as

| LRFD | ASD |
|---|---|
| $\varphi_v R_{nv}$ = design shear strength $\geq V_r$<br>where $\varphi_v$ = resistance factor = 0.75<br>$V_r$ = required shear strength using LRFD load combinations | $R_{nv}/\Omega_v$ = allowable shear strength $\geq V_r$<br>where $\Omega_v$ = safety factor = 2.00<br>$V_r$ = required shear strength using ASD load combinations |

American Institute of Steel Construction, *Steel Construction Manual* (AISC Manual)[5] Table 7-1 provides the available shear capacities of bolts with diameters of 5/8 in to 1½ in.

## Bearing Strength

The bearing strength of connected parts is specified in ASD Section J3.10 and is based on the projected area $A_{pb}$ of the nominal bolt diameter $d$ and the thickness of the critical connected part $t$, and on the clear distance $\ell_c$ between holes or between a hole and the edge of the material. Bearing strength is also limited by whether deformation of the hole is a design consideration. Limitations are also imposed on the use of long-slotted holes.

### Deformation of the Hole at Service Load Is a Design Consideration

When deformation of the hole at service load is a design consideration, independent of the direction of loading, the nominal bearing capacity of the connected material is given by AISC 360 Eq. (J3-6a) as

$$R_n = 1.2\ell_c tF_u \text{ ... when tear out strength governs}$$

$$\leq 2.4dtF_u \text{ ... when bearing strength governs}$$

where $F_u$ = tensile strength of the critical connected part

$\ell_c$ = clear distance, in the direction of force, between the edge of the hole and the edge of the adjacent hole or edge of the connected part

= $s - (d + 1/16)$ ... between holes

$(d + 1/16)$ = nominal hole diameter given in AISC 360 Table J3.3

$s$ = bolt center-to-center spacing

Hence, to ensure that tear out does not occur, the clear distance between adjacent holes or between the edge of a hole and the edge of the connected part is obtained from AISC 360 Eq. (J3-6a) as

$$\ell_c \geq 2d$$

Similarly, the minimum distance between the centers of standard, oversized, or slotted holes to ensure that tear out does not occur is

$$s \geq 3d + 1/16 \text{ in}$$

The available bearing strength is obtained from AISC 360, Sec. J3.10 as

| LRFD | ASD |
|---|---|
| $\varphi R_n$ = design bearing strength<br>$\geq P_r$<br>where $\varphi$ = resistance factor<br>= 0.75<br>$P_r$ = required bearing strength using LRFD load combinations | $R_n/\Omega$ = allowable bearing strength<br>$\geq P_r$<br>where $\Omega$ = safety factor<br>= 2.00<br>$P_r$ = required bearing strength using ASD load combinations |

The available bearing capacity, considering deformation of the connected parts, is given in AISC Manual Table 7-4 for various bolt center-to-center spacings for bolt diameters of 5/8 in to 1½ in. The available bearing capacity, considering deformation of the connected parts, is given in AISC 360 Table 7-5 for various bolt edge distances, measured from center of bolt to edge of connected part, for bolt diameters of 5/8 in to 1½ in.

**Example 10.1.** Bolts in Shear and Bearing with Deformation a Design Consideration
The connection shown in Fig. 10.7 consists of four, grade A490, ¾-in-diameter bolts. The bolts are snug-tight and threads are excluded from the shear planes. Deformation around the bolt holes is a design consideration and the bolt spacing is as indicated. The angles and gusset plate are fabricated from A36 steel. Assuming that the angles and gusset plate are satisfactory, determine the shear force that may be applied to the bolts in the connection.

The bolt edge distance perpendicular to the direction of the tensile force is obtained from Fig. 10.7 as

$$\ell_e = 1.75 \text{ in}$$

$$> 1 \text{ in ... complies with AISC 360 Table J3.4}$$

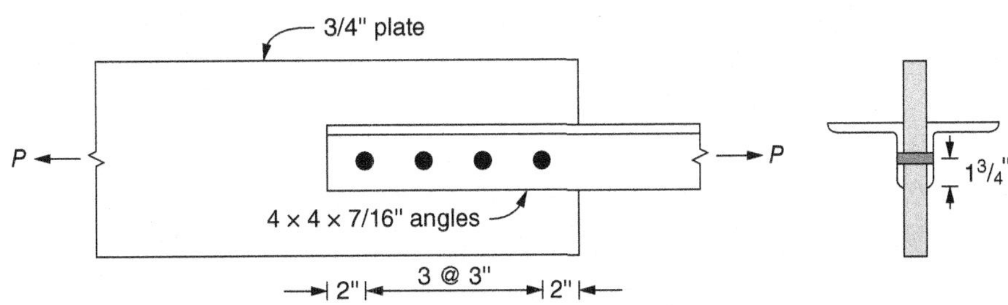


**FIGURE 10.7** Details for Example 10.1.

The bolt edge distance in the direction of the tensile force is obtained from Fig. 10.7 as

$$\ell_e = 2 \text{ in}$$

and $\ell_c$ = clear distance, in the direction of force, between the edge of the hole and the edge of the 3/4-in gusset plate

$$= \ell_e - (d + 1/16)/2$$
$$= 2 - (3/4 + 1/16)/2$$
$$= 1.59 \text{ in}$$
$$> 2d \text{ ... full bearing capacity possible on end bolt}$$

The bolt pitch is obtained from Fig. 10.7 as

$$s = 3 \text{ in}$$
$$= 4d$$
$$> 3d + 1/16 \text{ in ... full bearing capacity possible on all bolts}$$

Hence, tear out does not govern and a nominal bearing stress of $2.4F_u$ is applicable. The available bearing capacity of the four bolts on the 3/4-in plate material is

| LRFD | ASD |
|---|---|
| The design bearing strength is | The allowable bearing strength is |
| $\varphi R_n = \varphi n \times 2.4dtF_u$ | $R_n/\Omega = n \times 2.4dtF_u/\Omega$ |
| $= 0.75 \times 4 \times 2.4 \times 0.75 \times 0.75 \times 58$ | $= 4 \times 2.4 \times 0.75 \times 0.75 \times 58/2$ |
| $= 235$ kips | $= 157$ kips |
| Alternatively, from AISC Manual Table 7-4: | Alternatively, from AISC Manual Table 7-4: |
| $\varphi R_n = nt(\varphi r_n)$ | $R_n/\Omega = nt(r_n/\Omega)$ |
| $= 4 \times 0.75 \times 78.3$ | $= 4 \times 0.75 \times 52.2$ |
| $= 235$ kips | $= 157$ kips |

The double shear capacity of the four, grade A490, 3/4-in-diameter bolts with threads excluded from the shear planes is

| LRFD | ASD |
|---|---|
| The design shear strength is | The allowable shear strength is |
| $\varphi_v R_{nv} = 2n\varphi_v F_{nv} A_b$ | $R_{nv}/\Omega_v = 2nF_{nv}A_b/\Omega_v$ |
| $= 2 \times 4 \times 0.75 \times 84 \times 0.442$ | $= 2 \times 4 \times 84 \times 0.442/2$ |
| $= 223$ kips ... governs | $= 149$ kips ... governs |
| $< 235$ kips | $< 157$ kips |
| Alternatively, from AISC Manual Table 7-1: | Alternatively, from AISC Manual Table 7-1: |
| $\varphi_v R_{nv} = n(\varphi r_n)$ | $R_{nv}/\Omega_v = n(r_n/\Omega)$ |
| $= 4 \times 55.7$ | $= 4 \times 37.1$ |
| $= 223$ kips | $= 148$ kips |

### Deformation of the Hole at Service Load Is Not a Design Consideration

When deformation of the hole is not a design consideration, the nominal bearing capacity of each bolt is given by AISC 360 Eq. (J3-6b) as

$$R_n = 1.5\ell_c tF_u \text{ ... when tear out strength governs}$$
$$\leq 3dtF_u \text{ ... when bearing strength governs}$$

The minimum clear distance between adjacent holes, or between the edge of a hole and the edge of the connected part, to ensure that tear out does not occur with deformation not a design consideration is obtained from AISC 360 Eq. (J3-6b) as

$$\ell_c \geq 2d$$

The minimum distance between the centers of holes to ensure that tear out does not occur, with deformation not a design consideration, is given by

$$s \geq 3d + 1/16 \text{ in}$$

**Example 10.2.** Bolts in Shear and Bearing with Deformation Not a Design Consideration
The connection shown in Fig. 10.7 consists of four, grade A490, 3/4-in-diameter bolts. The bolts are snug-tight and threads are excluded from the shear planes. Deformation around the bolt holes is not a design consideration and the bolt spacing is as indicated. The angles and gusset plate are fabricated from A36 steel. Assuming that the angles and gusset plate are satisfactory, determine the shear force that may be applied to the bolts in the connection.

The bolt edge distance perpendicular to the direction of the tensile force is obtained from Fig. 10.7 as

$$\ell_e = 1.75 \text{ in}$$
$$> 1 \text{ in ... complies with AISC 360 Table J3.4}$$

The bolt edge distance in the direction of the tensile force is obtained from Fig. 10.7 as

$$\ell_e = 2 \text{ in}$$

and $\ell_c$ = clear distance, in the direction of force, between the edge of the hole and the edge of the 3/4-in gusset plate

$$= \ell_e - (d + 1/16)/2$$
$$= 2 - (3/4 + 1/16)/2$$
$$= 1.59 \text{ in}$$
$$> 2d \text{ ... full bearing capacity possible on end bolt}$$

The bolt pitch is obtained from Fig. 10.7 as

$$s = 3 \text{ in}$$
$$= 4d$$
$$> 3d + 1/16 \text{ in ... full bearing capacity possible on all bolts}$$

Hence, tear out does not govern and a nominal bearing stress of $3F_u$ is applicable. The available bearing capacity of the four bolts on the 3/4-in plate material is

| LRFD | ASD |
|---|---|
| The design bearing strength is | The allowable bearing strength is |
| $\varphi R_n = \varphi n \times 3dtF_u$ | $R_n/\Omega = n \times 3dtF_u/\Omega$ |
| $= 0.75 \times 4 \times 3 \times 0.75 \times 0.75 \times 58$ | $= 4 \times 3 \times 0.75 \times 0.75 \times 58/2$ |
| $= 294$ kips | $= 196$ kips |

The double shear capacity of the four, grade A490, 3/4-in-diameter bolts with threads excluded from the shear planes is obtained from Example 10.1 as

| LRFD | ASD |
|---|---|
| $\varphi_v R_{nv} = 223$ kips ... governs | $R_{nv}/\Omega_v = 149$ kips ... governs |
| $<294$ kips | $<196$ kips |

## Long-Slotted Holes

For long-slotted holes with the axis of the slot perpendicular to the direction of the tensile force, the nominal bearing capacity of each bolt is given by AISC 360 Eq. (J3-6c) as

$$R_n = 1.0\ell_c tF_u \text{ ... when tear out strength governs}$$

$$\leq 2.0dtF_u \text{ ... when bearing strength governs}$$

Hence, to ensure that tear out does not occur, the clear distance between adjacent holes or between the edge of a hole and the edge of the connected part is obtained from AISC 360 Eq. (J3-6c) as

$$\ell_c \geq 2d$$

Similarly, the distance between the centers of holes to ensure that tear out does not occur is

$$s \geq 3d + 1/16 \text{ in}$$

**Example 10.3.** Bolts in Long-slotted Holes

The connection shown in Fig. 10.7 consists of four, grade A490, 3/4-in-diameter bolts with long-slotted holes transverse to the direction of force in the gusset plate. The bolts are snug-tight and threads are excluded from the shear planes. The angles and gusset plate are fabricated from A36 steel. Assuming that the angles and gusset plate are satisfactory, determine the shear force that may be applied to the bolts in the connection.

The double shear capacity of the four, grade A490, 3/4-in-diameter bolts with threads excluded from the shear planes is obtained from Example 10.1 as

| LRFD | ASD |
|---|---|
| $\varphi_v R_{nv} = 223$ kips | $R_{nv}/\Omega_v = 149$ kips |

The bolt edge distance perpendicular to the direction of the tensile force is obtained from Fig. 10.7 as

$$\ell_e = 1.75 \text{ in}$$

The minimum permissible distance is obtained from AISC 360 Tables J3.4 and J3.5 as

$$\ell_{e,\,min} = 1 \text{ in} + C_2$$

$$= 1 \text{ in} + 0.75d \text{ ... long axis perpendicular to edge}$$

$$= 1.56 \text{ in}$$

$$< 1.75 \text{ in ... satisfactory}$$

The bolt edge distance in the direction of the tensile force is obtained from Fig. 10.7 as

$$\ell_e = 2 \text{ in}$$

and $\ell_c$ = clear distance, in the direction of force, between the edge of the hole and the edge of the 3/4-in gusset plate

$$= \ell_e - (d + 1/16)/2$$
$$= 2 - (3/4 + 1/16)/2$$
$$= 1.59 \text{ in}$$
$$> 2d \text{ ... full bearing capacity possible on end bolt}$$

The bolt pitch is obtained from Fig. 10.7 as

$$s = 3 \text{ in}$$
$$= 4d$$
$$> 3d + 1/16 \text{ in ... full bearing capacity possible on all bolts}$$

Hence, tear out does not govern and a nominal bearing stress of $2F_u$ is applicable. The available bearing capacity of the four bolts on the 3/4-in gusset plate is

| LRFD | ASD |
|---|---|
| The design bearing strength is | The allowable bearing strength is |
| $\varphi R_n = \varphi n \times 2dtF_u$ | $R_n/\Omega = n \times 2dtF_u/\Omega$ |
| $= 0.75 \times 4 \times 2 \times 0.75 \times 0.75 \times 58$ | $= 4 \times 2 \times 0.75 \times 0.75 \times 58/2$ |
| $= 196$ kips ... governs | $= 131$ kips ... governs |
| $< 223$ kips | $< 149$ kips |

## 10.3 Snug-Tight Bolts in Shear and Tension

### Bolts in Tension Only

Common bolts and high-strength Group A bolts installed in the snug-tight condition are permitted to be used in connections subject to tensile forces.

Nominal tensile stress for bolts is based on the unthreaded nominal cross-sectional area of the bolts and is given in AISC 360 Table J3.2 and in Table 10.2.

The available bolt tensile strength is obtained from AISC 360, Sec. J3.6 as

| LRFD | ASD |
|---|---|
| $\varphi_t R_{nt}$ = design tensile strength | $R_{nt}/\Omega_t$ = allowable shear strength |
| $\geq T_u$ | $\geq T_a$ |
| where $\varphi_t$ = resistance factor | where $\Omega_t$ = safety factor |
| $= 0.75$ | $= 2.00$ |
| $T_u$ = required tensile strength using LRFD load combinations | $T_a$ = required tensile strength using ASD load combinations |

The available tensile loads on bolts are tabulated in AISC Manual Table 7-2.

| Type of Bolt | A307 | Group A | Group B |
|---|---|---|---|
| Nominal tensile stress, ksi | 45 | 90 | 113 |

**TABLE 10.2** Bolt Nominal Tensile Stress

## Bolts in Combined Tension and Shear

Common bolts and high-strength Group A bolts installed in the snug-tight condition are permitted to be used in connections subject to combined tension and shear forces.

When a bearing-type connector is subjected to combined shear and tension, AISC 360 Sec. J3.7 specifies that the nominal tensile stress is reduced while the nominal shear stress is unaffected. The required shear stress $f_{rv}$ must not exceed the available shear stress $F_{cv}$.

The available tensile capacity of a bolt subjected to combined shear and tension, is given by AISC 360 Sec. J3.7 as

| LRFD | ASD |
|---|---|
| $\varphi_t R_{nt}$ = design tensile strength | $R_{nt}/\Omega_t$ = allowable shear strength |
| $= \varphi_t F'_{nt} A_b$ | $= F'_{nt} A_b/\Omega_t$ |
| where $F'_{nt}$ = modified nominal tensile stress | where $F'_{nt}$ = modified nominal tensile stress |
| $= 1.3F_{nt} - f_{rv}F_{nt}/\varphi_v F_{nv}$ | $= 1.3F_{nt} - \Omega_v f_{rv} F_{nt}/F_{nv}$ |
| $A_b$ = nominal area of bolt | $A_b$ = nominal area of bolt |
| $R_{nt}$ = nominal tensile capacity, kips | $R_{nt}$ = nominal tensile capacity, kips |
| $\varphi_v$ = resistance factor | $\Omega_v$ = safety factor |
| $= 0.75$ | $= 2.00$ |
| $f_{rv}$ = required shear stress using LRFD load combinations | $f_{rv}$ = required shear stress using ASD load combinations |
| $F_{nt}$ = nominal tensile stress from AISC 360 Table J3.2 | $F_{nt}$ = nominal tensile stress from AISC 360 Table J3.2 |
| $F_{nv}$ = nominal shear stress from AISC 360 Table J3.2 | $F_{nv}$ = nominal shear stress from AISC 360 Table J3.2 |

When either $f_{rv} \le 0.3F_{cv}$ or $f_{rt} \le 0.3F_{ct}$, where $F_c$ is the available stress, the effects of combined stress need not be investigated.

**Example 10.4.** Bolts in Combined Shear and Tension

All bolts in the connection of the WT gusset to the girder flange shown in Fig. 10.8 are grade A325, 3/4-in-diameter snug-tight bolts. The bolts are in standard holes, deformation around the bolt holes is a design consideration, and threads are excluded from the shear planes. The WT gusset and the girder are grade 50 steel. The service load $P$ consists of dead load $P_D$ = 20 kips and live load $P_L$ = 60 kips. Assuming that the angles and the connection of the angles to the WT gusset are satisfactory, determine if the connection of the WT gusset to the girder is adequate to support the load. Prying action may be neglected.

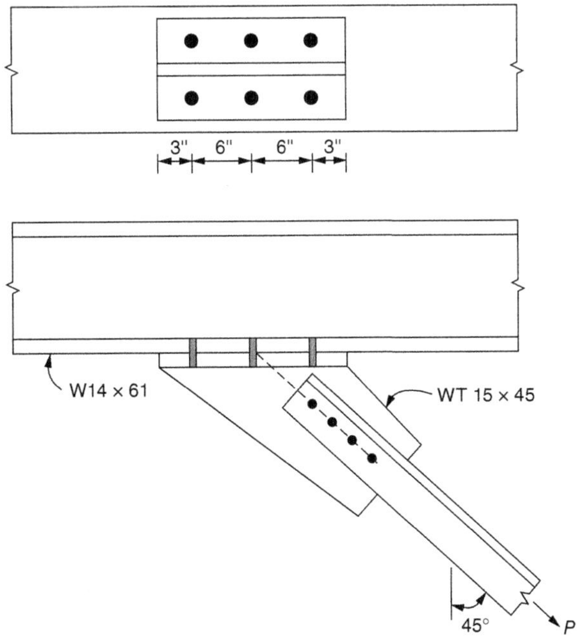


**Figure 10.8** Details for Example 10.4.

Applying American Society of Civil Engineers, *Minimum Design Loads for Buildings and Other Structures* (ASCE 7)[6] Secs. 2.3 and 2.4 gives the factored loads as

| LRFD | ASD |
|---|---|
| From ASCE 7 Sec. 2.3.2 combination 2: | From ASCE 7 Sec. 2.4.1 combination 2: |
| $P_u$ = factored load | $P_a$ = factored load |
| $= 1.2P_D + 1.6P_L$ | $= P_D + P_L$ |
| $= 1.2 \times 20 + 1.6 \times 60$ | $= 20 + 60$ |
| = 120 kips | = 80 kips |
| $V_u$ = factored shear | $V_a$ = factored shear |
| $= 0.707P_u$ | $= 0.707P_a$ |
| = 84.8 kips | = 56.6 kips |
| $T_u$ = factored tension | $T_a$ = factored tension |
| $= 0.707P_u$ | $= 0.707P_a$ |
| = 84.8 kips | = 56.6 kips |

The shear stress on each bolt is

| LRFD | ASD |
|---|---|
| $\varphi_v F_{nv}$ = design shear strength from Table 10.1 | $F_{nv}/\Omega_v$ = allowable shear stress from Table 10.1 |
| $= 0.75 \times 68$ | $= 68/2$ |
| $= 51$ kips | $= 34$ ksi |
| $f_v$ = shear stress on each bolt | $f_v$ = shear stress on each bolt |
| $= V_u/6A_b$ | $= V_a/6A_b$ |
| $= 84.8/(6 \times 0.442)$ | $= 56.6/(6 \times 0.442)$ |
| $= 31.98$ ksi | $= 21.34$ ksi |
| $< \varphi F_{nv}$ ... satisfactory | $< F_{nv}/\Omega$ ... satisfactory |
| $> 0.3\ \varphi F_{nv}$ | $> 0.3F_{nv}/\Omega$ |

The tensile stress on each bolt is

| LRFD | ASD |
|---|---|
| $\varphi_t F_{nt}$ = design tensile stress from Table 10.2 | $F_{nt}/\Omega_t$ = allowable tensile stress from Table 10.2 |
| $= 0.75 \times 90$ | $= 90/2$ |
| $= 67.5$ ksi | $= 45$ ksi |
| $f_t$ = tensile stress on each bolt | $f_t$ = tensile stress on each bolt |
| $= T_u/6A_b$ | $= T_a/6A_b$ |
| $= 84.8/(6 \times 0.442)$ | $= 56.6/(6 \times 0.442)$ |
| $= 31.98$ ksi | $= 21.34$ ksi |
| $< \varphi_t F_{nt}$ ... satisfactory | $< F_{nt}/\Omega_t$ ... satisfactory |
| $> 0.3\varphi_t F_{nt}$ | $> 0.3F_{nt}/\Omega_t$ |

Hence, it is necessary to investigate the effects of the combined shear and tensile stress. The modified nominal tensile stress $F'_{nt}$ of a bolt, subjected to combined shear and tension, is given by AISC 360 Eq. (J3-3) as

| LRFD | ASD |
|---|---|
| $F'_{nt} = 1.3F_{nt} - f_v F_{nt}/\varphi_v F_{nv}$ | $F'_{nt} = 1.3F_{nt} - \Omega_v f_v F_{nt}/F_{nv}$ |
| $= 1.3 \times 90 - 31.98 \times 90/51$ | $= 1.3 \times 90 - 2 \times 21.34 \times 90/68$ |
| $= 60.6$ ksi | $= 60.5$ ksi |
| $\varphi_t F'_{nt}$ = design tensile stress | $F'_{nt}/\Omega_t$ = allowable tensile stress |
| $= 0.75 \times 60.6$ | $= 60.5/2$ |
| $= 45.5$ ksi | $= 30.3$ ksi |
| $> f_t$ ... satisfactory | $> f_t$ ... satisfactory |

The bolt pitch is obtained from Fig. 10.8 as

$$s = 6 \text{ in}$$
$$= 8d_b$$
$$> 3.0d_b + 1/16 \text{ in ... full bearing capacity is possible}$$

Hence, tear out does not govern and a nominal bearing stress of $2.4F_u$ is applicable. The available bearing capacity of the six bolts on the 0.61-in-thick flange of the WT gusset is

| LRFD | ASD |
|---|---|
| The design bearing strength is | The allowable bearing strength is |
| $\varphi R_n = \varphi n \times 2.4dtF_u$ | $R_n/\Omega = n \times 2.4dtF_u/\Omega$ |
| $= 0.75 \times 6 \times 2.4 \times 0.75 \times 0.61 \times 65$ | $= 6 \times 2.4 \times 0.75 \times 0.61 \times 65/2$ |
| $= 321$ kips ... bearing does not govern | $= 214$ kips ... bearing does not govern |
| $> V_u$ | $> V_a$ |

The connection is adequate.

## 10.4 Slip-Critical Bolts in Shear and Tension

### Bolts in Shear Only

Slip-critical bolts are high strength Group A or Group B bolts, which are pretensioned to the values specified in AISC 360 Table J3.1 and in Table 10.3. The pretension is intended to produce a sufficiently high clamping force between the parts to transfer the shear load from one connected part to another. Undesirable slip at service loads is precluded and the shear force between the connected plates is transferred by friction. At the strength limit state, the connection may slip sufficiently to place the bolts in bearing. Hence, in accordance with AISC 360 Sec. J3.10, slip-critical connections must also comply with the design requirements of snug-tight connections.

Connections with standard holes or short-slotted holes perpendicular to the direction of the load may slip a maximum of 1/16 in before the bolts bear against the sides of the holes. This amount of slip will not result in unacceptable deformations in the structure. Hence, a smaller safety factor is acceptable in this situation. However, as indicated by Schlafly and Muir[7] connections with long-slotted holes parallel to the direction of the load may slip sufficiently to result in failure of the structure. An example of this is the possible slip in a splice in a flat roof truss causing excessive deflection and failure by ponding. A larger factor of safety is necessary in this situation. An intermediate case is a connection with oversized holes or short-slotted holes parallel to the direction of the load. For all three cases, the available strength is determined using the safety factor or resistance factor appropriate to each case.

| Bolt Size | Group A | Group B |
|---|---|---|
| 3/4 in | 28 kips | 35 kips |
| 7/8 in | 39 kips | 49 kips |
| 1 in | 51 kips | 64 kips |

**TABLE 10.3** Bolt Pretension, $T_b$

The frictional resistance developed in a slip-critical connection is dependant on the condition of the faying surfaces. Values of the mean slip coefficient μ for two types of surface condition are given in AISC 360 Sec. J3.8. The two conditions are defined as

- Class A surface conditions—unpainted clean mill scale surfaces or blast-cleaned surfaces with class A coatings
- Class B surface conditions—unpainted blast-cleaned surfaces or blast-cleaned surfaces with class B coatings

Surface coatings are qualified as Class A or Class B in accordance with the test procedure given in RCSC App. A.

A slip-critical connection using a single filler plate is shown in Fig. 10.9. Connections with multiple filler plates have a reduced slip resistance and the filler factor $h_f$ compensates for this. The surface of fills must be prepared to the same or higher slip coefficient than the other faying surfaces in the connection.

Because of the variability of pretension achieved in a bolt, the parameter $D_u$ derived from statistical analysis is used to reflect the possible reduced slip resistance.

The nominal slip resistance is given by AISC 360 Eq. (J3-4) as

$$R_n = \mu D_u h_f T_b n_s$$

where μ = mean slip coefficient for the applicable surfaces
= 0.30 ... for a Class A surface
= 0.50 ... for a Class B surface
$D_u$ = a multiplier that reflects the ratio of the mean installed bolt tension to the specified minimum bolt pretension
= 1.13
$h_f$ = modification factor for fillers
= 1.00 ... where bolts are added to distribute loads in the filler
= 1.00 ... for one filler between connected parts
= 0.85 ... for two or more fillers between connected parts
$T_b$ = minimum bolt pretension given in AISC 360 Table J3.1
$n_s$ = number of slip planes required to permit the connection to slip

The available slip resistance is determined from

- For standard size and short-slotted holes perpendicular to the direction of the load:

$$\varphi = 1.00 \text{ (LRFD)} \qquad \Omega = 1.50 \text{ (ASD)}$$

- For oversized and short-slotted holes parallel to the direction of the load:

$$\varphi = 0.85 \text{ (LRFD)} \qquad \Omega = 1.76 \text{ (ASD)}$$

- For long-slotted holes:

$$\varphi = 0.70 \text{ (LRFD)} \qquad \Omega = 2.14 \text{ (ASD)}$$

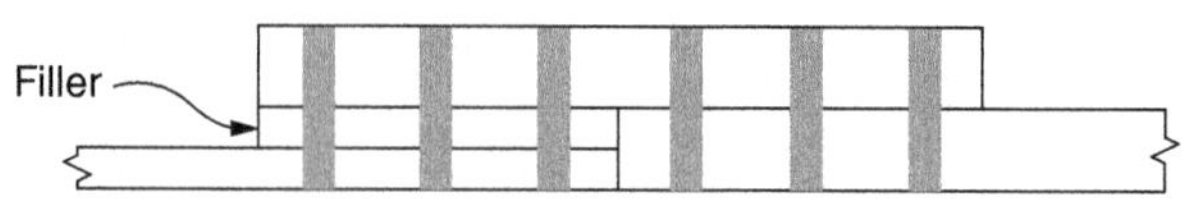


**FIGURE 10.9** Connection with single filler plate.

Values of available slip-critical shear strength for Group A and Group B bolts with a Class A faying surface are given in AISC Manual Table 7-3. These values are multiplied by 1.67 for a Class B faying surface.

**Example 10.5.** Slip-Critical Bolts in Shear
The slip-critical connection shown in Fig. 10.10 consists of four grade A490 3/4-in-diameter bolts, in standard holes, with threads excluded from the shear planes. The angles and gusset plate have Class A faying surfaces and are fabricated from steel with a tensile strength of $F_u = 65$ ksi. Assuming that the angles and gusset plate are satisfactory, determine the shear force that may be applied to the bolts in the connection.

For a connection with a Class A faying surface, the mean slip coefficient is

$$\mu = 0.30$$

For a connection without fillers, the modification factor is

$$h_f = 1$$

The multiplier that reflects the ratio of the mean installed bolt tension to the specified minimum bolt pretension is

$$D_u = 1.13$$

The minimum bolt pretension is given in Table 10.3 as

$$T_b = 35 \text{ kips}$$

The number of slip planes is

$$n_s = 2$$

The nominal slip resistance for one bolt is given by AISC 360 Eq. (J3-4) as

$$\begin{aligned} R_n &= \mu D_u h_f T_b n_s \\ &= 0.30 \times 1.13 \times 1.00 \times 35 \times 2 \\ &= 23.7 \text{ kips} \end{aligned}$$

Alternatively, from AISC Manual Table 7-3, $R_n = 23.7$ kips

The nominal slip resistance for 4 bolts is

$$\begin{aligned} R_n &= 4 \times 23.7 \\ &= 94.8 \text{ kips} \end{aligned}$$

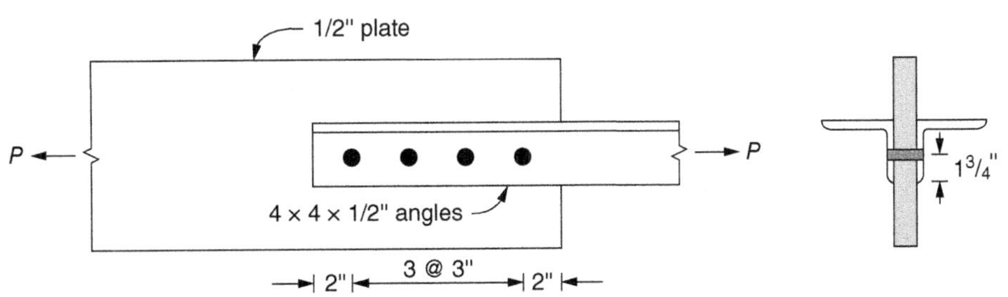


**Figure 10.10** Details for Example 10.5.

For a connection with standard holes, the available slip resistance is

| LRFD | ASD |
|---|---|
| The design slip resistance is | The allowable slip resistance is |
| $\varphi R_n = 1.00 \times 94.8$ | $R_{nv}/\Omega_v = 94.8/1.5$ |
| $= 94.8$ kips | $= 63.2$ kips |

Should the connectors slip into bearing, the 1/2-in-thick gusset plate is the critical part, and the available bearing capacity of the four bolts on the gusset plate is

| LRFD | ASD |
|---|---|
| The design bearing strength is | The allowable bearing strength is |
| $\varphi R_n = \varphi n \times 2.4dtF_u$ | $R_n/\Omega = n \times 2.4dtF_u/\Omega$ |
| $= 0.75 \times 4 \times 2.4 \times 0.75 \times 0.50 \times 65$ | $= 4 \times 2.4 \times 0.75 \times 0.50 \times 65/2$ |
| $= 176$ kips | $= 117$ kips |
| $>94.8$ kips ... satisfactory | $>63.2$ kips ... satisfactory |

## Bolts in Combined Shear and Tension

Nominal tensile stress for a bolt is based on the unthreaded nominal cross-sectional area of the bolt and is given in AISC 360 Table J3.2 and in Table 10.2. The available tensile stress is independent of the pretension in the bolt. The required tensile load on a bolt is the sum of the external applied load and any tension resulting from prying action due to the deformation of the connected parts.

When a slip-critical connection is subjected to combined shear and direct tension, as shown in Fig. 10.11, the allowable tensile force in a connector is unaffected. However, the clamping force on the faying surface is reduced and this results in a reduction in the slip resistance. Hence, the available shear capacity must be reduced in proportion to the loss of pretension.

In accordance with AISC 360 Sec. J3.9, the available slip resistance given in AISC 360 Sec. J3.8 is multiplied by the reduction factor:

$$k_{sc} = 1 - T_r/D_uT_bn_b \text{ ... for LRFD load combinations}$$

$$k_{sc} = 1 - 1.5T_r/D_uT_bn_b \text{ ... for ASD load combinations}$$

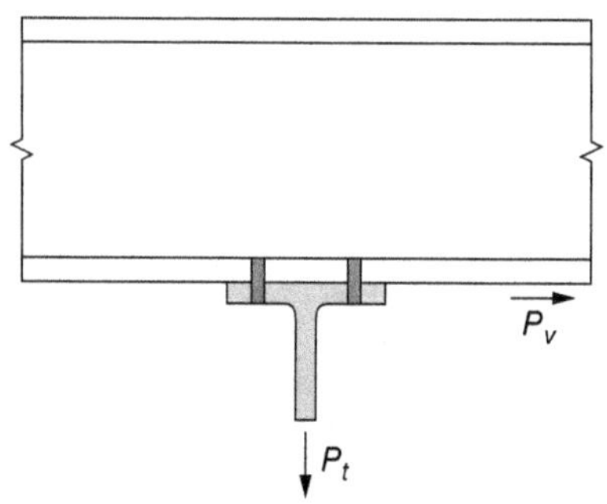


**FIGURE 10.11** Shear and tension on a slip critical connection.

where $T_r$ = required tensile force as appropriate for ASD or LRFD load combinations
$T_b$ = pretension force specified in AISC 360 Table J3.1
$n_b$ = number of bolts carrying the applied tension

However, when the tensile force in the connector is due to a bending moment applied in a plane perpendicular to the faying surface, as shown in Fig. 10.6, no reduction is necessary since the increase in compressive force at the bottom of the connection compensates for the tension produced in the connectors at the top of the connection.

**Example 10.6.** Slip-critical Bolts in Combined Shear and Tension
All bolts in the connection of the WT gusset to the girder flange shown in Fig. 10.8 are grade A325 3/4-in-diameter slip-critical bolts in standard holes with threads excluded from the shear planes. The WT gusset and the girder are grade 50 steel with a Class A faying surface. The service load $P$ consists of dead load $P_D$ = 9 kips and live load $P_L$ = 27 kips. Assuming that the angles and the connection of the angles to the WT gusset are satisfactory, determine if the connection of the WT gusset to the girder is adequate to support the load. Prying action may be neglected.

From ASCE 7 Secs. 2.3 and 2.4 the factored loads are

| LRFD | ASD |
|---|---|
| From ASCE 7 Sec. 2.3.2 combination 2: | From ASCE 7 Sec. 2.4.1 combination 2: |
| $P_u$ = factored load | $P_a$ = factored load |
| $= 1.2P_D + 1.6P_L$ | $= P_D + P_L$ |
| = 54 kips | = 36 kips |
| $V_u$ = factored shear | $V_a$ = factored shear |
| $= 0.707P_u/6$ | $= 0.707P_a/6$ |
| = 6.36 kips/bolt | = 4.24 kips/bolt |
| $T_u$ = factored tension | $T_a$ = factored tension |
| $= 0.707P_u/6$ | $= 0.707P_a/6$ |
| = 6.36 kips/bolt | = 4.24 kips/bolt |

Check the tensile stress on each bolt:

| LRFD | ASD |
|---|---|
| $\varphi_t F_{nt}$ = design tensile stress from Table 10.2 | $F_{nt}/\Omega_t$ = allowable tensile stress from Table 10.2 |
| $= 0.75 \times 90$ | $= 90/2$ |
| = 67.5 ksi | = 45 ksi |
| $f_t$ = required tensile stress on each bolt | $f_t$ = required tensile stress on each bolt |
| $= T_u/A_b$ | $= T_a/A_b$ |
| = 6.36/0.442 | = 4.24/0.442 |
| = 14.39 ksi | = 9.60 ksi |
| $< \varphi_t F_{nt}$ ... satisfactory in tension | $< F_{nt}/\Omega_t$ ... satisfactory in tension |

The combined tension and shear coefficient is given by AISC 360 Eq. (J3-5) as

| LRFD | ASD |
|---|---|
| $k_{sc} = 1 - T_u/D_u T_b n_b$ | $k_{sc} = 1 - 1.5T_a/D_u T_b n_b$ |
| $= 1 - 6.36/(1.13 \times 28 \times 1.0)$ | $= 1 - 1.5 \times 4.24/(1.13 \times 28 \times 1.0)$ |
| = 0.80 | = 0.80 |

The reduced nominal slip resistance for one bolt is

$$k_{sc}R_n = k_{sc}\mu D_u h_f T_b n_s$$
$$= 0.80 \times 0.30 \times 1.13 \times 1 \times 28 \times 1$$
$$= 7.59 \text{ kips/bolt}$$

For a connection with standard holes, the available slip resistance is

| LRFD | ASD |
|---|---|
| The design slip resistance is | The allowable slip resistance is |
| $\varphi R_n = 1.00 \times 7.59$ | $R_n/\Omega = 7.59/1.5$ |
| $= 7.59$ kips/bolt | $= 5.06$ kips/bolt |
| $> T_u$ ... satisfactory | $> T_a$ ... satisfactory |

Should the connectors slip into bearing, the 0.61 in thick flange of the WT gusset is the critical part, and the available bolt bearing capacity on the gusset plate is obtained from Example 10.4 as

| LRFD | ASD |
|---|---|
| The design bearing strength is | The allowable bearing strength is |
| $\varphi R_n = \varphi n \times 2.4dtF_u$ | $R_n/\Omega = n \times 2.4dtF_u/\Omega$ |
| $= 0.75 \times 1 \times 2.4 \times 0.75 \times 0.61 \times 65$ | $= 1 \times 2.4 \times 0.75 \times 0.61 \times 65/2$ |
| $= 53.5$ kips/bolt ... satisfactory | $= 35.7$ kips/bolt ... satisfactory |
| $> V_u$ | $> V_a$ |

The connection is adequate.

## 10.5 Prying Action

The double angle hanger, shown in Fig. 10.12, is fabricated from angles that are relatively rigid and are not subjected to distortion. The forces acting on the connection are as shown, where $2T$ is the required strength of the connection. For an applied load on each angle of $T$, the force in each bolt, exclusive of initial tightening and prying force, is $T/n$, where $n$ = number of bolts in each angle.

The double angle hanger, shown in Fig. 10.13, is fabricated from angles that are relatively flexible and these are distorted as shown, producing the prying force $Q$ at the tips of the angles. This produces bending moments $M_1$ and $M_2$ in the angle legs and increases the force in a bolt to

$$B_r = (T + Q)/n$$

The bolt must be designed for the total force $B_r$. AISC Manual Part 15 Table 15-2 provides a preliminary selection table for hanger type connections and a detailed design technique is presented in AISC Manual Part 9.

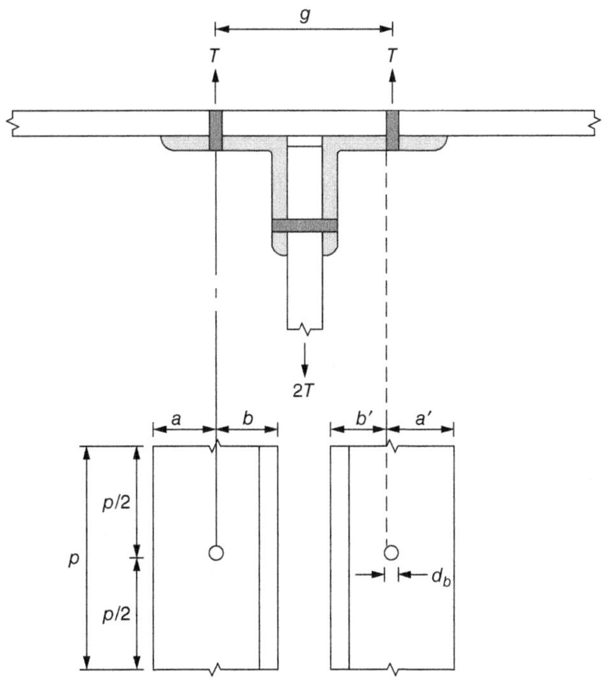


FIGURE 10.12 Double angle hanger without prying action.

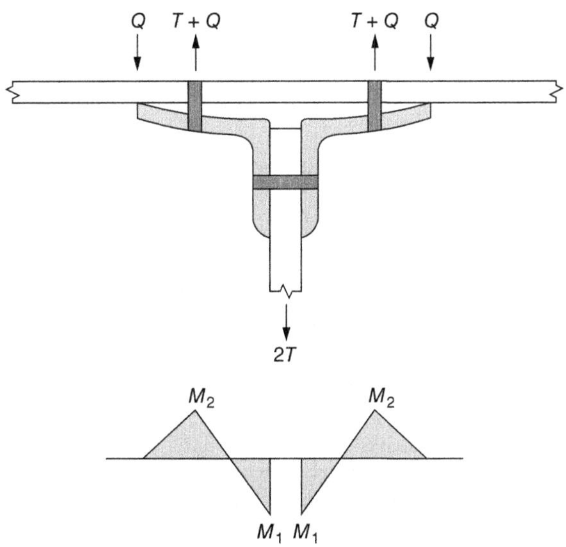


FIGURE 10.13 Double angle hanger with prying action.

The design procedure begins with estimating the number of bolts required and choosing an angle with a suitable leg thickness $t$. Locating the bolts on the angle gage enables the dimension $b$ shown in Fig. 10.12, to be determined and from AISC Manual Part 15 Table 15-2 the estimated allowable load $q$ in kips/in of angle leg is obtained. The required length of angle leg tributary to each bolt is then

$$p \approx T/q \leq 2b \leq \text{bolt spacing, } s$$

The length of each angle required is

$$\begin{aligned} L &= np \\ &= nT/q \end{aligned}$$

Also, from Fig. 10.12, the distance, $a$, from bolt centerline to edge of angle leg may be obtained and

$$a \leq 1.25b$$

Using the notation in AISC Manual Chap. 9, required angle leg thickness may now be determined from AISC Manual Eq. (9-23) as

$$t_{req} = [\eta Tb'/pF_u(1 + \delta\alpha')]^{0.5}$$

where $\eta = 4/\varphi = 4.44$ ... for LRFD load combinations with $\varphi = 0.90$
$= 4\Omega = 6.68$ ... for ASD load combinations with $\Omega = 1.67$
$b' = b - d_b/2$
$d_b$ = bolt diameter
$\delta$ = net area at bolt line/gross area at face of angle leg
$= 1 - d'/p$
$d'$ = width of bolt hole along length of fitting
$\alpha'$ = value of $\alpha$ for which $t_{req}$ is a minimum
$= 1$ ... for $\beta \geq 1$
or $\alpha' = \beta/\delta(1 - \beta)$ ... for $\beta < 1$
$\leq 1$
$\beta = (B/T - 1)/\rho$
$B$ = available tension per bolt, $\varphi r_n$ or $r_n/\Omega$, kips
$\rho = b'/a'$
$a' = a + d_b/2 \leq 1.25b + d_b/2$

The thickness of the angle leg to eliminate prying action may be determined from AISC Manual Eq. (9-20)

$$t_{min} = (\eta Tb'/pF_u)^{0.5}$$

**Example 10.7.** Prying Action in Bolted Connection
The four bolts in the WT section hanger connection, shown in Fig. 10.14, are grade A325, 3/4-in-diameter bolts in standard holes. The WT hanger is grade 50 steel. The service load $T$ consists of dead load $P_D$ = 8 kips and live load $P_L$ = 24 kips. Assuming that the angles and the connection of the angles to the WT hanger are satisfactory, determine if the connection of the WT hanger to the girder is adequate to support the load, allowing for prying action.

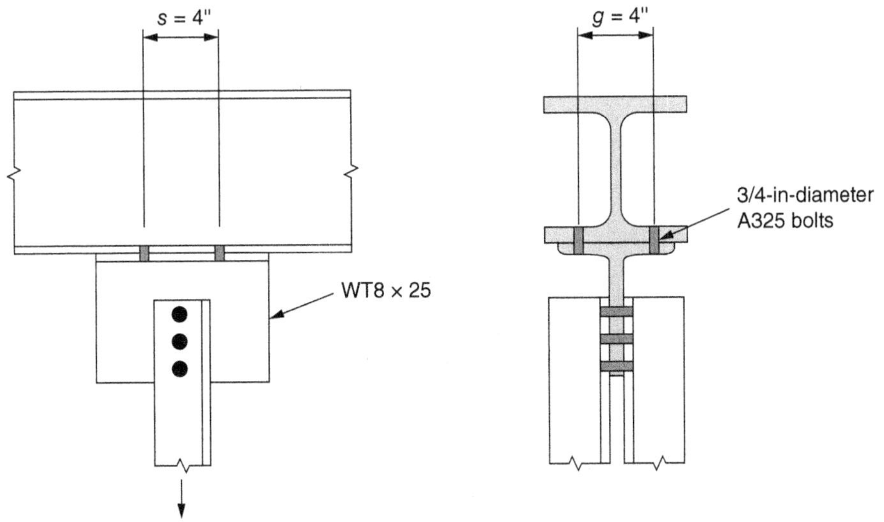


**FIGURE 10.14** Details for Example 10.7.

Applying ASCE 7 Secs. 2.3 and 2.4 gives the factored loads as

| LRFD | ASD |
|---|---|
| From ASCE 7 Sec. 2.3.2 combination 2: | From ASCE 7 Sec. 2.4.1 combination 2: |
| $P$ = factored load<br>$= 1.2P_D + 1.6P_L$<br>$= 1.2 \times 8 + 1.6 \times 24$<br>$= 48$ kips | $P$ = factored load<br>$= P_D + P_L$<br>$= 8 + 24$<br>$= 32$ kips |
| Load on one bolt neglecting prying force is | Load on one bolt neglecting prying force is |
| $T = P/4$<br>$= 12$ kips | $T = P/4$<br>$= 8$ kips |
| From AISC Manual Table 7-2 the available tensile load on one bolt is | From AISC Manual Table 7-2 the available tensile load on one bolt is |
| $B = 29.8$ kips<br>$> T$ ... without prying effects bolt is satisfactory | $B = 19.9$ kips<br>$> T$ ... without prying effects bolt is satisfactory |

For the WT8 × 25 hanger

$$t_f = 0.63 \text{ in}$$

$$t_w = 0.38 \text{ in}$$

Hence,

$$b = (g - t_w)/2$$
$$= (4 - 0.38)/2$$
$$= 1.81 \text{ in}$$
$$b' = b - d_b/2$$
$$= 1.81 - 0.75/2$$
$$= 1.44$$
$$p = 2b = 2 \times 1.81 = 3.62 \text{ in} < s \text{ ... satisfactory}$$

The required thickness of the flange to eliminate prying action is given by AISC Manual Eq. (9-20) as

| LRFD | ASD |
|---|---|
| $t_{min} = (\eta Tb'/pF_u)^{0.5}$<br>$= [4.44 \times 12 \times 1.44/(3.62 \times 65)]^{0.5}$<br>$= 0.57$ in<br>$< t_f$ ... satisfactory | $t_{min} = (\eta Tb'/pF_u)^{0.5}$<br>$= [6.68 \times 8 \times 1.44/(3.62 \times 65)]^{0.5}$<br>$= 0.57$ in<br>$< t_f$ ... satisfactory |

The connection is adequate.

## 10.6 Bolt Group Eccentrically Loaded in Plane of Faying Surface

### Elastic Unit Area Method

Eccentrically loaded bolt groups of the type shown in Fig. 10.15 may be conservatively designed by means of the elastic unit area method.

The moment of inertia of the bolt group about the $x$-axis is

$$I_x = \Sigma y^2$$

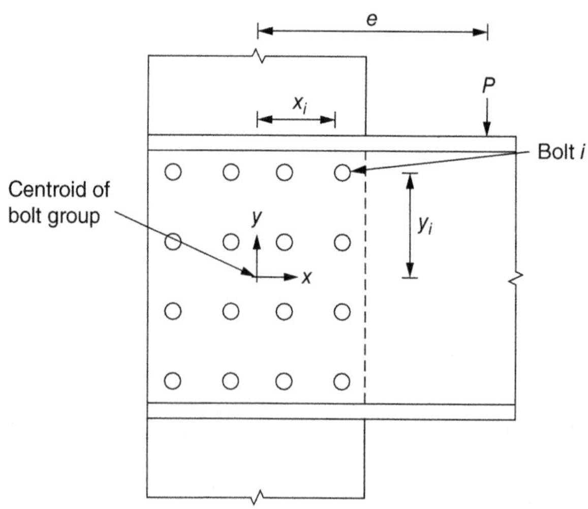


**FIGURE 10.15** Eccentrically loaded bolt group.

The moment of inertia of the bolt group about the $y$-axis is

$$I_y = \Sigma x^2$$

The polar moment of inertia of the bolt group about the centroid is

$$I_o = I_x + I_y$$

The vertical force on bolt i due to the applied load $P$ is

$$V_P = P/n$$

The vertical force on bolt $i$ due to the eccentricity $e$ is

$$V_e = Pex_i/I_o$$

The horizontal force on bolt $i$ due to the eccentricity $e$ is

$$H_e = Pey_i/I_o$$

The resultant force on bolt $i$ is

$$R = [(V_P + V_e)^2 + (H_e)^2]^{0.5}$$

The elastic unit area method may be readily applied to unusual as well as common bolt groups.

**Example 10.8.** Bolt Group Loaded in Plane of Faying Surface: Elastic Method
Determine the adequacy of the 3/4-in-diameter A325 bolts, in standard holes, in the bolted bracket shown in Fig. 10.16. The bolts are snug-tight and threads are excluded from the shear planes. The W-shape is grade 50 steel and the channel is grade A36 steel. The service load $P$ consists of dead load $P_D = 4$ kips and live load $P_L = 12$ kips. Use the elastic unit area method.

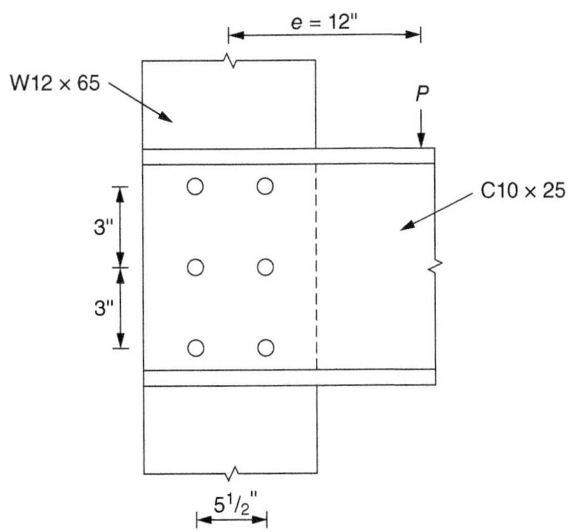


**FIGURE 10.16** Details for Example 10.8.

Applying ASCE 7 Secs. 2.3 and 2.4 gives the factored loads as

| LRFD | ASD |
|---|---|
| From ASCE 7 Sec. 2.3.2 combination 2: | From ASCE 7 Sec. 2.4.1 combination 2: |
| $P_u$ = factored load | $P_a$ = factored load |
| $= 1.2P_D + 1.6P_L$ | $= P_D + P_L$ |
| $= 1.2 \times 4 + 1.6 \times 12$ | $= 4 + 12$ |
| = 24 kips | = 16 kips |

The geometrical properties of the bolt group are obtained by applying the unit area method. The moment of inertia about the $x$-axis is

$$\begin{aligned} I_x &= \Sigma y^2 \\ &= 4 \times 3^2 \\ &= 36 \text{ in}^4/\text{in}^2 \end{aligned}$$

The moment of inertia about the $y$-axis is

$$\begin{aligned} I_y &= \Sigma x^2 \\ &= 6 \times 2.75^2 \\ &= 45.38 \text{ in}^4/\text{in}^2 \end{aligned}$$

The polar moment of inertia about the centroid is

$$\begin{aligned} I_o &= I_x + I_y \\ &= 81.38 \text{ in}^4/\text{in}^2 \end{aligned}$$

The top right bolt is the most heavily loaded and the co-existent forces on this bolt are

| LRFD | ASD |
|---|---|
| $V_p$ = Vertical force due to applied load | $V_p$ = Vertical force due to applied load |
| $= P_u/n$ | $= P/\text{n}$ |
| $= 24/6$ | $= 16/6$ |
| = 4 kips | = 2.67 kips |
| $V_e$ = Vertical force due to eccentricity | $V_e$ = Vertical force due to eccentricity |
| $= P_u e x_i / I_o$ | $= Pex_i/I_o$ |
| $= 24 \times 12 \times 2.75/81.38$ | $= 16 \times 12 \times 2.75/81.38$ |
| = 9.73 kips | = 6.49 kips |
| $H_e$ = Horizontal force due to eccentricity | $H_e$ = Horizontal force due to eccentricity |
| $= P_u e y_i / I_o$ | $= Pey_i/I_o$ |
| $= 24 \times 12 \times 3/81.38$ | $= 16 \times 12 \times 3/81.38$ |
| = 10.62 kips | = 7.08 kips |
| $R$ = resultant force | $R$ = resultant force |
| $= [(V_p + V_e)^2 + (H_e)^2]^{0.5}$ | $= [(V_p + V_e)^2 + (H_e)^2]^{0.5}$ |
| $= [(4 + 9.73)^2 + 10.62^2]^{0.5}$ | $= [(2.67 + 6.49)^2 + 7.08^2]^{0.5}$ |
| = 17.36 kips | = 11.58 kips |

Bolt shear controls and from AISC 360 Table J3.2, the available shear strength of a 3/4-in-diameter A325X bolt in a standard hole in single shear is

| LRFD | ASD |
|---|---|
| The design shear strength is<br>$\varphi_v R_{nv} = 0.75 \times 68 \times 0.442$<br>$= 22.54$<br>$> 17.36$ kips ... satisfactory | The allowable shear strength is<br>$R_{nv}/\Omega_v = 68 \times 0.442/2$<br>$= 15.03$ kips<br>$> 11.58$ kips ... satisfactory |

## Instantaneous Center of Rotation Method

The instantaneous center of rotation method of analyzing eccentrically loaded bolt groups affords a more realistic estimate of a bolt group's capacity. The eccentrically loaded member is assumed to rotate and translate as a rigid body and this may be reduced to a pure rotation about a point designated the instantaneous center of rotation. AISC Manual Part 7 details the method.

As shown in Fig. 10.17, the bolt that is farthest from the instantaneous center will experience the greatest deformation, and will be the first bolt to fail, as the applied load is increased. For 3/4-in-diameter A325 bolts, the maximum deformation at failure and the corresponding ultimate shear force have been obtained experimentally as

$$\Delta_{max} = 0.34 \text{ in}$$
$$R_{ult} = 74 \text{ kips}$$

The deformations in the remaining bolts in the group are proportional to their distances from the instantaneous center and the shear force in bolt i is given by AISC Manual Eq. (7-1)

$$R_i = R_{ult}(1 - e^{-10\Delta})^{0.55}$$

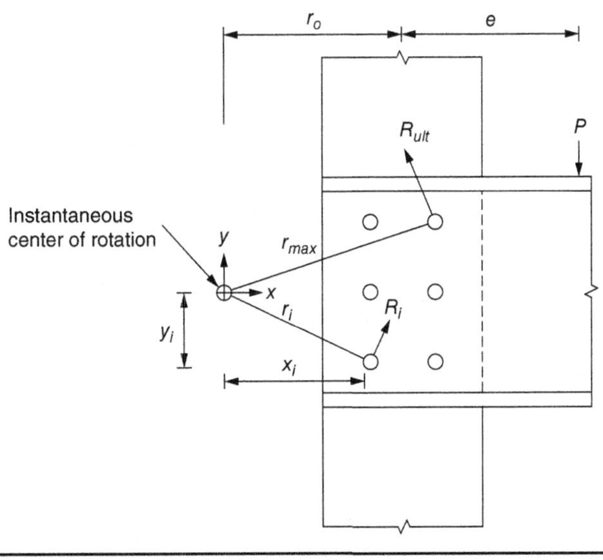


**FIGURE 10.17** Instantaneous center of rotation method.

where $\Delta$ = deformation in bolt $i$
$= \Delta_{max} r_i / r_{max}$
$e$ = base of natural logarithm
$r_i$ = distance of bolt $i$ from the instantaneous center of rotation
$r_{max}$ = distance of farthest bolt from the instantaneous center of rotation

The method is applied by assuming an initial location for the instantaneous center of rotation and determining the shear force in each bolt. Then, equating horizontal forces provides the expression

$$\Sigma R_i y_i / r_i = 0$$

Equating vertical forces gives

$$\Sigma R_i x_i / r_i = P$$

Equating moments about the instantaneous center of rotation gives

$$\Sigma R_i r_i = P(e + r_0)$$

The correct location of the instantaneous center of rotation is determined when these three equations of equilibrium are satisfied.

AISC Manual Tables 7-6 to 7-13 provide a convenient means of designing common bolt group patterns by this method. For a particular bolt geometry, the coefficient $C$ is obtained from the tables where

$$C = P_r / R_c$$

and $P_r$ = externally applied eccentric load

$R_c$ = available strength of one bolt

The method may be applied to bearing-type and slip-critical bolts, of any diameter, in single or double shear connections.

**Example 10.9.** Bolt Group Loaded in Plane of Faying Surface: Instantaneous Center Method
Determine the adequacy of the 3/4-in-diameter A325 bolts, in standard holes, in the bolted bracket shown in Fig. 10.16. The bolts are snug-tight and threads are excluded from the shear planes. The W-shape is grade 50 steel and the channel is grade A36 steel. The service load $P$ consists of dead load $P_D = 4$ kips and live load $P_L = 12$ kips. Use the instantaneous center method.

From Example 10.8 the factored loads and the available strength of one bolt are

| LRFD | ASD |
|---|---|
| $P_u$ = factored load | $P_a$ = factored load |
| = 24 kips | = 16 kips |
| $\varphi_v R_{nv}$ = design shear strength of one bolt | $R_{nv}/\Omega_v$ = allowable shear strength of one bolt |
| = 22.54 kips | = 15.03 kips |

From Fig. 10.16

$n$ = number of bolts in one vertical row
= 3

$s$ = bolt pitch
= 3 in

$D$ = lateral distance between bolts
= 5½ in

$e$ = eccentricity
= 12 in

From AISC Manual Table 7-8 for $\theta = 0°$ the coefficient $C$ is given as

$$C = 1.55$$

The allowable factored load on the connection is

| LRFD | ASD |
|---|---|
| $P_u = C\varphi_v R_{nv}$ | $P_a = CR_{nv}/\Omega_v$ |
| $= 1.55 \times 22.54$ | $= 1.55 \times 15.03$ |
| $= 34.9$ kips | $= 23.3$ |
| $>24$ kips ... satisfactory | $>16$ kips ... satisfactory |

## 10.7 Bolt Group Eccentrically Loaded Normal to the Faying Surface

Eccentrically loaded bolt groups of the type shown in Fig. 10.18 may be conservatively designed by assuming that the neutral axis is located at the centroid of the bolt group and that a plastic stress distribution is produced in the bolts. The tensile force in each

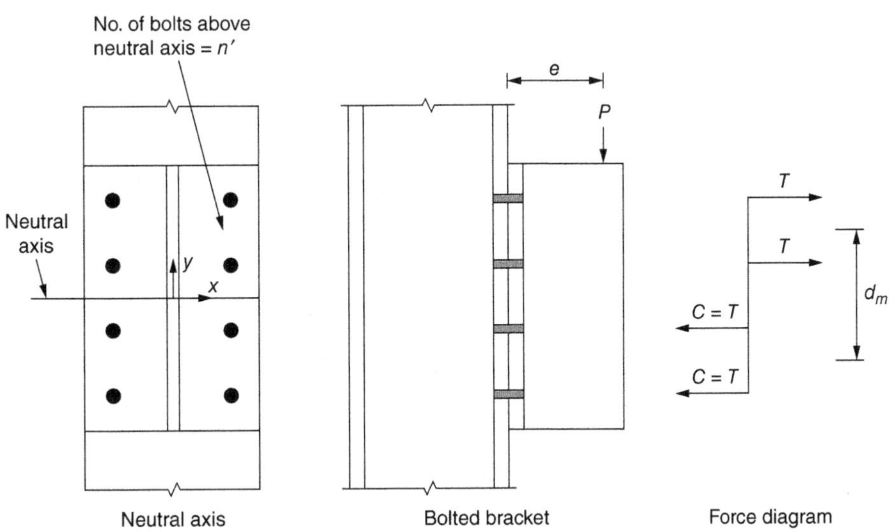


**FIGURE 10.18** Bolt group eccentrically loaded normal to the faying surface.

bolt above the neutral axis due to the eccentricity of the factored load about the bolt group is given by AISC Manual Eq. (7-14) as

$$T = Pe/n'd_m$$

where $P$ = applied load
$e$ = eccentricity of applied load about the bolt group
$n'$ = number of bolts above the neutral axis
$d_m$ = moment arm between resultant tensile and compressive forces in the bolts

The shear force in each bolt due to the factored applied load is given by AISC Manual Eq. (7-13) as

$$V_r = P/n$$

Bolts above the neutral axis are subject to shear force, tensile force, and the effects of prying action. Bolts below the neutral axis are subject to shear force only.

**Example 10.10.** Bolt Group Loaded Normal to Faying Surface
Determine the adequacy of the 3/4-in-diameter A325 bolts, in standard holes, in the bolted bracket shown in Fig. 10.19. The bolts are snug-tight and threads are not excluded from the shear planes. The service load P consists of dead load $P_D$ = 12 kips and live load $P_L$ = 36 kips. Prying action may be neglected.

Applying ASCE 7 Secs. 2.3 and 2.4 gives the factored loads as

| LRFD | ASD |
|---|---|
| From ASCE 7 Sec. 2.3.2 combination 2: | From ASCE 7 Sec. 2.4.1 combination 2: |
| $P_u$ = factored load | $P_a$ = factored load |
| $= 1.2P_D + 1.6P_L$ | $= P_D + P_L$ |
| $= 1.2 \times 12 + 1.6 \times 36$ | $= 12 + 36$ |
| = 72 kips | = 48 kips |

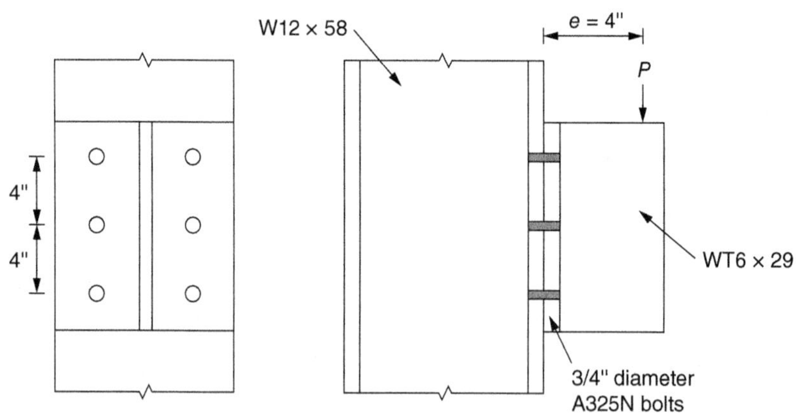


**Figure 10.19** Details for Example 10.10.

Assuming the neutral axis occurs at the centroid of the bolt group, the forces on the bolts are

| LRFD | ASD |
|---|---|
| $M_u$ = applied moment on the bolt group | $M_a$ = applied moment on the bolt group |
| $= P_u e$ | $= P_a e$ |
| $= 72 \times 4$ | $= 48 \times 4$ |
| $= 288$ kip-in | $= 192$ kip-in |
| $T_u$ = tensile force on each top bolt | $T_a$ = tensile force on each top bolt |
| $= P_u e / n' d_m$ | $= P_a e / n' d_m$ |
| $= 288/(2 \times 8)$ | $= 192/(2 \times 8)$ |
| $= 18$ kips | $= 12$ kips |
| $V_u$ = shear force on each bolt | $V_a$ = shear force on each bolt |
| $= P_u/6$ | $= P_a/6$ |
| $= 72/6$ | $= 48/6$ |
| $= 12$ kips | $= 8$ kips |

Bearing is not critical and the shear stress on each grade A325N bolt is

| LRFD | ASD |
|---|---|
| $\varphi_v F_{nv}$ = design shear strength from Table 10.1 | $F_{nv}/\Omega_v$ = allowable shear stress from Table 10.1 |
| $= 0.75 \times 54$ | $= 54/2$ |
| $= 40.5$ kips | $= 27$ ksi |
| $f_v$ = shear stress on each bolt | $f_v$ = shear stress on each bolt |
| $= V_u/A_b$ | $= V_a/A_b$ |
| $= 12/0.442$ | $= 8/0.442$ |
| $= 27.2$ ksi | $= 18.1$ ksi |
| $< \varphi F_{nv}$ ... satisfactory | $< F_{nv}/\Omega$ ... satisfactory |
| $> 0.3\varphi F_{nv}$ | $> 0.3F_{nv}/\Omega$ |

The tensile stress on each top bolt is

| LRFD | ASD |
|---|---|
| $\varphi_t F_{nt}$ = design tensile stress from Table 10.2 | $F_{nt}/\Omega_t$ = allowable tensile stress from Table 10.2 |
| $= 0.75 \times 90$ | $= 90/2$ |
| $= 67.5$ ksi | $= 45$ ksi |
| $f_t$ = tensile stress on each bolt | $f_t$ = tensile stress on each bolt |
| $= T_u/A_b$ | $= T_a/A_b$ |
| $= 18/0.442$ | $= 12/0.442$ |
| $= 40.7$ ksi | $= 27.1$ ksi |
| $< \varphi_t F_{nt}$ ... satisfactory | $< F_{nt}/\Omega_t$ ... satisfactory |
| $> 0.3\varphi_t F_{nt}$ | $> 0.3F_{nt}/\Omega_t$ |

Hence, it is necessary to investigate the effects of the combined shear and tensile stress. The modified nominal tensile stress $F'_{nt}$ of a bolt, subjected to combined shear and tension, is given by AISC 360 Eq. (J3-3) as

| LRFD | ASD |
|---|---|
| $F'_{nt} = 1.3F_{nt} - f_v F_{nt}/\varphi_v F_{nv}$ | $F'_{nt} = 1.3F_{nt} - \Omega_v f_v F_{nt}/F_{nv}$ |
| $= 1.3 \times 90 - 27.2 \times 90/40.5$ | $= 1.3 \times 90 - 2 \times 18.1 \times 90/54$ |
| $= 56.6$ ksi | $= 56.7$ ksi |
| $\varphi_t F'_{nt}$ = design tensile stress | $F'_{nt}/\Omega_t$ = allowable tensile stress |
| $= 0.75 \times 56.6$ | $= 56.7/2$ |
| $= 42.5$ ksi | $= 28.4$ ksi |
| $> f_t$ ... satisfactory | $> f_t$ ... satisfactory |

## References


1. Research Council on Structural Connections (RCSC). 2009. *Specification for Structural Joints Using High-Strength Bolts.* AISC, Chicago, IL.
2. Kulak, G. 2003. *High Strength Bolts, a Primer for Structural Engineers.* Design Guide No. 17. AISC, Chicago, IL.
3. Shaw, R. E. 2000. "New Bolting Specification for Steel Structures." *Structural Engineer.* Vol. 1, No. 8, September, 2000.
4. American Institute of Steel Construction (AISC). 2010. *Specification for Structural Steel Buildings* (AISC 360-10). AISC, Chicago, IL.
5. American Institute of Steel Construction (AISC). 2011. *Steel Construction Manual,* 14th edition. AISC, Chicago, IL.
6. American Society of Civil Engineers (ASCE). 2010. *Minimum Design Loads for Buildings and Other Structures* (ASCE 7-10). ASCE, Reston, VA.
7. Schlafly, T. and Muir, L. 2006. "Slip-Critical Bolts: New Available Strength Values." *Modern Steel Construction,* Vol. 46, No. 10, August 2006.


## Problems

**10.1** *Given:* The connection shown in Fig. 10.20 consists of two, grade A307, 5/8-in-diameter bolts in standard holes. The bolts are snug-tight and threads are excluded from the shear planes. Deformation around the bolt holes is a design consideration and the bolt spacing is as indicated. The plate material is A36 steel.

*Find:* Using allowable stress level design (ASD), the available strength of the connection.

**10.2** *Given:* The connection shown in Fig. 10.20 consists of two, grade A307, 5/8-in-diameter bolts in standard holes. The bolts are snug-tight and threads are excluded from the shear planes. Deformation around the bolt holes is a design consideration and the bolt spacing is as indicated. The plate material is A36 steel.

*Find:* Using strength level design (LRFD), the available strength of the connection.

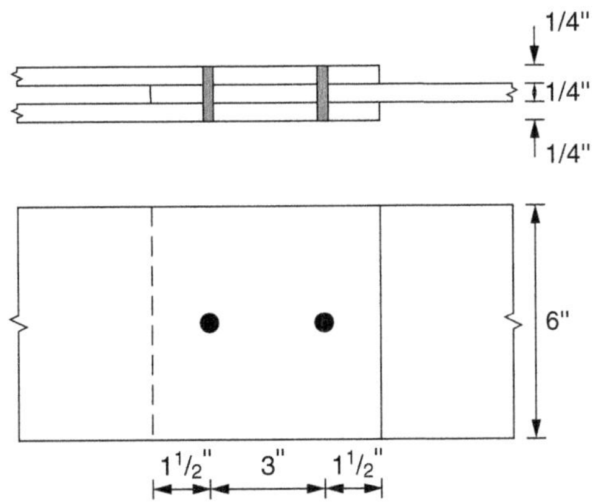


**FIGURE 10.20** Details for Problem 10.1.

**10.3** *Given:* The connection shown in Fig. 10.20 consists of two, grade A325, 5/8-in-diameter bolts in long-slotted holes transverse to the direction of force. The bolts are snug-tight and threads are not excluded from the shear planes. Deformation around the bolt holes is a design consideration and the bolt spacing is as indicated. The plate material is A36 steel.

*Find:* Using allowable stress level design (ASD), the available strength of the connection.

**10.4** *Given:* The connection shown in Fig. 10.20 consists of two, grade A325, 5/8-in-diameter bolts in long-slotted holes transverse to the direction of force. The bolts are snug-tight and threads are not excluded from the shear planes. Deformation around the bolt holes is a design consideration and the bolt spacing is as indicated. The plate material is A36 steel.

*Find:* Using strength level design (LRFD), the available strength of the connection.

**10.5** *Given:* The slip-critical connection shown in Fig. 10.20 consists of two grade A490 3/4-in-diameter bolts, in standard holes, with threads excluded from the shear planes. The plates are A36 steel with Class A faying surfaces.

*Find:* Using allowable stress level design (ASD), the available strength of the connection.

**10.6** *Given:* The slip-critical connection shown in Fig. 10.20 consists of two grade A490 3/4-in-diameter bolts, in standard holes, with threads excluded from the shear planes. The plates are A36 steel with Class A faying surfaces.

*Find:* Using strength level design (LRFD), the available strength of the connection.

**10.7** *Given:* All bolts in the connection of the WT gusset to the girder flange shown in Fig. 10.21 are grade A307, 1-in-diameter snug-tight bolts. The bolts are in standard holes and deformation around the bolt holes is a design consideration. The WT gusset and the girder are grade 50 steel.

*Find:* Using allowable stress level design (ASD), the available strength of the connection. Assume that the angles and the connection of the angles to the WT gusset are satisfactory. Prying action may be neglected.

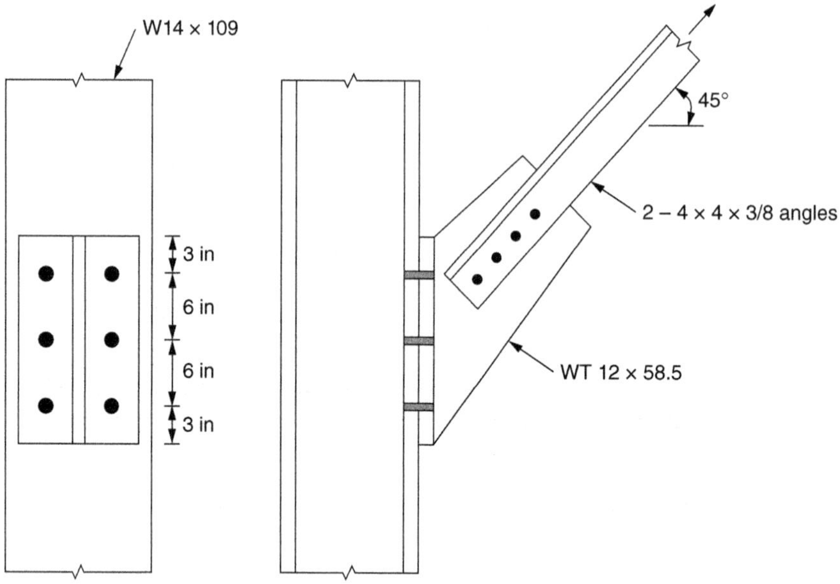


**FIGURE 10.21** Details for Problem 10.7.

**10.8** *Given:* All bolts in the connection of the WT gusset to the girder flange shown in Fig. 10.21 are grade A307, 1-in-diameter snug-tight bolts. The bolts are in standard holes and deformation around the bolt holes is a design consideration. The WT gusset and the girder are grade 50 steel.

*Find:* Using strength level design (LRFD), the available strength of the connection. Assume that the angles and the connection of the angles to the WT gusset are satisfactory. Prying action may be neglected.

**10.9** *Given:* The four bolts in the double angle hanger connection, shown in Fig. 10.22, are grade A307, 7/8-in-diameter bolts in standard holes. The angles are grade 36 steel. The service load $T$ consists of dead load $P_D = 12$ kips and live load $P_L = 36$ kips.

*Find:* Using allowable stress level design (ASD), the minimum size of 4 × 4 angle that will eliminate prying action.

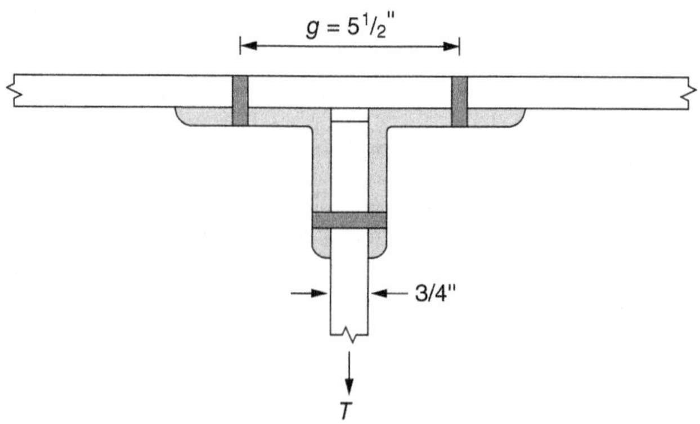


**FIGURE 10.22** Details for Problem 10.9.

**10.10** *Given:* The four bolts in the double angle hanger connection, shown in Fig. 10.22, are grade A307, 7/8-in-diameter bolts in standard holes. The angles are grade 36 steel. The service load $T$ consists of dead load $P_D$ = 12 kips and live load $P_L$ = 36 kips.

*Find:* Using strength level design (LRFD), the minimum size of 4 × 4 angle that will eliminate prying action.

**10.11** *Given:* The bolted bracket shown in Fig. 10.23 is connected to the column with 7/8-in-diameter A490 slip-critical bolts in standard holes. The bolt threads are excluded from the shear planes and faying surfaces are Class A. The W-shape is grade 50 steel and the channel is grade A36 steel. The service load $P$ consists of dead load $P_D$ = 4 kips and live load $P_L$ = 12 kips.

*Find:* Using allowable stress level design (ASD) if the connection is adequate. Use the elastic unit area method.

**10.12** *Given:* The bolted bracket shown in Fig. 10.23 is connected to the column with 7/8-in-diameter A490 slip-critical bolts in standard holes. The bolt threads are excluded from the shear planes and faying surfaces are Class A. The W-shape is grade 50 steel and the channel is grade A36 steel. The service load P consists of dead load $P_D$ = 4 kips and live load $P_L$ = 12 kips.

*Find:* Using strength level design (LRFD) if the connection is adequate. Use the elastic unit area method.

**10.13** *Given:* The bolted bracket shown in Fig. 10.23 is connected to the column with 7/8-in-diameter A490 slip-critical bolts in standard holes. The bolt threads are excluded from the shear planes and faying surfaces are Class A. The W-shape is grade 50 steel and the channel is grade A36 steel. The service load $P$ consists of dead load $P_D$ = 4 kips and live load $P_L$ = 12 kips.

*Find:* Using allowable stress level design (ASD) if the connection is adequate. Use the instantaneous center of rotation method.

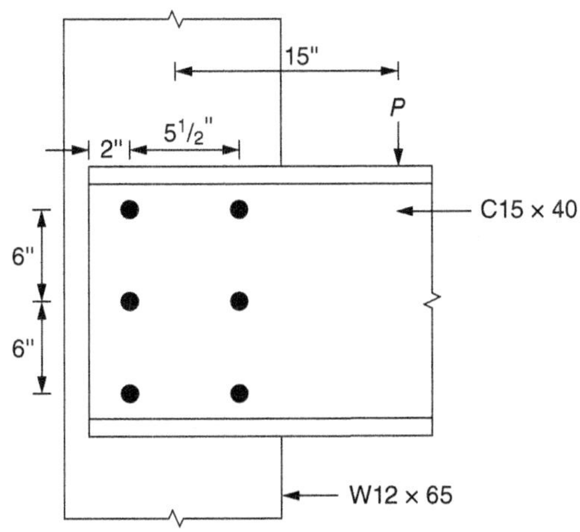


**FIGURE 10.23** Details for Problem 10.11.

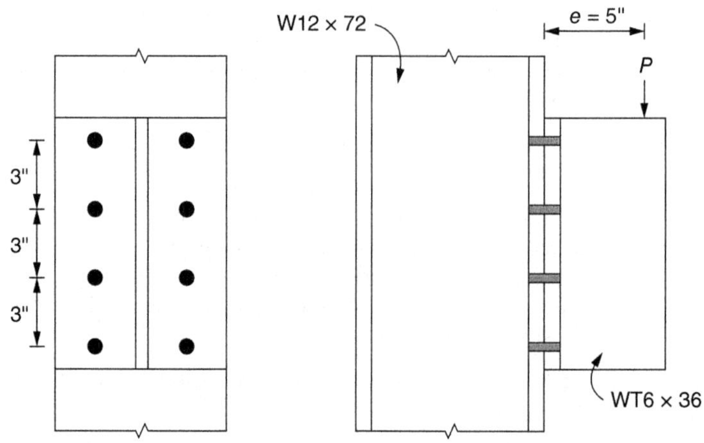


**FIGURE 10.24** Details for Problem 10.15.

**10.14** *Given:* The bolted bracket shown in Fig. 10.23 is connected to the column with 7/8-in-diameter A490 slip-critical bolts in standard holes. The bolt threads are excluded from the shear planes and faying surfaces are Class A. The W-shape is grade 50 steel and the channel is grade A36 steel. The service load $P$ consists of dead load $P_D = 4$ kips and live load $P_L = 12$ kips.

*Find:* Using strength level design (LRFD) if the connection is adequate. Use the instantaneous center of rotation method.

**10.15** *Given:* The bolted bracket shown in Fig. 10.24 is connected to the column with A325 bolts, in standard holes. The bolts are snug-tight and threads are not excluded from the shear planes. The service load P consists of dead load $P_D = 15$ kips and live load $P_L = 45$ kips.

*Find:* Using allowable stress level design (ASD), the required bolt diameter. Prying action may be neglected.

**10.16** *Given:* The bolted bracket shown in Fig. 10.24 is connected to the column with A325 bolts, in standard holes. The bolts are snug-tight and threads are not excluded from the shear planes. The service load $P$ consists of dead load $P_D = 15$ kips and live load $P_L = 45$ kips.

*Find:* Using strength level design (LRFD), the required bolt diameter. Prying action may be neglected.

# CHAPTER 11
# Design of Welded Connections

## 11.1 Introduction

### The Welding Process

Several different processes are utilized in welding structural steel and many of these are listed by Miller.[1] The two most commonly adopted are the shielded metal arc welding (SMAW) process and the submerged arc welding (SAW) process. In both processes, heat is generated by an electric arc produced between a welding electrode and the base metals being joined. The base metals and the electrode itself are melted and the weld metal from the electrode is deposited in the joint. During the process, the weld pool must be shielded from the atmosphere to prevent the formation of oxides that cause porosity and embrittlement. This is accomplished in the shielded metal arc process by a coating on the electrode that decomposes as the electrode melts producing a gaseous shield. In the submerged arc process, the shielding is produced by a previously deposited powdered flux that covers the welding zone. The shielded metal arc technique is a manual process and the submerged arc technique is an automated process.

### Welding Applications

The welding process has several advantages over bolting for the fabrication of steel structures and these include

- The process produces an airtight and watertight joint.
- Splice plates may be eliminated.
- Design changes are readily implemented in the field.
- There is no reduction in area for bolt holes.
- Connections to pipes and hollow structural sections are simplified.
- Strengthening of existing structures is facilitated.

## Quality Assurance

To ensure the production of satisfactory welds, stringent regulations must be followed and these are provided by American Welding Society, *AWS D1.1-08, Structural Welding Code—Steel* (AWS D1.1).[2] The principal requirements are

- All welding must be performed by a certified welder.
- A qualified welding inspector must be present on site.
- In accordance with AWS D1.1 Provision 5.5, a written welding procedure specification must be prepared for all welds. The written specification is required for both prequalified and qualified welds.
- Finished welds are subject to inspection to check for defects.

Most welding is carried out using prequalified welding procedure specifications. These are listed in American Institute of Steel Construction, *Steel Construction Manual* (AISC Manual)[3] Table 8-2. Prequalified welding procedure specifications are for welds that traditionally have proved satisfactory. The prequalified welding procedure provides written details of the required base metal material, approved welding process, joint geometry limitations, and necessary electrical procedure. A weld that is not prequalified must be qualified by preparing a test specimen for mechanical and nondestructive testing. If the test specimen proves satisfactory, a specific welding procedure may be written for the weld.

Visual inspection is the most commonly required inspection process. AISC Manual Part 8 provides details of the following additional methods:

- Penetrant testing involves the application of a red dye to locate cracks in the weld.
- Magnetic particle testing reveals defects near the surface by applying a magnetic powder to the magnetized joint.
- Ultrasonic testing detects voids in a weld by recording the amplitude of a high-frequency sound wave transmitted through the material.
- Radiographic testing is used to record cracks on an X-ray film.

## Weld Metal Strength

The strength of a welded connection depends on both the strength of the base metal and the strength of the weld metal. Weld metal that matches, or has properties comparable with, the base metal is used for complete-penetration groove weld connections in which the weld metal is in tension. For other types of welded connections, the weld metal strength may be one classification (10 ksi) lower than the matching weld metal. The relevant nominal strength values are given in American Institute of Steel Construction, *Specification for Structural Steel Buildings* (AISC 360)[4] Table J2.5 and Sec. J2.6. For A36 and A572 Grade 50 steels, the matching weld metal has a tensile strength of 70 ksi. This strength is provided in the shielded metal arc welding process by an electrode with the classification E70XX. Here the first letter E indicates an electrode, 70 specifies the tensile strength in ksi, and the postscript symbols indicate the position in which the electrode may be used, the coating type, and welding current.

## 11.2 Weld Types

### Complete Joint Penetration Groove Welds

Examples of single-sided and double-sided complete joint penetration (CJP) groove welds are shown in Fig. 11.1. Single-sided welds are typically used when access to the joint is possible from only one side. Single-sided welds require backing to retain the molten weld metal and they tend to produce more angular distortion than double-sided welds. Double-sided welds require the root weld to be backgouged from the opposite side, to remove slag and defects, before completing the second weld. In accordance with AISC 360 Table J2.5, the nominal strength in tension, shear, and compression is governed by the base metal and computation of the strength of the weld is not required.

### Partial Joint Penetration Groove Welds

Examples of partial joint penetration (PJP) groove welds are shown in Fig. 11.2. Partial joint penetration groove welds are used when it is not necessary to develop the full strength of the base metal. The effective throat thickness $t_e$ for shielded metal arc welds is defined in AISC 360 Table J2.1 and shown in Fig. 11.2.

The minimum permitted effective throat thickness is specified in AISC 360 Table J2.3, and is given in Table 11.1.

As specified in AISC 360 Table J2.5, compressive stress need not be considered at column splices and at column to base plate connections when the column is designed to bear in accordance with AISC 360 Sec. J1.4(1).

When connections other than columns are designed to bear in accordance with AISC 360 Sec. J1.4(2), the nominal strength of the weld metal in compression is given by AISC 360 Eq. (J2-3) as

$$R_n = F_{nw}A_{we}$$

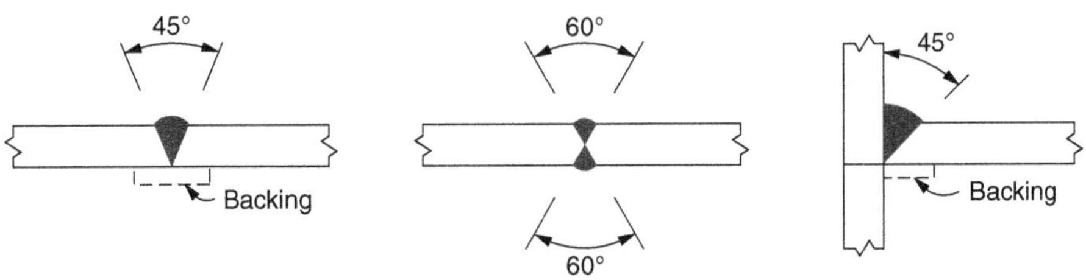


**FIGURE 11.1** Complete joint penetration groove welds.

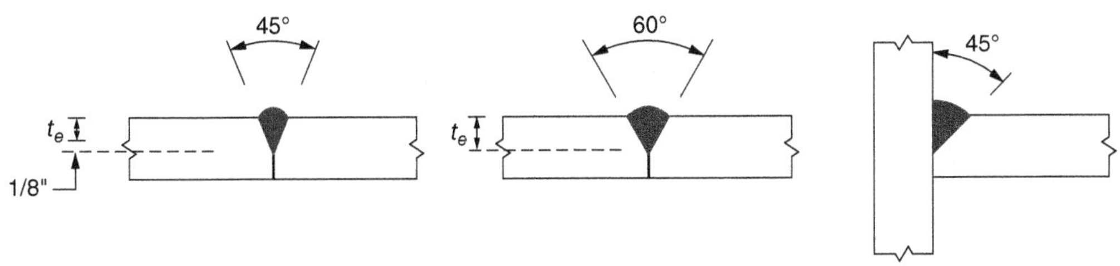


**FIGURE 11.2** Partial joint penetration groove weld: SMAW.

| Thickness of Thinner Part Joined, in | Minimum Effective Throat, in |
|---|---|
| $t \leq 1/4$ | 1/8 |
| $1/4 < t \leq 1/2$ | 3/16 |
| $1/2 < t \leq 3/4$ | 1/4 |
| $3/4 < t \leq 1\frac{1}{2}$ | 5/16 |

**TABLE 11.1** Minimum Effective Throat Thickness of Partial-Penetration Groove Welds

where $A_{we}$ = effective area of the weld metal
$= t_e \ell_e$
$\ell_e$ = effective length of weld
$F_{nw}$ = nominal stress of the weld metal
$= 0.6F_{EXX}$
$F_{EXX}$ = filler metal classification strength

The nominal strength of the base metal in compression is given by AISC 360 Eq. (J2-2) as

$$R_n = F_{nBM} A_{BM}$$

where $A_{BM}$ = effective area of the base metal
$F_{nBM}$ = nominal stress of the base metal
$= F_y$

The available compression strength of the weld metal and the base metal is obtained from AISC 360, Table J2.5 as the lesser of

| LRFD | ASD |
|---|---|
| $\varphi R_n = \varphi F_{nw} A_{we}$ ... weld metal | $R_n/\Omega = F_{nw} A_{we}/\Omega$ ... weld metal |
| $= 0.80 \times 0.60 F_{EXX} A_{we}$ | $= 0.60 F_{EXX} A_{we}/1.88$ |
| $\varphi R_n = \varphi F_{nBM} A_{BM}$ ... base metal | $R_n/\Omega = F_{nBM} A_{BM}/\Omega$ ... base metal |
| $= 0.90 F_y A_{BM}$ | $= F_y A_{BM}/1.67$ |

When connections are not designed to bear, the available compression strength of the weld metal and the base metal is obtained from AISC 360, Table J2.5 as the lesser of

| LRFD | ASD |
|---|---|
| $\varphi R_n = \varphi F_{nw} A_{we}$ ... weld metal | $R_n/\Omega = F_{nw} A_{we}/\Omega$ ... weld metal |
| $= 0.80 \times 0.90 F_{EXX} A_{we}$ | $= 0.90 F_{EXX} A_{we}/1.88$ |
| $\varphi R_n = \varphi F_{nBM} A_{BM}$ ... base metal | $R_n/\Omega = F_{nBM} A_{BM}/\Omega$ ... base metal |
| $= 0.90 F_y A_{BM}$ | $= F_y A_{BM}/1.67$ |

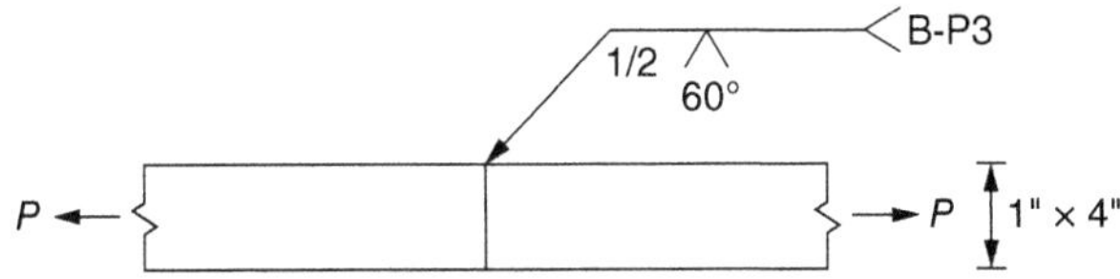


**FIGURE 11.3** Details for Example 11.1.

The available capacity in tension of the weld metal and the base metal is obtained from AISC 360, Table J2.5 as the lesser of

| LRFD | ASD |
|---|---|
| $\varphi R_n = \varphi F_{nw} A_{we}$ ... weld metal | $R_n/\Omega = F_{nw} A_{we}/\Omega$ ... weld metal |
| $= 0.80 \times 0.60 F_{EXX} A_{we}$ | $= 0.60 F_{EXX} A_{we}/1.88$ |
| $\varphi R_n = \varphi F_{nBM} A_{BM}$ ... base metal | $R_n/\Omega = F_{nBM} A_{BM}/\Omega$ ... base metal |
| $= 0.75 F_u A_{BM}$ | $= F_u A_{BM}/2.00$ |

**Example 11.1.** Partial Joint Penetration Groove Weld
The two 4 × 1 in plates shown in Fig. 11.3 are connected with a V-groove weld with a 1/2-in penetration as indicated. The plates are Grade A36 steel and the electrodes are E70XX. Determine the available tensile capacity of the connection.

For an included angle of 60°, the effective throat thickness equals the depth of the weld which is specified as 1/2 in. The total effective area of the weld metal is

$$A_{we} = 4 \times 0.5$$
$$= 2 \text{ in}^2$$

The total effective area of the base metal is

$$A_{we} = 4 \times 0.5$$
$$= 2 \text{ in}^2$$

The available tensile capacity, normal to the effective area, is given by the lesser of

| LRFD | ASD |
|---|---|
| $\varphi R_n = \varphi F_{nBM} A_{BM}$ ... base metal | $R_n/\Omega = F_{nBM} A_{BM}/\Omega$ ... base metal |
| $= 0.75 \times 58 \times 2$ | $= 58 \times 2/2.00$ |
| $= 87$ kips | $= 58$ kips |
| $\varphi R_n = \varphi F_{nw} A_{we}$ ... weld metal | $R_n/\Omega = F_{nw} A_{we}/\Omega$ ... weld metal |
| $= 0.80 \times 0.60 \times 70 \times 2$ | $= 0.60 \times 70 \times 2/1.88$ |
| $= 67.2$ kips ... governs | $= 44.7$ kips ... governs |

## Fillet Welds

The leg length $w$ of a fillet weld is used to designate the nominal size of the weld. The effective throat thickness, in accordance with AISC 360 Sec. J2.2, is illustrated in

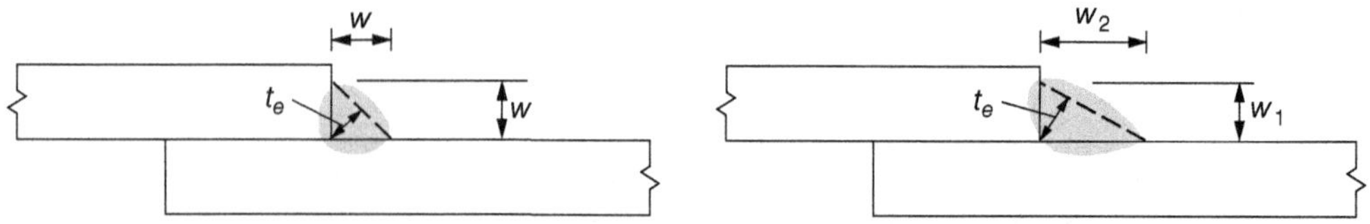


FIGURE 11.4 Fillet weld leg length.

Fig. 11.4. As indicated, the effective throat thickness is the shortest distance between the root and the weld face and, for a weld with equal leg lengths, is given by

$$t_e = 0.707w$$

where $w$ is leg length.

For a weld with unequal leg lengths, the effective throat thickness is given by

$$t_e = w_1 w_2 / (w_1^2 + w_2^2)^{0.5}$$

The force acting on a fillet weld is always a shear force. However, the force acting on the base elements may be shear, tension, or compression depending on the orientation of the element. These effects are defined in AISC 360 Sec. J4. As shown in Fig. 11.5, the vertical element of the tee-joint is subject to tension and the horizontal element is subject to double shear. The available shear strength of the weld metal is obtained from AISC 360, Table J2.5 as

| LRFD | ASD |
|---|---|
| $\varphi R_n = \varphi F_{nw} A_{we}$ ... weld metal | $R_n/\Omega = F_{nw} A_{we}/\Omega$ ... weld metal |
| $= 0.75 \times 0.60 F_{EXX} A_{we}$ | $= 0.60 F_{EXX} A_{we}/2.00$ |

The minimum permitted size of fillet welds is specified in AISC 360 Sec. J2.2b and is given in Table 11.2. This limitation is due to the quench effect of the base metal on

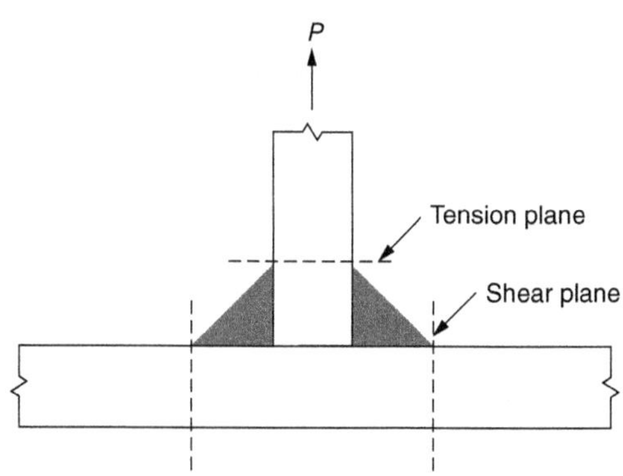


FIGURE 11.5 Fillet weld at T joint.

| Thickness of Thinner Part Joined, in | Minimum Size of Fillet Weld, in |
|---|---|
| $t \le 1/4$ | $w \ge 1/8$ |
| $1/4 < t \le 1/2$ | $w \ge 3/16$ |
| $1/2 < t \le 3/4$ | $w \ge 1/4$ |
| over 3/4 | $w \ge 5/16$ |

**Table 11.2** Minimum Size of Fillet Welds

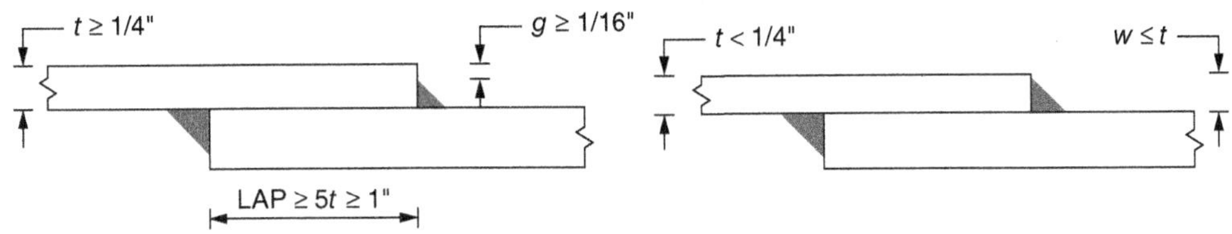


**Figure 11.6** Fillet weld along edge of material.

small welds resulting in rapid cooling and loss of ductility. The minimum weld size is determined by the thinner of the two parts joined.

Along the edge of thick material at a lap joint, it is possible for a welder to melt away the upper corner if the weld is too large. Hence, as shown in Fig. 11.6, it is necessary for the weld leg length to be less than the edge thickness. For material less than 1/4 in thick, this limitation is not required. For lap joints in tension, the minimum permissible lap is five times the thickness of the thinner part joined, but not less than 1 in so as to avoid undue rotation of the joint. When transverse welds are used, the welds must be applied to the ends of both lapped parts.

The maximum size of fillet welds, permitted along the edges of connected parts, is specified in AISC 360 Sec. J2.2b and is given in Table 11.3.

As specified in AISC 360 Sec. J2.2b and shown in Fig. 11.7, the minimum permissible length of a fillet weld is four times the nominal weld size. In addition, when longitudinal fillet welds are used alone in a connection, the length of each fillet weld shall be not less than the perpendicular distance between them because of shear lag.

| Thickness of Edge, in | Maximum Size of Fillet Weld, in |
|---|---|
| $t < 1/4$ | $w \le t$ |
| $t \ge 1/4$ | $w \le t - 1/16$ |

**Table 11.3** Maximum Size of Fillet Welds

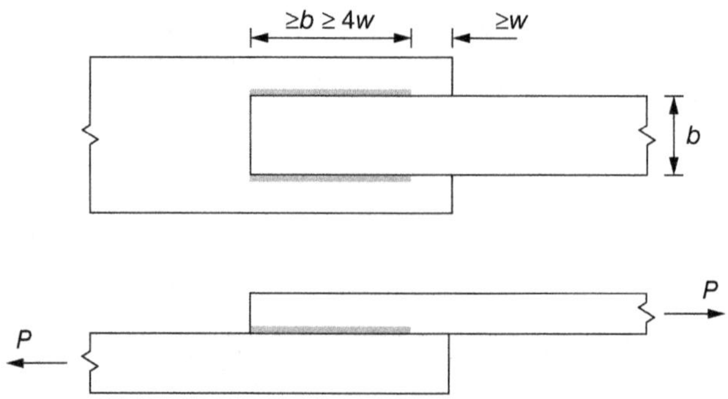


FIGURE 11.7 Fillet weld minimum dimensions.

For design purposes, it is convenient to determine the available strength of a 1/16-in fillet weld per inch run of E70XX Grade electrodes and this is given by

| LRFD | ASD |
|---|---|
| $q_u = \varphi F_{nw} A_{we}$ | $q = F_{nw} A_{we} / \Omega$ |
| $= 0.75 \times 0.60 \times 70 \times 0.707 \times 1/16 \times 1.0$ | $= 0.60 \times 70 \times 0.707 \times 1/16 \times 1.0/2$ |
| $= 1.392$ kips/in/1/16 in | $= 0.928$ kip/in/1/16 in |

If $D$ denotes the number of 1/16-in in weld size, the available capacity of a weld is

$$Q_u = Dq_u \text{ kips/in ... for LRFD load combinations}$$

$$Q = Dq \text{ kips/in ... for ASD load combinations}$$

In accordance with AISC 360 Commentary Sec. C-J2.4, the design of a welded connection is governed by the capacity of the weakest shear plane which is either through the weld or through the base material. Increasing the size of the weld to provide a strength in

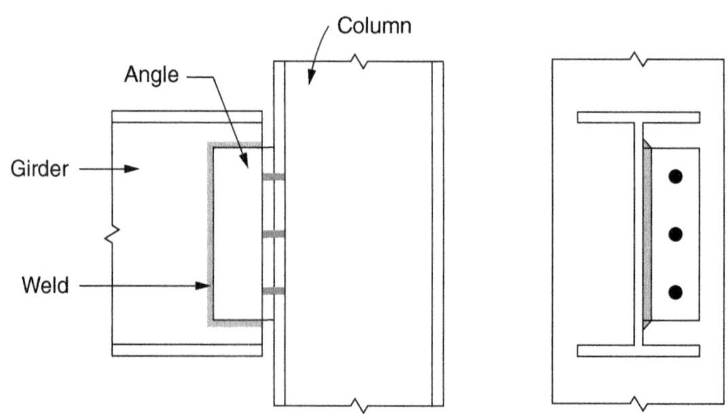


FIGURE 11.8 Fillet weld on one side of girder web.

excess of the rupture strength of the base material will not increase the capacity of the connection. Figure 11.8 shows a girder web, of thickness $t$, with an angle fillet welded to one side. The available shear capacity of the weld and the base material per linear inch is

| LRFD | ASD |
|---|---|
| $Q_u = Dq_u$ ... weld metal | $Q = Dq$ kips/in ... weld metal |
| $= 1.392D$ kips/in | $= 0.928D$ kip/in |
| $Q_{uBM} = \varphi F_{BM}t$ ... base metal | $Q_{BM} = F_{BM}t/\Omega$ ... base metal |
| $= 0.75 \times 0.60F_u t$ | $= 0.60F_u t/2$ |
| $= 0.75 \times 0.60 \times 65t$ ... Grade 50 steel | $= 0.60 \times 65t/2$ ... Grade 50 steel |
| $= 29.25t$ kips/in | $= 19.50t$ kips/in |
| The largest effective weld size is given by | The largest effective weld size is given by |
| $Q_u = Q_{uBM}$ | $Q = Q_{BM}$ |
| $D = 21.01t$ sixteenths | $D = 21.01t$ sixteenths |

For A36 steel with a tensile strength of 58 ksi, the corresponding value is

$$D = 18.75t \text{ sixteenths ... weld on one side of web plate only}$$

Figure 11.9 shows a girder web, of thickness $t$, with angles fillet welded to both sides. The largest effective weld size to match the strength of the web is similarly obtained as

$$D = 10.51t \text{ sixteenths ... Grade 50 steel with } F_u = 65 \text{ ksi}$$

$$D = 9.38t \text{ sixteenths ... A36 steel with } F_u = 58 \text{ ksi}$$

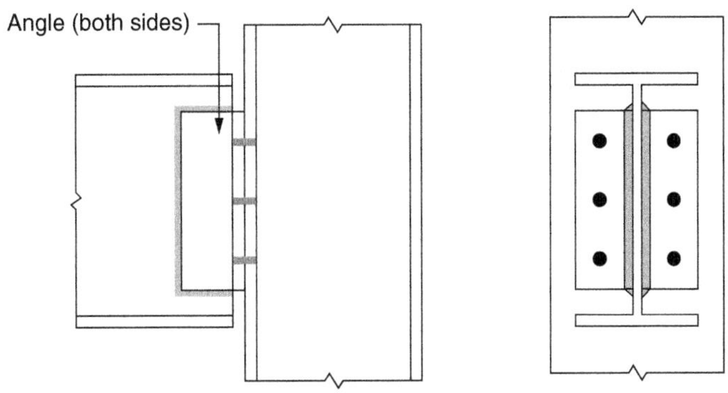


**FIGURE 11.9** Fillet weld on both sides of girder web.

## 11.3 Available Strength of Fillet Welds

### Summary

Three methods are presented in AISC 360 Sec. J2.4 for determining the strength of fillet weld groups. These are

- A linear weld group with a uniform leg size loaded through the center of gravity as specified in AISC 360 Sec. J2.4(a).
- Weld group analyzed using the instantaneous center of rotation method given in AISC 360 Sec. J2.4(b).
- Weld group with concentric loading with a uniform leg size and with elements oriented either longitudinally or transversely to the direction of the applied load, as specified in AISC 360 Sec. J2.4(c).

### Linear Weld Group Loaded through the Center of Gravity

A linear weld group is one in which all the elements are in line or are parallel. Fillet welds, loaded perpendicular to the weld axis as in the case of a transverse weld, have a nominal strength approximately 50 percent greater than that of a longitudinal weld. AISC 360 Sec. J2.4(a) accounts for the angle of inclination of the applied loading to the longitudinal axis of the weld. The nominal strength in shear is given by AISC 360 Eq. (J2-4) as

$$R_n = F_{nw}A_{we}$$

where $F_{nw}$ = nominal strength of the weld metal given by AISC 360 Eq. (J2-5)
$= 0.60F_{EXX}(1.0 + 0.50\sin^{1.5}\theta)$
$A_{we}$ = effective area of the weld
$F_{EXX}$ = filler metal classification strength
$\theta$ = angle of inclination of loading measured from the weld longitudinal axis

The term $(1.0 + 0.50\sin^{1.5}\theta)$ is an amplification factor that accounts for the orientation of the load with respect to the longitudinal axis of the weld.

**Example 11.2.** Fillet Weld with Inclined Load
The bracket shown in Fig. 11.10 supports a load acting at 45° to the axis of the weld. The supporting beam is a Grade 50 W14 × 61 beam, with a tensile strength of 65 ksi. The bracket is of A36 steel and

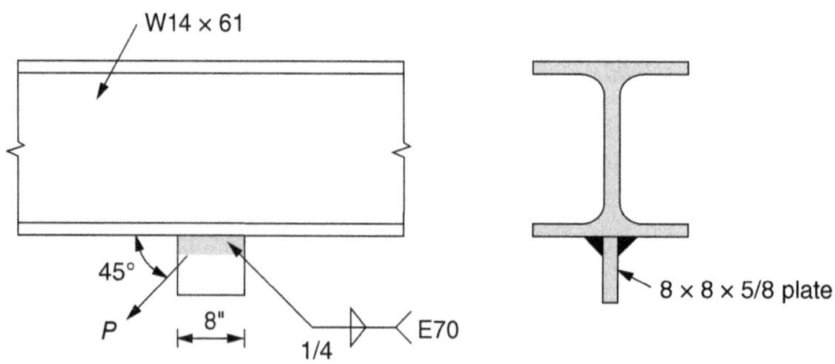


**FIGURE 11.10** Details for Example 11.2.

the electrodes are E70XX. The bracket is welded to the beam flange with 1/4 in fillet welds. Determine the maximum load that can be applied to the bracket.

The flange thickness of the W14 × 61 is

$$t_f = 0.645 \text{ in}$$
$$> 5/8 \text{ in}$$

Hence, the thickness of the bracket governs and from Table 11.2 the minimum size of fillet weld allowed is

$$w = 1/4 \text{ in}$$

The size of weld provided is

$$w = 1/4 \text{ in} \ldots \text{satisfactory}$$

The angle of inclination of the applied load measured from the axis of the weld is

$$\theta = 45°$$

The corresponding amplification of the weld shear strength is given by AISC 360 Eq. (J2-5) as

$$\varepsilon = (1.0 + 0.50\sin^{1.5}\theta)$$
$$= 1.0 + 0.50 \times 0.707^{1.5}$$
$$= 1.30$$

The available shear capacity of the weld and the bracket per linear inch is

| LRFD | ASD |
|---|---|
| $Q_{uBM} = \varphi F_{BM} t$ ... bracket | $Q_{BM} = F_{BM} t/\Omega$ ... bracket |
| $= 0.75 \times 0.60 \times 58 \times 0.625$ | $= 0.60 \times 58 \times 0.625/2$ |
| $= 16.31$ kips/in | $= 10.88$ kips/in |
| $Q_{uw} = 2Dq_u \times \varepsilon$ ... weld metal | $Q_w = 2Dq \times \varepsilon$ ... weld metal |
| $= 2 \times 4 \times 1.392 \times 1.30$ | $= 2 \times 4 \times 0.928 \times 1.3$ |
| $= 14.48$ kips/in ... governs | $= 9.65$ kips/in ... governs |
| $P_u = 8Q_{uw}$ ... design load | $P = 8Q_w$ ... allowable load |
| $= 8 \times 14.48$ | $= 8 \times 9.65$ |
| $= 116$ kips | $= 77$ kips |

## Weld Group with Concentric Loading

A weld group with concentric loading and with uniformly sized elements oriented both longitudinally and transversely to the direction of the applied load may be analyzed as specified in AISC 360 Sec. J2.4(c). Longitudinal welds are more ductile than transverse welds. Hence, both welds do not share the applied load equally. To address this issue the combined nominal strength of the weld group is given by the greater of

$$R_n = R_{nw\ell} + R_{nwt} \text{ ... AISC 360 Eq. (J2.10a)}$$

or

$$R_n = 0.85R_{nw\ell} + 1.5R_{nwt} \text{ ... AISC 360 Eq. (J2.10b)}$$

where $R_{nw\ell}$ = total nominal strength of longitudinally loaded fillet welds, as determined in accordance with AISC 360 Table J2.5

$R_{nwt}$ = total nominal strength of transversely loaded fillet welds, as determined in accordance with AISC 360 Table J2.5 without the amplification of the weld shear strength given by AISC 360 Eq. (J2-5)

When the length of the longitudinal welds is not less than 3.33 times the length of the transverse welds, Eq. (J2.10a) governs.

**Example 11.3.** Fillet Welds Oriented Longitudinally and Transversely to the Load
The two 4×4×3/8 angles shown in Fig. 11.11 are connected by 1/4-in E70XX fillet welds to a 5/8-in-thick gusset plate as indicated. The angles and gusset plate are fabricated from A36 steel. Determine the tensile force that may be applied to the connection. The applied load is concentric with the weld group.

From Table 11.2, the minimum size of fillet weld for the 5/8-in-thick gusset plate is

$$w_{min} = 1/4 \text{ in}$$

From Table 11.3, the maximum size of fillet weld for the 3/8-in-thick angles is

$$w_{max} = 3/8 - 1/16$$
$$= 5/16 \text{ in}$$

The minimum thickness of gusset plate to develop the full required strength of the 1/4-in fillet welds opposite to each other on both sides of the plate is

$$t = 4/9.38$$
$$= 0.43 \text{ in}$$
$$< 5/8 \text{ in}$$

Hence, a 1/4-in fillet weld is satisfactory.

The total length of longitudinally loaded weld is

$$\ell_{w\ell} = 4 \times 6$$
$$= 24 \text{ in}$$

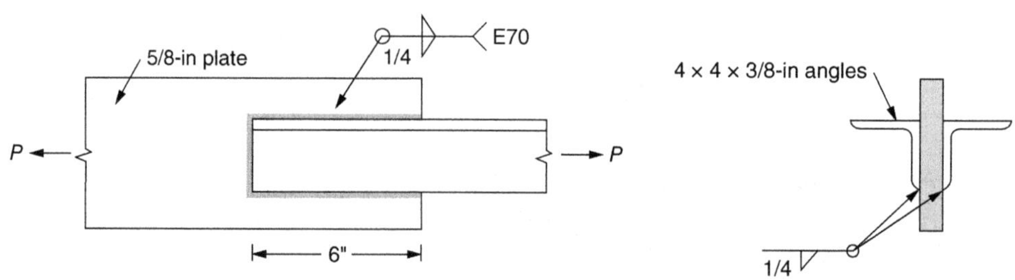


**FIGURE 11.11** Details for Example 11.3.

The total length of transversely loaded weld is

$$\ell_{wt} = 2 \times 4$$
$$= 8 \text{ in}$$

The nominal strength of one linear in of 1/4-in fillet weld is

$$F_{nw}A_{we} = 0.60 \times 70 \times 0.707 \times 0.25$$
$$= 7.42 \text{ kips/in}$$

Applying AISC 360 Eq. (J2-10a), the nominal strength of the connection is

$$R_n = R_{nw\ell} + R_{nwt}$$
$$= 24 \times 7.42 + 8 \times 7.42$$
$$= 237 \text{ kips}$$

Applying AISC 360 Eq. (J2-10b), the nominal strength of the connection is

$$R_n = 0.85R_{nw\ell} + 1.5R_{nwt}$$
$$= 0.85 \times 24 \times 7.42 + 1.5 \times 8 \times 7.42$$
$$= 240 \text{ kips ... governs}$$
$$> 237 \text{ kips}$$

The available shear capacity of the weld is

| LRFD | ASD |
|---|---|
| $\varphi R_n = 0.75 \times 240$ | $R_n/\Omega = 240/2$ |
| $= 180$ kips | $= 120$ kips |

## 11.4 Weld Group Eccentrically Loaded in Plane of Faying Surface

### Elastic Vector Analysis

Eccentrically loaded weld groups of the type shown in Fig. 11.12 may be conservatively designed by means of the elastic vector analysis technique assuming unit size of weld. To facilitate the process, AISC Manual Fig. 8-6 provides the geometrical properties of typical weld segments.

The analysis commences by first determining the location of the centroid of the weld group. Then, the polar moment of inertia of the weld group about the centroid is

$$I_o = I_x + I_y$$

where $I_x$ is moment of inertia of the weld group about the $x$-axis and $I_y$ is moment of inertia of the weld group about the $y$-axis.

For a total length of weld $L$, the vertical force per linear inch of weld due to the applied load $P$ is given by AISC Manual Eq. (8-5) as

$$V_P = P/L$$

The vertical force at point $i$ due to the eccentricity $e$ is given by AISC Manual Eq. (8-10) as

$$V_e = Pex_i/I_o$$

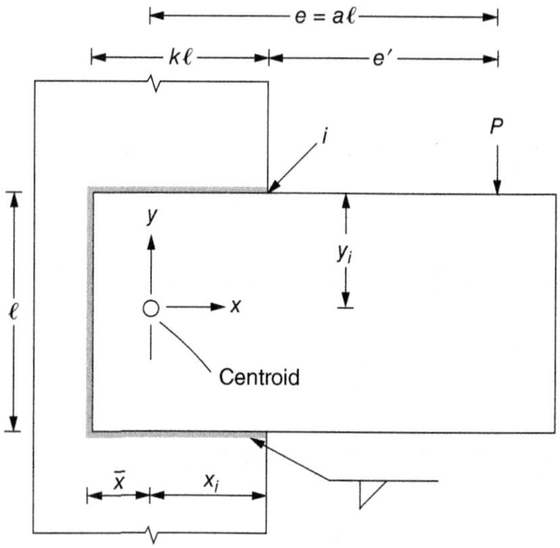


**FIGURE 11.12** Weld group eccentrically loaded in plane of faying surface.

The horizontal force at point $i$ due to the eccentricity $e$ is given by AISC Manual Eq. (8-9) as

$$H_e = Pey_i/I_o$$

The resultant force at point $i$ is

$$R = [(V_p + V_e)^2 + (H_e)^2]^{0.5}$$

The elastic vector analysis technique may be readily applied to unusual as well as common weld groups.

**Example 11.4.** Weld Group Eccentrically Loaded in Plane of Faying Surface
Determine the size of E70XX fillet weld required in the welded bracket shown in Fig. 11.13. The W-shape is Grade 50 steel and the channel is Grade A36 steel. The service load $P$ consists of dead load $P_D = 5$ kips and live load $P_L = 15$ kips. Use the elastic vector analysis method.

Applying American Society of Civil Engineers, *Minimum Design Loads for Buildings and Other Structures* (ASCE 7)[5] Secs. 2.3 and 2.4 gives the factored loads as

| LRFD | ASD |
|---|---|
| From ASCE 7 Sec. 2.3.2 combination 2: | From ASCE 7 Sec. 2.4.1 combination 2: |
| $P_u$ = factored load | $P_a$ = factored load |
| $= 1.2P_D + 1.6P_L$ | $= P_D + P_L$ |
| $= 1.2 \times 5 + 1.6 \times 15$ | $= 5 + 15$ |
| = 30 kips | = 20 kips |

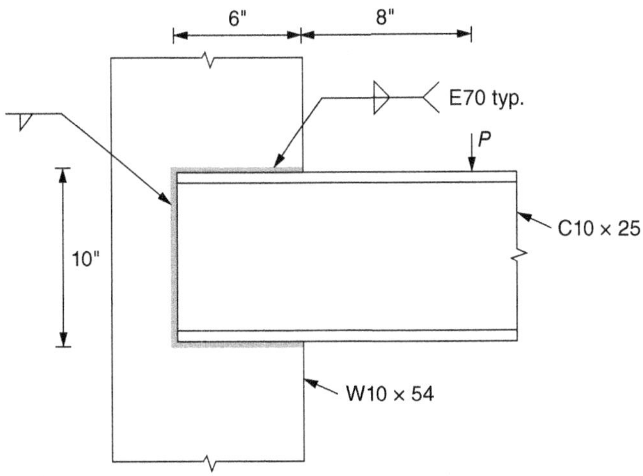


**FIGURE 11.13** Details for Example 11.4.

Assuming unit size of weld, the properties of the weld group are obtained by applying the elastic vector technique. The total length of the weld is

$$
\begin{aligned}
L &= \ell + 2k\ell \\
&= 10 + 2 \times 6 \\
&= 22 \text{ in}
\end{aligned}
$$

The centroid location is given by

$$
\begin{aligned}
\bar{x} &= 2(k\ell)^2/2L \\
&= 6^2/22 \\
&= 1.64 \text{ in}
\end{aligned}
$$

Moment of inertia about the $x$-axis is obtained from AISC Manual Fig. 8-6 as

$$
\begin{aligned}
I_x &= \ell^3/12 + 2(k\ell)(\ell/2)^2 \\
&= 10^3/12 + 2 \times 6 \times 5^2 \\
&= 383 \text{ in}^4/\text{in}
\end{aligned}
$$

Moment of inertia about the $y$-axis is

$$
\begin{aligned}
I_y &= 2(k\ell)^3/12 + 2(k\ell)(k\ell/2 - \bar{x})^2 + \ell\bar{x}^2 \\
&= 2 \times 6^3/12 + 2 \times 6 \times (1.36)^2 + 10 \times (1.64)^2 \\
&= 85 \text{ in}^4/\text{in}
\end{aligned}
$$

The polar moment of inertia is

$$
\begin{aligned}
I_o &= I_x + I_y \\
&= 383 + 85 \\
&= 468 \text{ in}^4/\text{in}
\end{aligned}
$$

The eccentricity of the applied load about the centroid of the weld profile is

$$e = e' + k\ell - \bar{x}$$
$$= 8 + 6 - 1.64$$
$$= 12.36 \text{ in}$$

The top right hand corner of the weld profile is the most highly stressed and the coordinates of point $i$ are

$$y_i = \ell/2$$
$$= 10/2$$
$$= 5 \text{ in}$$
$$x_i = k\ell - \bar{x}$$
$$= 6 - 1.64$$
$$= 4.36 \text{ in}$$

The coexistent forces acting at point $i$ in the $x$-direction and $y$-direction are

| LRFD | ASD |
|---|---|
| $V_p$ = vertical force due to applied load | $V_p$ = vertical force due to applied load |
| $= P_u/L$ | $= P/L$ |
| $= 30/22$ | $= 20/22$ |
| $= 1.36$ kips/in | $= 0.91$ kip/in |
| $V_e$ = vertical force due to eccentricity | $V_e$ = vertical force due to eccentricity |
| $= P_u e x_i/I_o$ | $= Pex_i/I_o$ |
| $= 30 \times 12.36 \times 4.36/468$ | $= 20 \times 12.36 \times 4.36/468$ |
| $= 3.45$ kips/in | $= 2.30$ kips/in |
| $H_e$ = horizontal force due to eccentricity | $H_e$ = horizontal force due to eccentricity |
| $= P_u e y_i/I_o$ | $= Pey_i/I_o$ |
| $= 30 \times 12.36 \times 5/468$ | $= 20 \times 12.36 \times 5/468$ |
| $= 3.96$ kips/in | $= 2.64$ kips/in |
| $R$ = resultant force | $R$ = resultant force |
| $= [(V_p + V_e)^2 + (H_e)^2]^{0.5}$ | $= [(V_p + V_e)^2 + (H_e)^2]^{0.5}$ |
| $= [(1.36 + 3.45)^2 + 3.96^2]^{0.5}$ | $= [(0.91 + 2.30)^2 + 2.64^2]^{0.5}$ |
| $= 6.23$ kips/in | $= 4.16$ kips/in |

The required fillet weld size per 1/16 in is

| LRFD | ASD |
|---|---|
| $D = R/q_u$ | $D = R/q$ |
| $= 6.23/1.392$ | $= 4.16/0.928$ |
| $= 4.48$ sixteenths | $= 4.48$ sixteenths |

Hence,

$$w = 4.48/16$$
$$= 0.28$$
$$= 5/16 \text{ in ... to nearest } 1/16 \text{ in}$$

The flange thickness of the W10 × 54 is

$$t_f = 0.615 \text{ in}$$
$$\approx 5/8 \text{ in}$$

From Table 11.2, the minimum size of fillet weld for the W10 × 54 flange is

$$w_{min} = 1/4 \text{ in}$$
$$< w \text{ ... satisfactory}$$

The web thickness of the C10 × 25 is

$$t_w = 0.526 \text{ in}$$
$$\approx 1/2 \text{ in}$$

From Table 11.3, the maximum size of fillet weld for the channel web is

$$w_{max} = 1/2 - 1/16$$
$$= 7/16 \text{ in}$$
$$> w \text{ ... satisfactory}$$

Hence, a 5/16-in weld is required.

## Instantaneous Center of Rotation Method

The instantaneous center of rotation method of analyzing eccentrically loaded weld groups affords a more realistic estimate of the capacity of a weld group (Fig. 11.14). As

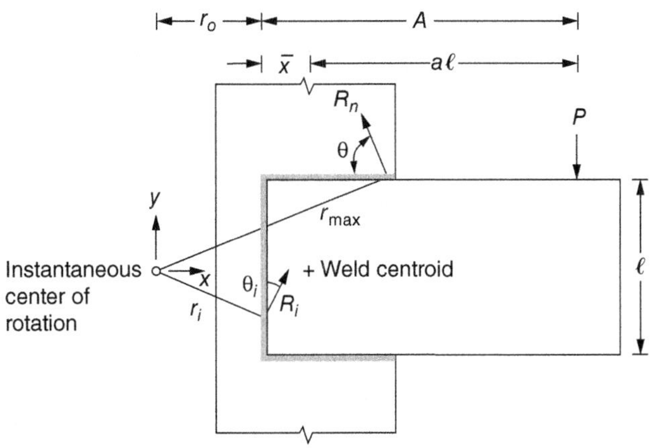


**FIGURE 11.14** Instantaneous center of rotation method.

shown in Fig. 11.14, the eccentrically loaded member is assumed to rotate and translate as a rigid body and this may be reduced to a pure rotation about a point designated the instantaneous center of rotation. AISC 360 Sec. J2.4(b) and its Commentary detail the method.

The deformation and shear force on an element of weld are proportional to the angle $\theta$ that the elemental force makes with the axis of the weld element. The maximum deformation of the critical element, when fracture is imminent, and the corresponding nominal strength are given by AISC Manual Eq. (8-4) as

$$\Delta_u = 1.087w(\theta + 6)^{-0.65}$$
$$\leq 0.17w$$

and

$$R_n = 0.60F_{EXX}t_e(1.0 + 0.50\sin^{1.5}\theta)$$

where $\theta$ is angle between elemental shear force and axis of weld element in degrees.

The deformations of other weld elements vary linearly with distance from the instantaneous center and are given by

$$\Delta_i = \Delta_u r_i / r_{crit}$$

The corresponding nominal strength of the weld element is given by AISC Manual Eq. (8-3) as

$$R_i = R_n[p_i(1.9 - 0.9p_i)]^{0.3}$$

where $r_i$ = distance of element $i$ from the instantaneous center of rotation
$r_{crit}$ = distance from the instantaneous center of rotation to the critical element
= distance from the instantaneous center of rotation to the element with the minimum $\Delta_u / r_i$ ratio
$p_i$ = ratio of element $i$ deformation to its deformation at maximum stress
= $\Delta_i / \Delta_{mi}$
$\Delta_{mi}$ = deformation of element $i$ at maximum stress
= $0.209w(\theta_i + 2)^{-0.32}$

The method is applied by assuming an initial location for the instantaneous center of rotation and determining the design strength of the elements. Then, equating horizontal forces provides the expression

$$\Sigma R_i y_i / r_i = 0$$

Equating vertical forces gives

$$\Sigma R_i x_i / r_i = P$$

Equating moments about the instantaneous center of rotation gives

$$\Sigma R_i r_i = P(A + r_0)$$

where $A$ is distance of applied load from the reference weld and $r_o$ is distance of instantaneous center from the reference weld. The correct location of the instantaneous center of rotation is determined when these three equations of equilibrium are satisfied.

| Electrode | E60 | E70 | E80 | E90 | E100 |
|---|---|---|---|---|---|
| $C_1$ | 0.857 | 1.0 | 1.03 | 1.16 | 1.21 |

**TABLE 11.4** Correction Factor for Electrode Strength

AISC Manual Tables 8-4 to 8-11 provide a convenient means of designing common weld group patterns by this method. For a particular weld geometry, the nominal strength is obtained as

$$R_n = CC_1D\ell$$

where $C$ = value tabulated in the tables
$C_1$ = correction factor for electrode strength given in Table 11.4
$D$ = the number of 1/16 in weld size
$\ell$ = length of reference weld

**Example 11.5.** Instantaneous Center of Rotation Method
Determine the maximum load that may be supported by the welded bracket shown in Fig. 11.15. All welds are 1/4-in E70XX fillet welds and the bracket and column may be considered adequate. Use the instantaneous center of rotation method.
From Fig. 11.15

$$\ell = 10 \text{ in}$$
$$e' = 8 \text{ in}$$
$$k\ell = 6 \text{ in}$$

and

$$k = k\ell/\ell$$
$$= 6/10$$
$$= 0.6$$

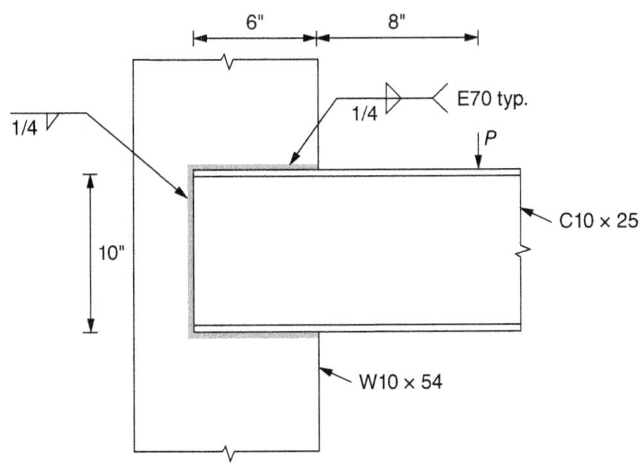


**FIGURE 11.15** Details for Example 11.5.

From AISC Manual Table 8-8 for $\theta = 0°$ and $k = 0.6$

$$\begin{aligned}\bar{x} &= 0.164\ell \\ &= 0.164 \times 10 \\ &= 1.64 \text{ in}\end{aligned}$$

and

$$\begin{aligned}a &= (e' + k\ell - \bar{x})/\ell \\ &= (8 + 6 - 1.64)/10 \\ &= 1.24\end{aligned}$$

From AISC Manual Table 8-8 for values of $a = 1.24$ and $k = 0.6$, the coefficient $C$ is given as

$$C = 1.59$$

For E70XX electrodes, the correction factor for electrode strength is obtained from Table 11.4 as

$$C_1 = 1.0$$

Hence, the nominal strength is obtained as

$$\begin{aligned}R_n &= CC_1D\ell \\ &= 1.59 \times 1.0 \times 4 \times 10 \\ &= 63.60 \text{ kips}\end{aligned}$$

The available capacity of the connection is

| LRFD | ASD |
|---|---|
| $\varphi R_n = 0.75 \times 63.6$<br>$= 47.7$ kips | $R_n/\Omega = 63.6/2$<br>$= 31.8$ kips |

## 11.5 Weld Group Eccentrically Loaded Normal to Faying Surface

### Elastic Vector Analysis

Eccentrically loaded weld groups of the type shown in Fig. 11.16 may be conservatively designed by means of the elastic vector analysis technique assuming unit size of weld.

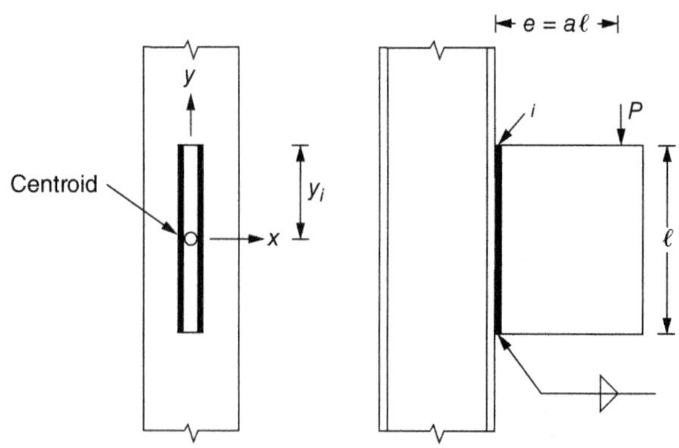


**FIGURE 11.16** Weld group eccentrically loaded normal to the faying surface.

For a total length of weld $L$, the vertical force per linear inch of weld due to the factored load $P$ is

$$V_p = P/L$$
$$= P/2\ell$$

Moment of inertia about $x$-axis is

$$I_x = 2\ell^3/12$$
$$= \ell^3/6$$

The horizontal force at point $i$ due to the eccentricity $e$ is

$$H_e = Pey_i/I_x$$
$$= 3Pe/\ell^2$$

The resultant force at point $i$ is

$$R = [(V_p)^2 + (H_e)^2]^{0.5}$$

The elastic vector analysis technique may be readily applied to unusual as well as common weld groups.

**Example 11.6.** Weld Group Eccentrically Loaded Normal to Faying Surface
Determine the size of E70XX fillet weld required in the welded bracket shown in Fig. 11.17. The W-shape is Grade 50 steel and the plate is Grade A36 steel. The service load $P$ consists of dead load $P_D = 8$ kips and live load $P_L = 24$ kips. Use the elastic vector analysis method.

Applying ASCE 7 Secs. 2.3 and 2.4 gives the factored loads as

| LRFD | ASD |
|---|---|
| From ASCE 7 Sec. 2.3.2 combination 2: | From ASCE 7 Sec. 2.4.1 combination 2: |
| $P_u$ = factored load | $P_a$ = factored load |
| $= 1.2P_D + 1.6P_L$ | $= P_D + P_L$ |
| $= 1.2 \times 8 + 1.6 \times 24$ | $= 8 + 24$ |
| $= 48$ kips | $= 32$ kips |

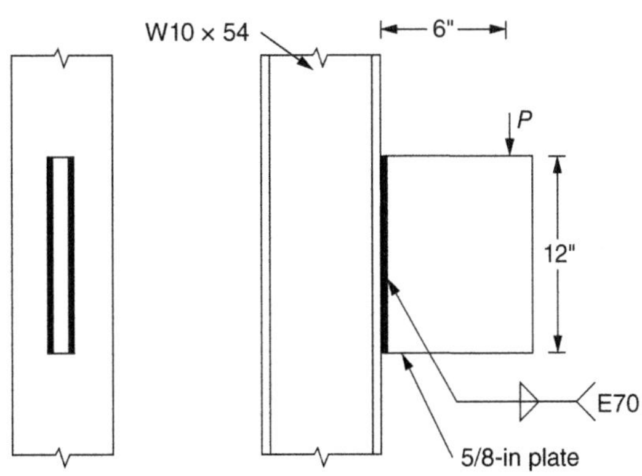


**FIGURE 11.17** Details for Example 11.6.

Assuming unit size of weld, the properties of the weld group are obtained by applying the elastic vector technique. The total length of the weld is

$$L = 2\ell$$
$$= 2 \times 12$$
$$= 24 \text{ in}$$

Moment of inertia about the $x$-axis is

$$I_x = 2\ell^3/12$$
$$= 2 \times 12^3/12$$
$$= 288 \text{ in}^4/\text{in}$$

The top corner $i$ of the weld profile is the most highly stressed and the coexistent forces acting at point $i$ in the $x$-direction and $y$-direction are

| LRFD | ASD |
|---|---|
| $V_P$ = vertical force due to applied load | $V_P$ = vertical force due to applied load |
| $= P_u/L$ | $= P/L$ |
| $= 48/24$ | $= 32/24$ |
| $= 2.00$ kips/in | $= 1.33$ kips/in |
| $H_e$ = horizontal force due to eccentricity | $H_e$ = horizontal force due to eccentricity |
| $= P_u e y_i/I_x$ | $= Pey_i/I_x$ |
| $= 48 \times 6 \times 6/288$ | $= 32 \times 6 \times 6/288$ |
| $= 6.00$ kips/in | $= 4.00$ kips/in |
| $R$ = resultant force | $R$ = resultant force |
| $= [(V_P)^2 + (H_e)^2]^{0.5}$ | $= [(V_P)^2 + (H_e)^2]^{0.5}$ |
| $= [(2.00)^2 + (6.00)^2]^{0.5}$ | $= [(1.33)^2 + (4.00)^2]^{0.5}$ |
| $= 6.32$ kips/in | $= 4.22$ kips/in |

The required fillet weld size per 1/16 in is

| LRFD | ASD |
|---|---|
| $D = R/q_u$ | $D = R/q$ |
| $= 6.32/1.392$ | $= 4.22/0.928$ |
| $= 4.54$ sixteenths | $= 4.55$ sixteenths |

Hence,

$$w = 4.55/16$$
$$= 0.28$$
$$= 5/16 \text{ in ... to nearest } 1/16 \text{ in}$$

The thickness of the plate is

$$t = 5/8 \text{ in}$$

The flange thickness of the W10 × 54 is

$$t_f = 0.615 \text{ in ... thinnest part governs}$$
$$\approx 5/8 \text{ in}$$

From Table 11.2, the minimum size of fillet weld for the W10 × 54 flange is

$$w_{min} = 1/4 \text{ in}$$
$$< w \text{ ... satisfactory}$$

The minimum thickness of the A36 plate to develop the full required strength of the 5/16-in fillet welds opposite to each other on both sides of the plate is

$$t = 4.55/9.38$$
$$= 0.49 \text{ in}$$
$$< 5/8 \text{ in ... satisfactory}$$

Hence, the 5/16-in fillet weld is adequate.

## Instantaneous Center of Rotation Method

The instantaneous center of rotation method may also be applied to welds of the type shown in Fig. 11.18 that are eccentrically loaded normal to the faying surface. AISC Manual Table 8-4 is applicable with the value $k = 0$.

**Example 11.7.** Instantaneous Center of Rotation Method
Determine the maximum load that may be supported by the welded bracket shown in Fig. 11.19. All welds are 1/4-in E70XX fillet welds and the bracket and column may be considered adequate. Use the instantaneous center of rotation method.
From Fig. 11.19

$$\ell = 12 \text{ in}$$
$$e = a\ell$$
$$= 6 \text{ in}$$

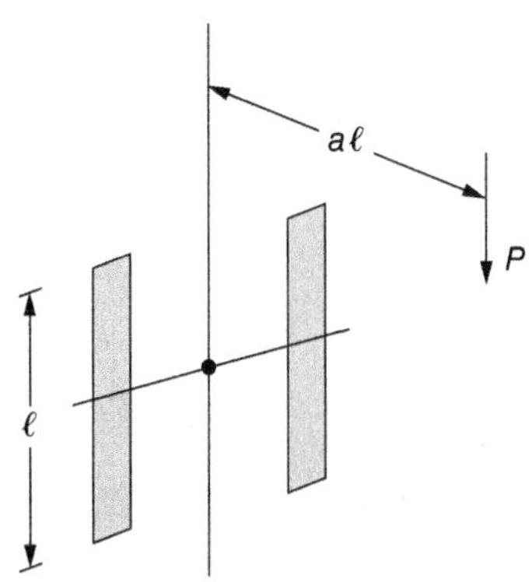


**FIGURE 11.18** Instantaneous center of rotation method load normal to faying surface.

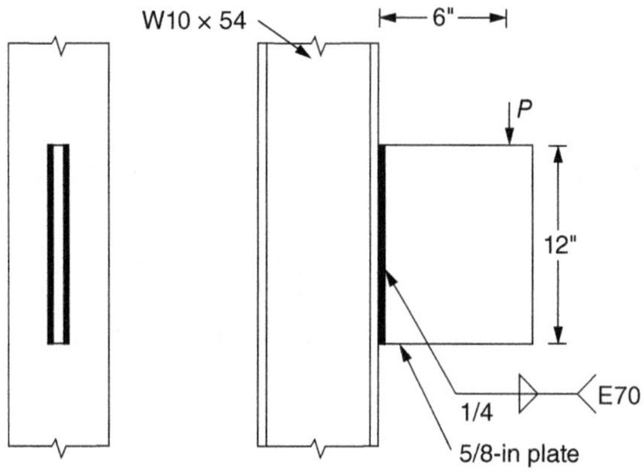


**FIGURE 11.19** Details for Example 11.7.

and
$$a = a\ell/\ell = 6/12 = 0.5$$

From AISC Manual Table 8-4 for values of $a = 0.5$ and $k = 0$, the coefficient $C$ is given as

$$C = 2.30$$

For E70XX electrodes, the correction factor for electrode strength is obtained from Table 11.4 as

$$C_1 = 1.0$$

Hence, the nominal strength is obtained as

$$R_n = CC_1D\ell = 2.30 \times 1.0 \times 4 \times 12 = 110 \text{ kips}$$

The available capacity of the weld is

| LRFD | ASD |
|---|---|
| $\varphi R_n = 0.75 \times 110$<br>$= 82.5$ kips | $R_n/\Omega = 110/2$<br>$= 55$ kips |

## References


1. Miller, K. M. 2006. *Welded Connections—A Primer for Engineers.* Design Guide No. 21. AISC, Chicago, IL.
2. American Welding Society (AWS). 2008. *AWS D1.1-08, Structural Welding Code—Steel* (AWS D1.1). AWS, Miami, FL.
3. American Institute of Steel Construction (AISC). 2011. *Steel Construction Manual,* 14th edition. AISC, Chicago, IL.

4. American Institute of Steel Construction (AISC). 2010. *Specification for Structural Steel Buildings* (AISC 360-10), AISC, Chicago, IL.
5. American Society of Civil Engineers (ASCE). 2010. *Minimum Design Loads for Buildings and Other Structures* (ASCE 7-10), ASCE, Reston, VA.

# Problems

**11.1** *Given:* The connection shown in Fig. 11.20 consists of two 3 × 1½-in plates connected with a double V-groove weld with a 1/2-in penetration at each face, as indicated. The plates are Grade A36 steel and the electrodes are E70XX.

*Find:* Using allowable stress level design (ASD), the available tensile capacity of the connection.

**11.2** *Given:* The connection shown in Fig. 11.20 consists of two 3 × 1½-in plates connected with a double V-groove weld with a 1/2-in penetration at each face, as indicated. The plates are Grade A36 steel and the electrodes are E70XX.

*Find:* Using strength level design (LRFD), the available tensile capacity of the connection.

**11.3** *Given:* The lap connection shown in Fig. 11.21 consists of two 3 × 1/2-in plates connected with transverse fillet welds as indicated. The plates are Grade A36 steel and the electrodes are E70XX.

*Find:* Using allowable stress level design (ASD), the size of fillet weld to develop the full capacity of the plates.

**11.4** *Given:* The lap connection shown in Fig. 11.21 consists of two 3 × 1/2-in plates connected with transverse fillet welds as indicated. The plates are Grade A36 steel and the electrodes are E70XX.

*Find:* Using strength level design (LRFD), the size of fillet weld to develop the full capacity of the plates.

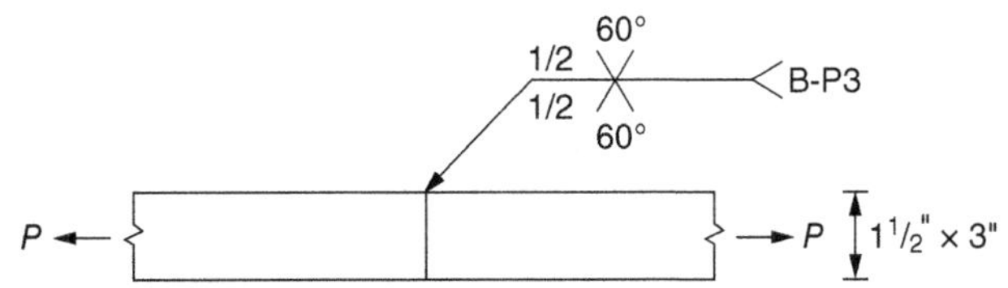


**FIGURE 11.20** Details for Problem 11.1.

**FIGURE 11.21** Details for Problem 11.3.

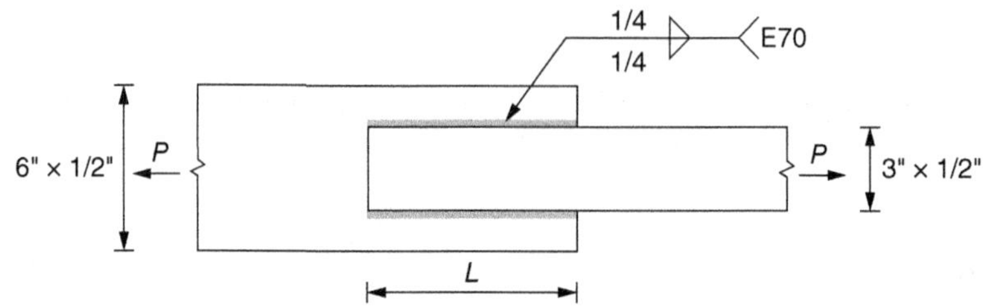


**FIGURE 11.22** Details for Problem 11.5.

**11.5** *Given:* The lap connection shown in Fig. 11.22 consists of a 3 × 1/2-in plate connected to a 1/2-in gusset plate with 1/4-in longitudinal fillet welds as indicated. The plates are Grade A36 steel and the electrodes are E70XX.

*Find:* Using allowable stress level design (ASD), the lap length $L$ to develop the full capacity of the plates.

**11.6** *Given:* The lap connection shown in Fig. 11.22 consists of a 3 × 1/2-in plate connected to a 1/2-in gusset plate with 1/4-in longitudinal fillet welds as indicated. The plates are Grade A36 steel and the electrodes are E70XX.

*Find:* Using strength level design (LRFD), the lap length $L$ to develop the full capacity of the plates.

**11.7** *Given:* The lap connection shown in Fig. 11.23 consists of a 3 × 1/2-in plate connected to a 1/2-in gusset plate with 3/16-in fillet welds as indicated. The plates are Grade A36 steel and the electrodes are E70XX.

*Find:* Using allowable stress level design (ASD), the lap length $L$ to develop the full capacity of the plates.

**11.8** *Given:* The lap connection shown in Fig. 11.23 consists of a 3 × 1/2-in plate connected to a 1/2-in gusset plate with 3/16-in fillet welds as indicated. The plates are Grade A36 steel and the electrodes are E70XX.

*Find:* Using strength level design (LRFD), the lap length $L$ to develop the full capacity of the plates.

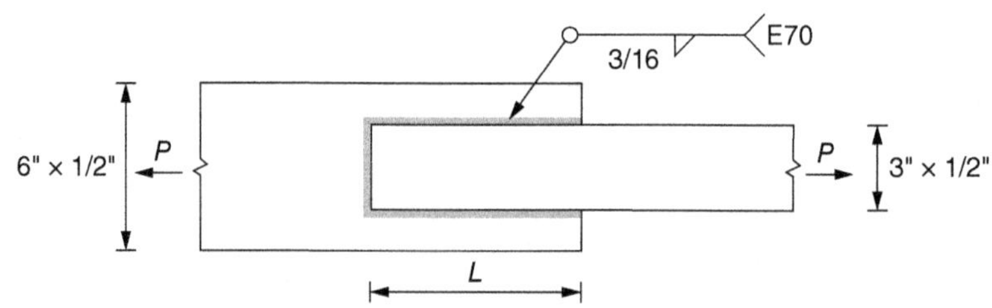


**FIGURE 11.23** Details for Problem 11.7.

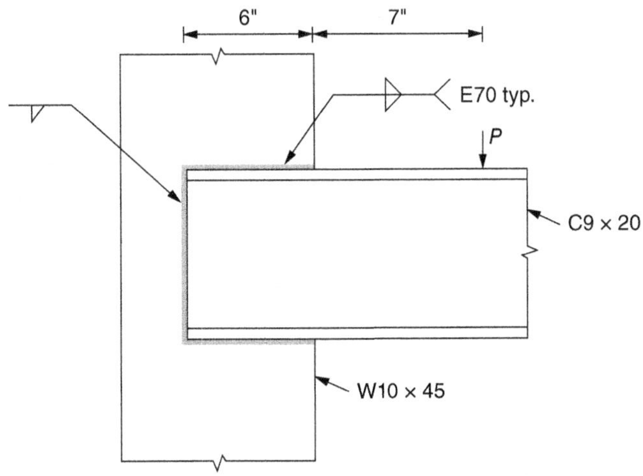


**FIGURE 11.24** Details for Problem 11.9.

**11.9** *Given:* The welded bracket shown in Fig. 11.24 consists of a C9 × 20 connected to a W10 × 45 column. The W-shape is Grade 50 steel, the channel is Grade A36 steel, and the electrodes are E70XX. The service load $P$ consists of dead load $P_D = 4$ kips and live load $P_L = 12$ kips.

*Find:* Using allowable stress level design (ASD), the size of fillet weld required to support the load. Use the elastic vector analysis method.

**11.10** *Given:* The welded bracket shown in Fig. 11.24 consists of a C9 × 20 connected to a W10 × 45 column. The W-shape is Grade 50 steel, the channel is Grade A36 steel, and the electrodes are E70XX. The service load $P$ consists of dead load $P_D = 4$ kips and live load $P_L = 12$ kips.

*Find:* Using strength level design (LRFD), the size of fillet weld required to support the load. Use the elastic vector analysis method.

**11.11** *Given:* The welded bracket shown in Fig. 11.24 consists of a C9 × 20 connected to a W10 × 45 column. The W-shape is Grade 50 steel, the channel is Grade A36 steel, and the electrodes are E70XX. The service load $P$ consists of dead load $P_D = 6$ kips and live load $P_L = 18$ kips.

*Find:* Using allowable stress level design (ASD), the size of fillet weld required to support the load. Use the instantaneous center of rotation method.

**11.12** *Given:* The welded bracket shown in Fig. 11.24 consists of a C9 × 20 connected to a W10 × 45 column. The W-shape is Grade 50 steel, the channel is Grade A36 steel, and the electrodes are E70XX. The service load $P$ consists of dead load $P_D = 6$ kips and live load $P_L = 18$ kips.

*Find:* Using strength level design (LRFD), the size of fillet weld required to support the load. Use the instantaneous center of rotation method.

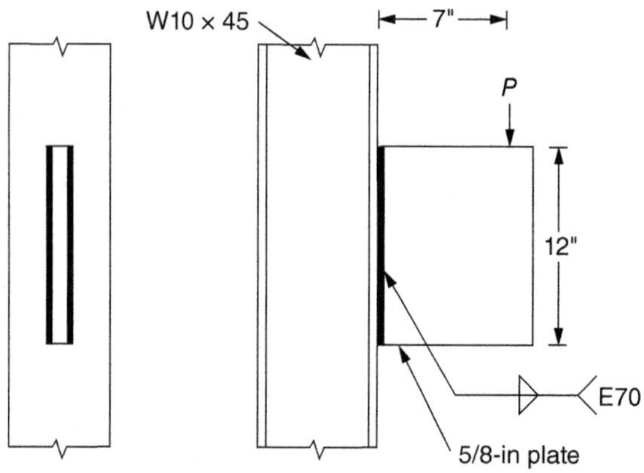


**Figure 11.25** Details for Problem 11.13.

**11.13** *Given:* The welded bracket shown in Fig. 11.25 consists of a 5/8-in plate connected to a W10 × 45 column. The W-shape is Grade 50 steel, the plate is Grade A36 steel, and the electrodes are E70XX. The service load $P$ consists of dead load $P_D = 8$ kips and live load $P_L = 24$ kips.

*Find:* Using ASD, the size of fillet weld required to support the load. Use the elastic vector analysis method.

**11.14** *Given:* The welded bracket shown in Fig. 11.25 consists of a 5/8-in plate connected to a W10 × 45 column. The W-shape is Grade 50 steel, the plate is Grade A36 steel, and the electrodes are E70XX. The service load $P$ consists of dead load $P_D = 8$ kips and live load $P_L = 24$ kips.

*Find:* Using LRFD, the size of fillet weld required to support the load. Use the elastic vector analysis method.

**11.15** *Given:* The welded bracket shown in Fig. 11.25 consists of a 5/8-in plate connected to a W10 × 45 column. The W-shape is Grade 50 steel, the plate is Grade A36 steel, and the electrodes are E70XX. The service load $P$ consists of dead load $P_D = 8$ kips and live load $P_L = 24$ kips.

*Find:* Using ASD, the size of fillet weld required to support the load. Use the instantaneous center of rotation method.

**11.16** *Given:* The welded bracket shown in Fig. 11.25 consists of a 5/8-in plate connected to a W10 × 45 column. The W-shape is Grade 50 steel, the plate is Grade A36 steel, and the electrodes are E70XX. The service load $P$ consists of dead load $P_D = 8$ kips and live load $P_L = 24$ kips.

*Find:* Using LRFD, the size of fillet weld required to support the load. Use the instantaneous center of rotation method.

# CHAPTER 12

# Plate Girders

## 12.1 Introduction

The typical plate girder shown in Fig. 12.1 is fabricated by welding together three steel plate elements. These consist of two heavy flange plates and a slender web plate. To increase the shear resistance of the slender web, intermediate stiffeners are provided and, at the location of concentrated loads, bearing stiffeners are used to prevent vertical buckling. Plate girders are usually deeper than the largest W-section that can be produced by a rolling mill and can support greater loads over longer spans. Plate girders may be used for long span floors in buildings, for crane girders in industrial structures, and for bridge superstructures.

To ensure maximum economy, plate girders are relatively deep with depth-to-span ratios of 1/10 to 1/12. This not only increases the flexural capacity but also increases the moment of inertia thus reducing deflection. The web plate may also be fabricated from steel with a strength lower than that used for the flanges and the lower strength steel is also lower in cost. Hence, the web may be made thicker without unduly increasing cost, thus reducing the slenderness ratio and increasing the shear capacity. This type of girder is known as a hybrid plate girder.

A plate girder has higher fabrication costs than a rolled section but the advantages provided by a plate girder include

- The girder is designed specifically for its required purpose.
- The girder has a high strength to weight ratio.
- The deflection is reduced.
- As the moment demand on the girder reduces, the flange plates may be correspondingly reduced in size.

As an example of the efficiency of a plate girder, a comparison may be made with the W-section which has the largest flexural capacity. This is W36 × 800 which has a nominal flexural resistance of 15,200 kip-ft, a moment of inertia of 64,700 in$^4$, and a weight of 800 lb/ft. By comparison, a 65-in-deep plate girder with 24 × 2.5 in flanges and a 60 × 5/16-in web, all in Grade 50 steel, has a nominal flexural resistance of 15,550 kip-ft, a moment of inertia of 123,000 in$^4$, and a weight of 470 lb/ft.

Some alternative forms of plate girders are shown in Fig. 12.2. These are the singly symmetric type (*a*) used typically for crane girders and type (*b*) used for composite sections. The box section shown at (*c*) is typically used for bridge construction and where torsion is a design consideration. This chapter deals only with doubly symmetric I-shaped built-up members.

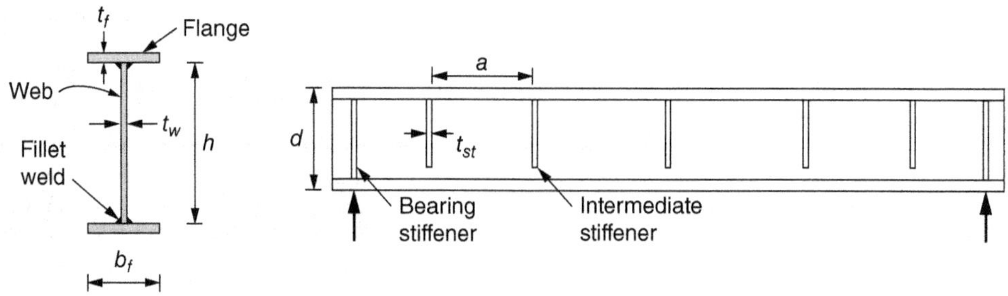


**FIGURE 12.1** Typical plate girder.

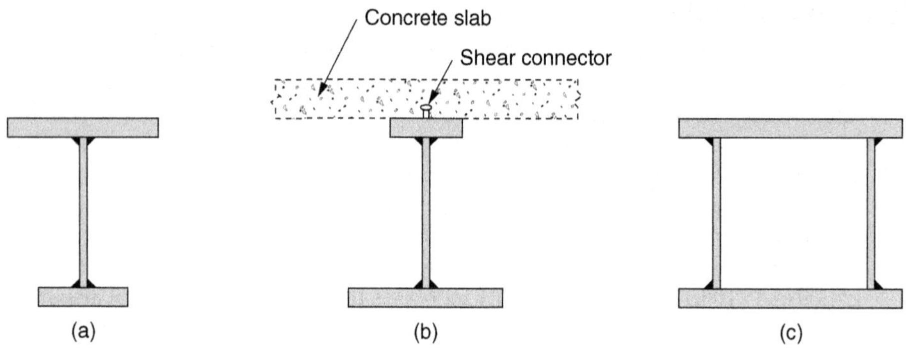


**FIGURE 12.2** Alternative types of plate girder.

## 12.2 Girder Proportions

Design aids are available (Nasados)[1] that provide available flexural and shear strengths for a wide range of plate girders. If these design aids do not provide an acceptable solution, an initial estimate of girder proportions may be obtained by assuming that the flanges carry all the moment and the web carries all the shear.

### Girder Depth

The overall girder depth $d$ will usually be in the range $L/10$ to $L/12$, where $L$ is the span length. A deeper girder is preferred for heavier loads and smaller deflections and a shallower girder is preferred for lighter loads.

### Flange Area

The required flange plate area is approximately

$$A_f = M_u/[(h + t_f)F_y\varphi_b] \text{ ... for LRFD load combinations}$$

$$\approx M_u/(hF_y\varphi_b)$$

or

$$A_f = \Omega M_a/[(h + t_f)F_y] \text{ ... for ASD load combinations}$$

$$\approx \Omega M_a/hF_y$$

### Flange Width

The flange width $b_f$ will usually be in the range $h/3$ to $h/5$, where $h$ is the clear depth between flanges. The wider width is preferred for shallower girders.

### Flange Thickness

To provide a compact compression flange, the slenderness parameter $b_{fc}/2t_f$ is limited by American Institute of Steel Construction, *Specification for Structural Steel Buildings* (AISC 360)[2] Table B4.1b, Case 11 to

$$\lambda_p = 0.38(E/F_y)^{0.5}$$
$$= 10.8 \ldots \text{for } F_y = 36 \text{ ksi}$$
$$= 9.2 \ldots \text{for } F_y = 50 \text{ ksi}$$

Provided this limit is not exceeded, and lateral bracing is adequate, the critical compression flange stress equals $F_y$.

### Web Thickness

Limitations are placed on the slenderness of the web in order to prevent vertical buckling of the compression flange. Intermediate transverse stiffeners may be required, on one or both sides of the web, to increase the shear capacity of the web. Additional shear strength may also be obtained by designing for postbuckling behavior, or tension field action, of the web. An estimate of the required web thickness may be obtained by using American Institute of Steel Construction, *Steel Construction Manual* (AISC Manual)[3] Tables 3-16 and 3-17. These tables provide available shear stress for various values of $a/h$ and $h/t_w$, where $a$ is the clear distance between stiffeners. Values of available shear strength are provided for webs both with and without tension field action.

### Intermediate Transverse Stiffeners

In accordance with AISC 360 Sec. G2.2, intermediate stiffeners need not be provided when the required shear strength does not exceed the available shear capacity of the web or the web depth-to-thickness ratio does not exceed

$$h/t_w = 2.46(E/F_y)^{0.5}$$
$$= 70 \ldots \text{for } F_y = 36 \text{ ksi}$$
$$= 59 \ldots \text{for } F_y = 50 \text{ ksi}$$

As specified in AISC 360 Sec. F13.2, the web depth-to-thickness ratio of an unstiffened web, to prevent vertical buckling of the compression flange, may not exceed the value

$$h/t_w = 260$$

In addition, the ratio of the web area to the compression flange area may not exceed 10.

When intermediate stiffeners are provided, the web depth-to-thickness ratio is governed by the panel aspect ratio $a/h$. When $a/h$ exceeds 1.5, the web depth-to-thickness ratio is limited by AISC 360 Eq. (F13-4) to a maximum value of

$$h/t_w = 0.40E/F_y$$
$$= 322 \text{ ... for } F_y = 36 \text{ ksi}$$
$$= 232 \text{ ... for } F_y = 50 \text{ ksi}$$

When $a/h$ is less than or equal to 1.5, the web depth-to-thickness ratio is limited by AISC 360 Eq. (F13-3) to a maximum value of

$$h/t_w = 12(E/F_y)^{0.5}$$
$$= 341 \text{ ... for } F_y = 36 \text{ ksi}$$
$$= 289 \text{ ... for } F_y = 50 \text{ ksi}$$

To utilize postbuckling behavior of the web, the panel aspect ratio is limited by AISC 360 Sec. G3.1 to a maximum value given by the lesser of

$$a/h = 3$$

and

$$a/h = [260/(h/t_w)]^2$$

## 12.3 Postbuckling Strength of the Web

The panels in the web of a plate girder are bounded top and bottom by the flanges and on either side by transverse stiffeners. Low shear loads are resisted by the development of a uniform shear stress in the web of the girder. At the onset of web buckling, a panel, stiffened with transverse stiffeners, does not collapse but does exhibit slight lateral web displacements that are not of structural significance. However, the panel can continue to support a substantial increase in load in the postbuckling range due to development of the tension field mechanism. As shown in Fig. 12.3, the shear load is now carried by a diagonal tensile membrane stress field in the web. This tension field is anchored by

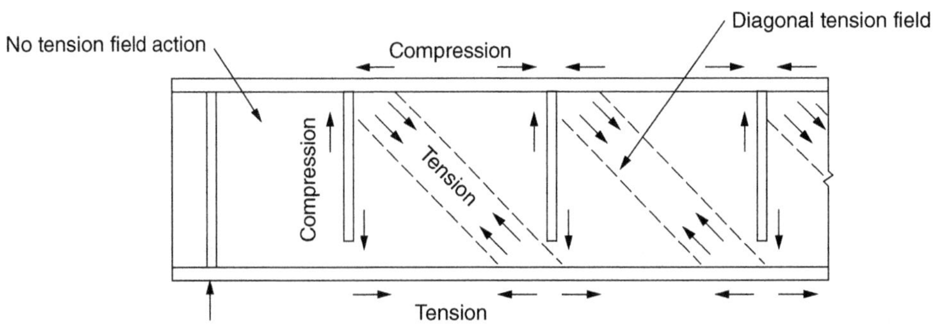


**Figure 12.3** Tension field action.

transverse stiffeners acting as struts and by the top and bottom flanges acting as chords, thus forming an effective Pratt or N-truss. In the postbuckling range, the resistance offered by the web plates is equivalent to that of the diagonal tie bars in a truss.

The web panels, on either side of an intermediate stiffener, provide anchorage for the diagonal tension field. No such anchorage is available in a web panel containing a large hole. Hence, in this panel and the adjacent panels, design using tension field action is not permitted and the shear capacity is determined from AISC 360 Eq. (G2-1). In the case of end panels, there is a panel on one side only and anchorage is not available for the diagonal tension field. Hence, design using tension field action is not permitted in end-panels. At large panel aspect ratios, tension field action is inhibited and design using tension field action is not permitted when $a/h$ exceeds 3.0 or $[260/(h/t_w)]^2$. Tension field action is also not permitted, in accordance with AISC Sec. G3.1, when

$$2A_w/(A_{fc} + A_{ft}) > 2.5$$

$$h/b_{fc} > 6.0$$

$$h/b_{ft} > 6.0$$

where $A_w$ = area of the web
$= dt_w$
$d$ = overall depth of section
$t_w$ = web thickness
$A_{fc}$ = area of compression flange
$A_{ft}$ = area of tension flange
$b_{fc}$ = width of compression flange
$b_{ft}$ = width of tension flange

There are three possible approaches to the shear design of plate girders. These are

- Design using an unstiffened web
- Design using transverse stiffeners without utilizing tension field action
- Design using transverse stiffeners with tension field action utilized

## 12.4 Design for Shear with Unstiffened Web

The nominal shear capacity of a plate girder with unstiffened web depends on the slenderness parameter $\lambda = h/t_w$. As the slenderness parameter is increased, AISC 360 Sec. G2.1(b) indicates that web failure occurs by either

- Plastic yielding of the web in girders with a compact web and $\lambda \leq 1.10(k_vE/F_y)^{0.5}$
- Inelastic buckling of the web in girders with a noncompact web and $1.10(k_vE/F_y)^{0.5} < \lambda \leq 1.37(k_vE/F_y)^{0.5}$
- Elastic buckling of the web in girders with a slender web and $\lambda > 1.37(k_vE/F_y)^{0.5}$

The shear buckling coefficient for an unstiffened web is given by AISC 360 Sec. G2.1(b) as

$$k_v = 5$$

Hence, to produce elastic buckling of the web, the minimum slenderness parameter is

$$\begin{aligned}\lambda &= 1.37(k_vE/F_y)^{0.5}\\ &= 3.06(E/F_y)^{0.5}\\ &= 87 \text{ ... for } F_y = 36 \text{ ksi}\\ &= 74 \text{ ... for } F_y = 50 \text{ ksi}\end{aligned}$$

Plate girders may be expected to exceed these limits and, hence, the nominal shear strength is given by AISC 360 Eq. (G2-1) as

$$\begin{aligned}V_n &= 0.6F_yA_wC_v\\ &= 0.6F_yA_w[7.55E/F_y(h/t_w)^2]\\ &= 4.53A_wE/(h/t_w)^2\end{aligned}$$

where $C_v = 1.51k_vE/F_y(h/t_w)^2$
$= 7.55E/F_y(h/t_w)^2$

**Example 12.1.** Unstiffened Web

Determine the available shear capacity of the welded plate girder, of Grade 50 steel, shown in Fig. 12.4. No intermediate stiffeners are provided.

The area of the web is

$$\begin{aligned}A_w &= dt_w\\ &= 75 \times 0.375\\ &= 28.13 \text{ in}^2\end{aligned}$$

The web is unstiffened and the web plate buckling coefficient is given by AISC 360 Sec. G2.1(b) as

$$k_v = 5$$

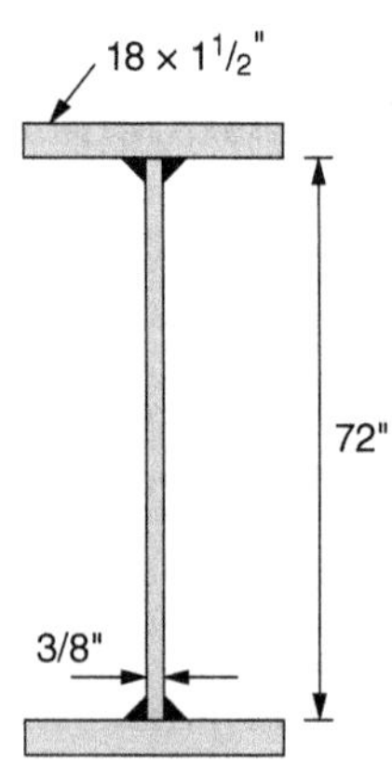


**FIGURE 12.4** Details for Example 12.1.

The web depth-to-thickness ratio is

$$h/t_w = 72/0.375$$
$$= 192$$
$$< 260$$

Hence, from AISC 360 Sec. F13.2, stiffeners are not mandatory.

Also,
$$h/t_w > 1.37(k_vE/F_y)^{0.5}$$
$$= 74$$

Hence, from AISC 360 Sec. G2.1(b), elastic buckling of the web governs and the nominal shear strength is given by AISC 360 Eq. (G2-1) as

$$V_n = 0.6F_yA_wC_v$$
$$= 0.6F_yA_w[7.55E/F_y(h/t_w)^2]$$
$$= 4.53A_wE/(h/t_w)^2$$
$$= 4.53 \times 28.13 \times 29{,}000/192^2$$
$$= 100.25 \text{ kips}$$

The available shear strength of the web is obtained from AISC 360 Sec. G1 as

| LRFD | ASD |
|---|---|
| $\varphi V_n$ = design shear capacity | $V_n/\Omega$ = allowable shear capacity |
| $= 0.90 \times 100.25$ | $= 100.25/1.67$ |
| = 90 kips | = 60 kips |

## 12.5 Design for Shear with Stiffened Web: Tension Field Action Excluded

The nominal shear capacity of a plate girder with stiffened web depends on the web depth-to-thickness ratio $h/t_w$, the panel aspect ratio $a/h$, and the shear buckling coefficient $k_v$. The shear buckling coefficient for a stiffened web is given by AISC 360 Eq. (G2-6) as

$$k_v = 5 + 5/(a/h)^2$$

The nominal shear strength for girders with transverse stiffeners is given by AISC 360 Eq. (G2-1) as

$$V_n = 0.6F_yA_wC_v$$

For shear yielding of the web with $h/t_w \le 1.1(k_vE/F_y)^{0.5}$, the shear coefficient is given by AISC 360 Eq. (G2-3) as

$$C_v = 1.0$$

For inelastic buckling of the web with $1.1(k_vE/F_y)^{0.5} < h/t_w \le 1.37(k_vE/F_y)^{0.5}$, the shear coefficient is given by AISC 360 Eq. (G2-4) as

$$C_v = 1.1(k_vE/F_y)^{0.5}/(h/t_w)$$

For elastic buckling of the web with $h/t_w > 1.37(k_vE/F_y)^{0.5}$, the shear coefficient is given by AISC 360 Eq. (G2-5) as

$$C_v = 1.51Ek_v/F_y(h/t_w)^2$$

**Example 12.2.** Stiffened Web with Tension Field Action Excluded
Determine the available shear capacity of the welded plate girder, of Grade 50 steel, shown in Fig. 12.5. Tension field action is not utilized. Intermediate stiffeners are provided at a clear spacing of 80 in. Check the solution using AISC Manual Table 3-17a.

The area of the web is

$$\begin{aligned} A_w &= dt_w \\ &= 75 \times 0.375 \\ &= 28.13 \text{ in}^2 \end{aligned}$$

The panel aspect ratio is

$$\begin{aligned} a/h &= 80/72 \\ &= 1.11 \end{aligned}$$

The web is stiffened and the web plate buckling coefficient is given by AISC 360 Sec. G2.1(b) as

$$\begin{aligned} k_v &= 5 + 5/(a/h)^2 \\ &= 5 + 5/1.11^2 \\ &= 9.06 \end{aligned}$$

The web depth-to-thickness ratio is

$$\begin{aligned} h/t_w &= 72/0.375 \\ &= 192 \\ &> 1.37(k_vE/F_y)^{0.5} \\ &= 99 \end{aligned}$$

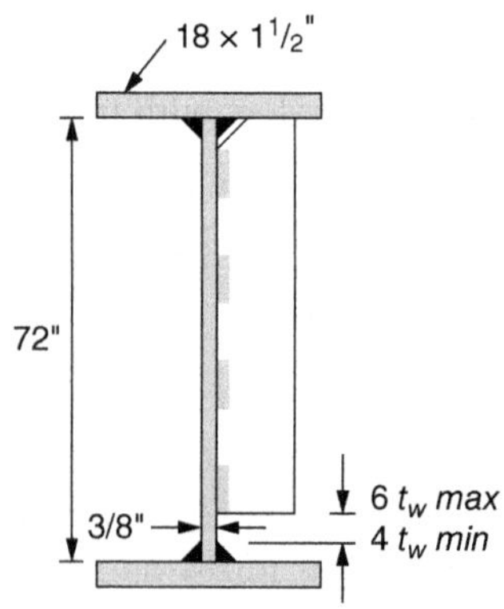


**FIGURE 12.5** Details for Example 12.2.

Hence, the shear coefficient is given by AISC 360 Eq. (G2-5) as

$$C_v = 1.51Ek_v/F_y(h/t_w)^2$$
$$= 1.51 \times 29{,}000 \times 9.06/50(192)^2$$
$$= 0.22$$

The nominal shear strength is given by AISC 360 Eq. (G2-1) as

$$V_n = 0.6F_yA_wC_v$$
$$= 0.6 \times 50 \times 28.13 \times 0.22$$
$$= 185.66 \text{ kips}$$

The available shear strength of the web is obtained from AISC 360 Sec. G1 as

| LRFD | ASD |
|---|---|
| $\varphi V_n$ = design shear capacity<br>$= 0.90 \times 185.66$<br>$= 167$ kips | $V_n/\Omega$ = allowable shear capacity<br>$= 185.66/1.67$<br>$= 111$ kips |
| From AISC Manual Table 3-17a, for $h/t_w = 192$ and $a/h = 1.11$, the design web shear capacity is | From AISC Manual Table 3-17a, for $h/t_w = 192$ and $a/h = 1.11$, the allowable web shear capacity is |
| $\varphi V_n = 6.0A_w$<br>$= 6.0 \times 28.13$<br>$= 169$ kips | $\varphi V_n = 4.0A_w$<br>$= 4.0 \times 28.13$<br>$= 113$ kips |

## 12.6 Design for Shear with Stiffened Web: Tension Field Action Included

When tension field action is included in the design of the web, the nominal shear stress is determined in accordance AISC 360 Sec. G3.2. The shear buckling coefficient for a stiffened web is given by AISC 360 Sec. G2.1(b) as

$$k_v = 5 + 5/(a/h)^2$$

For inelastic buckling of the web with $1.1(k_vE/F_y)^{0.5} < h/t_w \leq 1.37(k_vE/F_y)^{0.5}$, the shear coefficient is given by AISC 360 Eq. (G2-4) as

$$C_v = 1.1(k_vE/F_y)^{0.5}/(h/t_w)$$

For elastic buckling of the web with $h/t_w > 1.37(k_vE/F_y)^{0.5}$, the shear coefficient is given by AISC 360 Eq. (G2-5) as

$$C_v = 1.51Ek_v/F_y(h/t_w)^2$$

For shear yielding of the web with $h/t_w \leq 1.1(k_vE/F_y)^{0.5}$, the nominal shear strength is given by AISC 360 Eq. (G3-1) as

$$V_n = 0.60F_yA_w$$

For a girder with $h/t_w > 1.10(k_vE/F_y)^{0.5}$, the nominal shear strength is given by AISC 360 Eq. (G3.2) as

$$V_n = 0.6F_yA_w\{C_v + (1 - C_v)/1.15[1 + (a/h)^2]^{0.5}\}$$

Values of available shear stress are given in AISC Manual Tables 3-16b and 3-17b for a range of values of $h/t_w$ and $a/h$ for steel with a yield stress of 36 ksi and 50 ksi.

**Example 12.3.** Stiffened Web with Tension Field Action Included
Determine the available shear capacity of the welded plate girder, of Grade 50 steel, shown in Fig. 12.5. Tension field action is utilized. Intermediate stiffeners are provided at a clear spacing of 80 in. Check the solution using AISC Manual Table 3-17b.

From Example 12.2

$$A_w = 28.13 \text{ in}^2$$
$$a/h = 1.11$$
$$k_v = 9.06$$
$$h/t_w = 192$$
$$> 1.37(k_vE/F_y)^{0.5}$$
$$= 99$$
$$C_v = 0.22$$

For a girder with $h/t_w > 1.10(k_vE/F_y)^{0.5}$, the nominal shear strength is given by AISC 360 Eq. (G3.2) as

$$V_n = 0.6F_yA_w\{C_v + (1 - C_v)/1.15[1 + (a/h)^2]^{0.5}\}$$
$$= 0.6 \times 50 \times 28.13\{0.22 + (1 - 0.22)/1.15[1 + (1.11)^2]^{0.5}\}$$
$$= 569 \text{ ksi}$$

The available shear strength of the web is obtained from AISC 360 Sec. G1 as

| LRFD | ASD |
|---|---|
| $\varphi V_n$ = design shear capacity | $V_n/\Omega$ = allowable shear capacity |
| $= 0.90 \times 569$ | $= 569/1.67$ |
| = 512 kips | = 341 kips |
| From AISC Manual Table 3-17b, for $h/t_w = 192$ and $a/h = 1.11$, the design web shear capacity is | From AISC Manual Table 3-17b, for $h/t_w = 192$ and $a/h = 1.11$, the allowable web shear capacity is |
| $\varphi V_n = 18.5A_w$ | $\varphi V_n = 12.3A_w$ |
| $= 18.5 \times 28.13$ | $= 12.3 \times 28.13$ |
| = 520 kips | = 346 kips |

## 12.7 Design of Transverse Stiffeners

### Tension Field Action Excluded

The stiffeners must be of adequate rigidity perpendicular to the plane of the web to prevent lateral deflection throughout their height for loads up to the shear buckling load. The intermediate stiffeners may consist of pairs of stiffeners, provided on opposite sides of the web, or single stiffeners on one side of the web. The required moment of inertia of a single stiffener about the face in contact with the web plate or of a pair of stiffener plates about the web center line is specified by AISC 360 Eq. (G2-7) as

$$I_{st} = bt_w^3 j$$

where $j$ = transverse stiffener factor given by AISC 360 Eq. (G2-8)
$= 2.5/(a/h)^2 - 2$
$\geq 0.5$
$b$ = smaller of dimensions $a$ and $h$

Since the tension flange is self-aligning, the stiffeners can be stopped short of the tension flange. This avoids potential fatigue induced cracking occurring if the stiffener is welded to the tension flange. As specified in AISC 360 Sec. G2.2 and shown in Fig. 12.5, the stiffener may be stopped short of the tension flange with welding terminated a distance from the near toe of the web-to-flange weld between $4t_w$ and $6t_w$. When single stiffeners are used, they must be attached to the compression flange to resist any twisting of the flange.

**Example 12.4.** Transverse Stiffener with Tension Field Action Excluded
Design the intermediate stiffeners for the welded plate girder of Example 12.2. The stiffeners are of A36 steel and are provided at a clear spacing of 80 in on one side of the web only as shown in Fig. 12.5. Tension field action is not utilized.

From Example 12.2

$$a = 80 \text{ in}$$

$$h = 72 \text{ in}$$

$$a/h = 1.11$$

$$t_w = 0.375 \text{ in}$$

$$h/t_w = 192$$

The transverse stiffener factor given by AISC 360 Eq. (G2-8) is

$$j = 2.5/(a/h)^2 - 2$$

$$= 2.5/(1.11)^2 - 2$$

$$= 0.03$$

use $j = 0.5$ ... minimum

The smaller of the dimensions $a$ and $h$ is

$$b = h$$

$$= 72 \text{ in}$$

The required moment of inertia of a stiffener plate about the face of the web plate is given by AISC 360 Eq. (G2-7) as

$$I_{st} = bt_w^3 j$$

$$= 72(0.375)^3 \times 0.5$$

$$= 1.90 \text{ in}^4$$

Selecting a 3 × 1/4-in plate, the moment of inertia provided is

$$
\begin{aligned}
I_{st} &= t_{st} b_{st}^{3}/3 \\
&= 0.25 \times 3^3/3 \\
&= 2.25 \text{ in}^4 \\
&> 1.90 \text{ in}^4 \ldots \text{ satisfactory}
\end{aligned}
$$

In accordance with AISC 360 Sec. G3.3, the limiting width-to-thickness ratio of a stiffener subject to tension field action is

$$
\begin{aligned}
b_{st}/t_{st} &= 0.56(E/F_{yst})^{0.5} \\
&= 0.56(29{,}000/36)^{0.5} \\
&= 15.90 \ldots \text{ conservative for a stiffener not subject to tension field action}
\end{aligned}
$$

The width-to-thickness ratio of the stiffener is

$$
\begin{aligned}
b_{st}/t_{st} &= 3/0.25 \\
&= 12 \\
&< 15.90 \ldots \text{ satisfactory}
\end{aligned}
$$

The minimum permissible fillet weld size, connecting the 1/4-in stiffener to the 3/8-in web, is given by AISC 360 Table J2.4 as

$$w = 1/8 \text{ in}$$

Allowing for a 1-in corner clip, provide 12 runs of 1/8-in E70XX intermittent fillet welds 3 in long, spaced 3 in apart on both sides of the stiffener and terminating 2 in from the tension flange. In accordance with AISC 360 Sec. G2.2, the weld must terminate from the bottom flange a minimum distance of

$$
\begin{aligned}
g_{min} &= 4t_w + (\text{flange weld leg length}) \\
&= 4 \times 0.375 + 0.313 \ldots \text{ assuming a 5/16-in flange weld} \\
&= 1.81 \text{ in}
\end{aligned}
$$

and a maximum distance of

$$
\begin{aligned}
g_{max} &= 6t_w + (\text{flange weld leg length}) \\
&= 6 \times 0.375 + 0.313 \\
&= 2.56 \text{ in}
\end{aligned}
$$

The weld, and the stiffener provided, terminates 2 in from the bottom flange and is satisfactory.

### Tension Field Action Included

In addition to the rigidity required to prevent lateral deflection of the web, the stiffener must resist the vertical component of the tension field force. The limiting width-to-thickness ratio of a stiffener subject to tension field action is given by AISC 360 Eq. (G3-3) as

$$b_{st}/t_{st} = 0.56(E/F_{yst})^{0.5}$$

The minimum required moment of inertia of a single stiffener about the face in contact with the web plate or of a pair of stiffener plates about the web center line is specified by AISC 360 Eq. (G3-4) as

$$I_{st} = I_{st1} + (I_{st2} - I_{st1})(V_r - V_{c1})/(V_{c2} - V_{c1})$$

where $I_{st1}$ = moment of inertia as defined by AISC 360 Eq. (G2-7)

$I_{st2}$ = moment of inertia required for development of the web shear buckling resistance plus web tension field resistance

$= (h^4\rho_{st}^{1.3}/40)(F_{yw}/E)^{1.5}$

$\rho_{st}$ = the larger of $F_{yw}/F_{yst}$ and 1.0

$V_{c1}$ = smaller of the available shear strengths in the adjacent panels with $V_n$ as defined in AISC 360 Sec. G2.1

$V_{c2}$ = smaller of the available shear strengths in the adjacent panels with $V_n$ as defined in AISC 360 Sec. G3.2

$V_r$ = larger of the required shear strengths in the adjacent panels using LRFD or ASD load combinations

$F_{yw}$ = yield stress of the web material

**Example 12.5** Transverse Stiffener with Tension Field Action Included

Design the intermediate stiffeners for the welded plate girder of Example 12.3. The stiffeners are of A36 steel and are provided at a clear spacing of 80 in on one side of the web only as shown in Fig. 12.5. Tension field action is utilized and the larger of the required shear strengths in the adjacent web panels is 450 kips (LRFD load combinations) and 300 kips (ASD load combinations).

From Examples 12.2, 12.3, and 12.4

$$h = 72 \text{ in}$$

$$I_{st1} = 1.90 \text{ in}^4$$

$$V_{c1} = 167 \text{ kips ... LRFD load combinations}$$

$$V_{c1} = 111 \text{ kips ... ASD load combinations}$$

$$V_{c2} = 512 \text{ kips ... LRFD load combinations}$$

$$V_{c2} = 341 \text{ kips ... ASD load combinations}$$

$$F_{yw}/F_{yst} = 50/36$$

$$= 1.39$$

$$= \rho_{st}$$

Hence,

$$I_{st2} = (h^4\rho_{st}^{1.3}/40)(F_{yw}/E)^{1.5}$$

$$= (72^4 \times 1.39^{1.3}/40)(50/29{,}000)^{1.5}$$

$$= 73.8 \text{ in}^4$$

Then,

$$I_{st} = I_{st1} + (I_{st2} - I_{st1})(V_r - V_{c1})/(V_{c2} - V_{c1})$$

$$= 1.9 + (73.8 - 1.9)(450 - 167)/(512 - 167) \text{ ... for LRFD load combinations}$$

$$= 1.9 + 71.9 \times 283/345$$

$$= 60.9 \text{ in}^4$$

Selecting a 7.5 × 1/2-in plate, the moment of inertia provided is

$$\begin{aligned} I_{st} &= t_{st}b_{st}^{3}/3 \\ &= 0.5 \times 7.5^{3}/3 \\ &= 70.3 \text{ in}^{4} \\ &> 60.9 \text{ in}^{4} \text{ ... satisfactory} \end{aligned}$$

In accordance with AISC 360 Eq. (G3-3), the limiting width-to-thickness ratio is

$$\begin{aligned} b_{st}/t_{st} &= 0.56(E/F_{yst})^{0.5} \\ &= 0.56(29{,}000/36)^{0.5} \\ &= 15.90 \end{aligned}$$

The width-to-thickness ratio of the stiffener is

$$\begin{aligned} b_{st}/t_{st} &= 7.5/0.5 \\ &= 15 \\ &< 15.90 \text{ ... satisfactory} \end{aligned}$$

## 12.8 Flexural Design of Plate Girders

The web of a doubly symmetric, nonhybrid, I-shaped girder is classified as slender by AISC 360 Table B4.1b, Case 15, when the depth-to-thickness ratio of the web exceeds the value

$$h/t_w = 5.70(E/F_y)^{0.5}$$

where $h$ is clear distance between flanges for a welded built-up section and $t_w$ is web thickness. For this situation, AISC 360 Sec. F5 is applicable.

The applicable limit states in flexure are compression flange yielding, lateral-torsional buckling, compression flange local buckling, and tension flange yielding. The available flexural capacity is given by AISC 360 Sec. F1 as

$$\begin{aligned} M_c &= \varphi_b M_n \text{ ... for LRFD load combinations} \\ &= M_n/\Omega_b \text{ ... for ASD load combinations} \end{aligned}$$

where $\varphi_b$ = resistance factor for flexure
= 0.90
$\Omega_b$ = safety factor for flexure
= 1.67

The following presentation applies to doubly symmetric sections.

### Compression Flange Yielding

The nominal flexural strength of a plate girder with a slender web cannot exceed the yield moment. In addition, the lateral buckling of a slender web results in a loss of compressive capacity in the web, since an increase in strain does not result in a proportional increase in stress. To compensate for this, the nominal flexural capacity is reduced to the value given by AISC 360 Eq. (F5-1) as

$$M_n = R_{pg}F_yS_{xc}$$

where $S_{xc}$ = elastic section modulus referred to the compression flange
$R_{pg}$ = bending strength reduction factor given by AISC 360 Eq. (F5-6)
$= 1 - a_w[h/t_w - 5.70(E/F_y)^{0.5}]/(1200 + 300a_w)$ ... for a doubly symmetrical section
$\leq 1.0$
$a_w$ = ratio of web area to compression flange area defined by AISC 360 Eq. (F4-11)
$= ht_w/A_{fc}$ ... for a doubly symmetrical section
$A_{fc}$ = compression flange area
$= b_{fc}t_{fc}$
$b_{fc}$ = compression flange width
$t_{fc}$ = compression flange thickness

## Lateral-Torsional Buckling

Lateral-torsional buckling is influenced by unbraced length as illustrated in Fig. 12.6. The nominal flexural capacity depends on the critical compression flange stress $F_{cr}$ and the nominal capacity is given by AISC 360 Eq. (F5-2) as

$$M_n = R_{pg}F_{cr}S_x$$

For $L_b \leq L_p$ lateral torsional buckling does not occur and the critical compression flange stress is

$$F_{cr} = F_y$$

where $L_b$ = length of a beam segment between points of lateral restraint
$L_p$ = maximum unbraced length for the limit state of yielding as given by AISC 360 Eq. (F4-7)
$= 1.1r_t(E/F_y)^{0.5}$
$r_t$ = radius of gyration, about the $y$-axis, of the compression flange plus one-sixth of the total area of the web
$\approx b_f/[12(1 + a_w/6)]^{0.5}$ ... for a symmetrical section

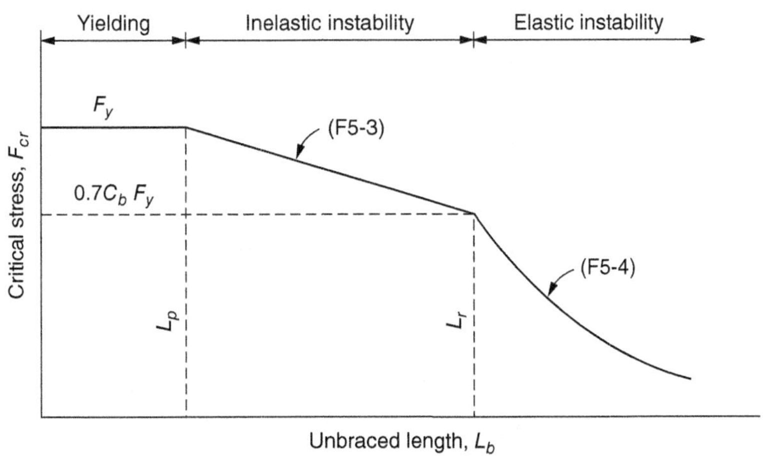


**FIGURE 12.6** Variation of critical stress with lateral support.

For inelastic lateral-torsional buckling with $L_p < L_b \leq L_r$ the critical compression flange stress is given by AISC 360 Eq. (F5-3) as

$$F_{cr} = C_b F_y[1 - 0.3(L_b - L_p)/(L_r - L_p)]$$
$$\leq F_y$$

where $C_b$ = bending coefficient given by AISC 360 Eq. (F1-1) as
$= 12.5M_{max}/(2.5M_{max} + 3M_A + 4M_B + 3M_C)$
$M_A$ = absolute value of the bending moment at the quarter point of an unbraced segment
$M_B$ = absolute value of the bending moment at the centerline of an unbraced segment
$M_C$ = absolute value of the bending moment at the three-quarter point of an unbraced segment
$M_{max}$ = absolute value of the maximum bending moment in the unbraced segment
$L_r$ = maximum unbraced length for the limit state of inelastic lateral-torsional buckling as given by AISC 360 Eq. (F5-5)
$= \pi r_t(E/0.7F_y)^{0.5}$

For elastic lateral-torsional buckling with $L_b > L_r$, the critical compression flange stress is given by AISC 360 Eq. (F5- 4) as

$$F_{cr} = C_b\pi^2E/(L_b/r_t)^2$$
$$\leq F_y$$

## Compression Flange Local Buckling

Local buckling of the flange is influenced by the flange slenderness ratio as illustrated in Fig. 12.7. The nominal flexural capacity depends on the critical compression flange stress $F_{cr}$ and the nominal capacity is given by AISC 360 Eq. (F5-7) as

$$M_n = R_{pg}F_{cr}S_{xc}$$

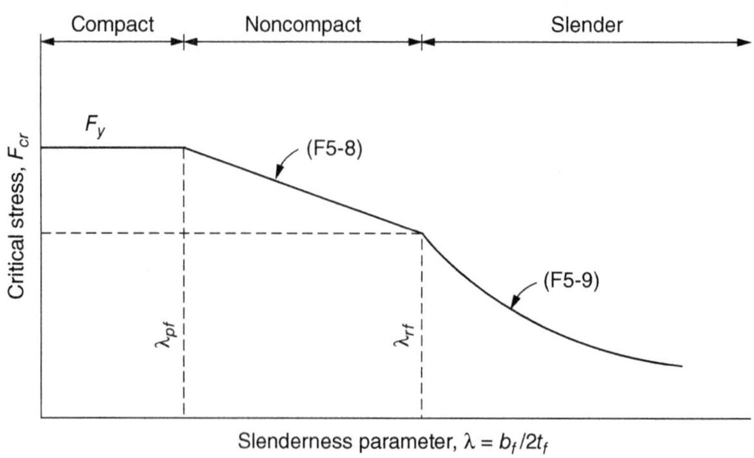


**FIGURE 12.7** Variation of critical stress with flange slenderness ratio.

For $\lambda \leq \lambda_{pf}$, the flange is compact, local buckling does not occur, and the critical compression flange stress is

$$F_{cr} = F_y$$

where $\lambda$ = slenderness parameter for the flange of a plate girder specified by AISC 360 Sec. F5.3
$= b_{fc}/2t_{fc}$
$\lambda_{pf}$ = limiting slenderness parameter, to prevent compression flange local buckling, given
by AISC 360 Table B4.1b, Case 11
$= 0.38(E/F_y)^{0.5}$
$= 10.8$ ... for $F_y = 36$ ksi
$= 9.2$ ... for $F_y = 50$ ksi

For $\lambda_p < \lambda \leq \lambda_r$, the flange is noncompact, inelastic local buckling occurs, and the critical compression flange stress is given by AISC 360 Eq. (F5-8) as

$$F_{cr} = F_y[1 - 0.3(\lambda - \lambda_{pf})/(\lambda_{rf} - \lambda_{pf})]$$

where $\lambda_{rf}$ = limiting slenderness parameter, to prevent elastic buckling of the flange, given by AISC 360 Table B4.1b, Case 11
$= 0.95(k_c E/0.7F_y)^{0.5}$
$k_c = 4/(h/t_w)^{0.5}$
$\geq 0.35$
$\leq 0.76$

For $\lambda > \lambda_{rf}$, the flange is slender, elastic local buckling occurs, and the critical compression flange stress is given by AISC 360 Eq. (F5-9) as

$$F_{cr} = 0.9Ek_c/\lambda^2$$

## Tension Flange Yielding

Tension flange yielding does not occur in a doubly symmetric I-section.

**Example 12.6.** Plate Girder Flexural Capacity
Determine the design flexural capacity of the welded plate girder, of Grade 50 steel, shown in Fig. 12.4. No intermediate stiffeners are provided. The girder supports a uniformly distributed load over a simple span of 54 ft, with lateral bracing to the compression flange at third points.

The relevant section properties are

$A_{fc}$ = compression flange area
$= 18 \times 1.5$
$= 27 \text{ in}^2$
$ht_w = 72 \times 0.375$
$= 27 \text{ in}^2$
$a_w$ = ratio of $ht_w$ to compression flange area
$= ht_w/A_f$
$= 27/27$
$= 1.0$

$I_x$ = moment of inertia

$= 84{,}600 \text{ in}^4$

$S_{xc}$ = section modulus

$= 2260 \text{ in}^3$

$r_t$ = radius of gyration of the compression flange plus one-sixth of the web

$= b_f/[12(1 + a_w/6)]^{0.5}$

$= 18/[12(1 + 1/6)]^{0.5}$

$= 4.81 \text{ in}$

$h/t_w$ = web depth-to-thickness ratio

$= 72/0.375$

$= 192$

$> 5.70(E/F_y)^{0.5}$ ... web is slender according to AISC 360 Table B4.1b, Case 15 and AISC 360 Sec. F5 applies

The limiting slenderness parameter for a compact flange is given by AISC 360 Table B4.1b, Case 11 as

$$\lambda_p = 0.38(E/F_y)^{0.5}$$
$$= 0.38(29{,}000/50)^{0.5}$$
$$= 9.2$$

The actual flange slenderness provided is given by AISC 360 Sec. F5.3 as

$$b_{fc}/2t_{fc} = 18/(2 \times 1.5)$$
$$= 6$$
$$< 9.2$$

Hence, the critical stress for flange local buckling is given by

$$F_{cr} = F_y$$
$$= 50 \text{ ksi}$$

The limiting unbraced length of the beam for the limit state of yielding is given by AISC 360 Eq. (F4-7) as

$$L_p = 1.1r_t(E/F_y)^{0.5}$$
$$= 1.1 \times 4.81(29{,}000/50)^{0.5}$$
$$= 127 \text{ in}$$
$$= 10.6 \text{ ft}$$

The limiting unbraced length for the limit state of inelastic lateral-torsional buckling is given by AISC 360 Eq. (F5-5) as

$$L_r = \pi r_t(E/0.7F_y)^{0.5}$$
$$= 3.14 \times 4.81(29{,}000/0.7 \times 50)^{0.5}$$
$$= 435 \text{ in}$$
$$= 36.2 \text{ ft}$$

The actual unbraced length is

$$L_b = 54/3$$
$$= 18 \text{ ft}$$

Hence, $L_p < L_b \leq L_r$ and AISC 360 Eq. (F5-3) applies.

The bending coefficient is given by AISC Manual Table 3-1 as

$$C_b = 1.0$$

Hence, the critical stress based on lateral torsional stability criteria and AISC 360 Eq. (F5-3) is

$$F_{cr} = C_b F_y [1 - 0.3(L_b - L_p)/(L_r - L_p)]$$
$$= 1.0 \times 50[1 - 0.3(18 - 10.6)/(36.2 - 10.6)]$$
$$= 45.66 \text{ ksi ... governs}$$
$$< F_y$$

The plate girder bending strength reduction factor is given by AISC 360 Eq. (F5-6) as

$$R_{pg} = 1 - a_w[h/t_w - 5.7(E/F_y)^{0.5}]/(1200 + 300a_w)$$
$$= 1 - 1.0[192 - 5.7(29{,}000/50)^{0.5}]/(1200 + 300 \times 1.0)$$
$$= 0.964$$

The available flexural strength is obtained from AISC 360 Sec. F1 as

| LRFD | ASD |
|---|---|
| $\varphi_b M_n = \varphi_b S_{xc} R_{pg} F_{cr}$ | $M_n/\Omega_b = S_{xc} R_{pg} F_{cr}/\Omega_b$ |
| $= 0.9 \times 2260 \times 0.964 \times 45.66/12$ | $= 2260 \times 0.964 \times 45.66/(12 \times 1.67)$ |
| $= 7461$ kip-ft | $= 4964$ kip-ft |

## 12.9 Design of Bearing Stiffeners

Bearing stiffeners are required in a plate girder when a concentrated load, applied to either the top or bottom flange, exceeds the local yielding, crippling, or sidesway buckling capacity of the web. These capacities are given by AISC 360 Eqs. (J10-2) to (J10-7).

As shown in Fig. 12.8, stiffeners are placed in pairs at the location of the load. In accordance with AISC 360 Sec. J10.8, the stiffener is designed as an axially loaded cruciform column with a section composed of the two stiffener plates plus a strip of web having a width of $25t_w$ at interior stiffeners and $12t_w$ at end stiffeners. The effective length factor is given by AISC 360 Sec. J10.8 as $K = 0.75$. The slenderness ratio is

$$KL/r = Kh/r$$

where $r$ is radius of gyration of the column about the axis of the web and $h$ is clear distance between flanges. When the slenderness ratio of the column is $KL/r \leq 25$, the nominal strength is given by AISC 360 Eq. (J4-6) as

$$P_n = F_y A_g$$

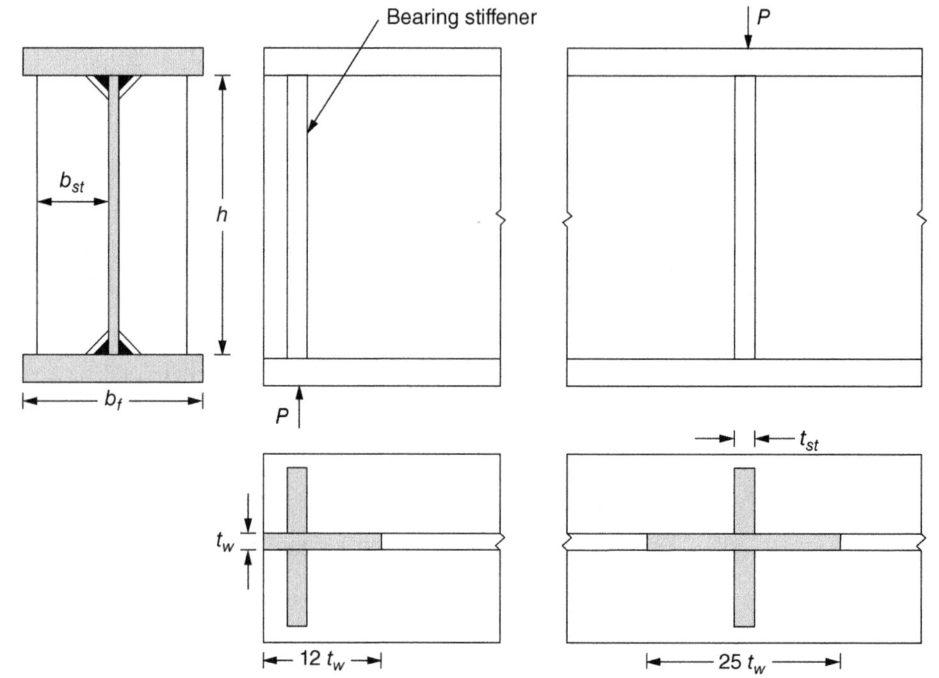


**Figure 12.8** Details of bearing stiffeners.

The available compression capacity is given by AISC 360 Sec. J4.4(a) as

$$P_c = \varphi P_n \text{ ... for LRFD load combinations}$$
$$= P_n/\Omega \text{ ... for ASD load combinations}$$

where $\varphi$ = resistance factor for compression
= 0.90
$\Omega$ = safety factor for compression
= 1.67

When the slenderness ratio of the column is $KL/r > 25$, the nominal strength is obtained from AISC 360 Chap. E.

Load-bearing stiffeners must extend for the full height of the web and provide close bearing on, or be welded to, the loaded flange. Where a concentrated load is directly over a support, this requirement applies to both top and bottom flanges. The stiffener plates should extend, approximately, to the edge of the flanges and the limiting width-thickness ratio of each plate, in accordance with AISC 360 Table B4.1a Case 3 is

$$b_{st}/t_{st} = 0.45(E/F_y)^{0.5}$$
$$= 12.8 \text{ ... for } F_y = 36 \text{ ksi}$$
$$= 10.8 \text{ ... for } F_y = 50 \text{ ksi}$$

The nominal bearing strength on the area of the stiffener plate in contact with the loaded flange is specified by AISC 360 Eq. (J7-1) as

$$R_n = 1.8F_yA_{pb}$$

The available bearing capacity is given by AISC 360 Sec. J7 as

$$R_c = \varphi R_n \text{ ... for LRFD load combinations}$$
$$= R_n/\Omega \text{ ... for ASD load combinations}$$

where $\varphi$ = resistance factor for bearing
= 0.75
$\Omega$ = safety factor for bearing
= 2.00

The capacity of the fillet weld between the stiffener plate and the web must be sufficient to transmit the applied compression force.

**Example 12.7.** Bearing Stiffener
The end bearing stiffener for the welded plate girder of Example 12.3 consists of a pair of 8½ × 3/4-in A36 plates. The end reaction consists of dead load $P_D = 70$ kips and live load $P_L = 210$ kips. Determine if the stiffener is adequate.

From Example 12.2:

$$h = 72 \text{ in}$$
$$t_w = 0.375 \text{ in}$$

The width-thickness ratio for each plate is

$$b_{st}/t_{st} = 8.5/0.75$$
$$= 11.3$$
$$< 12.8 \text{ ... satisfies AISC 360 Table B4.1a Case 3}$$

The effective column consisting of the two stiffener plates plus a strip of web having a width of $12t_w$ has a moment of inertia of

$$I \approx t_{st}(2b_{st} + t_w)^3/12$$
$$= 0.75 \times (17.375)^3/12$$
$$= 327.83 \text{ in}^4$$

The area of the effective column is

$$A = 2b_{st}t_{st} + 12t_w^2$$
$$= (2 \times 8.5 \times 0.75) + (12 \times 0.375^2)$$
$$= 14.44 \text{ in}^2$$

The radius of gyration of the effective column is

$$r = (I/A)^{0.5}$$
$$= (327.83/14.44)^{0.5}$$
$$= 4.77 \text{ in}$$

For an effective length factor of $K = 0.75$ the slenderness ratio of the effective column is

$$KL/r = 0.75h/r$$
$$= 0.75 \times 72/4.77$$
$$= 11.32$$
$$< 25$$

Hence, the nominal compression strength is given by AISC 360 Eq. (J4-6) as

$$P_n = F_y A_g$$
$$= 36 \times 14.44$$
$$= 520 \text{ kips}$$

Applying American Society of Civil Engineers, *Minimum Design Loads for Buildings and Other Structures* (ASCE 7)[4] Secs. 2.3 and 2.4 gives the factored loads as

| LRFD | ASD |
|---|---|
| From ASCE 7 Sec. 2.3.2 combination 2: | From ASCE 7 Sec. 2.4.1 combination 2: |
| $P_u$ = factored load | $P_a$ = factored load |
| $= 1.2P_D + 1.6P_L$ | $= P_D + P_L$ |
| $= 1.2 \times 70 + 1.6 \times 210$ | $= 70 + 210$ |
| $= 420$ kips | $= 280$ kips |
| The design compression strength is | The allowable compression strength is |
| $\varphi P_n = 0.9 \times 520$ | $P_n/\Omega = 520/1.67$ |
| $= 468$ kips | $= 311$ kips |
| $> P_u$ ... satisfactory | $> P_a$ ... satisfactory |

Allowing for a 1-in corner clip on the stiffeners to clear the flange-to-web fillet weld, the stiffener bearing area is

$$A_{pb} = 2t_{st}(b_{st} - 1)$$
$$= 2 \times 0.75(8.5 - 1)$$
$$= 11.25 \text{ in}^2$$

The nominal bearing strength is given by AISC 360 Eq. (J7-1) as

$$R_n = (1.8F_y A_{pb})$$
$$= 1.8 \times 36 \times 11.25$$
$$= 729 \text{ kips}$$

| LRFD | ASD |
|---|---|
| The design bearing strength is | The allowable bearing strength is |
| $\varphi R_n = 0.75 \times 729$ | $R_n/\Omega = 729/2$ |
| $= 547$ kips | $= 365$ kips |
| $> P_u$ ... satisfactory | $> P_a$ ... satisfactory |

The web thickness is 3/8 in and the minimum allowable fillet weld size, connecting the stiffeners to the web, is given by AISC 360 Table J2.4 as

$$w = 3/16 \text{ in}$$

Allowing for a 1-in corner clip at each end, provide 7 runs of 3/16-in E70XX intermittent fillet welds 4 in long spaced 7 in apart on both sides of both stiffeners. The total length of weld is

$$L = 4 \times 4 \times 7$$
$$= 112 \text{ in}$$

| LRFD | ASD |
|---|---|
| The design weld strength is | The allowable weld strength is: |
| $\varphi R_n = LDq_u$ | $R_n/\Omega = LDq$ |
| $= 112 \times 3 \times 1.392$ | $= 112 \times 3 \times 0.928$ |
| $= 468$ kips | $= 312$ kips |
| $> P_u$ ... satisfactory | $> P_a$ ... satisfactory |

## References


1. Nasados, P. P. 2009. "Design Aids for Built-Up I-Shaped Beams with Slender Webs," *Engineering Journal*, 1st Quarter 2009.
2. American Institute of Steel Construction (AISC). 2010. *Specification for Structural Steel Buildings* (AISC 360-10), AISC, Chicago, IL.
3. American Institute of Steel Construction (AISC). 2011. *Steel Construction Manual*, 14th edition. AISC, Chicago, IL.
4. American Society of Civil Engineers (ASCE). 2010. *Minimum Design Loads for Buildings and Other Structures* (ASCE 7-10), ASCE, Reston, VA.


## Problems

**12.1** *Given:* The welded plate girder, of Grade 50 steel, shown in Fig. 12.9. The web plate is 84 × 3/8 in and the flanges 18 × 2 in. No intermediate stiffeners are provided.

*Find:* Using allowable stress level design (ASD), the allowable shear capacity of the welded plate girder.

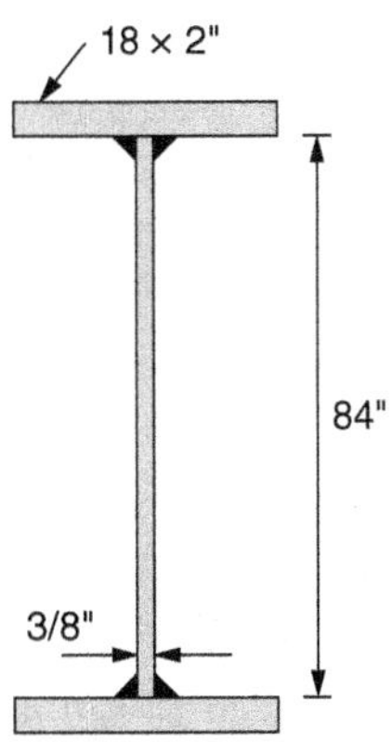


**Figure 12.9** Details for Problem 12.1.

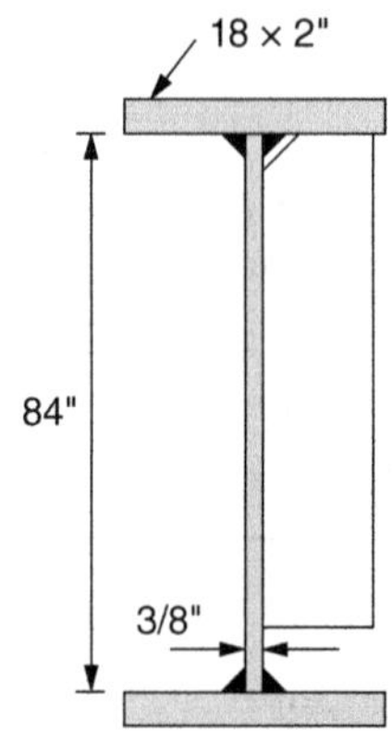


**FIGURE 12.10** Details for Problem 12.3.

**12.2** *Given:* The welded plate girder, of Grade 50 steel, shown in Fig. 12.9. The web plate is 84 × 3/8 in and the flanges 18 × 2 in. No intermediate stiffeners are provided.

*Find:* Using strength level design (LRFD), the design shear capacity of the welded plate girder.

**12.3** *Given:* The welded plate girder, of Grade 50 steel, shown in Fig. 12.10. The web plate is 84 × 3/8 in and the flanges 18 × 2 in. Tension field action is not utilized and intermediate stiffeners are provided at a clear spacing of 90 in on one side of the web only.

*Find:* Using allowable stress level design (ASD), the allowable shear capacity of the welded plate girder.

**12.4** *Given:* The welded plate girder, of Grade 50 steel, shown in Fig. 12.10. The web plate is 84 × 3/8 in and the flanges 18 × 2 in. Tension field action is not utilized and intermediate stiffeners are provided at a clear spacing of 90 in on one side of the web only.

*Find:* Using strength level design (LRFD), the allowable shear capacity of the welded plate girder.

**12.5** *Given:* The welded plate girder, of Grade 50 steel, shown in Fig. 12.10. The web plate is 84 × 3/8 in and the flanges 18 × 2 in. Tension field action is utilized and intermediate stiffeners are provided at a clear spacing of 90 in on one side of the web only.

*Find:* Using allowable stress level design (ASD), the allowable shear capacity of the welded plate girder.

**12.6** *Given:* The welded plate girder, of Grade 50 steel, shown in Fig. 12.10. The web plate is 84 × 3/8 in and the flanges 18 × 2 in. Tension field action is utilized and intermediate stiffeners are provided at a clear spacing of 90 in on one side of the web only.

*Find:* Using strength level design (LRFD), the allowable shear capacity of the welded plate girder.

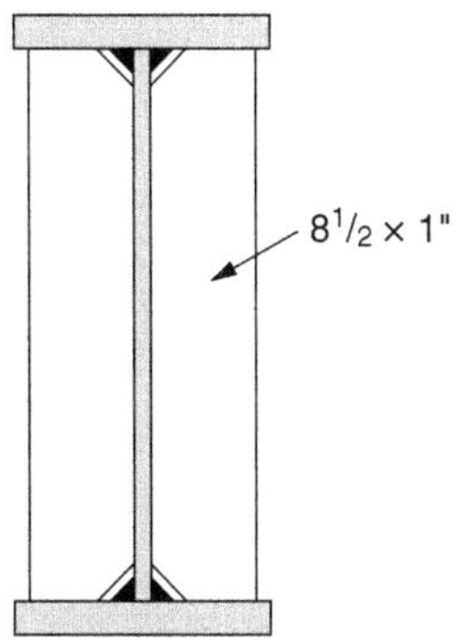


**FIGURE 12.11** Details for Problem 12.11.

**12.7** *Given:* The welded plate girder, of Grade 50 steel, shown in Fig. 12.10. The stiffeners are of A36 steel and are provided at a clear spacing of 90 in on one side of the web only. Tension field action is not utilized.

*Find:* Using allowable stress level design (ASD), a suitable stiffener size.

**12.8** *Given:* The welded plate girder, of Grade 50 steel, shown in Fig. 12.10. The stiffeners are of A36 steel and are provided at a clear spacing of 90 in on one side of the web only. Tension field action is not utilized.

*Find:* Using strength level design (LRFD), a suitable stiffener size.

**12.9** *Given:* The welded plate girder, of Grade 50 steel, shown in Fig. 12.10. The stiffeners are of A36 steel and are provided at a clear spacing of 90 in on one side of the web only. Tension field action is utilized.

*Find:* Using allowable stress level design (ASD), a suitable stiffener size.

**12.10** *Given:* The welded plate girder, of Grade 50 steel, shown in Fig. 12.10. The stiffeners are of A36 steel and are provided at a clear spacing of 90 in on one side of the web only. Tension field action is utilized.

*Find:* Using strength level design (LRFD), a suitable stiffener size.

**12.11** *Given:* As shown in Fig. 12.11, the end bearing stiffener for the welded plate girder of Problem 12.9 consists of a pair of 8½ × 1-in A36 plates. The stiffeners have a 1-in corner clip.

*Find:* Using allowable stress level design (ASD), the allowable end reaction.

**12.12** *Given:* As shown in Fig. 12.11, the end bearing stiffener for the welded plate girder of Problem 12.10 consists of a pair of 8½ × 1-in A36 plates. The stiffeners have a 1-in corner clip.

*Find:* Using strength level design (LRFD), the design end reaction.

# CHAPTER 13
# Composite Members

## 13.1 Introduction

In a composite member, the steel and concrete are interconnected so as to resist the applied loading as a whole. As shown in Fig. 13.1, a composite column may consist of

- A steel section completely encased in structural concrete
- A rectangular, square, or round hollow structural section or pipe filled with structural concrete

The advantage of a filled section is that the hollow structural section acts as formwork for the concrete. Since the concrete acts compositely with the steel column, the concrete provides additional load-carrying capacity and stiffness to the column. Because of the restraining effect of the concrete fill, the local buckling strength of the tube is increased. An additional economic advantage is that concrete has relatively low material costs.

The advantage of an encased section is that the concrete provides fireproofing and protection from corrosion to the steel section. Since the concrete acts compositely with the steel column, the concrete provides additional load-carrying capacity and stiffness to the column. Alternatively, the size of the steel section may be reduced. Cost savings are obtained because of the relative low material cost of concrete.

As shown in Fig. 13.2, a composite beam may consist of

- A steel shape completely encased in concrete
- A flat soffit concrete slab cast on a steel shape with the slab connected to the steel shape with steel anchors
- A concrete slab poured on a formed steel deck supported on a steel shape with steel anchors to connect the slab to the steel shape
- A concrete slab poured on a formed steel deck supported on an open web steel joist with steel anchors as described by Samuelson.[1]

A composite beam is more efficient than a concrete slab cast noncompositely on a steel beam. This results in a reduction in the size of steel beam required for a given imposed load or an increase in span length for a given beam size. The advantage of an encased section is that the concrete provides fireproofing and protection from corrosion to the steel section. The advantage of a composite steel joist is the ability to route services through the joist open webs. The advantage of using a formed steel deck is that

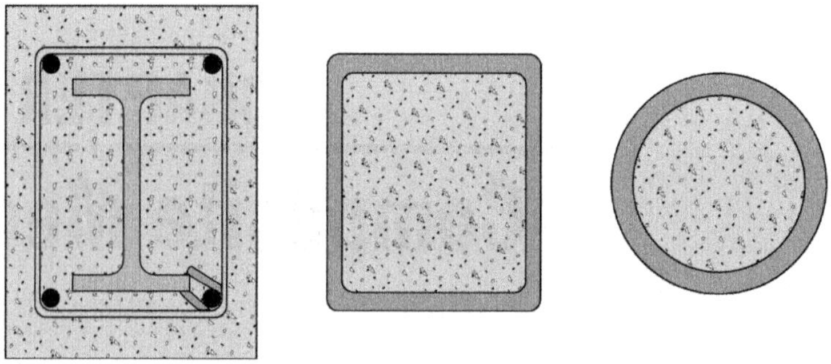

**FIGURE 13.1** Composite columns.

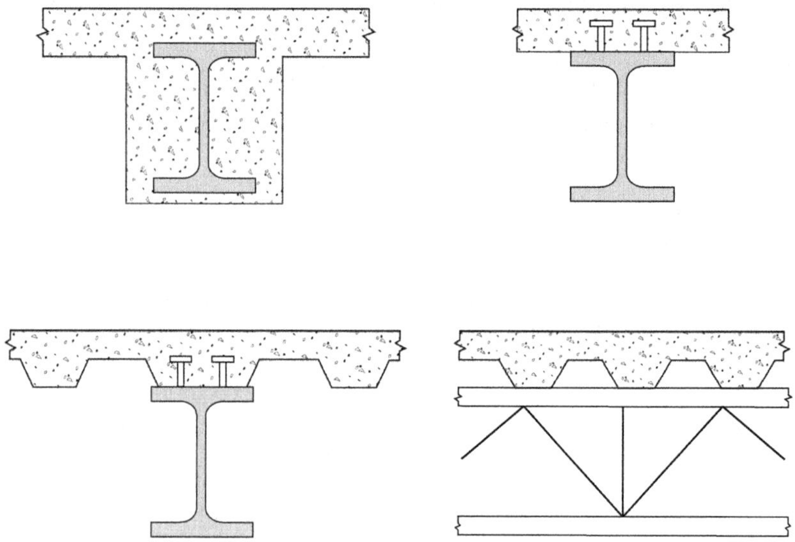

**FIGURE 13.2** Composite beams.

the deck provides formwork to support the poured concrete deck. A disadvantage of a composite floor is that the combination of longer span and reduced mass may produce a vibration serviceability problem.

In accordance with American Institute of Steel Construction, *Specification for Structural Steel Buildings* (AISC 360)[2] Sec. I1.2, design of composite sections may be either by the plastic stress distribution method or the strain-compatibility method.

For the plastic stress distribution method, the nominal strength is determined assuming that steel components have reached a stress of $F_y$ in either tension or compression and concrete components in compression have reached a stress of $0.85f'_c$. For round hollow structural sections filled with concrete, a stress of $0.95f'_c$ may be used to account for the effects of concrete confinement.

For irregular sections and beam-columns, the strain compatibility method is used. This assumes a linear distribution of strains across the section with the maximum concrete compressive strain equal to 0.003 in/in. The steel components are assumed to exhibit ideal elastic-plastic properties. A review of the method is given by Leon and Hajjar.[3]

## 13.2 Encased Composite Columns

### Limitations

As shown in Fig. 13.1, an encased composite column consists of a concrete encased rolled steel section reinforced with longitudinal and lateral reinforcing bars. In order to qualify as an encased composite column, AISC 360 Secs. I1.3 and I2.1a impose the following limitations:

- The minimum compressive strength of normal weight concrete is 3 ksi and the maximum is 10 ksi. The minimum compressive strength of lightweight concrete is 3 ksi and the maximum is 6 ksi.
- In determining the capacity of a composite column, the assumed yield stress of both the structural steel element and the reinforcing shall not exceed 75 ksi.
- The cross-sectional area of the structural steel element shall not be less than 1 percent of the total area of the composite section.
- The concrete encasement of a structural steel element shall be reinforced with continuous longitudinal bars and lateral ties or spirals.
- The cross-sectional area of the longitudinal reinforcing shall not be less than 0.4 percent of the gross area of the composite member and at least four continuous bars shall be used.
- Where lateral ties are used, a minimum of either a No. 3 bar spaced at a maximum of 12 in on center or a No. 4 bar spaced at a maximum of 16 in on center is required. To ensure good confinement of the concrete, the maximum spacing of lateral ties must not exceed 0.5 times the least dimension of the composite section.

### Compressive Strength

The development of the design provisions for composite columns is detailed by Leon et al.[4] The nominal axial compressive strength, or squash load, without consideration of length effects, is given by AISC 360 Eq. (I2-4) as the sum of the plastic capacities of the components

$$P_{no} = A_s F_y + A_{sr} F_{ysr} + 0.85 f'_c A_c$$

where $F_y$ = yield stress of steel section
$F_{ysr}$ = yield stress of longitudinal reinforcing bars
$A_s$ = area of steel section
$A_{sr}$ = area of continuous longitudinal reinforcing bars
$A_c$ = concrete area
$f'_c$ = concrete compressive strength

The elastic critical buckling load is given by AISC 360 Eq. (I2-5) as

$$P_e = \pi^2 EI_{eff}/(KL)^2$$

where $EI_{eff}$ = effective stiffness of composite section from AISC 360 Eq. (I2-6), kip-in$^2$
$= E_s I_s + 0.5 E_s I_{sr} + C_1 E_c I_c$
$E_s$ = modulus of elasticity of steel
= 29,000 ksi
$E_c$ = modulus of elasticity of concrete, ksi
$= w_c^{1.5}(f'_c)^{0.5}$
$w_c$ = unit weight of concrete, lb/ft$^3$
$I_c$ = moment of inertia of concrete section about the elastic neutral axis of the composite section, in$^4$
$I_s$ = moment of inertia of steel shape about the elastic neutral axis of the composite section, in$^4$
$I_{sr}$ = moment of inertia of reinforcing bars about the elastic neutral axis of the composite section, in$^4$
$C_1$ = coefficient for calculation of effective rigidity of an encased composite section from AISC 360 Eq. (I2-7)
$= 0.1 + 2A_s/(A_c + A_s)$
$\leq 0.3$
$K$ = effective length factor
$L$ = laterally unbraced length, in

For a short column with $P_{no} \leq 2.25P_e$, inelastic buckling governs and the nominal axial strength is given by AISC 360 Eq. (I2-2) as

$$P_n = P_{no}(0.658)^{\kappa}$$

where $\kappa$ is equal to $P_{no}/P_e$.

For a long column with $P_{no} > 2.25P_e$, elastic buckling governs and the nominal axial strength is given by AISC 360 Eq. (I2-3) as

$$P_n = 0.877P_e$$

The design compressive strength and the allowable compressive strength may be obtained from AISC 360, Sec. I2.1b as

$\varphi_c P_n$ = design compressive strength
$\geq P_u$ ... required compressive strength using LRFD load combinations

$P_n/\Omega_c$ = allowable compressive strength
$\geq P_a$ ... required compressive strength using ASD load combinations

where $\varphi_c$ = resistance factor for compression
= 0.75
$\Omega_c$ = safety factor for compression
= 2.00

**Example 13.1.** Encased Composite Column
Determine the available axial strength for a composite column, consisting of a W8 × 31 shape with 18 × 18-in concrete encasement. The steel section has a yield stress of $F_y = 50$ ksi and the concrete has a specified strength of $f'_c = 4$ ksi and a weight of 145 lb/ft³. Continuous longitudinal reinforcement is provided consisting of four No. 8 Grade 60 bars at 14 in on center. Adequate lateral reinforcement and load transfer is provided. The column is 24 ft high and is pinned at each end with the load applied directly to the steel section.

Gross area of the composite section is

$$A_g = 18 \times 18$$
$$= 324 \text{ in}^2$$
$$0.01A_g = 3.24 \text{ in}^2$$
$$0.004A_g = 1.30 \text{ in}^2$$

From American Institute of Steel Construction, *Steel Construction Manual* (AISC Manual)[5] Table 1-1, the relevant values are

$$A_s = \text{area of steel section}$$
$$= 9.13 \text{ in}^2$$
$$> 3.24 \text{ in}^2 \text{ ... satisfies AISC 360 Sec. I2.1a (1)}$$
$$I_s = \text{moment of inertia of steel section about } y\text{-axis}$$
$$= 37.1 \text{ in}^4$$

The reinforcement area is

$$A_{sr} = 3.16 \text{ in}^2$$
$$> 1.3 \text{ in}^2 \text{ ... satisfies AISC 360 Sec. I2.1a (3)}$$

The distance between the centroid of reinforcing bars is

$$y = 14 \text{ in}$$

The moment of inertia of the reinforcing bars is

$$I_{sr} = A_{sr}(y/2)^2 + 4\pi d_b^4/64$$
$$= 3.16 \times 7^2 + 4 \times 3.14 \times 1.0^4/64$$
$$= 155 \text{ in}^4$$

The concrete area is

$$A_c = A_g - A_s - A_{sr}$$
$$= 18 \times 18 - 9.13 - 3.16$$
$$= 312 \text{ in}^2$$

The nominal axial compressive stress without consideration of length effects is given by AISC 360 Eq. (I2-4) as

$$P_{no} = A_sF_y + A_{sr}F_{ysr} + 0.85f'_cA_c$$
$$= 9.13 \times 50 + 3.16 \times 60 + 0.85 \times 4 \times 312$$
$$= 1707 \text{ kips}$$

The concrete reduction factor is given by AISC 360 Eq. (I2-7) as

$$\begin{aligned} C_1 &= 0.1 + 2A_s/(A_c + A_s) \\ &= 0.1 + 2 \times 9.13/(312 + 9.13) \\ &= 0.157 \\ &< 0.3 \text{ ... satisfactory} \end{aligned}$$

The moment of inertia of the concrete section is

$$\begin{aligned} I_c &= I_g - I_s - I_{sr} \\ &= 18^4/12 - 37.1 - 155 \\ &= 8556 \text{ in}^4 \end{aligned}$$

The modulus of elasticity of the concrete is

$$\begin{aligned} E_c &= w_c^{1.5}(f'_c)^{0.5} \\ &= 145^{1.5}(4)^{0.5} \\ &= 3492 \text{ ksi} \end{aligned}$$

The effective stiffness of the composite section is given by AISC 360 Eq. (I2-6) as

$$\begin{aligned} EI_{eff} &= E_s I_s + 0.5E_s I_{sr} + C_1 E_c I_c \\ &= 29{,}000 \times 37.1 + 0.5 \times 29{,}000 \times 155 + 0.157 \times 3492 \times 8556 \\ &= 8{,}014{,}176 \text{ kip-in}^2 \end{aligned}$$

The elastic critical buckling stress is given by AISC 360 Eq. (I2-5) as

$$\begin{aligned} P_e &= \pi^2 EI_{eff}/(KL)^2 \\ &= 3.14^2 \times 8{,}014{,}176/(1 \times 24 \times 12)^2 \\ &= 953 \text{ kips} \\ P_{no} &= 1707 \text{ kips} \\ &< 2.25P_e \\ &= 2144 \text{ kips} \end{aligned}$$

Hence, AISC 360 Eq. (I2-2) is applicable and the nominal axial strength is

$$\begin{aligned} P_n &= P_{no}(0.658)^{\kappa} \\ &= 1707(0.658)^{1707/953} \\ &= 807 \text{ kips} \end{aligned}$$

The available axial strength of the composite column is obtained from AISC 360 Sec I2.1b as

| LRFD | ASD |
|---|---|
| $\varphi_c P_n = 0.75 \times 807$<br>$= 605$ kips | $P_n/\Omega_c = 807/2$<br>$= 404$ kips |

## Load Transfer

In encased composite members steel headed stud anchors or channel anchors are used to transfer longitudinal shear at the interface. The longitudinal shear is determined from the full plastic capacity of the two materials. When the entire external force $P_r$ is applied to the concrete encasement, the shear at the interface is given by AISC 360 Eq. (I6-2) as

$$
\begin{aligned}
V'_r &= P_r(A_sF_y/P_{no}) \\
&= \text{force in steel section at equilibrium} \\
&= P_r\,(\text{steel strength/member strength})
\end{aligned}
$$

When the entire external force is applied to the steel section, the shear at the interface is given by AISC 360 Eq. (I6-1) as

$$
\begin{aligned}
V'_r &= P_r(1 - A_sF_y/P_{no}) \\
&= \text{force in concrete at equilibrium} \\
&= P_r\,(\text{concrete strength /member strength})
\end{aligned}
$$

When the external force is applied concurrently to the two materials, the shear at the interface is given by AISC 360 Commentary Sec. I6.2 as

$$V'_r = P_s - P_r(A_sF_y/P_{no})$$

where $P_r$ = required external force applied to the composite member
$P_{no}$ = nominal axial compressive strength without consideration of length effects as determined by AISC 360 Eq. (I2-4) for an encased composite member
$P_s$ = portion of external force applied directly to the steel section

The portion of the external force applied to the steel section, $P_s$, is determined by considering the axial stiffness of the individual components of the composite member. Then,

$$P_s = P_rA_sE_s/(A_sE_s + A_cE_c + A_{sr}E_{sr})$$

where $A_c = A_g - A_s - A_{sr}$
$E_s$ = modulus of elasticity of steel section
= 29,000 ksi
$E_{sr}$ = modulus of elasticity of continuous longitudinal reinforcing bars
= 29,000 ksi
$E_c$ = modulus of elasticity of concrete, ksi
$= w_c^{1.5}(f'_c)^{0.5}$
$w_c$ = unit weight of concrete, lb/ft$^3$
$f'_c$ = concrete compressive strength
$A_s$ = area of steel section
$A_{sr}$ = area of continuous longitudinal reinforcing bars
$A_c$ = concrete area
$A_g$ = gross area of composite
$P_r$ = external force applied to composite member

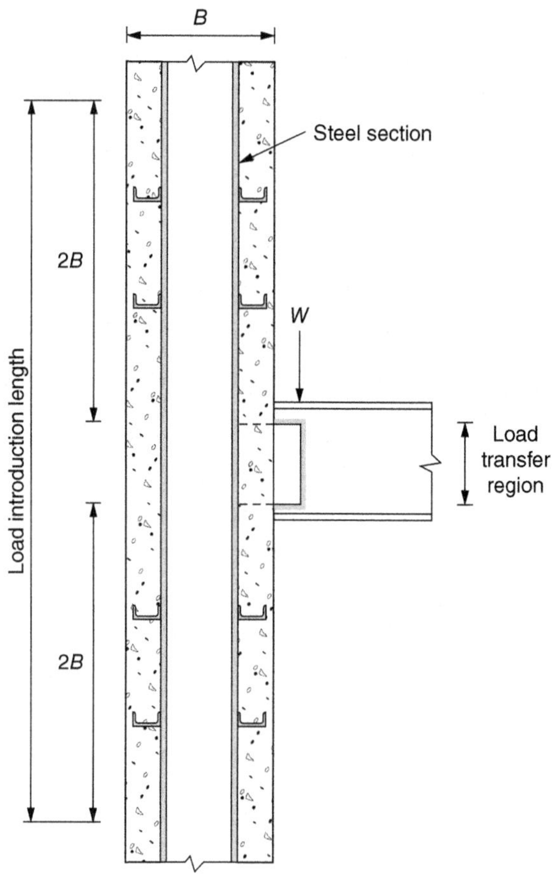


**Figure 13.3** Load introduction length for encased composite member.

To avoid overstressing the steel section or the concrete encasement, the transfer of the longitudinal shear is required by AISC 360 Sec. I6.4 to occur within the load introduction length. As shown in Fig. 13.3, this is assumed to extend a distance of twice the minimum dimension of the composite member both above and below the load transfer region.

The nominal shear strength of a steel headed stud anchor, when concrete breakout is precluded by the encasement, is given by AISC 360 Eq. (I8-3) as

$$Q_{nv} = A_{sc}F_u$$

where $A_{sc}$ is cross-sectional area of stud anchor and $F_u$ is tensile strength of stud anchor. The available shear capacity is obtained from AISC 360, Sec. I8.3a as

$$\begin{aligned}\varphi_v Q_{nv} &= \text{design shear capacity}\\ &= 0.65Q_{nv}\\ Q_{nv}/\Omega_v &= \text{allowable shear capacity}\\ &= Q_{nv}/2.31\end{aligned}$$

| Anchor diameter, in | 1/2 | 5/8 | 3/4 |
|---|---|---|---|
| Nominal shear strength $Q_{nv}$, kips | 12.8 | 20.0 | 28.7 |

**TABLE 13.1** Nominal Shear Strength of Steel Headed Stud Anchors

The nominal shear strength of steel headed stud anchors with a tensile strength of 65 ksi is given in Table 13.1.

Dimensional requirements for stud anchors are given in AISC 360 Secs. I8.3 and I8.3e. For normal weight concrete, stud anchors subjected to shear only are required to be five-stud diameters in length from the base of the stud to the top of the stud head after installation. For lightweight concrete, stud anchors subjected to shear only are required to be seven-stud diameters in length from the base of the stud to the top of the stud head after installation. A minimum of one inch lateral clear concrete cover is required to stud anchors. The minimum required spacing is four diameters in any direction and the maximum spacing is 32 times the shank diameter. To maintain composite action over the length of the column, stud anchors are required over the remainder of the composite member at a maximum spacing of 32 times the shank diameter.

The nominal shear strength of a steel channel anchor is given by AISC 360 Eq. (I8-2) as

$$Q_n = 0.3(t_f + 0.5t_w)\ell_c(f'_c E_c)^{0.5}$$

where $\ell_c$ = length of channel anchor
$t_f$ = flange thickness of channel anchor
$t_w$ = web thickness of channel anchor

The available shear capacity is obtained from AISC 360, Sec. I8.3d as

$$\begin{aligned}\varphi_s Q_n &= \text{design shear capacity}\\ &= 0.75Q_n\\ Q_n/\Omega_s &= \text{allowable shear capacity}\\ &= Q_n/2.00\end{aligned}$$

The nominal shear strength of a $3 \times 6$ steel channel anchor with a length of 1 in is given in Table 13.2.

A minimum of one inch lateral clear concrete cover is required to channel anchors. The maximum permitted spacing is 24 in. To maintain composite action over the length of the column, channel anchors are required over the remainder of the composite member at a maximum spacing 24 in.

| Concrete strength $f'_c$, ksi | 3 | 4 | 5 |
|---|---|---|---|
| Concrete modulus of elasticity $E_c$, ksi | 3024 | 3492 | 3904 |
| Nominal shear strength $Q_n$, kips | 12.9 | 16.0 | 18.9 |

**TABLE 13.2** Nominal Shear Strength of a C3 × 6 Steel Channel Anchor with a Length of 1 in

**Example 13.2.** Load Transfer in Encased Composite Column
The encased composite column of Example 13.1 has two beams framing into the top of the column opposite to the beam flanges. The load is applied directly to the steel section. Both beams have a depth of 12 in and the total applied load is 600 kips (LRFD load combinations) or 400 kips (ASD load combinations). Using C3 × 6 channel anchors determine the load transfer requirements.

From Example 13.1

$$A_s = 9.13 \text{ in}^2$$
$$F_y = 50 \text{ ksi}$$
$$B = \text{overall width of encasement}$$
$$= 18 \text{ in}$$
$$P_{no} = 1707 \text{ kips}$$

The external force is applied directly to the steel section and the shear at the interface is given by AISC 360 Eq. (I6-1) as

| LRFD | ASD |
|---|---|
| $V'_r = P_r(1 - A_sF_y/P_{no})$ | $V'_r = P_r(1 - A_sF_y/P_{no})$ |
| $= 600(1 - 9.13 \times 50/1707)$ | $= 400(1 - 9.13 \times 50/1707)$ |
| $= 440$ kips | $= 293$ kips |

Since the column extends only below the point of application of the external force, the transfer of the longitudinal shear is required by AISC 360 Sec. I6.4 to occur within the load introduction length, underneath the load transfer region, which is

$$L_i = 2B$$
$$= 2 \times 18$$
$$= 36 \text{ in}$$

Weld four C3 × 6 channel anchors with a length of 5 in at 10 in on center to both flanges of the W8 × 31, underneath the load transfer region, to provide an available shear capacity of

| LRFD | ASD |
|---|---|
| $\varphi R_n = 0.75 \times 8 \times 5 \times 16.0$ ... from Table 13.2 | $R_n/\Omega = 8 \times 5 \times 16/2$ ... from Table 13.2 |
| $= 480$ kips | $= 320$ kips |
| $> V'_r$ ... satisfactory | $> V'_r$ ... satisfactory |

## 13.3 Filled Composite Columns

### Limitations

As shown in Fig. 13.1, a filled composite column consists of a rectangular, square, or round hollow structural section filled with structural concrete. In order to qualify as a filled composite column, AISC 360 Secs. I1.3, I1.4, and I2.2a impose the following limitations:

- The minimum compressive strength of normal weight concrete is 3 ksi and the maximum is 10 ksi. The minimum compressive strength of lightweight concrete is 3 ksi and the maximum is 6 ksi.
- In determining the capacity of a composite column, the assumed yield stress of both the structural steel element and the reinforcing shall not exceed 75 ksi.
- The maximum width-to-thickness ratio of concrete filled rectangular and square hollow structural sections is

$$b/t = 5.00(E/F_y)^{0.5}$$
$$= 126 \ldots \text{for } F_y = 46 \text{ ksi}$$

- The maximum diameter-to-thickness ratio of concrete filled round hollow structural sections and pipes is

$$D/t = 0.31E/F_y$$
$$= 214 \ldots \text{for } F_y = 42 \text{ ksi}$$
$$= 257 \ldots \text{for } F_y = 35 \text{ ksi}$$

- The cross-sectional area of the structural steel element shall not be less than 1 percent of the total area of the composite section.

## Slenderness Limits

The slenderness limits for a concrete filled rectangular or square hollow structural section in axial compression are given in AISC 360 Table I1.1a. The limiting width-to-thickness ratio of a compact section is

$$b/t = \lambda_p$$
$$= 2.26(E/F_y)^{0.5}$$

The limiting width-to-thickness ratio of a noncompact section is

$$b/t = \lambda_r$$
$$= 3.00(E/F_y)^{0.5}$$

All square hollow structural sections are compact for axial compression with the exception of HSS7 × 7 × 1/8, HSS 8 × 8 × 1/8, HSS9 × 9 × 1/8 and HSS12 × 12 × 3/16 which are noncompact.

The slenderness limits for a concrete filled round section in axial compression are given in AISC 360 Table I1.1a. The limiting diameter-to-thickness ratio of a compact section is

$$D/t = \lambda_p$$
$$= 0.15E/F_y$$

The limiting diameter-to-thickness ratio of a noncompact section is

$$D/t = \lambda_r$$
$$= 0.19E/F_y$$

All round hollow structural sections are compact for axial compression.

## Compressive Strength

The full plastic strength in compression, or the squash load, is given by AISC 360 Eq. (I2-9b) as

$$P_p = A_sF_y + C_2f'_c(A_c + A_{sr}E_s/E_c)$$

where $C_2 = 0.85$ ... for a rectangular section
$= 0.95$ ... for a circular section

The elastic critical buckling load is given by AISC 360 Eq. (I2-5) as

$$P_e = \pi^2 EI_{eff}/(KL)^2$$

where $EI_{eff}$ = effective stiffness of composite section from AISC 360 Eq. (I2-12), kip-in²
$= E_sI_s + E_sI_{sr} + C_3E_cI_c$
$C_3$ = coefficient for calculation of effective rigidity of a filled composite section from AISC 360 Eq. (I2-13)
$= 0.6 + 2A_s/(A_c + A_s)$
$\leq 0.9$

The nominal axial compressive strength without consideration of length effects $P_{no}$ is determined separately for compact, noncompact, and slender steel elements as follows.

For a compact section with $\lambda \leq \lambda_p$ AISC 360 Eq. (I2-9a) gives

$$P_{no} = P_p$$

For a noncompact section with $\lambda_p < \lambda \leq \lambda_r$ AISC 360 Eq. (I2-9c) gives

$$P_{no} = P_p - (P_p - P_y)(\lambda - \lambda_p)^2/(\lambda_r - \lambda_p)^2$$

where $P_y = A_sF_y + 0.7f'_c(A_c + A_{sr}E_s/E_c)$ ... from AISC 360 Eq. (I2-9d).

For a slender section with $\lambda > \lambda_r$ AISC 360 Eq. (I2-9e) gives

$$P_{no} = A_sF_{cr} + 0.7f'_c(A_c + A_{sr}E_s/E_c)$$

For a rectangular section, AISC 360 Eq. (I2-10) gives

$$F_{cr} = 9E_s/(b/t)^2$$

For a round section, AISC 360 Eq. (I2-11) gives

$$F_{cr} = 0.72F_y/[(D/t)(F_y/E_s)]^{0.2}$$

The nominal axial compressive strength allowing for length effects $P_n$ is determined from AISC 360 Sec. I2.1b as follows.

For a short column with $P_{no} \leq 2.25P_e$, inelastic buckling governs and the nominal axial strength is given by AISC 360 Eq. (I2-2) as

$$P_n = P_{no}(0.658)^{\kappa}$$

where $\kappa$ is equal to $P_{no}/P_e$

For a long column with $P_{no} > 2.25P_e$, elastic buckling governs and the nominal axial strength is given by AISC 360 Eq. (I2-3) as

$$P_n = 0.877P_e$$

The available compressive strength is obtained from AISC 360, Sec. I2.1b as

$$\varphi_c P_n = \text{design compressive strength}$$
$$\geq P_u \text{ ... required compressive strength using LRFD load combinations}$$
$$P_n/\Omega_c = \text{allowable compressive strength}$$
$$\geq P_a \text{ ... required compressive strength using ASD load combinations}$$

where $\varphi_c$ = resistance factor for compression
$= 0.75$
$\Omega_c$ = safety factor for compression
$= 2.00$

Values of the available axial load are tabulated in AISC Manual Tables 4-17 to 4-20 for round hollow structural members and pipes with a yield stress of 42 ksi and 35 ksi and with concrete fill strengths of 4 ksi and 5 ksi. Values of the available axial load are tabulated in AISC Manual Tables 4-13 to 4-16 for rectangular and square hollow structural members with a yield stress of 46 ksi and with concrete fill strengths of 4 ksi and 5 ksi.

**Example 13.3.** Filled Composite Column
Determine the available axial strength for a composite column, consisting of a $6 \times 6 \times 1/4$-in square hollow structural member with a yield stress of 46 ksi, filled with 4 ksi concrete with a weight of 145 lb/ft$^3$. Adequate load transfer is provided. The column is 24 ft high and is pinned at each end.

From AISC Manual Table 1-12, the relevant properties of the square hollow structural section are

$$A_s = 5.24 \text{ in}^2$$
$$I_s = 28.6 \text{ in}^4$$
$$t = \text{design wall thickness}$$
$$= 0.233 \text{ in}$$
$$b/t = 22.8$$
$$< 2.26(E/F_y)^{0.5} \text{ ... section is compact}$$
$$b = \text{outside nominal side length}$$
$$= 6 \text{ in}$$

Neglecting the corner radius, the concrete area is given by

$$A_c = (b - 2t)^2$$
$$= (6 - 0.5)^2$$
$$= 30.25 \text{ in}^2$$
$$A_c + A_s = 30.25 + 5.24$$
$$= 35.49 \text{ in}^2$$

and
$$A_s/(A_c + A_s) = 5.25/35.49$$
$$= 0.148$$
$$> 0.01 \text{ ... conforms to AISC 360 Sec. I2.2a}$$

Neglecting the corner radius, the moment of inertia of the concrete section is given by

$$I_c = (b - 2t)^4/12$$
$$= (6 - 0.5)^4/12$$
$$= 76.26 \text{ in}^4$$

The full plastic strength in compression without consideration of length effects is given by AISC 360 Eq. (I2-9b) as

$$P_p = A_s F_y + C_2 f'_c A_c$$
$$= 5.24 \times 46 + 0.85 \times 4 \times 30.25$$
$$= 344 \text{ kips}$$
$$= P_{no} \text{ ... for a compact section}$$
$$C_3 = \text{coefficient for calculation of effective rigidity of a filled composite section from AISC 360 Eq. (I2-13)}$$
$$= 0.6 + 2A_s/(A_c + A_s)$$
$$= 0.6 + 2 \times 0.148$$
$$= 0.896$$
$$< 0.9 \text{ ... satisfactory}$$

The modulus of elasticity of the concrete is

$$E_c = w_c^{1.5}(f'_c)^{0.5}$$
$$= 145^{1.5}(4)^{0.5}$$
$$= 3492 \text{ ksi}$$

The effective stiffness of the composite section is given by AISC 360 Eq. (I2-12) as

$$EI_{eff} = E_s I_s + C_3 E_c I_c$$
$$= 29{,}000 \times 28.6 + 0.896 \times 3492 \times 76.26$$
$$= 1{,}068{,}005 \text{ kip-in}^2$$

The elastic critical buckling stress is given by AISC 360 Eq. (I2-5) as

$$\begin{aligned}P_e &= \pi^2 EI_{eff}/(KL)^2\\ &= 3.14^2 \times 1{,}068{,}005/(1 \times 24 \times 12)^2\\ &= 127 \text{ kips}\\ P_{no} &= 344 \text{ kips}\\ &> 2.25P_e\\ &= 286 \text{ kips}\end{aligned}$$

Hence, AISC 360 Eq. (I2-3) is applicable and the nominal axial strength is

$$\begin{aligned}P_n &= 0.877P_e\\ &= 0.877 \times 127\\ &= 111.4 \text{ kips}\end{aligned}$$

The available axial strength of the composite column is obtained from AISC 360 Sec I2.1b as

| LRFD | ASD |
|---|---|
| $\varphi_c P_n = 0.75 \times 111.4$ | $P_n/\Omega_c = 111.4/2$ |
| $= 83.6$ kips | $= 55.7$ kips |
| From AISC Manual Table 4-15: | From AISC Manual Table 4-15: |
| $\varphi_c P_n = 83.9$ kips | $P_n/\Omega_c = 55.9$ kips |

## Load Transfer

In filled composite members, because of the confinement provided by the steel encasement, direct bond interaction may be utilized to transfer longitudinal shear at the interface. The longitudinal shear is determined from the full plastic capacity of the two materials. When the entire external force is applied to the concrete fill, the shear at the interface is given by AISC 360 Eq. (I6-2) as

$$\begin{aligned}V'_r &= P_r(A_sF_y/P_{no})\\ &= \text{force in steel at equilibrium}\\ &= P_r \text{ (steel strength/member strength)}\end{aligned}$$

When the entire external force is applied to the steel section, the shear at the interface is given by AISC 360 Eq. (I6-1) as

$$\begin{aligned}V'_r &= P_r(1 - A_sF_y/P_{no})\\ &= \text{force in concrete at equilibrium}\\ &= P_r \text{ (concrete strength/member strength)}\end{aligned}$$

When the external force is applied concurrently to the two materials, the shear at the interface is given by AISC 360 Commentary Sec. I6.2 as

$$V'_r = P_s - P_r(A_sF_y/P_{no})$$

where $P_r$ = required external force applied to the composite member
$P_{no}$ = nominal axial compressive strength without consideration of length effects as determined by AISC 360 Eq. (I2-9a) for a filled composite member
$P_s$ = portion of external force applied directly to the steel section

The portion of the external force applied to the steel section, $P_s$, is determined by considering the axial stiffness of the individual components of the composite member. Then,

$$P_s = P_r A_s E_s / (A_s E_s + A_c E_c + A_{sr} E_{sr})$$

where $A_c = A_g - A_s - A_{sr}$
$E_s$ = modulus of elasticity of steel section
= 29,000 ksi
$E_{sr}$ = modulus of elasticity of continuous longitudinal reinforcing bars
= 29,000 ksi
$E_c$ = modulus of elasticity of concrete, ksi
$= w_c^{1.5}(f'_c)^{0.5}$
$w_c$ = unit weight of concrete, lb/ft$^3$
$f'_c$ = concrete compressive strength
$A_s$ = area of steel section
$A_{sr}$ = area of continuous longitudinal reinforcing bars
$A_c$ = concrete area
$A_g$ = gross area of composite
$P_r$ = external force applied to composite member

To avoid overstressing the steel section or the concrete fill, the transfer of the longitudinal shear is required by AISC 360 Sec. I6.4 to occur within the load introduction length. As shown in Fig. 13.4, this is assumed to extend a distance of twice the minimum dimension of the composite member both above and below the load transfer region.

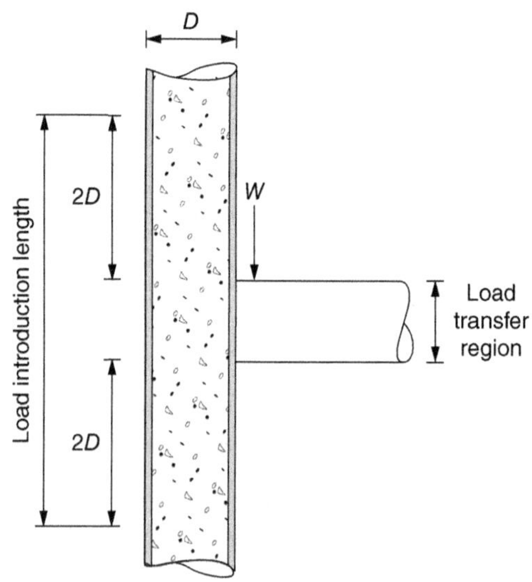


**FIGURE 13.4** Load introduction length for filled composite member.

When a connecting beam frames into one side of the composite member, it is assumed that bond interaction occurs on one face of a rectangular filled composite column or one-quarter of the perimeter of a round filled composite column. Hence the nominal bond capacity for a rectangular filled composite is given by AISC 360 Eq. (I6-5) as

$$R_n = B^2 C_{in} F_{in}$$

The nominal bond capacity for a round filled composite is given by AISC 360 Eq. (I6-6) as

$$R_n = 0.25\pi D^2 C_{in} F_{in}$$

where $B$ = overall width of rectangular steel section along face transferring load
$D$ = outside diameter of round steel section
$C_{in}$ = 2 if the composite member extends to one side of the point of load transfer
= 4 if the composite member extends both sides of the point of load transfer
$F_{in}$ = nominal bond stress
= 0.06 ksi

If connecting elements frame into the composite column from multiple sides, the bond interaction capacity may be increased accordingly.

The design bond capacity and the allowable bond capacity may be obtained from AISC 360, Sec. I6.3c as

$$\varphi R_n = \text{design bond capacity}$$
$$= 0.45 R_n$$
$$R_n/\Omega = \text{allowable bond capacity}$$
$$\geq R_n/3.33$$

When the external load is applied to the concrete of a filled composite member through a bearing plate, the confinement provided by the steel encasement allows the maximum bearing strength permitted by AISC 360 Eq. (J8-2) to be applied. Hence, the nominal bearing strength is

$$P_p = 1.7 f'_c A_1$$

where $A_1$ is loaded area of concrete.

The design bearing capacity and the allowable bearing capacity may be obtained from AISC 360, Sec. J8 as

$$\varphi_c P_p = \text{design bearing capacity}$$
$$= 0.65 P_p$$
$$P_p/\Omega_c = \text{allowable bearing capacity}$$
$$= P_p/2.31$$

**Example 13.4.** Load Transfer in Filled Composite Column

The filled composite column of Example 13.3 has four beams framing into the top of the column with each attached to the steel section. All beams have a depth of 12 in and the applied load on each beam is 6 kips (LRFD load combinations) or 4 kips (ASD load combinations). Determine if the load transfer requirements are satisfied.

From Example 13.3

$$A_s = 5.24 \text{ in}^2$$

$$F_y = 46 \text{ ksi}$$

$$B = \text{overall width of rectangular steel section along face transferring load}$$
$$= 6 \text{ in}$$

$$P_{no} = 344 \text{ kips}$$

The external force is applied directly to the steel section and, for each face of the steel section, the shear at the interface is given by AISC 360 Eq. (I6-1) as

| LRFD | ASD |
|---|---|
| $V'_r = P_r(1 - A_sF_y/P_{no})$<br>$= 6(1 - 5.24 \times 46/344)$<br>$= 1.80$ kips | $V'_r = P_r(1 - A_sF_y/P_{no})$<br>$= 4(1 - 5.24 \times 46/344)$<br>$= 1.20$ kips |

The composite member extends below the point of load transfer and

$$C_{in} = 2$$

The overall width of the steel section is

$$B = 6 \text{ in}$$

The nominal bond stress for a filled composite member is

$$F_{in} = 0.06 \text{ ksi}$$

The nominal bond capacity on each face of the steel section is given by AISC 360 Eq. (I6-5) as

$$R_n = B^2C_{in}F_{in}$$
$$= 6^2 \times 2 \times 0.06$$
$$= 4.32 \text{ kips}$$

The available bond capacity is given by AISC 360, Sec. I6.3c as

| LRFD | ASD |
|---|---|
| $\varphi R_n = 0.45 \times 4.32$<br>$= 1.94$ kips<br>$> V'_r$ ... satisfactory | $R_n/\Omega = 4.32/3.33$<br>$= 1.30$ kips<br>$> V'_r$ ... satisfactory |

## 13.4 Encased Composite Beams

The concrete encasement effectively inhibits local and flexural buckling of the steel section and AISC 360 Sec. I3.3 permits the following three alternative design methods for determination of the nominal flexural strength:

- The superposition of elastic stresses on the composite section, based on the first yield of the tension flange. Shear anchors are not required as natural bond is considered adequate to connect the steel beam to the concrete.

- The plastic moment strength of the steel section alone. Shear anchors are not required.
- The strength of the composite section obtained from the plastic stress distribution method or the strain-compatibility method. Shear anchors are required to ensure composite action after the breakdown of natural bond.

The available flexural capacity is obtained from AISC 360, Sec. I3.3 as

$$\varphi_b M_n = \text{design flexural capacity}$$
$$= 0.90M_n$$
$$M_n/\Omega_b = \text{allowable flexural capacity}$$
$$= M_n/1.67$$

**Example 13.5.** Fully Encased Composite Beam
The W24 × 62 beam of Grade 50 steel, shown in Fig. 13.5, is fully encased in concrete and is fabricated without shear anchors. Determine the maximum available moment that the beam can support.
From AISC Manual Table 3-2, the plastic modulus of the W24 × 62 beam is

$$Z_x = 153 \text{ in}^3$$

The beam nominal flexural strength is

$$M_n = Z_x F_y$$
$$= 153 \times 50/12$$
$$= 638 \text{ kip-ft}$$

The available flexural capacity is given by AISC 360, Sec. I3.3 as

| LRFD | ASD |
|---|---|
| $\varphi_b M_n = 0.90 \times 638$<br>$= 574$ kip-ft | $M_n/\Omega_b = 638/1.67$<br>$= 382$ kip-ft |

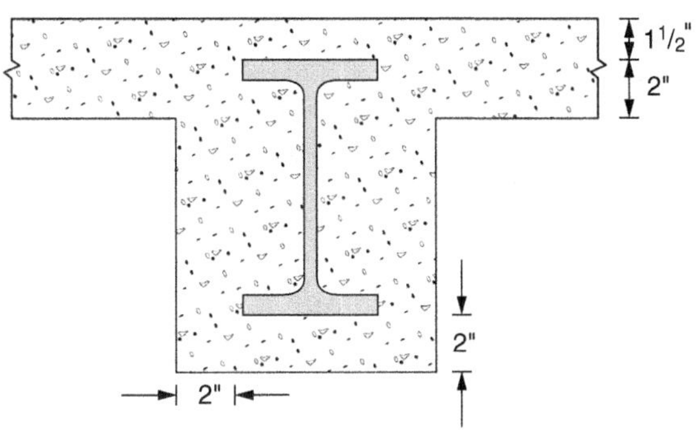


**FIGURE 13.5** Details for Example 13.5.

## 13.5 Composite Beam with Flat Soffit Concrete Slab

Composite action is ensured by the provision of concrete anchors. The nominal flexural strength of a composite beam is determined by assuming a fully plastic stress distribution across the section. At the ultimate load, the location of the plastic neutral axis may be either within the depth of the concrete slab, within the flange of the steel beam, or within the web of the steel beam. It is assumed in AISC 360 Sec. I1.2a that the steel beam has attained a yield stress of $F_y$ in tension below the plastic neutral axis and a yield stress of $F_y$ in compression above the plastic neutral axis. The portion of the concrete slab that is in compression is assumed uniformly stressed in compression to $0.85f'_c$. It is further assumed that any portion of the concrete slab below the plastic neutral axis has zero stress and is neglected. For a positive bending moment, the concrete anchors effectively inhibit local and flexural buckling of the compression flange of the steel section.

### Effective Slab Width

The composite system, shown in Fig. 13.6, consists of a flat soffit concrete slab acting integrally with steel beams to resist the applied moment. Steel anchors, welded to the top flange of the steel beams, ensure composite action between the concrete slab and the beams. Because of shear lag in the composite section, the compressive stress in the concrete slab is not uniform but varies in intensity over the width of the slab. To simplify computation, an effective width of slab is defined with an assumed equivalent uniform stress. In accordance with AISC 360 Sec. I3.1a, the maximum effective width of the concrete slab, on either side of the beam centerline, is given as

$$\begin{aligned} b &= \text{one-eighth of the beam span} \\ &\le \text{one-half of the beam spacing} \\ &\le \text{the distance to edge of the slab} \end{aligned}$$

### Nominal Strength

As shown in Fig. 13.7*c*, when the location of the plastic neutral axis is within the depth of the steel beam, the concrete slab is fully utilized in compression and the depth of the concrete stress block is $a = t_c$. As shown in Fig. 13.7*b*, when the location of the plastic neutral axis is within the depth of the concrete slab, the steel beam is fully utilized in tension and the depth of the concrete stress block is $a < t_c$. It is assumed in AISC 360 Sec. I1.2a that a rectangular stress block is formed in the effective concrete compression zone with a stress of

$$f_u = 0.85f'_c$$

where $f'_c$ is compressive strength of the concrete.

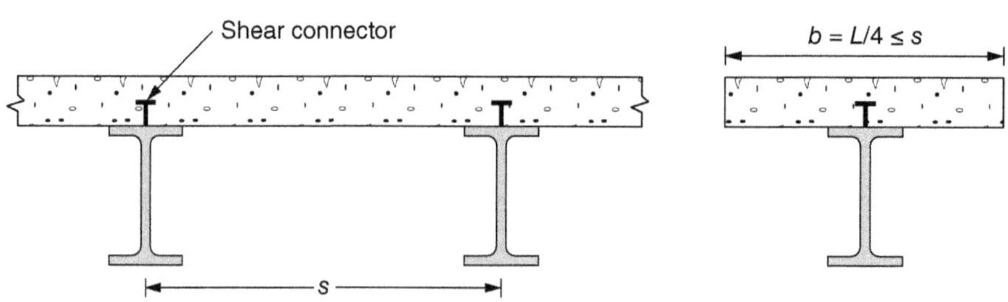


**FIGURE 13.6** Effective slab width.

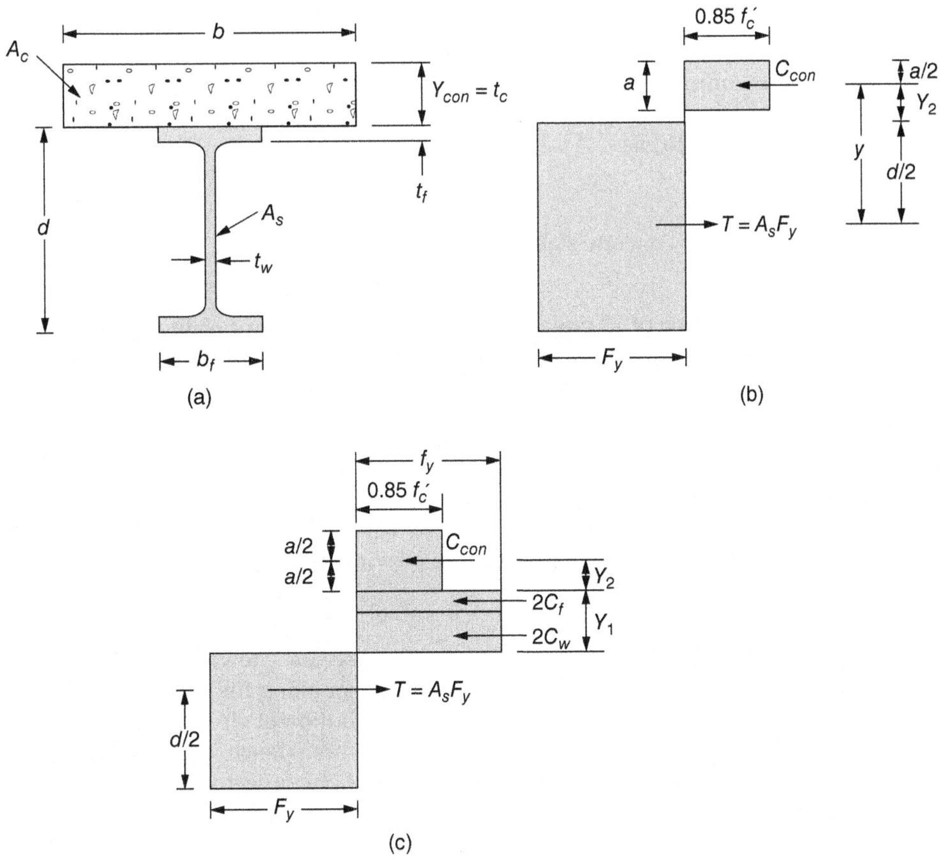


**FIGURE 13.7** Fully composite beam with flat soffit concrete slab.

It is further assumed that the portion of the steel beam below the plastic neutral axis is stressed in tension to the yield stress $F_y$ and that the portion of the steel beam above the plastic neutral axis is stressed in compression to the yield stress $F_y$. Concrete tensile stress is neglected.

## Fully Composite and Partially Composite Beams

In order to achieve composite action between the steel section and the concrete slab, steel anchors are provided to transfer the horizontal shear force across the interface. The number of connectors required to provide full composite action is derived from ultimate strength concepts.

When the plastic neutral axis is in the concrete slab, the steel beam has fully yielded in tension. Then the shear force developed at the interface between the point of maximum positive moment and the point of zero moment is given by AISC 360 Eq. (I3-1b) as

$$V' = F_yA_s$$

where $A_s$ is cross-sectional area of steel section. When the plastic neutral axis is within the steel beam, the full depth of the concrete slab is stressed to its maximum capacity in compression. Then the shear force developed at the interface between the point of maximum positive moment and the point of zero moment is given by AISC 360 Eq. (I3-1a) as

$$V' = 0.85f'_cA_c$$

where $A_c$ = area of concrete slab within effective width
$= bt_c$

The minimum value of $V'$ governs and, to achieve full composite action, it is necessary to provide steel anchors between the point of maximum positive moment and the point of zero moment with a total strength of

$$\Sigma Q_n = \text{lesser value of } V'$$

where $Q_n$ is nominal strength of one steel anchor.

The number of steel anchors required between the point of maximum positive moment and the point of zero moment to provide full composite action is

$$n = V'/Q_n$$

In accordance with AISC 360 Sec. I8.2d, it is not necessary to vary the spacing of connectors to accommodate the varying intensity of shear along the beam since the flexibility and ductility of the connectors allow the redistribution of stress from the more heavily loaded connectors to the lightly loaded connectors. Hence, for a beam supporting a uniformly distributed load, the required number of connectors may be uniformly distributed between the point of maximum positive moment and the adjacent points of zero moment.

When fewer steel anchors are provided than are required for full composite action, the nominal flexural strength is governed by the horizontal shear force developed across the interface and this is given by AISC 360 Eq. (I3-1c) as

$$V' = \Sigma Q_n$$

It is generally considered that in order to prevent excessive deflection and limited ductility at the nominal strength of the member, sufficient anchors are required to provide a minimum of 50 percent composite action.

## Nominal Strength of Fully Composite Beam with PNA in Concrete Slab

For the composite beam shown in Fig. 13.7*b*, the steel beam has fully yielded in tension before the full capacity of the concrete slab has been mobilized. Hence,

$$F_yA_s < 0.85f'_cA_c$$

Equating horizontal forces gives

$$T = C_{con}$$

or

$$F_yA_s = 0.85f'_cba$$

Hence, the depth of the stress block in a fully composite beam is

$$a = F_y A_s / 0.85 f'_c b$$

The distance from top of steel beam to top of concrete slab is

$$Y_{con} = t_c$$

The distance from top of steel beam to compression force in concrete slab is

$$Y_2 = Y_{con} - a/2$$

The moment arm between centroids of the tensile force and the compressive force is

$$y = Y_2 + d/2$$

The nominal flexural capacity is then

$$\begin{aligned} M_n &= Ty \\ &= T(Y_2 + d/2) \end{aligned}$$

The available flexural capacity is obtained from AISC 360, Sec. I3.2a as

$$\begin{aligned} \varphi_b M_n &= \text{design flexural capacity} \\ &= 0.90 M_n \\ M_n/\Omega_b &= \text{allowable flexural capacity} \\ &= M_n/1.67 \end{aligned}$$

**Example 13.6.** Fully Composite Beam with Flat Soffit Concrete Slab
The fully composite section shown in Fig. 13.8 consists of W18 × 50 beams of Grade 50 steel spaced at 11 ft centers and spanning 40 ft. The 5-in-thick solid slab is of normal weight concrete with a compressive strength of 4 ksi. Determine the available strength of the fully composite section.

The relevant properties of the W18 × 50 beam are

$$A_s = 14.7 \text{ in}^2$$

$$d = 18 \text{ in}$$

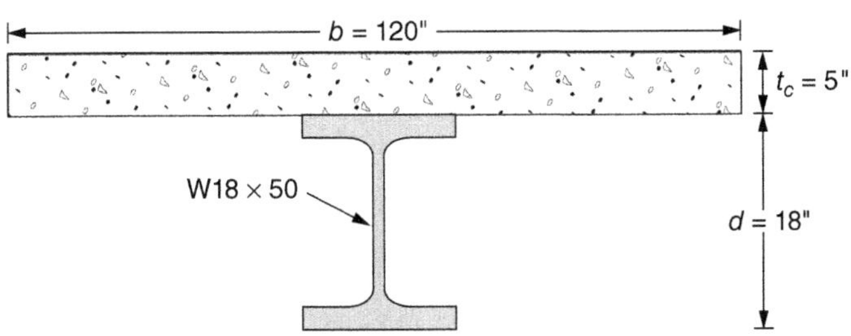


**FIGURE 13.8** Details for Example 13.6.

The effective width of the concrete slab is the lesser of

$$\begin{aligned} b &= s \\ &= 11 \times 12 \\ &= 132 \text{ in} \end{aligned}$$

or

$$\begin{aligned} b &= L/4 \\ &= 40 \times 12/4 \\ &= 120 \text{ in ... governs} \end{aligned}$$

The area of concrete slab within the effective width is

$$\begin{aligned} A_c &= bt_c \\ &= 120 \times 5 \\ &= 600 \text{ in}^2 \end{aligned}$$

The compression capacity of the concrete slab assuming full utilization is

$$\begin{aligned} 0.85f'_cA_c &= 0.85 \times 4 \times 600 \\ &= 2040 \text{ kips} \end{aligned}$$

The tensile capacity of the steel beam assuming it has fully yielded in tension is

$$\begin{aligned} T &= F_yA_s \\ &= 50 \times 14.7 \\ &= 735 \text{ kips} \\ &< 0.85f'_cA_c \end{aligned}$$

Hence, the depth of the concrete slab exceeds the depth of the stress block and, for a fully composite beam

$$\begin{aligned} C_{con} &= T \\ &= 735 \text{ kips} \end{aligned}$$

The depth of the stress block is given by

$$\begin{aligned} a &= C_{con}/0.85f'_cb \\ &= 735/(0.85 \times 4 \times 120) \\ &= 1.8 \text{ in} \end{aligned}$$

The distance from top of steel beam to compressive force in the concrete slab is

$$\begin{aligned} Y_2 &= Y_{con} - a/2 \\ &= 5 - 1.8/2 \\ &= 4.1 \text{ in} \end{aligned}$$

The moment arm between the centroids of the tensile force and the compressive force is

$$\begin{aligned} y &= Y_2 + d/2 \\ &= 4.1 + 18/2 \\ &= 13.1 \end{aligned}$$

The nominal flexural capacity is

$$M_n = Ty$$
$$= 735 \times 13.1$$
$$= 9629 \text{ kip-in}$$
$$= 802 \text{ kip-ft}$$

The available flexural capacity is given by AISC 360, Sec. I3.2a as

| LRFD | ASD |
|---|---|
| $\varphi_b M_n = 0.90 \times 802$ <br> $= 722$ kip-ft | $M_n/\Omega_b = 802/1.67$ <br> $= 480$ kip-ft |

## Design Tables

To facilitate the determination of section properties, composite beam selection tables are provided in AISC Manual Table 3-19. For a range of W-shapes, with steel anchors providing full composite action, and with known values of $\Sigma Q_n$ and $Y_2$ the tables provide values of $M_n/\Omega_b$, and $\varphi_b M_n$, where

$\Sigma Q_n = 0.85 f'_c A_c$ ... when plastic neutral axis is located in the steel beam

or, $\Sigma Q_n = F_y A_s$ ... when plastic neutral axis is located in the concrete slab

$Y_2$ = distance from top of steel beam to compressive force in the concrete slab

The Table also lists the value of

$Y_1$ = distance from top of steel beam to plastic neutral axis in the steel beam

= 0 ... for full composite action with plastic neutral axis in the concrete slab

The Table may also be used to determine the available flexural strength for partially composite members where

$\Sigma Q_n$ = horizontal shear force developed across the interface

= sum of nominal shear strengths of steel anchors between the point of maximum positive moment and the point of zero moment

**Example 13.7.** Fully Composite Beam with Flat Soffit Concrete Slab, Tabular Solution
The fully composite section shown in Fig. 13.8 consists of W18 × 50 beams of Grade 50 steel spaced at 11 ft centers and spanning 40 ft. The 5-in-thick solid slab is of normal weight concrete with a compressive strength of 4 ksi. Using AISC Manual Table 3-19, determine the available strength of the fully composite section.

From Example 13.6, the depth of the concrete slab exceeds the depth of the stress block and, for a fully composite beam

$$Y_2 = 4.1 \text{ in}$$
$$\Sigma Q_n = F_y A_s$$
$$= 735 \text{ kips}$$

From AISC Manual Table 3-19, the available flexural capacity is

| LRFD | ASD |
|---|---|
| $\varphi_b M_n = 720$ kip-ft | $M_n/\Omega_b = 479$ kip-ft |

**Example 13.8.** Partially Composite Beam with Flat Soffit Concrete Slab
The composite section shown in Fig. 13.8 consists of W18 × 50 beams of Grade 50 steel spaced at 11 ft centers and spanning 40 ft. The 5-in-thick solid slab is of normal weight concrete with a compressive strength of 4 ksi. Determine the available strength of the composite section if the horizontal shear force developed across the interface by the steel anchors is $\Sigma Q_n = 520$ kips.

The relevant properties of the W18 × 50 beam are

$$A_s = 14.7 \text{ in}^2$$
$$d = 18 \text{ in}$$
$$b_f = 7.5 \text{ in}$$
$$t_f = 0.57 \text{ in}$$

From Example 13.6

$$b = 120 \text{ in}$$
$$T = F_y A_s$$
$$= 735 \text{ kips}$$
$$> 520 \text{ kips}$$

and,

$$C_{con} = \Sigma Q_n$$
$$= 520 \text{ kips}$$

Hence, the full strength of the steel beam can not be developed and the depth of the concrete stress block is given by

$$a = C_{con}/0.85f'_c b$$
$$= 520/(0.85 \times 4 \times 120)$$
$$= 1.27 \text{ in}$$

The distance from top of steel beam to compressive force in the concrete slab is

$$Y_2 = Y_{con} - a/2$$
$$= 5 - 1.27/2$$
$$= 4.37 \text{ in}$$

Assuming that the plastic neutral axis is in the flange of the steel beam a distance $Y_1$ below the top of the flange and that the compression developed in the flange is $C_s$, the plastic stress distribution is shown in Fig. 13.9.

Equating horizontal forces gives

$$T = C_{con} + 2C_s$$
$$C_s = (735 - 520)/2$$
$$= 107.5 \text{ kips}$$

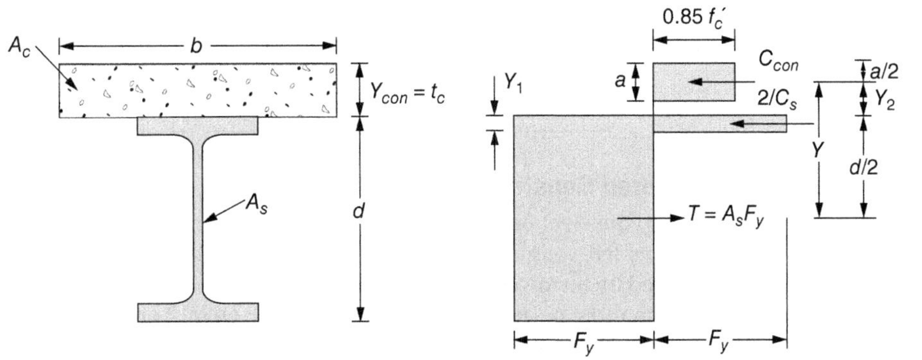


**FIGURE 13.9** Plastic stress distribution, partially composite beam.

The distance from the top of the steel beam to the plastic neutral axis is

$$\begin{aligned} Y_1 &= C_s/F_yb_f \\ &= 107.5/(50 \times 7.5) \\ &= 0.29 \text{ in} \\ &< t_f \ldots \text{ plastic neutral axis is in the flange} \end{aligned}$$

Taking moments about top of flange gives

$$\begin{aligned} M_n &= C_{con}Y_2 + Td/2 - 2C_sY_1/2 \\ &= 520 \times 4.37 + 735 \times 18/2 - 107.5 \times 0.29 \\ &= 8856 \text{ kip-in} \\ &= 738 \text{ kip-ft} \end{aligned}$$

The available flexural capacity is given by AISC 360, Sec. I3.2a as

| LRFD | ASD |
|---|---|
| $\varphi_b M_n = 0.90 \times 738$ | $M_n/\Omega_b = 738/1.67$ |
| $= 664$ kip-ft | $= 442$ kip-ft |

**Example 13.9.** Partially Composite Beam with Flat Soffit Concrete Slab, Tabular Solution
The composite section shown in Fig. 13.8 consists of W18 × 50 beams of Grade 50 steel spaced at 11 ft centers and spanning 40 ft. The 5-in-thick solid slab is of normal weight concrete with a compressive strength of 4 ksi. Using AISC Manual Table 3-19, determine the available strength of the composite section if the horizontal shear force developed across the interface by the steel anchors is 520 kips.

From Example 13.8

$$Y_2 = 4.37 \text{ in}$$
$$\Sigma Q_n = 520 \text{ kips}$$

From AISC Manual Table 3-19, the available flexural capacity for $\Sigma Q_n = 520$ kips and $Y_2 = 4.37$ in is

| LRFD | ASD |
|---|---|
| $\varphi_b M_n = 662$ kip-ft | $M_n/\Omega_b = 440$ kip-ft |

## Shored and Unshored Construction

During construction, the steel beam is initially placed in position and is subjected to the moment caused by its self-weight. If the beam is unshored, it then has to support the construction loads caused by formwork, pouring the concrete slab, workmen, and equipment. Hence, the steel beam must be designed for its self-weight, weight of formwork, weight of concrete slab, and construction loads. Adequate bracing must be provided to prevent lateral-torsional buckling as the slab is poured. Deflection of the steel beam during pouring of the concrete slab may cause an inadvertent increase in slab thickness and dead load. Providing camber to the beam and pouring the slab to a constant thickness may prevent this eventuality. Dead load and live load, imposed after the concrete has attained 75 percent of its required strength, are considered supported by the composite section and produce bending moments in the composite section. Because of stress redistribution at ultimate loads, the adequacy of the composite section is determined as though all loads are applied to the composite section. Hence, the composite section must be designed for steel beam self-weight, weight of concrete slab, and imposed dead load and live load.

In shored construction, temporary shores are employed to support the steel beam during casting of the concrete slab. The steel beam is initially placed in position and is subjected to the moment caused by its self-weight. The next stages in the construction sequence are locating the props under the beam, placing formwork, and pouring the concrete slab. Negligible stresses are produced by these operations. Removing the props after the concrete has attained 75 percent of its required strength causes the weight of the concrete slab to be supported by the composite section and produces a bending moment in the composite section. Subsequent application of the superimposed dead load and live load causes additional bending moments in the composite section. Because of stress redistribution at ultimate loads, the adequacy of the composite section is determined as though all loads are applied to the composite section. Hence, the composite section is designed for steel beam self-weight, weight of concrete slab, and imposed dead load and live load.

**Example 13.10.** Applied Loading

The fully composite section shown in Fig. 13.8 consists of W18 × 50 beams of Grade 50 steel spaced at 11 ft centers and spanning 40 ft. The 5-in-thick solid slab is of normal weight concrete with a compressive strength of 4 ksi. The construction load = 20 lb/ft$^2$, imposed dead load = 15 lb/ft$^2$, and imposed live load = 125 lb/ft$^2$. Reinforcement in the slab gives an equivalent concrete weight of 150 lb/ft$^3$. During construction, the steel beams are laterally braced and are not shored. Determine if the beam is adequate.

At the *construction stage* the loads on the steel beam are

$$
\begin{aligned}
&\text{Beam self-weight} && = 50 \text{ lb/ft} \\
&\text{Concrete slab} = 150 \times 11 \times 5/12 && = \underline{688} \text{ lb/ft} \\
&\text{Total dead load} = w_D && = 738 \text{ lb/ft} \\
&\text{Construction live load} = w_L && = 20 \times 11 \\
& && = 220 \text{ lb/ft}
\end{aligned}
$$

Applying American Society of Civil Engineers, *Minimum Design Loads for Buildings and Other Structures* (ASCE 7)[6] Secs. 2.3 and 2.4 gives the factored loads as

| LRFD | ASD |
|---|---|
| From ASCE 7 Sec. 2.3.2 combination 2:<br>$w_u$ = factored load<br>$= 1.2w_D + 1.6w_L$<br>$= 1.2 \times 738 + 1.6 \times 220$<br>$= 1237$ lb/ft | From ASCE 7 Sec. 2.4.1 combination 2:<br>$w_a$ = factored load<br>$= w_D + w_L$<br>$= 738 + 220$<br>$= 958$ lb/ft |

The applied bending moment is

| LRFD | ASD |
|---|---|
| $M_u = 1.237 \times 40^2/8$<br>$= 247$ kip-ft<br>Beam is compact and fully braced. Hence,<br>$\varphi M_n = 0.9 \times F_y Z_x$ ... AISC 360 Eq. (F2-1)<br>$= 0.9 \times 50 \times 101/12$<br>$= 379$ kip-ft<br>$> M_u$ ... satisfactory | $M_a = 0.958 \times 40^2/8$<br>$= 192$ kip-ft<br>Beam is compact and fully braced. Hence,<br>$M_n/\Omega = F_y Z_x/1.67$ ... AISC 360 Eq. (F2-1)<br>$= 50 \times 101/(1.67 \times 12)$<br>$= 252$ kip-ft<br>$> M_a$ ... satisfactory |

The applied shear force is

| LRFD | ASD |
|---|---|
| $V_u = 1.237 \times 40/2$<br>$= 24.7$ kips<br>AISC 360 Eq. (G2-1) applies and hence,<br>$\varphi V_n = 1.0 \times 0.6 F_y A_w$<br>$= 1.0 \times 0.6 \times 50 \times 18 \times 0.355$<br>$= 192$ kips<br>$> V_u$ ... satisfactory | $V_a = 0.958 \times 40/2$<br>$= 19.2$ kips<br>AISC 360 Eq. (G2-1) applies and hence,<br>$V_n/\Omega = 0.6 F_y A_w/1.50$<br>$= 0.6 \times 50 \times 18 \times 0.355/1.50$<br>$= 128$ kips<br>$> V_a$ ... satisfactory |

Deflection of the steel beam due to beam self-weight and concrete slab is

$$\delta = 5w_D L^4/384EI$$
$$= 5 \times 0.738 \times 40^4 \times 1728/(384 \times 29{,}000 \times 800)$$
$$= 1.8 \text{ in ... camber beam to facilitate placement of the concrete slab}$$

At the *composite stage*, the loads on the composite beam are

| | |
|---|---|
| Steel beam self-weight | = 50 lb/ft |
| Concrete slab = $150 \times 11 \times 5/12$ | = 688 lb/ft |
| Imposed dead load = $15 \times 11$ | = 165 lb/ft |
| Total dead load = $w_D$ | = 903 lb/ft |
| Imposed live load = $w_L$ | = $125 \times 11$ |
| | = 1375 lb/ft |

Applying ASCE-07 Secs. 2.3 and 2.4 gives the factored loads as

| LRFD | ASD |
|---|---|
| From ASCE 7 Sec. 2.3.2 combination 2: | From ASCE 7 Sec. 2.4.1 combination 2: |
| $w_u$ = factored load | $w_a$ = factored load |
| $= 1.2w_D + 1.6w_L$ | $= w_D + w_L$ |
| $= 1.2 \times 903 + 1.6 \times 1375$ | $= 903 + 1375$ |
| $= 3284$ lb/ft | $= 2278$ lb/ft |

The applied bending moment is

| LRFD | ASD |
|---|---|
| $M_u = 3.284 \times 40^2/8$ | $M_a = 2.278 \times 40^2/8$ |
| $= 657$ kip-ft | $= 456$ kip-ft |
| From Example 13.6: | From Example 13.6: |
| $\varphi M_n = 722$ kip-ft | $M_n/\Omega = 480$ kip-ft |
| $> M_u$ ... satisfactory | $> M_a$ ... satisfactory |

The applied shear force is

| LRFD | ASD |
|---|---|
| $V_u = 3.284 \times 40/2$ | $V_a = 2.278 \times 40/2$ |
| $= 65.7$ kips | $= 45.6$ kips |
| AISC 360 Eq. (G2-1) applies and hence, | AISC 360 Eq. (G2-1) applies and hence, |
| $\varphi V_n = 1.0 \times 0.6F_yA_w$ | $V_n/\Omega = 0.6F_yA_w/1.50$ |
| $= 1.0 \times 0.6 \times 50 \times 18 \times 0.355$ | $= 0.6 \times 50 \times 18 \times 0.355/1.50$ |
| $= 192$ kips | $= 128$ kips |
| $> V_u$ ... satisfactory | $> V_a$ ... satisfactory |

## Composite Beam Deflection

The moment of inertia of a composite beam may be determined using the transformed section properties shown in Fig. 13.10*a*. The effective section is converted to an equivalent homogenous section, termed the transformed section, with the transformed slab width given by

$$b/n = b/(E_s/E_c)$$

where $E_s$ = modulus of elasticity of the steel beam
$E_c$ = modulus of elasticity of the concrete slab
$n$ = modular ratio
$= E_s/E_c$

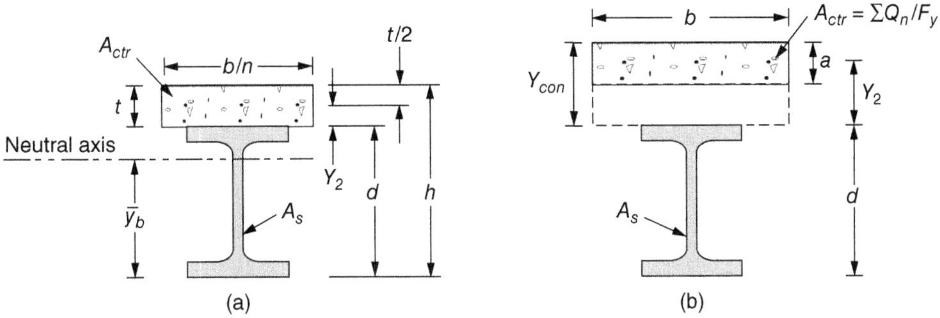


**FIGURE 13.10** Composite beam moment of inertia.

The area of the concrete slab in compression is then treated as an equivalent area of steel and the concrete transformed area in compression is given by

$$A_{ctr} = bt/n$$

To simplify the determination of deflections, AISC Manual Table 3-20 provides values of a lower bound moment of inertia for a range of values of $Y_1$ and $Y_2$. This moment of inertia is derived using the effective concrete area at the factored load which is given by

$$A_{ctr} = \Sigma Q_n / F_y$$

where $\Sigma Q_n$ is summation of the nominal shear resistance of all shear connectors between the points of maximum and zero moment.

As shown in Fig. 13.10*b*, this technique includes only the portion of the concrete slab used to balance $\Sigma Q_n$ and neglects the area of concrete below. In effect, the lower bound moment of inertia is the moment of inertia at the strength level which is smaller than that at the service level.

**Example 13.11.** Composite Beam Deflection
The fully composite section shown in Fig. 13.8 consists of W18 × 50 beams of Grade 50 steel spaced at 11 ft centers and spanning 40 ft. The 5-in-thick solid slab is of normal weight concrete with a compressive strength of 4 ksi. The imposed live load on the composite section = 125 lb/ft$^2$. Determine the live load deflection of the composite section.

From Example 13.6 the following details are obtained

$$C_{con} = 735 \text{ kips}$$
$$= \Sigma Q_n$$
$$Y_2 = 4.1 \text{ in}$$
$$Y_1 = 0$$

From AISC Manual Table 3-20, the lower bound moment of inertia is obtained as

$$I_{LB} = 2060 \text{ in}^4$$

The imposed live load is

$$w_L = 125 \times 11$$
$$= 1375 \text{ lb/ft}$$

Deflection of the composite beam due to imposed live load is

$$\delta = 5w_L L^4/384EI_{LB}$$
$$= 5 \times 1.375 \times 40^4 \times 1728/(384 \times 29{,}000 \times 2060)$$
$$= 1.33 \text{ in}$$
$$= L/360 \text{ ... satisfactory}$$

## Negative Flexural Strength

In the negative moment regions of continuous composite beams, the concrete slab is in tension and is considered not to contribute to the properties of the composite section. However, when adequate shear connection is provided and the beam is compact and adequately braced, AISC 360 Sec. I3.2b allows the longitudinal reinforcing steel, within the effective width of the slab, to be considered as acting compositely with the steel beam. The slab reinforcement must be properly developed over the negative moment region. The plastic stress distribution in the composite member is shown in Fig. 13.11 and the tensile force $T_r$ in the reinforcing steel is given in AISC 360 Commentary Eqs. (C-I3-11) and (C-I3-12) as the lesser of

$$T_r = F_{yr} A_r$$

or

$$T_r = \Sigma Q_n$$

where $A_r$ = total area of longitudinal reinforcing steel at an interior support located within the effective flange width

$F_{yr}$ = minimum yield stress of the longitudinal reinforcing steel

$\Sigma Q_n$ = sum of the nominal strengths of shear connectors between an interior support and the adjacent point of contraflexure

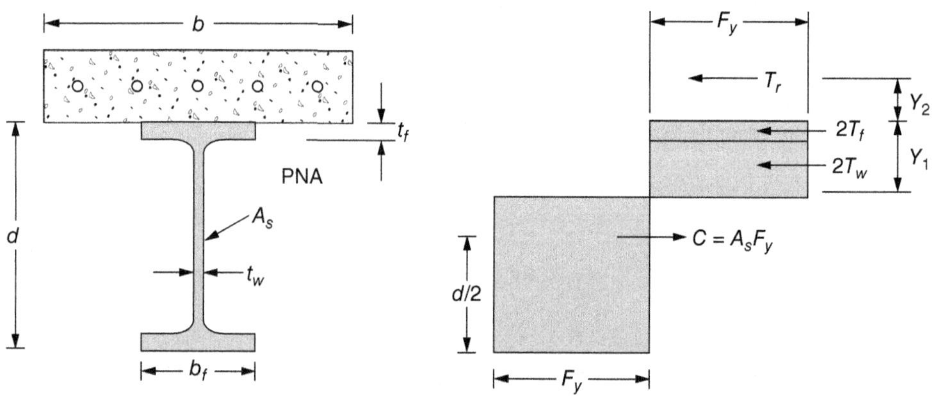


**FIGURE 13.11** Plastic stress distribution in negative moment region.

The distance from the top of the steel beam to the tensile force in the reinforcing steel is $Y_2$. Assuming the plastic neutral axis is located in the web of the steel beam and equating horizontal forces gives

$$C = T_r + 2T_f + 2T_w$$

$$F_yA_s = T_r + 2F_yb_ft_f + 2F_yt_w(Y_1 - t_f)$$

Hence, $$Y_1 = t_f + (A_s/2 - b_ft_f - T_r/2F_y)/t_w$$

Taking moments about the plastic neutral axis gives the nominal flexural capacity as

$$M_n = T_r(Y_1 + Y_2) + 2T_f(Y_1 - t_f/2) + T_w(Y_1 - t_f) + C(d/2 - Y_1)$$
$$= T_r(Y_1 + Y_2) + 2F_yb_ft_f(Y_1 - t_f/2) + F_yt_w(Y_1 - t_f)^2 + F_yA_s(d/2 - Y_1)$$

**Example 13.12.** Negative Flexural Strength
The fully composite section shown in Fig. 13.12 consists of W18 × 50 beams of Grade 50 steel spaced at 11 ft centers and spanning 40 ft. The 5-in-thick solid slab is of normal weight concrete with a compressive strength of 4 ksi. The slab is reinforced, over its effective width of 120 in, with eight No. 8 reinforcing bars with a yield stress of 60 ksi. If the bottom flange of the steel beam is adequately braced, determine the design negative moment capacity of the fully composite section.

The area of longitudinal reinforcing steel provided is

$$A_r = 8 \times 0.79$$
$$= 6.32 \text{ in}^2$$

The tensile force in the reinforcing steel is

$$T_r = F_{yr}A_r \text{ ... full composite action}$$
$$= 60 \times 6.32$$
$$= 379 \text{ kips}$$

The distance from top of steel beam to the tensile force in the reinforcing is

$$Y_2 = t_c/2$$
$$= 5/2$$
$$= 2.5 \text{ in}$$

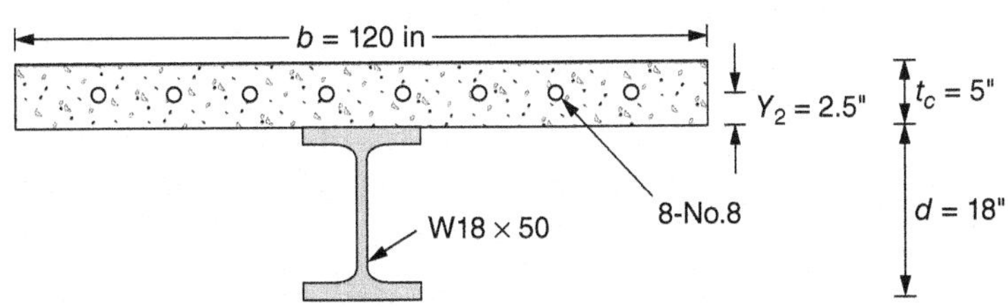


**FIGURE 13.12** Details for Example 13.12.

From AISC Manual Table 3-19, the available negative moment capacity of the fully composite section for $Y_2 = 2.5$ in and $\Sigma Q_n = 379$ kips is

| LRFD | ASD |
|---|---|
| $\varphi_b M_n = 559$ kip-ft | $M_n/\Omega_b = 372$ kip-ft |

The total horizontal shear to be resisted by steel anchors between the point of maximum negative moment and the point of zero moment on either side is given by AISC 360 Eq. (C-I3-12) as

$$V' = T_r$$
$$= 379 \text{ kips}$$

## Steel-Headed Stud Anchors in Composite Beam with Flat Soffit Concrete Slab

Steel anchors are provided at the interface of a composite beam to transfer the horizontal shear force across the interface and prevent separation of the slab from the steel beam. The most commonly used anchor is the steel headed stud which is welded to the top flange of the steel beam by means of a stud welding gun. The capacity of a stud anchor is limited by the shear strength of the stud shank, the strength of the weld, and the crushing and pryout strength of the concrete as described by Ollgaard et al.[7]

The nominal strength of one steel headed stud anchor embedded in a flat soffit slab is given by AISC 360 Eq. (I8-1) as

$$Q_n = 0.5A_{sa}(f'_c E_c)^{0.5}$$
$$\leq R_g R_p A_{sa} F_u$$

where $A_{sa}$ = cross-sectional area of a steel headed stud anchor, in$^2$
$f'_c$ = specified compressive strength of concrete, ksi
$F_u$ = minimum specified tensile strength of steel headed stud anchor, ksi
= 65 ksi for Type B steel headed stud anchors made from ASTM A108 material
$E_c$ = modulus of elasticity of concrete, ksi
$= w_c^{1.5}(f'_c)^{0.5}$
$w_c$ = unit weight of concrete, lb/ft$^3$
= 145 lb/ft$^3$ ... for normal weight concrete

Hence, $Q_n = 0.5A_{sa}(w_c f'_c)^{0.75}$
$\leq R_g R_p A_{sa} F_u$

where $R_g$ = anchor group coefficient
= 1.0 ... for flat soffit, solid slabs, with the steel headed stud anchor welded directly to the girder flange
$R_p$ = anchor position coefficient
= 0.75 ... for flat soffit, solid slabs, with the steel headed stud anchor welded directly to the girder flange

Steel headed stud anchors are the usual type of connector used for composite beams with the stud welded directly to the girder flange, the nominal shear strengths are given in Table 13.3.

For lightweight concrete, reduced values apply and Table 13.4 provides values for 3/4-in steel-headed stud anchors in flat soffit concrete slabs with a unit weight of 110 lb/ft$^3$.

| Anchor | Specified Concrete Compressive Strength, $f'_c$ | | | |
|---|---|---|---|---|
| | 3.0 ksi | 3.5 ksi | 4.0 ksi | 4.5 ksi |
| 1/2 in diameter | 9.33 | 9.56 | 9.56 | 9.56 |
| 5/8 in diameter | 14.6 | 15.0 | 15.0 | 15.0 |
| 3/4 in diameter | 21.1 | 21.6 | 21.6 | 21.6 |
| 7/8 in diameter | 28.6 | 29.3 | 29.3 | 29.3 |

**TABLE 13.3** Nominal Shear Strength, in kips, of Steel-Headed Stud Anchors in Normal Weight Concrete

| Connector | Specified Concrete Compressive Strength, $f'_c$ | | | |
|---|---|---|---|---|
| | 3.0 ksi | 3.5 ksi | 4.0 ksi | 4.5 ksi |
| 3/4 in diameter | 17.1 | 19.2 | 21.2 | 21.2 |

**TABLE 13.4** Nominal Shear Strength, in kips, of Steel-Headed Stud Anchors in Light Weight Concrete

To prevent concrete breakout, AISC 360 Sec. I8.2d requires anchors to have a minimum 1-in lateral concrete cover in a direction perpendicular to the shear force, except for anchors installed in the ribs of formed steel decks. The minimum distance required from the center of an anchor to a free edge in the direction of the shear force is 8 in for normal weight concrete and 10 in for lightweight concrete. For normal weight concrete, stud anchors are required to be five stud diameters in length from the base of the stud to the top of the stud head after installation. For lightweight concrete, stud anchors are required to be seven stud diameters in length from the base of the stud to the top of the stud head after installation. Additional dimensional requirements are given in Fig. 13.13.

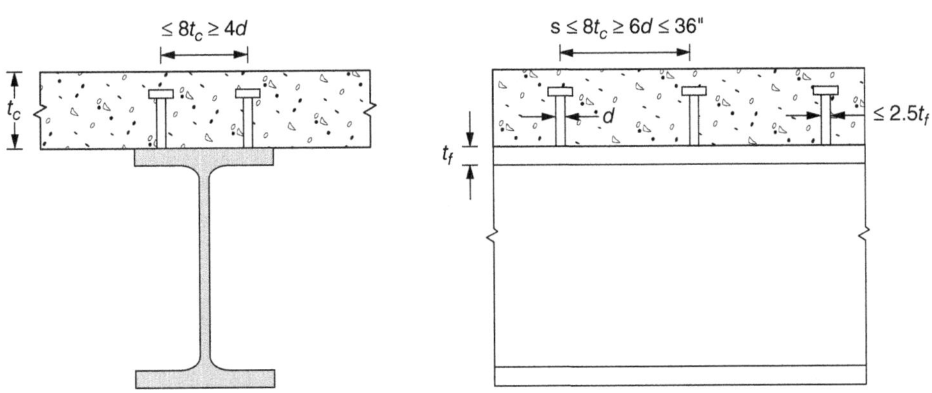


**FIGURE 13.13** Dimensional requirements for steel headed stud anchors in flat soffit slabs.

**Example 13.13.** Steel Headed Stud Anchors in Fully Composite Section
The fully composite section shown in Fig. 13.8 consists of W18 × 50 beams of Grade 50 steel spaced at 11 ft centers and spanning 40 ft. The 5-in-thick solid slab is of normal weight concrete with a compressive strength of 4 ksi. The imposed load is uniformly distributed. Determine the number of 5/8-in-diameter steel-headed stud anchors required to provide full composite action.

The total horizontal shear to be resisted by the anchors is the lesser value obtained from AISC 360 Sec. I3.2d(1) as

$$V' = 0.85f'_cA_c$$
$$= 0.85 \times 4 \times 120 \times 5$$
$$= 2040 \text{ kips}$$

or

$$V' = F_yA_s$$
$$= 50 \times 14.7$$
$$= 735 \text{ kips ... governs}$$

The nominal area of a 5/8-in-diameter steel-headed stud anchor is

$$A_{sa} = 0.3068 \text{ in}^2$$

The nominal shear strength of a 3¼-in long, 5/8-in-diameter steel-headed stud anchor in normal weight 4 ksi concrete is the lesser value obtained from AISC 360 Eq. (I8-1) as

$$Q_n = 0.5A_{sa}(w_cf'_c)^{0.75}$$
$$= 0.5 \times 0.3068(145 \times 4)^{0.75}$$
$$= 18.1 \text{ kips}$$

or

$$Q_n = R_gR_pA_{sa}F_u$$
$$= 1.0 \times 0.75 \times 0.3068 \times 65$$
$$= 15.0 \text{ kips ... governs}$$

Hence, for full composite action, the total number of anchors required on the beam is given by

$$2n = 2V'/Q_n$$
$$= 2 \times 735/15$$
$$= 98$$

Providing 98 studs in pairs, equally spaced, with end distances of 8 in gives a spacing of

$$s = (12 \times 40 - 2 \times 8)/(98/2 - 1)$$
$$\approx 9\tfrac{5}{8} \text{ in}$$

The minimum spacing requirement is

$$s = 6d$$
$$= 6 \times 5/8$$
$$= 3.75 \text{ in}$$
$$< 9\tfrac{5}{8} \text{ in ... satisfactory}$$

The maximum spacing requirement is

$$\begin{aligned} s &= 8t_c \\ &= 8 \times 5 \\ &= 40 \text{ in} \\ &> 9\tfrac{5}{8} \text{ in ... satisfactory} \end{aligned}$$

**Example 13.14.** Steel Headed Stud Anchors in Partially Composite Section
The composite section shown in Fig. 13.8 consists of W18 × 50 beams of Grade 50 steel spaced at 11 ft centers and spanning 40 ft. The 5-in-thick solid slab is of normal weight concrete with a compressive strength of 4 ksi. The imposed dead load = 15 lb/ft$^2$, the imposed live load = 125 lb/ft$^2$. Using both ASD and LRFD methods, determine the minimum number of 5/8-in-diameter steel-headed stud anchors required to provide the required moment capacity.

From Example 13.10, the required number of anchors must provide an available moment of

$$\varphi M_n = 657 \text{ kip-ft}$$

or

$$M_n/\Omega = 456 \text{ kip-ft}$$

| LRFD | ASD |
|---|---|
| Provide a total of 69 anchors to give | Provide a total of 80 anchors to give |
| $\Sigma Q_n = nQ_n$ | $\Sigma Q_n = nQ_n$ |
| $= 69 \times 15/2$ | $= 80 \times 15/2$ |
| $= 518$ kips | $= 600$ kips |
| $= 0.85 f'_c ab$ | $= 0.85 f'_c ab$ |
| $a = 518/(0.85 \times 4 \times 120)$ | $a = 600/(0.85 \times 4 \times 120)$ |
| $= 1.27$ in | $= 1.47$ in |
| $Y_2 = t_c - a/2$ | $Y_2 = t_c - a/2$ |
| $= 5 - 1.27/2$ | $= 5 - 1.47/2$ |
| $= 4.37$ in | $= 4.27$ in |
| From AISC Manual Table 3-19 with $\Sigma Q_n = 518$ and $Y_2 = 4.37$, the available moment is | From AISC Manual Table 3-19 with $\Sigma Q_n = 600$ and $Y_2 = 4.27$, the available moment is |
| $\varphi M_n = 663$ kip-ft | $M_n/\Omega = 456$ kip-ft |
| $> 657$ kip-ft ... satisfactory | = required value ... satisfactory |
| Hence, a total of 69 anchors are required. | Hence, a total of 80 anchors are required. |

## Steel-Headed Stud Anchors in Composite Section with Concentrated Loads

As required by AISC 360 Sec. I8.2c, when a concentrated load is applied to a composite section, the number of connectors provided between the support and the concentrated load must be sufficient to develop the bending moment at the load.

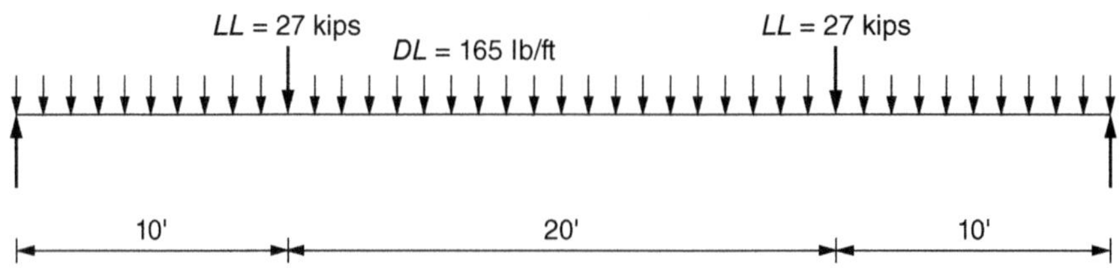


**Figure 13.14** Details for Example 13.15.

**Example 13.15.** Steel Headed Stud Anchors in Composite Section with Concentrated Load
The composite section shown in Fig. 13.8 consists of W18 × 50 beams of Grade 50 steel spaced at 11 ft centers and spanning 40 ft. The 5-in-thick solid slab is of normal weight concrete with a compressive strength of 4 ksi. The imposed dead load = 15 lb/ft$^2$, the imposed live load consists of two concentrated loads each of 27 kips as indicated in Fig. 13.14. Determine the minimum number of 5/8-in-diameter steel-headed stud anchors required between the supports and the loads and the number required in the central 20 ft of the span.

From Example 13.10, the total distributed dead load on the composite beam is

$$w_D = 903 \text{ lb/ft}$$

Applying ASCE-07 Secs. 2.3 and 2.4 gives the factored loads as

| LRFD | ASD |
|---|---|
| From ASCE 7 Sec. 2.3.2 combination 2: | From ASCE 7 Sec. 2.4.1 combination 2: |
| $w_u$ = factored distributed load | $w_a$ = factored distributed load |
| $= 1.2w_D$ | $= w_D$ |
| $= 1.2 \times 903$ | $= 903$ lb/ft |
| $= 1084$ lb/ft | |
| $W_u$ = factored concentrated load | $W_a$ = factored concentrated load |
| $= 1.6 \times 27$ | $= 27$ kips |
| $= 43.2$ kips | |

The bending moment at the location of the concentrated load is

| LRFD | ASD |
|---|---|
| $M_u = 1.084(20 \times 10 - 10^2/2) + 43.2 \times 10$ | $M_a = 0.903(20 \times 10 - 10^2/2) + 27 \times 10$ |
| $= 595$ kip-ft | $= 405$ kip-ft |

The nominal shear strength of a 3¼-in long, 5/8-in-diameter steel-headed stud anchor in normal weight 4 ksi concrete is obtained from Table 13.3 as

$$Q_n = 15 \text{ kips}$$

| LRFD | ASD |
|---|---|
| Provide 22 anchors between support and load | Provide 24 anchors between support and load |
| $\Sigma Q_n = nQ_n$ | $\Sigma Q_n = nQ_n$ |
| $= 22 \times 15$ | $= 24 \times 15$ |
| $= 330$ kips | $= 360$ kips |
| $= 0.85f'_c ab$ | $= 0.85f'_c ab$ |
| $a = 330/(0.85 \times 4 \times 120)$ | $a = 360/(0.85 \times 4 \times 120)$ |
| $= 0.81$ in | $= 0.88$ in |
| $Y_2 = t_c - a/2$ | $Y_2 = t_c - a/2$ |
| $= 5 - 0.81/2$ | $= 5 - 0.88/2$ |
| $= 4.60$ in | $= 4.56$ in |
| From AISC Manual Table 3-19 with $\Sigma Q_n = 330$ and $Y_2 = 4.60$, the available moment is | From AISC Manual Table 3-19 with $\Sigma Q_n = 360$ and $Y_2 = 4.56$, the available moment is |
| $\varphi M_n = 600$ kip-ft | $M_n/\Omega = 406$ kip-ft |
| $> 595$ kip-ft ... satisfactory | $> 405$ kip-ft ... satisfactory |

The bending moment at mid span of the composite section is

| LRFD | ASD |
|---|---|
| $M_u = 1.084 \times 40^2/8 + 43.2 \times 10$ | $M_a = 0.903 \times 40^2/8 + 27 \times 10$ |
| $= 649$ kip-ft | $= 451$ kip-ft |

From Example 13.14

| LRFD | ASD |
|---|---|
| Using a total of $2n = 69$ anchors over the whole beam provides an available moment: | Using a total of $2n = 80$ anchors over the whole beam provides an available moment: |
| $M_n = 663$ kip-ft | $M_n/\Omega = 456$ kip-ft |
| $> 649$ kip-ft ... satisfactory | $> 451$ kip-ft ... satisfactory |
| Hence, number of anchors required in central 20 ft of beam is | Hence, number of anchors required in central 20 ft of beam is |
| $n' = 69 - 2 \times 22$ | $n' = 80 - 2 \times 24$ |
| $= 25$ anchors | $= 32$ anchors |

## 13.6 Formed Steel Deck with Ribs Perpendicular to Beams

### Requirements

Formed steel deck generally consists of steel sheet formed into a trapezoidal profile. The advantage of using a formed steel deck in the construction of composite beams is that the deck provides formwork to support the poured concrete slab and provides the

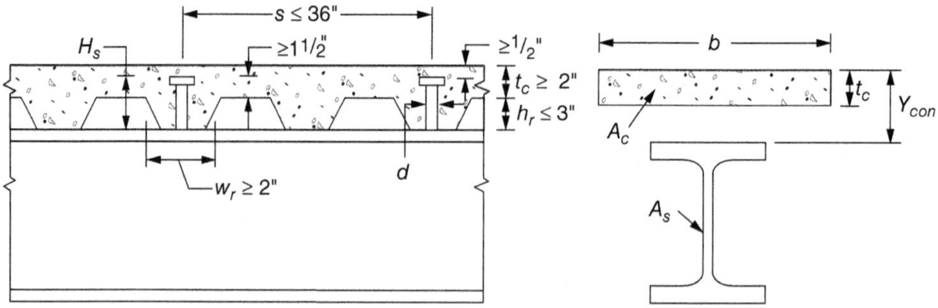


**FIGURE 13.15** Formed steel deck with ribs perpendicular to beam.

tensile reinforcement for the slab. In addition, the slab is lighter in weight than a flat soffit slab and, in most cases, provides the same strength.

The requirements for the use of formed steel deck, with ribs perpendicular to the steel beams, are detailed in AISC 360 Sec. I3.2c and are illustrated in Fig. 13.15. Welded stud shear anchors, not exceeding 3/4 in in diameter, may be used to connect the concrete slab and the steel beams. The studs may be welded directly to the steel beams or through the steel deck and, to prevent uplift, the steel deck must be anchored to the steel beams at a maximum spacing of 18 in. This anchorage may be provided by a combination of stud anchors and puddle welds. When determining the composite section properties or in calculating the value of $V'$, the concrete below the top of the steel deck is neglected. Detailed design procedures are given by Vogel.[8]

**Example 13.16.** Formed Steel Deck with Ribs Perpendicular to Beam
The composite section shown in Fig. 13.16 consists of W18 × 50 beams of Grade 50 steel spaced at 11 ft centers and spanning 40 ft. The 3-in concrete topping is of normal weight concrete with a compressive strength of 4 ksi. The 18-gage, 2-in-deep, formed steel deck has an average rib width of 6 in spaced at 12 in. The ribs are oriented perpendicular to the steel beam. Determine the properties of the composite section assuming full composite action.

The effective width of the concrete slab is the lesser of

$$
\begin{aligned}
b &= s \\
&= 11 \times 12 \\
&= 132 \text{ in}
\end{aligned}
$$

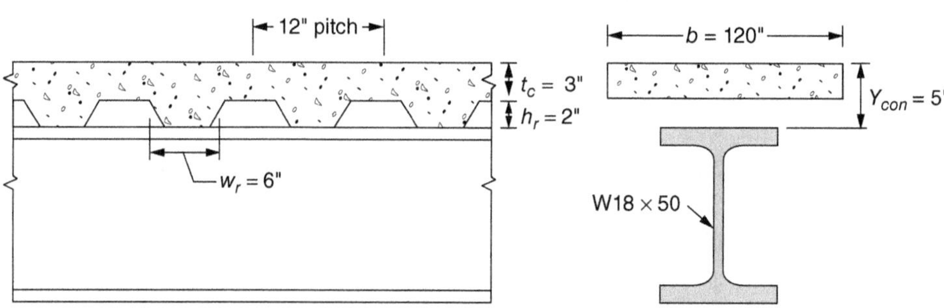


**FIGURE 13.16** Details for Example 13.16.

or $$b = L/4$$

$$= 40 \times 12/4$$

$$= 120 \text{ in ... governs}$$

The area of concrete slab within the effective width is

$$A_c = bt_c$$

$$= 120 \times 3$$

$$= 360 \text{ in}^2$$

The compressive capacity of the concrete slab assuming full utilization is

$$0.85f'_cA_c = 0.85 \times 4 \times 360$$

$$= 1224 \text{ kips}$$

The tensile capacity of the steel beam assuming it has fully yielded in tension is

$$F_yA_s = 50 \times 14.7$$

$$= 735 \text{ kips}$$

$$< 0.85f'_cA_c$$

Hence, the depth of the concrete slab exceeds the depth of the stress block and, for a fully composite beam, the depth of the stress block is given by

$$a = F_yA_s/0.85f'_cb$$

$$= 735/(0.85 \times 4 \times 120)$$

$$= 1.80 \text{ in}$$

The distance from top of steel beam to concrete flange force is

$$Y_2 = Y_{con} - a/2$$

$$= 5 - 1.80/2$$

$$= 4.10 \text{ in}$$

$$Y_1 = 0$$

$$\Sigma Q_n = F_yA_s \text{ ... depth of concrete slab exceeds depth of stress block}$$

$$= 735 \text{ kips}$$

From AISC Manual Table 3-19, for $Y_2 = 4.10$ and $Y_1 = 0$ the available flexural capacity is

| LRFD | ASD |
|---|---|
| $\varphi_bM_n = 720$ kip-ft | $M_n/\Omega_b = 479$ kip-ft |

## Steel-Headed Stud Anchors in Formed Steel Deck with Ribs Perpendicular to Beam

The nominal strength of stud anchors in formed steel deck depends on the location of the anchor in the rib and on the number of anchors in a group positioned on each beam. As shown in Fig. 13.17, with one stud positioned in the rib, the group coefficient is $R_g = 1.0$. With two studs in a group, the group coefficient is $R_g = 0.85$. With three studs in a group, the group coefficient is $R_g = 0.7$.

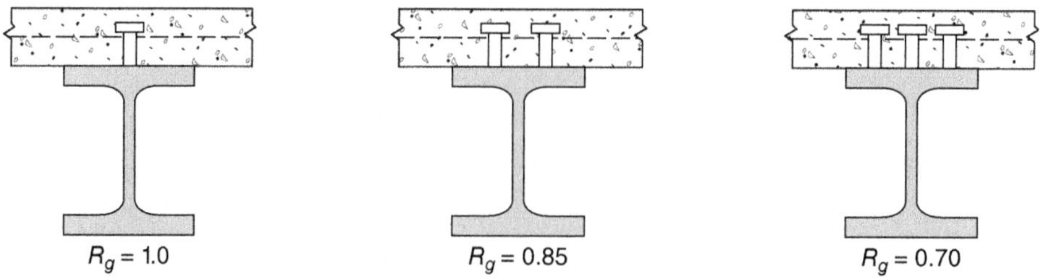


**FIGURE 13.17** Stud group coefficient, ribs perpendicular to beam.

Trapezoidal formed steel deck is usually manufactured with a stiffener in the center of the bottom flange. As shown in Fig. 13.18, this necessitates welding the stud off center, to one side or other of the stiffener, and this affects the strength of the stud. The shear force on the stud acts in the direction from the beam center to the beam support. As described by Easterling et al.,[9] a stud located on the side of the stiffener nearest to the beam support has a greater volume of concrete between the stud and the web of the deck in the direction of the shear force, and is in the strong location, with a position coefficient of $R_p = 0.75$. A stud located on the side of the stiffener nearest to the location of maximum moment is in the weak location with a position coefficient of $R_p = 0.60$.

When deck ribs are perpendicular to the steel beam, the nominal strength of one stud shear connector is given by AISC 360 Eq. (I8-1) as

$$Q_n = 0.5A_{sa}(f'_cE_c)^{0.5}$$
$$\leq R_gR_pA_{sa}F_u$$

where $R_g$ = stud group coefficient, as shown in Fig. 13.17
= 1.0 ... for one stud welded in a steel deck rib
= 0.85 ... for two studs welded in a steel deck rib
= 0.70 ... for three or more studs welded in a steel deck rib

$R_p$ = stud position coefficient, as shown in Fig. 13.18
= 0.75 ... for studs welded in a steel deck rib with $e_{mid\text{-}ht} \geq 2$ in (strong position)
= 0.60 ... for studs welded in a steel deck rib with $e_{mid\text{-}ht} < 2$ in (weak position)

$e_{mid\text{-}ht}$ = distance from the edge of the stud shank to the steel deck web, measured at midheight of the deck rib, in the direction of maximum moment for a simply supported beam

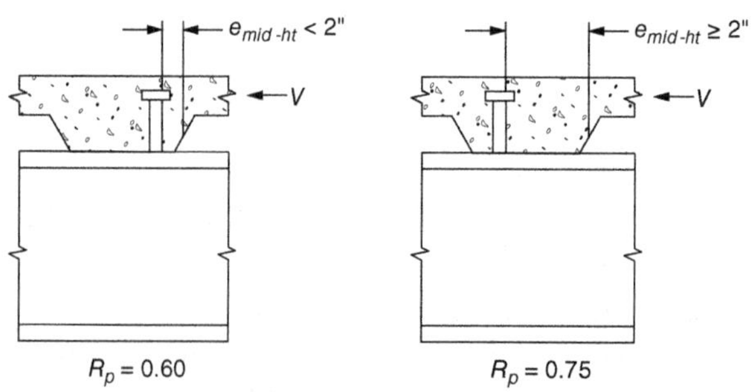


**FIGURE 13.18** Stud position coefficient, ribs perpendicular to beam.

**Example 13.17.** Applied Loading on Formed Steel Deck with Ribs Perpendicular to Beam
The composite section shown in Fig. 13.16 consists of W18 × 50 beams of Grade 50 steel spaced at 11 ft centers and spanning 40 ft. The 3-in concrete topping is of normal weight concrete with a compressive strength of 4 ksi. The 18-gage, 2-in-deep, formed steel deck has an average rib width of 6 in spaced at 12 in. The ribs are oriented perpendicular to the steel beam. The imposed dead load = 15 lb/ft², the imposed live load = 125 lb/ft². Reinforcement in the slab gives an equivalent concrete weight of 150 lb/ft³. Determine the minimum number of 5/8-in-diameter steel-headed stud anchors, placed in the strong position, required to provide the required moment capacity.

At the composite stage the loads on the composite beam are

| | |
|---|---|
| Steel beam self-weight | = 50 lb/ft |
| Metal deck = 2.7 × 11 | = 30 lb/ft |
| Concrete slab = 150 × 11(3 + 2/2)/12 | = 550 lb/ft |
| Imposed dead load = 15 × 11 | = 165 lb/ft |
| Total dead load = $w_D$ | = 795 lb/ft |
| Imposed live load = 125 × 11 = $w_L$ | = 1375 lb/ft |

Assume two stud anchors per rib. Then

$$R_g = 0.85$$

Assume all stud anchors are located in the strong position. Then

$$R_p = 0.75$$

Hence, the nominal shear strength of a 3½-in-long, 5/8-in-diameter steel-headed stud anchor in normal weight 4 ksi concrete is the lesser value obtained from AISC 360 Eq. (I8-1) as

$$Q_n = 0.5A_{sa}(w_c f'_c)^{0.75}$$
$$= 0.5 \times 0.3068(145 \times 4)^{0.75}$$
$$= 18.1 \text{ kips}$$

or

$$Q_n = R_g R_p A_{sa} F_u$$
$$= 0.85 \times 0.75 \times 0.3068 \times 65$$
$$= 12.7 \text{ kips ... governs}$$

Applying ASCE-07 Secs. 2.3 and 2.4 gives the factored loads as

| LRFD | ASD |
|---|---|
| From ASCE 7 Sec. 2.3.2 combination 2: | From ASCE 7 Sec. 2.4.1 combination 2: |
| $w_u$ = factored load | $w_a$ = factored load |
| $= 1.2w_D + 1.6w_L$ | $= w_D + w_L$ |
| $= 1.2 \times 795 + 1.6 \times 1375$ | $= 795 + 1375$ |
| = 3154 lb/ft | = 2170 lb/ft |

The applied bending moment is

| LRFD | ASD |
|---|---|
| $M_u = 3.154 \times 40^2/8$ | $M_a = 2.170 \times 40^2/8$ |
| = 631 kip-ft | = 434 kip-ft |

The required number of stud anchors is obtained as

| LRFD | ASD |
|---|---|
| Provide a total of 70 anchors to give | Provide a total of 80 anchors to give |
| $\Sigma Q_n = nQ_n$ | $\Sigma Q_n = nQ_n$ |
| $= 70 \times 12.7/2$ | $= 80 \times 12.7/2$ |
| $= 445$ kips | $= 508$ kips |
| $= 0.85f'_c ab$ | $= 0.85f'_c ab$ |
| $a = 445/(0.85 \times 4 \times 120)$ | $a = 508/(0.85 \times 4 \times 120)$ |
| $= 1.09$ in | $= 1.25$ in |
| $Y_2 = t_c - a/2$ | $Y_2 = t_c - a/2$ |
| $= 5 - 1.09/2$ | $= 5 - 1.25/2$ |
| $= 4.46$ in | $= 4.38$ in |
| From AISC Manual Table 3-19 with $\Sigma Q_n = 445$ and $Y_2 = 4.46$, the available moment is | From AISC Manual Table 3-19 with $\Sigma Q_n = 508$ and $Y_2 = 4.38$, the available moment is |
| $\varphi M_n = 634$ kip-ft | $M_n/\Omega = 438$ kip-ft |
| $> 631$ kip-ft ... satisfactory | $> 434$ kip-ft ... satisfactory |
| Hence, a total of 70 anchors are required. | Hence, a total of 80 anchors are required. |

## 13.7 Formed Steel Deck with Ribs Parallel to Beams

### Requirements

The requirements for the use of formed steel deck, with ribs parallel to the steel beams, are detailed in AISC 360 Sec. I3.2c and are illustrated in Fig. 13.19. Welded stud shear anchors, not exceeding 3/4 in in diameter, may be used to connect the concrete slab and the steel beams. The studs may be welded directly to the steel beams or through the steel deck and, to prevent uplift, the steel deck must be anchored to the steel beams at a maximum spacing of 18 in. This anchorage may be provided by a combination of stud anchors and puddle welds.

When determining the composite section properties, the concrete below the top of the steel deck may be neglected. However, when calculating the value of $V'_r$ the concrete below the top of the steel deck must be included. The ribs of the steel decks over

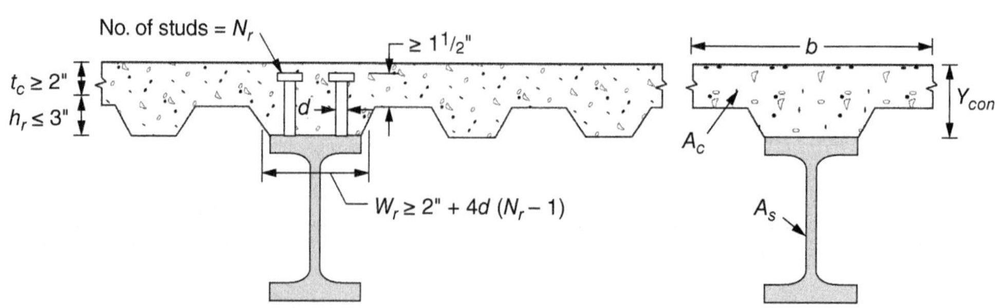


**Figure 13.19** Formed steel deck with ribs parallel to beam.

supporting beams may be split longitudinally and separated to form a concrete haunch. When the depth of the steel deck is 1½ in or greater, the average width $w_r$ of the supported haunch or rib shall not be less than 2 in for the first stud anchor in a row plus four stud diameters for each additional stud anchor in a transverse row.

**Example 13.18.** Formed Steel Deck with Ribs Parallel to Beam
The composite section shown in Fig. 13.20 consists of W18 × 50 beams of Grade 50 steel spaced at 11 ft centers and spanning 40 ft. The 3-in concrete topping is of normal weight concrete with a compressive strength of 4 ksi. The 18-gage, 2-in-deep, formed steel deck has an average rib width of 6 in spaced at 12 in. The ribs are oriented parallel to the steel beam. Determine the properties of the composite section assuming full composite action.

The concrete below the top of the steel deck may be neglected in determining the section properties and the effective section is shown in Fig. 13.20. This is identical with the composite section analyzed in Example 13.16. Hence, the available flexural capacity is

| LRFD | ASD |
|---|---|
| $\varphi_b M_n = 720$ kip-ft | $M_n/\Omega_b = 479$ kip-ft |

## Steel-Headed Stud Anchors in Formed Steel Deck with Ribs Parallel to Beam

The nominal strength of stud anchors in formed steel deck depends on the profile of the steel deck. As shown in Fig. 13.21, a wide shallow rib provides an increased stud strength.

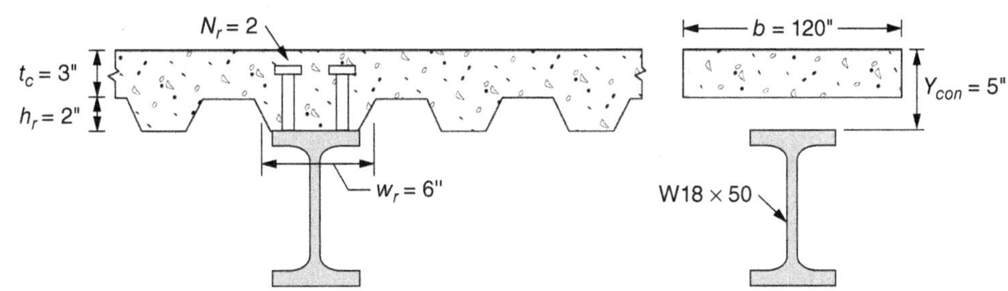


**FIGURE 13.20** Details for Example 13.18.

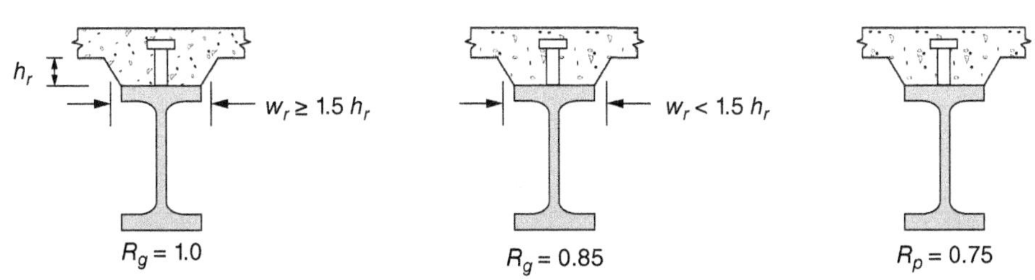


**FIGURE 13.21** Stud group and position coefficients, ribs parallel to beam.

When the deck ribs are parallel to the steel beam, the nominal strength of one stud shear connector is given by AISC 360 Eq. (I8-1) as

$$Q_n = 0.5A_{sa}(f'_c E_c)^{0.5}$$
$$\leq R_g R_p A_{sa} F_u$$

where $R_g$ = stud group coefficient, as shown in Fig. 13.21
= 1.0 ... for any number of studs welded in a row through the steel deck when $w_r \geq 1.5h_r$
= 0.85 ... for one stud welded through the steel deck when $w_r < 1.5h_r$
$R_p$ = stud position coefficient, as shown in Fig. 13.21
= 0.75 ... for studs welded through the steel deck

**Example 13.19.** Applied Loading on Formed Steel Deck with Ribs Parallel to Beam
The composite section shown in Fig. 13.22 consists of W18 × 50 beams of Grade 50 steel spaced at 11 ft centers and spanning 40 ft. The 3-in concrete topping is of normal weight concrete with a compressive strength of 4 ksi. The 18-gage, 2-in-deep, formed steel deck has an average rib width of 6 in spaced at 12 in. The ribs are oriented parallel to the steel beam. The imposed dead load = 15 lb/ft², the imposed live load = 125 lb/ft². Reinforcement in the slab gives an equivalent concrete weight of 150 lb/ft³. Determine the minimum number of 5/8-in-diameter steel-headed stud anchors required to provide the required moment capacity.

As shown in Example 13.16, the depth of the concrete slab exceeds the depth of the stress block and the concrete below the top of the steel deck may be neglected. The ratio of the average rib width to the rib depth is

$$w_r/h_r = 6/2$$
$$= 3$$
$$> 1.5$$

Hence, the stud group coefficient is

$$R_g = 1.0$$

The stud position coefficient, for deck oriented parallel to the steel beam, is

$$R_p = 0.75$$

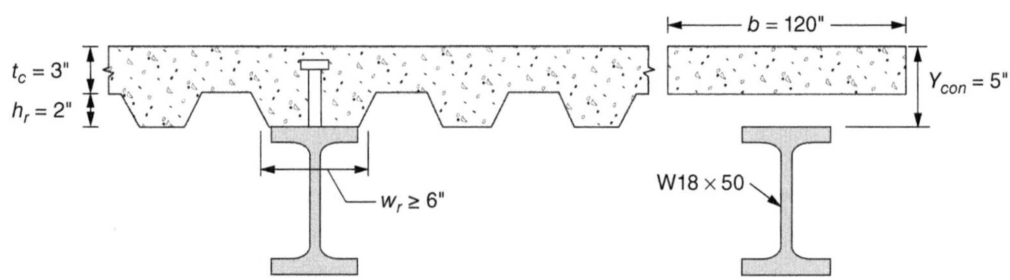


**Figure 13.22** Details for Example 13.19.

Hence, the nominal shear strength of a 3½-in-long, 5/8-in-diameter steel-headed stud anchor in normal weight 4 ksi concrete is the lesser value obtained from AISC 360 Eq. (I8-1) as

$$Q_n = 0.5A_{sa}(w_c f'_c)^{0.75}$$
$$= 0.5 \times 0.3068(145 \times 4)^{0.75}$$
$$= 18.1 \text{ kips}$$

or

$$Q_n = R_g R_p A_{sa} F_u$$
$$= 1.00 \times 0.75 \times 0.3068 \times 65$$
$$= 15.0 \text{ kips ... governs}$$

From Example 13.17, the applied bending moment is

| LRFD | ASD |
|---|---|
| $M_u = 3.154 \times 40^2/8$<br>$= 631$ kip-ft | $M_a = 2.170 \times 40^2/8$<br>$= 434$ kip-ft |

The required number of stud anchors is obtained as

| LRFD | ASD |
|---|---|
| Provide a total of 56 anchors to give | Provide a total of 68 anchors to give |
| $\Sigma Q_n = nQ_n$<br>$= 56 \times 15.0/2$<br>$= 420$ kips<br>$= 0.85f'_c ab$<br>$a = 420/(0.85 \times 4 \times 120)$<br>$= 1.0$ in<br>$Y_2 = t_c - a/2$<br>$= 5 - 1.0/2$<br>$= 4.5$ in | $\Sigma Q_n = nQ_n$<br>$= 68 \times 15.0/2$<br>$= 510$ kips<br>$= 0.85f'_c ab$<br>$a = 510/(0.85 \times 4 \times 120)$<br>$= 1.25$ in<br>$Y_2 = t_c - a/2$<br>$= 5 - 1.25/2$<br>$= 4.38$ in |
| From AISC Manual Table 3-19 with $\Sigma Q_n = 420$ and $Y_2 = 4.5$, the available moment is | From AISC Manual Table 3-19 with $\Sigma Q_n = 510$ and $Y_2 = 4.38$, the available moment is |
| $\varphi M_n = 632$ kip-ft<br>$> 631$ kip-ft ... satisfactory | $M_n/\Omega = 439$ kip-ft<br>$> 434$ kip-ft ... satisfactory |
| Hence, a total of 56 anchors are required. | Hence, a total of 68 anchors are required. |

## References

1. Samuelson, D. 2002. "Composite Steel Joists," *Engineering Journal*, 3rd Quarter 2002.
2. American Institute of Steel Construction (AISC). 2010. *Specification for Structural Steel Buildings* (AISC 360-10), AISC, Chicago, IL.
3. Leon, R. T. and Hajjar, J. F. 2008. "Limit State Response of Composite Columns and Beam-Columns Part II: Application of Design Provisions for the 2005 AISC Specification," *Engineering Journal*, 1st Quarter 2008.
4. Leon, R. T., Kim, D. K., and Hajjar, J. F. 2007. "Limit State Response of Composite Columns and Beam-Columns Part I: Formulation of Design Provisions for the 2005 AISC Specification," *Engineering Journal*, 4th Quarter 2007.
5. American Institute of Steel Construction (AISC). 2011. *Steel Construction Manual*, 14th edition. AISC, Chicago, IL.
6. American Society of Civil Engineers (ASCE). 2010. *Minimum Design Loads for Buildings and Other Structures* (ASCE 7-10), ASCE, Reston, VA.
7. Ollgaard, J. G., Slutter, R. G., and Fisher, J. W. 1971. "Strength of Shear Studs in Steel Deck on Composite Beams and Joists," *Engineering Journal*, 2nd Quarter 1971.
8. Vogel, R. 1991. *LRFD-Composite Beam Design with Metal Deck*. Steel Committee of California, El Monte, CA.
9. Easterling, W. S., Gibbings, D. R., and Murray, T. M. 1993. "Strength of Shear Studs in Steel Deck on Composite Beams and Joists," *Engineering Journal*, 2nd Quarter 1993.

## Problems

**13.1** *Given:* The composite column shown in Fig. 13.23 consists of a W10 × 49 shape with 20- × 20-in concrete encasement. The steel section has a yield stress of $F_y = 50$ ksi and the concrete has a specified strength of $f'_c = 4$ ksi and a weight of 145 lb/ft$^3$. Continuous longitudinal reinforcement is provided consisting of four No. 9 Grade 60 bars at 16 in on center. Adequate lateral reinforcement and load transfer is provided. The column is 20 ft high and is pinned at each end with the load applied directly to the steel section.

*Find:* Using allowable stress level design (ASD), the available axial strength of the composite column.

**13.2** *Given:* The composite column shown in Fig. 13.23 consists of a W10 × 49 shape with 20 × 20-in concrete encasement. The steel section has a yield stress of $F_y = 50$ ksi and the concrete has a specified strength of $f'_c = 4$ ksi and a weight of 145 lb/ft$^3$. Continuous longitudinal reinforcement is provided consisting of four No. 9 Grade 60 bars at 16 in on center. Adequate lateral reinforcement and load transfer is provided. The column is 20 ft high and is pinned at each end with the load applied directly to the steel section.

*Find:* Using load and resistance factor design (LRFD), the available axial strength of the composite column.

**13.3** *Given:* The composite column, shown in Fig. 13.24 consisting of a HSS6.000 × 0.250 round hollow structural member with a yield stress of 42 ksi, filled with 4 ksi concrete with a weight of 145 lb/ft$^3$. Adequate load transfer is provided. The column is 20 ft high and is pinned at each end.

*Find:* Using allowable stress level design (ASD), the available axial strength of the composite column.

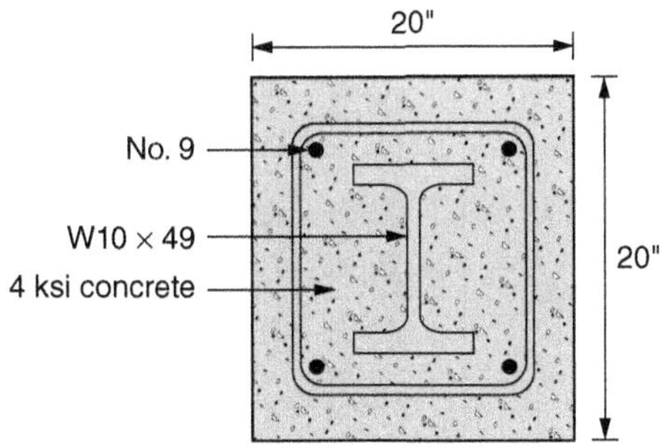


**FIGURE 13.23** Details for Problem 13.1.

**13.4** *Given:* The composite column, shown in Fig. 13.24 consisting of a HSS6.000 × 0.250 round hollow structural member with a yield stress of 42 ksi, filled with 4 ksi concrete with a weight of 145 lb/ft$^3$. Adequate load transfer is provided. The column is 20 ft high and is pinned at each end.

*Find:* Using load and resistance factor design (LRFD), the available axial strength of the composite column.

**13.5** *Given:* The W21 × 101 beam of Grade 50 steel, shown in Fig. 13.25, is fully encased in 4 ksi concrete and is fabricated without shear anchors.

*Find:* Using allowable stress level design (ASD), the maximum available moment that the beam can support.

**13.6** *Given:* The W21 × 101 beam of Grade 50 steel, shown in Fig. 13.25, is fully encased in 4 ksi concrete and is fabricated without shear anchors.

*Find:* Using load and resistance factor design (LRFD), the maximum available moment that the beam can support.

**13.7** *Given:* The composite section shown in Fig. 13.26 consists of W14 × 26 beams of Grade 50 steel spaced at 10 ft centers and spanning 30 ft. The 5-in-thick solid slab is of normal weight concrete with a compressive strength of 4 ksi.

*Find:* Using allowable stress level design (ASD), the available strength of the fully composite section.

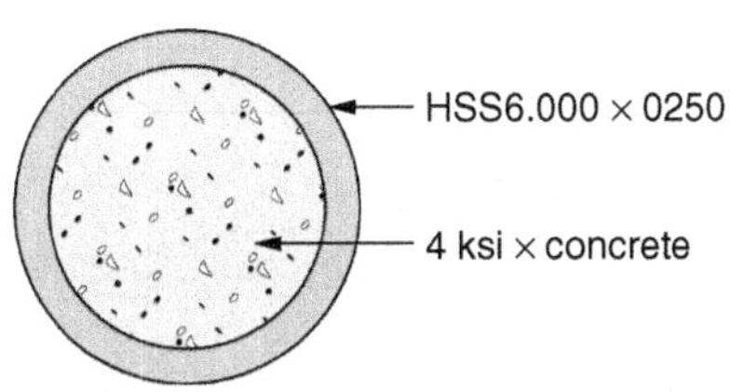


**FIGURE 13.24** Details for Problem 13.3.

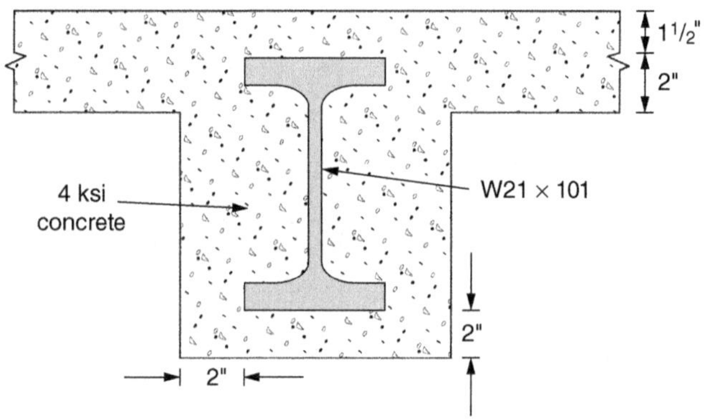


**Figure 13.25** Details for Problem 13.5.

**13.8** *Given:* The composite section shown in Fig. 13.26 consists of W14 × 26 beams of Grade 50 steel spaced at 10 ft centers and spanning 30 ft. The 5-in-thick solid slab is of normal weight concrete with a compressive strength of 4 ksi.

*Find:* Using load and resistance factor design (LRFD), the available strength of the fully composite section.

**13.9** *Given:* The composite section shown in Fig. 13.26 consists of W14 × 26 beams of Grade 50 steel spaced at 10 ft centers and spanning 30 ft. The 5-in-thick solid slab is of normal weight concrete with a compressive strength of 4 ksi. The construction load = 20 lb/ft$^2$, imposed dead load = 15 lb/ft$^2$, and imposed live load = 65 lb/ft$^2$. Reinforcement in the slab gives an equivalent concrete weight of 150 lb/ft$^3$. During construction, the steel beams are laterally braced and are not shored.

*Find:* Using allowable stress level design (ASD):

(i) If the steel beam is adequate during construction.
(ii) The number of 5/8-in-diameter steel-headed stud anchors required to support the total load on the composite section.

**13.10** *Given:* The composite section shown in Fig. 13.26 consists of W14 × 26 beams of Grade 50 steel spaced at 10 ft centers and spanning 30 ft. The 5-in-thick solid slab is of normal weight concrete with a compressive strength of 4 ksi. The construction load = 20 lb/ft$^2$, imposed dead load = 15 lb/ft$^2$, and imposed live load = 65 lb/ft$^2$. Reinforcement in the slab gives an equivalent concrete weight of 150 lb/ft$^3$. During construction, the steel beams are laterally braced and are not shored.

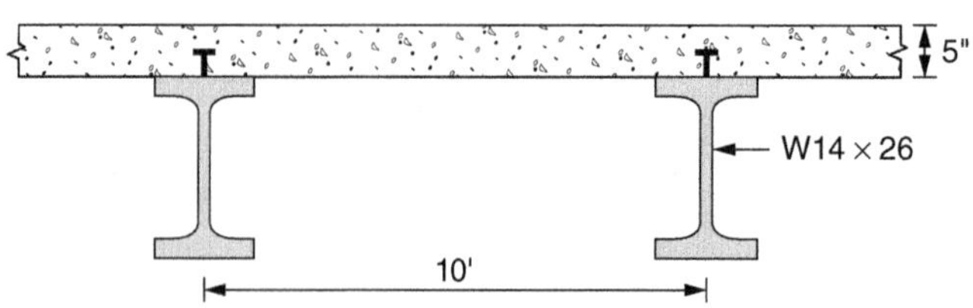


**Figure 13.26** Details for Problem 13.7.

*Find:* Using load and resistance factor design (LRFD):

(i) If the steel beam is adequate during construction.
(ii) The number of 5/8-in-diameter steel-headed stud anchors required to support the total load on the composite section.

**13.11** *Given:* The composite section shown in Fig. 13.26 consists of W14 × 26 beams of Grade 50 steel spaced at 10 ft centers and spanning 30 ft. The 5-in-thick solid slab is of normal weight concrete with a compressive strength of 4 ksi. The imposed dead load = 15 lb/ft$^2$, the imposed live load consists of two concentrated loads each of 7 kips as indicated in Fig. 13.27.

*Find:* Using allowable stress level design (ASD), the minimum number of 5/8-in-diameter steel-headed stud anchors required between the supports and the concentrated loads and the number required in the central 10 ft of the span.

**13.12** *Given:* The composite section shown in Fig. 13.26 consists of W14 × 26 beams of Grade 50 steel spaced at 10 ft centers and spanning 30 ft. The 5-in-thick solid slab is of normal weight concrete with a compressive strength of 4 ksi. The imposed dead load = 15 lb/ft$^2$, the imposed live load consists of two concentrated loads each of 7 kips as indicated in Fig. 13.27.

*Find:* Using load and resistance factor design (LRFD), the minimum number of 5/8-in-diameter steel-headed stud anchors required between the supports and the concentrated loads and the number required in the central 10 ft of the span.

**13.13** *Given:* A composite section consisting of W14 × 26 beams of Grade 50 steel spaced at 10 ft centers and spanning 30 ft. The 3-in concrete topping is of normal weight concrete with a compressive strength of 4 ksi. The 18-gage, 2-in-deep, formed steel deck has an average rib width of 6 in spaced at 12 in. The ribs are oriented perpendicular to the steel beam.

*Find:* Using allowable stress level design (ASD), the properties of the composite section assuming full composite action.

**13.14** *Given:* A composite section consisting of W14 × 26 beams of Grade 50 steel spaced at 10 ft centers and spanning 30 ft. The 3-in concrete topping is of normal weight concrete with a compressive strength of 4 ksi. The 18-gage, 2-in-deep, formed steel deck has an average rib width of 6 in spaced at 12 in. The ribs are oriented perpendicular to the steel beam.

*Find:* Using load and resistance factor design (LRFD), the properties of the composite section assuming full composite action.

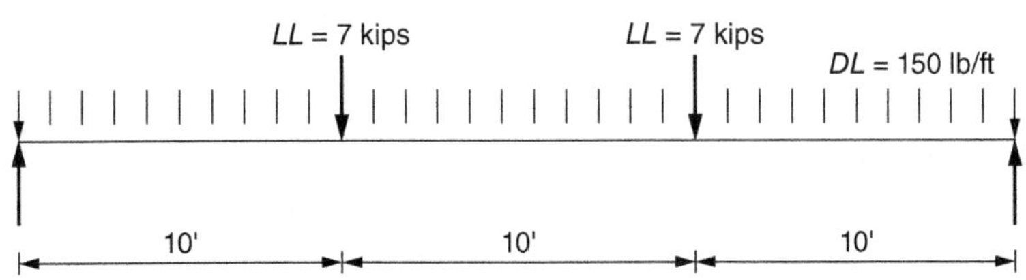


**FIGURE 13.27** Details for Problem 13.11.

**13.15** *Given:* A composite section consisting of W14 × 26 beams of Grade 50 steel spaced at 10 ft centers and spanning 30 ft. The 3-in concrete topping is of normal weight concrete with a compressive strength of 4 ksi. The 18-gage, 2-in-deep, formed steel deck has an average rib width of 6 in spaced at 12 in. The ribs are oriented perpendicular to the steel beam.

*Find:* Using allowable stress level design (ASD), the number of 5/8-in-diameter steel-headed stud anchors to provide full composite action.

**13.16** *Given:* A composite section consisting of W14 × 26 beams of Grade 50 steel spaced at 10 ft centers and spanning 30 ft. The 3-in concrete topping is of normal weight concrete with a compressive strength of 4 ksi. The 18-gage, 2-in-deep, formed steel deck has an average rib width of 6 in spaced at 12 in. The ribs are oriented perpendicular to the steel beam.

*Find:* Using load and resistance factor design (LRFD), the number of 5/8-in-diameter steel-headed stud anchors to provide full composite action.

**13.17** *Given:* A composite section consisting of W14 × 26 beams of Grade 50 steel spaced at 10 ft centers and spanning 30 ft. The 3-in concrete topping is of normal weight concrete with a compressive strength of 4 ksi. The 18-gage, 2-in-deep, formed steel deck has an average rib width of 6 in spaced at 12 in. The ribs are oriented parallel to the steel beam.

*Find:* Using allowable stress level design (ASD), the properties of the composite section assuming full composite action.

**13.18** *Given:* A composite section consisting of W14 × 26 beams of Grade 50 steel spaced at 10 ft centers and spanning 30 ft. The 3-in concrete topping is of normal weight concrete with a compressive strength of 4 ksi. The 18-gage, 2-in-deep, formed steel deck has an average rib width of 6 in spaced at 12 in. The ribs are oriented parallel to the steel beam.

*Find:* Using load and resistance factor design (LRFD), the properties of the composite section assuming full composite action.

**13.19** *Given:* A composite section consisting of W14 × 26 beams of Grade 50 steel spaced at 10 ft centers and spanning 30 ft. The 3-in concrete topping is of normal weight concrete with a compressive strength of 4 ksi. The 18-gage, 2-in-deep, formed steel deck has an average rib width of 6 in spaced at 12 in. The ribs are oriented parallel to the steel beam.

*Find:* Using allowable stress level design (ASD), the number of 5/8-in-diameter steel-headed stud anchors to provide full composite action.

**13.20** *Given:* A composite section consisting of W14 × 26 beams of Grade 50 steel spaced at 10 ft centers and spanning 30 ft. The 3-in concrete topping is of normal weight concrete with a compressive strength of 4 ksi. The 18-gage, 2-in-deep, formed steel deck has an average rib width of 6 in spaced at 12 in. The ribs are oriented parallel to the steel beam.

*Find:* Using load and resistance factor design (LRFD), the number of 5/8-in-diameter steel headed stud anchors to provide full composite action.

**13.21** *Given:* A composite section consisting of W14 × 26 beams of Grade 50 steel spaced at 10 ft centers and spanning 30 ft. The 3-in concrete topping is of normal weight concrete with a compressive strength of 4 ksi. The 18-gage, 2-in-deep, formed steel deck has an average rib width of 6 in spaced at 12 in. The ribs are oriented parallel to the steel beam.

*Find:* Using allowable stress level design (ASD), the number of 5/8-in-diameter steel-headed stud anchors to provide for a required bending moment of $M_a = 180$ kip-ft.

**13.22** *Given:* A composite section consisting of W14 × 26 beams of Grade 50 steel spaced at 10 ft centers and spanning 30 ft. The 3-in concrete topping is of normal weight concrete with a compressive strength of 4 ksi. The 18-gage, 2-in-deep, formed steel deck has an average rib width of 6 in spaced at 12 in. The ribs are oriented parallel to the steel beam.

*Find:* Using load and resistance factor design (LRFD), the number of 5/8-in-diameter steel-headed stud anchors to provide for a required bending moment of $M_u = 270$ kip-ft.

# CHAPTER 14

# Design of Steel Structures for Lateral Loads

## 14.1 Introduction

The *International Building Code*[1] (IBC) requires the seismic design of steel structures to be in accordance with IBC Sec. 2205.2. Buildings assigned to seismic design categories B or C must be designed in accordance with the requirements of *Seismic Provisions for Structural Steel Buildings*[2] (AISC 341). The response modification coefficient, $R$, listed in *Minimum Design Loads for Buildings and Other Structures*[3] (ASCE 7) Table 12.2-1, governs the magnitude of the lateral force for which the structure must be designed. An exception is permitted for "Steel systems not specifically detailed for seismic resistance, excluding cantilever column systems" listed in ASCE 7, Table 12.2-1 with a response modification coefficient of 3.0. These buildings are designed and detailed for the earthquake load effects using *Specification for Structural Steel Buildings*[4] (AISC 360). It is assumed that by adopting a value of $R = 3.0$, a nominally elastic response is produced by the seismic loads and the special detailing required by AISC 341 to resist inelastic response is unnecessary. This may result in a more economical building.

For buildings assigned to seismic design categories D, E, or F, the provisions of AISC 341 are mandatory. In situations where wind effects exceed seismic effects, the building elements must still be detailed in accordance with AISC 341 provisions so as to accommodate the inelastic demand required.

Nonbuilding structures assigned to seismic design categories D, E, or F must also be designed in accordance with the requirements of AISC 341. An exception is permitted for "Steel ordinary concentrically braced frame with unlimited height" listed in ASCE 7 Table 15.4-1 with a response modification coefficient of 1.5 and "Steel ordinary moment frames with unlimited height" listed in ASCE 7 Table 15.4-1 with a response modification coefficient of 1.0. These buildings are designed and detailed for the earthquake load effects using AISC 360.

Buildings assigned to seismic design category A are exempt from the seismic design requirements of ASCE 7 Chap. 12. General structural integrity is ensured by compliance with ASCE 7 Sec. 1.4. These provisions ensure that a building in seismic design category A is provided with a lateral force resisting system with sufficient strength to provide stability under nominal lateral forces.

## 14.2 General Considerations

In designing a structure, using the provisions of AISC 341, it is necessary to ensure that the structure can withstand large inelastic deformations without structural collapse. To achieve this requires the careful selection of materials and components and adherence to the detailing provisions of the standard.

### Ductility

Ductility in the steel used is necessary to ensure that the members of a structure can deform sufficiently to accommodate large inelastic cyclic displacements without precipitate collapse. This allows a redundant structure to continuously deform and absorb energy as plastic hinges progressively form in the structure until a collapse mechanism is produced. Energy absorption in a steel structure is typically by yielding of the members in bending, shear, tension, or compression. Examples of collapse mechanisms are shown in Fig. 14.1.

Steels permitted for use in seismic applications are selected based on their inelastic properties and weldability. In general, these steels possess the following characteristics:

- A pronounced stress-strain plateau at the yield stress
- A large inelastic strain capability
- Good weldability

Wide flange shapes are generally of ASTM A992 material. This has a specified maximum yield stress and maximum yield-to-tensile ratio, to ensure ductility, and a limit on the carbon equivalent to ensure weldabiblity. In general, the specified yield stress must not exceed 50 ksi for members in which inelastic behavior is expected. This limit is increased to 55 ksi for the following systems:

- Ordinary moment frames
- Ordinary concentrically braced frames
- Composite ordinary moment frames
- Composite ordinary braced frames
- Composite ordinary shear walls

These limits may be exceeded when it can be shown that this is justified.

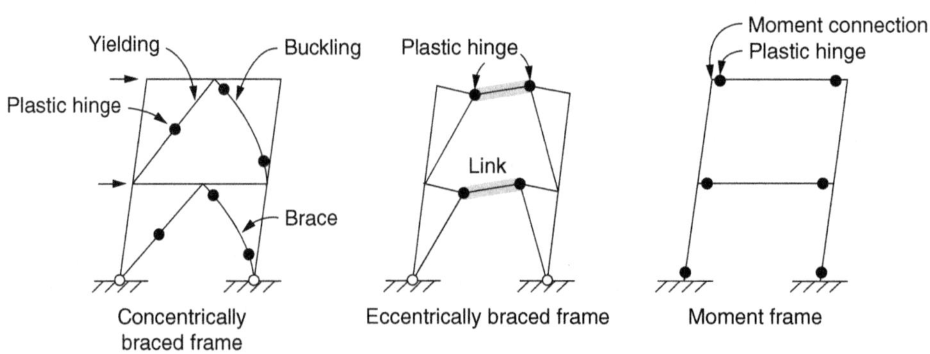


**FIGURE 14.1** Collapse mechanisms.

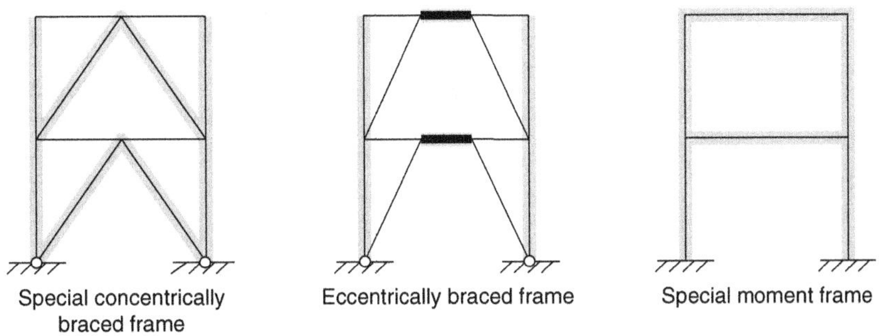


**Figure 14.2** Highly ductile members.

Members of the seismic force resisting system that undergo inelastic deformation are classified as either moderately ductile or highly ductile members. Members of the structural system that undergo extremely large plastic rotations in excess of 0.04 radian under the design earthquake are designated as highly ductile members. These members generally act as fuses in the system, through the formation of plastic hinges, and limit the load transferred to other members of the system which remain essentially elastic. The links in an eccentrically braced frame act as fuses as also do the diagonal braces in a concentrically braced frame. In a special moment frame, plastic hinges occur at the ends of the girders near, but away from, their connections to the columns. The final hinges, prior to collapse, form at the base of the columns. Examples of highly ductile members are shown shaded in Fig. 14.2.

Elements of the structural system that undergo moderate plastic rotations not exceeding 0.02 radian under the design earthquake are designated as moderately ductile members. The *Seismic Design Manual*[5] (SDM) Table 1-2 provides a comprehensive list of ductility requirements for the members of different systems. Table 14.1 lists the ductility requirements for an abridged number of systems.

| System | Highly Ductile | Moderately Ductile |
|---|---|---|
| **Ordinary concentrically braced frames** | | |
| Diagonal braces | | ✓ |
| **Special concentrically braced frames** | | |
| Diagonal braces | ✓ | |
| Beams | | ✓ |
| Columns | ✓ | |
| **Eccentrically braced frames** | | |
| Diagonal braces | | ✓ |
| Columns | ✓ | |
| Link beams | ✓ | |
| Beams outside the link | | ✓ |
| **Special moment frames** | | |
| Beams | ✓ | |
| Columns | ✓ | |

**Table 14.1** Member Ductility Requirements

| Element | Width-to-Thickness Ratio | Limiting Width-to-Thickness Ratio | |
|---|---|---|---|
| | | Moderately Ductile, $\lambda_{md}$ | Highly Ductile, $\lambda_{mh}$ |
| Round HSS | $D/t$ | $0.044E/F_y^*$ | $0.038E/F_y$ |
| Rectangular HSS | $b/t$ | $0.64(E/F_y)^{0.5\dagger}$ | $0.55(E/F_y)^{0.5\ddagger}$ |
| Angles | $b/t$ | $0.38(E/F_y)^{0.5}$ | $0.30(E/F_y)^{0.5}$ |
| Flanges of I-shaped members and channels | $b/t$ | $0.38(E/F_y)^{0.5}$ | $0.30(E/F_y)^{0.5}$ |
| Webs of I-shaped sections used as beams or columns¶ | $h/t_w$ | $3.76(1 - 2.75C_a)(E/F_y)^{0.5}$ ... for $C_a \leq 0.125$ | $2.45(1 - 0.93C_a)(E/F_y)^{0.5}$ ... for $C_a \leq 0.125$ |
| | | $1.12(2.33 - C_a)(E/F_y)^{0.5} \geq 1.49(E/F_y)^{0.5}$ ... for $C_a > 0.125$ | $0.77(2.93 - C_a)(E/F_y)^{0.5} \geq 1.49(E/F_y)^{0.5}$ ... for $C_a > 0.125$ |
| Webs of I-shaped sections used as diagonal braces | $h/t_w$ | $1.49(E/F_y)^{0.5}$ | $1.49(E/F_y)^{0.5}$ |

*The limiting diameter-to-thickness ratio of round HSS members used as beams or columns shall not exceed $0.07E/F_y$.

†The limiting width-to-thickness ratio of walls of rectangular HSS members used as beams or columns shall not exceed $1.12(E/F_y)^{0.5}$.

‡The limiting width-to-thickness ratio of flanges of rectangular HSS members used as columns shall not exceed $0.6(E/F_y)^{0.5}$.

¶For I-shaped beams in SMF systems, where $C_a$ is less than or equal to 0.125, the limiting ratio $h/t_w$ shall not exceed $2.45(E/F_y)^{0.5}$. For I-shaped beams in IMF systems, where $C_a$ is less than or equal to 0.125, the limiting width-to-thickness ratio shall not exceed $3.76(E/F_y)^{0.5}$. $C_a = P_u/\varphi_c P_y$ ... LRFD, $C_a = \Omega_c P_u/P_y$ ... ASD, SMF = special moment frames, IMF = intermediate moment frames.

**TABLE 14.2** Limiting Width-to-Thickness Ratios of Structural Sections

Local buckling and lateral-torsional buckling failure are nonductile failure modes and must be prevented from occurring in a seismic force resisting system. Members in the system that are designated moderate or highly ductile members, have severe restrictions placed on their width-to-thickness ratios to prevent local buckling occurring as plastic hinges develop. These restrictions are tabulated in AISC 341 Table D1.1 and in an abridged form in Table 14.2.

Structural sections that satisfy the moderately or highly ductile width-to-thickness requirements for specific seismic force resisting systems are listed in SDM Tables 1-3 to 1-7.

To prevent lateral-torsional buckling of beams, AISC 341 Sec. D2 specifies the maximum unbraced length for moderate or highly ductile members. For moderately ductile beams, the maximum unbraced length is given by AISC 341 Eq. (D1-2) as

$$L_b = 0.17r_y E/F_y$$

Both flanges of the beam must be laterally braced or the beam cross section must be torsionally braced. For highly ductile beams, the maximum unbraced length is given by AISC 341 Sec. D1.2b as

$$L_b = 0.086r_y E/F_y$$

Both flanges of the beam must be laterally braced or the beam cross section must be torsionally braced. Bracing must be located adjacent to expected plastic hinge locations.

## Capacity Design and Expected Material Strength

For the design of special concentrically braced frames, eccentrically braced frames, and special moment frames, the seismic provisions adopts a capacity design methodology. One element in the system is designated as the structural fuse and the required strength of the remaining members is determined from the expected strength of the fuse. The links in an eccentrically braced frame act as fuses, as do the braces in a concentrically braced frame and the plastic hinges that occur at the ends of the girders of a special moment frame adjacent to their connections to the columns. Inelastic ductile behavior is confined to the fuse and the remaining members in the system are designed to resist the force produced by the fuse so as to remain essentially elastic.

Steel sections usually have a yield stress and a tensile strength greater than the specified minimum values. An accurate estimate of the fuse capacity requires an accurate estimate of the expected yield stress and tensile strength and these are given by

$$R_yF_y = \text{expected yield stress}$$

$$R_tF_u = \text{expected tensile strength}$$

where $R_y$ = ratio of the expected yield stress to the specified minimum yield strength
$R_t$ = ratio of the expected tensile strength to the specified minimum tensile strength
$F_y$ = specified minimum yield stress of the type of steel used
$F_u$ = specified minimum tensile strength of the type of steel used

Values of $R_y$ and $R_t$ are tabulated in AISC 341 Table A3.1 and in an abridged form in Table 14.3.

## Protected Zones

Discontinuities produced by fabrication and erection procedures or from other attachments introduced into plastic hinging regions produce stress concentrations that may lead to fracture. For each seismic force resisting system, the seismic provisions designate

| Application | Grade | $F_y$ | $F_u$ | $R_y$ | $R_t$ |
|---|---|---|---|---|---|
| Hot-rolled structural shapes, and bars | A36 | 36 | 58 | 1.5 | 1.2 |
| | A992 | 50 | 65 | 1.1 | 1.1 |
| | A572 Gr. 50 | 50 | 65 | 1.1 | 1.1 |
| Hollow structural sections | A500 Gr. B, rectangular | 46 | 58 | 1.4 | 1.3 |
| | A500 Gr. B, round | 42 | 58 | 1.4 | 1.3 |
| Pipes | A53 Gr. B | 35 | 60 | 1.6 | 1.2 |
| Plates, strips, and sheets | A36 | 36 | 58 | 1.3 | 1.2 |
| | A572 Gr. 50 | 50 | 65 | 1.1 | 1.2 |

**TABLE 14.3** Values of $R_y$ and $R_t$

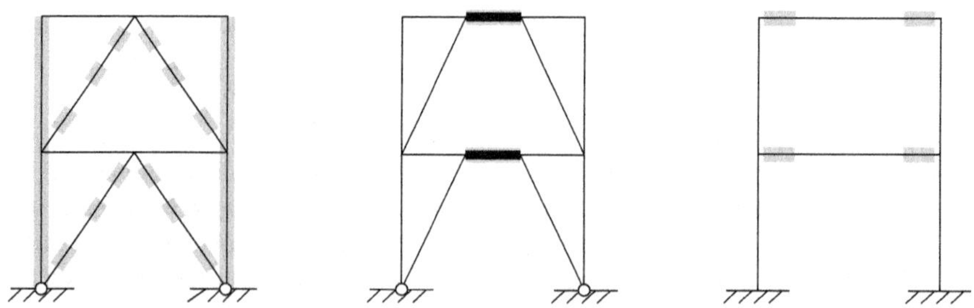


**FIGURE 14.3** Examples of protected zones.

protected zones in which discontinuities are prohibited. Examples of protected zones are shown shaded in Fig. 14.3.

The discontinuities that are prohibited are listed in AISC 341 Sec. I2.1 and consist of

- Holes, tack welds, erection aids, air-arc gouging, and unspecified thermal cutting
- Steel-headed stud anchors and decking attachments that penetrate the beam flange shall not be placed on beam flanges; arc spot welds as required to secure decking are permitted
- Welded, bolted, screwed or shot-in attachments for perimeter edge angles, exterior facades, partitions, duct work, piping, or other construction

Exceptions are permitted for steel-headed stud anchors when designated in AISC 358[6] or as otherwise determined with a prequalified connection.

## Demand Critical Welds

Welds are designated as demand critical based on the expected inelastic strain demand and the consequence of failure. The location of demand critical welds are specified in AISC 341 and AISC 358. These welds must be made with filler metals that have twice the Charpy V-notch toughness values of filler metals commonly used. Details of filler metal requirements are given in AISC 341 Sec. A3.4b. Other requirements such as the use of backing bars and run-off tabs are also given in the seismic provisions.

## Loads and Load Combinations

For load combinations comprising seismic loads, IBC Sec. 1605.1 adopts ASCE 7 Secs. 12.4.2 and 12.4.3. For LRFD design, the prescribed load combinations are given in ASCE 7 Sec. 12.4.2.3. Load combination 5 is applicable when the effects of gravity and seismic loads are additive and is

$$(1.2 + 0.2S_{DS})D + \rho Q_E + 0.5L + 0.2S$$

where $0.5L$ is replaced with $L$ for garages, places of public assembly, and areas where live load is greater than 100 lb/ft$^2$. Load combination 7 is applicable when the effects of gravity and seismic loads counteract and is

$$(0.9 - 0.2S_{DS})D + \rho Q_E$$

For ASD design, the prescribed load combinations are given in ASCE 7 Sec. 12.4.2.3. Load combinations 5 and 6b are applicable when the effects of gravity and seismic loads are additive and are

$$(1.0 + 0.14S_{DS})D + 0.7\rho Q_E$$

$$(1.0 + 0.105S_{DS})D + 0.525\rho Q_E + 0.75L + 0.75S$$

Load combination 8 is applicable when the effects of gravity and seismic loads counteract and is

$$(0.6 - 0.14S_{DS})D + 0.7\rho Q_E$$

where $Q_E$ = effect of horizontal seismic forces
$S_{DS}$ = 5 percent damped, design spectral response acceleration, for a period of 0.2 second
$D$ = effect of dead load
$\rho$ = redundancy factor
$L$ = effect of live load
$S$ = effect of snow load

For LRFD design, the prescribed load combinations with overstrength factor are given by load combinations 5 and 6 of ASCE 7 Sec. 12.4.3.2. Load combination 5 is applicable when the effects of gravity and seismic loads are additive and is

$$(1.2 + 0.2S_{DS})D + \Omega_0 Q_E + 0.5L + 0.2S$$

where $0.5L$ is replaced with $L$ for garages, places of public assembly, and areas where live load is greater than 100 lb/ft$^2$. Load combination 6 is applicable when the effects of gravity and seismic loads counteract and is

$$(0.9 - 0.2S_{DS})D + \Omega_0 Q_E$$

For ASD design, the prescribed load combinations with overstrength factor are given in ASCE 7 Sec. 12.4.3.2. Load combinations 5 and 6b are applicable when the effects of gravity and seismic loads are additive and are

$$(1.0 + 0.14S_{DS})D + 0.7\Omega_0 Q_E$$

$$(1.0 + 0.105S_{DS})D + 0.525\Omega_0 Q_E + 0.75L + 0.75S$$

When the effects of gravity and seismic loads counteract, load combination 8 is applicable, which is

$$(0.6 - 0.14S_{DS})D + 0.7\Omega_0 Q_E$$

where $\Omega_0$ = overstrength factor given in ASCE 7 Table 12.2-1
$\Omega_0 Q_E = E_{mh}$
= effect of horizontal seismic forces, including overstrength

For some structural systems where capacity design is used, the value of $E_{mh}$ is defined in AISC 341 as the amplified load and this is substituted in the above equations.

## 14.3 Concentrically Braced Frames

As shown in Fig. 14.4, concentrically braced frames[7,8] are basically truss systems where the centerlines of the members coincide or nearly coincide thus, minimizing flexural action. Lateral forces are resisted primarily by truss action with members subjected to axial forces in the elastic range. The bracing members and their connections are subject to cyclic tension and compression. During a severe earthquake, bracing members and their connections may undergo significant inelastic deformations and are specially detailed to avoid premature failure.

Concentrically braced frames are usually more economical to construct than moment resisting frames and eccentrically braced frames. However, concentrically braced frames impose restrictions on floor layout because of the location of the braces. Concentrically braced frames are subdivided into two types, ordinary concentrically braced frames and special concentrically braced frames. The response modification coefficient for an ordinary concentrically braced frame is approximately half the value of that for a special concentrically braced frame. Hence, an ordinary concentrically braced frame is designed for roughly twice the seismic load of the corresponding special concentrically braced frame and the members of an ordinary concentrically braced frame remain essentially elastic during seismic loading. This is indicated by the ductility requirements shown in Table 14.1 which shows that the braces of an ordinary concentrically braced frame are required to be only moderately ductile.

K-braced frames are not permitted in either ordinary or special concentrically braced frames. This bracing, framing into columns between diaphragms, produces lateral deflection of the columns leading to instability and collapse. Tension-only frames, in which the brace compression resistance is neglected and the brace is designed for tension force only, are permitted in ordinary concentrically braced frames but not in special concentrically braced frames.

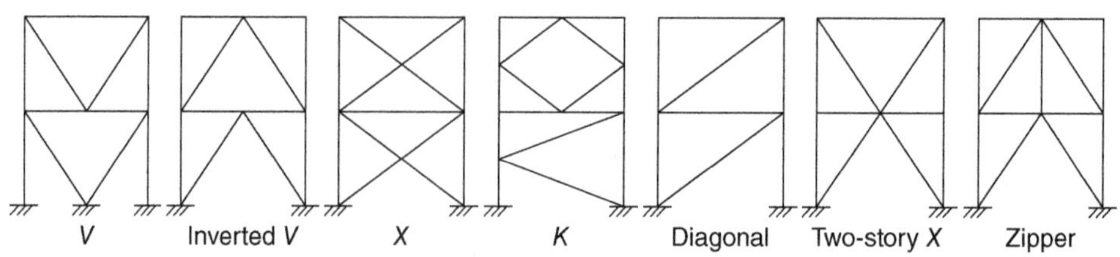


**FIGURE 14.4** Concentrically braced frame system configurations.

## 14.4 Ordinary Concentrically Braced Frames

Ordinary concentrically braced frames are specified in AISC 341 Sec. F1, and may be used in all seismic design categories with the exception of F. As specified in ASCE 7 Table 12.2-1, $R = 3.25$, $\Omega_0 = 2.0$ and no limitation is imposed on building height in seismic design categories A, B, and C. The maximum height permitted in seismic design categories D and E is 35 ft. An exception is permitted in seismic design categories D and E for penthouse structures and in single story buildings up to a height of 60 ft when the dead load of the roof does not exceed 20 lb/ft$^2$. Ordinary concentrically braced frames are not permitted in seismic design category F with the exception of single story buildings up to a height of 60 ft when the dead load of the roof does not exceed 20 lb/ft$^2$.

Ordinary concentrically braced frames are designed for the design seismic force of the applicable building code. Because of the low value of $R$ assigned to the system, ordinary concentrically braced frames are designed using a high seismic force level and are expected to remain essentially elastic under a seismic event. In accordance with AISC 341 Comm. Sec. F1.3, because of the limited inelastic demands on the system, an elastic analysis is considered sufficient when supplemented with the design of the brace connections using the amplified seismic load. By precluding failure in the connection, yielding of the tension brace acts as the ductile fuse in the system. Ordinary concentrically braced frames are an economical choice for smaller buildings and nonbuilding structures because of their simple design and construction procedures.

Diagonal braces are required to be moderately ductile members with no ductility conditions imposed on the other members in the system. There are no protected zones specified.

### Design of Braces

To reduce the likelihood of buckling causing large unbalanced forces in the floor beam, AISC 341 Sec. F1.5b requires bracing members in a V or inverted V configuration to be designed with a slenderness ratio not exceeding

$$KL/r = 4(E/F_y)^{0.5}$$

where $L$ is the length of the bracing member and $r$ is the governing radius of gyration.

The required values of $KL/r$ are given in Table 14.4.

This slenderness limitation does not apply to two-story X-braced frames or zipper column configurations as an unbalanced force does not develop in these systems.

| Member | $F_y$, ksi | $KL/r$ |
|---|---|---|
| Pipe | 35 | 115 |
| Hot-rolled shapes | 36 | 114 |
| Rectangular HSS | 42 | 105 |
| Round HSS | 46 | 100 |
| Hot-rolled shapes | 50 | 96 |

**Table 14.4** Limiting $KL/r$ Values for V and Inverted V Braces

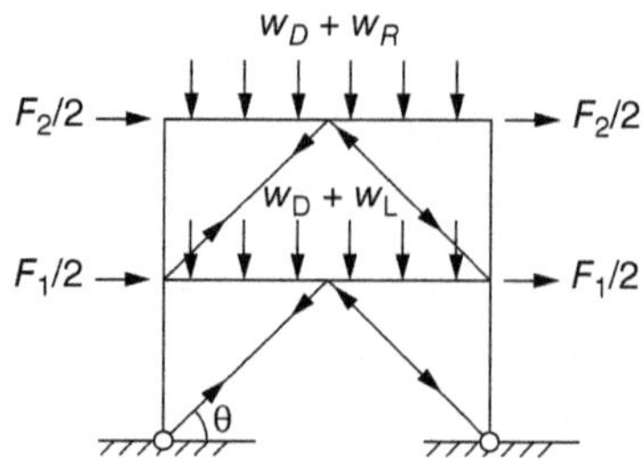


**FIGURE 14.5** Loading on an inverted V-braced frame.

For these systems and for configurations, other than systems with V or inverted V braces, the recommended brace slenderness ratio is given by AISC 360 Sec. E2 user note as

$$KL/r \leq 200$$

The design force in a diagonal brace is obtained as shown in Fig. 14.5. The indicated loads are applied to the frame which is analyzed as a pin-jointed truss, with the exception of the beams which are continuous at brace connections. The seismic loads are applied symmetrically on either side of the frame. The force in a brace in the bottom story due to the seismic load is

$$Q_E = (F_1 + F_2)/2\cos\theta$$

The force in a brace in the bottom story due to the gravity loads is

$$D = 0.625w_D l/2\sin\theta \text{ ... for dead load}$$
$$L = 0.625w_L l/2\sin\theta \text{ ... for live load}$$

The compression strength of a brace is

$$P_n = F_{cr} A_g$$

where $F_{cr}$ is the critical stress given by AISC 360 Sec. E3 using the specified yield stress of the brace, $F_y$, and $l$ is the width of bay.

Values of $\varphi_c P_n$ and $P_n/\Omega_c$ are given in AISC Manual[9] Tables 4-1 through 4-20.

**Example 14.1.** Design of a Brace in an Inverted V-Brace Frame

The single bay ordinary concentrically braced steel frame shown in Fig. 14.6 forms part of the building frame system of a structure in seismic design category D with a redundancy factor of $\rho = 1.0$ and a design response acceleration of $S_{DS} = 1.0g$. The seismic lateral load of $F = 20$ kips is applied symmetrically on either side of the frame giving two loads of 10 kips each. The indicated dead load on the beam of $w_D = 0.2$ kip/ft includes the weight of the beam. The live load is $w_L = 0.1$ kip/ft. Determine a suitable standard steel pipe section for the brace. Neglect second-order effects and the self-weight of the pipe.

The length of the brace between workpoints is

$$\begin{aligned} l &= h/\sin\theta \\ &= 15/\sin 45° \\ &= 21.21 \text{ ft} \end{aligned}$$

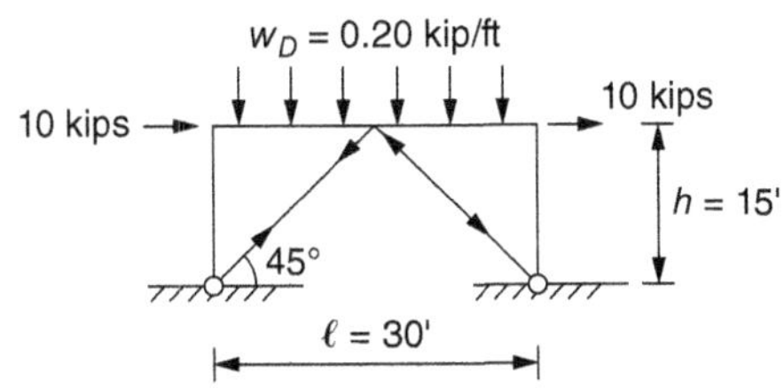


**FIGURE 14.6** Details for Example 14.1.

Allowing for a connection length of 18 in at each end, the actual brace length is

$$L = 21.21 - 3$$
$$18.21 \text{ ft}$$

The effective length factor for the pin-ended brace is obtained from AISC 360 Table C-A-7.1 as

$$K = 1.0$$

The effective length of the brace is

$$KL = 18.21 \text{ ft}$$

From Table 14.4, the limiting $KL/r$ ratio for a standard pipe section used as an inverted V brace is

$$KL/r = 115$$

Hence the minimum required radius of gyration for the brace is

$$\begin{aligned} r &= KL/115 \\ &= 18.21 \times 12/115 \\ &= 1.90 \text{ in} \end{aligned}$$

The forces acting on a brace are obtained from Fig. 14.6. The force in a brace due to the seismic load is

$$\begin{aligned} Q_E &= F/2\cos\theta \\ &= 20/2\cos 45° \\ &= 14.14 \text{ kips} \end{aligned}$$

The force in each brace due to the gravity load is

$$\begin{aligned} D &= 0.625 w_D l/2\sin\theta \\ &= 0.625 \times 0.20 \times 30/(2 \times \sin 45°) \\ &= 2.65 \text{ kips} \\ L &= 2.65/2 \\ &= 1.33 \text{ kips} \end{aligned}$$

For a value of the design response acceleration of $S_{DS} = 1.0g$

$$(1.2 + 0.2S_{DS}) = 1.4$$
$$(1.0 + 0.14S_{DS}) = 1.14$$

| LRFD | ASD |
|---|---|
| From ASCE 7 Sec. 12.4.2.3 combination 5:<br>$P_u$ = factored axial load<br>$= (1.2 + 0.2S_{DS})D + \rho Q_E + 0.5L + 0.2S$<br>$= (1.4)(2.65) + 1.0 \times 14.14 + 0.5 \times 1.33 + 0$<br>$= 18.52$ kips | From ASCE 7 Sec. 12.4.2.3 combination 5:<br>$P_a$ = factored axial load<br>$= (1.0 + 0.14S_{DS})D + 0.7\rho Q_E$<br>$= (1.0 + 0.14 \times 1.0)(2.65) + 0.7 \times 14.14$<br>$= 12.92$ kips |
| From AISC Manual Table 4-6 for a value of $KL = 18.21$ ft and a value of $F_y = 35$ kips/in$^2$, a steel pipe 5 std has a design strength of<br>$\varphi_c P_n = 63.3$ kips<br>$> 18.52$ kips ... satisfactory | From AISC Manual Table 4-6 for a value of $KL = 18.21$ ft and a value of $F_y = 35$ kips/in$^2$, a steel pipe 5 std has an allowable strength of<br>$P_n/\Omega_c = 42.2$ kips<br>$> 12.92$ kips ... satisfactory |
| It also has a radius of gyration of<br>$r = 1.88$ in<br>$\approx 1.90$ in ... satisfactory | It also has a radius of gyration of<br>$r = 1.88$ in<br>$\approx 1.90$ in ... satisfactory |

The brace is designated a moderately ductile member and requires the minimum width-to-thickness ratio specified in AISC 341 Table D1.1 in order to prevent local buckling occurring. As indicated in SDM Table 1-7, a steel pipe 5 std meets these requirements. Hence, the steel pipe 5 std is adequate.

## Design of Brace Connections and Gusset Plates

Gusset plates for V-type and inverted V-type braces may either be flat bar gussets[10] connecting each brace to the beam individually or a combined gusset[11] in which a single plate is used to connect both braces to the beam. The two types are shown in Fig. 14.7.

Bracing connections are designed for forces of sufficient magnitude to ensure that the braces will fail prior to a connection failure. In accordance with AISC 341 Sec. F1.6a and shown in Fig. 14.8*a*, the required strength of the brace connection is the load effect based on the amplified seismic load. There are three exceptions to this requirement:

1. The required strength of the connection in tension, as shown in Fig. 14.8*b*, need not exceed the value

$$R_u = R_y F_y A_g \text{ ... for LRFD}$$
$$R_a = R_y F_y A_g / 1.5 \text{ ... for ASD}$$

2. The required strength of the connection in compression, as shown in Fig. 14.8*c*, need not exceed the lesser value of

$$P_u = R_y F_y A_g \text{ ... for LRFD}$$
$$\le 1.14 F_{cre} A_g$$
$$P_a = R_y F_y A_g / 1.5 \text{ ... for ASD}$$
$$\le 1.14 F_{cre} A_g / 1.5$$

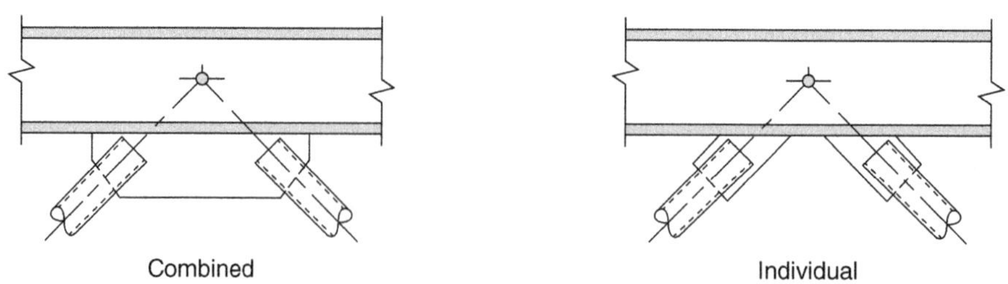


**Figure 14.7** Types of gusset plates.

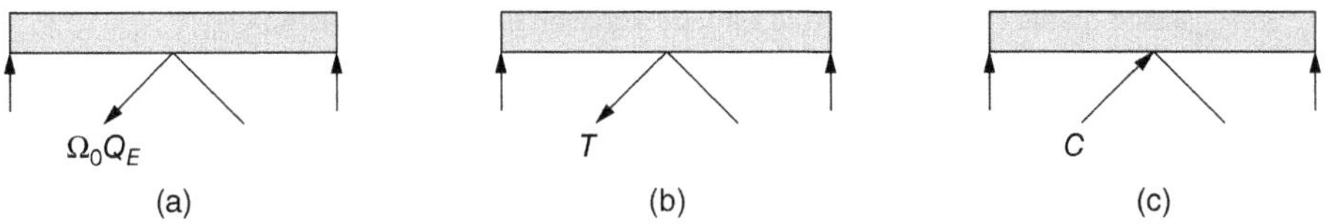


**FIGURE 14.8** Brace connection design loads.

When $KL/r \leq 4.71(E/R_yF_y)^{0.5}$, AISC 360 Eq. (E3-2) governs and the critical stress is

$$F_{cre} = (0.658^{\kappa})R_yF_y$$

where $\kappa = R_yF_y/F_e$

$F_{cre}$ = critical stress given by AISC 360 Sec. E3 using the expected yield stress of the brace, $R_yF_y$, instead of the specified yield stress, $F_y$

$F_e$ = elastic critical buckling stress

$= \pi^2E/(KL/r)^2$ ... from AISC 360 Eq. (E3-4)

$\geq R_yF_y/2.25$

When $KL/r > 4.71(E/R_yF_y)^{0.5}$, AISC 360 Eq. (E3-3) governs and the critical stress is:

$$F_{cre} = 0.877F_e$$

The brace length used for the determination of $F_{cre}$ must not exceed the distance from brace end to brace end.

3. When oversized holes are used, the available slip resistance of the connection need not exceed the force in the connection determined from the design seismic loads, not including the amplified seismic load. Bolt slip does not constitute connection failure and the associated energy dissipation can serve to reduce seismic response.

**Example 14.2.** Determination of Brace Connection Force
Determine the appropriate brace connection forces for the single bay ordinary concentrically braced steel frame shown in Fig. 14.6 and analyzed in Example 14.1. A welded connection is proposed.

From Example 14.1, the seismic force in a brace is

$$Q_E = 14.14 \text{ kips}$$

The force in a brace due to the gravity load is

$$D = 2.65 \text{ kips}$$
$$L = 1.33 \text{ kips}$$

From ASCE 7 Table 12.2-1, the overstrength factor for an ordinary concentrically braced frame is

$$\Omega_0 = 2.0$$
$$\Omega_0 Q_E = 2.0 \times 14.14$$
$$= 28.28 \text{ kips}$$

Based on the amplified seismic load, the force on the brace connection is

| LRFD | ASD |
|---|---|
| From ASCE 7 Sec. 12.4.3.2 combination 5:<br>$P_u$ = factored axial load<br>$= (1.2 + 0.2S_{DS})D + \Omega_0 Q_E + 0.5L + 0.2S$<br>$= (1.4)(2.65) + 28.28 + 0.5 \times 1.33 + 0$<br>= 32.67 kips ... compression<br>$P_u = (0.9 - 0.2S_{DS})D - \Omega_0 Q_E$<br>$= (0.9 - 0.2 \times 1.0)(2.65) - 28.28$<br>= −26.43 kips ... tension | From ASCE 7 Sec. 12.4.3.2 combination 5:<br>$P_a$ = factored axial load<br>$= (1.0 + 0.14S_{DS})D + 0.7\Omega_0 Q_E$<br>$= (1.0 + 0.14 \times 1.0)(2.65) + 0.7 \times 28.28$<br>= 22.82 kips ... compression<br>$P_a = (0.6 - 0.14S_{DS})D - 0.7\Omega_0 Q_E$<br>$= (0.6 - 0.14 \times 1.0)(2.65) - 0.7 \times 28.28$<br>= −18.58 kips ... tension |

The relevant properties of the steel pipe 5 std are

$$F_y = 35 \text{ kips/in}^2$$
$$R_y = 1.6$$
$$A_g = 4.01 \text{ in}^2$$
$$R_y F_y = 1.6 \times 35$$
$$= 56.00 \text{ kips/in}^2$$

In accordance with AISC 341 Sec. F1.6a, the required strength of the connection in tension need not exceed the value

| LRFD | ASD |
|---|---|
| $R_u = R_y F_y A_g$<br>$= 56 \times 4.01$<br>= 225 kips ... does not govern<br>> 26.43 kips | $R_a = R_y F_y A_g / 1.5$<br>$= 56 \times 4.01/1.5$<br>= 150 kips ... does not govern<br>> 18.58 kips |

From Example 14.1, the slenderness ratio of the pipe 5 std brace is

$$KL/r = 1 \times 18.21 \times 12/1.88$$
$$= 116$$

The elastic buckling stress is given by AISC 360 Eq. (E3-4) as

$$F_e = \pi^2 E/(KL/r)^2$$
$$= 3.14^2 \times 29{,}000/116^2$$
$$= 21.25 \text{ kips/in}^2$$
$$4.71(E/R_y F_y)^{0.5} = 4.71(29{,}000/56)^{0.5}$$
$$= 107.2$$
$$< KL/r$$

Hence, AISC 360 Eq. (E3-3) governs and the critical stress is

$$F_{cre} = 0.877 F_e$$
$$= 0.877 \times 21.25$$
$$= 18.64 \text{ kips/in}^2$$

In accordance with AISC 341 Sec. F1.6a, the required strength of the connection in compression need not exceed the lesser value of

| LRFD | ASD |
|---|---|
| $P_u = R_y F_y A_g$<br>$= 56 \times 4.01$<br>$= 225$ kips ... does not govern<br>or $P_u = 1.14 F_{cre} A_g$<br>$= 1.14 \times 18.64 \times 4.01$<br>$= 85$ kips ... does not govern<br>$< 225$ kips<br>$> 32.67$ kips | $P_a = R_y F_y A_g / 1.5$<br>$= 56 \times 4.01/1.5$<br>$= 150$ kips ... does not govern<br>or $P_a = 1.14 F_{cre} A_g / 1.5$<br>$= 1.14 \times 18.64 \times 4.01/1.5$<br>$= 57$ kips ... does not govern<br>$< 150$ kips<br>$> 22.82$ kips |

The required brace connection forces are

| LRFD | ASD |
|---|---|
| $P_u = 32.67$ kips ... compression<br>$= 26.43$ kips ... tension | $P_a = 22.82$ kips ... compression<br>$= 18.58$ kips ... tension |

## Design of Beams in V-Braced and Inverted V-Braced Configurations

When the compression brace buckles in a V-type or inverted V-type bracing system, unbalanced vertical and horizontal loads occur in the beam. The beam must be designed for the unbalanced forces produced when the tension brace yields and the compression brace buckles. For this reason, AISC 341 Sec. F1.4a requires beams to be continuous between columns and it is assumed that the beam supports all gravity loads without any support from the braces. In addition, both flanges of the beam must be provided with lateral braces at the point of intersection of the braces to ensure the stability of the beam. The residual strength of the compression brace after buckling is given as

$$0.3P_n = 0.3F_{cr}A_g$$

where $A_g$ = gross area of member
$F_{cr}$ = critical stress given by AISC 360 Sec. E3 using the specified yield stress $F_y$
$P_n$ = axial strength

The yield strength of the tension brace is given as

$$R_u = R_y F_y A_g$$

where $R_y$ is the ratio of the expected yield stress to the minimum specified yield strength and is tabulated in Table 14.3, and $F_y$ is the specified minimum yield stress of the type of steel used.

In designing the beam, three load cases are considered. In all three, the force in the compression brace is assumed to be $0.3P_n$.

As shown in Fig. 14.9, the force in the tension brace is assumed to be the least of the following:

1. The expected yield strength, as shown in Fig. 14.9*a*, which is

$$R_u = R_y F_y A_g$$

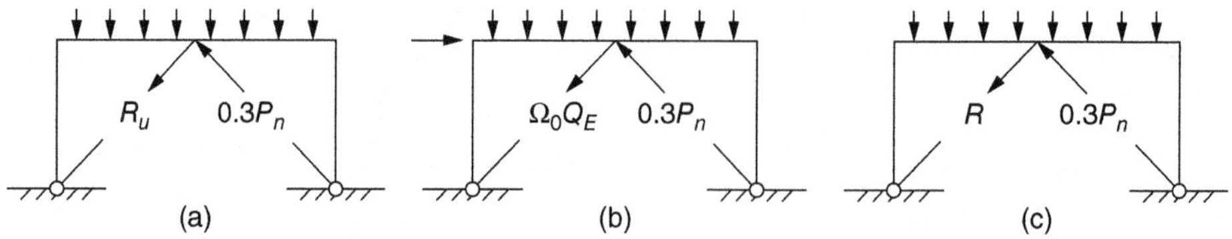


**FIGURE 14.9** Beam loading cases.

The seismic load effects are determined as shown in Fig. 14.10. The unbalanced vertical force is

$$E_v = (R_u - 0.3P_n)\sin\theta$$

The unbalanced horizontal force is

$$E_h = (R_u + 0.3P_n)\cos\theta$$

2. The force developed in the brace, as shown in Fig. 14.9*b*, when the amplified seismic load is applied to the system, which is

$$R_\Omega = \Omega_0 Q_E$$

where $\Omega_0$ is the overstrength factor given in ASCE 7 Table 12.2-1, and $Q_E$ is the effect of horizontal seismic forces.

The seismic load effects are determined as shown in Fig. 14.11. The unbalanced vertical force is

$$E_v = (\Omega_0 Q_E - 0.3P_n)\sin\theta$$

The unbalanced horizontal force is

$$E_h = (\Omega_0 Q_E + 0.3P_n)\cos\theta$$

3. The maximum force that can be developed by the system.

At system failure, the force in the tension brace is $R$ and the seismic load effects are

$$E_v = (R - 0.3P_n)\sin\theta$$

$$E_h = (R + 0.3P_n)\cos\theta$$

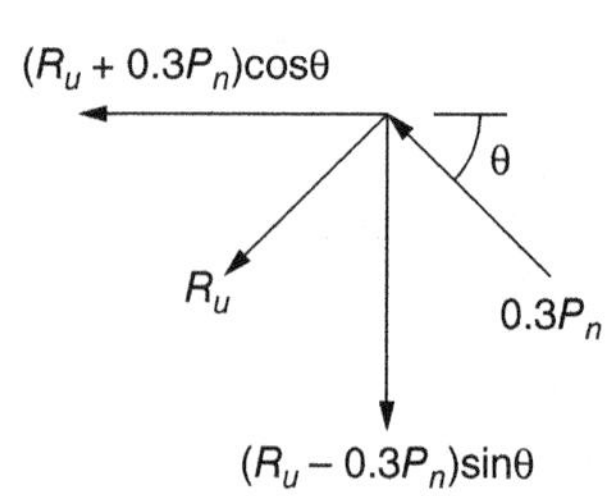


**FIGURE 14.10** Unbalanced forces for loading case 1.

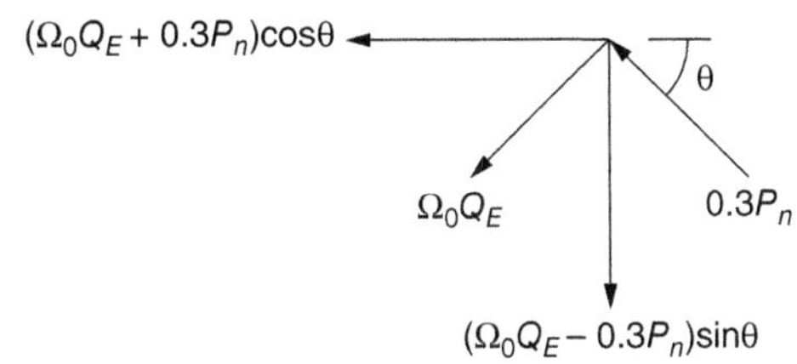


**FIGURE 14.11** Unbalanced forces for loading case 2.

**Example 14.3.** Design of a Beam in an Inverted V-Brace Configuration
Select a suitable W14 section, with a yield stress of 50 kips/in$^2$, for the beam in the single bay ordinary concentrically braced steel frame shown in Fig. 14.6 and analyzed in Example 14.1. Lateral braces are provided at the midpoint of the beam. The force in the brace is not governed by the maximum force that can be developed by the system. Neglect second-order effects.

The beam is considered pinned at each end and resists all gravity loads without any support from the braces.

From Examples 14.1 and 14.2

$$\begin{aligned} l &= \text{beam span} \\ &= 30 \text{ ft} \end{aligned}$$

For the brace,

$$\begin{aligned} R_u &= 225 \text{ kips} \\ \Omega_0 Q_E &= 28.28 \text{ kips} \\ \varphi_c P_n &= 63.3 \text{ kips} \\ 0.3P_n &= 0.3\varphi_c P_n/\varphi_c \\ &= 0.3 \times 63.3/0.90 \\ &= 21.1 \text{ kips} \end{aligned}$$

The gravity load on the beam is

$$\begin{aligned} D &= w_D \\ &= 0.20 \text{ kip/ft} \\ L &= 0.10 \text{ kip/ft} \end{aligned}$$

By inspection load case 1 governs.

For load case 1, the unbalanced vertical load is

$$\begin{aligned} E_v &= (R_u - 0.3P_n)\sin\theta \\ &= (225 - 21.1)\sin 45° \\ &= 144 \text{ kips} \end{aligned}$$

The unbalanced horizontal load, which is carried equally by the two halves of the beam, is

$$\begin{aligned} E_h &= 0.5(R_u + 0.3P_n)\cos\theta \\ &= 0.5(225 + 21.1)\cos 45° \\ &= 87 \text{ kips} \end{aligned}$$

The loads acting on the beam are shown in Fig. 14.12.

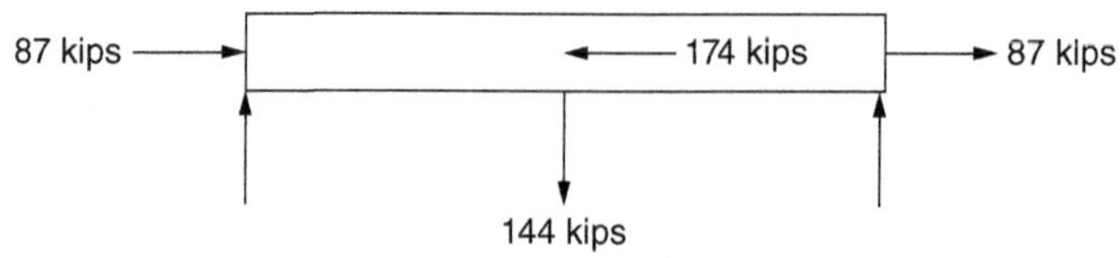


**FIGURE 14.12** Beam loads for Example 14.3.

For vertical loads, the beam is considered simply supported over a span of 30 ft. Applying the factored loads to the beam gives

| LRFD | ASD |
|---|---|
| From ASCE 7 Sec. 12.4.3.2 combination 5, the required flexural strength is<br>$M_r = (1.2 + 0.2S_{DS})Dl^2/8 + E_v l/4 + 0.5Ll^2/8 + 0.2S$<br>$= (1.4)(0.2)(30)^2/8 + 144 \times 30/4 + 0.5(0.1)30^2/8 + 0$<br>$= 1118$ kip-ft<br>From ASCE 7 Sec. 12.4.3.2 combination 5, the required axial strength is<br>$P_r = (1.2 + 0.2S_{DS})D + E_h + 0.5L + 0.2S$<br>$= (1.2 + 0.2 \times 1.0) \times 0 + 87 + 0 + 0$<br>$= 87$ kips ... compression | From ASCE 7 Sec. 12.4.3.2 combination 5, the required flexural strength is<br>$M_r = (1.0 + 0.14S_{DS})Dl^2/8 + 0.7E_v l/4$<br>$= (1.0 + 0.14 \times 1.0)\ (0.2)(30)^2/8 + 0.7 \times 144 \times 30/4$<br>$= 782$ kip-ft<br>From ASCE 7 Sec. 12.4.3.2 combination 5, the required axial strength is<br>$P_r = (1.0 + 0.14S_{DS})D + 0.7E_h$<br>$= (1.0 + 0.14 \times 1.0) \times 0 + 0.7 \times 87$<br>$= 61$ kips ... compression |

Select a W14 × 193 for the beam which satisfies all seismic width-to-thickness requirements and requires lateral bracing only at mid span as indicated in SDM Table 1-3. The beam is pinned at each end, giving a value for the effective length factor of $K = 1.0$ about both axes, and is braced about the $y$-axis at mid span. From AISC Manual Table 6-1, obtain the value

$$r_x/r_y = 1.6$$

The effective length about the $y$-axis is

$$KL_y = 15 \text{ ft}$$

The effective length about the $x$-axis is

$$KL_x = 30 \text{ ft}$$

The equivalent effective length about the $x$-axis with respect to the $y$-axis is

$$\begin{aligned}(KL_y)_{equiv} &= (KL_x)/(r_x/r_y)\\ &= 30/1.6\\ &= 18.75 \text{ ft ... governs}\\ &> KL_y\end{aligned}$$

The equivalent effective length about the $y$-axis for axial compression is 18.75 ft and the unbraced length for flexure is 30 ft. The adequacy of the W14 × 193 beam to sustain combined compression and flexure is determined by the interaction equations given in AISC 360 Sec. H1.1. From AISC Manual Table 6-1, the following information is obtained

| LRFD | ASD |
|---|---|
| $p = 0.490 \times 10^{-3}$ (kips)$^{-1}$<br>$b_x = 0.737 \times 10^{-3}$ (kip-ft)$^{-1}$<br>$P_r/\varphi_c P_n = pP_r = 0.490 \times 10^{-3} \times 87$<br>$= 0.043 < 0.2$ ... Eq. (H1-1b) applies<br>$pP_r/2 + (9/8)b_x M_{rx} \le 1.0$<br>$0.000490 \times 87/2 + (9/8)0.000737 \times 1118 = 0.95$<br>$< 1.0$ ... satisfactory | $p = 0.737 \times 10^{-3}$ (kips)$^{-1}$<br>$b_x = 1.11 \times 10^{-3}$ (kip-ft)$^{-1}$<br>$\Omega_c P_r/P_n = pP_r = 0.737 \times 10^{-3} \times 61$<br>$= 0.045 < 0.2$ ... Eq. (H1-1b) applies<br>$pP_r/2 + (9/8)b_x M_{rx} \le 1.0$<br>$0.000737 \times 61/2 + (9/8)0.00111 \times 782 = 0.999$<br>$< 1.0$ ... satisfactory |

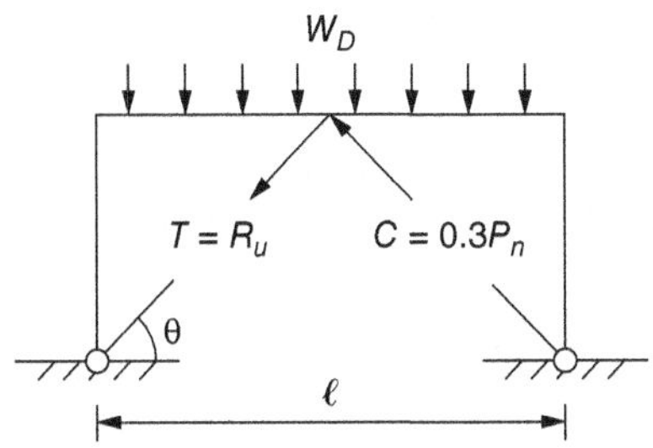


**FIGURE 14.13** Determination of column loads.

## Design of Columns

A column in a V- or inverted V-braced frame may be designed by assuming that the tension brace has yielded at the expected yield stress and the compression brace has buckled simultaneously. As shown in Fig. 14.13, the residual strength of the compression brace is

$$0.3P_n = 0.3F_{cr}A_g$$

The expected strength of the tension brace is

$$R_u = R_yF_yA_g$$

The total axial load on the column due to the seismic load is

$$P = (1/2)(R_u - 0.3P_n)\sin\theta$$

where $A_g$ = gross area of member
$F_{cr}$ = critical stress given by AISC 360 Sec. E3 using the specified yield stress $F_y$
$P_n$ = axial strength
$R_y$ = ratio of the expected yield stress to the minimum specified yield strength and is tabulated in Table 14.3
$F_y$ = specified minimum yield stress of the type of steel used

For the two-story X-braced frame shown in Fig. 14.14, the loads on the columns in the bottom story, due to the applied seismic load, are

$$P = (R_{u2} + R_{u1} + 0.3P_{n2})\sin\theta \text{ ... compression}$$
$$P = (R_{u2} + 0.3P_{n2} + 0.3P_{n1})\sin\theta \text{ ... tension}$$

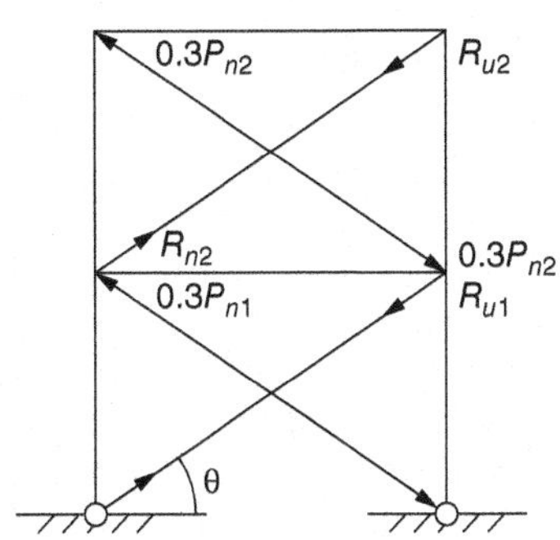


**FIGURE 14.14** Two-story X-braced frame.

**Example 14.4.** Design of a Column in an Inverted V-Brace Configuration
Select a suitable W14 section, with a yield stress of 50 kips/in$^2$, for the column in the single bay ordinary concentrically braced steel frame shown in Fig. 14.6 and analyzed in Example 14.1. Neglect second-order effects and the self-weight of the column.

The column is considered pinned at each end with an effective length about both axes of

$$Kh = 15 \text{ ft}$$

From Example 14.3 for the brace,

$$R_u = 225 \text{ kips}$$
$$0.3P_n = 21.1 \text{ kips}$$

The seismic force in a brace is

$$\begin{aligned} E &= (1/2)(R_u - 0.3P_n)\sin\theta \\ &= (1/2)(225 - 21.1)\sin 45^\circ \\ &= 72 \text{ kips} \end{aligned}$$

The force in a column due to the gravity load is

$$\begin{aligned} D &= w_D l/2 \\ &= 0.2 \times 30/2 \\ &= 3 \text{ kips} \\ L &= 1.5 \text{ kips} \end{aligned}$$

Applying the factored loads to the column gives

| LRFD | ASD |
|---|---|
| From ASCE 7 Sec. 12.4.3.2 combination 5, the required axial strength is<br>$P_r = (1.2 + 0.2S_{DS})D + E + 0.5L + 0.2S$<br>$= (1.2 + 0.2 \times 1.0) \times 3 + 72 + 1.5/2 + 0$<br>$= 77$ kips ... compression<br>From AISC Manual Table 4-1 for a value of $KL$ = 15 ft and a value of $F_y = 50$ kips/in$^2$, a W14 × 48 has a design strength of<br>$\varphi_c P_n = 331$ kips<br>$> 77$ kips ... satisfactory | From ASCE 7 Sec. 12.4.3.2 combination 5, the required axial strength is:<br>$P_r = (1.0 + 0.14S_{DS})D + 0.7E$<br>$= (1.0 + 0.14 \times 1.0) \times 3 + 0.7 \times 72$<br>$= 53.82$ kips ... compression<br>From AISC Manual Table 4-1 for a value of $KL$ = 15 ft and a value of $F_y = 50$ kips/in$^2$, a W14 × 48 has an available strength of<br>$P_n/\Omega_0 = 221$ kips<br>$> 53.82$ kips ... satisfactory |

As shown in SDM Table 1-3, a W14 × 48 satisfies all seismic width-to-thickness requirements.

## 14.5 Special Concentrically Braced Frames

Special concentrically braced frames[12] are specified in AISC 341 Sec. F2, and may be used in all seismic design categories. As specified in ASCE 7 Table 12.2-1, $R = 6.0$, $\Omega_0 = 2.0$ and no limitation is imposed on building height in seismic design categories A, B, and C. The maximum height permitted in seismic design categories D and E is 160 ft, and in seismic design category F is 100 ft.

Special concentrically braced frames may be utilized in dual systems with special moment frames in all seismic design categories. In accordance with ASCE 7 Table 12.2-1, $R = 7.0$, $\Omega_0 = 2.5$ and no limitation is imposed on the building height.

Special concentrically braced frames may be utilized in dual systems with intermediate moment frames, in all seismic design categories with the exceptions of E and F,

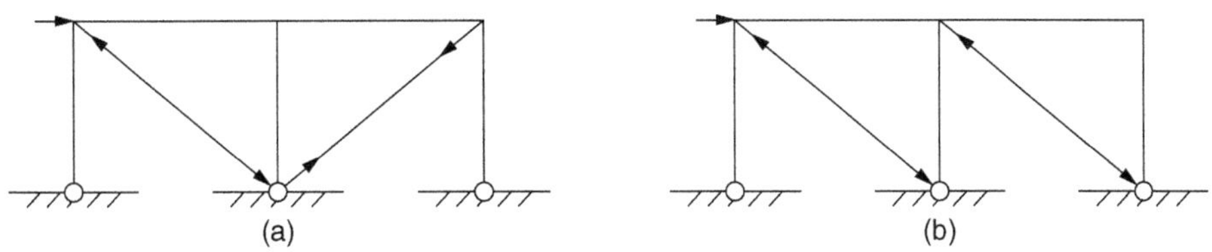


**FIGURE 14.15** Brace orientation.

using values of $R = 6.0$, $\Omega_0 = 2.5$. No limitation is imposed on the building height in seismic design categories A, B, and C. The maximum height permitted in seismic design category D is 35 ft.

Beams are required to be moderately ductile members and columns and braces required to be highly ductile members. Protected zones are specified for gusset plates and for brace ends and central region.

In accordance with AISC 341 Sec. F2.4a, braces must be deployed in alternate directions, as shown in Fig. 14.15*a*, to ensure that at least 30 percent but no more than 70 percent of the horizontal force along any line of bracing is resisted by braces in tension. Figure 14.15*b* shows a bracing configuration in which the horizontal force is resisted by braces deployed in a single direction. The post-buckling strength of a brace is considerably less than its tensile strength. Hence, during cyclic loading, there is an accumulation of inelastic drift in the direction corresponding to compression in the braces and this may lead to instability. This requirement is relaxed provided that the braces are designed to resist the compression force produced by the amplified seismic force applied to the frame.

## Mechanism Analysis

Special concentrically braced frames provide a stable and ductile response to cyclic loading by compression brace buckling and tension brace yielding. Local buckling of the compression brace must be prevented by limiting its width-to-thickness ratios. In the elastic range, the frame behaves essentially as a vertical truss and the required strength of braces is determined, in accordance with AISC 341 Comm. Sec. F2.3, by analyzing the frame with the seismic loads applied to the system. The required strength of the remaining elements in the system is determined from a plastic-mechanism analysis using the load combinations in the applicable building code that include the amplified seismic load. The amplified seismic load, $E_{mh}$, is defined as the greater force determined from

- An analysis in which all braces are assumed to have reached their maximum forces corresponding to their expected strength in compression or in tension, as shown in Fig. 14.16*a*
- An analysis in which all braces in tension are assumed to have reached their maximum forces corresponding to their expected strength and all braces in compression are assumed to have reached their minimum expected post-buckling strength, as shown in Fig. 14.16*b*

When determining the brace forces, the effects of gravity loads are neglected.

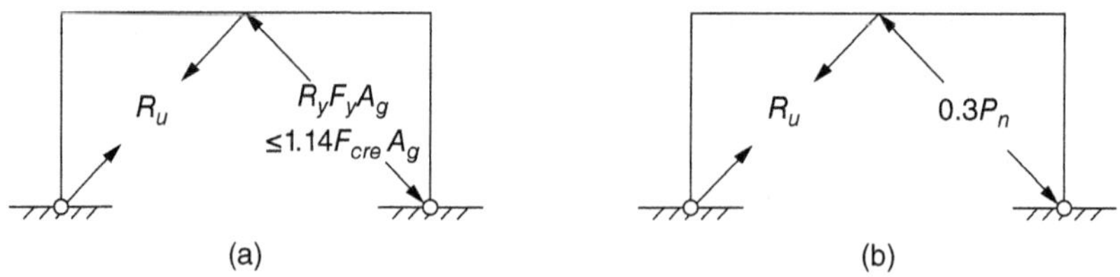


**FIGURE 14.16** Determination of amplified seismic load.

The expected brace strength in tension is

$$R_u = R_yF_yA_g$$

The expected brace strength in compression need not exceed the lesser of

$$P_u = R_yF_yA_g$$
$$\leq 1.14F_{cre}A_g$$

When $KL/r \leq 4.71(E/R_yF_y)^{0.5}$, AISC 360 Eq. (E3-2) governs and the critical stress is

$$F_{cre} = (0.658^{\kappa})R_yF_y$$

where $\kappa = R_yF_y/F_e$
$F_{cre}$ = critical stress given by AISC 360 Sec. E3 using the expected yield stress of the brace, $R_yF_y$, instead of the specified yield stress, $F_y$
$F_e$ = elastic critical buckling stress
$= \pi^2E/(KL/r)^2$ ... from AISC 360 Eq. (E3-4)
$\geq R_yF_y/2.25$

When $KL/r > 4.71(E/R_yF_y)^{0.5}$, AISC 360 Eq. (E3-3) governs and the critical stress is

$$F_{cre} = 0.877F_e$$

The residual strength of the compression brace after buckling is given as

$$0.3P_n = 0.3F_{cr}A_g$$

where $A_g$ = gross area of member
$F_{cr}$ = critical stress given by AISC 360 Sec. E3 using the specified yield stress $F_y$
$P_n$ = axial strength

A conservative value of $0.3F_{cr}$ = 2.1 kips/in$^2$ may be assumed as this is the value for a brace with the maximum permitted slenderness ratio of 200. Alternatively, a value of $0.3F_{cr}$ = 0 kip/in$^2$ may be assumed. The brace length used for the determination of $F_{cre}$ must not exceed the distance from brace end to brace end. Braces shall be determined to be in compression or tension neglecting the effects of gravity loads.

## Design of Braces

In accordance with AISC 341 Comm. Sec. F2.3, the required strength of braces is determined by analyzing the frame with the seismic loads applied to the system. The compression brace is intended to buckle in a severe earthquake. To prevent local buckling

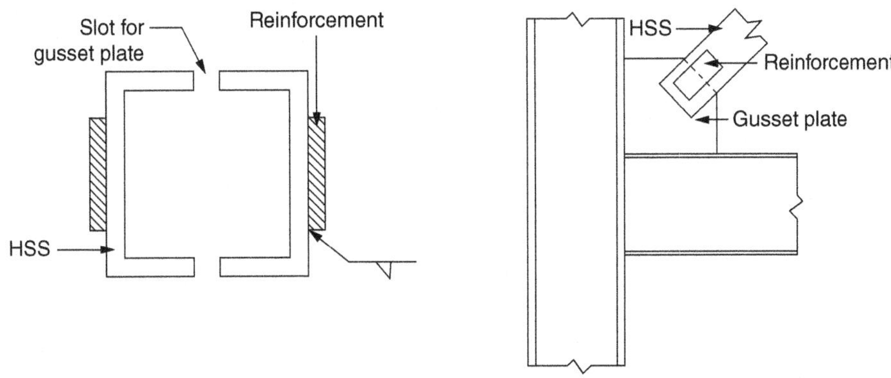


**FIGURE 14.17** Reinforcement at a connection.

from occurring when plastic hinges form in the brace, AISC 341 Sec. F2.5a designates the brace as a highly ductile member with severe restrictions on the width-to-thickness ratio. These ratios are listed in AISC 341 Table D1.1, and AISC 341 Tables 1-3 through 1-7 indicate the sections that are suitable for each seismic resisting system. The brace slenderness ratio is limited to a maximum value of

$$KL/r = 200$$

To prevent net section rupture of a brace at the connection to a gusset plate, the effective net area of the brace must not be less than the gross area of the brace. As shown in Fig. 14.17 for a knife-plate connection, this requires the addition of reinforcement at the ends of the brace to compensate for the slot cut in the brace to accommodate the gusset plate. In accordance with AISC 341 Sec. F2.5b(3), the specified minimum yield strength of the reinforcement may not be less than the specified minimum yield strength of the brace. In addition, the connection of the reinforcement to the brace shall have sufficient strength to develop the expected reinforcement strength on each side of the reduced section.

In accordance with AISC 341 Sec. F2.5c, the protected zones for braces extend over the center one-quarter of the brace length and adjacent to each connection for a length equal to the brace depth in the plane of buckling. In order to prevent individual buckling of elements in built-up bracing members, stitch spacing closer than normal is required. As specified in AISC 341 Sec. F2.5b(2), a minimum of two stitches is required and bolted stitches may not be located within the central quarter of the clear brace length. The total design shear strength of the stitches shall at least be equal to the design tensile strength of each element. The slenderness ratio of the individual members between stitches may not exceed 40 percent of the governing slenderness ratio of the built-up member.

**Example 14.5.** Design of a Brace in an Inverted V-Brace Frame
The single bay special concentrically braced steel frame shown in Fig. 14.18 forms part of the building frame system of a structure in seismic design category D with a redundancy factor of $\rho = 1.0$ and a design response acceleration of $S_{DS} = 1.0g$. The seismic lateral load of $F = 20$ kips is applied symmetrically on either side of the frame giving two loads of 10 kips each. The indicated dead load on the beam of $w_D = 0.2$ kip/ft includes the weight of the beam. The live load is 0.1 kip/ft. Determine a suitable standard steel pipe section for the brace. Neglect second-order effects and the self-weight of the pipe. Include the effects of gravity loads.

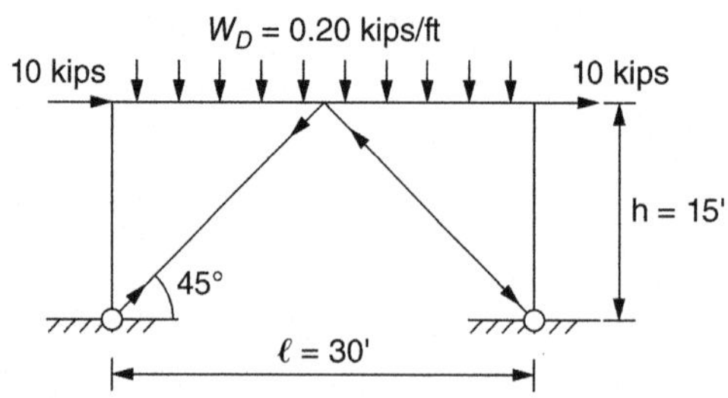


**FIGURE 14.18** Details for Example 14.5.

The length of the brace between workpoints is

$$l = h/\sin\theta$$
$$= 15/\sin 45°$$
$$= 21.21 \text{ ft}$$

Allowing for a connection length of 18 in at each end, the actual brace length is

$$L = 21.21 - 3$$
$$18.21 \text{ ft}$$

The effective length factor for the pin-ended brace is obtained from AISC 360 Table C-A-7.1 as

$$K = 1.0$$

The effective length of the brace is

$$KL = 18.21 \text{ ft}$$

From AISC 341 Sec. F2.5b, the limiting $KL/r$ ratio for a standard pipe section used as an inverted V brace in a special concentric braced frame is

$$KL/r = 200$$

Hence, the minimum required radius of gyration for the brace is

$$r = KL/115$$
$$= 18.21 \times 12/200$$
$$= 1.09 \text{ in}$$

The forces acting on a brace are obtained from Fig. 14.18. The force in a brace due to the seismic load is

$$Q_E = F/2\cos\theta$$
$$= 20/2\cos 45°$$
$$= 14.14 \text{ kips}$$

The force in each brace due to the gravity load is

$$D = 0.625 w_D l/2\sin\theta$$
$$= 0.625 \times 0.20 \times 30/(2 \times \sin 45°)$$
$$= 2.65 \text{ kips}$$
$$L = 2.65/2$$
$$= 1.33 \text{ kips/in}$$

| LRFD | ASD |
|---|---|
| From ASCE 7 Sec. 12.4.2.3 combination 5:<br>$P_u$ = factored axial load<br>$= (1.2 + 0.2S_{DS})D + \rho Q_E + 0.5L + 0.2S$<br>$= (1.4)(2.65) + 1.0 \times 14.14 + 1.33/2 + 0$<br>$= 18.52$ kips<br>From AISC Manual Table 4-6 for a value of $KL$ = 18.21 ft and a value of $F_y = 35$ kips/in², a steel pipe 3½ XS has a design strength of<br>$\varphi_c P_n = 27.9$ kips<br>$> 18.52$ kips ... satisfactory<br>It also has a radius of gyration of<br>$r = 1.31$ in<br>$> 1.09$ in ... satisfactory | From ASCE 7 Sec. 12.4.2.3 combination 5:<br>$P_a$ = factored axial load<br>$= (1.0 + 0.14S_{DS})D + 0.7\rho Q_E$<br>$= (1.0 + 0.14 \times 1.0)(2.65) + 0.7 \times 14.14$<br>$= 12.92$ kips<br>From AISC Manual Table 4-6 for a value of $KL$ = 18.21 ft and a value of $F_y = 35$ kips/in², a steel pipe 3½ XS has an allowable strength of<br>$P_n/\Omega_c = 18.6$ kips<br>$> 12.92$ kips ... satisfactory<br>It also has a radius of gyration of<br>$r = 1.31$ in<br>$> 1.09$ in ... satisfactory |

The steel pipe 3½ XS satisfies all seismic width-to-thickness requirements as indicated in SDM Table 1-7.

## Design of Brace Connections and Gusset Plates

Bracing connections are designed to ensure that the braces will fail prior to a connection failure. In accordance with AISC 341 Sec. F2.6c. the required strength of the brace connection must meet the following criteria.

1. The expected strength of the brace in tension is

$$R_u = R_y F_y A_g \text{ ... for LRFD}$$
$$R_a = R_y F_y A_g/1.5 \text{ ... for ASD}$$

   but need not exceed.

2. The maximum force that can be developed in the brace by the system.
3. The expected strength of the brace in compression which is

$$1.1 \times P_u = 1.1 R_y F_y A_g \text{ ... for LRFD}$$
$$\leq 1.1 \times 1.14 F_{cre} A_g$$
$$1.1 \times P_a = 1.1 R_y F_y A_g/1.5 \text{ ... for ASD}$$
$$\leq 1.1 \times 1.14 F_{cre} A_g/1.5$$

   When $KL/r \leq 4.71(E/R_y F_y)^{0.5}$, AISC 360 Eq. (E3-2) governs and the critical stress is

$$F_{cre} = (0.658^{\kappa}) R_y F_y$$

   where $\kappa = R_y F_y / F_e$
   $F_{cre}$ = critical stress given by AISC 360 Sec. E3 using the expected yield stress of the brace, $R_y F_y$, instead of the specified yield stress, $F_y$
   $F_e$ = elastic critical buckling stress
   $= \pi^2 E/(KL/r)^2$ ... from AISC 360 Eq. (E3-4)
   $\geq R_y F_y / 2.25$

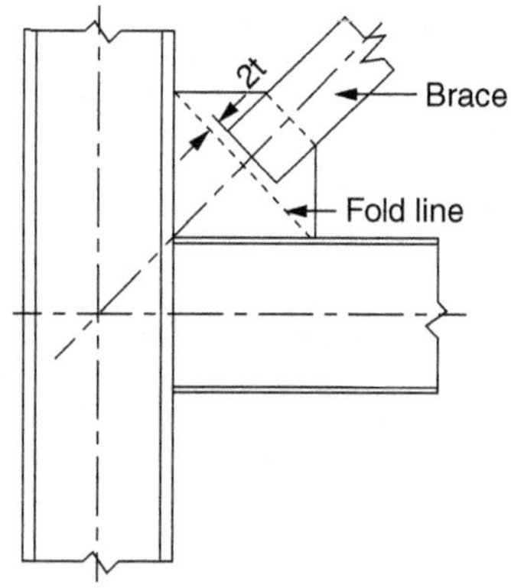


**FIGURE 14.19** Brace-to-gusset plate requirements.

When $KL/r > 4.71(E/R_yF_y)^{0.5}$, AISC 360 Eq. (E3-3) governs and the critical stress is

$$F_{cre} = 0.877F_e$$

The effects of gravity loads are neglected in determining the brace loads.

Brace connections must accommodate the cyclic buckling of the diagonal braces. To prevent fracture of the connection, resulting from brace rotations, bracing connections must either have sufficient strength to confine inelastic rotation to the bracing member or sufficient ductility to accommodate brace end rotations. For brace buckling in the plane of the gusset plates, the end connections should be designed to resist the expected flexural strength of the brace, which is

$$M_u = 1.1R_yF_yZ_b \text{ ... for LRFD}$$

$$M_a = 1.1R_yF_yZ_b/1.5 \text{ ... for ASD}$$

where $Z_b$ is the plastic section modulus of the brace.

For brace buckling out of the plane of single plate gussets, weak-axis bending in the gusset is induced by member end rotation. As shown in Fig. 14.19, the brace is terminated on the gusset plate a minimum of twice the gusset plate thickness from a line about which the gusset plate can bend unrestrained by the column or beam.

**Example 14.6.** Determination of Brace Connection Force
Determine the appropriate brace connection forces for the single bay special concentrically braced steel frame shown in Fig. 14.18 and analyzed in Example 14.5. The maximum force that can be transferred to the brace by the system does not govern.

The relevant properties of the steel pipe 3½ XS are

$$F_y = 35 \text{ kips/in}^2$$

$$R_y = 1.6$$

$$A_g = 3.43 \text{ in}^2$$

$$r = 1.31 \text{ in}$$

$$R_yF_y = 1.6 \times 35$$

$$= 56.00 \text{ kips/in}^2$$

In accordance with AISC 341 Sec. F2.6c(1), the required strength of the connection in tension need not exceed the value.

| LRFD | ASD |
|---|---|
| $R_u = R_yF_yA_g$<br>$= 56 \times 3.43$<br>$= 192$ kips | $R_a = R_yF_yA_g/1.5$<br>$= 56 \times 3.43/1.5$<br>$= 128$ kips |

From Example 14.5, the slenderness ratio of the pipe 3½ XS brace is

$$\begin{aligned} KL/r &= 1 \times 18.21 \times 12/1.31 \\ &= 167 \end{aligned}$$

The elastic buckling stress is given by AISC 360 Eq. (E3-4) as

$$\begin{aligned} F_e &= \pi^2E/(KL/r)^2 \\ &= 3.14^2 \times 29{,}000/167^2 \\ &= 10.25 \text{ kips/in}^2 \\ 4.71(E/R_yF_y)^{0.5} &= 4.71(29{,}000/56)^{0.5} \\ &= 107.2 \\ &< KL/r \end{aligned}$$

Hence, AISC 360 Eq. (E3-3) governs and the critical stress is

$$\begin{aligned} F_{cre} &= 0.877F_e \\ &= 0.877 \times 10.25 \\ &= 9.00 \text{ kips/in}^2 \end{aligned}$$

In accordance with AISC 341 Sec. F2.6c(2), the required strength of the connection in compression, with a factor of 1.1 to compensate for the conservative column curves, need not exceed the lesser value of

| LRFD | ASD |
|---|---|
| $1.1 \times P_u = R_yF_yA_g$<br>$= 1.1 \times 56 \times 3.43$<br>$= 211$ kips<br>or $1.1 \times P_u = 1.1 \times 1.14F_{cre}A_g$<br>$= 1.1 \times 1.14 \times 9.00 \times 3.43$<br>$= 38.7$ kips ... governs<br>$< 211$ kips | $1.1 \times P_a = R_yF_yA_g/1.5$<br>$= 1.1 \times 56 \times 3.43/1.5$<br>$= 141$ kips<br>or $1.1 \times P_a = 1.1 \times 1.14F_{cre}A_g/1.5$<br>$= 1.1 \times 1.14 \times 9.00 \times 3.43/1.5$<br>$= 25.8$ kips ... governs<br>$< 141$ kips |

The required brace connection forces are

| LRFD | ASD |
|---|---|
| $P_u = 192$ kips ... tension<br>$= 38.7$ kips ... compression | $P_a = 128$ kips ... tension<br>$= 25.8$ kips ... compression |

## Design of Beams in V-Braced and Inverted V-Braced Configurations

The beam must be designed for the unbalanced forces produced when the tension brace yields and the compression brace buckles. AISC 341 Sec. F2.4b requires beams

**FIGURE 14.20** Beam loading cases.

to be continuous between columns and it is assumed that the beam supports all gravity loads without any support from the braces. In addition, both flanges of the beam must be provided with lateral braces at the point of intersection of the braces to ensure the stability of the beam. The beam is classified as a moderately ductile member and requires lateral bracing, as specified by AISC 341 Eq. (D1-2), at a maximum spacing of

$$L_b = 0.17r_yE/F_y$$

The required strength of the lateral bracing is given by AISC 360 Eq. (A-6-5) as

$$P_{rb} = 0.02M_rC_d/h_o$$

where $M_r = R_yF_yZ$ ... for LRFD
$= R_yF_yZ/1.5$ ... for ASD
$C_d = 1.0$ ... for single curvature
$h_o$ = distance between flange centroids

In accordance with AISC 341 Sec. F2.3, the beam is designed for the expected loads in the braces with the more critical of the two loading cases shown in Fig. 14.20*a* or *b* governing.

The expected brace strength in tension is

$$R_u = R_yF_yA_g$$

The expected brace strength in compression need not exceed the lesser of

$$P_u = R_yF_yA_g$$
$$\leq 1.14F_{cre}A_g$$

The residual strength of the compression brace after buckling is given as

$$0.3P_n = 0.3F_{cr}A_g$$

**Example 14.7.** Design of a Beam in an Inverted V-Brace Configuration
Select a suitable W14 section, with a yield stress of 50 kips/in$^2$, for the beam in the single bay special concentrically braced steel frame shown in Fig. 14.18 and analyzed in Example 14.5. Lateral braces are provided at the midpoint of the beam. Neglect second-order effects.

The beam is considered pinned at each end and resists all gravity loads without any support from the braces.

From Examples 14.5 and 14.6

$$l = \text{beam span} = 30 \text{ ft}$$

For the brace,

$$KL = 18.2 \text{ ft}$$
$$r = 1.31 \text{ in}$$
$$R_u = 192 \text{ kips}$$
$$P_u = 38.7/1.1 = 35.2 \text{ kips}$$
$$\varphi_c P_n = 27.9 \text{ kips}$$
$$0.3P_n = 0.3\varphi_c P_n/\varphi_c = 0.3 \times 27.9/0.90 = 9.30 \text{ kips}$$

The gravity load on the beam is

$$D = w_D = 0.20 \text{ kip/ft}$$
$$L = 0.10 \text{ kip/ft}$$

By inspection load case (a) governs.

For load case (a), the unbalanced vertical load is

$$E_v = (R_u - 0.3P_n)\sin\theta = (192 - 9.30)\sin 45° = 129 \text{ kips}$$

The unbalanced horizontal load, which is carried equally by the two halves of the beam, is

$$E_h = 0.5(R_u + 0.3P_n)\cos\theta = 0.5(192 + 9.30)\cos 45° = 71 \text{ kips}$$

The loads acting on the beam are shown in Fig. 14.21.

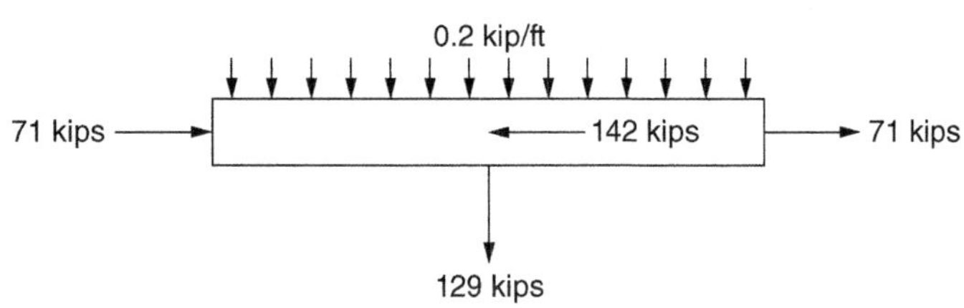


**FIGURE 14.21** Beam loads for Example 14.7.

For vertical loads, the beam is considered simply supported over a span of 30 ft. Applying the factored loads to the beam gives

| LRFD | ASD |
|---|---|
| From ASCE 7 Sec. 12.4.3.2 combination 5, the required flexural strength is<br>$M_r = (1.2 + 0.2S_{DS})Dl^2/8 + E_v l/4 + 0.5Ll^2/8 + 0.2S$<br>$= (1.4)(0.2)(30)^2/8 + 129 \times 30/4 + (0.5)(0.1)30^2/8 + 0$<br>$= 1005$ kip-ft | From ASCE 7 Sec. 12.4.3.2 combination 5, the required flexural strength is<br>$M_r = (1.0 + 0.14S_{DS})Dl^2/8 + 0.7E_v l/4$<br>$= (1.0 + 0.14 \times 1.0)\ (0.2)(30)^2/8 + 0.7 \times 129 \times 30/4$<br>$= 677$ kip-ft |
| From ASCE 7 Sec. 12.4.3.2 combination 5, the required axial strength is<br>$P_r = (1.2 + 0.2S_{DS})D + E_h + 0.5L + 0.2S$<br>$= (1.2 + 0.2 \times 1.0) \times 0 + 71 + 0 + 0$<br>$= 71$ kips ... compression | From ASCE 7 Sec. 12.4.3.2 combination 5, the required axial strength is:<br>$P_r = (1.0 + 0.14S_{DS})D + 0.7E_h$<br>$= (1.0 + 0.14 \times 1.0) \times 0 + 0.7 \times 71$<br>$= 50$ kips ... compression |

Select a W14 × 176 for the beam which satisfies all seismic width-to-thickness requirements and requires lateral bracing only at mid span as indicated in SDM Table 1-3. The beam is pinned at each end, giving a value for the effective length factor of $K = 1.0$ about both axes, and is braced about the $y$-axis at mid span. From AISC Manual Table 6-1 obtain the value

$$r_x/r_y = 1.6$$

The effective length about the $y$-axis is

$$KL_y = 15 \text{ ft}$$

The effective length about the $x$-axis is

$$KL_x = 30 \text{ ft}$$

The equivalent effective length about the $x$-axis with respect to the $y$-axis is

$$\begin{aligned}(KL_y)_{equiv} &= (KL_x)/(r_x/r_y)\\ &= 30/1.6\\ &= 18.75 \text{ ft ... governs}\\ &> KL_y\end{aligned}$$

The equivalent effective length about the $y$-axis for axial compression is 18.75 ft and the unbraced length for flexure is 30 ft. The adequacy of the W14 × 176 beam to sustain combined compression and flexure is determined by the interaction equations given in AISC 360 Sec. H1.1. From AISC Manual Table 6-1, the following information is obtained

| LRFD | ASD |
|---|---|
| $p = 0.539 \times 10^{-3}$ (kips)$^{-1}$<br>$b_x = 0.826 \times 10^{-3}$ (kip-ft)$^{-1}$<br>$P_r/\varphi_c P_n = pP_r = 0.539 \times 10^{-3} \times 71$<br>$= 0.038 < 0.2$ ... Eq. (H1-1b) applies<br>$pP_r/2 + (9/8)b_x M_{rx} \le 1.0$<br>$0.038/2 + (9/8)0.000826 \times 1005 = 0.95$<br>$< 1.0$ ... satisfactory | $p = 0.811 \times 10^{-3}$ (kips)$^{-1}$<br>$b_x = 1.24 \times 10^{-3}$ (kip-ft)$^{-1}$<br>$\Omega_c P_r/P_n = pP_r = 0.811 \times 10^{-3} \times 50$<br>$= 0.041 < 0.2$ ... Eq. (H1-1b) applies<br>$pP_r/2 + (9/8)b_x M_{rx} \le 1.0$<br>$0.041/2 + (9/8)0.00124 \times 677 = 0.96$<br>$< 1.0$ ... satisfactory |

## Design of Columns

In accordance with AISC 341 Sec. F2.3, the column is designed for the expected loads in the braces with the more critical of the two loading cases shown in Fig. 14.22*a* or *b* governing.

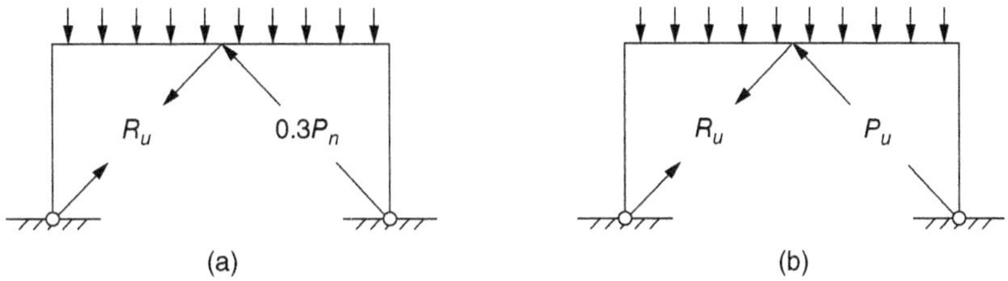


**FIGURE 14.22** Column loading cases.

The expected brace strength in tension is

$$R_u = R_y F_y A_g$$

The expected brace strength in compression need not exceed the lesser of

$$P_u = R_y F_y A_g$$
$$\leq 1.14 F_{cre} A_g$$

In accordance with AISC 342 Sec. F2.3, the brace length used for the determination of $F_{cre}$ must not exceed the distance from brace end to brace end.

The residual strength of the compression brace after buckling is given as

$$0.3P_n = 0.3F_{cr}A_g$$

As specified in AISC 341 Sec. F2.3(1), flexural forces on a column due to story drift may be neglected and $P$-$\Delta$ effects need not be considered.

Figure 14.23 shows a three-story single-bay frame constructed with inverted V bracing with the same brace size at every level. The maximum compressive force in a column in the bottom story, due to the brace forces, is given by the most critical of load case (a) and load case (b). For load case (a), the applied seismic load is

$$E_v = 3[(R_u - 0.3P_n)\sin\theta]/2$$

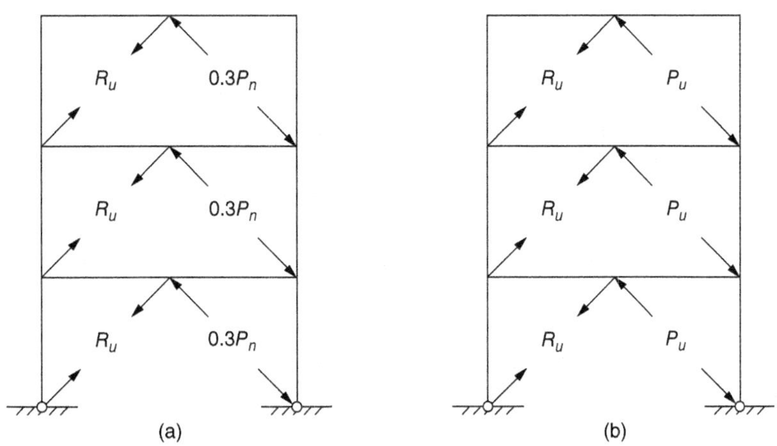


**FIGURE 14.23** Determination of column loads for an inverted V-braced frame.

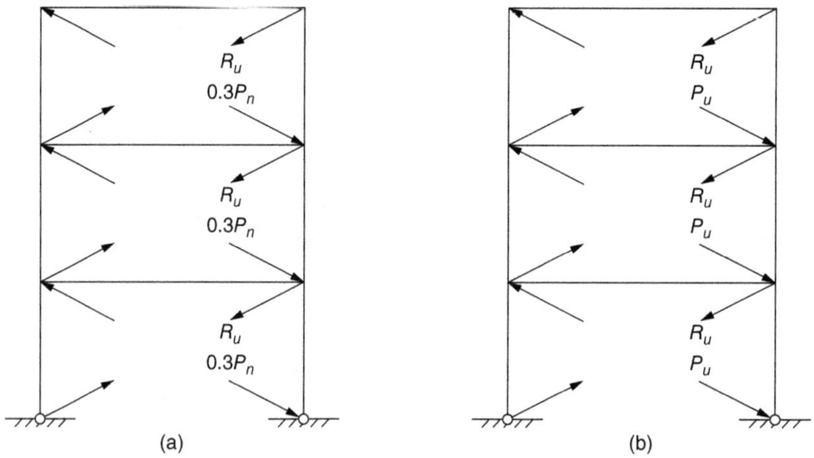


**FIGURE 14.24** Determination of column loads for an X-braced frame.

For load case (b), the applied seismic load is

$$E_v = 3[(R_u - P_u)\sin\theta]/2$$

Figure 14.24 shows a three-story single-bay frame constructed with X-bracing with the same brace size at every level. The maximum compressive force in a column in the bottom story, due to the brace forces, is given by the most critical of load case (a) and load case (b). For load case (a), the applied seismic load is

$$E_v = 3(R_u\sin\theta) + 2(0.3P_n\sin\theta)$$

For load case (b), the applied seismic load is

$$E_v = 3(R_u\sin\theta) + 2(P_u\sin\theta)$$

As an exception to this design methodology, AISC 341 Sec. F2.3(2) specifies that the required strength of columns need not exceed the least of the following:

- The forces resulting from the amplified seismic load, applied to the frame with all compression braces removed, as shown in Fig. 14.25
- The forces corresponding to the resistance of the foundation to overturning uplift
- The forces determined from a nonlinear analysis

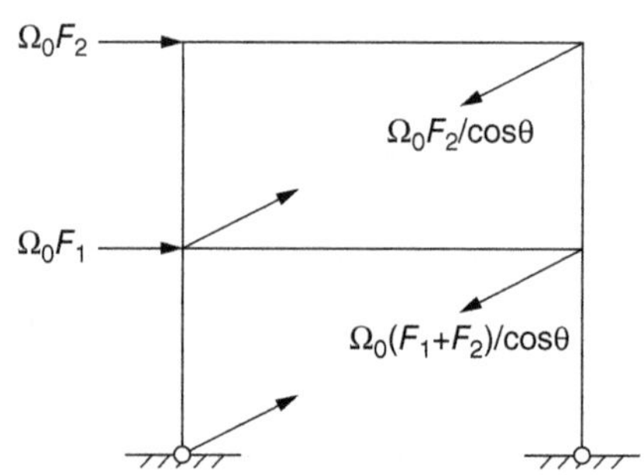


**FIGURE 14.25** Determination of column loads with compression brace removed.

**Example 14.8.** Design of a Column in an Inverted V-Brace Configuration
Determine the appropriate column design force for the single bay special concentrically braced steel frame shown in Fig. 14.18 and analyzed in Example 14.5. The maximum force that can be transferred to the column by the system does not govern nor does the force determined from a nonlinear analysis. Neglect second-order effects and the self-weight of the column.

The properties of the brace are determined in Example 14.6 as

$$R_u = 192 \text{ kips}$$
$$P_u = 35.2 \text{ kips}$$
$$0.3P_n = 9.30 \text{ kips}$$

For load case (a), the applied seismic load is

$$\begin{aligned} E_v &= [(R_u - 0.3P_n)\sin\theta]/2 \\ &= [(192 - 9.30)\sin 45°]/2 \\ &= 65 \text{ kips} \end{aligned}$$

For load case (b), the applied seismic load is

$$\begin{aligned} E_v &= [(R_u - P_u)\sin\theta]/2 \\ &= [(192 - 35.2)\sin 45°]/2 \\ &= 55 \text{ kips} \ldots \text{ does not govern} \\ &< 65 \text{ kips} \end{aligned}$$

Figure 14.26 shows the amplified seismic load applied to the frame with the compression brace removed. The force in the tension brace is

$$\begin{aligned} R_\Omega &= 40/\cos 45° \\ &= 56.57 \text{ kips} \end{aligned}$$

The compression force in the column is

$$\begin{aligned} E_v &= (R_\Omega \sin 45°)/2 \\ &= 20 \text{ kips} \end{aligned}$$

According to SDM Sec. 5.3, it is not prudent to use this value for the exterior columns of a braced frame. Hence, use the value

$$E_v = 65 \text{ kips}$$

The force in a column due to the gravity load is

$$\begin{aligned} D &= w_D l/2 \\ &= 0.2 \times 30/2 \\ &= 3 \text{ kips} \\ L &= 1.5 \text{ kips} \end{aligned}$$

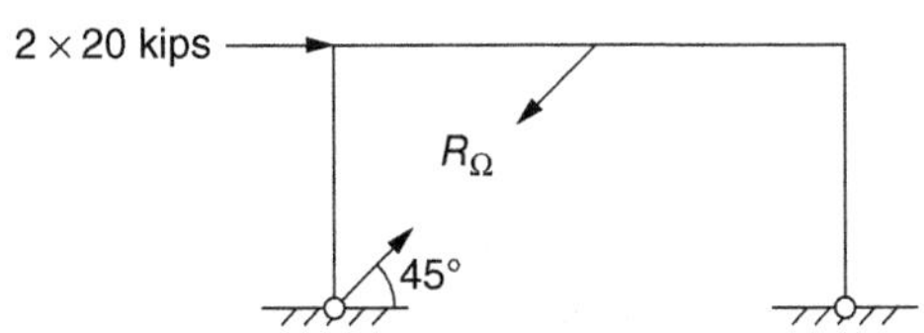


**FIGURE 14.26** Frame with compression brace removed.

Applying the factored loads to the column gives

| LRFD | ASD |
|---|---|
| From ASCE 7 Sec. 12.4.3.2 combination 5, the required axial strength is<br>$P_r = (1.2 + 0.2S_{DS})D + E_v + 0.5L + 0.2S$<br>$= 1.4 \times 3 + 65 + 1.5/2 + 0$<br>$= 70$ kips ... compression | From ASCE 7 Sec. 12.4.3.2 combination 5, the required axial strength is<br>$P_r = (1.0 + 0.14S_{DS})D + 0.7E_v$<br>$= (1.0 + 0.14 \times 1.0) \times 3 + 0.7 \times 65$<br>$= 48.9$ kips ... compression |

## 14.6 Eccentrically Braced Frames

Eccentrically braced frames[8,13] are specified in AISC 341 Sec. F3, and may be used in all seismic design categories. As specified in ASCE 7 Table 12.2-1, $R = 8.0$, $\Omega_0 = 2.0$, $C_d = 4$ and no limitation is imposed on building height in seismic design categories A, B, and C. The maximum height permitted in seismic design categories D and E is 160 ft, and in seismic design category F is 100 ft.

Eccentrically braced frames may be utilized in dual systems with special moment frames in all seismic design categories. In accordance with ASCE 7 Table 12.2-1, $R = 8.0$, $\Omega_0 = 2.5$, $C_d = 4$ and no limitation is imposed on the building height.

Braces and beams outside the link are required to be moderately ductile members and columns and link beams are required to be highly ductile members. The link beam is specified as a protected zone in AISC 341 Sec. 3.5c.

Types of eccentrically braced frame configurations are shown in Fig. 14.27. Eccentrically braced frames provide a high degree of stiffness in the elastic range that is comparable to that of concentrically braced frames, particularly when short link lengths are used. They also provide excellent ductility and energy dissipation capacity comparable to special moment frames in the inelastic range when the links are not too short. When the ratio of link length to frame length, $e/L$, approaches zero, the stiffness of an eccentrically braced frame approaches that of a concentrically braced frame. When the ratio $e/L$ approaches one, the stiffness of an eccentrically braced frame approaches that of a special moment frame.

As shown in Fig. 14.27, the bracing member is connected to the beam so as to form a short link between the brace and the column or between two opposing braces. The link acts as a fuse to prevent other elements in the frame from being overstressed.

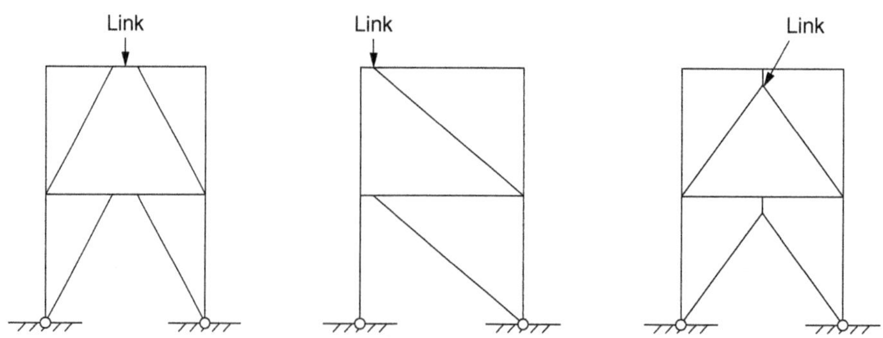


**Figure 14.27** Eccentrically braced frame configurations.

## Mechanism Analysis

Eccentrically braced frames provide a stable and ductile response to seismic loading by cyclic yielding in the links. Stable behavior of the link is ensured by limiting its width-to-thickness ratios and by the provision of lateral bracing and adequate stiffeners to I-shaped cross sections. Links constructed with built-up box sections[14] may eliminate the need for lateral bracing and stiffeners. In accordance with AISC 341 Comm. Sec. F3.3, links are designed for the forces produced by the design level earthquake as determined by an elastic analysis. The required strength of the remaining elements in the system is determined from a plastic-mechanism analysis using the load combinations in the applicable building code that include the amplified seismic load. The amplified seismic load, $E_{mh}$, is defined as the force developed by the fully yielded and strain hardened link.

## Link Characteristics

The forces acting on a link are shown in Fig. 14.28 and the nominal strength of the link is given by the lesser of

$V_n = 2M_p/e$ ... flexural yielding of the link occurs, AISC Eq. (F3-7)

$V_n = V_p$ ... shear yielding of the link occurs, AISC 342 Eq. (F3-1)

The length of the link determines whether the plastic strength of the link will be determined by shear yielding or flexural yielding.

When $P_r/P_c \leq 0.15$, the effect of axial force on the link moment capacity need not be considered and flexural yielding occurs, in accordance with AISC 341 Eq. (F3-8), when the moment on the link is

$$M = M_p = F_y Z$$

where $Z$ = plastic section modulus of the link
$F_y$ = minimum specified yield stress of the link
$e$ = length of the link
$P_r$ = required axial strength of the link
= $P_u$ ... for LRFD
= $P_a$ ... for ASD
$P_c$ = available axial strength
= $P_y$ ... for LRFD
= $P_y/1.5$ ... for ASD
$P_y$ = nominal axial yield strength given by AISC 341 Eq. (F3-6) as
= $F_y A_g$
$A_g$ = gross area of link
$V_p$ = nominal shear strength of an active link

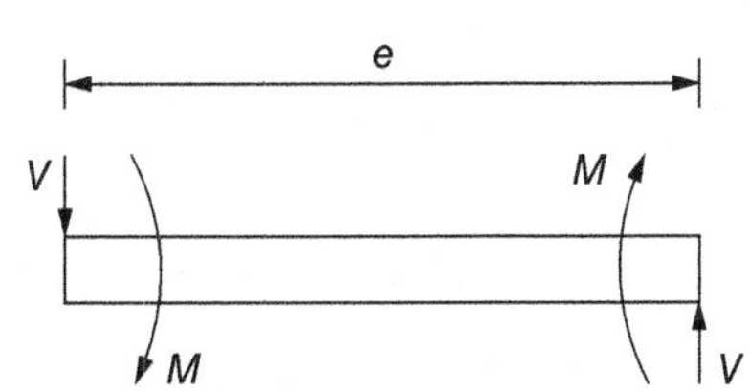


**Figure 14.28** Forces acting on the link.

When $P_r/P_c > 0.15$, the effect of axial force on the link moment capacity must be considered and flexural yielding occurs, in accordance with AISC 341 Eq. (F3-9), when the moment on the link is

$$M = M_p = F_y Z(1 - P_r/P_c)/0.85$$

When $P_r/P_c \le 0.15$, the effect of axial force on the link shear capacity need not be considered and shear yielding occurs, in accordance with AISC 341 Eq. (F3-1), when the shear on the link is

$$V = V_n = V_p$$

where $V_p = 0.60F_y A_{lw}$ ... AISC 341 Eq. (F3-2)
$A_{lw}$ = web area of the link
$= (d_b - 2t_f)t_w$ ... for I-shaped link sections
$d_b$ = depth of link
$t_f$ = flange thickness
$t_w$ = web thickness

The moment produced at the ends of the link is

$$M_r = V_p e/2$$

When $P_r/P_c > 0.15$, the effect of axial force on the link shear capacity must be considered and shear yielding occurs, in accordance with AISC 341 Eq. (F3-3), when the shear on the link is

$$V = V_p = 0.6F_y A_{lw}[1 - (P_r/P_c)^2]^{0.5}$$

In addition, for these higher axial loads, the length of the link is limited by AISC 341 Sec. F3.5b(3) so as to ensure a more reliable shear failure mode. The required length of the link is given by

$$e \le 1.6M_p/V_p \text{ ... for } \rho' \le 0.5$$
$$e \le [1.15 - 0.3\rho']1.6M_p/V_p \text{ ... for } \rho' > 0.5$$

where $\rho' = (P_r/P_c)/(V_r/V_c)$
$V_r$ = required shear strength
$= V_u$ ... for LRFD
$= V_a$ ... for ASD
$V_c$ = available shear strength
$= V_y$ ... for LRFD
$= V_y/1.5$ ... for ASD
$V_y$ = nominal shear yield strength given by AISC 341 Eq. (F3-13) as
$= 0.6F_y A_{lw}$

When $P_r/P_c \le 0.15$, there is no upper limitation on link length.

Shear and flexural yielding occur simultaneously when $V = V_p$ and $M = M_p$ at a link length of

$$e = 2M_p/V_p$$

Shear yielding governs when

$$e < 2M_p/V_p$$

Flexural yielding governs when

$$e > 2M_p/V_p$$

## Link Rotation

The link rotation angle is the inelastic angle between the link and the beam outside of the link due to the total story drift produced by the design level earthquake forces. The link rotation angle is determined by assuming the frame deforms as a rigid plastic mechanism as shown in Fig. 14.29 and AISC 341 Comm. Sec. F3.4a gives the value

$$\begin{aligned}\gamma_p &= L\Delta_p/he \\ &= L\theta_p/e\end{aligned}$$

where $L$ = frame length between column centers
$\Delta_p$ = plastic story drift
$h$ = story height
$e$ = length of link
$\theta_p$ = plastic story drift angle
$\gamma_p$ = plastic link rotation angle

The total drift of the single-story frame, elastic plus plastic, is given by ASCE 7 as

$$\delta_x = C_d\delta_{xe}/I_e$$

where $C_d$ = deflection amplification factor given in ASCE 7 Table 12.2-1
= 4 for an eccentrically braced frame
$\delta_{xe}$ = theoretical displacement, caused by the design level earthquake forces, as determined by an elastic analysis
$I_e$ = seismic importance factor

Hence, the plastic story drift is

$$\begin{aligned}\Delta_p &= \delta_x - \delta_{xe} \\ &= \delta_{xe}(C_d/I_e - 1)\end{aligned}$$

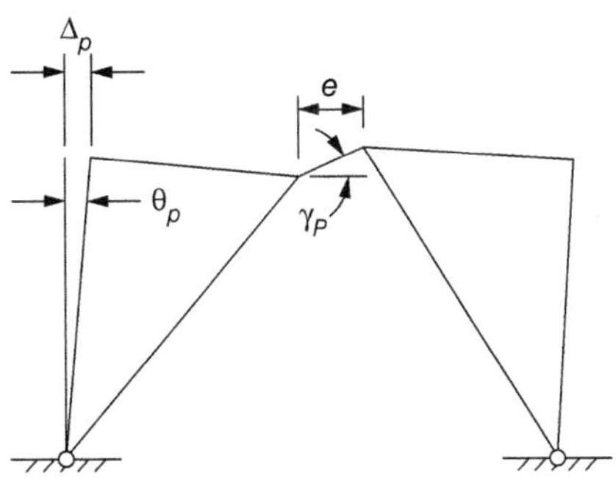


**FIGURE 14.29** Link rotation.

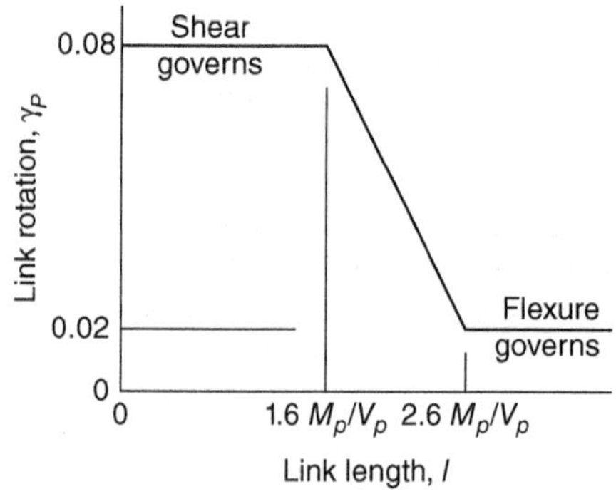


**FIGURE 14.30** Link rotation angle.

As illustrated in Fig. 14.30, for values of $e \leq 1.6M_p/V_p$ the link behavior is governed by shear and for values of $e \geq 2.6M_p/V_p$ the link behavior is governed by flexure. The link rotation angle is limited by AISC 341 Sec. F3.4a to the following values:

$$\gamma_p \leq 0.080 \text{ radian} \ldots \text{ for short links of length } e \leq 1.6M_p/V_p$$
$$\gamma_p \leq 0.020 \text{ radian} \ldots \text{ for long links of length } e \geq 2.6M_p/V_p$$

Linear interpolation may be used for intermediate link lengths.

## Design of Links

In accordance with AISC 341 Comm. Sec. F3.3, links are designed for the forces produced by the design level earthquake as determined by an elastic analysis. The link is designed to deform inelastically and provide an energy absorbing capacity similar to that of a steel special moment frame.

**Example 14.9.** Design of the Link in an Eccentrically Braced Frame
The eccentrically braced bottom story of a three-story residential building is shown in Fig. 14.31. The structure is assigned to seismic design category D with a redundancy factor of $\rho = 1.0$ and a design response acceleration of $S_{DS} = 1.0g$. The structure has a seismic importance factor of $I_e = 1.0$. The connection of a brace to the beam may be considered fixed and the beam is simply supported at its ends. The seismic lateral load of $F = 200$ kips is applied symmetrically on either side of the frame giving two loads of 100 kips each. The preliminary member sizes are indicated in Fig. 14.31 and all members have a yield strength of $F_y = 50$ kips/in$^2$. Determine if the W14 × 82 is a suitable section for the link. Neglect second-order effects.

Applying the dead load to the frame produces the following forces in the link at the first floor as determined by a first-order elastic analysis

$$P_D = 6.4 \text{ kips} \qquad V_D = 1.2 \text{ kips} \qquad M_D = 10 \text{ kip-ft}$$

Applying the live load to the frame produces the following forces in the link as determined by a first-order elastic analysis

$$P_L = 4.3 \text{ kips} \qquad V_L = 0.8 \text{ kip} \qquad M_L = 6 \text{ kip-ft}$$

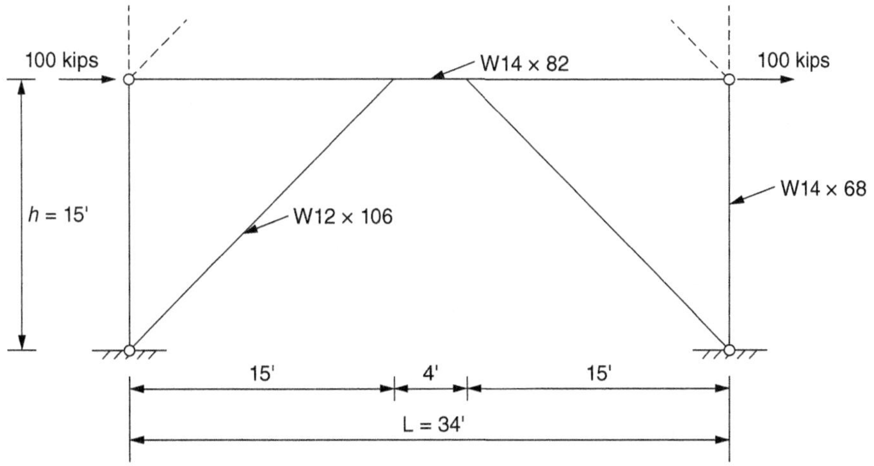


**Figure 14.31** Details for Example 14.9.

Applying the design level seismic load to the frame produces the following forces in the link as determined by a first-order elastic analysis

$$P_E = 0 \text{ kip} \qquad V_E = 90 \text{ kips} \qquad M_E = 180 \text{ kip-ft}$$

The drift of the bottom story of the frame, caused by the design level earthquake forces, as determined by a first-order elastic analysis is

$$\begin{aligned}
\delta_{xe} &= 0.30 \text{ in} \\
I_e &= \text{seismic importance factor} \\
&= 1 \\
C_d &= \text{deflection amplification factor} \\
&= 4 \ldots \text{ from AISC 7 Table 12.2-1 for an eccentrically braced frame}
\end{aligned}$$

Hence, the plastic story drift is

$$\begin{aligned}
\Delta_p &= \delta_x - \delta_{xe} \\
&= \delta_{xe}(C_d/I_e - 1) \\
&= 0.30(4/1 - 1) \\
&= 0.90 \text{ in}
\end{aligned}$$

The link rotation angle is

$$\begin{aligned}
\gamma_p &= L\Delta/he \\
&= 34 \times 0.90/(15 \times 48) \\
&= 0.043 \text{ radian}
\end{aligned}$$

For a value of the design response acceleration of $S_{DS} = 1.0g$

$$(1.2 + 0.2S_{DS}) = 1.4$$

$$(1.0 + 0.14S_{DS}) = 1.14$$

Applying the factored loads to the link gives

| LRFD | ASD |
|---|---|
| From ASCE 7 Sec. 12.4.2.3 combination 5:<br>$P_u$ = factored axial load<br>$= (1.2 + 0.2S_{DS})D + \rho Q_E + 0.5L + 0.2S$<br>$= 1.4 \times 6.4 + 1.0 \times 0 + 0.5 \times 4.3 + 0$<br>$= 11.11$ kips<br>$V_u$ = factored shear load<br>$= (1.2 + 0.2S_{DS})D + \rho Q_E + 0.5L + 0.2S$<br>$= 1.4 \times 1.2 + 1.0 \times 90 + 0.5 \times 0.8 + 0$<br>$= 92$ kips<br>$M_u$ = factored moment<br>$= (1.2 + 0.2S_{DS})D + \rho Q_E + 0.5L + 0.2S$<br>$= 1.4 \times 10 + 1.0 \times 180 + 0.5 \times 6 + 0$<br>$= 197$ kip-ft | From ASCE 7 Sec. 12.4.2.3 combination 5:<br>$P_a$ = factored axial load<br>$= (1.0 + 0.14S_{DS})D + 0.7\rho Q_E$<br>$= 1.14 \times 6.4 + 0.7 \times 1 \times 0$<br>$= 7.30$ kips<br>$V_a$ = factored shear load<br>$= (1.0 + 0.14S_{DS})D + 0.7\rho Q_E$<br>$= 1.14 \times 1.2 + 0.7 \times 1 \times 90$<br>$= 64$ kips<br>$M_a$ = factored moment<br>$= (1.0 + 0.14S_{DS})D + 0.7\rho Q_E$<br>$= 1.14 \times 10 + 0.7 \times 1 \times 180$<br>$= 137$ kip-ft |

The relevant properties of the W14 × 82 are

$$d = 14.3 \text{ in} \quad b_f = 10.1 \text{ in} \quad t_w = 0.51 \text{ in} \quad t_f = 0.855 \text{ in} \quad A_g = 24.0 \text{ in}^2$$
$$h_o = 13.4 \text{ in} \quad b_f/2t_f = 5.92 \quad h/t_w = 22.4 \quad Z_x = 139 \text{ in}^3 \quad k_{det} = 1.69 \text{ in}$$

From SDM Table 1-3, a W14 × 82 satisfies the seismic width-to-thickness requirements for the link of an eccentrically braced frame.

The nominal axial yield strength of the link is given by AISC 341 Equation (F3-6) as

$$\begin{aligned} P_y &= F_y A_g \\ &= 50 \times 24 \\ &= 1200 \text{ kips} \end{aligned}$$

The available axial strength is given by AISC 341 Sec. F3.5b(2) as

| LRFD | ASD |
|---|---|
| $P_c = P_y$<br>$= 1200$ kips<br>$P_r/P_c = 11.11/1200$<br>$= 0.009$ | $P_c = P_y/1.5$<br>$= 800$ kips<br>$P_r/P_c = 7.30/800$<br>$= 0.009$ |

The web area of the link is

$$\begin{aligned} A_w &= (d - 2t_f)t_w \\ &= (14.3 - 2 \times 0.855)0.51 \\ &= 6.42 \text{ in}^2 \end{aligned}$$

The nominal shear strength of the link for $P_r/P_c < 0.15$ is given by AISC 341 Eq. (F3-2) as

$$\begin{aligned} V_n &= V_y = V_p \\ &= 0.60F_y A_w \\ &= 0.60 \times 50 \times 6.42 \\ &= 193 \text{ kips} \end{aligned}$$

The nominal plastic flexural strength of the link for $P_r/P_c < 0.15$ is given by AISC 341 Eq. (F3-8) as

$$\begin{aligned} M_p &= Z_x F_y \\ &= 139 \times 50/12 \\ &= 579 \text{ kip-ft} \\ 1.6M_p/V_p &= 1.6 \times 579/193 \\ &= 4.80 \\ &> e \ldots \text{ link behavior is governed by shear yielding} \end{aligned}$$

For values of $e < 1.6M_p/V_p$, the link rotation angle is limited by AISC 341 Sec. F3.4a to a maximum value of $\gamma_p = 0.080$ radian. The actual link rotation angle is

$$\begin{aligned} \gamma_p &= 0.043 \text{ radian} \\ &< 0.080 \text{ radian} \ldots \text{ satisfactory} \end{aligned}$$

The effective length factor of the link may be taken as $K = 1$ and the effective length of the link is

$$\begin{aligned} KL &= 1 \times 4 \\ &= 4 \text{ ft} \end{aligned}$$

The available compressive strength of the link is

| LRFD | ASD |
|---|---|
| From AISC Manual Table 4-1 for a value of $KL$ = 4 ft and a value of $F_y$ = 50 kips/in$^2$, a W14×82 has a design axial strength of<br>$\varphi_c P_n = 1040$ kips<br>$> P_u = 11.11$ kips ... satisfactory | From AISC Manual Table 4-1 for a value of $KL$ = 4 ft and a value of $F_y$ = 50 kips/in$^2$, a W14 × 82 has an available axial strength of<br>$P_n/\Omega_c = 690$ kips<br>$> P_a = 7.30$ kips ... satisfactory |

The available shear strength of the link is

| LRFD | ASD |
|---|---|
| The nominal shear strength is<br>$V_n = 193$ kips<br>The design shear strength is<br>$\varphi_v V_n = 1 \times 193$<br>$= 193$ kips<br>$> V_u = 92$ kips ... satisfactory | The nominal shear strength is<br>$V_n = 193$ kips<br>The available shear strength is<br>$V_n/\Omega_v = 193/1.5$<br>$= 129$ kips<br>$> V_a = 64$ kips ... satisfactory |

The available flexural strength of the link is

| LRFD | ASD |
|---|---|
| The nominal flexural strength is<br>$M_n = 579$ kip-ft<br>The design flexural strength is<br>$\varphi_b M_n = 0.9 \times 579$<br>$= 521$ kip-ft<br>$> M_u = 197$ kip-ft ... satisfactory | The nominal flexural strength is<br>$M_n = 579$ kip-ft<br>The available flexural strength is<br>$M_n/\Omega_b = 579/1.67$<br>$= 347$ kip-ft<br>$> M_a = 137$ kip-ft ... satisfactory |

The adequacy of the W14 × 193 link to sustain combined compression and flexure is determined by the interaction equations given in AISC 360 Sec. H1.1.

| LRFD | ASD |
|---|---|
| $P_u/\varphi_c P_n = 11.11/1040$<br>$= 0.011 < 0.2$ ... Eq. (H1-1a) applies<br>$P_u/2\varphi_c P_n + M_u/\varphi_b M_n \le 1.0$<br>$0.011/2 + 197/521 = 0.38$<br>$< 1.0$ ... satisfactory | $\Omega_c P_a/P_n = 7.3/690$<br>$= 0.011 < 0.2$ ... Eq. (H1-1a) applies<br>$\Omega_c P_a/2P_n + \Omega_b M_a/M_n \le 1.0$<br>$0.011/2 + 137/347 = 0.40$<br>$< 1.0$ ... satisfactory |

## Design of Link Stiffeners and Bracing

For I-shaped links, stiffeners are required to provide stability and a typical detail is shown in Fig. 14.32.

AISC 341 Sec. F3.5b(4) requires double-sided full depth stiffeners at each end of the link. The minimum required width of each stiffener is

$$b_{st} = (b_f - 2t_w)/2$$

The minimum required thickness is

$$t_{st} = 0.75t_w \\ \ge 3/8 \text{ in}$$

where $b_f$ is the link flange width and $t_w$ is the link web thickness.

To ensure full development of the shear capacity of the link, intermediate stiffeners are required:

a. For a link length of $1.6M_p/V_p$ or less the required intermediate stiffener spacing is

$$s \le 30t_w - d/5 \text{ ... for } \gamma_p = 0.08 \text{ radian}$$
$$\le 52t_w - d/5 \text{ ... for } \gamma_p \le 0.02 \text{ radian}$$

Linear interpolation may be used for intermediate link rotations.

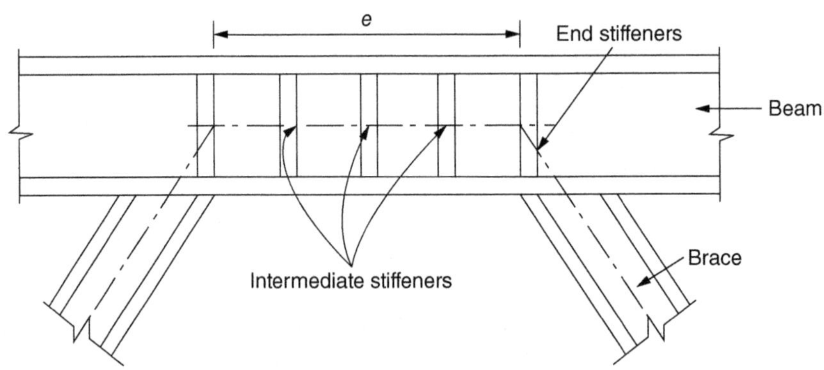


**FIGURE 14.32** Link details.

b. For a link length defined by $2.6M_p/V_p \leq e < 5M_p/V_p$ intermediate stiffeners are required at a distance of $1.5b_f$ from each end of the link.

c. For a link length defined by $1.6M_p/V_p < e < 2.6M_p/V_p$ intermediate stiffeners are required to satisfy the requirements of both (a) and (b).

d. For a link length greater than $5M_p/V_p$ intermediate stiffeners are not required.

Intermediate web stiffeners are required to be full depth. Provided the link depth is less than 25 in, single-sided full-depth intermediate stiffeners are permitted. The required width of single-sided intermediate stiffeners is

$$b_{st} \geq b_f/2 - t_w$$

and the required thickness is

$$\begin{aligned} t_{st} &= t_w \\ &\geq 3/8 \text{ in} \end{aligned}$$

When the link depth is 25 in or greater, similar intermediate stiffeners are required on both sides of the web.

The required strength of the fillet weld between a stiffener and the link web is

$$\begin{aligned} P_w &= A_{st}F_y \text{ ... for LRFD} \\ &= A_{st}F_y/1.5 \text{ ... for ASD} \end{aligned}$$

The required strength of the fillet weld between a stiffener and the link flange is

$$\begin{aligned} P_w &= A_{st}F_y/4 \text{ ... for LRFD} \\ &= A_{st}F_y/(4 \times 1.5) \text{ ... for ASD} \end{aligned}$$

where $A_{st}$ is the area of stiffener $= b_{st}t_{st}$.

As specified in AISC 341 Sec. F3.4b, lateral bracing must be provided to both the top and bottom flanges at the ends of the link. The required strength of each brace is given by AISC 341 Eq. (D1-4a) as

$$\begin{aligned} P_r &= 0.06R_yF_yZ/h_o \text{ ... for LRFD} \\ &= (0.06/1.5)R_yF_yZ/h_o \text{ ... for ASD} \end{aligned}$$

The required stiffness of each brace is given by AISC 341 Eq. (D1-6a) as

$$\begin{aligned} M_r &= R_yF_yZ \text{ ... for LRFD} \\ &= R_yF_yZ/1.5 \text{ ... for ASD} \end{aligned}$$

where $R_y$ = ratio of the expected yield stress to the minimum specified yield stress
$Z$ = plastic section modulus of the link
$h_o$ = distance between flange centroids

**Example 14.10.** Design of the Link Stiffeners

Design the stiffeners for the link of the eccentrically braced frame shown in Fig. 14.31 using A36 steel.

In accordance with AISC 341 Sec. F3.5b(4), double-sided, full-depth web stiffeners are required at each end of the link. The minimum required width of each stiffener is

$$\begin{aligned} b_{st} &= (b_f - 2t_w)/2 \\ &= (10.1 - 2 \times 0.51)/2 \\ &= 4.54 \text{ in} \ldots \text{ use } 4\tfrac{1}{2} \text{ in} \end{aligned}$$

The required thickness is given by the greater of $t_{st} = 3/8$ in or

$$\begin{aligned} t_{st} &= 0.75t_w \\ &= 0.75 \times 0.51 \\ &= 0.38 \text{ in} \ldots \text{ use } \tfrac{1}{2} \text{ in} \end{aligned}$$

Single-sided, full-depth web intermediate stiffeners are permitted because the link depth is less than 25 in. The minimum required width is

$$\begin{aligned} b_{st} &= b_f/2 - t_w \\ &= 10.1/2 - 0.51 \\ &= 4.54 \text{ in} \ldots \text{ use } 4\tfrac{1}{2} \text{ in} \end{aligned}$$

The required thickness is given by the greater of $t_{st} = 3/8$ in or

$$\begin{aligned} t_{st} &= t_w \\ &= 0.51 \text{ in} \ldots \text{ use } \tfrac{1}{2} \text{ in} \end{aligned}$$

For a link length of e $< 1.6M_p/V_p$ and a link rotation of $\gamma_p = 0.08$ radian, the required intermediate stiffener spacing is

$$\begin{aligned} s &= 30t_w - d/5 \\ &= 30 \times 0.51 - 14.3/5 \\ &= 12.4 \text{ in} \end{aligned}$$

For a link length of $e < 1.6M_p/V_p$ and a link rotation of $\gamma_p = 0.02$ radian, the required intermediate stiffener spacing is

$$\begin{aligned} s &= 52t_w - d/5 \\ &= 52 \times 0.51 - 14.3/5 \\ &= 23.7 \text{ in} \end{aligned}$$

For a link length of $e < 1.6M_p/V_p$ and a link rotation of $\gamma_p = 0.043$ radian, the intermediate stiffener spacing is obtained by interpolation as

$$\begin{aligned} s &= 43.6t_w - d/5 \\ &= 43.6 \times 0.51 - 14.3/5 \\ &= 19.4 \text{ in} \end{aligned}$$

Provide two intermediate stiffeners and two end stiffeners to give a spacing of

$$\begin{aligned} s &= 48/3 \\ &= 16 \text{ in} \\ &< 19.4 \text{ in} \ldots \text{ satisfactory} \end{aligned}$$

As required by AISC 341 Sec. F3.5b(1), the weld between a stiffener and the web is required to develop the force

$$
\begin{aligned}
P_w &= A_{st}F_y \ \ldots \text{ for LRFD} \\
&= 4.5 \times 0.5 \times 36 \\
&= 81.0 \text{ kips} \\
P_w &= A_{st}F_y/1.5 \ldots \text{ for ASD} \\
&= 54.0 \text{ kips}
\end{aligned}
$$

In accordance with AISC 341 Comm. Sec. F3.5b(4), welding within 1 in of the *k*-area of the link should be avoided. This gives the length of the weld along the web of

$$
\begin{aligned}
l_w &= d - 2(k_{det} + 1.5) \\
&= 14.3 - 2(1.69 + 1.5) \\
&= 7.9 \text{ in}
\end{aligned}
$$

The required weld size of the double-sided fillet weld per 1/16 in is

| LRFD | ASD |
|---|---|
| $D = P_w/(2 \times 1.39 l_w)$<br>$= 81/(2 \times 1.39 \times 7.9)$<br>$= 3.7$ sixteenths<br>Use $w = ¼$ in ... satisfactory | $D = P_w/(2 \times 1.39 l_w)$<br>$= 54/(2 \times 0.93 \times 7.9)$<br>$= 3.7$ sixteenths<br>Use $w = ¼$ in ... satisfactory |

From AISC 360 Table J2.4, the minimum size of weld permitted for the ½ in stiffener thickness is 3/16 in.

As required by AISC 341 Sec. F3.5b(1), the weld between a stiffener and the flange is required to develop the force

$$
\begin{aligned}
P_w &= A_{st}F_y/4 \ \ldots \text{ for LRFD} \\
&= 4.5 \times 0.5 \times 36/4 \\
&= 20.3 \text{ kips} \\
P_w &= A_{st}F_y/(4 \times 1.5) \ldots \text{ for ASD} \\
&= 13.5 \text{ kips}
\end{aligned}
$$

Provide a 1-in corner clip on the stiffener to clear the corner fillet. This gives the length of the weld along the flange of

$$
\begin{aligned}
l_w &= b_{st} - 1.0 \text{ in} \\
&= 4.5 - 1.0 \\
&= 3.5 \text{ in}
\end{aligned}
$$

The required weld size of the double-sided fillet weld per 1/16 in is

| LRFD | ASD |
|---|---|
| $D = P_w/(2 \times 1.39 l_w)$<br>$= 20.3/(2 \times 1.39 \times 3.5)$<br>$= 2.1$ sixteenths<br>Use $w = 3/16$ in ... satisfactory | $D = P_w/(2 \times 1.39 l_w)$<br>$= 13.5/(2 \times 0.93 \times 3.5)$<br>$= 2.1$ sixteenths<br>Use $w = 3/16$ in ... satisfactory |

From AISC 360 Table J2.4, the minimum size of weld permitted for the ½ in stiffener thickness is 3/16 in.

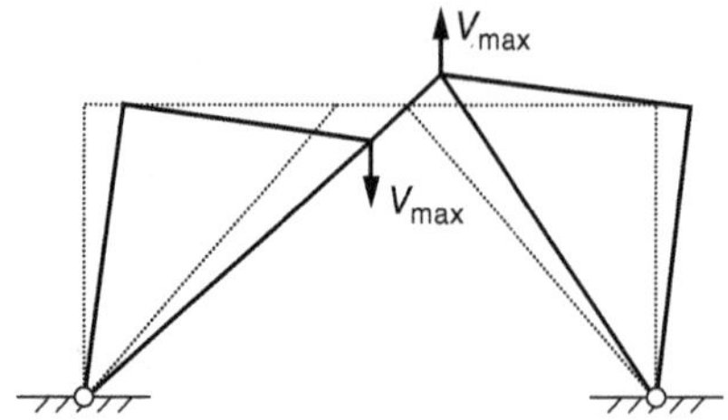


**FIGURE 14.33** Amplified seismic load.

## Design of Members in an Eccentrically Braced Frame Other Than the Link

The required strength of elements, other than the link, is determined from a plastic-mechanism analysis using the load combinations in the applicable building code that include the amplified seismic load. The amplified seismic load, $E_{mh}$, is defined in AISC 341 Sec. F3.3 as the force generated by the adjusted link shear strength, $V_{max}$, and is shown in Fig. 14.33. Because of strain hardening of the link and of the anticipated higher yield strength, the adjusted link yield strength for I-shaped links is

$$V_{max} = 1.25R_yV_n$$

where $R_y$ = ratio of the expected yield stress to the minimum specified yield strength of the link
1.25 = strain hardening factor for I-shaped links
$V_n$ = nominal shear capacity of the link

Because of the strengthening provided to the beam by a composite floor slab, the beam is designed for only 88 percent of the adjusted link forces. For columns in frames with three or more stories of bracing it is unlikely that all links will strain-harden simultaneously and columns are designed for 88 percent of the adjusted link forces. Braces must be designed for the maximum adjusted link forces.

## Design of the Beam Outside the Link

AISC 341 Sec. F3.3 and Comm. Sec. F3.3 specify the design requirements for the beam outside the link. The required axial, flexural, and shear capacities of the beam are determined from the load combinations of ASCE 7 Sec. 12.4.3.2 with the amplified seismic force, $E_{mh}$, replaced by 88 percent of the forces developed by the adjusted link shear strength.

In accordance with AISC 341 Sec. F3.5a, *User Note*, the design capacity of the beam is determined using the expected yield strength of $R_yF_y$ since the beam is the same member as the link.

Where required, the beam shall be provided with lateral support at both the top and bottom flanges. The beam is classified as a moderately ductile member and requires lateral bracing, as specified by AISC 341 Eq. (D1-2), at a maximum spacing of

$$L_b = 0.17r_yE/F_y$$

The required strength of the lateral bracing is given by AISC 360 Eq. (A-6-7) as

$$P_{rb} = 0.02MrC_d/h_o$$

where $M_r = R_y F_y Z$ ... for LRFD
$= R_y F_y Z/1.5$ ... for ASD
$C_d = 1.0$ ... for single curvature
$h_o$ = distance between flange centroids

**Example 14.11.** Design of the Beam in an Eccentrically Braced Frame
Determine the required forces for the beam in the bottom story of the eccentrically braced building shown in Fig. 14.31. Neglect second-order effects.

Applying the dead load independently to the frame produces the following forces in the beam as determined by a first-order elastic analysis.

$$P_D = 0.08 \text{ kip} \qquad V_D = 5.4 \text{ kips} \qquad M_D = 13 \text{ kip-ft}$$

Applying the live load independently to the frame produces the following forces in the beam as determined by a first-order elastic analysis.

$$P_L = 0.05 \text{ kip} \qquad V_L = 3.6 \text{ kips} \qquad M_L = 8.7 \text{ kip-ft}$$

Applying the design level seismic load to the frame produces the following forces in the beam as determined by a first-order elastic analysis

$$P_E = 100 \text{ kips} \qquad V_E = 6 \text{ kips} \qquad M_E = 95 \text{ kip-ft}$$

The forces produced in the beam by the amplified seismic load, consisting of 88 percent of the adjusted link shear strength, are determined by multiplying the listed beam forces due to the design level seismic load by an amplification factor. The amplification factor is

$$\alpha = 0.88V_{max}/V_{Elink}$$

where $V_{max} = 1.25R_y V_n$
$R_y$ = ratio of the expected yield stress to the minimum specified yield strength of the link
= 1.1 ... for grade 50 steel
$V_n$ = nominal shear strength of the link
= 193 kips ... from Example 14.9
$V_{Elink}$ = force in the link produced by the design level seismic load
= 90 kips ... from Example 14.9

Then,
$$V_{max} = 1.25R_y V_n = 1.25 \times 1.1 \times 193 = 265 \text{ kips}$$
$$\alpha = 0.88V_{max}/V_{Elink} = 0.88 \times 265/90 = 2.60$$

The forces produced in the beam by the amplified seismic load are

$$P_{Emh} = \alpha P_E = 2.60 \times 100 = 260 \text{ kips}$$
$$V_{Emh} = \alpha V_E = 2.60 \times 6 = 15.6 \text{ kips}$$
$$M_{Emh} = \alpha M_E = 2.6 \times 95 = 247 \text{ kip-ft}$$

For a value of the design response acceleration of $S_{DS} = 1.0g$

$$(1.2 + 0.2S_{DS}) = 1.4$$
$$(1.0 + 0.14S_{DS}) = 1.14$$

Applying the factored loads to the beam gives the following required strengths:

| LRFD | ASD |
|---|---|
| From ASCE 7 Sec. 12.4.3.2 combination 5:<br>$P_u$ = factored axial load<br>$= (1.2 + 0.2S_{DS})D + E_{mh} + 0.5L + 0.2S$<br>$= 1.4 \times 0.08 + 260 + 0.5 \times 0.05 + 0$<br>= 260 kips<br>$V_u$ = factored shear load<br>$= (1.2 + 0.2S_{DS})D + E_{mh} + 0.5L + 0.2S$<br>$= 1.4 \times 5.4 + 15.6 + 0.5 \times 3.6 + 0$<br>= 25 kips<br>$M_u$ = factored moment<br>$= (1.2 + 0.2S_{DS})D + E_{mh} + 0.5L + 0.2S$<br>$= 1.4 \times 13 + 247 + 0.5 \times 8.7 + 0$<br>= 270 kip-ft | From ASCE 7 Sec. 12.4.3.2 combination 5:<br>$P_a$ = factored axial load<br>$= (1.0 + 0.14S_{DS})D + 0.7E_{mh}$<br>$= 1.14 \times 0.08 + 0.7 \times 260$<br>= 182 kips<br>$V_a$ = factored shear load<br>$= (1.0 + 0.14S_{DS})D + 0.7E_{mh}$<br>$= 1.14 \times 5.4 + 0.7 \times 15.6$<br>= 17 kips<br>$M_a$ = factored moment<br>$= (1.0 + 0.14S_{DS})D + 0.7E_{mh}$<br>$= 1.14 \times 13 + 0.7 \times 247$<br>= 188 kip-ft |

The W14 × 82 satisfies all seismic width-to-thickness requirements as indicated in SDM Table 1-3.

## Design of the Diagonal Brace

As required in AISC 341 Sec. F3.3, the brace is designed for the factored gravity loads plus the effects of the maximum adjusted link yield strength as shown in Fig. 14.33. The braces are designated as moderately ductile members and are intended to remain essentially elastic under the amplified seismic load. Because of the large axial force and bending moment on the brace it is designed as a beam-column using AISC 360 Sec. H1.1.

**Example 14.12.** Design of the Diagonal Brace in an Eccentrically Braced Frame
Determine if a W12 × 106 is a suitable section for the diagonal brace in the bottom story of the eccentrically braced building shown in Fig. 14.31. Neglect second-order effects.

Applying the dead load independently to the frame produces the following forces in the brace as determined by a first-order elastic analysis.

$$P_D = 9.2 \text{ kips} \qquad V_D = 0.2 \text{ kip} \qquad M_D = 3.3 \text{ kip-ft}$$

Applying the live load independently to the frame produces the following forces in the brace as determined by a first-order elastic analysis.

$$P_L = 6.2 \text{ kips} \qquad V_L = 0.1 \text{ kip} \qquad M_L = 2.2 \text{ kip-ft}$$

Applying the design level seismic load to the frame produces the following forces in the brace as determined by a first-order elastic analysis

$$P_E = 139 \text{ kips} \qquad V_E = 4.8 \text{ kips} \qquad M_E = 81 \text{ kip-ft}$$

The forces produced in the brace by the amplified seismic load, consisting of 100 percent of the adjusted link shear strength, are determined by multiplying the listed brace forces due to the design level seismic load by the amplification factor

$$\alpha = V_{max}/V_{Elink}$$

where $V_{max} = 1.25R_yV_n$
= 265 kips ... from Example 14.11
$V_{Elink}$ = force in the link produced by the design level seismic load
= 90 kips ... from Example 14.9

Then, $\alpha = V_{max}/V_{Elink}$
$= 265/90$
$= 2.94$

The forces produced in the brace by the amplified seismic load are

$$P_{Emh} = \alpha P_E = 2.94 \times 139 = 409 \text{ kips}$$

$$V_{Emh} = \alpha V_E = 2.94 \times 4.8 = 14 \text{ kips}$$

$$M_{Emh} = \alpha M_E = 2.94 \times 81 = 238 \text{ kip-ft}$$

For a value of the design response acceleration of $S_{DS} = 1.0g$

$$(1.2 + 0.2S_{DS}) = 1.4$$

$$(1.0 + 0.14S_{DS}) = 1.14$$

Applying the factored loads to the brace gives the following required strengths

| LRFD | ASD |
|---|---|
| From ASCE 7 Sec. 12.4.3.2 combination 5:<br>$P_u$ = factored axial load<br>$= (1.2 + 0.2S_{DS})D + E_{mh} + 0.5L + 0.2S$<br>$= 1.4 \times 9.2 + 409 + 0.5 \times 6.2 + 0$<br>$= 425$ kips<br>$V_u$ = factored shear load<br>$= (1.2 + 0.2S_{DS})D + E_{mh} + 0.5L + 0.2S$<br>$= 1.4 \times 0.2 + 14 + 0.5 \times 0.1 + 0$<br>$= 14$ kips<br>$M_u$ = factored moment<br>$= (1.2 + 0.2S_{DS})D + E_{mh} + 0.5L + 0.2S$<br>$= 1.4 \times 3.3 + 238 + 0.5 \times 2.2 + 0$<br>$= 244$ kip-ft | From ASCE 7 Sec. 12.4.3.2 combination 5:<br>$P_a$ = factored axial load<br>$= (1.0 + 0.14S_{DS})D + 0.7E_{mh}$<br>$= 1.14 \times 9.2 + 0.7 \times 409$<br>$= 297$ kips<br>$V_a$ = factored shear load<br>$= (1.0 + 0.14S_{DS})D + 0.7E_{mh}$<br>$= 1.14 \times 0.2 + 0.7 \times 14$<br>$= 10$ kips<br>$M_a$ = factored moment<br>$= (1.0 + 0.14S_{DS})D + 0.7E_{mh}$<br>$= 1.14 \times 3.3 + 0.7 \times 238$<br>$= 170$ kip-ft |

The work point-to-work point length of the brace is

$$l = h/\sin\theta = 15/\sin 45^\circ = 21.21 \text{ ft}$$

The effective length factor for a member in a braced frame is

$$K = 1.0$$

Allowing 1.21 ft for the depth of members and a gusset plate, the effective length about the $y$-axis for axial compression is

$$KL = 18 \text{ ft}$$

The unbraced length for flexure is 18 ft. The adequacy of the W12 × 106 brace to sustain combined compression and flexure is determined by the interaction equations given in AISC 360 Sec. H1.1. From AISC Manual Table 6-1, the following information is obtained

| LRFD | ASD |
|---|---|
| $p = 1.01 \times 10^{-3}$ (kips)$^{-1}$<br>$b_x = 1.55 \times 10^{-3}$ (kip-ft)$^{-1}$<br>$P_r/\varphi_c P_n = pP_r = 1.01 \times 10^{-3} \times 425$<br>$= 0.429 > 0.2$ ... Eq. (H1-1a) applies<br>$pP_r + b_x M_{rx} \leq 1.0$<br>$0.429 + 0.00155 \times 244 = 0.81$<br>$< 1.0$ ... satisfactory | $p = 1.52 \times 10^{-3}$ (kips)$^{-1}$<br>$b_x = 2.33 \times 10^{-3}$ (kip-ft)$^{-1}$<br>$\Omega_c P_r/P_n = pP_r = 1.52 \times 10^{-3} \times 297$<br>$= 0.451 > 0.2$ ... Eq. (H1-1a) applies<br>$pP_r + b_x M_{rx} \leq 1.0$<br>$0.451 + 0.00233 \times 170 = 0.85$<br>$< 1.0$ ... satisfactory |

The W12 × 106 satisfies all seismic width-to-thickness requirements as indicated in SDM Table 1-3.

## Design of Columns

As required in AISC 341 Sec. F3.3, the column is designed for the factored gravity loads plus the effects of the adjusted link yield strength at each floor above the top of the column as shown in Fig. 14.34. When the frames consist of three or more stories of bracing above the level considered, columns are designed for 88 percent of the adjusted link forces.

In accordance with AISC Sec. F3.3, flexural forces resulting from seismic drift are neglected when designing the column. The required strength of columns need not exceed the lesser of the following two conditions:

- Forces corresponding to the resistance of the foundation to overturning uplift
- Forces determined from nonlinear analysis

Welds at a column splice and at column-to-base plate connections are designated as demand critical in AISC 341 Sec. F3.6a. In accordance with AISC 341 Sec. F3.6d, welds at a splice must be complete joint-penetration groove welds, and the splice must be designed to develop at least 50 percent of the lesser available flexural strength of the connected columns. The required shear strength of the joint is

$$V_r = \Sigma F_y Z_c / H_c \text{ ... for LRFD}$$
$$= \Sigma F_y Z_c / (1.5 H_c) \text{ ... for ASD}$$

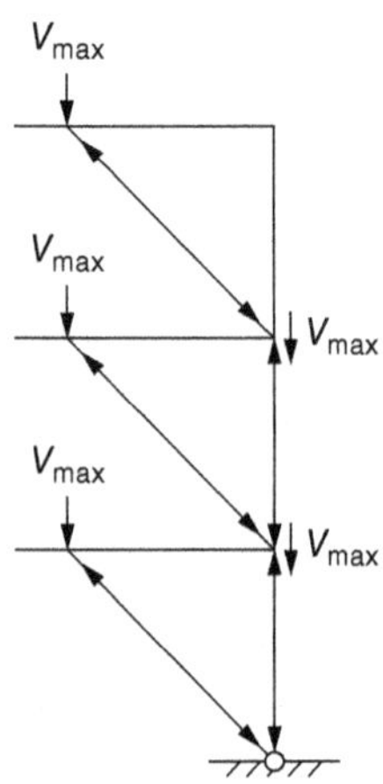


**FIGURE 14.34** Column loads due to link shear.

where $Z_c$ = plastic section modulus of a column
$\Sigma F_y Z_c$ = sum of the nominal plastic flexural strengths of the columns above and below the splice
$H_c$ = clear height of the column between beam connections

**Example 14.13.** Design of the Column in an Eccentrically Braced Frame
Determine if a W14 × 68 is a suitable section for the column in the bottom story of the eccentrically braced building shown in Fig. 14.31. The section properties of the members are identical at all stories. Neglect second-order effects.

Applying the dead load independently to the frame produces the following forces in the column as determined by a first-order elastic analysis.

$$P_D = 31 \text{ kips} \qquad V_D = 0 \text{ kip} \qquad M_D = 0 \text{ kip-ft}$$

Applying the live load independently to the frame produces the following forces in the column as determined by a first-order elastic analysis.

$$P_L = 21 \text{ kips} \qquad V_L = 0 \text{ kip} \qquad M_L = 0 \text{ kip-ft}$$

The adjusted link yield strength for the I-shaped link is obtained from Example 14.11 as

$$\begin{aligned} V_{max} &= 1.25R_yV_n \\ &= 265 \text{ kips} \end{aligned}$$

The frame has less than three stories of bracing above the level of the column and no reduction in $V_{max}$ is permitted. As shown in Fig. 14.34, the adjusted link yield strength at the third floor and the roof contribute to the axial force in the column in the bottom story and the amplified seismic load is

$$\begin{aligned} P_{emh} &= 2V_{max} \\ &= 2 \times 265 \\ &= 530 \text{ kips} \end{aligned}$$

For a value of the design response acceleration of $S_{DS} = 1.0g$

$$(1.2 + 0.2S_{DS}) = 1.4$$

$$(1.0 + 0.14S_{DS}) = 1.14$$

Applying the factored loads to the brace gives the following required strengths:

| LRFD | ASD |
|---|---|
| From ASCE 7 Sec. 12.4.3.2 combination 5:<br>$P_u$ = factored axial load<br>$= (1.2 + 0.2S_{DS})D + E_{mh} + 0.5L + 0.2S$<br>$= 1.4 \times 31 + 530 + 0.5 \times 21 + 0$<br>$= 584$ kips | From ASCE 7 Sec. 12.4.3.2 combination 5:<br>$P_a$ = factored axial load<br>$= (1.0 + 0.14S_{DS})D + 0.7E_{mh}$<br>$= 1.14 \times 31 + 0.7 \times 530$<br>$= 406$ kips |

The effective length factor for a member in a braced frame is

$$K = 1.0$$

Allowing 1.0 ft for the depth of members and a gusset plate, the effective length about the $y$-axis for axial compression is

$$\begin{aligned} KL &= 1(15 - 1) \\ &= 14 \text{ ft} \end{aligned}$$

| LRFD | ASD |
|---|---|
| From AISC Manual Table 4-1 for a value of $KL = 14$ ft and a value of $F_y = 50$ kips/in$^2$, a W14 × 68 has a design strength of<br>$\varphi_c P_n = 640$ kips<br>$> 584$ kips ... satisfactory | From AISC Manual Table 4-1 for a value of $KL = 14$ ft and a value of $F_y = 50$ kips/in$^2$, a W14 × 68 has an allowable strength of<br>$P_n/\Omega_c = 426$ kips<br>$> 406$ kips ... satisfactory |

A W14 × 68 satisfies all seismic width-to-thickness requirements as indicated in SDM Table 1-3.

## 14.7 Special Moment Frames

Steel special moment frames[15] are specified in AISC 341 Sec. E3 and may be used in all seismic design categories. As specified in ASCE 7 Table 12.2-1, $R = 8$, $\Omega_0 = 3.0$ and $C_d = 5.5$. No limitation is imposed on building height. Steel special moment frames may also be used in dual systems, combined with braced frames or shear walls, with the moment frame required to resist at least 25 percent of the prescribed seismic forces. The system requirements for the different types of dual systems are specified in ASCE Table 12.2-1.

In contrast to braced frame and shear wall systems, moment frame systems provide open bays that allow for greater architectural freedom in design. However, moment frames may be more costly than braced frame or shear wall structures because of the heavier member sections required and the more expensive beam-to-column connections used. The system has a low lateral stiffness which may result in considerable drift with significant damage to architectural features and nonstructural components in the event of a severe earthquake. Hence, the design of special moment frames is often governed by code requirements to limit story drift and *P*-delta effects rather than to provide structural strength. The system is highly redundant and has a correspondingly high value of the response modification coefficient, *R*, resulting in a low lateral design force.

A steel special moment frame provides resistance to seismic forces by means of flexure in the beams and columns. During a severe earthquake, the structure experiences large inelastic deformations concentrated at the beam-to-column connections and the structure relies on the ductility of the system and the integrity of the moment resisting connections to prevent collapse. Ductility is developed by flexural yielding of the beams and columns and by shear yielding of column panel zones. Plastic hinges form adjacent to the connections and preferably in the beams, as shown in Fig. 14.1, so as to avoid a soft story collapse in the structure.

### Beam-to-Column Connections

The prescriptive beam-to-column connection prior to 1994 is shown in Fig. 14.35 and consisted of a welded flange and bolted web moment connection. The design criteria was limited to a requirement that the connection must be capable of developing the strength of the connected members, and this was considered a deemed-to-comply detail. The Northridge earthquake that occurred in the San Fernando Valley on January 17, 1994 had a magnitude of 6.8 on the Richter scale. After the earthquake, extensive failures were observed in these prescriptive connections[16] and it was determined that the following factors contributed to the damage:

- High stress concentrations at the beam bottom flange weld
- The use of weld filler metals with inherent low toughness

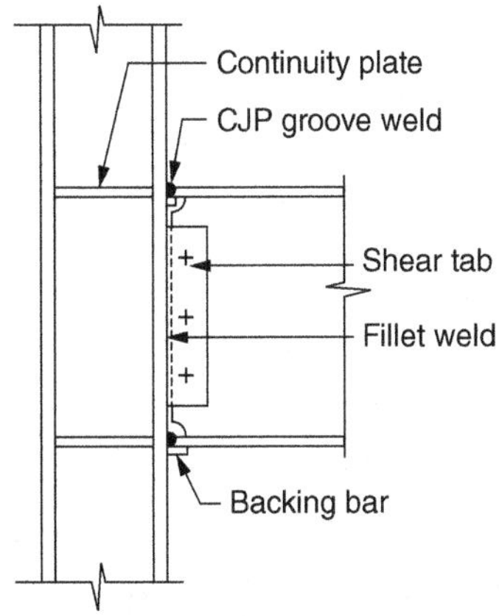


**FIGURE 14.35** Pre-Northridge moment connection.

- Variations of member strength from prescribed values
- The use of larger beams than had previously been tested
- Inadequate quality control
- Fatigue failure at the joint

As a result of an extensive research program, requirements are now detailed in AISC 341 and AISC 358 to eliminate these defects. In accordance with AISC 341 Sec. E3.6b, beam-to-column connections are required to be capable of developing an interstory drift angle of at least 0.04 radian with a residual moment capacity of 80 percent of the nominal plastic moment capacity of the beam.

A number of joint assemblies has been demonstrated by extensive testing and analysis to meet these criteria and has been approved by the connection prequalification review panel. These joint assemblies are designated prequalified connections and may be categorized into the seven main types: reduced beam section connections; bolted stiffened and unstiffened extended end-plates; bolted flange plate connections; welded unreinforced flange-welded web connections; Kaiser bolted bracket connections; ConX-tech CONXL connections; and SidePlate connections. These prequalified connections are fully described and detailed in AISC 358 and when adopted for a specific project are acceptable to local authorities. In accordance with AISC 358 Sec. 1.3, prequalified connections must be designed by the LRFD procedure only. For each of the prequalified connections, AISC 358 provides a detailed step-by-step procedure for the design of the connection.

Several proprietary connection systems are also available[17] and these have their own library of test data and evaluation reports and are widely used and accepted. If no prequalified connection or proprietary connection is suitable for a particular design situation, AISC 341 Sec. E3.6c requires project-specific testing of the proposed detail and the approval of the connection prequalification review panel.

Beams and columns are required by AISC 341 Sec. E3.5a to be highly ductile members. The location of protected zones and demand critical welds are designated in AISC 358 or as determined in a connection prequalification program in accordance with AISC 341 Sec. K1 or a qualification testing program in accordance with AISC 341 Sec. K2.

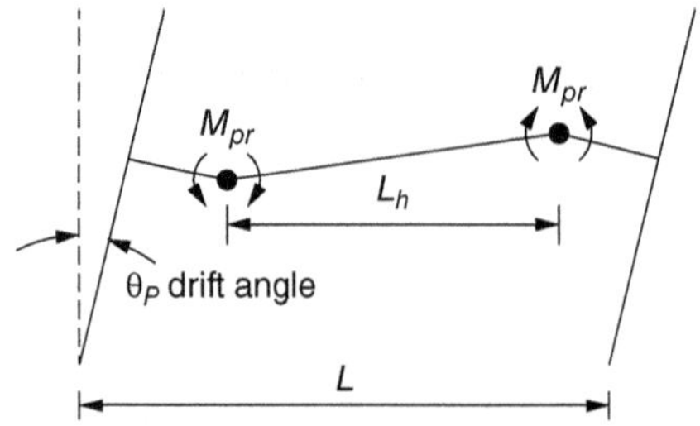


FIGURE 14.36 Plastic hinge locations.

## Design Concepts

In order to eliminate the high stress concentrations at the beam bottom flange weld, beam-to-column connections typically seek to move the plastic hinge location away from the face of the column as shown in Fig. 14.36.

The design of the connection is formulated on an expected strength basis using the probable strengths of the members. The probable beam plastic hinge moment, allowing for overstrength of the steel, the difference in yield strengths of the beam flanges and web, and the estimated strain hardening is given by AISC 358 Eq. (2.4.3-1) as

$$M_{pr} = C_{pr}R_yZ_eF_y$$

where $R_y$ = overstrength coefficient given in Table 14.3
= ratio of the expected yield stress to the minimum specified yield strength of the member
$Z_e$ = plastic section modulus of the section at the location of the plastic hinge
$F_y$ = minimum specified yield stress of the member
$C_{pr}$ = peak connection strength coefficient to account for the peak connection strength, including strain hardening, local restraint, additional reinforcement, and other connection conditions. For most connection types, $C_{pr}$ is defined by AISC 358 Eq. (2.4.3-2) as
$C_{pr} = (F_y + F_u)/2F_y \leq 1.2$
= 1.15 ... for ASTM A992 steel with $F_y$ = 50 ksi and $F_u$ = 65 ksi
= 1.4 .... for unreinforced flange-welded web connections, from AISC 358 Sec. 8.7
$F_u$ = minimum specified tensile strength of the beam

The connection design must ensure that the ratio of flexural demand to plastic hinge capacity is a maximum within the beam span. This is obtained either by reinforcing the flanges at the ends of the beam or by reducing the beam section some distance from the face of the column by trimming the beam flanges. Two exceptions to these practices are the welded unreinforced flange-welded web connection and the slotted web connection.[18] These two connections eliminate the seismic shear in the beam flanges and reduce the high stress and strain concentrations and gradients in the beam flange welds that contributed to the failure of the pre-Northridge connection.

When a sufficient number of plastic hinges develops in a frame, a mechanism is formed and the frame can deform laterally, as shown in Fig. 14.1, resulting in the dissipation of a significant amount of energy. The formation of hinges in columns, as shown

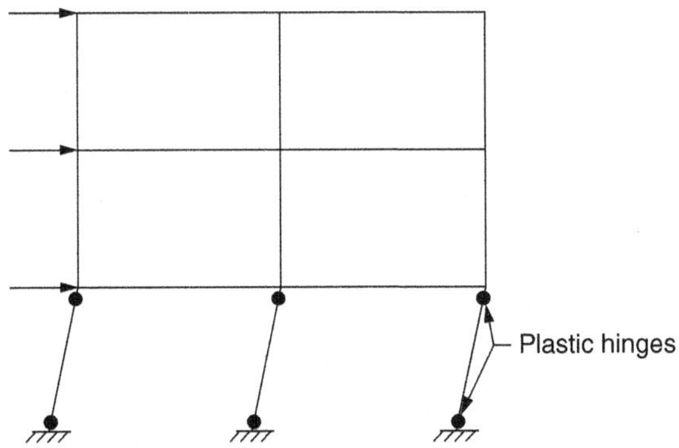


FIGURE 14.37 Soft story mechanism.

in Fig. 14.37, is undesirable as this may result in the formation of a soft story mechanism with few columns participating, little energy dissipation, and potential collapse.

## Beam Requirements

To prevent local flange buckling in a highly ductile member, AISC 341 Table D1.1 specifies the use of sections with a maximum flange width-to-thickness ratio of

$$b_f/2t_f = 0.30(E/F_y)^{0.5}$$

where $b_f$ is the flange width, and $t_f$ is the flange thickness.

In accordance with AISC 358 Sec. 5.3.1(6), this ratio may be determined, in reduced beam section connections, at the ends of the center two-thirds of the reduced section of the beam, unless gravity loading moves the hinge point significantly from the center of the reduced section. If the width-to-thickness ratio of the full flange is satisfactory, the width-to-thickness ratio of the reduced flange must also be satisfactory.

To provide adequate web stability for a highly ductile member with $C_a \approx 0$, AISC 341 Table D1.1 gives a maximum height-to-thickness ratio of

$$h/t_w = 2.45(E/F_y)^{0.5}$$

where $h$ = clear distance between flanges less the corner radius
$t_w$ = web thickness
$C_a$ = ratio of required axial strength to available axial strength

Lateral bracing is necessary, as specified in AISC 341 Sec. D1.2, to the top and bottom flanges of the beam to prevent instability. Bracing is required adjacent to expected plastic hinge locations and at a maximum spacing of

$$L_b = 0.086r_y E/F_y$$

When the beam supports a concrete slab along its whole length, lateral bracing is not required to the top flange.

The required strength of the lateral bracing is given by AISC 360 Eq. (A-6-7) as

$$P_{rb} = 0.02M_r C_d/h_o$$

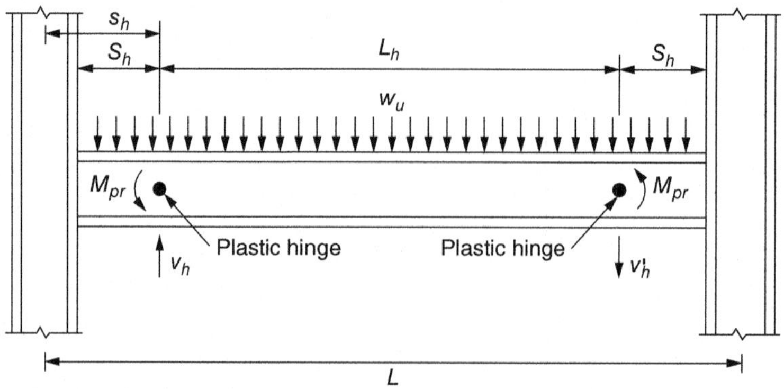


**FIGURE 14.38** Shear at plastic hinges.

where $M_r = R_y F_y Z$ ... for LRFD
$= R_y F_y Z/1.5$ ... for ASD
$C_d$ = bracing coefficient
= 1.0 ... for single curvature
$h_o$ = distance between flange centroids

In order to design the beam and beam-to-column connection, it is necessary to determine the shear and flexural strength demands at each critical section. These demands are calculated by reference to Figs. 14.38 and 14.39 for a beam with a uniformly distributed factored dead load plus live load of $w_u$ lb/ft.

From Fig. 14.38 and AISC 358 Eq. (C-5.8-1), the shear force at the plastic hinge is

$$V_h = 2M_{pr}/L_h + w_u L_h/2$$
$$V_h' = -2M_{pr}/L_h + w_u L_h/2$$

where $w_u$ = factored gravity load on the beam, from ASCE 7 Sec. 12.4.2.3
$= 1.2D + 0.5L + 0.2S$ ... for LRFD with $L \leq 100$ lb/ft²
$= 1.0D + 0.75L + 0.75S$ ... for ASD
$L_h$ = length between plastic hinges

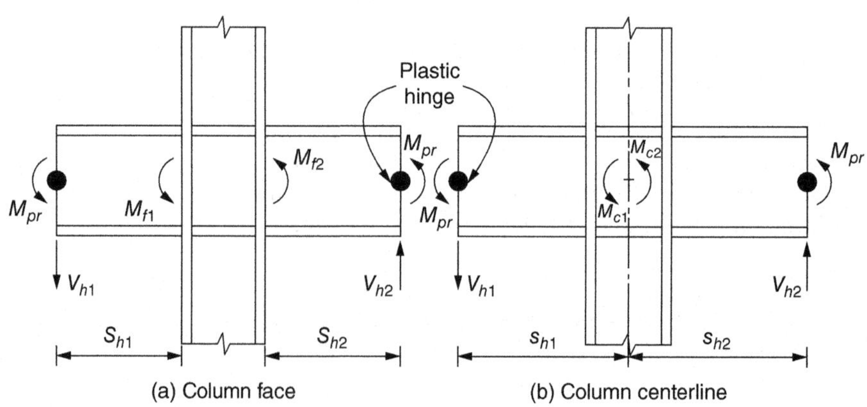


**FIGURE 14.39** Moment demand at critical sections.

Neglecting the gravity load on the length, $S_h$, the resulting bending moment at the face of the column is given by Fig. 14.39*a* and AISC 358 Eq. (5.8-6) as

$$M_f = M_{pr} + V_h S_h$$

where $S_h$ is the distance from face of column to centroid of plastic hinge.

For the column indicated, the resulting bending moment at the center of the column is given by Fig. 14.39*b* as

$$M_{pb}^* = \Sigma(M_{pr} + V_h s_h)$$

where $s_h$ is the distance from center of column to centroid of plastic hinge.

**Example 14.14.** Special Moment Frame Beam-to-Column Connection
Determine if the reduced beam section moment connection shown in Fig. 14.40 is adequate. The span between column centers is $L = 26$ ft and story height is 15 ft. The LRFD factored gravity load on the beam is $w_u = 2$ kips/ft, and the axial load on the column due to the LRFD load combination including the overstrength factor, $\Omega_0$, is $P_{uc} = 240$ kips. The beam supports a concrete slab that is connected with welded shear connectors spaced at a maximum of 12 in between protected zones. All material has a yield stress of 50 ksi and a tensile strength of 65 ksi. The frame satisfies story drift, stability, and strength requirements under the governing load combination from ASCE 7 Sec. 12.4.2.3.

The relevant section properties of the W14 × 145 column are

$$Z_c = 260 \text{ in}^3 \qquad F_{yc} = 50 \text{ ksi} \qquad F_{uc} = 65 \text{ ksi} \qquad d_c = 14.8 \text{ in}$$
$$t_{cf} = 1.09 \text{ in} \qquad b_{cf} = 15.5 \text{ in} \qquad t_{cw} = 0.680 \text{ in} \qquad h_c/t_{cw} = 16.8$$
$$b_{cf}/2t_{cf} = 7.11 \qquad k_1 = 1.56 \text{ in} \qquad k_{des} = 1.69 \text{ in} \qquad k_{det} = 2.38 \text{ in}$$
$$A_g = 42.7 \text{ in}^2$$

The relevant section properties of the W21 × 93 beam are

$$Z_b = 221 \text{ in}^3 \qquad F_y = 50 \text{ ksi} \qquad d_b = 21.6 \text{ in} \qquad t_{bf} = 0.93 \text{ in}$$
$$b_{bf} = 8.42 \text{ in} \qquad t_{bw} = 0.58 \text{ in} \qquad h_b/t_{bw} = 32.3 \qquad b_{bf}/2t_{bf} = 4.53$$
$$h_{bo} = 20.7 \text{ in}$$

In accordance with AISC 358 Sec. 1.3, prequalified connections must be designed by the LRFD procedure only.

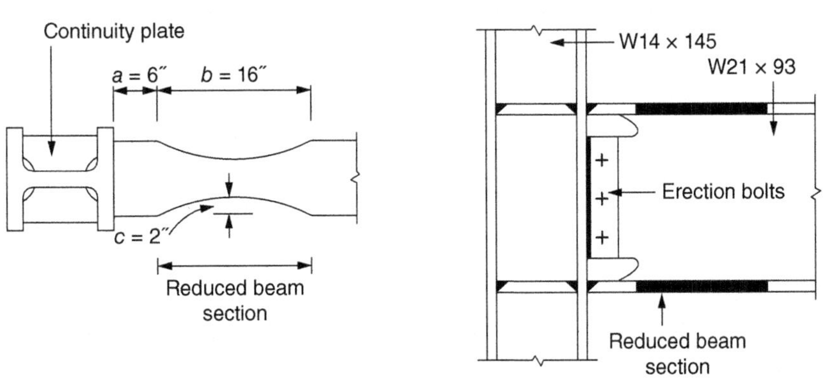


**FIGURE 14.40** Details for Example 14.14.

AISC 358 Sec. 5.8 restricts the dimensions of a reduced section to

$$a = (0.5 \text{ to } 0.75)b_{bf}$$
$$b = (0.65 \text{ to } 0.85)d_b$$
$$c = (0.1 \text{ to } 0.25)b_{bf}$$
$$r = (4c^2 + b^2)/8c$$

The dimensions provided are

$$a = 6 \text{ in}$$
$$= 0.71b_{bf} \ldots \text{ satisfactory}$$
$$b = 16 \text{ in}$$
$$= 0.74d_b \ldots \text{ satisfactory}$$
$$c = 2 \text{ in}$$
$$= 0.24b_{bf} \ldots \text{ satisfactory}$$
$$r = (4 \times 2^2 + 16^2)/(8 \times 3)$$
$$= 11.3 \text{ in}$$

The plastic section modulus of the beam at the center of the reduced section is given by AISC 358 Eq. (5.8-4) as

$$Z_{RBS} = Z_b - 2ct_{bf}(d_b - t_{bf})$$
$$= 221 - 2 \times 2 \times 0.93(21.6 - 0.93)$$
$$= 144 \text{ in}^3$$

The distance of the center of the reduced beam section from the face of the column is obtained from Fig. 14.40 as

$$S_h = a + b/2$$
$$= 6 + 16/2$$
$$= 14 \text{ in}$$

The distance between the centers of the reduced beam sections is obtained from Fig. 14.38 as

$$L_h = L - 2S_h - d_c$$
$$= 26 \times 12 - 2 \times 14 - 14.8$$
$$= 269.2 \text{ in}$$

The probable beam plastic moment is given by AISC 358 Eq. (2.4.3-1) as

$$M_{pr} = C_{pr}R_yZ_{RBS}F_y$$
$$= 1.15 \times 1.1 \times 144 \times 50$$
$$= 9108 \text{ kip-in}$$

The shear force at the center of the reduced beam section is given by AISC 358 Eq. (5.8-9) as

$$V_h = 2M_{pr}/L_h + w_uL_h/2$$
$$= 2 \times 9108/269.2 + (2/12)269.2/2$$
$$= 90 \text{ kips}$$

The design shear stress of the beam is obtained from AISC Manual Table 3-6 as

$$\varphi V_n = 376 \text{ kips}$$
$$> V_h \ldots \text{ satisfactory}$$

Neglecting the gravity load on the length $S_h$, the probable maximum moment at the face of the column is given by AISC 358 Eq. (5.8-6) and Fig. 14.39 as

$$\begin{aligned} M_f &= M_{pr} + V_h S_h \\ &= 9108 + 90 \times 14 \\ &= 10{,}368 \text{ kip-in} \end{aligned}$$

The plastic moment of the beam based on the expected yield stress $R_y F_y$ is obtained from SDM Table 4-2 as

$$\begin{aligned} M_{pe} &= 1010 \times 12 \\ &= 12{,}120 \text{ kip-in} \end{aligned}$$

The ductile resistance factor is given by AISC 358 Sec. 2.4.1 as

$$\varphi_d = 1.0$$

and

$$\begin{aligned} \varphi_d M_{pe} &= 1.0 \times 12{,}120 \\ &= 12{,}120 \text{ kip-in} \\ &> M_f \ldots \text{ satisfies AISC 358 Eq. (5.8-8)} \end{aligned}$$

The beam satisfies all seismic width-to-thickness requirements as indicated in SDM Table 1-3. The clear span-to-depth ratio is

$$\begin{aligned} L_{clear}/d_b &= (L - d_c)/d_b \\ &= (26 \times 12 - 14.8)/21.6 \\ &= 13.8 \\ &> 7 \ldots \text{ satisfies AISC 358 Sec. 5.3.1} \end{aligned}$$

The beam depth is

$$\begin{aligned} d_b &= 21.6 \text{ in} \\ &< \text{W36} \ldots \text{ satisfies AISC 358 Sec. 5.3.1} \end{aligned}$$

The beam weight is

$$\begin{aligned} w &= 93 \text{ lb/ft} \\ &< 300 \text{ lb/ft} \ldots \text{ satisfies AISC 358 Sec. 5.3.1} \end{aligned}$$

The flange thickness is

$$\begin{aligned} t_{bf} &= 0.93 \text{ in} \\ &< 1.75 \text{ in} \ldots \text{ satisfies AISC 358 Sec. 5.3.1} \end{aligned}$$

Because of the concrete slab, supplemental lateral bracing is not required at the reduced section. As indicated in SDM Table 4-2, lateral bracing is required to the bottom flange at a maximum spacing of 7.65 ft.

In accordance with AISC 358 Sec. 5.5, the beam flanges are connected to the column flanges with complete-joint-penetration groove welds that are designated demand critical welds. As required by AISC 358 Sec. 3.3.2, the backing bar to the bottom flange weld must be removed, backgouged to sound weld metal, and backwelded with a 5/16-in reinforcing fillet. The backing bar to the top flange weld may remain in place and is attached to the column flange with a continuous 5/16-in fillet weld as shown in Fig. 14.41.

The dimensions of the weld access holes are specified in AISC 360 Sec. J1.6. The required minimum length is 1½ in and the minimum depth ¾ in. A 3/8-in single plate shear tab is required by AISC 358 Sec. 5.6 to extend between the access holes to support erection loads. The shear tab is used as backing

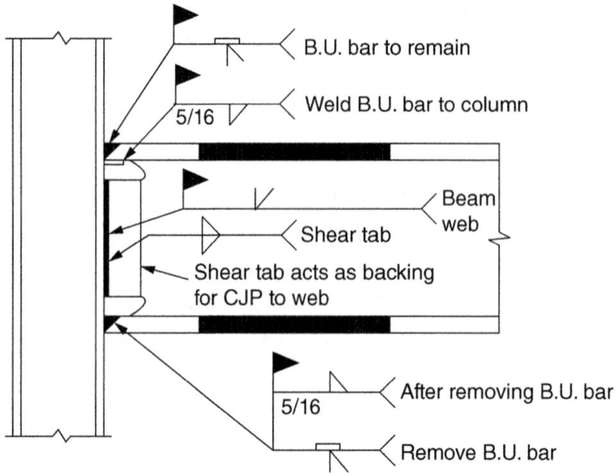


**FIGURE 14.41** Connection details.

for the complete-joint-penetration groove welds connecting the beam web to the column flange. The web depth between access holes is

$$\begin{aligned} l_w &= d_b - (2 \times t_{bf}) - (2 \times ¾ \text{ in}) \\ &= 21.6 - (2 \times 0.93) - (2 \times ¾ \text{ in}) \\ &= 18.24 \text{ in} \end{aligned}$$

For the W21 × 93 beam, the nominal strength of the beam web to column flange connection is given by AISC 360 Sec. G2.1 as

$$\begin{aligned} V_n &= 0.6F_y t_{bw} l_w C_v \\ &= 0.6 \times 50 \times 0.58 \times 18.24 \times 1.0 \\ &= 317 \text{ kips} \end{aligned}$$

The design strength of the beam web to column flange connection is

$$\begin{aligned} \varphi V_n &= 1.00 \times V_n \\ &= 317 \text{ kips} \\ &> V_h \ldots \text{ satisfactory} \end{aligned}$$

## Strong-Column Weak-Beam Concept

A strong-column weak-beam design requirement is adopted to ensure that inelastic drift is uniformly distributed over the height of the building. Plastic hinges are formed at each end of the beams, as shown in Fig. 14.1, and the dissipation of a significant amount of energy occurs before a collapse mechanism is produced. The formation of hinges in columns, as shown in Fig. 14.37, is undesirable as this may result in the formation of a soft story mechanism with few columns participating, little energy dissipation, and potential collapse due to *P*-delta effects.

AISC 341 Section E3.4a(a) relaxes the strong column-weak beam requirement for columns when the required axial strength of the column, $P_{rc}$, is less than 30 percent of the nominal axial strength of the column, $P_c$, and the column is either (1) located in a one-story building or in the top story of a multistory building where the formation of a soft story is not significant, or (2) located in a column line in which the available shear strength of all exempted columns is less than 33 percent of the total available shear strength of the column line, and the available shear strength of all exempted columns in the story is less than 20 percent of the total available shear strength of the story. This provides reasonable flexibility for large transfer girders.

The nominal axial strength of the column is given by AISC 341 Sec. E3.4a as

$$P_c = F_y A_g \ldots \text{ for LRFD}$$
$$= F_y A_g / 1.5 \ldots \text{ for ASD}$$

The required axial strength of the column is given by AISC 341 Sec. E3.4a as

$$P_{rc} = P_{uc} \ldots \text{ for LRFD}$$
$$P_{rc} = P_{ac} \ldots \text{ for ASD}$$

AISC 341 Sec. E3.4a(b) also provides an exemption for a column located in a story with a ratio of available shear strength to required shear strength 50 percent greater than that of the story above because column yielding at the stronger level would be unlikely to occur.

The strong column-weak beam design requirement may be achieved in accordance with AISC 341 Eq. (E3-1) by ensuring that

$$\Sigma M^*_{pc} / \Sigma M^*_{pb} > 1.0$$

where $\Sigma M^*_{pc}$ = the sum of the projections of the nominal flexural strengths of the columns above and below the joint to the beam centerline with a reduction for the axial force in the column as given by

$\Sigma M^*_{pc} = \Sigma Z_c(F_{yc} - P_{uc}/A_g) \ldots$ AISC 341 Eq. (E3-2a) LRFD

$= \Sigma Z_c(F_{yc} - 1.5P_{ac}/A_g) \ldots$ AISC 341 Eq. (E3-2b) ASD

$\Sigma M^*_{pb}$ = the sum of the projections of the expected flexural strengths of the beams at the plastic hinge locations to the column centerline

$= \Sigma(M_{pr} + V_h s_h)$

**Example 14.15.** Special Moment Frame Strong-Column Weak-Beam Concept

For the special moment frame analyzed in Example 14.14, determine if the strong-column weak-beam concept is satisfied.

The distance of the center of the reduced beam section from the center of the column is obtained from Fig. 14.40 as

$$s_h = a + b/2 + d_c/2$$
$$= 6 + 16/2 + 14.8/2$$
$$= 21.4 \text{ in}$$

The bending moment at the center of the column is

$$M^*_{pb} = M_{pr} + V_h s_h$$
$$= 9108 + 90 \times 21.4$$
$$= 11{,}034 \text{ kip-in}$$

The sum of the nominal flexural strengths of the column above and below the joint at the beam centerline, with a reduction for the factored axial force in the column, is given by AISC 341 Eq. (E3-2a) as

$$\Sigma M^*_{pc} = \Sigma Z_c(F_{yc} - P_{uc}/A_g)$$
$$= 2 \times 260(50 - 240/42.7)$$
$$= 23{,}077 \text{ kip-in}$$

The ratio of column moments to beam moment is

$$\begin{aligned}\Sigma M_{pc}^*/\Sigma M_{pb}^* &= 23{,}077/11{,}034\\ &= 2.1\\ &> 1.0 \ldots \text{ satisfactory}\end{aligned}$$

## Web Panel Zone Design

As shown in Fig. 14.42 the column panel zone[19] is the web area bounded by the top and bottom girder flanges. Panel zone shear results from the combined effects of the beam flange force and the story shear.

The beam flange force resulting from the probable maximum moments at the face of the column is

$$\Sigma(M_f/h_o) = \Sigma[(M_{pr} + V_h S_h)/h_o]$$

where $M_f$ = the probable maximum moment at the face of the column
$= M_{pr} + V_h S_h$
$M_{pr}$ = probable beam plastic moment
$V_h$ = shear force at the center of the reduced beam section
$S_h$ = distance of the plastic hinge centroid from the face of the column
$h_{bo}$ = distance between beam flange centroids

It is assumed that a point of inflection occurs at the midheight of each column in the frame as shown in Fig. 14.42. Then, the shear produced in the story is

$$V_{story} = \Sigma M_f/H$$

where $H$ is the distance between points of inflection.

The net shear acting on the panel zone is

$$\begin{aligned}V_{pz} &= \Sigma(M_f/h_{bo}) - V_{story}\\ &= \Sigma(M_f/h_{bo}) - \Sigma M_f/H\\ &= \text{required shear strength of panel zone}\end{aligned}$$

When adequate ductility is provided to accommodate deformations in the inelastic range, the post-yield panel zone strength may be utilized. Hence, when the effect of

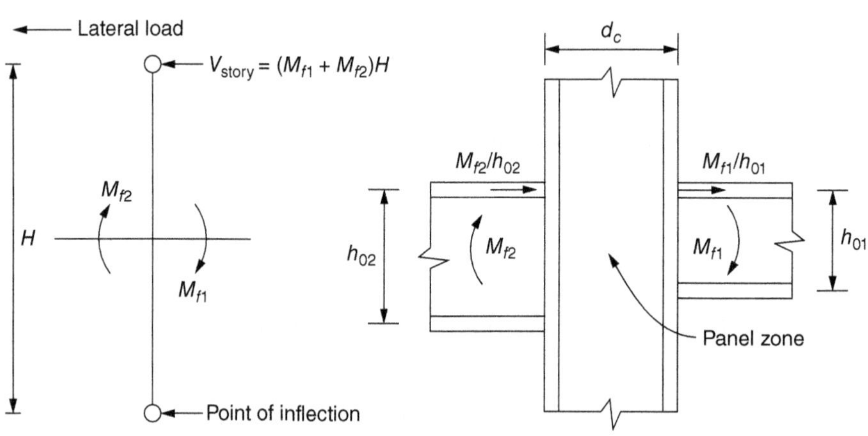


FIGURE 14.42 Panel zone shear.

panel zone deformation on frame stability is considered, the nominal shear strength of the panel zone is given by AISC 360 Eqs. (J10-11) and (J10-12) as

$$R_n = 0.60F_y d_c t_{cw}[1 + 3b_{cf}t_{cf}^2/(d_b d_c t_{cw})] \ldots \text{for } P_r \leq 0.75P_c$$

$$R_n = 0.60F_y d_c t_{cw}[1 + 3b_{cf}t_{cf}^2/(d_b d_c t_{cw})]\,[1.9 - (1.2P_r)/P_c] \ldots \text{for } P_r > 0.75P_c$$

where $t_{cw}$ = column web thickness
$b_{cf}$ = width of column flange
$t_{cf}$ = thickness of column flange
$d_c$ = column depth
$d_b$ = girder depth
$F_y$ = yield stress of the column web
$P_c = P_y$ for LRFD load combinations
$= 0.6P_y$ for ASD load combinations
$P_r$ = required axial strength using LRFD or ASD load combinations
$P_y$ = axial yield strength of the column
$= F_y A_g$
$A_g$ = column cross-sectional area

The available shear strength of the panel zone is obtained from AISC 341 Sec. E3.6e(1) as

$$\varphi R_n = \text{design shear strength for LRFD}$$

where φ is the resistance factor = 1.00,

or $$R_n/\Omega = \text{allowable shear strength for ASD}$$

where Ω is the safety factor = 1.50.

Equation (J10-11) may be rewritten as

$$\varphi R_n = \varphi R_{v1} + \varphi R_{v2}/d_b \ \ldots \text{for LRFD}$$

or $$R_n/\Omega = R_{v1}/\Omega + R_{v2}/\Omega d_b \ \ldots \text{for ASD}$$

Values of $\varphi R_{v1}$, $\varphi R_{v2}$, $R_{v1}/\Omega$, and $R_{v2}/\Omega$ are tabulated in SDM Table 4-2.

## Doubler Plates

When the available web shear strength is inadequate, a web doubler plate, as shown in Fig. 14.43, may be welded to the column web to carry the additional force. Any required doubler plate must extend over the full area of the panel zone and provide the additional thickness necessary to resist the difference between the required strength and the available strength. To prevent shear buckling under cyclic loading, the web thickness and the doubler thickness, without the aid of intermediate plug welds, must each not be less than the value given by AISC 341 Eq. (E3-7) as

$$t_{min} = (d_z + w_z)/90$$

where $d_z = d_b - 2t_{bf}$ of deeper beam at connection, and $w_z$ is panel zone width between column flanges.

Alternatively, when a minimum of four plug welds are used to connect the doubler and the column web, the total panel zone thickness must satisfy AISC 341 Eq. (E3-7). When the doubler plate is placed against the column web, it must be

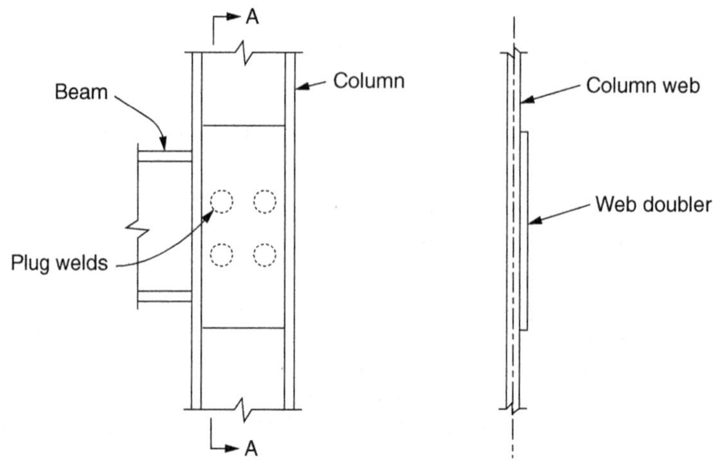


**FIGURE 14.43** Web doubler plates.

welded to the column flanges to develop the available strength of the full doubler plate thickness. The doubler plate must be either complete-joint-penetration groove welded or fillet welded to the column flanges. When continuity plates are not used, the doubler plate must be fillet welded top and bottom to develop the proportion of the total force that is transmitted to the doubler plate. For this situation, the doubler plates must extend a minimum of 6 in above and below the top and bottom of the deeper beam. When continuity plates are used, the doubler plate must be welded to the continuity plates to develop the proportion of the total force that is transmitted to the doubler plate. When the doubler plates are placed away from the column web, they must be placed symmetrically in pairs and complete-joint-penetration groove welded to continuity plates to develop the pro rata share of the total force transmitted to the doubler plate.

**Example 14.16.** Doubler Plate

For the special moment frame analyzed in Example 14.14, determine if doubler plates are required.

The minimum column web thickness to prevent shear buckling is given by AISC 341 Eq. (E3-7) as

$$
\begin{aligned}
t_{min} &= (d_z + w_z)/90 \\
&= (d_b - 2t_{bf} + d_c - 2t_{cf})/90 \\
&= (21.6 - 2 \times 0.93 + 14.8 - 2 \times 1.09)/90 \\
&= 0.360 \text{ in}
\end{aligned}
$$

The actual column web thickness is

$$
\begin{aligned}
t_{cw} &= 0.680 \text{ in} \\
&> t_{min} \ldots \text{ satisfactory}
\end{aligned}
$$

From AISC 341 Sec. E3.6e(1), the required shear strength of the panel zone is

$$
\begin{aligned}
V_{pz} &= \Sigma(M_f/h_{bo}) - \Sigma M_f/H \\
&= 10{,}368/20.7 - 10{,}368/(15 \times 12) \\
&= 443 \text{ kips}
\end{aligned}
$$

Frame stability and panel zone deformation is considered in the analysis and the available shear strength of the panel zone depends on the ratio of required axial strength, $P_{uc}$, to available axial strength, $P_c$, of the column.

$$\begin{aligned} 0.75P_c &= 0.75P \\ &= 0.75F_y A_g \\ &= 0.75 \times 50 \times 42.7 \\ &= 1601 \text{ kips} \\ > P_{uc} &= 240 \text{ kips} \ldots \text{AISC 360 Eq. (J10-11) applies} \end{aligned}$$

The available shear strength of the column panel zone is given by AISC 360 Eq. (J10-11) and SDM Table 4-2 as

$$\begin{aligned} \varphi R_n &= \varphi R_{v1} + \varphi R_{v2}/d_b \\ &= 302 + 1660/21.6 \\ &= 379 \text{ kips} \\ < V_{pz} &= 443 \text{ kips} \ldots \text{unsatisfactory} \end{aligned}$$

A doubler plate is required to provide a combined thickness of web plus doubler plate of

$$\begin{aligned} t_{cw} + t_p &= [V_{pz} - 0.60F_y(3b_{cf}t_{cf}^2/d_b)][1/(0.60F_y d_c)] \\ &= [443 - 0.60 \times 50(3 \times 15.5 \times 1.09^2/21.6)][1/(0.60 \times 50 \times 14.8)] \\ &= 0.825 \text{ in} \end{aligned}$$

The required doubler plate thickness is

$$\begin{aligned} t_p &= t_{cw} + t_p - t_{cw} \\ &= 0.825 - 0.680 \\ &= 0.145 \text{ in} \\ < t_{min} &= 0.360 \text{ in} \ldots \text{plug weld doubler plate to the column web} \end{aligned}$$

Use a 3/16-in plate connected with complete-joint-penetration groove welds to the column flanges. The proportion of the panel shear transmitted to the doubler plate is

$$\begin{aligned} V_p &= V_{pz}t_p/(t_{cw} + t_p) \\ &= (443 \times 0.145)/0.825 \\ &= 78 \text{ kips} \end{aligned}$$

Continuity plates are provided at the top and bottom of the panel zone and the doubler plate is connected to the continuity plates with fillet welds to resist $V_p$. The total length of fillet weld available at the top and bottom of the doubler plate is

$$\begin{aligned} l_w &= 2(d_c - 2t_{cf}) \\ &= 2(14.8 - 2 \times 1.09) \\ &= 25.24 \text{ in} \end{aligned}$$

The design fillet weld strength per 1/16 in of E70XX electrodes is given by AISC 360 Table J2.5 as

$$\begin{aligned} q_u &= 0.6\varphi F_{EXX}A_w \\ &= 0.6 \times 0.75 \times 70 \times 0.707 \times 1/16 \\ &= 1.39 \text{ kips/in} \end{aligned}$$

The required weld size per 1/16 in is

$$D = V_p / l_w q_u$$
$$= 78/(25.24 \times 1.39)$$
$$= 2.2 \text{ sixteenths}$$

Hence, the required weld size is

$$w = 2.2/16$$

Provide $w = 3/16$ in

$> 1/8$ in … satisfies AISC 360 Table J2.4

## Continuity Plates

Continuity plates, as shown in Fig. 14.44, are required in accordance with AISC 341 Sec. E3.6f except when the column flange thickness satisfies both of the following equations:

$$t_{cf} \geq 0.4(1.8 b_{bf} t_{bf} F_{yb} R_{yb} / F_{yc} R_{yc})^{0.5} \text{ … AISC 341 Eq. (E3-8)}$$

and

$$t_{cf} \geq b_{bf}/6 \text{ … AISC 341 Eq. (E3-9)}$$

where $t_{cf}$ = minimum required thickness of column flange when no continuity plates are provided
$b_{bf}$ = beam flange width
$t_{bf}$ = beam flange thickness
$F_{yb}$ = minimum specified yield stress of the beam flange
$F_{yc}$ = minimum specified yield stress of the column flange
$R_{yb}$ = ratio of the expected yield strength of the beam material to the minimum specified yield strength
$R_{yc}$ = ratio of the expected yield strength of the column material to the minimum specified yield strength

The minimum continuity plate thickness specified by AISC 341 Sec. E3.6f(2) is

$$t_{st} = t_{bf} \text{ … for two-sided connections}$$

and

$$t_{st} = t_{bf}/2 \text{ … for one-sided connections}$$

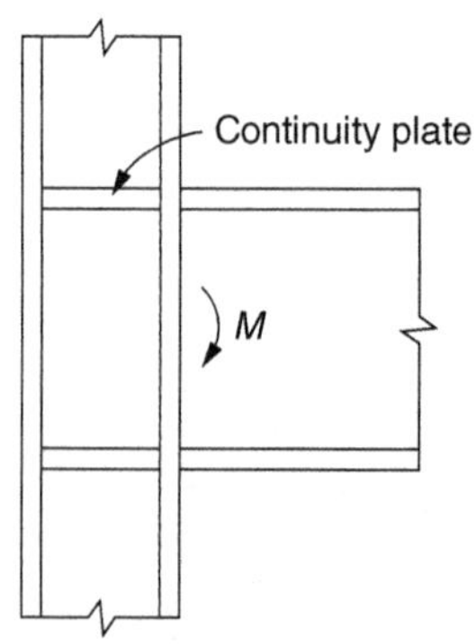


**Figure 14.44** Continuity plates.

The minimum width of a continuity plate is typically selected to match the beam flange with a maximum width not to exceed the column flange.

Continuity plates are welded to the column flange using complete-joint-penetration groove welds. Continuity plates are clipped as detailed in AISC 358 Sec. 3.6 to avoid the column k-area. The plates are welded to column webs using complete-joint-penetration groove welds or fillet welds. The required strength of the sum of the welded joints of the continuity plate to the column web is the smallest of the following:

1. The sum of the design strengths in tension of the contact areas of the continuity plates to the column flanges, which is

$$R_u = 2\varphi F_{yst} t_{st}(b_{st} - k_1 - 0.5 \text{ in})$$

2. The design strength in shear of the contact area of the continuity plates with the column web, which is

$$R_u = 2 \times 0.60\varphi F_{yst} t_{st} L_{net}$$

where $L_{net}$ = net length of continuity plate

$= d_c - 2(k_{det} + 1.5 \text{ in})$ … from AISC 358 Sec. 3.6

$k_{det}$ = distance from outer face of column flange to web toe of fillet

3. The design strength in shear of the column panel zone, which is given by AISC 360 Eq. (J10-11) as

$$R_u = 0.60\varphi F_y d_c t_p(1 + 3b_{cf} t_{cf}^2/d_b d_c t_p) \ \ldots \text{ for } P_r \le 0.75P_c$$

where $\varphi = 1.0$ ... from AISC 341 Sec. E3.6c

$t_p = t_{cw} + d_{dbl}$

$d_{dbl}$ = doubler plate thickness

4. The sum of the expected yield strengths of the beam flanges transmitting force to the continuity plates, which is

$$R_u = M_{pr} / h_{bo}$$

where $h_{bo}$ = distance between beam flange centroids

**Example 14.17.** Continuity Plate

For the special moment frame analyzed in Example 14.14, design the required continuity plates.

Applying the requirements of AISC 341 Sec. E3.6f

$$\begin{aligned} b_{bf}/6 &= 8.42/6 \\ &= 1.40 \text{ in} \\ &> t_{cf} \ldots \text{ continuity plates are required} \end{aligned}$$

The minimum continuity plate thickness specified by AISC 341 Sec. E3.6f(2) is

$$\begin{aligned} t_{st} &= t_{bf}/2 \ \ldots \text{ for one-sided connections} \\ &= 0.93/2 \end{aligned}$$

Use $t_{st} = 0.5$ in

The minimum continuity plate width is typically selected to match the beam flange which gives

$$\begin{aligned} b_{st} &= (b_{bf} - t_{cw})/2 \\ &= (8.42 - 0.68)/2 \\ &= 3.87 \text{ in} \end{aligned}$$

To provide adequate contact with the column flange use

$$b_{st} = 6.5 \text{ in}$$

As specified in AISC 358 Sec. 3.6f(3), the continuity plates are welded to the column flange using complete-joint-penetration groove welds.

The length of the continuity plate in contact with the column flange is given by AISC 358 Sec. 3.6 as

$$\begin{aligned} L_{flange} &= b_{st} - k_1 - 0.5 \text{ in} \\ &= 6.5 - 1.56 - 0.5 \text{ in} \\ &= 4.44 \text{ in} \end{aligned}$$

The length of the continuity plate in contact with the column web is given by AISC 358 Sec. 3.6 as

$$\begin{aligned} L_{web} &= d_c - 2(k_{det} + 1.5 \text{ in}) \\ &= 14.8 - 2(2.38 + 1.5 \text{ in}) \\ &= 7.04 \text{ in} \end{aligned}$$

As specified in AISC 358 Sec. 3.6f(3), the required strength of the sum of the welded joints of the continuity plates to the column web is the smallest of the following:

1. The sum of the design strengths in tension of the contact areas of both continuity plates to the column flanges, which is

$$\begin{aligned} R_u &= 2 \times 2\varphi F_{yst} t_{st} L_{flange} \\ &= 2 \times 2 \times 0.9 \times 50 \times 0.5 \times 4.44 \\ &= 400 \text{ kips} \end{aligned}$$

2. The design strength in shear of the contact area of both continuity plates with the column web, which is

$$\begin{aligned} R_u &= 2 \times 0.60\ \varphi F_{yst} t_{st} L_{web} \\ &= 2 \times 0.6 \times 0.9 \times 50 \times 0.5 \times 7.04 \\ &= 190 \text{ kips} \ldots \text{ governs} \end{aligned}$$

3. The design strength in shear of the column panel zone, which is

$$R_u = 443 \text{ kips} \ldots \text{ from Example 14.16 for the combined thickness } t_{cw} + t_p$$

4. The sum of the expected yield strengths of the beam flanges transmitting force to the continuity plates, which is

$$\begin{aligned} R_u &= M_{pr}/h_{bo} \\ &= 9108/20.7 \\ &= 440 \text{ kips} \end{aligned}$$

Connect the continuity plate and the column web with fillet welds. The total length of fillet weld available using fillet welds on both sides of both continuity plates is

$$\begin{aligned} l_w &= 2 \times 2L_{web} \\ &= 2 \times 2 \times 7.04 \\ &= 28.16 \text{ in} \end{aligned}$$

The design fillet weld strength per 1/16 in of E70XX electrodes is given by AISC 360 Table J2.5 as

$$\begin{aligned} q_u &= 0.6\varphi F_{EXX} A_w \\ &= 0.6 \times 0.75 \times 70 \times 0.707 \times 1/16 \\ &= 1.39 \text{ kips/in} \end{aligned}$$

The required weld size per 1/16 in is

$$\begin{aligned} D &= R_u / l_w q_u \\ &= 190/(28.16 \times 1.39) \\ &= 4.9 \text{ sixteenths} \end{aligned}$$

Hence, the required weld size is

$$\begin{aligned} w &= 4.9/16 \\ &= 5/16 \text{ in} \ldots \text{ to the nearest } 1/16 \text{ in} \end{aligned}$$

The thickness of the column web is ½ in $< t_{cw} <$ ¾ in and the minimum allowable fillet weld size connecting the continuity plates to the web is given by AISC 360 Table J2.4 as

$$\begin{aligned} w_{min} &= 1/4 \text{ in} \\ &< w \ldots \text{ satisfactory} \end{aligned}$$

The minimum thickness of web required to match the shear rupture strength of the welds on opposite sides of the web is given by AISC Manual Part 9 Eq. (9-3) as

$$\begin{aligned} t_{cw,min} &= 6.19D/F_{uc} \\ &= 6.19 \times 4.9/65 \\ &= 0.47 \text{ in} \\ &< t_{cw} = 0.68 \text{ in} \ldots \text{ satisfactory} \end{aligned}$$

Panel zone details are shown in Fig. 14.45.

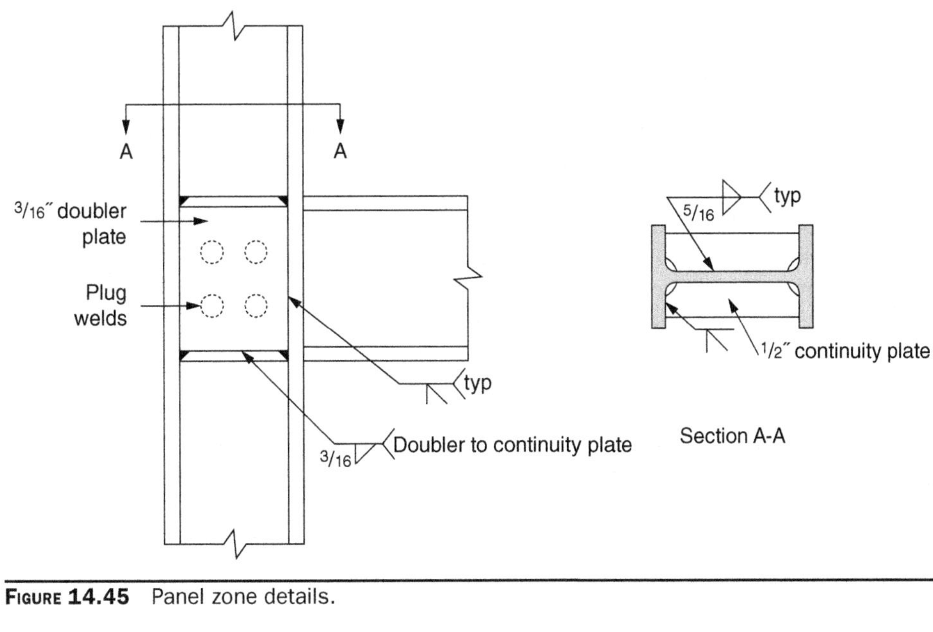


**FIGURE 14.45** Panel zone details.

## References

1. International Code Council (ICC). 2015. *International Building Code*, 2015 edition. ICC, Falls Church, VA.
2. American Institute of Steel Construction (AISC). 2010. *Seismic Provisions for Structural Steel Buildings* (AISC 341-10), AISC, Chicago, IL.
3. American Society of Civil Engineers (ASCE). 2010. *Minimum Design Loads for Buildings and Other Structures* (ASCE 7-10), ASCE, Reston, VA.
4. American Institute of Steel Construction (AISC). 2010. *Specification for Structural Steel Buildings* (AISC 360-10), AISC, Chicago, IL.
5. American Institute of Steel Construction (AISC). 2011. *Seismic Design Manual*. 2nd edition. AISC, Chicago, IL.
6. American Institute of Steel Construction. (AISC). 2010. *Prequalified Connections for Special and Intermediate Steel Moment Frames for Seismic Applications*, including Supplements No. 1 and No. 2. (AISC 358-10). AISC, Chicago, IL.
7. Sabol, T. A. 2013. *Seismic Design Manual and Application of the 2010 Seismic Provisions*. AISC, Chicago, IL.
8. Sabelli, R. 2011. "Mechanism Analysis in the 2010 Seismic Provisions." *NASCC Steel Conference Proceedings*. AISC. Chicago, IL.
9. American Institute of Steel Construction (AISC). 2011. *Steel Construction Manual*. 14th edition. AISC, Chicago, IL.
10. Fortney, P. J. 2015. *Analysis and Design of Chevron Brace Connections with Flat Bar Gussets*. AISC, Chicago, IL.
11. Fortney, P. J. 2015. "The Chevron Effect-Not an Isolated Problem." *Engineering Journal*, Vol. 52, No. 2, Second Quarter 2015.
12. Sabelli, R., Roeder, C. W., and Hajjar, J. F. 2013. "Seismic Design of Steel Special Concentrically Braced Frame Systems." *NEHRP Seismic Design Technical Brief No. 8*. National Institute of Standards and Technology. Gaithersburg, MD.
13. Lindsay, S. D. 2003. *Eccentric Braced Steel Frames for Wind and Low-to-Moderate Seismic Loads*. American Institute of Steel Construction. Chicago, IL.
14. Berman, J. W., and Bruneau, M. 2013. "Overview of the Development of Design Recommendations for Eccentrically Braced Frame Links with Built-Up Sections." *Engineering Journal*, Vol. 50, No. 1, First Quarter 2013.
15. Hamburger, R. O., Krawinkler, H., Malley, J. O., and Adan, S. M. 2009. "Seismic Design of Steel Special Moment Frames: A Guide for Practicing Engineers." *NEHRP Seismic Design Technical Brief No.* 2. National Institute of Standards and Technology. Gaithersburg, MD.
16. Sabol, T. A. 1994. "Damage to Ductile Steel Frames in the Northridge Earthquake." *Proceedings of the AISC Special Task Committee on the Northridge Earthquake*. American Institute of Steel Construction, Chicago, March 1994.
17. Cordova, P. P., and Hamburger, R. O. 2011. "Steel Connections: Proprietary or Public Domain." *Modern Steel Construction*. October 2011.
18. International Code Council. 2011. *Slotted Web Beam-to-Column Steel Moment Frame Connection*. ICC-ES Evaluation Report, ESR-1093. Washington, DC, 2011.
19. Adan, S. M. 2011. "Design Parameters for Steel Special Moment Frame Connections." *SEAOC* 2011 *Convention Proceedings*. SEAOC, Sacramento, CA, 2011.

## Problems

**14.1** *Given:* The single bay ordinary concentrically braced steel frame shown in Fig. 14.46 forms part of an office building assigned to seismic design category D with a redundancy factor of $\rho = 1.0$ and a design response acceleration of $S_{DS} = 1.0g$. The seismic lateral loads and the gravity loads at each level are indicated. The dead loads include the self weights of the members. Neglect second-order effects.

*Find:* Using allowable stress level design (ASD), a suitable round HSS member for the brace in the bottom story.

**14.2** *Given:* The single bay ordinary concentrically braced steel frame shown in Fig. 14.46 forms part of an office building assigned to seismic design category D with a redundancy factor of $\rho = 1.0$ and a design response acceleration of $S_{DS} = 1.0g$. The seismic lateral loads and the gravity loads at each level are indicated. The dead loads include the self weights of the members. Neglect second-order effects.

*Find:* Using load and resistance factor design (LRFD), a suitable round HSS member for the brace in the bottom story.

**14.3** *Given:* The single bay ordinary concentrically braced steel frame shown in Fig. 14.46 forms part of an office building assigned to seismic design category D with a redundancy factor of $\rho = 1.0$ and a design response acceleration of $S_{DS} = 1.0g$. The seismic lateral loads and the gravity loads at each level are indicated. The dead loads include the self weights of the members. Neglect second-order effects.

*Find:* Using allowable stress level design (ASD), the lightest weight W24 section of ASTM 992 steel suitable for the beam in the bottom story.

**14.4** *Given:* The single bay ordinary concentrically braced steel frame shown in Fig. 14.46 forms part of an office building assigned to seismic design category D with a redundancy factor of $\rho = 1.0$ and a design response acceleration of $S_{DS} = 1.0g$. The seismic lateral loads and the gravity loads at each level are indicated. The dead loads include the self weights of the members. Neglect second-order effects.

*Find:* Using load and resistance factor design (LRFD), the lightest weight W24 section of ASTM 992 steel suitable for the beam in the bottom story.

**14.5** *Given:* The single bay ordinary concentrically braced steel frame shown in Fig. 14.46 forms part of an office building assigned to seismic design category D with a redundancy factor of $\rho = 1.0$ and a design response acceleration of $S_{DS} = 1.0g$. The seismic lateral loads and the gravity loads at each level are indicated. The dead loads include the self weights of the members. Neglect second-order effects.

*Find:* Using allowable stress level design (ASD), the lightest weight W14 section of ASTM 992 steel suitable for the column in the bottom story.

**14.6** *Given:* The single bay ordinary concentrically braced steel frame shown in Fig. 14.46 forms part of an office building assigned to seismic design category D with a redundancy factor of $\rho = 1.0$ and a design response acceleration of $S_{DS} = 1.0g$. The seismic lateral loads and the gravity loads at each level are indicated. The dead loads include the self weights of the members. Neglect second-order effects.

*Find:* Using load and resistance factor design (LRFD), the lightest weight W14 section of ASTM 992 steel suitable for the column in the bottom story.

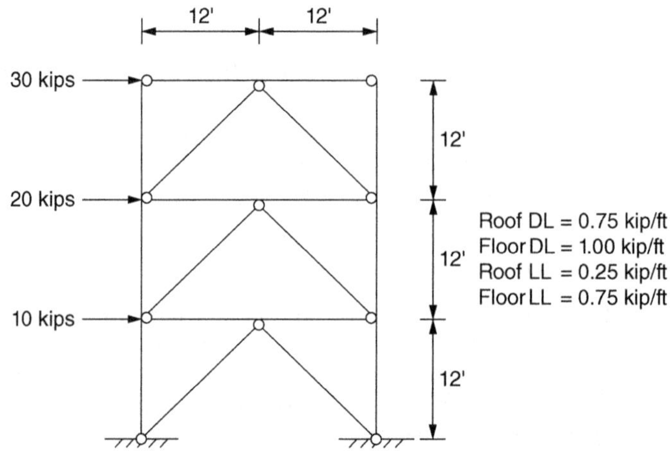


**FIGURE 14.46** Details for Problem 14.1.

**14.7** *Given:* The single bay special concentrically braced steel frame shown in Fig. 14.46 forms part of an office building assigned to seismic design category D with a redundancy factor of $\rho = 1.0$ and a design response acceleration of $S_{DS} = 1.0g$. The seismic lateral loads and the gravity loads at each level are indicated. The dead loads include the self weights of the members. Neglect second-order effects.

*Find:* Using allowable stress level design (ASD), a suitable round HSS member for the brace in the bottom story.

**14.8** *Given:* The single bay special concentrically braced steel frame shown in Fig. 14.46 forms part of an office building assigned to seismic design category D with a redundancy factor of $\rho = 1.0$ and a design response acceleration of $S_{DS} = 1.0g$. The seismic lateral loads and the gravity loads at each level are indicated. The dead loads include the self weights of the members. Neglect second-order effects.

*Find:* Using load and resistance factor design (LRFD), a suitable round HSS member for the brace in the bottom story.

**14.9** *Given:* The single bay special concentrically braced steel frame shown in Fig. 14.46 forms part of an office building assigned to seismic design category D with a redundancy factor of $\rho = 1.0$ and a design response acceleration of $S_{DS} = 1.0g$. The seismic lateral loads and the gravity loads at each level are indicated. The dead loads include the self weights of the members. Neglect second-order effects.

*Find:* Using allowable stress level design (ASD), the lightest weight W24 section of ASTM 992 steel suitable for the beam in the bottom story.

**14.10** *Given:* The single bay special concentrically braced steel frame shown in Fig. 14.46 forms part of an office building assigned to seismic design category D with a redundancy factor of $\rho = 1.0$ and a design response acceleration of $S_{DS} = 1.0g$. The seismic lateral loads and the gravity loads at each level are indicated. The dead loads include the self weights of the members. Neglect second-order effects.

*Find:* Using load and resistance factor design (LRFD), the lightest weight W24 section of ASTM 992 steel suitable for the beam in the bottom story.

**14.11** *Given:* The single bay special concentrically braced steel frame shown in Fig. 14.46 forms part of an office building assigned to seismic design category D with a redundancy factor of $\rho = 1.0$ and a design response acceleration of $S_{DS} = 1.0g$. The seismic lateral loads and the gravity loads at each level are indicated. The dead loads include the self weights of the members. Neglect second-order effects.

*Find:* Using allowable stress level design (ASD), the lightest weight W14 section of ASTM 992 steel suitable for the column in the bottom story.

**14.12** *Given:* The single bay special concentrically braced steel frame shown in Fig. 14.46 forms part of an office building assigned to seismic design category D with a redundancy factor of $\rho = 1.0$ and a design response acceleration of $S_{DS} = 1.0g$. The seismic lateral loads and the gravity loads at each level are indicated. The dead loads include the self weights of the members. Neglect second-order effects.

*Find:* Using load and resistance factor design (LRFD), the lightest weight W14 section of ASTM 992 steel suitable for the column in the bottom story.

**14.13** *Given:* The single bay eccentrically braced steel frame shown in Fig. 14.47 forms part of a residential building assigned to seismic design category D with a redundancy factor of $\rho = 1.0$ and a design response acceleration of $S_{DS} = 1.0g$. The seismic lateral loads at each level are indicated. The preliminary member sizes are indicted in Fig. 14.47 and all members have a yield stress of 50 kips/in$^2$. The connection of a brace to the beam may be considered fixed and the beam is simply supported at its ends. From a first-order analysis of the frame, the forces in the link and the drift in the bottom story are determined as shown in the table.

| Load | Axial Force, kips | Shear Force, kips | Moment, kip-ft | Drift, in |
|---|---|---|---|---|
| Dead | 4.0 | 0.7 | 6.0 | – |
| Live | 3.0 | 0.5 | 4.0 | – |
| Seismic | 0 | 54 | 110 | 0.2 |

*Find:* Using allowable stress level design (ASD), determine if the W12 × 96 member is suitable for the link in the bottom story. Neglect second-order effects.

**14.14** *Given:* The single bay eccentrically braced steel frame shown in Fig. 14.47 forms part of a residential building assigned to seismic design category D with a redundancy factor of $\rho = 1.0$ and a design response acceleration of $S_{DS} = 1.0g$. The seismic lateral loads at each level are indicated. The preliminary member sizes are indicted in Fig. 14.47 and all members have a yield stress of 50 kips/in$^2$. The connection of a brace to the beam may be considered fixed and the beam is simply supported at its ends. From a first-order analysis of the frame, the forces in the link and the drift in the bottom story are determined as shown in the table.

| Load | Axial Force, kips | Shear Force, kips | Moment, kip-ft | Drift, in |
|---|---|---|---|---|
| Dead | 4.0 | 0.7 | 6.0 | – |
| Live | 3.0 | 0.5 | 4.0 | – |
| Seismic | 0 | 54 | 110 | 0.2 |

*Find:* Using load and resistance factor design (LRFD), determine if the W12 × 96 member is suitable for the link in the bottom story. Neglect second-order effects.

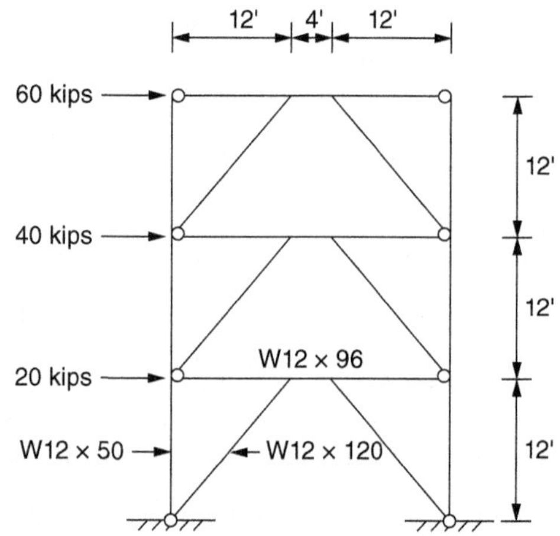


**FIGURE 14.47** Details for Problem 14.13.

**14.15** *Given:* The single bay eccentrically braced steel frame shown in Fig. 14.47 forms part of a residential building assigned to seismic design category D with a redundancy factor of $\rho = 1.0$ and a design response acceleration of $S_{DS} = 1.0g$. The seismic lateral loads at each level are indicated. The preliminary member sizes are indicted in Fig. 14.47 and all members have a yield stress of 50 kips/in$^2$. The connection of a brace to the beam may be considered fixed and the beam is simply supported at its ends.

*Find:* Using allowable stress level design (ASD), design the stiffeners for the link in the bottom story.

**14.16** *Given:* The single bay eccentrically braced steel frame shown in Fig. 14.47 forms part of a residential building assigned to seismic design category D with a redundancy factor of $\rho = 1.0$ and a design response acceleration of $S_{DS} = 1.0g$. The seismic lateral loads at each level are indicated. The preliminary member sizes are indicted in Fig. 14.47 and all members have a yield stress of 50 kips/in$^2$. The connection of a brace to the beam may be considered fixed and the beam is simply supported at its ends.

*Find:* Using load and resistance factor design (LRFD), design the stiffeners for the link in the bottom story.

**14.17** *Given:* The single bay eccentrically braced steel frame shown in Fig. 14.47 forms part of a residential building assigned to seismic design category D with a redundancy factor of $\rho = 1.0$ and a design response acceleration of $S_{DS} = 1.0g$. The seismic lateral loads at each level are indicated. The preliminary member sizes are indicted in Fig. 14.47 and all members have a yield stress of 50 kips/in$^2$. The connection of a brace to the beam may be considered fixed and the beam is simply supported at its ends. From a first-order analysis of the frame, the forces in the beam in the bottom story are determined as shown in the table.

| Load | Axial Force, kips | Shear Force, kips | Moment, kip-ft |
|---|---|---|---|
| Dead | 0.05 | 3.5 | 8.0 |
| Live | 0.03 | 2.5 | 5.0 |
| Seismic | 60 | 4.0 | 58 |

*Find:* Using allowable stress level design (ASD), determine if the W12 × 96 member is suitable for the beam in the bottom story. Neglect second-order effects.

**14.18** *Given:* The single bay eccentrically braced steel frame shown in Fig. 14.47 forms part of a residential building assigned to seismic design category D with a redundancy factor of $\rho = 1.0$ and a design response acceleration of $S_{DS} = 1.0g$. The seismic lateral loads at each level are indicated. The preliminary member sizes are indicted in Fig. 14.47 and all members have a yield stress of 50 kips/in$^2$. The connection of a brace to the beam may be considered fixed and the beam is simply supported at its ends. From a first-order analysis of the frame, the forces in the beam in the bottom story are determined as shown in the table.

| Load | Axial Force, kips | Shear force, kips | Moment, kip-ft |
|---|---|---|---|
| Dead | 0.05 | 3.5 | 8.0 |
| Live | 0.03 | 2.5 | 5.0 |
| Seismic | 60 | 4.0 | 58 |

*Find:* Using load and resistance factor design (LRFD), determine if the W12 × 96 member is suitable for the beam in the bottom story. Neglect second-order effects.

**14.19** *Given:* The single bay eccentrically braced steel frame shown in Fig. 14.47 forms part of a residential building assigned to seismic design category D with a redundancy factor of $\rho = 1.0$ and a design response acceleration of $S_{DS} = 1.0g$. The seismic lateral loads at each level are indicated. The preliminary member sizes are indicted in Fig. 14.47 and all members have a yield stress of 50 kips/in$^2$. The connection of a brace to the beam may be considered fixed and the beam is simply supported at its ends. From a first-order analysis of the frame, the forces in the diagonal brace in the bottom story are determined as shown in the table.

| Load | Axial Force, kips | Shear Force, kips | Moment, kip-ft |
|---|---|---|---|
| Dead | 5.5 | 0.1 | 2.0 |
| Live | 3.7 | 0.1 | 2.0 |
| Seismic | 83 | 2.9 | 49 |

*Find:* Using allowable stress level design (ASD), determine if the W12 × 120 member is suitable for the brace in the bottom story. Neglect second-order effects.

**14.20** *Given:* The single bay eccentrically braced steel frame shown in Fig. 14.47 forms part of a residential building assigned to seismic design category D with a redundancy factor of $\rho = 1.0$ and a design response acceleration of $S_{DS} = 1.0g$. The seismic lateral loads at each level are indicated. The preliminary member sizes are indicted in Fig. 14.47 and all members have a yield stress of 50 kips/in$^2$. The connection of a brace to the beam may be considered fixed and the beam is simply supported at its ends. From a first-order analysis of the frame, the forces in the diagonal brace in the bottom story are determined as shown in the table.

| Load | Axial Force, kips | Shear Force, kips | Moment, kip-ft |
|---|---|---|---|
| Dead | 5.5 | 0.1 | 2.0 |
| Live | 3.7 | 0.1 | 2.0 |
| Seismic | 83 | 2.9 | 49 |

*Find:* Using load and resistance factor design (LRFD), determine if the W12 × 120 member is suitable for the brace in the bottom story. Neglect second-order effects.

**14.21** *Given:* The single bay eccentrically braced steel frame shown in Fig. 14.47 forms part of a residential building assigned to seismic design category D with a redundancy factor of $\rho = 1.0$ and a design response acceleration of $S_{DS} = 1.0g$. The seismic lateral loads at each level are indicated. The preliminary member sizes are indicted in Fig. 14.47 and all members have a yield stress of 50 kips/in$^2$. The connection of a brace to the beam may be considered fixed and the beam is simply supported at its ends. From a first-order analysis of the frame, the forces in the column in the bottom story are determined as shown in the table.

| Load | Axial Force, kips | Shear Force, kips | Moment, kip-ft |
|---|---|---|---|
| Dead | 19 | 0 | 0 |
| Live | 13 | 0 | 0 |

*Find:* Using allowable stress level design (ASD), determine if the W12 × 50 member is suitable for the column in the bottom story. Neglect second-order effects.

**14.22** *Given:* The single bay eccentrically braced steel frame shown in Fig. 14.47 forms part of a residential building assigned to seismic design category D with a redundancy factor of $\rho = 1.0$ and a design response acceleration of $S_{DS} = 1.0g$. The seismic lateral loads at each level are indicated. The preliminary member sizes are indicted in Fig. 14.47 and all members have a yield stress of 50 kips/in$^2$. The connection of a brace to the beam may be considered fixed and the beam is simply supported at its ends. From a first-order analysis of the frame, the forces in the column in the bottom story are determined as shown in the table.

| Load | Axial Force, kips | Shear Force, kips | Moment, kip-ft |
|---|---|---|---|
| Dead | 19 | 0 | 0 |
| Live | 13 | 0 | 0 |

*Find:* Using load and resistance factor design (LRFD), determine if the W12 × 50 member is suitable for the column in the bottom story. Neglect second-order effects.

**14.23** *Given:* Figure 14.48 shows the members selected for the third floor of a single bay special moment frame that forms part of a residential building assigned to seismic design category D with a redundancy factor of $\rho = 1.0$ and a design response acceleration of $S_{DS} = 1.0g$. The story height is 12 ft. The LRFD factored gravity load on the beam is $w_u = 2$ kips/ft, and the axial load on the column due to the LRFD load combination including the overstrength factor, $\Omega_0$, is $P_{uc} = 300$ kips. The beam supports a concrete slab that is connected with welded shear connectors spaced at a maximum of 12 in between protected zones. All material has a yield stress of 50 ksi and a tensile strength of 65 ksi. The frame satisfies story drift, stability, and strength requirements under the governing load combination from ASCE 7 Sec. 12.4.2.3.

*Find:* Select suitable dimensions for the cut in a reduced beam flange moment connection and determine if the W24 × 131 beam is adequate.

**14.24** *Given:* Figure 14.48 shows the members selected for the third floor of a single bay special moment frame that forms part of a residential building assigned to seismic design category D with a redundancy factor of $\rho = 1.0$ and a design response acceleration of $S_{DS} = 1.0g$. The story height is 12 ft. The LRFD factored gravity load on the beam is $w_u = 2$ kips/ft, and the axial load on the column due to the LRFD load combination including the overstrength factor, $\Omega_0$, is $P_{uc} = 300$ kips. The beam supports a concrete slab that is connected with welded shear connectors spaced at a maximum of 12 in between protected zones. All material has a yield stress of 50 ksi and a tensile strength of 65 ksi. The frame satisfies story drift, stability, and strength requirements under the governing load combination from ASCE 7 Sec. 12.4.2.3.

*Find:* For the reduced beam section connection designed in Problem 14.23, determine if the strong-column weak-beam concept is satisfied.

**14.25** *Given:* Figure 14.48 shows the members selected for the third floor of a single bay special moment frame that forms part of a residential building assigned to seismic design category D with a redundancy factor of $\rho = 1.0$ and a design response acceleration of $S_{DS} = 1.0g$. The story height is 12 ft. The LRFD factored gravity load on the beam is $w_u = 2$ kips/ft, and the axial load on the column due to the LRFD load combination including the overstrength factor, $\Omega_0$, is $P_{uc} = 300$ kips. The beam supports a concrete slab that is connected with welded shear connectors spaced at a maximum of 12 in between protected zones. All material has a yield stress of 50 ksi and a tensile strength of 65 ksi. The frame satisfies story drift, stability, and strength requirements under the governing load combination from ASCE 7 Sec. 12.4.2.3.

*Find:* For the reduced beam section connection designed in Problem 14.23, design suitable doubler plates, if required.

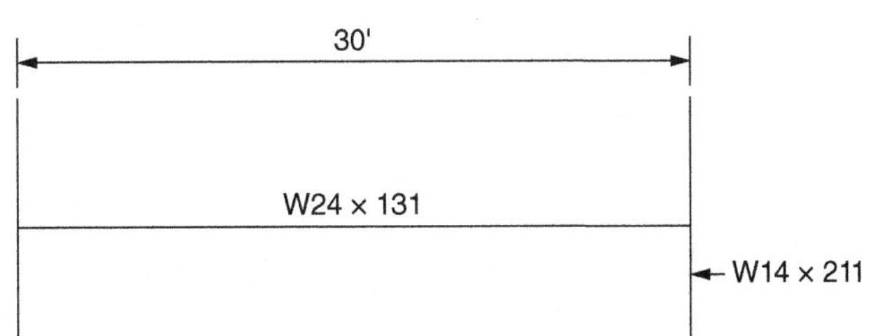


**FIGURE 14.48** Details for Problem 14.23.

**14.26** *Given:* Figure 14.48 shows the members selected for the third floor of a single bay special moment frame that forms part of a residential building assigned to seismic design category D with a redundancy factor of $\rho = 1.0$ and a design response acceleration of $S_{DS} = 1.0g$. The story height is 12 ft. The LRFD factored gravity load on the beam is $w_u = 2$ kips/ft, and the axial load on the column due to the LRFD load combination including the overstrength factor, $\Omega_o$, is $P_{uc} = 300$ kips. The beam supports a concrete slab that is connected with welded shear connectors spaced at a maximum of 12 in between protected zones. All material has a yield stress of 50 ksi and a tensile strength of 65 ksi. The frame satisfies story drift, stability, and strength requirements under the governing load combination from ASCE 7 Sec. 12.4.2.3.

*Find:* For the reduced beam section connection designed in Problem 14.23, design suitable continuity plates, if required.

# Index

*Note:* Page numbers followed by *f* denote figures; page numbers followed by *t* denote tables.

## E

## G

## M

## Q



## R



## S

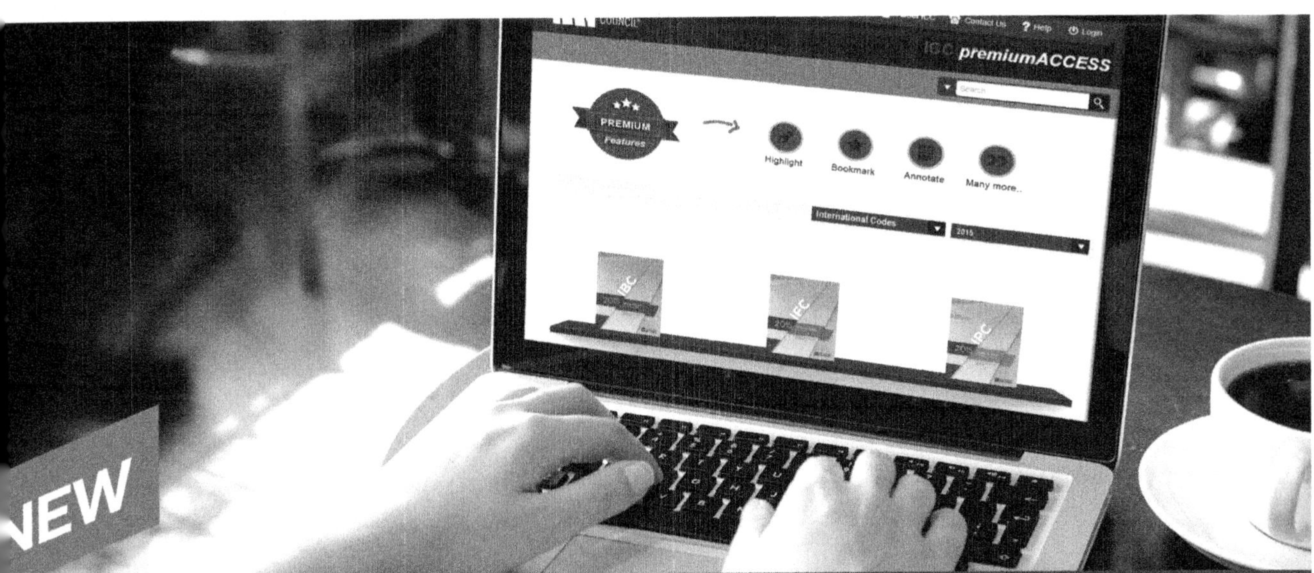
premiumACCESS
PREMIUM
Features
Highlight
Bookmark
Annotate
Many more..
International Codes
2015
NEW

CPSIA information can be obtained
at www.ICGtesting.com
Printed in the USA
LVOW04*1929060817
544044LV00004B/7/P

9 781259 588013